Quantitative Methods in Reservoir Engineering

Quantitative Methods in Reservoir Engineering

Second Edition

Wilson C. Chin
Stratamagnetic Software, LLC
Houston, Texas and Beijing, China

ELSEVIER

AMSTERDAM • BOSTON • HEIDELBERG • LONDON
NEW YORK • OXFORD • PARIS • SAN DIEGO
SAN FRANCISCO • SINGAPORE • SYDNEY • TOKYO
Gulf Professional Publishing is an imprint of Elsevier

Gulf Professional Publishing is an imprint of Elsevier
50 Hampshire Street, 5th Floor, Cambridge, MA 02139, United States
The Boulevard, Langford Lane, Kidlington, Oxford, OX5 1GB, United Kingdom

Library of Congress Cataloging-in-Publication Data
A catalog record for this book is available from the Library of Congress

British Library Cataloguing-in-Publication Data
A catalogue record for this book is available from the British Library

ISBN: 978-0-12-810518-4

For information on all Gulf Professional publications
visit our website at https://www.elsevier.com/

Publisher: Joe Hayton
Acquisition Editor: Katie Hammon
Editorial Project Manager: Kattie Washington
Production Project Manager: Vijayaraj Purushothaman
Cover Designer: Greg Harris

Typeset by SPi Global, India

Contents

11. Effective Properties in Single and Multiphase Flows

12. Modeling Stochastic Heterogeneities

13. Real and Artificial Viscosity

14. Borehole Flow Invasion, Lost Circulation and Time Lapse Logging

15. Horizontal, Deviated and Modern Multilateral Well Analysis

About the Author

Wilson C. Chin earned his PhD from the Massachusetts Institute of Technology and his MSc from the California Institute of Technology.

He has authored more than one hundred papers, over three dozen patents, and fifteen books in petroleum engineering, many with Elsevier. His interests include formation testing, reservoir engineering, acoustic and resistivity logging, MWD design, and managed pressure drilling.

Professional Interests

Contact information
Email: wilsonchin@aol.com
United States cell: (832) 483-6899

Scientific Book Publications

Fifteen original books, describing personal research in reservoir engineering, electromagnetics, formation evaluation, Measurement-While-Drilling, sensor design, and drilling and cementing rheology, namely,

- Resistivity Modeling: Propagation, Laterolog and Micro-Pad Analysis (John Wiley & Sons, 2016)
- Reservoir Engineering in Modern Oilfields: Vertical, Deviated, Horizontal and Multilateral Well Systems (John Wiley & Sons, 2016)
- Quantitative Methods in Reservoir Engineering, 2nd Edition—*with New Topics in Formation Testing and Multilateral Well Flow Analysis* (Elsevier, 2016)
- Formation Testing: Low Mobility Pressure Transient Analysis (*with CNOOC*, John Wiley & Sons, 2015)
- Wave Propagation in Drilling, Well Logging and Reservoir Applications (John Wiley, 2014)
- Measurement-While-Drilling Signal Analysis, Optimization and Design (*with CNPC*, John Wiley, 2014)
- Electromagnetic Well Logging: Models for MWD/LWD Interpretation and Tool Design (John Wiley, 2014)
- Formation Testing Pressure Transient and Contamination Analysis (*with CNOOC*, John Wiley, 2014)
- Managed Pressure Drilling: Modeling, Strategy and Planning (Elsevier, 2012)
- Quantitative Methods in Reservoir Engineering (Elsevier, 2002)
- Computational Rheology for Pipeline and Annular Flow (Elsevier, 2001)
- Formation Invasion, *with Applications to Measurement-While-Drilling, Time Lapse Analysis and Formation Damage* (Gulf Publishing, 1995)
- Wave Propagation in Petroleum Engineering, *with Applications to Drillstring Vibrations, Measurement-While-Drilling, Swab-Surge and Geophysics* (Gulf Publishing, 1994)
- Modern Reservoir Flow and Well Transient Analysis (Gulf Publishing, 1993)
- Borehole Flow Modeling in Horizontal, Deviated and Vertical Wells (Gulf Publishing, 1992)

United States Patents

- U.S. Patent No. 7,243,537. Methods for Measuring a Formation Supercharge Pressure, with M. Proett, J. Beique, J. Hardin, J. Fogal, D. Welshans, and G. Gray, Jul. 17, 2007
- U.S. Patent No. 7,224,162. System and Methods for Upscaling Petrophysical Data, with M. Proett, J. Fogal, and P. Aadireddy, May 29, 2007
- U.S. Patent No. 7,082,078. Magneto-Rheological Fluid Controlled Mud Pulser, with M. Fripp and N. Skinner, Jul. 25, 2006
- U.S. Patent No. 7,059,179. Multi-Probe Pressure Transient Analysis for Determination of Horizontal Permeability, Anisotropy and Skin in an Earth Formation, with M. Proett, Jun. 13, 2006
- U.S. Patent No. 6,327,538. Method and Apparatus for Evaluating Stoneley Waves, and for Determining Formation Parameters in Response Thereto, Dec. 4, 2001
- U.S. Patent No. 5,969,638. Multiple Transducer MWD Surface Signal Processing, Oct. 19, 1999
- U.S. Patent No. 5,831,177. Fluid Driven Siren Flowmeter, with J. Anders, M. Proett, and M. Waid, Nov. 3, 1998
- U.S. Patent No. 5,787,052. Snap Action Rotary Pulser, with W. Gardner, Jul. 28, 1998
- U.S. Patent No. 5,740,126. Turbosiren Signal Generator for Measurement While Drilling Systems, with T. Ritter, Apr. 14, 1998
- U.S. Patent No. 5,703,286. Method of Formation Testing, with M. Proett and C. Chen, Dec. 30, 1997
- U.S. Patent No. 5,672,819. Formation Evaluation Using Phase Shift Periodic Pressure Pulse Testing, with M. Proett, Sept. 30, 1997
- U.S. Patent No. 5,644,076. Wireline Formation Tester Supercharge Correction Method, with M. Proett and M. Waid, Jul. 1, 1997
- U.S. Patent No. 5,586,083. Turbosiren Signal Generator for Measurement While Drilling Systems, with T. Ritter, Dec. 17, 1996
- U.S. Patent No. 5,583,827. Measurement-While-Drilling System and Method, Dec. 10, 1996
- U.S. Patent No. 5,535,177. MWD Surface Signal Detector Having Enhanced Acoustic Detection Means, with K. Hamlin, Jul. 9, 1996
- U. S. Patent No. 5,515,336. MWD Surface Signal Detector Having Bypass Loop Acoustic Detection Means, with W. Gardner and M. Waid, May 7, 1996
- U. S. Patent No. 5,459,697. MWD Surface Signal Detector Having Enhanced Acoustic Detection Means, with K. Hamlin, Oct. 17, 1995
- U. S. Patent No. 4,785,300. Pressure Pulse Generator, with J. Trevino, Nov. 15, 1988

Recent Patent Applications

- Formation Evaluation Using Phase Shift Periodic Pressure Pulse Testing in Anisotropic Media, Jun. 2015
- Multiple Drawdown Pressure Transient Analysis for Low Mobility Formation Testing, Jun. 2015
- High Signal Strength Mud Siren for MWD Telemetry, with K. Iftikhar, Jan. 2016
- High Speed Telemetry Signal Processing, with Y. Jiang, May 2016

International and Domestic Patents

- CA2556937, 9/21/2010, Methods for measuring a formation supercharge pressure
- BRPI0508357, 7/24/2007, Método para determinar a pressão de supercarga em uma formação interceptada por um furo de sondagem
- US7243537, 7/17/2007, Methods for measuring a formation supercharge pressure
- US7224162, 5/29/2007, System and methods for upscaling petrophysical data
- CA2156224, 10/17/2006, MWD surface signal detector having bypass loop acoustic detection means
- CA2156223, 8/1/2006, MWD surface signal detector having bypass loop acoustic detection means
- US7082078, 7/25/2006, Magneto-rheological fluid controlled mud pulser
- US7059179, 6/13/2006, Multi-probe pressure transient analysis for determination of horizontal permeability, anisotropy and skin in an earth formation
- WO2005084332, 9/15/2005, Methods for measuring a formation supercharge pressure
- WO2005036338, 4/21/2005, System and methods for upscaling petrophysical data
- WO2005017301, 2/24/2005, Electroactive fluid controlled mud pulser
- US06327538, 12/04/2001, Method and apparatus for evaluating Stoneley waves, and for determining formation parameters in response thereto
- EP00936477A3, 12/13/2000, Evaluating Stoneley waves and formation parameters
- EP1053488, 11/22/2000, Multiple transducer MWD surface signal processing
- NO20003826A, 09/26/2000, Behandling av signal fra multippel MWD-transducer p overflaten
- NO20003826A0, 07/26/2000, Behandling av signal fra multippel MWD-transducer p overflaten
- EP00747571B1, 02/02/2000, Downhole pressure pulse generator
- EP0950795, 10/20/1999, Tool for and method of geological formation evaluation testing

- US5969638, 10/19/1999, Multiple transducer MWD surface signal processing
- NO00990872A, 10/18/1999, Verktoey og fremgangsmte for geologisk formasjonsevaluering og testing
- EP936477A2, 8/18/1999, Evaluating Stoneley waves and formation parameters
- NO00990615A, 8/18/1999, Evaluering av Stoneley-boelger og formasjonsparametre
- WO09938032, 7/29/1999, Multiple transducer MWD surface signal processing
- NO00990872A0, 02/24/1999, Verkt y og fremgangsmte for geologisk formasjonsevaluering og testing
- NO00990615A0, 02/09/1999, Evaluering av Stoneley-boelger og formasjonsparametre
- US05831177, 11/03/1998, Fluid driven siren flowmeter
- US05787052, 07/28/1998, Snap action rotary pulser
- US05740126, 04/14/1998, Turbo-siren signal generator for measurement while drilling systems
- US05703286, 12/30/1997, Method of formation testing
- EP0747571, 12/11/1996, Downhole pressure pulse generator
- US05672819, 09/30/1997, Formation evaluation using phase shift periodic pressure pulse testing
- EP00697499A3, 07/30/1997, Apparatus for detecting an acoustic signal in drilling mud
- US05644076, 07/01/1997, Wireline formation tester supercharge correction method
- US05586083, 12/17/1996, Turbo siren signal generator for measurement while drilling systems
- EP00747571A2, 12/11/1996, Downhole pressure pulse generator
- US05583827, 12/10/1996, Measurement-while-drilling system and method
- US05535177, 07/09/1996, MWD surface signal detector having enhanced acoustic detection means
- US05515336, 05/07/1996, MWD surface signal detector having bypass loop acoustic detection means
- EP00697498, 02/21/1996, Apparatus for detecting pressure pulses in a drilling fluid supply
- EP00697499A2, 02/21/1996, Apparatus for detecting an acoustic signal in drilling mud
- EP00697498A2, 02/21/1996, Apparatus for detecting pressure pulses in a drilling fluid supply
- NO00953224A, 02/19/1996, Anordning til paavisning av trykkpulser i en ledning for tilfoersel av borevaeske
- NO00953223A, 02/19/1996, Overflate-signaldetektor for maaling I loepet av boringen med forsterket akustisk detektorinnretning

- CA02156224AA, 02/18/1996, MWD surface signal detector having bypass loop acoustic detection means
- CA02156223AA, 02/18/1996, MWD surface signal detector having enhanced acoustic detection means
- US05459697, 10/17/1995, MWD surface signal detector having enhanced acoustic detection means
- NO00953224A0, 08/16/1995, Anordning for aa detektere trykkpulser I en borefluid
- NO00953223A0, 08/16/1995, Overflate-signaldetektor for maaling I loepet av boringen med forsterket akustisk detektorinnretning
- US4785300, 11/15/1988, Pressure pulse generator
- CA1228909, 11/03/1987, Pressure pulse generator
- BRPI8405278, 8/27/1985, Gerador de pulsos de pressao
- EP0140788, 5/8/1985, Pressure pulse generator
- NO00844026A, 04/25/1985, Trykkpulsgenerator

Journal Articles and Conference Publications

Single-authored unless noted otherwise.

- Exact Three-Dimensional Electromagnetic Model: MWD/LWD Anisotropic Prediction for R_h and R_v, *Journal of Sustainable Energy Engineering*, 2015
- High-Data-Rate Measurement-While-Drilling System for Very Deep Wells, with Y. Su, L. Sheng, L. Li, H. Bian and R. Shi, *Well Logging Technology Journal*, Xi'an, China, Dec. 2014
- Strategies in High-Speed MWD Mud Pulse Telemetry, with Y. Su, L. Sheng, L. Li, H. Bian, R. Shi and X. Zhuang, *Journal of Sustainable Energy Engineering*, Dec. 2014
- Formation Testing: New Methods for Rapid Mobility and Pore Pressure Prediction, with Y. Zhou, Z. Hao, Y. Feng and Q. Yu, Paper OTC-24890-MS, *2014 Offshore Technology Conference Asia (OTC Asia)*, Kuala Lumpur, Malaysia, Mar. 25–28, 2014
- Formation Testing: New Methods for Rapid Mobility and Pore Pressure Prediction, with Y. Zhou, L. Zhao, Y. Feng and Q. Yu, Paper 17214, *7th International Petroleum Technology Conference (IPTC)*, Doha, Qatar, Jan. 19–22, 2014
- Formation Tester Flow Analysis in Anisotropic Media With Flowline Storage and Skin at Arbitrary Dip, *Well Logging Technology Journal*, Xi'an, China, Feb. 2013
- Advances in Swab-Surge Modeling for Managed Pressure Drilling, with X. Zhuang, Paper OTC-21115-PP, *2011 Offshore Technology Conference*, Houston, TX, May 2–5, 2011

- Effect of Rotation on Flowrate and Pressure Gradient in Eccentric Holes, with X. Zhuang, Paper AADE-11-NTCE-45, *AADE 2011 National Technical Conference and Exhibition*, Houston, TX, Apr. 12–14, 2011
- Advances in Swab-Surge Modeling for Managed Pressure Drilling, with X. Zhuang, Paper AADE-11-NTCE-46, *AADE 2011 National Technical Conference and Exhibition*, Houston, TX, Apr. 12–14, 2011
- Transient, Multiphase, Three-Dimensional Pumping Models for Cementing and Drilling, with X. Zhuang, Paper AADE-11-NTCE-72, *AADE 2011 National Technical Conference and Exhibition*, Houston, TX, Apr. 12–14, 2011
- Comprehensive Annular Flow Models for Drilling and Completions, with X. Zhuang, Paper AADE-11-NTCE-73, *AADE 2011 National Technical Conference and Exhibition*, Houston, TX, Apr. 12–14, 2011
- High-Data-Rate Measurement-While-Drilling System for Very Deep Wells, with Y. Su, L. Sheng, L. Li, H. Bian and R. Shi, Paper AADE-11-NTCE-74, *AADE 2011 National Technical Conference and Exhibition*, Houston, TX, Apr. 12–14, 2011
- Flow Simulation Methods for Managed Pressure Drilling and Cementing, *Drilling and Completing Trouble Zones Conference*, Galveston, TX, Oct. 19–21, 2010
- Modeling and Simulation of Managed Pressure Drilling for Improved Design, Risk Assessment, Training and Operations, *RPSEA Ultra-Deepwater Technology Conference*, Houston, TX, Jun. 22–23, 2010
- Exact Non-Newtonian Flow Analysis of Yield Stress Fluids in Highly Eccentric Borehole Annuli with Pipe or Casing Translation and Rotation, with X. Zhuang, SPE Paper 131234-PP, *CPS/SPE International Oil & Gas Conference and Exhibition*, Beijing, China, Jun. 8–10, 2010
- Displacement of Viscoplastic Fluids in Eccentric Annuli: Numerical Simulation and Experimental Validation, with T. Deawwanich, J.C. Liew, Q.D. Nguyen, M. Savery and N. Tonmukayakul, *Chemeca 2008 Conference*, Newcastle, Australia, Sept. 28 to Oct. 1, 2008
- Laminar Displacement of Viscoplastic Fluids in Eccentric Annuli—Numerical Simulation and Experimental Validations, with M. Savery, P. Tonmukayakul, T. Deawwanich, J. Liew and Q. Dzuy Nguyen, *XXII International Congress of Theoretical and Applied Mechanics (ICTAM 2008)*, Adelaide, Australia, Aug. 24–29, 2008
- Flow Visualization and Numerical Simulation of Viscoplastic Fluid Displacements in Eccentric Annuli, with Q.D. Nguyen, T. Deawwanich, N. Tonmukayakul and M.R. Savery, *XVth International Congress on Rheology (ICR 2008)*, Society of Rheology 80th Annual Meeting, Monterey, CA, Aug. 3–8, 2008
- Modeling Cement Placement Using a New Three-Dimensional Flow Simulator, with M. Savery, *AADE 2008 Fluids Technology Conference*, Houston, Apr. 8–9, 2008

- Formation Tester Inverse Permeability Interpretation for Liquids in Anisotropic Media with Flowline Storage and Skin at Arbitrary Dip, with X. Zhuang, *48th Annual SPWLA Meeting*, Austin, TX, Jun. 3–6, 2007
- Modeling Fluid Interfaces During Cementing Using a Three-Dimensional Mud Displacement Simulator, with M. Savery and R. Darbe, OTC Paper 18513, *2007 Offshore Technology Conference (OTC)*, Houston, TX, Apr. 30 to May 3, 2007
- Formation Tester Immiscible and Miscible Flow Modeling for Job Planning Applications, with M. Proett, *46th Annual SPWLA Meeting*, New Orleans, LA, Jun. 26–29, 2005
- MWD Siren Pulser Fluid Mechanics, *Petrophysics*, Journal of the Society of Petrophysicists and Well Log Analysts (SPWLA), Vol. 45, No. 4, Jul.–Aug. 2004, pp. 363–379
- Formation Testing in the Dynamic Drilling Environment, with M. Proett, D. Seifert, S. Lysen, and P. Sands, *SPWLA 45th Annual Logging Symposium*, Noordwijk, The Netherlands, Jun. 6–9, 2004
- Job Planning Simulators for Immiscible and Miscible Flow, *SPWLA Spring Topical Conference, Formation Testing: Applications and Practices*, Taos, NM, Mar. 28 to Apr. 1, 2004
- Mudcake Growth, Invasion, and Dynamic Coupling with Reservoir Flow: Experiment and Theory, *SPWLA Spring Topical Conference, Formation Testing: Applications and Practices*, Taos, NM, Mar. 28 to Apr. 1, 2004
- Sample Quality Prediction with Integrated Oil and Water-based Mud Invasion Modeling, with M. Proett, D. Belanger, M. Manohar, and J. Wu, *PetroMin Magazine*, Sept. 2003
- Improved Rheology Model and Hydraulics Analysis for Tomorrow's Wellbore Fluid Applications, with R. Morgan, T. Becker, and J. Griffith, SPE Paper 82415, *SPE Production and Operations Symposium*, Oklahoma City, OK, Mar. 2003
- Sample Quality Prediction with Integrated Oil and Water-based Mud Invasion Modeling, with M. Proett, D. Belanger, M. Manohar, and J. Wu, SPE Paper 77964, *SPE Asia Pacific Oil & Gas Conference and Exhibition (APOGCE)*, Melbourne, Australia, Oct. 2002
- Multiple Factors That Influence Wireline Formation Tester Pressure Measurements and Fluid Contacts Estimates, with M. Proett, M. Manohar, R. Sigal, and J. Wu, SPE Paper 71566, *SPE Annual Technical Conference and Exhibition*, New Orleans, LA, Oct. 2001
- Comprehensive Look at Factors Influencing Wireline Formation Tester Pressure Measurements and Fluid Contacts, with M. Proett, M. Manohar, and R. Sigal, *42nd SPWLA Annual Symposium, Society of Professional Well Log Analysts*, Houston, TX, Jun. 2001
- New Wireline Formation Testing Tool With Advanced Sampling Technology, with M. Proett, G. Gilbert, and M. Monroe, *SPE Reservoir Evaluation and Engineering*, Apr. 2001

- Advanced Permeability and Anisotropy Measurements While Testing and Sampling in Real-Time Using a Dual Probe Formation Tester, with M. Proett, SPE Paper 64650, *Seventh International Oil & Gas Conference and Exhibition*, Beijing, China, Nov. 2000
- Modern Flow Assurance Methods, Part III: Coupled Velocity and Temperature Fields in Bundled Pipelines, *Offshore*, Nov. 2000
- Modern Flow Assurance Methods, Part II: Detailed Physical Properties and Engineering Application, *Offshore*, Oct. 2000
- Modern Flow Assurance Methods, Part I: Clogged Pipelines, Wax Deposition, and Hydrate Plugs, *Offshore*, Sept. 2000
- Advanced Permeability and Anisotropy Measurements While Testing and Sampling in Real Time Using a Dual Probe Formation Tester, with M. Proett, SPE Paper 62919, *2000 SPE Annual Technical Conference and Exhibition*, Dallas, TX, Oct. 2000
- General Three-Dimensional Electromagnetic Model for Nondipolar Transmitters in Layered Anisotropic Media With Dip, *Well Logging Technology Journal*, Xi'an, China, Aug. 2000
- New Dual Probe Wireline Formation Testing and Sampling Tool Enables Real-Time Permeability and Anisotropy Measurements, with M. Proett, *41st SPWLA Annual Symposium, Society of Professional Well Log Analysts*, Dallas, TX, Jun. 2000
- Clogged Pipe, Non-Newtonian Fluid, and Coupled Solids Deposition Flow Modeling, Final Technical Report, Brown & Root Energy Services, Houston, TX, May 2000
- Irregular Grid Generation and Rapid 3D Color Display Algorithm, Final Technical Report, DOE Grant No. DE-FG03-99ER82895, United States Department of Energy, May 2000
- New Dual-Probe Wireline Formation Testing and Sampling Tool Enables Real-Time Permeability and Anisotropy Measurements, with M. Proett, M. Manohar, G. Gilbert, and M. Monroe, SPE Paper 59701, *SPE Permian Basin Oil & Gas Recovery Conference*, Midland, TX, Mar. 2000. Also presented at *SPE Rocky Mountain Regional Low Permeability Reservoirs Symposium and Exhibition*, Denver, CO, Mar. 2000
- New Wireline Formation Testing Tool with Advanced Sampling Technology, with M. Proett, G. Gilbert, and M. Monroe, SPE Paper 56711, *1999 SPE Annual Technical Conference and Exhibition*, Houston, TX
- Exact Spherical Flow Solution with Storage for Early-Time Test Interpretation, with M. Proett, *SPE Journal of Petroleum Technology*, Nov. 1998
- New Exact Spherical Flow Solution with Storage *and Skin* for Early-Time Interpretation, with Applications to Wireline Formation and Early-Evaluation Drillstem Testing, with M. Proett, SPE Paper 49140, *1998 SPE Annual Technical Conference and Exhibition*, New Orleans, LA, Sept. 1998

- New Exact Spherical Flow Solution for Early-Time Well Test Interpretation with Applications to Wireline Formation Testing and Early-Evaluation Drillstem Testing, with M. Proett, SPE Paper 39915, *SPE Rocky Mountain Regional Meeting/Low Permeability Reservoirs Symposium*, Apr. 1998
- New Exact Spherical Flow Solution for Early-Time Well Test Interpretation with Applications to Wireline Formation Testing and Early-Evaluation Drillstem Testing, with M. Proett, SPE Paper 39768, *SPE Permian Basin Oil and Gas Recovery Conference*, Midland, TX, Mar. 1998
- Pressure Test Validity Shown by Comparing Field Tests and Simulations, Part II: Formation Pressure Test Tools, with N. Skinner, M. Proett, P. Ringgenberg, and R. Aadireddy, *Oil & Gas Journal*, Jan. 12, 1998
- Testing System Combines Advantages of Wireline and Drillstem Testers, Part I: Formation Pressure Test Tools, with N. Skinner, M. Proett, P. Ringgenberg, and R. Aadireddy, *Oil & Gas Journal*, Jan. 5, 1998
- New Early Formation Pressure System Field Test Results and Advances in Early Time Pressure Buildup Analysis, with N. Skinner, M. Proett, P. Ringgenberg, K. Manke, H. Smith, and R. Aadireddy, SPE Paper 38648, *1997 SPE Annual Technical Conference and Exhibition*, San Antonio, TX, Oct. 1997
- Permeability Prediction from Formation Tester "Phase Delay" and "Sonic Pulse" Analysis, with M. Proett, *GRI—SPWLA Research Forum on Permeability Logging*, Houston, TX, Feb. 1997
- Permeability Prediction from Stoneley Waveform Data, *GRI—SPWLA Research Forum on Permeability Logging*, Houston, TX, Feb. 1997
- Supercharge Pressure Compensation Using a New Wireline Testing Method and Newly Developed Early Time Spherical Flow Model, with M. Proett, SPE Paper 36524, *1996 Annual Technical Conference and Exhibition of the Society of Petroleum Engineers*, Denver, CO, Oct. 1996
- Supercharge Pressure Compensation with New Wireline Formation Testing Method, with M. Proett, *1996 Annual Meeting of the Society of Professional Well Log Analysts*, New Orleans, LA, Jun. 1996
- Supercharge Pressure Compensation with New Wireline Formation Testing Method, with M. Proett, *SPE Paper 35178, Permian Basin Oil and Gas Recovery Conference*, Midland, TX, Mar. 1996
- Modeling Complex Horizontal Wells in Heterogeneous Formations, *Offshore*, Oct. 1993
- Eccentric Annular Flow Modeling for Highly Deviated Boreholes, *Offshore*, Aug. 1993
- Model Offers Insight Into Spotting Fluid Performance, *Offshore*, Feb. 1991
- Annular Flow Model Explains Conoco's Borehole Cleaning Success, *Offshore*, Oct. 1990
- Exact Cuttings Transport Correlations Developed for High Angle Wells, *Offshore*, May 1990
- Advances in Annular Borehole Flow Modeling, *Offshore*, Feb. 1990

- Comparative Studies in Dual Porosity Reservoir Simulation, *BP Exploration Reservoir Technology Report No. H090.0005*, Houston, TX, Jan. 1990
- Automating the Acquisition Planning Process at the Johnson Space Center (using Artificial Intelligence Methods), with J. Golej, MITRE Corporation Report MTR-88D00065, NASA/JSC Contract No. NAS9-18057, Sept. 1988
- Simulating Horizontal Well Fracture Flow, *Offshore*, Aug. 1988
- Why Drill Strings Fail at the Neutral Point, *Petroleum Engineer International*, May 1988
- Fatal Tubular Bending Motions Difficult to Detect Uphole, *Offshore*, Apr. 1988
- Formation Evaluation Using Repeated MWD Logging Measurements, with A. Suresh, P. Holbrook, L. Affleck, and H. Robertson, *SPWLA 27th Annual Logging Symposium*, Houston, TX, Jun. 9–13, 1986
- Inviscid Steady Flow Past Turbofan Mixer Nozzles, *ASME J. Applied Mechanics*, Dec. 1984
- Comment on "PAN AIR Applications to Aero-Propulsion Integration," *J. Aircraft*, Nov. 1984
- Thin Airfoil Theory for Planar Inviscid Shear Flow, *ASME J. Applied Mechanics*, Mar. 1984
- Engine Power Simulation for Transonic Flow-Through Nacelles, *AIAA J.*, Oct. 1983
- An Axisymmetric Nacelle and Turboprop Inlet Analysis Including Power Simulation, with D. Golden and T. Barber, *J. Aircraft*, Jun. 1983
- Superpotential Solution for Jet Engine External Potential and Internal Rotational Flow Interaction, *ASME J. Applied Mechanics*, Jun. 1983
- An Axisymmetric Nacelle and Turboprop Inlet Analysis with Flow-Through and Power Simulation Capabilities, with D. Golden and T. Barber, AIAA Paper 82-0256, *AIAA 20th Aerospace Sciences Meeting*, Orlando, FL, Jan. 1982
- Direct Approach to Aerodynamic Design Problems, *ASME J. Applied Mechanics*, Dec. 1981
- Harmonic Analysis of Unsteady Transonic Flow, *AIAA J.*, Feb. 1981
- Kinematic Barrier for Gravity Waves on Variable Currents, *J. Hydronautics*, Jan. 1981
- Optimal Coordinates for Squire's Jet, *AIAA J.*, Jan. 1981
- Transonic Nacelle Inlet Analyses, with W. Presz, D. Ives, D. Paris and D. Golden, *NASA Lewis Workshop on Application of Advanced Computational Methods*, Nov. 1980
- Wave Focusing and Hydraulic Jump Formation, *J. Hydronautics*, Jul. 1980
- Inviscid Parallel Flow Stability with Mean Profile Distortion, *J. Hydronautics*, Jul. 1980
- Class of Shockfree Airfoils Producing the Same Surface Pressure, *J. Aircraft*, Apr. 1980

- Effect of Dissipation and Dispersion on Slowly Varying Wavetrains, *AIAA J.*, Feb. 1980
- Airfoil Design in Subcritical and Supercritical Flow, with D. Rizzetta, *ASME J. Applied Mechanics*, Dec. 1979
- Kinematic Wave Approach to Hydraulic Jumps with Waves, *J. Hydronautics*, Oct. 1979
- Effect of Frequency in Unsteady Transonic Flow, with D. Rizzetta, *AIAA J.*, Jul. 1979
- Stability of Inviscid Shear Flow Over Flexible Membranes, *AIAA J.*, Jun. 1979
- Some Exact Solutions to Guderley's Equation, *AIAA J.*, Apr. 1979
- On the Design of Thin Subsonic Airfoils, *ASME J. Applied Mechanics*, Mar. 1979
- Algorithm for Inviscid Flow Using the Viscous Transonic Equation, *AIAA J.*, Aug. 1978
- Type-Independent Solutions for Mixed Compressible Flows, *AIAA J.*, Aug. 1978
- Similar Solutions for Unsteady Transonic Flow, *AIAA J.*, Jun. 1978
- Nonlinear Formulation for Low-Frequency Transonic Flow, *AIAA J.*, Jun. 1978
- Supersonic Wave Drag of Planar Singularity Distributions, *AIAA J.*, May 1978
- Some Singular Aspects of Three-Dimensional Transonic Flow, *AIAA J.*, Mar. 1978
- Pseudo-Transonic Equation with a Diffusion Term, *AIAA J.*, Jan. 1978
- Goethert's Rule with an Improved Boundary Condition, *AIAA J.*, Oct. 1977
- Numerical Solution for Viscous Transonic Flow, *AIAA J.*, Sept. 1977
- Supersonic Wave Drag for Nonplanar Singularity Distributions, *AIAA J.*, Jun. 1977

Acknowledgments

The author wishes to express his appreciation to his colleagues at Schlumberger, Halliburton, Baker Hughes, British Petroleum, GE Oil & Gas, and other companies for their insights and suggestions over the years and for shaping his approach to reservoir engineering and simulator development. He is especially grateful to Katie Hammon, Senior Acquisitions Editor, for providing him with the platform to communicate important ideas in a highly technical volume. Here content is developed that enables users to solve important and modern problems using the latest modeling technologies available. Moreover, the methods are not "me too" in nature; they are unique, e.g., singular integral equation, advanced finite difference methods, specialized conformal mappings, boundary-conforming grid generation techniques, and so on, but no attempt is made to be all things to all people. It is often said that "one does not understand, until one can compute it." Few recent oilfield achievements have approached areas like horizontal and multilateral well design, formation testing, and fracture flow in heterogeneous reservoirs in economic and practical significance, the subject of this Second Edition—and with the methodologies presented here, we believe that we have closed an important gap in petroleum engineering by simultaneously advancing simulation methods while reducing planning and modeling costs by several orders of magnitude. For the opportunity to present these ideas, many new and developed only within the past several years, to a broad audience, I thank Elsevier, and again, Katie. And I look forward to constructive feedback from readers both young and experienced, important feedback necessary to improve upon our methods and upon which all new technologies thrive.

Preface

Most reservoir flow analysis books provide the basic equations, such as Darcy's law, single-phase radial flow solutions, simple well test models, and the usual descriptions of relative permeability and capillary pressure, and explain elementary concepts in finite difference methods and modeling before referring readers to commercial simulators and industry case studies. These books, and the courses that promote them, are useful in introducing students to fundamental methodologies and company practices. However, few develop the physical and mathematical insight needed to create the next generation of models or to evaluate the limitations behind existing simulation tools. Many analysis techniques and computational approaches employed, in fact, are incorrect, despite their common use in reservoir evaluation. Rigor is noticeably absent.

I earned my PhD at MIT and an earlier MSc degree from Caltech. My major areas were high-speed aerodynamics and wave propagation, which are synonymous with applied math and nonlinear differential equations—specialties that focus on rigorous solutions to practical problems. From MIT, I joined Boeing's prestigious computational fluid dynamics group in Seattle and, three years later, headed up engine flow analysis at United Technologies' Pratt and Whitney, the company that develops the world's most powerful jet engines.

But the thrill of the hunt lost its allure, despite the thrill of being published in journals and attending high-tech conferences. Like all of you, I was attracted to the petroleum industry because of its excitement and the opportunities it offered. That was just five years into my career, as I joined a new industry undergoing rapid change—a transition requiring me to learn anew the fluid dynamics of flows as far underground as my prior learning was above ground. Since then, three decades have elapsed, in which I actively engaged in oil field research and development. In that time, for example, with leading operating and service companies like British Petroleum, Schlumberger and Halliburton, I was fortunate to have been continuously challenged by new problems, both mathematical and operational.

This reservoir flow analysis and simulation book is unique because it brings several decades of perspectives and experience to the fluid mechanics of Darcy flows. Many commonly accepted "recipes" for flow evaluation are critiqued and incorrect underlying assumptions are noted. This volume aims at a rigorous and scientifically correct approach to reservoir simulation. In each of dozens of difficult problems surveyed, the state of the art is examined, and analytical or numerical solutions are offered, with the exact physical assumptions always stated precisely. Industry "common sense" approaches are avoided: once the correct model is formulated, the entire arsenal of modern analysis tools is brought to bear—we then focus on ways to extract formation information using the new solution or clever means to exploit the physics just uncovered.

Fortunately, this book does not require advanced mathematics or numerical analysis to understand. Great care was undertaken to explain and develop very advanced methods in simple terms that undergraduates can comprehend. For example, "conformal mapping" usually requires a background course in complex variables, and complementary subjects like stream functions and streamline tracing in homogeneous media are typically taught within this framework—one that could be intimidating and esoteric. Quite to the contrary, our special derivations require just simple calculus but more generally apply to anisotropic, heterogeneous media, and not necessarily those described using rectangular coordinate systems. This book addresses "difficult" flows, such as liquid and gas flows from fractures, general flows past shales, production from multilateral horizontal wells, multiple well interactions, rigorous approaches to effective properties, and so on, problems not often treated in the literature but relevant to modern petroleum engineering. In doing so, we strive to avoid the simplistic "recipe" approaches our industry often encourages.

Every effort is made to define and formulate the mathematical problem precisely and then to solve it as exactly as modern analysis methods will allow. These include classic differential equation models as well as modern singular integral equation approaches, all of which were unavailable to Morris Muskat when he wrote his lasting monographs on Darcy flow analysis. Our techniques go beyond purely analytical ones. For example, the problem of accurately modeling flow from interacting multilateral drain holes in anisotropic heterogeneous layered media—despite the inefficiencies imposed by nonneighboring grid point connections—is formulated and solved in Chapter 15 (the groundwork for this research won a Chairman's Innovation Award at BP Exploration in 1991 and led to a substantial "Education for the Energy Industry" contract funded by ExxonMobil, ChevronTexaco and Shell in 2000–3).

Or consider boundary-conforming, curvilinear grids in Chapter 8. Fast and accurate mesh generation algorithms are developed in this book, which are cleverly applied to the solution of complicated reservoir flows. Suppose a "Houston well" produces from a "Texas-shaped" reservoir. This geometry is associated with an elementary function as unique as the logarithm is to radial coordinates. Its "analogous logarithm" permits us to instantly write the solution to all liquid and gas flows for any set of pressure-pressure and pressure-rate boundary conditions. What if the Texas-shaped reservoir is produced by a Dallas well? Ah yes, the "Dallas logarithm"—we'll address that too. This work won a prestigious Small Business Innovation Research Award from the United States Department of Energy in 2000, the author's first of five D.O.E. contracts.

Other areas addressed include opened fractures, curved shales, fractured holes, general heterogeneities, formation invasion, and time-lapse well logging using drilling data. In terms of techniques, we introduce modern ideas in singular integral equations, improper integrals, advanced conformal mapping, perturbation methods, numerical grid generation, artificial viscosity, moving boundary value problems, ADI and relaxation methods, and so on, developing these in context with the physics of the problem at hand. These methods, used by aerodynamicists and workers in theoretical elasticity, can be intimidating.

However, the presentation style adopted is far from difficult: while not exactly easy reading, there is nothing in this book that could not be grasped by a student who has taken basic freshman calculus. Whenever possible, Fortran source code is presented, so that students can test and evaluate ideas old and new without the trials and tribulations of debugging (free Fortran compilers are available online for unrestricted downloading from various third-party sources).

New approaches to old problems are emphasized. For example, how do mathematical aerodynamicists turned petroleum engineers view the physical world? Stare up the back end of a rocket lifting off: *Is that a fuselage with stabilizer fins, or is it a circular wellbore with radial fractures?* Pry open the maintenance box of your typical jet engine: *Are those cascades of airfoil blades, or are they distributions of stochastic shales? Can the solutions that describe brittle failure be repackaged to model arrays of fractures, say, the natural fracture systems that spur horizontal drilling?* Very often, the problems (inaccurately) crunched by our fastest computers can be solved (accurately) using closed-form analytical solutions found in other scientific disciplines.

In this Preface to the Second Edition, it is important to describe the additional content developed which extends the scope of the earlier work. All of the original work remains correct—additional perspectives and extensions are added from a since broader exposure to client activities. However, two subject areas are completely new, namely, "formation testing" and "multilateral well reservoir modeling." We shall give a brief synopsis of this material next.

Formation testers are borehole well logging instruments that, when pressed against the sandface, extract reservoir fluids into sealed chambers for examination at the surface. During this extraction, pressure transients in the flowline are recorded for analysis. Some "testers" consist of a single source or pumping probe, while others may contain two (or more) probes. Often, these additional probes are passive observation probes. In the 1960s, interpretation methods were developed to interrogate measured pressure transients—from recorded steady-state drawdown values, effective permeabilities can be obtained from single probe tools. Later in the 1990s, dual probe tools were used to provide k_h and k_v individually, again using steady state pressure drops. These methods were popular, not only because downhole sampling provided laboratory analysis that allowed direct assessment of petroleum fluid properties, but because measurements for permeability (which affect production planning and cash flow) could be obtained as a by-product of the survey. For these reasons, formation testing remains a powerful tool in petroleum exploration, and not surprisingly, a strong source of profits for oil service companies.

However, as new exploration fields continually decreased in mobility, the times required to achieve steady-state pressures increased substantially. Quite often, wait times of hours and sometimes days instead of minutes were required. This increased drilling and logging costs, and also, raised the risk of stuck tools lost in the well. Thus, improvements to formation testing methodologies were clearly needed. In 2004, the United States Department of Energy, through its Small Business Innovation Research (SBIR) program, awarded about two hundred contracts for all fields of energy, e.g., wind, ocean, geothermal, air conditioning, plasma and nuclear physics, and so on. Four awards were allotted to fossil fuels. Of these four, two were won by our Stratamagnetic Software LLC and both in the area of formation testing. Since then, this author has published two books in formation testing, and developed (patent pending) methods to provide k_h and k_v in real-time using early-time (seconds) data collected from low to very-low mobility formations. These methods, which have reduced both risks and costs, were developed by extending the cylindrical flow principles described in our prior book edition to the ellipsoidal flows now encountered in pumping from transversely isotropic reservoirs.

In Chapter 15 of the first edition, problems and limitations associated with horizontal and multilateral well analysis were discussed. A general and very versatile formulation was provided and successful but preliminary numerical results were described. Since then, extensive improvements allow both rapid computing speed on low-end hardware and excellent numerical stability. The methodology applies to general well systems in layered, anisotropic and heterogeneous media producing liquids or gases, where wells can act as producers or injectors or change roles or shut-in during simulation. Importantly, pressure and rate constraints can be altered during simulation, and complicated wells can be added to the reservoir if desired. "Drilling while simulating" is introduced for the first time.

In 2000, three major operating companies funded Houston's well known "E.E.I." or "Education for the Energy Industry" program to introduce petroleum focused careers to inner city students from Kindergarten to Grade 12 at the Aldine Independent School District. Stratamagnetic Software LLC won the competitive bidding over eighty more established companies with a simple but irresistible proposition: develop a learning platform and curriculum allowing teachers and students to understand horizontal and multilateral well production and simulate the most complicated flows using our custom-developed software. Our reservoir flow models have since been commercialized by several organizations. The additional content provided in this Second Edition summarizes key capabilities of the new approach.

More efficient technologies are now more important than ever in the present economic environment—companies are downsizing, research budgets are being reduced, and experienced staff are driven to focus on more practical and immediate activities. The author hopes that the new work will help the industry explore more efficiently and cost-effectively, and that it will convince new engineers joining the petroleum engineering industry that innovation continues and thrives despite the doom and the gloom. Our new approaches to horizontal and multilateral well modeling will be essential to economic planning for unconventional resources; our new approaches to formation tester pressure transient interpretation will hopefully pave the way for further development in petrophysics. And we trust that our original contributions in fracture modeling will enable more efficient fracturing programs and reduced needs for water resources. In short, we're here for the long haul, and we won't be "turning out the lights" any time soon.

Wilson C. Chin, PhD, MIT
Houston, Texas and Beijing, China
Email: wilsonchin@aol.com
Cell: (832) 483-6899

Chapter 1

Motivating Ideas and Governing Equations

It is no accident that the industry's first math models for fluid flow in petroleum reservoirs were developed by analogy to problems in electrostatics and heat transfer (Muskat, 1937; van Everdingen and Hurst, 1949; Carslaw and Jaeger, 1946, 1959). These solutions reflect well on the investigators; they did not fall prey to the maxim that "those who refuse to learn from history are doomed to repeat it." That the equations for single phase flow are identical to the classical equations of elliptic and parabolic type facilitated the initial progress; these similarities also assisted with the design and scaling of experiments, particularly, those based on electrical and temperature analogies.

To practitioners in reservoir engineering and well test analysis, the state-of-the-art has bifurcated into two divergent paths. The first searches for simple closed-form solutions. These are naturally restricted to simplified geometries and boundary conditions, but analytical solutions, many employing "method of images" techniques, nonetheless involve cumbersome infinite series. More recent solutions for transient pressure analysis, given in terms of Laplace and Fourier transforms, tend to be more computational than analytical: they require complicated and often inaccurate numerical inversion, and hence, shed little insight on the physics.

It seems, very often, that all of the analytical solutions that can be derived, have been derived. Thus, the second path noted above falls largely in the realm of supercomputers, high-powered workstations, and brute force numerical analysis: it is the science, or more appropriately the art, that we call reservoir simulation, requiring industry-wide "comparison projects" for validation. There has been no middle ground for smart solutions that solve difficult problems, that is, for solutions that provide physical insight and are in themselves useful, models that can be used for calibration purposes "to keep numerical solutions honest." This dearth of truly useful real world examples lends credence to the often-stated belief that high-speed machines, the marvels that they are in this day and age, only allow engineers to err more quickly and in much greater volume.

Quantitative Methods in Reservoir Engineering. http://dx.doi.org/10.1016/B978-0-12-810518-4.00001-3

Despite numerous computational researches purporting to model transient flows from line fractures, say, there is still no analytical solution encompassing the simpler steady-state limit satisfying practical boundary conditions. Transient solutions, consequently, are sometimes obtained incorrectly by assuming steady state asymptotic conditions that are physically inconsistent. And in spite of wide interest in reservoir heterogeneities, there are still no closed form solutions for flows past single-shale lenses or through mineralized faults. Clearly, there is a need for more work, deeper thought and fundamental investigation.

Like the earlier research of Muskat, Hurst and others, the solutions given in this book are drawn from related outside disciplines, in particular, aerodynamics and theoretical elasticity. A vast library of interesting and immediately useful analytical solutions can be constructed without difficulty; these apply to flows past impermeable shales, flows from fractures and through real-world faults, that, where applicable, satisfy variable flow rate and pressure boundary conditions. In the realm of numerical analysis, it is possible to formulate problems more elegantly, circumventing "brute force" approaches, by using modern methods in curvilinear grid generation. The resulting models provide improved physical resolution where needed and minimize computer storage and data processing requirements; they are especially important, for example, in numerical well testing, where the exact treatment of fracture and stratigraphic boundaries is crucial. These "new" techniques, almost three decades old, were developed in the aerospace industry, and only recently are being applied to problems in petroleum engineering.

But the methods produce more than smart numerics. Computational methods can be combined with analytical ones to form pseudo-analytical approaches that increase accuracy while minimizing hardware requirements. Intelligent personal computer based models founded on these techniques can produce solutions superior to those obtained from existing large scale models, and numerous such hybrid models are developed here. We shall show, for example, how the classic "log r" pressure distribution for radial flow can be generalized to arbitrary reservoirs. As an illustration, a "Texas-shaped" oil field is used and its "elementary solutions" for wells located in different cities are obtained by simple computation. This in turn is used to solve a super-set of reservoir engineering problems analytically and in closed form, that is, for different classes of mixed pressure and flow rate boundary conditions, for liquids and gases having different thermodynamic profiles, and so on.

This book introduces classes of steady-state solutions that the interested reader can extend and generalize. They are particularly meaningful to reservoirs that produce under near-steady conditions at high rates, typical of many oil fields outside the United States. The solutions are useful in studies related to flow heterogeneities, hydraulic fractures, nonlinear gas flows, horizontal drilling, infill drilling, and formation evaluation. The analytical techniques used are described in detail, applied to nontrivial flow problems, and extensions are outlined in the "Problems and Exercises" sections at the end of each chapter.

EXAMPLES OF INCORRECT FORMULATIONS

In this book, mathematical formulations are posed for well-defined physical problems and solved using rigorous solution methods. Assumptions are clearly stated and ad hoc analysis methods are avoided. In order to appreciate this philosophy, it is necessary to understand the proliferation of incorrect models and solutions in existing literature and software. A review of these formulations is imperative to teaching sound mathematics. But it is also important to petroleum engineers who purchase software responsible for field development and major financial commitments. For these reasons, we shall explain common errors and their ramifications. To be entirely fair, only those problems that are actually solved in this book are listed and discussed here.

Velocity Singularities

Flows from natural and massive induced hydraulic fractures are economically important, as are the effects of flow impediments due to shales and mineralized faults. Commercial simulators are available to model these effects, which consume hours of computing time to provide the accuracy needed for economic projections. However, this accuracy is illusory and mere perception. The velocities at fracture tips (when constant pressure boundary conditions apply) are "singular," that is, infinite in speed, although the integrals used to compute total flow rate are finite and bounded. Also, shale tip flows are marked by analogous singularities. Any attempt to model tip flows correctly by fine mesh discretization will promote numerical instability, originating at points with the highest flow gradients—this apparent instability *is* large because the physical velocities (as shown in Chapters 2 and 3) *are* truly large. The benefits of excessive computing are illusory and self-defeating, but accurate fracture and shale flow solutions *can* be developed using simple analytical methods.

Fracture Flows

In many simulators, fracture flows are modeled using rows of discrete point sources. The results are crude at best: incorrect end singularities, "lumpy" flowfields, and fictitious through-flow between widely separated source points appear in the results, in many cases worsened by typically large distances between grid points. These undesirable effects are easily eliminated by using continuous line source distributions for fracture flows. The resulting formulation can be solved analytically using integral equation methods that have been available for decades. Discrete singularity methods, such as the point sources just described, were originally used in aerodynamics a century ago and have been obsolete since then.

Uniform Flux Fractures

As noted above, velocities at fracture tips are singular for constant fracture pressure boundary conditions—this is an exact consequence of the pressure model applied at the fracture plane itself (pressures are constant because equilibration is instantaneous in steady-state flows). Various workers have introduced a complementary "uniform flux" or "uniform velocity" model, that is, one where constant perpendicular velocity holds (its integral over fracture area would give total flow rate). However, this model does not apply to "clean fractures" without proppants. A "uniform flux" (that is, normal velocity) model is often used to model nonideal proppant effects. The solution to this problem is also developed in closed form in Chapter 2.

Mudcake Buildup

During oil well drilling, high-pressure "drilling mud" is used to contain the formation, safely reducing the risk of dangerous "blowouts." As fluid penetrates the formation, a filter-cake or mudcake is left at the borehole face that grows with time and continuously reduces flow. The invading and reservoir fluids possess different flow and conductive properties. Accurate electromagnetic log interpretation requires precise knowledge of front position so that rock properties in the faraway zone can be predicted. Thus, a key element in this process is the use of an accurate mudcake filtration model, because it controls the salient physical features of the flow. But simulators typically invoke a $\sqrt{t}$ law which is not universal: it applies only in linear core (as opposed to radial) flow and, then, only to single-phase flows when the formation is much more permeable than the cake. Thus, the computations are useless, for example, in slimhole applications (where cake will also not grow indefinitely with time) and in tight or low-permeability formations which exert a strong back-influence on cake evolution. To correctly model the physics, the Darcy flows within the growing cake and the invaded reservoir must be dynamically coupled and solved as an integrated system. This effort is pursued later.

Geometric Gridding

The importance of curvilinear grid systems (using "corner-point" methods) that capture geometric details in oil fields is well understood in reservoir simulation. These general mappings typically introduce nonnegligible second-derivative cross-terms in the transformed flow equations, which, unfortunately, are deliberately and completely ignored by many matrix inversion routines because they introduce numerical inefficiencies and instability. In many applications, the ideal structure of the governing coefficient matrix is destroyed by real-world and gridding constraints, but these constraints are disregarded for computational expediency. Thus, reservoir engineering departments of at least several oil companies issue warnings to users, noting that corner-point results are suspect and probably incorrect. Proper use of boundary-conforming curvilinear meshes, developed here, avoids these problems.

Averaging Methods

Simple equivalent resistance calculations in electric circuits are based on appropriate use of lumped or averaged properties. For instance, the series and parallel equivalent resistor formulas used in freshman physics assume steady, single-directional conditions, e.g., see Fig. 1.1. Similar results are desired in reservoir engineering to increase computing speed, but in several widely used simulators, averaging techniques are systematically abused. Formulas derived for linear (versus cylindrical or spherical) flow under constant density, single-phase, steady, identical-block-size assumptions are indiscriminately used to process intermediate results in transient, compressible, multiphase, variable grid runs, leading to questionable results.

Upscaling Techniques

In electric circuits, equivalent resistance depends on the arrangement of the resistors and the location of the voltage source. In Fig. 1.1, the identical resistor arrangements possess different equivalent resistances depending on the *parallel* or *series* nature of the current flow—in short, the directions and paths of the current flow(s) must be known a priori. Use of related formulas when available significantly reduces computing times and are desirable for numerical expediency.

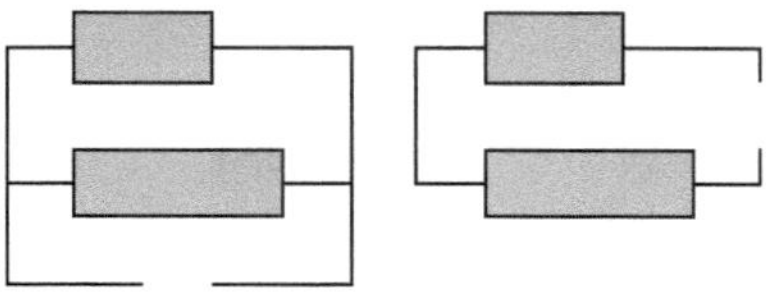

FIG. 1.1 Electrical resistance.

Similar upscaling techniques, motivated by the need to reduce grid block number, are important in practice. But the equivalent permeabilities within any reservoir will change if the reservoir is produced by different arrangements or patterns of production and/or injection wells, because the parallel and serial nature of the flow will have changed. Upscaled quantities are not properties of isolated formations but are also related to the production method. Moreover, it is well known in petrophysical logging that several heterogeneous flows that are isotropic locally may behave anisotropically "in the large" (this is clearly demonstrated in Chapter 7 in an example focusing on cross-bedded sands). However, several simulators compute fixed upscaled properties as if they were true properties and incorrectly use them in contrasting production scenarios.

Wells in Layered Media

Consider a layered reservoir produced by a general well, for simplicity, neglecting borehole friction and gravity. Production is controlled by one physical condition only: the same constant pressure acts at the sandface along the entire length of the well, whether it is vertical, horizontal, deviated, or multilateral with arbitrary drainholes. Pressure itself may be prescribed, in which case the well is pressure constrained. When the well is rate constrained, the total volume flow rate is specified but subject to the constancy of a single *unknown* pressure along the entire wellbore, whose value must be determined as part of the complete solution for pressure.

This mathematical description is exact, but such rate-constrained wells pose special challenges. While contributions to flow arise from the Darcy pressure gradient perpendicular to the sand face, the boundary value problem is not a standard Neumann formulation, where local normal (perpendicular) pressure gradients $\partial p/\partial n$ are prescribed. Instead, it must be the *integral* of all $\partial p/\partial n$ values that is specified, taken around the well, then along its entire length to include complicated tip end effects, and considering all branches if the well is multilateral, subject to the constancy of an unknown sand face pressure whose value is sought as part of the solution (obvious changes apply when friction and gravity cannot be neglected). This naturally destroys any symmetries in the governing coefficient matrix that may be needed for efficient solution—quite obviously because the driller or reservoir engineer is free to specify *any* well geometry or topology he sees fit. No wonder that rigorous solution has been avoided for decades—instead, un-validated, incorrect "permeability allocation" methods are used which prescribe local influx rates in a manner proportional to local bed "permeability-thickness" products—promising to provide approximate solutions in order to avoid the mathematical difficulties noted.

Rigorous analysis requires a special algorithm, developed in Chapter 15, since the general nonneighboring connections relating different portions of a general well destroy the symmetries required in many fast matrix inversion algorithms. Many simulators instead proportion the amount of flow that enters the well from layers according to their local permeability-thickness, kh, products. This appears to be a reasonable start, since it is obviously correct when k vanishes; however, there is no real mathematical basis for the so-called kh allocation methods. Many simulators also rule out interlayer flow at the outset, an a priori assumption that is untrue except for the most impermeable layers. For computational simplicity, these layers are also typically assumed to extend indefinitely from the well, which is rarely the case in practice.

Wellbore Models

Reservoir simulators employ large grid blocks that are thousands of feet across. Input properties are based on selected core samples whose sizes range from inches to several feet. Single computational grid blocks may contain multiple wells, with typical diameters no greater than one foot. Consequently, special "well models" are used to *mimic* real wells, augmented by "productivity indexes" that account for skin damage and perforations. The industry is still mired in rectangular grid systems that do not provide resolution anywhere, when curvilinear grids that accommodate multiple wells (that is, "airfoils" in a computational fluids context) are readily available in the aerospace industry, e.g., see Fig. 9.12 for a two-well grid. Local wellbore imperfections can be modeled with detailed local simulations. Small and large-scale flows can be fully integrated using rigorous "inner and outer matched asymptotic expansions" as discussed in the classic book by van Dyke (1964). Detailed discussion is offered at the end of our revised Chapter 15.

Formation Tester Multiphase Flow

In reservoir engineering, the effects of capillary pressure are initially unimportant if flow rates are high in a dimensionless sense. This well-known Buckley-Leverett limit does not otherwise apply. In formation evaluation, flow rates are high only initially when drilling mud invades an oil reservoir, since mudcake buildup rapidly slows the invading flow, typically within minutes. Thus, capillary pressure effects are important almost immediately for invasion modeling and must be considered in any immiscible two-phase model. In addition, while mudcake (being much less permeable than the formation) very often controls the overall filtration rate of the flow, this is not necessarily so in tight zones and certain two-phase flows with extremely low relative permeabilities. For such problems, immiscible Darcy flows in the formation and within the mudcake are dynamically coupled, and a combined boundary value problem formulation with "moving boundaries" (that is, cake-to-mud interfaces) must be solved. Capillary pressure and mudcake coupling are generally ignored because of their complexity. However, the general problem is solved in this book, whose numerical solutions are relevant to formation tester "clean-up" applications that predict pumpout times required to retrieve in situ fluid samples free of mud filtrate contamination. Formation testers, or so-called "testers," which extract reservoir fluid samples from the well, provide a wealth of new research topics, including interpretation methods for pressure transients used to predict pressure, which we discuss next.

Formation Tester Pressure Transient Interpretation

As indicated in the Preface, single probe formation testers in the 1960s provided effective permeability, or more precisely, "spherical permeability" predictions for $k_h^{2/3} k_v^{1/3}$ by evaluating steady-state pressure drop formulas using measured data. In the 1990s, service companies developed dual probe tools, and the author introduced similar steady-state equations to predict k_h and k_v individually.

These quantities are important to infill drilling planning, hydraulic fracturing, wellbore stability and other reservoir applications. As exploration targets diminished in permeability or mobility, waiting times required for steady-state equilibrium increased significantly, often exceeding hours and days. This added to drilling and rig rental costs, not to mention increasing the risk of stuck and lost tools. For single-probe tools, Halliburton in the late 1990s introduced pressure prediction techniques targeted specifically at low mobility reservoirs. Its GeoTap procedure, co-developed by the present author, at first employed empirical math models which, in conjunction with standard drawdown-buildup procedures, provided spherical permeability in just seconds. However, why the method worked was not initially clear. This author later placed the technique on sound analytical footing by solving the complete spherical flow transient formulation in closed analytical form and showing how an "exponential equation" arose as a specific limit of a more complete solution. Moreover, the single solution removed uncertainties commonly encountered in well testing, and notions like "early, intermediate and late" time, which depend on the unknown background mobility itself, no longer presented confusion to the log analyst. This early success motivated many more, which are discussed in detail in two recent formation testing books, namely, Chin (2014a,b, 2015a,b). New and important results are introduced and summaries are provided in a completely rewritten and updated Chapter 18 focused on formation testing developments.

Sweep Efficiency and Streamline Tracing

The saturation equation with capillary pressure terms removed describes the movement of single-phase fluids in heterogeneous reservoirs under ideal non-diffusive conditions. Computed results are useful in understanding basic ideas on reservoir connectivity and sweep efficiency. Within this framework, fluid never mixes: if initial portions of the fluid are arbitrarily dyed red and blue, mathematical proofs (given later) require that red fluid stays red, and blue fluid stays blue. In one commercial simulator, an entire spectrum of colors emerges that dazzles the user, presenting color results that are everything but physically meaningful. Mixing, in the absence of true diffusion, is the result of often large truncation error.

Book Objectives Recapitulated

This book addresses the inefficiencies just pointed out, bringing the power of singular integral equations, linear superposition, conformal mapping, modern curvilinear grid generation, moving boundary value problems, regular perturbation theory, advanced source and sink methods, and so on, to bear upon issues that have prohibited accurate solution: real-world problems are formulated and solved. But before studying these methods, it is important to understand the fundamental reservoir flow equations and their analogies in other branches of the physical sciences. Only by doing so can we exploit the wealth of techniques and solutions already available in the interdisciplinary literature, and then, in a manner that enhances our physical insight into the physics of petroleum reservoirs. A broad understanding of the interdisciplinary literature requires significant time and academic commitment, that is, more courses and homework than most graduate engineers can afford. It means the study of advanced mathematics, not to mention esoteric areas like aerodynamics, differential geometry and topology. Given these obstacles, it might appear that technology transfer is at best optimistic. But, it is not so: this book aims at "translating" the existing state of the art into practical terms relevant to complicated reservoir flows. We shall consider only physically significant problems, then develop the basic motivating ideas, taking care to introduce in a simple, readable, down-to-earth way *only* those mathematical notions that are absolutely essential. Wherever possible, Fortran source code is provided to guide the implementation of key algorithms, so that the models developed here provide immediate value.

DARCY'S EQUATIONS FOR FLOW IN POROUS MEDIA

Physical phenomena in science and engineering satisfy partial differential equations (PDEs), which relate changes in measurable quantities, like pressure or velocity, through partial derivatives taken in space and time. Unlike ordinary differential equations (ODEs), whose integration constants are fixed by specifying values of the function and its derivatives at one or more points, PDEs require, in addition to functional information on curved boundaries, the specification of initial conditions for equations of evolution. Boundaries, we emphasize, may be external or internal, stationary or moving. The exact manner in which auxiliary conditions apply depends on the physical nature of the problem and is reflected in the "type" classification of the PDE studied.

Differential Equations and Boundary Conditions

PDEs are classified accordingly as elliptic (e.g., $\partial^2U/\partial x^2 + \partial^2U/\partial y^2 = 0$), parabolic (e.g., $\partial^2U/\partial x^2 + \partial^2U/\partial y^2 = \partial U/\partial t$), or hyperbolic (wavelike, as in $\partial^2U/\partial x^2 + \partial^2U/\partial y^2 = \partial^2U/\partial t^2$). An equation that is parabolic, for instance, the pressure equation used in well testing, is sometimes referred to as "the heat equation" for historical reasons; as noted, well test methods were originally developed by diffusive heat transfer analogy. The reader is assumed to be familiar with, or at least cognizant of, these classifications and their auxiliary data requirements (Hildebrand, 1948; Tychonov and Samarski, 1964, 1967; Garabedian, 1964).

Elliptic equations are solved with boundary conditions related to the function itself (say, the pressure p for $\partial^2p/\partial x^2 + \partial^2p/\partial y^2 = 0$) or its normal derivative (e.g., $\partial p/\partial n$, which is proportional to the normal flow velocity) along prescribed curves. In the former case, we have a pure Dirichlet boundary value problem, while in the latter, the formulation is of the Neumann type. Mixed problems containing combined pressures and flow rates are also possible; for example, a flow-rate-constrained production well may act under the action of a nearby injector, in a reservoir partially opened to a large water aquifer modeled as a prescribed-pressure boundary. In petroleum engineering, elliptic equations describe general constant density flows and steady-state flows of compressible gases. Care must be taken to pose boundary conditions properly: an improper formulation that does *not* conserve mass *can* converge numerically and produce incorrect, misleading information.

The flow domain is singly connected when there are no "holes" anywhere, say, the Darcy flow in a heterogeneous sand without wells. Doubly and multiply connected domains contain one or more holes such as wells (e.g., a "donut" is doubly connected). Boundary conditions must be prescribed along all exterior and interior boundaries. Time may appear explicitly in elliptic problems for constant density fluids through boundary conditions, as in flows containing free surfaces, in wells with time-varying production rates, or while drilling under constant pressure with growing mudcake. The explicit presence of t does not mean that the problem is parabolic, hyperbolic, or compressible: in formation invasion, unsteadiness is associated with fronts that move much slower than the sound speed and the governing equations are often elliptic.

Parabolic equations describe transient compressible effects, including modeling well test buildup and drawdown. These also require boundary conditions as described previously; in addition, they must be solved together with initial conditions. For example, how active is the reservoir at the outset? Do transients dominate the physics? Or is it flowing at steady state? Perhaps it is static, held at constant uniform pressure? Boundary conditions may be related to skin resistance and storage in wells and flowlines. We shall formulate these auxiliary constraints generally and offer exact pressure diffusion solutions later.

Hyperbolic equations, like parabolic ones, are also equations of evolution. Although steady flows in aerodynamics, for example, can be both elliptic and hyperbolic (representing, respectively, subsonic and supersonic flow), the latter are not as often found in petroleum reservoir simulation except for certain immiscible two-phase flows dominated by inertia. They describe seismic wave propagation in the earth, but this subject, being entirely different, is not discussed in this book. These equation classifications are mathematical ones that apply to the equation only. Seismic waves and well test transients excited by periodic disturbances, such as thumpers and oscillating pistons, which are respectively hyperbolic and parabolic in the time domain, satisfy elliptic equations when the governing equations are expressed in the frequency domain.

This book, while it does approach mathematics rigorously, does not treat PDEs comprehensively. It does not attempt to catalog the broad range of solution techniques available for boundary value problem analysis. Instead, it describes reservoir flow problems in precise terms when the physics allows and offers rigorous solutions obtained from advanced analysis without introducing the ad hoc assumptions common to industry models. At the same time, we emphasize that many problems are *not* solved from scratch. We shall draw upon physical and mathematical analogies in heat transfer, electrodynamics and aerodynamics, taking advantage of existing solutions and techniques. Advanced methods are explained logically, but once discussed, the requisite solutions are summarized with detailed derivations omitted for brevity. Because numerous classes of flow problems are considered, it is impossible to follow one specific set of typographical conventions. Upper and lower case letters, and Greek and italicized letters are used to represent, at various times, different dimensional and dimensionless quantities, variables of integration, physical parameters in rectangular or mapped coordinates, and so on. However, the conventions applied within particular sections will always be consistent and clear.

Darcy's Laws

The fundamental equations of motion governing fluid flow in petroleum reservoirs are given in several books (Muskat, 1937; Collins, 1961; Aziz and Settari, 1979). We shall not rederive them but instead refer the reader to the cited publications. In this section, we shall list these equations for reference, and discuss motivating observations that should be useful to our subsequent work. These go beyond mere summary: subtleties related to the Navier-Stokes equations and important aerodynamic analogies are given.

Let $k_x(x,y,z)$, $k_y(x,y,z)$, and $k_z(x,y,z)$ denote heterogeneous anisotropic permeabilities in the x, y, and z rectangular directions, respectively. If a Newtonian fluid having constant viscosity μ flows under a superimposed pressure field $p(x,y,z,t)$, its velocities can be obtained by taking partial derivatives with respect x, y, and z:

$$u(x,y,z,t) = -(k_x/\mu)\,\partial p(x,y,z,t)/\partial x \qquad (1.1)$$

$$v(x,y,z,t) = -(k_y/\mu)\,\partial p(x,y,z,t)/\partial y \qquad (1.2)$$

$$w(x,y,z,t) = -(k_z/\mu)\,\partial p(x,y,z,t)/\partial z \qquad (1.3)$$

Here, u, v, and w are velocities in the x, y, and z directions, respectively, and Eqs. (1.1)–(1.3) are known as Darcy's equations, after the French engineer Henri Darcy, who discovered them empirically. They are Eulerian velocities at fixed points in space and not Lagrangian velocities following fluid elements. In a petroleum reservoir, the pressure and motion at each point (x,y,z) affects every other point, and vice versa. The dynamics of such flows are coupled by PDEs for $p(x,y,z,t)$, which must be solved subject to flow-rate, pressure and auxiliary constraints known as "boundary" and "initial" conditions.

The preceding represent momentum equations only and do not describe the complete physical picture since mass conservation has not yet entered. Now let $\phi(x,y,z)$ denote the formation porosity and $c(x,y,z)$ the effective compressibility of the fluid and underlying rock matrix. When the fluid is a slightly compressible *liquid*, mass conservation requires that the transient flow satisfy the classical parabolic heat equation given by

$$\begin{aligned}&\partial\{k_x(x,y,z)\,\partial p/\partial x\}/\partial x+\partial\{k_y(x,y,z)\,\partial p/\partial y\}/\partial y\\&+\partial\{k_z(x,y,z)\,\partial p/\partial z\}/\partial z=\phi\mu c\,\partial p/\partial t\end{aligned}\tag{1.4}$$

If the liquid is incompressible or if the compressible liquid has reached steady state flow conditions, the time derivative term vanishes. Then, the governing equation is elliptic, that is,

$$\begin{aligned}&\partial\{k_x(x,y,z)\,\partial p/\partial x\}/\partial x+\partial\{k_y(x,y,z)\,\partial p/\partial y\}/\partial y\\&+\partial\{k_z(x,y,z)\,\partial p/\partial z\}/\partial z=0\end{aligned}\tag{1.5}$$

Gases behave differently from liquids and must be characterized by an equation of state. Polytropic processes are studied in thermodynamics (Saad, 1966). Essentially, $pv^n=$ constant, say C, where v is the specific volume; also the index n of the process may vary from $-\infty$ to $+\infty$. For constant pressure processes, $n=0$; for isothermal processes assuming perfect gases, $n=1$. For reversible adiabatic processes, $n=Cp/Cv$, where Cp is the specific heat at constant pressure and Cv is the value obtained at constant volume. Finally, for constant volume processes, $n=\infty$.

In Muskat (1937, 1949), the gas density ρ is proportional to the mth power of pressure, that is, $\rho=\gamma_0 p^m$ so that $p=\gamma_0^{-1/m}\rho^{1/m}$. These parameters can be related to those in the preceding paragraph. The equations $pv^n=C$ and $v=1/\rho$ imply that $p=C\rho^n$. Thus, $C=\gamma_0^{-1/m}$, while $m=1/n$. Muskat's m, used throughout this book in lieu of n, describes both the properties of the gas and the thermodynamic process. The equations for a compressible *gas* are similar to those given already. In this case, Eqs. (1.4), (1.5) are replaced by

$$\begin{aligned}&\partial\{k_x(x,y,z)\,\partial p^{m+1}/\partial x\}/\partial x+\partial\{k_y(x,y,z)\partial p^{m+1}/\partial y\}/\partial y\\&+\partial\{k_z(x,y,z)\partial p^{m+1}/\partial z\}/\partial z=\phi\mu c^*\,\partial p^{m+1}/\partial t\end{aligned}\tag{1.6}$$

for transient flow and

$$\begin{aligned}&\partial\{k_x(x,y,z)\,\partial p^{m+1}/\partial x\}/\partial x+\partial\{k_y(x,y,z)\,\partial p^{m+1}/\partial y\}/\partial y\\&+\partial\{k_z(x,y,z)\,\partial p^{m+1}/\partial z\}/\partial z=0\end{aligned}\tag{1.7}$$

for steady flow. In Eqs. (1.6), (1.7), Muskat's gas exponent m satisfies

$$\begin{aligned}m&=1,\ \text{for isothermal expansion}\\&=Cv/Cp,\ \text{for adiabatic expansion}\\&=0,\ \text{for constant volume processes}\\&=\infty,\ \text{for constant pressure processes}\end{aligned}\tag{1.8}$$

The quantity

$$c^*=m/p(x,y,z,t)\tag{1.9}$$

in the transient equation denotes a compressibility-like quantity; at least for numerical purposes, the functional form in Eq. (1.6) superficially resembles a linear equation for $p^{m+1}(x,y,z,t)$, and it does allow algorithms developed for linear flows to be extended nonlinearly with minimal effort.

Of course, Eq. (1.6) is nonlinear; its solutions *cannot* be superposed, except approximately, only if small linearized disturbances to a large mean pressure are considered by linearizing $c^*(p)$ about a constant mean pressure. Writing it in a form similar to Eq. (1.4), as we have done, will of course assist with its numerical integration. We shall consider Eq. (1.6) as the most general equation for single-phase flow in this book, where $m=0$, $c^*=c$ for liquids, while nonzero m's and $c^*=m/p(x,y,z,t)$ apply to gases. Eq. (1.6) is complicated for obvious reasons, and simplifications are usually made to render the mathematics tractable. When $k_x(x,y,z)=k_y(x,y,z)=k_z(x,y,z)=k(x,y,z)$, say, the formation is said to be isotropic, with the consequence that transient flows satisfy the nonlinear parabolic equation

$$\begin{aligned}&\partial\{k(x,y,z)\,\partial p^{m+1}/\partial x\}/\partial x+\partial\{k(x,y,z)\,\partial p^{m+1}/\partial y\}/\partial y\\&+\partial\{k(x,y,z)\,\partial p^{m+1}/\partial z\}/\partial z=\phi\mu c^*\,\partial p^{m+1}/\partial t\end{aligned}\qquad(1.10)$$

If $k(x,y,z)$ is *constant*, Eq. (1.10) additionally simplifies to

$$\partial^2p^{m+1}/\partial x^2+\partial^2p^{m+1}/\partial y^2+\partial^2p^{m+1}/\partial z^2=(\phi\mu c^*/k)\,\partial p^{m+1}/\partial t\qquad(1.11)$$

which is still nonlinear. Only in the liquid $m=0$ limit does Eq. (1.11) become linear, and then, only when compressibility and porosity are constant does it become amenable to classical analysis (e.g., using Laplace and Fourier transforms, separation of variables, or superposition via Duhamel's integral).

Steady-state flows of gases and constant density flows of liquids are somewhat less complicated. In steady flow, these equations become

$$\begin{aligned}&\partial\{k(x,y,z)\,\partial p^{m+1}/\partial x\}/\partial x+\partial\{k(x,y,z)\,\partial p^{m+1}/\partial y\}/\partial y\\&+\partial\{k(x,y,z)\,\partial p^{m+1}/\partial z\}/\partial z=0\end{aligned}\qquad(1.12)$$

$$\partial^2p^{m+1}/\partial x^2+\partial^2p^{m+1}/\partial y^2+\partial^2p^{m+1}/\partial z^2=0\qquad(1.13)$$

These equations are linear in p^{m+1}. Eq. (1.13), the simplest of all the mathematical models cited, is Laplace's linear equation for the pressure function $p^{m+1}(x,y,z)$. For simplicity, auxiliary conditions on pressure can be written in terms of p^{m+1} (as opposed to p). Again, the PDEs in this section must be augmented by appropriate boundary and initial conditions as necessary.

LOGARITHMIC SOLUTIONS AND BEYOND

All of the above pressure equations are complicated, and given that boundary conditions are typically prescribed on awkward near and farfield boundaries, it is no wonder that recourse to numerical models is often made. Analytical approaches typically stop at the classical logarithmic solution for pressure, which is restricted to purely radial flows, and progress no further. However, two simple solutions, introduced here, can be leveraged to produce large classes of solutions for flows past fractures and shales. In most books, the simple radial flow model for liquids in homogeneous, isotropic media is discussed. It is based on the parabolic and elliptic equations

$$\partial^2 p/\partial r^2 + (1/r)\partial p/\partial r = (\phi\mu c/k)\partial p/\partial t, \quad \text{for transient flow} \tag{1.14a}$$

$$= 0, \quad \text{for steady flow} \tag{1.14b}$$

For these two classical equations, there is no shortage of solutions (Carslaw and Jaeger, 1946, 1959). The best known solution is the steady state logarithmic solution

$$P(r) = A + B \log r \tag{1.15}$$

for Eq. (1.14b), where A and B are constants, and the coordinate r satisfies

$$r = \sqrt{\{x^2 + y^2\}} \tag{1.16}$$

Surely, radial coordinates "r" are "natural" to flows with radial symmetry. So it is "unnatural," at least for now, to reconsider Eq. (1.15) in the form

$$p(x, y) = A + B \log \sqrt{\{x^2 + y^2\}} \tag{1.17}$$

But it is precisely this unconventional thinking that reaps the greatest benefit in dealing with more general problems as shown starting in Chapter 2. And just as the logarithm is natural to radial flow, we can and will, via Cartesian coordinates as an intermediary vehicle, extend the utility of this logarithmic solution to general fracture and nonfracture flows.

Coordinate systems, therefore, form a central underlying theme in this book. When presented correctly, they help us understand what types of elementary solutions are available for modeling, what their properties are, and what their potential uses might be. In the next section, we shall introduce a new elementary solution, an "arc tangent" or "θ" model that is complementary to the above logarithm, to Laplace's equation in polar coordinates; this solution is important in modeling general flows past impermeable shales. Many authors suggest that "Laplace's equation is Laplace's equation," point to simple analogies, and conclude brief discussions with obvious exercises in separation of variables and linear transforms. However, the connection between aerodynamics and Darcy flow—and different solutions to Laplace's equation—is subtle and deserves further elaboration. A significant part of the work developed in this book stems from aerodynamics and understanding its basics will be fruitful to the reader in future work. However, the next section may be omitted for now without loss of continuity.

FUNDAMENTAL AERODYNAMIC ANALOGIES

Aerodynamic theory and Darcy flow modeling in porous media are similar in one respect only: both derive from the Navier-Stokes equations governing viscous flows (Milne-Thomson, 1958; Schlichting, 1968; Slattery, 1981). We emphasize this because the great majority of our new solutions derive from the classical aerodynamics literature, but in a subtle manner. Very often, the superficial claim is made that, because petroleum pressure potentials satisfy $\partial^2 p/\partial x^2 + \partial^2 p/\partial y^2 = 0$, the analogy to aerodynamic flowfields, which satisfy Laplace's equation $\partial^2 \phi/\partial x^2 + \partial^2 \phi/\partial y^2 = 0$ for a similar velocity potential ϕ, can be readily drawn. This is rarely the case, and let us examine the reasons why.

Navier-Stokes Equations

There are pitfalls in the preceding reasoning: while true as far as the equation is concerned, the types of elementary solutions used in applications are different. To understand why, it is necessary to learn some aerodynamics. To be sure, the Navier-Stokes equations for Newtonian viscous flows *do* apply to both, but different limit processes are at work. For clarity, consider steady, constant density, planar, liquid flows governed by

$$\rho(u\, \partial u/\partial x + v\, \partial u/\partial y) = -\partial p/\partial x + \mu \left(\partial^2 u/\partial x^2 + \partial^2 u/\partial y^2\right) \tag{1.18}$$

$$\rho(u\, \partial v/\partial x + v\, \partial v/\partial y) = -\partial p/\partial y + \mu \left(\partial^2 v/\partial x^2 + \partial^2 v/\partial y^2\right) \tag{1.19}$$

Here, u and v are Eulerian velocities in the x and y directions; μ and ρ are constant fluid viscosity and density. These equations contain three unknowns, u, v, and the pressure p. To determine them, the mass continuity equation

$$\partial u/\partial x + \partial v/\partial y = 0 \tag{1.20}$$

is required. These equations appear in dimensional and possibly misleading form. The usual practice is to introduce nondimensional variables $p' = p/\rho U^2$, $u' = u/U$, $v' = v/U$, $x' = x/L$, and $y' = y/L$, based on a suitable set of reference parameters: a length L, a flow speed U, and a dynamic head ρU^2. This rescaling leads to the dimensionless momentum equations

$$u'\, \partial u'/\partial x' + v'\, \partial u'/\partial y' = -\partial p'/\partial x' + Re^{-1} \left(\partial^2 u'/\partial x'^2 + \partial^2 u'/\partial y'^2\right) \tag{1.21}$$

$$u'\, \partial v'/\partial x' + v'\, \partial v'/\partial y' = -\partial p'/\partial y' + Re^{-1} \left(\partial^2 v'/\partial x'^2 + \partial^2 v'/\partial y'^2\right) \tag{1.22}$$

where a single nondimensional number, the so-called Reynolds number

$$Re = \rho UL/\mu \tag{1.23}$$

measures the ratio of inertial to viscous forces (Schlichting, 1968; Slattery, 1981). We now demonstrate how Laplace's equation (we emphasize, for different quantities) arises in different Reynolds number limits.

The Darcy Flow Limit

In reservoir engineering, Eqs. (1.1), (1.2), known as Darcy's equations, apply (Muskat, 1937). Historically, they were determined empirically by the French engineer Henri Darcy, who observed that the inviscid, high Reynolds number models then in vogue did not describe hydraulics problems. Darcy's laws do not follow immediately from Eqs. (1.21), (1.22), but they can be derived through an averaging process taken over many pore spaces and, then, only in the *low Reynolds number* limit (Batchelor, 1970). If Eqs. (1.1), (1.2) are substituted in Eq. (1.20) and if constant viscosities and constant isotropic permeabilities are further assumed, Laplace's equation $\partial^2 p/\partial x^2 + \partial^2 p/\partial y^2 = 0$ for reservoir pressure p(x,y) follows. Now let us derive the Laplace's equation used in aerodynamics.

The Aerodynamic Limit

Inviscid aerodynamics, the study of nonviscous flow, is obtained by contrast in the limit of *infinite Reynolds number*. In this limit, Eqs. (1.21), (1.22) become

$$u' \partial u'/\partial x' + v' \partial u'/\partial y' = -\partial p'/\partial x' \quad (1.24)$$

$$u' \partial v'/\partial x' + v' \partial v'/\partial y' = -\partial p'/\partial y' \quad (1.25)$$

In airfoil theory, the inviscid assumption (with several exceptions in stratified and compressible flows, e.g., see Yih, 1969) requires that all fluid elements that are initially "irrotational" will remain irrotational in the absence of viscosity; that is, they do not rotate about their axes as they would on account of viscous shearing forces. This kinematic requirement is expressed by either of

$$\partial u'/\partial y' - \partial v'/\partial x' = 0 \quad (1.26a)$$

$$\partial u/\partial y - \partial v/\partial x = 0 \quad (1.26b)$$

Eqs. (1.26a), (1.26b) apply to Darcy flows, too, for example, substitution of Eqs. (1.1), (1.2) into Eq. (1.26b) shows how "0=0" holds. Now, Eqs. (1.24), (1.26a) combine to give $\partial\{p' + ½ (u'^2 + v'^2)\}/\partial x = 0$ while Eqs. (1.25), (1.26a) yield $\partial\{p' + ½ (u'^2 + v'^2)\}/\partial y = 0$. The first result implies that the quantity in curly brackets is independent of y, while the second implies that it is independent of x. Hence $p' + ½ (u'^2 + v'^2)$ must be the same *constant* throughout the entire flowfield, one that is in turn determined from known upstream conditions. Returning to dimensional variables, we have the well-known Bernoulli equation relating pressure to velocity,

$$p(x, y) + ½ \rho \left\{ u(x, y)^2 + v(x, y)^2 \right\} = \text{constant} \quad (1.27)$$

which we emphasize does *not* apply to Darcy flows (again, the latter satisfy $\partial^2 p/\partial x^2 + \partial^2 p/\partial y^2 = 0$ or the generalizations listed earlier).

In so-called "analysis" problems when the airfoil shape is given, pressure is the quantity of interest used to calculate airfoil lift or turbomachinery torque once surface velocities are known. At this point, though, the velocities u and v are still unknown. Eq. (1.26b) suggests that we can write, without loss of generality,

$$u(x, y) = \partial\phi\,(x, y)/\partial x \tag{1.28}$$

$$v(x, y) = \partial\phi\,(x, y)/\partial y \tag{1.29}$$

since the substitution of Eqs. (1.28), (1.29) in Eq. (1.26b) leads to an acceptable 0=0. When Eqs. (1.28), (1.29) are substituted in Eq. (1.20), which describes mass conservation, we obtain the Laplace equation

$$\partial^2\phi/\partial x^2 + \partial^2\phi/\partial y^2 = 0 \tag{1.30}$$

for the velocity potential $\phi(x,y)$. Eqs. (1.28), (1.29) are not unlike Eqs. (1.1), (1.2), while Eq. (1.30) resembles the Darcy equation $\partial^2 p/\partial x^2 + \partial^2 p/\partial y^2 = 0$. One might be inclined to view p(x,y) as a pressure potential and draw obvious analogies, but the aerodynamics potential possesses very different properties. Before discussing them, we cite the results in three dimensions. Simple extensions yield $u = \partial\phi/\partial x$, $v = \partial\phi/\partial y$, $w = \partial\phi/\partial z$, and $\partial^2\phi/\partial x^2 + \partial^2\phi/\partial y^2 + \partial^2\phi/\partial z^2 = 0$, and $p + ½\,\rho\,\{u^2 + v^2 + w^2\} = \text{constant}$, assuming steady, constant density, irrotational flow. This system and suitable boundary conditions are used to model the lift and "induced drag" associated with the nonviscous flow component in low-speed incompressible flows.

Validity of Laplace's Equation

Since Laplace's equation, that is, Eq. (1.13) for the Darcy pressure and Eq. (1.30) for the inviscid aerodynamic potential, arise in both problems as a result of *different* physical limits, it is of interest to ask when the approximate models apply and why. This understanding is crucial to the translation process alluded to earlier, so that "fixes" used in aerodynamics, which may be inappropriate to Darcy flows, can be removed if and when they are present. It is especially important because the analogies presupposed by nonspecialists are sometimes not analogous at all.

Fig. 1.2 shows a typical streamline in the Darcy flow beneath a dam; the sketch is based on photographs of sand model experiments, with sheet pilings at "heel alone" and "heel plus toe" (e.g., see Muskat, 1937). The complete

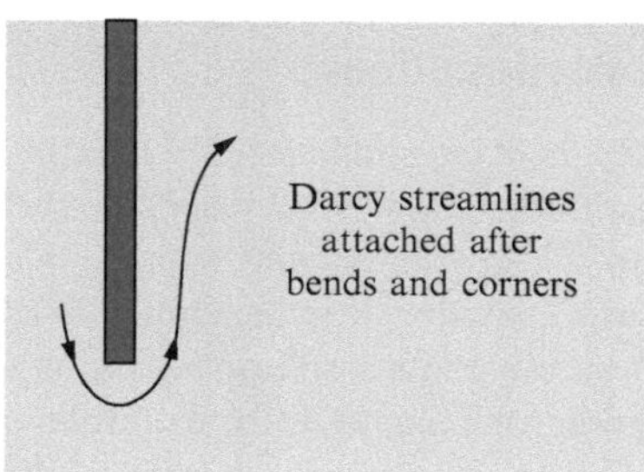

FIG. 1.2 Darcy flow streamline beneath dam (un-separated).

streamline pattern can be predicted quite well using the planar, liquid limit of Eq. (1.13), so that the solutions apply to all oncoming flow angles up to a sharp 180° (refer to Muskat's work for detailed drawings). Thus, Eq. (1.13) appears to be generally valid for all low Reynolds number flows. Now, Fig. 1.3 shows a flat plate airfoil at a not-so-small "angle of attack" or flow inclination relative to the oncoming fluid. The creation of eddies at the trailing edge, which increase in size with increasing angle, is indicated.

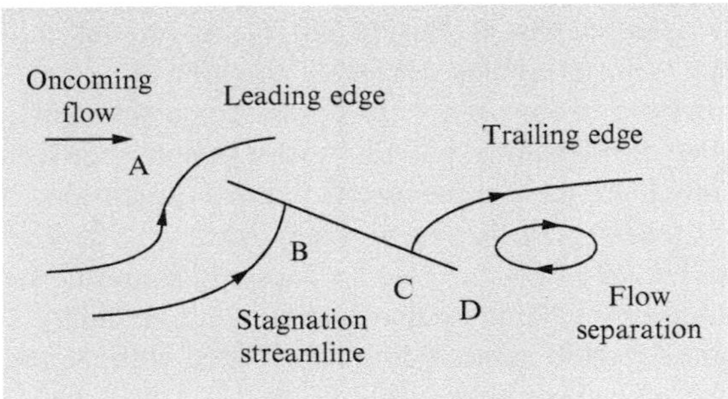

FIG. 1.3 Inviscid flow streamlines past thin airfoils (separated).

The viscous effects in Fig. 1.3 require an analysis using the full unsteady Navier-Stokes equations; they cannot be modeled using Eq. (1.30). But Eq. (1.30) is meaningful for small flow inclinations, say, less than 10 degrees. When this is the case, it admits an *infinity* of solutions, each corresponding to a different position of the aft stagnation point C. This position is fixed and the solution rendered unique by forcing C to coincide with the trailing edge location D. This so-called Kutta-Joukowski theorem allows Eq. (1.30) to *mimic* solutions of the more rigorous Navier-Stokes model. That solutions to Laplace's equation are *not* unique may not be well known to petroleum engineers, who are accustomed to dealing with log r solutions. This nonuniqueness is related to the existence of θ solutions, usually reserved for advanced math courses. These elementary solutions, discussed briefly next and in Chapter 3, are important to modeling impermeable flow barriers like shale lenses.

Different Physical Interpretations

Several important points must be emphasized when translating aerodynamics results into petroleum solutions. First, with respect to the preceding comments, the additional "circulatory" flow associated with the Kutta condition must be subtracted out before airfoil solutions can be applied to flows past impermeable shales. Second, not all aerodynamics solutions contain Kutta conditions; the results for fractures derived in Chapter 5, for example, are taken from slender body crossflow theory where circulatory solutions are not needed. Third, in aerodynamics, the airfoil surface is a streamline of the flow having a constant value of the streamfunction, supporting variable pressure; in Darcy fracture flows, the fracture surface is not a streamline, but pressure is (or may be) constant along it. On the other hand, shale surfaces do represent streamlines, although Kutta's condition does not apply. Careful attention to the physics is obviously required.

The Darcy pressure p and the aerodynamic potential ϕ appear to be similar, at least superficially, since velocities in both cases are obtained by taking derivatives; in this sense, they are, mathematically at least, potentials. But the key differences are significant: the potential ϕ is *not* a physical quantity like pressure or velocity; it is an abstract, generally multivalued entity, so defined in order to model the effects of lift. Second, pressures are obtained from Bernoulli's equation (see Eq. 1.27), which does *not* apply to Darcy flows. Other solutions used in this book contain similar pitfalls; for example, "analogous" heat transfer solutions with embedded insulators, which allow double-valued temperatures through thin surfaces, and bed interfaces in electrodynamics, which allow double-valued electric fields. The reader bent on studying the interdisciplinary literature should be aware that the translation process is not as straightforward as it may appear and that a detailed understanding of the physics is crucial. We now introduce the notion of double-valued functions in the aerodynamics context. Properties of the arc tan singularity, one that plays a role equally important to log r, are developed, which will be used extensively in Chapter 3 to model flows past shale distributions.

Meaning of Multivalued Solutions

We summarize our results thus far. First, the velocities in aerodynamics are obtained by solving Laplace's equation for the velocity potential, subject to kinematic "no flow through the surface" boundary conditions related to u, v, and $\phi(x,y)$, plus Kutta's condition at the trailing edge. Then, pressures are obtained from Eq. (1.27), where the integration constant is evaluated from known ambient conditions at infinity upstream. Aerodynamicists work with a dimensionless pressure coefficient

$$Cp = (p - p_\infty)/\left(1/2\,\rho U_\infty{}^2\right) \tag{1.31}$$

where U_∞ is the free stream speed; later, x is the coordinate parallel to the horizontal airfoil chord. The velocity potential $\phi(x,y)$ is usually expanded in a regular expansion (van Dyke, 1964) about free stream conditions, taking

$$\phi(x, y) = U_\infty x + \phi^{(0)}(x, y) + \text{higher order terms} \tag{1.32}$$

where $U_\infty x$ represents uniform flow effects. To leading order, for sufficiently thin airfoils, the disturbance potential and the pressure coefficient satisfy

$$\partial^2\phi^{(0)}/\partial x^2 + \partial^2\phi^{(0)}/\partial y^2 = 0 \tag{1.33}$$

$$Cp = (p - p_\infty)/\left(1/2\,\rho U_\infty{}^2\right) \approx -2\left\{\partial\phi^{(0)}(x,y)/\partial x\right\}/U_\infty \qquad (1.34)$$

Now, the lift, or upward force that raises airplanes off the ground, is proportional to the line integral of the pressure coefficient taken over *both* upper and lower airfoil surfaces. But $\int \partial\phi^{(0)}(x,y)/\partial x \, dx$ is just $\phi^{(0)}(x,y)$. And since the integration variable x traverses from left to right and then right to left, returning to the starting point, the integral must, one might prematurely conclude, vanish identically—implying incorrectly that lift is impossible!

One way to understand this inconsistency is to rewrite Eq. (1.27) in cylindrical coordinates, take $p + \frac{1}{2}\rho\left\{(\partial\phi/\partial r)^2 + 1/r^2\,(\partial\phi/\partial\theta)^2\right\} = \text{constant}$ and consider the lifting flow past a circle, which can be conveniently mapped into any airfoil. If ϕ is independent of the angle θ, then $\phi = \phi(r)$ only: the resulting flow symmetry implies that no resultant force acts. Thus, the key to modeling lift lies in using multivalued θ solutions to Laplace's equation. The original fallacy lies in the fact that the log solution (i.e., $\phi^{(0)}(x,y) = \log r$) as one might have for Darcy's radial flow equation, is *not* the only type of permissible solution. If we recognize that Eq. (1.33) can be equivalently expressed as

$$\partial^2\phi^{(0)}/\partial r^2 + (1/r)\,\partial\phi^{(0)}/\partial r + \left(1/r^2\right)\partial^2\phi^{(0)}/\partial\theta^2 = 0 \qquad (1.35)$$

in cylindrical coordinates, then it is obvious that another solution for the potential takes the form

$$\phi(0) = \theta = \text{arc tan}\,(y/x) \qquad (1.36)$$

This arc tan solution is multivalued: when the required line integral is taken, with x returning to its starting point, the value of the angle θ goes from 0, say, to 2π, resulting in the desired nonzero lift integral.

In summary, aerodynamicists and mathematicians employ superpositions using two types of elementary solutions, namely, *logarithms* and *arc tans*. The correct multiple of the latter function is determined by "Kutta's condition" (Milne-Thomson, 1958; Yih, 1969), simulating smooth flow from the trailing edge, as if the Navier-Stokes equations themselves had been solved. Thus, to use the results of so-called aerodynamic "analysis" models in Darcy pressure formulations, the level of "circulation" (that is, a suitable multiple of θ, which does not apply to Darcy flows) must be subtracted out.

Analogies From Inverse Formulations

In the preceding, the word "analysis" was added to emphasize aerodynamic analogies where airfoil geometry is specified and pressures are to be obtained. In addition to analysis problems, aerodynamicists work with inverse problems, where surface pressures are given and geometric shapes are desired. The governing dependent variable, the streamfunction Ψ, also satisfies Laplace's equation in the simplest limit, and as before, there are log r and arc tan solutions associated with inverse problems. To take advantage of the complete suite of formulations offered by modern aerodynamic theory, we need to understand how inverse solutions are constructed and how they can be undone for petroleum reservoir analysis.

Let us reconsider Eq. (1.30), and rewrite it in the conservation form $\partial(\partial\phi/\partial x)/\partial x + \partial(\partial\phi/\partial y)/\partial y = 0$. Then, it is clear that we can introduce a function Ψ satisfying $\partial\phi/\partial x = \partial\Psi/\partial y$ and $\partial\phi/\partial y = -\partial\Psi/\partial x$, since back substitution implies that $\partial^2\Psi/\partial x\partial y = \partial^2\Psi/\partial y\partial x$, which is always true. However, these definitions also imply that $\partial^2\Psi/\partial x^2 + \partial^2\Psi/\partial y^2 = 0$; thus, associated with every potential is a complementary streamfunction. A streamline is defined by a simple kinematic requirement: its slope is tangent to the local velocity vector, or $dy/dx = v/u$, where the right side can be written as $\partial\phi/\partial y/\partial\phi/\partial x$. This term is simply $-\partial\Psi/\partial x/\partial\Psi/\partial y$. Thus the "total derivative" satisfies $d\Psi = \partial\Psi/\partial x\, dx + \partial\Psi/\partial y\, dy = 0$, so that Ψ is constant along a streamline. Now, since the airfoil surface itself is a streamline, it must be a contour along which Ψ does not vary, and one might conclude that Ψ must be a log r type of single-valued function.

In many cases, this interpretation is correct, but numerous modeling advantages are possible by taking Ψ as a double-valued arc tan function. If this is pursued, the upper and lower surfaces of any calculated geometry must represent different streamlines: this is possible only if the trailing edge is opened and mass issues from it into the downstream flow. Of course, airfoils do not open and "spill" flow in practice, not intentionally anyway! But the mass that streams from this fictitious source does model the thick viscous wakes left behind by thick airfoils, or by thin airfoils at high angles of attack, which more often than not act as physical extensions of the airfoil (Chin, 1979, 1981, 1984; Chin and Rizzetta, 1979). Again, inverse aerodynamic solutions also provide a source for new petroleum results, but their nonuniquenesses must be carefully interpreted and exploited if physically meaningful reservoir engineering results are to be obtained.

PROBLEMS AND EXERCISES

1. Consider Laplace's equation $\partial^2F/\partial x^2 + \partial^2F/\partial y^2 = 0$ for F(x,y). Verify by differentiation that the following functions are solutions: (1) $\log \sqrt{(x^2+y^2)}$, (2) arc tan (y/x), (3) $\log \sqrt{\left\{(x-\xi)^2 + (y-\eta)^2\right\}}$, and (4) $\arctan\{(y-\eta)/(x-\xi)\}$, where ξ and η are constants. Explain the physical significance of these Greek constants. What are their mathematical properties? Why are these functions referred to as "elementary singularities"? Why are arbitrary linear combinations (or superpositions) of these also solutions? How can these be used in reservoir modeling applications?

2. Now consider the double integral $\iint f(\xi, \eta) \log \sqrt{\{(x\text{-}\xi)^2 + (y\text{-}\eta)^2\}}\, d\xi d\eta$, where constant limits are assumed. Verify that this integral is also a solution to Laplace's equation $\partial^2 F/\partial x^2 + \partial^2 F/\partial y^2 = 0$. How does this integral behave in the farfield? What are the implications of this result? How might this solution be used in modeling 3D flows containing localized fractures? Repeat this exercise with $\iint g(\xi, \eta) \arctan \{(y\text{-}\eta)/ (x\text{-}\xi)\}\, d\xi d\eta$.
3. Constant density, one-dimensional liquid flows in linear cores satisfy $d^2p/dx^2 = 0$. Now, consider the in-series flow through two linear cores having unequal lengths and different permeabilities. The left side is held at pressure p_{left}, while the far right is held at pressure p_{right}. The differential equations break down at the core interface because the first derivative of pressure is discontinuous. What matching conditions are required at this boundary to define a unique solution? What do these conditions mean physically? Write the solution for the complete boundary value problem, and explain its significance in reservoir engineering and its relevance to effective properties analysis. Define the effective permeability of the system, derive its value in analytical form, and state very clearly all of the assumptions used in the derivation. Would this effective value be useful in transient problems? Immiscible flow problems? Gas flows?
4. Rederive the solution in Exercise 3 to allow general continuous distributions of the permeability k(x) and show how the solutions correctly reduce to the constant permeability result.
5. Suppose, in the preceding problems, that the linear cores are characterized by two different porosities as well. What issues are important to effective properties when modeling production rate is important? When modeling tracer arrival times? What is the difference between the Eulerian velocity at a point versus the Lagrangian velocity following a particle?
6. Reconsider Exercises 3, 4, and 5 in the context of cylindrical radial flows satisfying $d^2p/dr^2 + 1/r\, dp/dr = 0$. Do the solutions and effective properties defined using them apply in transient compressible flows? Gas flows?

Chapter 2

Fracture Flow Analysis

Several new mathematical techniques are introduced to model flows from single straight-line fractures, using different boundary conditions. These methods are useful in the simulation of flow from massive hydraulic fractures and from horizontal wells that penetrate natural fracture systems. Fracture flows are often treated numerically. Here, we shall use rigorous singular integral equation methods to obtain closed-form solutions, first for liquids, and next for gases. Regular perturbation techniques are then introduced and used to extend thin fracture solutions to handle the effects of thickness. Elementary solutions are obtained, explained in clear terms, and generalized step-by-step. Flows from more complicated fracture systems will be discussed in Chapter 5.

EXAMPLE 2.1. SINGLE STRAIGHT-LINE FRACTURE IN AN ISOTROPIC CIRCULAR RESERVOIR CONTAINING INCOMPRESSIBLE FLUID

Given the geometric simplicity, it is surprising that closed-form solutions have not been available earlier, in view of the problem's practical significance. Here, a completely arbitrary specification of pressure along the fracture length is allowed. Thus, the results can be used to model less-than-ideal proppant-induced effects that may arise in stimulation by massive hydraulic fracturing or, perhaps, to model non-Darcy flows along mineralized fractures penetrated by horizontal and deviated wells. The analytical solutions importantly reveal velocity singularities at the fracture tips. These edge singularities, well known in aerodynamics and elasticity, reveal the complex nature of the flow and suggest caution with numerical schemes.

Formulation

We consider the flow of an incompressible liquid from (or into) a single straight-line fracture of length 2c, centered in a circular reservoir of radius $R \gg c$, as shown in Fig. 2.1A. The pressure P(X,Y) assumed along the fracture $-c \leq X \leq +c$, $Y = 0$ is the variable function $P_{ref}\ p_f(X/c)$, where P_{ref} is a reference level and p_f is dimensionless. The pressure at the farfield boundary

Quantitative Methods in Reservoir Engineering. http://dx.doi.org/10.1016/B978-0-12-810518-4.00002-5

is a constant P_R. For a uniform isotropic medium, P(X,Y) satisfies the Dirichlet boundary value problem

$$\partial^2 P/\partial X^2 + \partial^2 P/\partial Y^2 = 0 \tag{2.1}$$

$$P(X, 0) = P_{ref} p_f(X/c), \ -c \le X \le +c \tag{2.2}$$

$$P(X, Y) = P_R, \ X^2 + Y^2 = R^2 \tag{2.3}$$

for Laplace's equation. Since p_f (X/c) is variable, conformal mapping methods (see Chapter 5) are not convenient and alternative solutions are sought. In the stimulation literature, p_f(X/c) is often assumed to satisfy Darcy's law; in one dimension, for example, Darcy's governing equation for pressure reduces to $d^2p_f/dX^2 = 0$. This may well be true for certain flows. But to allow more general possibilities, we shall not restrict the dependence of p_f on X/c.

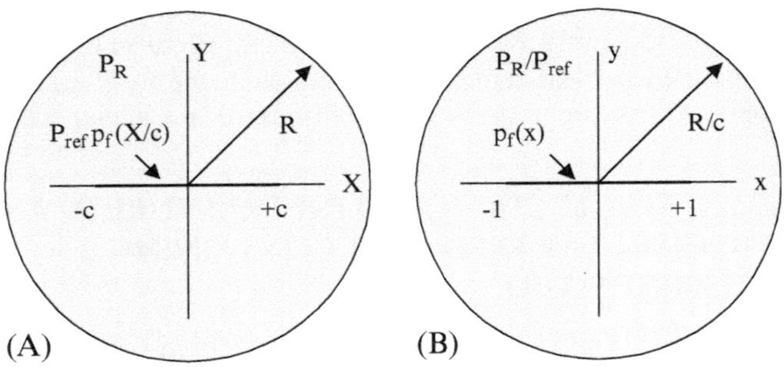

FIG. 2.1 Centered straight fracture formulations.

The use of infinitesimally thin "slits" for fracture flow modeling, due originally to Muskat (1937, 1949), describes the physics accurately. The aerodynamic analogy is thin airfoil theory, where boundary conditions are assigned along straight lines (Ashley and Landahl, 1965; Bisplinghoff et al., 1955). Slit models are appropriate to fracture analysis since typical fractures *are* thin. The model is actually less useful in airfoil theory, since local corrections must be used to account for blunt leading edges.

We shall consider the effects of nonzero thickness in Example 2.3, where high-order corrections for thickness (or more precisely, open fracture effects) are developed. The basic slit solution is considered here. For convenience, introduce the nondimensional variables x, y, and p, defined by

$$x = X/c \tag{2.4}$$

$$y = Y/c \tag{2.5}$$

$$P(X, Y) = P_{ref}\, p(x, y) \tag{2.6}$$

The dimensionless pressure p(x,y) is defined between the circle $x^2+y^2=(R/c)^2$ and the fracture $-1 \le x \le +1$, $y=0$ in Fig. 2.1B. Eqs. (2.1)–(2.3) become

$$\partial^2 p/\partial x^2 + \partial^2 p/\partial y^2 = 0 \tag{2.7}$$

$$p(x,0) = p_f(x),\ -1 \le x \le +1 \tag{2.8}$$

$$p(x,y) = P_R/P_{ref},\ x^2+y^2 = (R/c)^2 \tag{2.9}$$

Recourse to numerical methods is understandable, given the variability in $p_f(x)$ and the finite size of the reservoir. Fortunately, this is not necessary.

Singular Integral Equation Analysis

A closed form analytical solution can be obtained. We use results from thin airfoil theory (e.g., Ashley and Landahl, 1965) and singular integral equations (Muskhelishvili, 1953; Gakhov, 1966; Carrier et al., 1966). Now the standard log r source solution, centered at the origin $r=\sqrt{(x^2+y^2)}=0$ solves Eq. (2.7). Thus, $\log\sqrt{\{(x-\xi)^2+y^2\}}$ centered at $x=\xi$, $y=0$ also satisfies Laplace's equation, where ξ represents a shift in the choice of origin.

Now, ξ can be viewed as a general point source position over which the effects of numerous sources can be summed. But rather than examining multiple discrete point sources, we examine *continuous* line source distributions placed along the fracture to represent it. This is clearly the situation physically; many simulators model fractures using point sources, which allow fictitious flow between points. We consider the superposition

$$p(x,y) = \int_{-1}^{+1} f(\xi)\, \log\sqrt{\{(x-\xi)^2+y^2\}}\,d\xi + H \tag{2.10}$$

This integral satisfies Eq. (2.7) for pressure, since the governing equation is linear. Physically, Eq. (2.10) represents the pressure for a continuously distributed line source, where H is a crucial integration constant.

The problem reduces to finding solutions for H and the distributed source strength f(x) that yield pressures satisfying Eqs. (2.8), (2.9). Let us first combine Eqs. (2.8), (2.10) to obtain

$$\int_{-1}^{+1} f(\xi)\, \log|x-\xi|\, d\xi = p_f(x) - H \tag{2.11}$$

and, for the moment, assume that H is known. Then, when the fracture pressure is specified, Eq. (2.11) provides the desired singular integral equation for the unknown source strength f(x). It is important to note that, in the case of *discrete* wells with $p = A \log r + B$, it is not possible to evaluate the pressure at $r = 0$ without having it become singular. For *continuous* distributions of sources, however, the result is finite because of interference effects.

Unlike a partial differential equation such as Eq. (2.7), an integral equation involves unknown functions that fall within the integrand (Garabedian, 1964; Hildebrand, 1965; Mikhlin, 1965). And since the kernel (the function multiplying the unknown) in this case contains a logarithmic singularity at $x = \xi$, Eq. (2.11) is known as a singular integral equation.

The formulations posed by Eqs. (2.7), (2.11) are equivalent since superpositions of elementary solutions, which involve no additional assumptions, are used without loss of generality. Integral equation methods were not available to Muskat and his contemporaries; they were developed in aerodynamics and elasticity after the publication of his classic textbooks. The advantage in using Eq. (2.11) is a practical one: its completely analytical solution is available and is known as Carleman's formula (Carrier et al., 1966; Estrada and Kanwal, 1987). In fact, for the general equation

$$\int_{-1}^{+1} f(\xi) \log|x-\xi| \, d\xi = g(x) \tag{2.12}$$

Carleman, using complex variables methods, derives the exact solution

$$\begin{aligned} f(x) = [&\text{PV} \int_{-1}^{+1} \{g'(\xi)/(\xi-x)\} \sqrt{(1-\xi^2)} \, d\xi \\ &-(1/\log_e 2) \int_{-1}^{+1} g(\xi)/\sqrt{(1-\xi^2)} \, d\xi]/\{\pi^2 \sqrt{(1-x^2)}\} \end{aligned} \tag{2.13}$$

where $g'(\xi)$ is the derivative of $g(\xi)$ with respect to ξ.

The first integral in Eq. (2.13) is known as an improper or singular integral because it is infinite or singular at $\xi = x$. Its interpretation as a Cauchy principal value (hence, the PV prefix) is given in calculus textbooks (Thomas, 1960); we shall illustrate with examples later. The e subscript is retained in $\log_e 2$ for emphasis; all logarithms in this book are natural ones. The second integral, somewhat complicated, is a standard look-up integral.

Specializing Carleman's Results to Fracture Flow

In our particular problem, we identify, using Eqs. (2.11), (2.12) that

$$g(x) = p_f(x) - H \tag{2.14}$$

and immediately obtain the source strength

$$f(x) = [PV \int_{-1}^{+1} \{p_f'(\xi)/(\xi\text{-}x)\} \sqrt{(1\text{-}\xi^2)}\, d\xi$$
$$- (1/\log_e 2) \int_{-1}^{+1} p_f(\xi)/\sqrt{(1\text{-}\xi^2)}\, d\xi]/\{\pi^2 \sqrt{(1\text{-}x^2)}\} \tag{2.15}$$
$$+H/\{\pi \log_e 2 \sqrt{(1\text{-}x^2)}\}$$

This still contains the unknown constant H. To determine it, return to the complete expression for pressure in Eq. (2.10) and apply the farfield boundary condition of Eq. (2.9). We therefore have

$$p(x, y) = \int_{-1}^{+1} f(\xi) \log \sqrt{\{(x\text{-}\xi)^2 + y^2\}}\, d\xi + H$$
$$= \int_{-1}^{+1} f(\xi) \log \sqrt{\{x^2 + y^2\text{-}2x\xi + \xi^2\}}\, d\xi + H \tag{2.16a}$$

or, at distances large compared to the fracture length 2c, noting that $|\xi| \leq 1$ is bounded, the asymptotic result

$$p(x, y) = \int_{-1}^{+1} f(\xi) \log \sqrt{\{x^2 + y^2\}}\, d\xi + H$$
$$= \int_{-1}^{+1} f(\xi)\, d\xi \log r + H \tag{2.16b}$$

where $\sqrt{\{x^2 + y^2\}}$ is the conventional radial coordinate.

Eq. (2.16b) states that the flow behaves radially in the farfield as if it had been produced by a point source with the cumulative strength

$$\int_{-1}^{+1} f(\xi)\, d\xi \tag{2.16c}$$

of our distributed line source. If this collapses to a discrete point, $f(\xi)$ is proportional to the Dirac delta function $\delta(\xi)$. Since the integral

$$\int_{-1}^{+1} \delta(\xi)\, d\xi \tag{2.16d}$$

is exactly 1 (e.g., see Garabedian, 1964 or Lighthill, 1958), the integral in Eq. (2.16b) is a constant and, in particular, the one familiar from radial flow theory. This limit reproduces classical single well results. Now let us combine Eqs. (2.9), (2.16b). This leads to

$$P_R/P_{ref} = H + \{\log R/c\} \int_{-1}^{+1} f(x)\, dx \tag{2.17}$$

where we have changed the integration variable from ξ to x. Next substitute the expression for f(x) in Eq. (2.15) into Eq. (2.17). The result is

$$\begin{aligned} P_R/P_{ref} &= H + \{\log R/c\} \int f(x)\, dx \\ &= H + \{\log R/c\} \left[\int PV \int \{p_f'(\xi)/(\xi - x)\} \surd(1-\xi^2)\, d\xi / \{\pi^2 \surd(1-x^2)\}\, dx \right. \\ &\quad - \int (1/\log_e 2) \int p_f(\xi)/\surd(1-\xi^2)\, d\xi / \{\pi^2 \surd(1-x^2)\}\, dx \\ &\quad \left. + \{H/(\pi \log_e 2)\} \int dx/\surd(1-x^2) \right] \end{aligned} \tag{2.18}$$

where the limits (-1, +1) are omitted for clarity. Each of the double integrals in Eq. (2.18) represents a constant. To simplify our notation, we introduce

$$I_1 = \int PV \int \{p_f'(\xi)/(\xi - x)\} \surd(1-\xi^2) d\xi / \{\pi^2 \surd(1-x^2)\} dx \tag{2.19}$$

$$I_2 = \int (1/\log_e 2) \int p_f(\xi)/\surd(1-\xi^2)\, d\xi / \{\pi^2 \surd(1-x^2)\}\, dx \tag{2.20}$$

and evaluate the integral involving H to obtain

$$P_R/P_{ref} \;=\; H + (\log R/c)\, \{ I_1 - I_2 + H/\log_e 2 \} \tag{2.21}$$

Hence, it follows that

$$H = \{P_R/P_{ref} - (I_1 - I_2) \log R/c\}/\{1 + (\log R/c)/\log_e 2\} \tag{2.22}$$

The source strength f(x) and the constant H are now completely fixed. Note that H depends on *all* of the flow parameters, including the dimensionless ratios P_R/P_{ref} and R/c. Also note from Eq. (2.11) that the f(x) *cannot* be determined without H; that is, f(x) depends on the complete geometry of the reservoir and the pressure levels at its boundaries. The role of the constant H in Eq. (2.10), as we shall show, is no insignificant matter.

Physical Meaning of f(x)

We digress to consider several properties of f(x). An understanding of f(x) and its relationship to local velocity helps to improve numerical formulations for more complicated fracture geometries and assists in posing and solving fracture problems governed by other boundary conditions. Let us return to the expression for pressure in Eq. (2.10) and differentiate it with respect to the vertical coordinate y normal to the fracture.

$$\partial p(x,y)/\partial y = \partial/\partial y \left\{ \int_{-1}^{+1} f(\xi) \log \sqrt{\{(x-\xi)^2 + y^2\}}\, d\xi + H \right\} \tag{2.23}$$

$$= y \int_{-1}^{+1} f(\xi)/\{(x-\xi)^2 + y^2\}\, d\xi \tag{2.24}$$

Following the limit process in Yih (1969), introduce the change of coordinates

$$\eta = (\xi - x)/y \tag{2.25}$$

so that

$$\partial p(x,y)/\partial y = \int_{\eta-}^{\eta+} f(\xi)/(1+\eta^2)\, d\eta \tag{2.26}$$

Now for small positive y's, we find that on using Eq. (2.25) (written in the form $x = \xi - \eta y$), the vertical derivative satisfies

$$\partial p(x,0+)/\partial y = \int_{-\infty}^{+\infty} f(\xi)/(1+\eta^2) d\eta = \pi f(x) \tag{2.27}$$

Similarly, for small negative y's, we obtain

$$\partial p(x,0-)/\partial y = -\pi f(x) \tag{2.28}$$

Hence,

$$\partial p(x,0+)/\partial y - \partial p(x,0-)/\partial y = 2\pi f(x) \tag{2.29}$$

Eqs. (2.27), (2.28) are also easily combined to show that

$$\partial p(x,0+)/\partial y = - \partial p(x,0-)/\partial y \tag{2.30}$$

that is, the vertical Darcy velocities on either side of the slit are antisymmetric. This antisymmetry is a consequence of the physics: the velocities are equal and opposite, and the streamline pattern is therefore symmetric about the x axis. Eq. (2.30) shows how, when a distribution of logarithmic singularities is assumed as in Eq. (2.10), the normal derivative of the function "jumps" or is discontinuous through the slit. The pressure itself does not jump, because setting $y = 0$ in Eq. (2.10) provides a single-valued $p(x,0)$ on the x axis.

Since the derivative $\partial p(x,0+)/\partial y$, via Darcy's law, is proportional to the vertical velocity into the fracture, it goes without saying that the source strength f(x) is directly proportional to the y-component of velocity at $y=0$. This allows us to write a simple formula connecting the dimensionless source strength f(x) to the dimensional total volume flow rate Q issuing from (or into) the fracture. Let us now introduce the isotropic formation permeability k, the fluid viscosity μ, and the depth into the page D. Then, it is clear that

$$Q = D \int_{X=-c}^{X=+c} 2\,(-k/\mu)\,\partial P(X, Y=0+)/\partial Y\, dX \tag{2.31}$$

where the factor of two accounts for both sides of the fracture, the integration is performed over the entire length 2c, and $(-k/\mu)\ \partial P(X,Y=0+)/\partial Y$ is the Darcy velocity normal to the fracture. Using Eqs. (2.4)–(2.6), it follows that $\partial P/\partial Y = (P_{ref}/c)\ \partial p/\partial y$, and hence,

$$\begin{aligned} Q &= D \int_{x=-1}^{x=+1} 2\,(-k/\mu)\,(P_{ref}/c)\partial p/\partial y(x,0+)\,c\,dx \\ &= -P_{ref}(Dk/\mu) \int_{x=-1}^{x=+1} 2\,\partial p/\partial y\,(x,0+)dx = -2P_{ref}(Dk\pi/\mu) \int_{x=-1}^{x=+1} f(x)\,dx \end{aligned} \tag{2.32}$$

where we have employed Eq. (2.27).

An alternative problem might call for a prescribed total volume flow rate Q subject to constant pressure along the fracture. In this case, a series of problems would be initially solved to produce a parametric $Q=Q(P_f)$ relationship for later interpolation. Eq. (2.32) relates the dimensional volumetric flow rate to the dimensionless integral of source strength f(x) over fracture length, where f(x) is known from Eq. (2.13). Also, at the fracture, the relationship between the dimensional vertical Darcy velocity $V(X,Y=0+)$ and the source strength is

$$\begin{aligned} V(X, Y=0+) &= (-k/\mu)\ \partial P(X, Y=0+)/\partial Y \\ &= (-k/\mu)(P_{ref}/c)\,\partial p/\partial y(x,0+) \\ &= (-\pi k/\mu)(P_{ref}/c)\,f(x) \end{aligned} \tag{2.33}$$

where we have used Eq. (2.27).

Remark on Muskat's Solution

We have given the closed form solution, but several subtleties deserve further discussion. The first concerns volumetric calculations for Q using Eq. (2.32). We observe that the fracture half-length c and the farfield reservoir pressure P_R do not explicitly appear in that formula, but their effects do appear *implicitly* through our solution for the constant of integration H. That is, all nearfield and farfield effects are properly accounted for by using H and f(x) as determined, respectively, by Eqs. (2.22), (2.15).

Suppose that we had *not* allowed for the existence of H in Eq. (2.10). Then it is clear from Eq. (2.11) that the source strength f(x) would depend on the fracture pressure $p_f(x)$ only; that is, not on, say, c, since c does not appear at all in Eq. (2.11). It would follow, using Eq. (2.32), that the total flow rate Q can be obtained independently of the value of c. This incorrect result would be a consequence of not accounting for H. The loss of the required second degree of freedom in what is essentially a two-point boundary value problem, in fact, precludes farfield boundary conditions from being satisfied.

For example, one would *not* solve radial flow with $d^2P/dr^2 + 1/r\ dP/dr = 0$, $P(r_{well}) = P_{well}$ and $P(r_{farfield}) = P_{farfield}$ using $P(r) = A \log r$ only. The correct choice is $P(r) = A \log r + H$ where A *and* H are determined from coupled equations developed from both boundary conditions. Thus, it is not so much that one integration constant handles the nearfield, with the other handling farfield conditions; both are simultaneously required to handle nearfield and farfield interactions. The solution of Muskat (1937) is not valid in the foregoing sense. His pressure formula, which satisfied constant fracture pressures, was obtained as the real part of a complex analytic function of z (refer to Discussion 4.6 in Chapter 4). But farfield conditions were ignored, leading to an incomplete pressure solution and a volumetric flow rate that was independent of fracture length. This incorrect rate was then renormalized in an ad hoc manner to show some dependence on pressure drop. A pressure solution for large R must be handled as the limit of a two point boundary value problem; however, Muskat does correctly model clusters of discrete wells in a circular field.

Velocity Singularities at Fracture Tips

The result in Eq. (2.33) demonstrates that the vertical velocity at the fracture is proportional to the local source strength. This is important numerically. Since Eq. (2.15) yields a square root singularity in f(x) at both ends of the fracture, it follows that *both tip velocities are locally infinite*. The integral in Eq. (2.32) nevertheless exists because square root singularities are integrable (Thomas, 1960). Localized infinities do not necessarily cause integrals to diverge, e.g., the area under the curve $y = 1/\sqrt{x}$ (which "blows up" at $x = 0$) from $x = 0$ to $x = 1$ is *exactly* 2.

To be accurate, any numerical method must be capable of predicting this infinite pressure gradient (e.g., the computational solutions in Chapter 7); but this is impossible, since any such prediction is bound to cause numerical instability. Thus, one is torn between finely gridding both fracture tips to model reality, or doing the opposite to preserve stability. This dilemma means that an upper bound to accuracy limits the usefulness of computational schemes. This is also the case with transient flow simulations from fractures, important in well testing, where analogous edge singularities exist. These have not been discussed in the literature, let alone properly modeled. This edge singularity is known to aerodynamicists. In wing design, it does not really exist because leading edges are rounded (the radius of curvature is small but not negligible compared with the chord c). Local edge corrections, obtained using matched asymptotic expansions (van Dyke, 1964), are introduced to correct the fictitious singularity. But in reservoir flow, line fractures *do* exist and the singularity is real.

Streamline Orientation

We conclude with remarks on local streamline orientation. Our asymptotic log r expansion far away from the fracture shows that the flow behaves radially when $R \gg c$ (see Eq. 2.16b). In Chapter 4, we prove that in a uniform isotropic medium, streamlines and lines of constant pressure are orthogonal. Thus, when the fracture pressure is a prescribed constant, the flow at the fracture is everywhere perpendicular to it, except at the tips. However, when $p_f(x)$ varies with x, this orthogonality is lost; any scheme assuming local orthogonality is incorrect. Of course, the foregoing solutions can be used to predict local flow inclinations; such results apply qualitatively even in the presence of distant wells, fractures, and boundaries, effects considered later.

EXAMPLE 2.2. LINE FRACTURE IN AN ANISOTROPIC RESERVOIR WITH INCOMPRESSIBLE LIQUIDS AND COMPRESSIBLE GASES

Having demonstrated the power and elegance behind the use of distributed line sources and the use of singular integral equations, we now consider a slightly more complicated example involving incompressible liquids and compressible gases in anisotropic reservoirs under steady-state flow conditions. This second example will illustrate the flexibility of the thin airfoil technique. But it will also reveal the weaknesses inherent in analytical approaches and why a well formulated numerical method is necessary.

General Formulation

Consider the flow from (or into) a straight-line fracture of length 2c, centered in a circular reservoir of radius $R \gg c$, as in Fig. 2.1. The pressure P(X,Y) assumed on the fracture $-c \le X \le +c$, $Y = 0$ is the function $P_{ref}\, p_f(X/c)$, where P_{ref} is constant and p_f is dimensionless. The pressure at an assumed elliptical farfield boundary (see Eq. 2.36) is a constant P_R. For an anisotropic medium, P(X,Y) satisfies the Dirichlet problem

$$\partial\left(k_x\, \partial P^{m+1}/\partial X\right)/\partial X + \partial\left(k_y\, \partial P^{m+1}/\partial Y\right)/\partial Y = 0 \qquad (2.34)$$

$$P(X,0) = P_{ref}\, p_f(X/c),\ -c \le X \le +c \tag{2.35}$$

$$P(X,Y) = P_R,\ X^2 + (k_x/k_y)Y^2 = R^2 \tag{2.36}$$

where the permeabilities k_x and k_y parallel and perpendicular to the fracture depend on (X,Y). The ellipse is a requirement of the approach; there are more general techniques—which do *not* bear this requirement—that we shall discuss later. Again, $m = 0$ for liquids, while m takes on nonzero values for real gases.

To make progress, we consider constant permeabilities (a "log r" function is not available for heterogeneous reservoirs). This leads to the simpler equation

$$k_x\, \partial^2 P^{m+1}/\partial X^2 + k_y\, \partial^2 P^{m+1}/\partial Y^2 = 0 \tag{2.37}$$

For convenience, we introduce the nondimensional variables x, y, and p defined by

$$X = X/c \tag{2.38}$$

$$Y = \sqrt{(k_x/k_y)}\, Y/c \tag{2.39}$$

$$P(X,Y) = P_{ref}\, p(x,y) \tag{2.40}$$

Then, p(x,y) resides in the domain $x^2+y^2 < (R/c)^2$ external to the assumed fracture $-1 \le x \le +1$, $y = 0$ in Fig. 2.2. Eqs. (2.34)–(2.36) become

$$\partial^2 p^{m+1}/\partial x^2 + \partial^2 p^{m+1}/\partial y^2 = 0 \tag{2.41}$$

$$p(x,0) = p_f(x),\ -1 \le x \le +1 \tag{2.42}$$

$$p(x,y) = P_R/P_{ref},\ x^2+y^2 = (R/c)^2 \tag{2.43}$$

Eqs. (2.41)–(2.43) resemble Eqs. (2.7)–(2.9), except that m is nonzero. The scaling in Eqs. (2.38), (2.39) is chosen so that x remains between -1 and +1. This requirement is imposed so that existing results can be used without renormalization. This problem, while it involves powers of pressure, is *not* nonlinear; our pressure boundary conditions are easily rewritten in powers of pressure, resulting in a linear Dirichlet problem for p^{m+1}.

Singular Integral Equation Analysis

A closed-form analytical solution can be obtained. Observe that the standard source solution log r, centered at the origin $r = \sqrt{(x^2+y^2)} = 0$, solves Eq. (2.41) for p^{m+1}. Similarly, the expression $\log \sqrt{\{(x-\xi)^2+y^2\}}$ centered at

$x = \xi$, $y = 0$ satisfies Laplace's equation, with ξ being a constant. Now, ξ can be viewed as a generalized point source position, and it is of interest to consider line distributions of sources. In particular, we examine the superposition integral

$$p^{m+1}(x, y) = \int_{-1}^{+1} f(\xi) \log \sqrt{\left\{(x-\xi)^2 + y^2\right\}}\, d\xi + H \tag{2.44}$$

which also satisfies Eq. (2.41), since it is linear. Physically, Eq. (2.44) is viewed as a pressure equation corresponding to a continuously distributed line source, where f and H are not to be confused with their counterparts in Example 2.1. We emphasize that their physical dimensions are also different.

The problem reduces to finding the values of H and f(x) that yield pressure solutions satisfying Eqs. (2.42), (2.43). Following Example 2.1, let us first combine Eqs. (2.42), (2.44) to obtain

$$\int_{-1}^{+1} f(\xi) \log|x-\xi|\, d\xi = p_f^{\,m+1}(x) - H \tag{2.45}$$

and for now, assume that H is known. Thus, when the fracture pressure is specified, Eq. (2.45) provides an integral equation for the strength f(x). As before, we apply Carleman's formulas in Eqs. (2.12), (2.13), and set

$$g(x) = p_f^{\,m+1}(x) - H \tag{2.46}$$

to obtain

$$\begin{aligned} f(x) = {} & [PV \int_{-1}^{+1} \{p_f^{\,m+1,\prime}(\xi)/(\xi-x)\} \sqrt{(1-\xi^2)}\, d\xi \\ & - (1/\log_e 2) \int_{-1}^{+1} p_f^{\,m+1}(\xi)/\sqrt{(1-\xi^2)}\, d\xi] / \left\{\pi^2 \sqrt{(1-x^2)}\right\} \\ & + H/\left\{\pi \log_e 2 \sqrt{(1-x^2)}\right\} \end{aligned} \tag{2.47}$$

where $p_f^{\,m+1}(\xi)$ here denotes the *function* obtained by raising the fracture pressure $p_f(\xi)$ to the $(m+1)$th power. The (primed) expression $p_f^{\,m+1,\prime}(\xi)$ is the *first derivative of the function* defined, equal to $(m+1)p_f^{\,m}(\xi)\, dp_f/d\xi$.

Eq. (2.47) still contains the unknown constant H. To determine H, we return to Eq. (2.44) and evaluate it for distances that are large compared with the fracture length. Thus, the exact expression for pressure

$$p^{m+1}(x,y) = \int_{-1}^{+1} f(\xi)\log\sqrt{\{(x-\xi)^2 + y^2\}}\,d\xi + H$$
$$= \int_{-1}^{+1} f(\xi)\log\sqrt{\{x^2 + y^2 - 2x\xi + \xi^2\}}\,d\xi + H \tag{2.48a}$$

can be approximated by

$$p^{m+1}(x,y) \approx \int_{-1}^{+1} f(\xi)\log\sqrt{\{x^2+y^2\}}\,d\xi + H \approx \int_{-1}^{+1} f(\xi)\,d\xi\,\log r + H \tag{2.48b}$$

far from the fracture since $|\xi| \le 1$ is bounded. Observe that the dimensionless $r = \sqrt{\{x^2+y^2\}}$ describes the elliptical locus of points in Eq. (2.36). Now let us combine Eqs. (2.43), (2.48b). This leads to

$$(P_R/P_{ref})^{m+1} = H + \{\log R/c\}\int_{-1}^{+1} f(x)\,dx \tag{2.49}$$

where we have changed the integration variable from ξ to x. Next substitute f(x) from Eq. (2.47) into the integral of Eq. (2.49). The result is

$$(P_R/P_{ref})^{m+1} = H + \{\log R/c\}\int f(x)\,dx$$
$$= H + \{\log R/c\}[\int PV\int \{p_f^{\,m+1,}(\xi)/(\xi-x)\}\sqrt{(1-\xi^2)}d\xi/\{\pi^2\sqrt{(1-x^2)}\}\,dx$$
$$-\int (1/\log_e 2)\int p_f^{\,m+1}(\xi)/\sqrt{(1-\xi^2)}\,d\xi/\{\pi^2\sqrt{(1-x^2)}\}\,dx$$
$$+\{H/(\pi\log_e 2)\}\int dx/\sqrt{(1-x^2)}] \tag{2.50}$$

where the integration limits (-1, +1) are omitted for clarity. Each of the double integrals represents constants. To simplify the notation, we introduce

$$I_3 = \int PV\int \{p_f^{\,m+1,}(\xi)/(\xi-x)\}\sqrt{(1-\xi^2)}\,d\xi\,/\{\pi^2\sqrt{(1-x^2)}\}\,dx \tag{2.51}$$

$$I_4 = \int (1/\log_e 2)\int p_f^{\,m+1}(\xi)/\sqrt{(1-\xi^2)}\,d\xi\,/\{\pi^2\sqrt{(1-x^2)}\}\,dx \tag{2.52}$$

and evaluate the integral involving H to obtain

$$(P_R/P_{ref})^{m+1} = H + (\log R/c)\,\{I_3 - I_4 + H/\log_e 2\} \tag{2.53}$$

Hence, it follows that

$$H = \{(P_R/P_{ref})^{m+1} - (I_3 - I_4)\log R/c\}/\{1 + (\log R/c)/\log_e 2\} \qquad (2.54)$$

Now f(x) and H are fixed. Observe that H depends on *all* the flow parameters, including the dimensionless ratios P_R/P_{ref} and R/c. Also, from Eq. (2.45), f(x) cannot be determined without H: f(x) depends on the geometry of the reservoir and the pressure at its boundaries. The role of the integration constant H in Eq. (2.44) is as significant here as in Example 2.1.

The Physical Meaning of f(x)

We digress to consider several general properties of the pseudo-source strength f(x). An understanding of f(x) and its relationship to local velocity will help to improve numerical formulations for more complicated fracture geometries, as well as assist in posing and solving fracture flows problems governed by alternative boundary conditions. We return to the general expression for pressure in Eq. (2.44) and differentiate it with respect to the vertical coordinate y normal to the fracture. This gives

$$\begin{aligned}\partial p^{m+1}(x,y)/\partial y &= \partial/\partial y\{\int_{-1}^{+1} f(\xi)\log\surd\{(x-\xi)^2+y^2\}\,d\xi + H\} \\ &= y\int_{-1}^{+1} f(\xi)/\{(x-\xi)^2+y^2\}\,d\xi\end{aligned} \qquad (2.55)$$

Following the limiting process in Example 2.1, we again introduce

$$\eta = (\xi - x)/y \qquad (2.56)$$

$$\partial p^{m+1}(x,y)/\partial y = \int_{\eta-}^{\eta+} f(\xi)/(1+\eta^2)\,d\eta \qquad (2.57)$$

For small positive y's, we find on using Eq. (2.56) (in the form $x = \xi - \eta y$) that the vertical derivative satisfies

$$\partial p^{m+1}(x,0+)/\partial y = \int_{-\infty}^{+\infty} f(\xi)/(1+\eta^2)\,d\eta = \pi f(x) \qquad (2.58)$$

Similarly, for small negative y's, we obtain

$$\partial p^{m+1}(x,0-)/\partial y = -\pi f(x) \qquad (2.59)$$

Hence,

$$\partial p^{m+1}(x,0+)/\partial y - \partial p^{m+1}(x,0-)/\partial y = 2\pi f(x) \qquad (2.60)$$

If we now eliminate f(x) between Eqs. (2.58), (2.59), we have

$$\partial p^{m+1}(x,0+)/\partial y = -\partial p^{m+1}(x,0-)/\partial y \quad (2.61)$$

Carrying out the differentiation and canceling like powers of pressure leaves

$$\partial p(x,0+)/\partial y = -\partial p(x,0-)/\partial y \quad (2.62)$$

The normal derivatives of pressure, as in Example 2.1, are antisymmetric, meaning that the Darcy velocities perpendicular to the fracture are equal and opposite. This antisymmetry is physically the result of having streamlines symmetric about the x axis. Again the complete velocity vector is in general not perpendicular to the slit, if the prescribed fracture pressure is variable; this is due to the existence of a flow component along the slit.

Only in the case of liquids where $m = 0$ is the strength f(x) exactly proportional to the vertical Darcy velocity at the slit (see Example 2.1). In general, the differentiation suggested in Eq. (2.58) leads to

$$\partial p(x,0+)/\partial y = \pi f(x)/\{(m+1)\, p_f^{\,m}(x)\} \quad (2.63)$$

since $p(x,0) = p_f(x)$ from Eq. (2.42). The proportionality between $\partial p(x,0+)/\partial y$ and f(x) depends on the local fracture pressure; this generalizes the result obtained in Eq. (2.27).

We now relate the dimensionless strength f(x) to the dimensional volume flow rate Q issuing from (or into) the fracture. First, we introduce the permeability k_y, the viscosity μ, and the depth into the page D. Even though the streamlines adjacent to the fracture are not in general perpendicular to it, it is still the vertical velocity component that contributes to Q. Thus, it is clear that

$$Q = D \int_{X=-c}^{X=+c} 2\,(-k_y/\mu)\, \partial P(X, Y = 0+)/\partial Y\, dX \quad (2.64)$$

where the "2" accounts for both sides of the fracture, the integration is performed over 2c, and $(-k_y/\mu)\ \partial P(X,Y=0+)/\partial Y$ is the Darcy velocity normal to the fracture. Using Eqs. (2.38)–(2.40), $\partial P/\partial Y = (P_{ref}/c)\ \sqrt{(k_x/k_y)}\ \partial p/\partial y$, so

$$\begin{aligned} Q &= D \int_{x=-1}^{x=+1} 2\,(-k_y/\mu)(P_{ref}/c)\,\sqrt{(k_x/k_y)}\;\partial p/\partial y\,(x,0+)\,c\,dx \\ &= -2P_{ref}\,(Dk_y/\mu)\,\sqrt{(k_x/k_y)} \int_{x=-1}^{x=+1} \partial p/\partial y\,(x,0+)\,dx \\ &= -2P_{ref}\left(D\sqrt{(k_x k_y)}\,/\mu\right) \int_{x=-1}^{x=+1} \partial p/\partial y\,(x,0+)\,dx \\ &= -2P_{ref}\,D\sqrt{(k_x k_y)}\;\pi/\{(m+1)\mu\} \int_{x=-1}^{x=+1} f(x)/p_f^{\,m}(x)\,dx \end{aligned} \quad (2.65)$$

where we used Eq. (2.63). When $m=0$ and $k_x=k_y=k$, we recover Eq. (2.32) for constant density liquids in isotropic, homogeneous media. In Eq. (2.65), the $p_f^m(x)$ term can be moved across the integral if it is constant; if so, the boundary value problem prescribing Q subject to constant fracture pressure, as in Example 2.1, can be easily solved. Eq. (2.65) relates the dimensional volumetric flow rate to the dimensionless integral of f(x) over fracture length. Also, at the fracture, the relationship between the dimensional vertical Darcy velocity $V(X,Y=0+)$ and source strength, using Eq. (2.63), is

$$\begin{aligned} V(X,Y=0+) &= (-k_y/\mu)\,\partial P(X,Y=0+)/\partial Y \\ &= (-k_y/\mu)\,(P_{ref}/c)\sqrt{(k_x/k_y)}\,\partial p/\partial y(x,y=0+) \quad (2.66) \\ &= -\{P_{ref}\sqrt{(k_x k_y)}/(\mu c)\}\,\pi f(x)/\{(m+1)\,p_f{}^m(x)\} \end{aligned}$$

Velocity Singularities at Fracture Tips

Eq. (2.66), which extends Eq. (2.33), demonstrates that the vertical velocity at the fracture is again proportional to the local source strength. When the fracture pressure is a prescribed and bounded analytic function, Eq. (2.47) implies a square root singularity in velocity at both ends of the fracture; the expression $\sqrt{(1-x^2)}$ appears in the denominator, leading to *infinite tip velocities*. The integral in Eq. (2.65) nevertheless exists, of course, because weak singularities are integrable; local infinities do not necessarily cause integrals to diverge. Again, such singularities give reason for concern in numerical models. To be accurate, one must be able to predict this infinite pressure gradient, but this is impossible since such predictions will cause numerical instability. As in Example 2.1, one is torn between finely gridding both fracture tips to model reality, or doing the opposite to preserve stability. This dilemma means that an upper bound to accuracy limits the usefulness of numerical schemes. This is particularly the case with transient simulations of compressible flows from fractures, important to well test interpretation, where analogous edge singularities do exist and have not been properly modeled or discussed in the literature.

EXAMPLE 2.3. EFFECT OF NONZERO FRACTURE THICKNESS

We shall discuss methods for handling real fractures with general nonzero thickness and curved shapes later. For now, we demonstrate how fractures with thickness distributions that are symmetric about $y=0$ can be treated by a simple extension of the previous formalism. Consider essentially the same boundary value problem as in Example 2.2, with zero or nonzero m, either isotropic or anisotropic media, within the framework discussed there. However, assume that the pressure is applied along *the edges* of an open fracture with symmetric halves characterized by a dimensionless thickness function $\varepsilon\sqrt{(k_y/k_x)}\ T(X/c)$ where ε is a positive dimensionless number (see Eq. 2.68):

$$k_x\, \partial^2 P^{m+1}/\partial X^2 + k_y\, \partial^2 P^{m+1}/\partial Y^2 = 0 \tag{2.67}$$

$$P(X,0) = P_{ref} p_f(X/c),\ Y = \pm\, \varepsilon\sqrt{(k_y/k_x)}\, cT(X/c),\ -c \le X \le +c \tag{2.68}$$

$$P(X,Y) = P_R,\ X^2 + (k_x/k_y)\, Y^2 = R^2 \tag{2.69}$$

Eq. (2.68) is the only change to the problem in Example 2.2. For convenience, introduce the nondimensional variables x, y, and p, defined by Eqs. (2.38)–(2.40), with the result that

$$\partial^2 p^{m+1}/\partial x^2 + \partial^2 p^{m+1}/\partial y^2 = 0 \tag{2.70}$$

$$p^{m+1}(x,y) = p_f^{\ m+1}(x),\ y = \pm\, \varepsilon T(x),\ -1 \le x \le +1 \tag{2.71}$$

$$p^{m+1}(x,y) = (P_R/P_{ref})^{m+1},\ x^2 + y^2 = (R/c)^2 \tag{2.72}$$

Now assume that the fracture length greatly exceeds its thickness, that is, $\varepsilon \ll 1$. This is the thin airfoil limit in classical aerodynamics, or the thin fracture limit for fracture modeling. Then, the results of Example 2.2 apply to leading order. Our main task is to develop a mathematical formalism that captures this leading order description when ε vanishes but provides systematic, straightforward corrections when it is nonvanishing but small.

For this reason, we introduced ε explicitly in Eq. (2.71). Then the complete pressure boundary value problem can be represented as a power series in ε, with component problems defined by collecting coefficients of like powers of ε (this is plausible mathematically, but convergence to the exact solution must be rigorously established). Our approach is taken from perturbation theory, in particular, the use of "regular expansions," e.g., van Dyke (1964) and Ashley and Landahl (1965). Rather than considering expansions for p(x,y), it is simpler to consider p^{m+1} taken as a whole. Let us introduce the superscripted series shown in Eq. (2.73), where indexes greater than 0 are high-order corrections to the zeroth-order solution $p^{m+1(0)}(x,y)$ of Example 2.2, that is,

$$p^{m+1}(x,y) = p^{m+1(0)}(x,y) + \varepsilon p^{m+1(1)}(x,y) + \varepsilon^2 p^{m+1(2)}(x,y) + \ldots \tag{2.73}$$

Each of the superscripted functions, owing to the linearity of Eq. (2.70) in p^{m+1}, satisfies Laplace's equation. To each problem, we assign boundary conditions. We consider the fracture nearfield first. We expand the full p^{m+1} (without superscripts) near the fracture in a Taylor series in y about $y=0$,

$$p^{m+1}(x, \pm\, \varepsilon T(x)) = p^{m+1}(x,0) \pm \varepsilon T(x)\, \partial p^{m+1}(x, \pm 0)/\partial y + O(\varepsilon^2) \tag{2.74}$$

where the $O(\varepsilon^2)$ symbolically indicates high-order terms not considered here. Then, substitute the series in Eq. (2.73), retaining only the leading terms:

$$p^{m+1}(x, \pm \varepsilon T(x)) = p^{m+1(0)}(x,0) + \varepsilon p^{m+1(1)}(x,0) \pm \varepsilon T(x)\, \partial p^{m+1(0)}(x, \pm 0)/\partial y \tag{2.75}$$

If we compare Eq. (2.75) to the boundary condition $p^{m+1}(x,y) = p_f^{m+1}(x)$ in Eq. (2.71), we have on equating like powers of ε, the identities

$$p^{m+1(0)}(x,0) = p_f{}^{m+1}(x) \tag{2.76}$$

$$p^{m+1(1)}(x,0) = -\{\pm \varepsilon T(x)\}\, \partial p^{m+1(0)}(x, \pm 0)/\partial y \tag{2.77}$$

Eq. (2.76) corresponds to Eq. (2.42) for the zeroth-order problem. Eq. (2.77) may be confusing, but it is readily explained. Eq. (2.61) for the zeroth-order problem shows that the normal derivatives are antisymmetric; the ± in Eq. (2.77) accordingly renders the left side of Eq. (2.77) single-valued, as required, so that only one value of the pressure $p^{m+1(1)}$ can be assigned to any particular value of x.

The farfield boundary condition is similarly treated. The zeroth-order problem supports the entire pressure level P_R/P_{ref}, leaving "0" to the higher order corrections. When the coefficients of like powers in ε are collected, the following sequence of boundary value problems is identified. The first, that is,

$$\partial^2 p^{m+1(0)}/\partial x^2 + \partial^2 p^{m+1(0)}/\partial y^2 = 0 \tag{2.78}$$

$$p^{(0)}(x,y) = p_f(x),\ y = 0,\ -1 \le x \le +1 \tag{2.79}$$

$$p^{(0)}(x,y) = (P_R/P_{ref}),\ x^2 + y^2 = (R/c)^2 \tag{2.80}$$

is the problem treated in Example 2.2, while the second takes the form

$$\partial^2 p^{m+1(1)}/\partial x^2 + \partial^2 p^{m+1(1)}/\partial y^2 = 0 \tag{2.81}$$

$$p^{m+1(1)}(x,0) = -\{\pm \varepsilon T(x)\}\, \partial p^{m+1(0)}(x, \pm 0)/\partial y,\ -1 \le x \le +1 \tag{2.82}$$

$$p^{m+1(1)}(x,y) = 0,\ x^2 + y^2 = (R/c)^2 \tag{2.83}$$

This problem is similar to the first, and the same solution method applies.

Practical Algebraic Issues

The formulas here and in Example 2.2 are intimidating; e.g., source strength requires the evaluation of difficult integrals, as in Eqs. (2.15), (2.47). This is not to say that simple results are not easily generated; they have, for instance in van Dyke (1956), and specific solutions will be given later for several fracture pressures. But we emphasize that symbolic manipulation methods based on recursive algebras have become available on different computing platforms, e.g., *MathCad*, *Maple*, and *Mathematica*. They can be used to develop formulas analytically and then to evaluate and plot results. The primary benefit of analytical results is exactness and understanding. No numerical solution will ever uncover a singularity and its algebraic structure as beautifully as Eq. (2.47) does, but then, not all practical problems are amenable to elegant solution techniques. To obtain closed form results, limiting assumptions were invoked along the way; for instance, (i) circular or elliptical reservoir boundaries with $R \gg c$, (ii) uniform permeabilities, and (iii) the failure to handle mixed flow rate and pressure farfield boundary conditions. One cannot de-emphasize the importance of computational methods, but analytical solutions help us properly formulate numerical problems and assist in calibrating the computations.

EXAMPLE 2.4. FLOW RATE BOUNDARY CONDITIONS

In many production applications, the flow from fractures having prescribed pressure distributions is realistic; for example, the constant pressure along an infinitely conductive fracture is identical to that of the wellbore that drains it. This motivates the formulation in Example 2.2, in which source strength and flow rate are sought when pressures are prescribed. For other problems, the local flow rate normal to the fracture may be specified, as is often the case in hydraulic fracturing. This assumes that some control over the form of the normal velocity is available. Then, the mathematical problem determines pressure along the fracture and throughout the physical plane; this problem, too, is amenable to direct solution. For simplicity, return to the flow configuration in Example 2.1. The dimensionless pressure for the incompressible *liquid* under consideration again satisfies Eq. (2.7), that is,

$$\partial^2 p/\partial x^2 + \partial^2 p/\partial y^2 = 0 \tag{2.84}$$

As before, we assume a distributed line source having a pressure of the form

$$p(x, y) = \int_{-1}^{+1} f(\xi) \log \sqrt{\left\{(x-\xi)^2 + y^2\right\}}\, d\xi + G \tag{2.85}$$

as in Eq. (2.10), where G is a constant. Of course, Eq. (2.8) no longer applies; in fact, we must now determine p(x,y) and $p_f(x)$. To do this, recall from Eq. (2.27) that the normal derivative satisfies $\partial p(x,0+)/\partial y = \pi\, f(x)$, that is,

$$f(x) = (1/\pi)\, \partial p(x,0+)/\partial y \tag{2.86}$$

where the right side is known from the prescribed Darcy velocity (e.g., see Eq. 2.33). Thus, the solution to our problem takes the general form

$$p(x, y) = (1/\pi) \int_{-1}^{+1} \partial p(\xi,0+)/\partial y \log \sqrt{\left\{(x-\xi)^2 + y^2\right\}}\, d\xi + G \tag{2.87}$$

Following Eqs. (2.16a), (2.16b), we evaluate Eq. (2.87) for large distances away from the fracture, yielding

$$p(x, y) = (1/\pi) \int_{-1}^{+1} \partial p(\xi,0+)/\partial y \; d\xi \; \log r + G \qquad (2.88)$$

where $r = \sqrt{\{x^2 + y^2\}}$, and G is still undetermined. Applying the farfield boundary condition in Eq. (2.9), we have

$$P_R/P_{ref} = (1/\pi) \log (R/c) \int_{-1}^{+1} \partial p(\xi,0+)/\partial y \; d\xi + G, \qquad (2.89)$$

or

$$G = P_R/P_{ref} - (1/\pi) \log (R/c) \int_{-1}^{+1} \partial p(\xi,0+)/\partial y \; d\xi \qquad (2.90)$$

Since the integral in Eq. (2.90) is known from inputs, the flow rate problem is solved. The net result on combining Eqs. (2.87), (2.90) is

$$\begin{aligned} p(x, y) = (1/\pi) \int_{-1}^{+1} \partial p(\xi,0+)/\partial y \; \log \sqrt{\left\{(x-\xi)^2 + y^2\right\}} \; d\xi \\ + P_R/P_{ref} - (1/\pi) \log (R/c) \int_{-1}^{+1} \partial p(\xi,0+)/\partial y \; d\xi \end{aligned} \qquad (2.91)$$

EXAMPLE 2.5. UNIFORM VERTICAL VELOCITY ALONG THE FRACTURE

Example 2.4 solved the velocity flux problem for a general liquid inflow distribution $\partial p(x,0+)/\partial y$. In the form given, the integral on the first line of Eq. (2.91) describes the spatial variation of pressure, whereas the remainder of Eq. (2.91) supplies an overall pressure level that accounts for near and farfield interactions. How does fracture pressure behave with x for a simple flow rate distribution?

Evaluation of Singular Integrals

In Examples 2.1 and 2.2, we introduced singular or improper integrals whose integrands contain infinities. Most calculus books offer satisfactory discussions, e.g., refer to Thomas (1960). In this example, we evaluate typical integrals by assuming uniform velocity flux at the fracture, taking $\partial p(x,0+)/\partial y$ constant with x, setting

$$\partial p(x,0+)/\partial y = \beta \qquad (2.92)$$

From Eq. (2.91), the solution to this simplified problem is

$$p(x, y) = (\beta/\pi) \int_{-1}^{+1} \log \sqrt{\{(x-\xi)^2 + y^2\}}\, d\xi + P_R/P_{ref} - (2\beta/\pi) \log (R/c) \quad (2.93)$$

Along y = 0, the fracture pressure therefore satisfies

$$p(x, 0) = (\beta/\pi) \int_{-1}^{+1} \log |x-\xi|\, d\xi + P_R/P_{ref} - (2\beta/\pi) \log (R/c) \quad (2.94)$$

Then, it is clear from Eq. (2.94) that the singular integral of interest is

$$I(x) = \int_{-1}^{+1} \log |\xi-x|\, d\xi \quad (2.95)$$

Following Thomas (1960), I(x) is evaluated by first considering

$$I(x; \varepsilon) = \int_{\xi=-1}^{\xi=x-\varepsilon} \log (x-\xi)\, d\xi + \int_{\xi=x+\varepsilon}^{\xi=+1} \log (\xi-x)\, d\xi \quad (2.96)$$

and then taking the limit as ε vanishes (see Fig. 2.2). The logarithmic functions in Eq. (2.96) are available in standard tables.

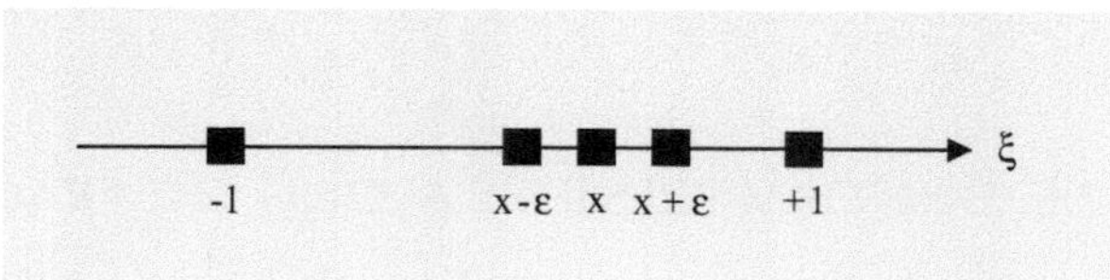

FIG. 2.2 Cauchy principal value limit process.

The subsequent limit process, which requires a single application of L'Hospital's Rule to evaluate the expression ε log ε, leads to

$$I(x) = -2 + \log\{(1+x)^{1+x}(1-x)^{1-x}\} \quad (2.97)$$

It is seen that Eq. (2.97) possesses the correct symmetries about x = 0; for example, replacing +x with -x leaves the result unchanged. Also, I(+1) = I(-1) = 2(-1 + log 2), which is finite. Combination of Eqs. (2.94), (2.96) yields

$$p(x, 0) = p_f(x) = (\beta/\pi) \left[-2 + \log\left\{(1+x)^{1+x}(1-x)^{1-x}\right\}\right] + P_R/P_{ref} - (2\beta/\pi) \log(R/c) \quad (2.98)$$

The function I(x) is tabulated in Fig. 2.3, where the even symmetry about x = 0 is apparent. For uniform velocity boundary conditions, pressure is therefore "well behaved" in that there are no singularities.

x	I (x)
- 0.99	- 0.62
- 0.95	- 0.85
- 0.90	- 1.01
- 0.80	- 1.26
- 0.70	- 1.46
- 0.60	- 1.61
- 0.50	- 1.74
- 0.40	- 1.84
- 0.30	- 1.91
- 0.20	- 1.96
- 0.10	- 1.99
- 0.00	- 2.00
.	.
.	.
+ 0.90	- 1.01
+ 0.95	- 0.85
+ 0.99	- 0.62

FIG. 2.3 The pressure function I(x).

We have spoken of constant pressure and uniform velocity fracture models. Where are these used? When the produced fluid in a fracture is "clean," its pressure equalizes almost instantaneously and the former applies. On the other hand, the latter describes injection problems in which a filter cake or extremely low permeability material is left at the sandface, which in turn controls local volume influx rate. The subject of formation invasion will be studied starting with Chapter 16.

EXAMPLE 2.6. UNIFORM PRESSURE ALONG THE FRACTURE

Here, we give a complementary solution to Example 2.5 and consider the flow of a constant density liquid into a fracture held at a constant pressure p_f. Thus, the derivative $p_f'(\xi)$ vanishes identically, and Eq. (2.15) reduces to

$$f(x) = [PV \int_{-1}^{+1} \{pf'(\xi)/(\xi - x)\} \sqrt{(1-\xi^2)}\, d\xi$$

$$- (1/\log_e 2) \int_{-1}^{+1} p_f(\xi)/\sqrt{(1-\xi^2)}\, d\xi] / \left\{\pi^2 \sqrt{(1-x^2)}\right\} + H/\{\pi \log_e 2 \sqrt{(1-x^2)}\}$$

$$= -(p_f/\log_e 2) \left\{ \int_{-1}^{+1} d\xi/\sqrt{(1-\xi^2)} \right\} / \left\{\pi^2 \sqrt{(1-x^2)}\right\} + H/\left\{\pi \log_e 2 \sqrt{(1-x^2)}\right\}$$

$$= (H - p_f)/\left\{\pi \log_e 2 \sqrt{(1-x^2)}\right\} \quad (2.99)$$

Note that the integral of f(x) over (-1, +1) equals $(H - p_f)/\log_e 2$, a result consistent with Eq. (2.113) when $C_0 = p_f$ and $C_2 = C_4 = 0$. Next, from Eqs. (2.19), (2.20), respectively, we have

$$I_1 = \int PV \int \{p_f'(\xi)/(\xi - x)\} \sqrt{(1-\xi^2)} d\xi / \{\pi^2 \sqrt{(1-x^2)}\} dx = 0 \qquad (2.100)$$

$$I_2 = \int (1/\log_e 2) \int p_f(\xi)/\sqrt{(1-\xi^2)}\, d\xi / \{\pi^2 \sqrt{(1-x^2)}\} dx = p_f/\log_e 2 \qquad (2.101)$$

Hence, from Eq. (2.22),

$$\begin{aligned} H &= \{P_R/P_{ref} - (I_1 - I_2) \log R/c\} / \{1 + (\log R/c)/\log_e 2\} \\ &= \{P_R/P_{ref} + (p_f/\log_e 2) \log R/c\} / \{1 + (\log R/c)/\log_e 2\} \end{aligned} \qquad (2.102)$$

The formula for the total volume flow rate Q is given in Eq. (2.114), where $C_0 = p_f$ and $C_2 = C_4 = 0$. Since H is now determined, the source strength in Eq. (2.99) is known. Eq. (2.27) shows that the Darcy velocity normal to the fracture is directly proportional to f(x) at the fracture; it is, as is evident from Eq. (2.99), singular at the tips $x = \pm 1$, although the flow rate integral over (-1, 1) itself exists. The complete pressure solution is obtained by substituting H and f(x) into Eq. (2.10), and that, in turn, in Eq. (2.6). The resulting pressure formula is evaluated numerically. Note that an expression for Q can be derived by combining Eqs. (2.65), (2.99); a simple equation relating p_f to Q follows, which allows problems that specify Q (subject to uniform fracture pressure) to be conveniently solved.

EXAMPLE 2.7. MORE GENERAL FRACTURE PRESSURE DISTRIBUTIONS

It is possible to obtain closed-form results for complicated fracture pressures $p_f(x)$ for the problems in Examples 2.1 and 2.2. This involves the tedious evaluation of Cauchy principal value (PV) integrals, as those who have attempted the operation in Eq. (2.96) know. Fortunately, this is not required, because the most useful singular integrals have already been identified, evaluated and cataloged in the aerospace industry. These are available in NASA publications; e.g., van Dyke (1956) lists over thirty commonly used integrals, while Gradshteyn and Ryzhik (1965) provide still more. From Eqs. (2.19), (2.20), (2.51), (2.52) of Examples 2.1 and 2.2, the integral pairs

$$I_5 = \int p_f^{m+1}(\xi) d\xi / \sqrt{(1-\xi^2)} \qquad (2.103)$$

$$I_6 = PV \int p_f^{m+1\prime}(\xi) \sqrt{(1-\xi^2)} d\xi / (\xi - x) \qquad (2.104)$$

are always encountered in pressure-pressure problems. Here, the limits of integration (-1, +1) are omitted for clarity. Simple solutions are obtained when $p_f^{m+1}(\xi)$ is taken in the polynomial form

$$p_f^{m+1}(\xi) = C_0\xi^0 + C_1\xi^1 + C_2\xi^2 + C_3\xi^3 + C_4\xi^4 \tag{2.105}$$

which provides for simple non-Darcy effects along fractures. It is a straightforward matter to show, using standard integral tables, that

$$I_5 = \pi(C_0 + 1/2C_2 + 3/8C_4) \tag{2.106}$$

The principal value integral in Eq. (2.104) requires more labor. However, we can easily verify that

$$PV\int \xi^0 \sqrt{(1-\xi^2)}\,d\xi/(x-\xi) = \pi x \tag{2.107}$$

$$PV\int \xi^1 \sqrt{(1-\xi^2)}\,d\xi/(x-\xi) = \pi\left(x^2-1/2\right) \tag{2.108}$$

$$PV\int \xi^2 \sqrt{(1-\xi^2)}\,d\xi/(x-\xi) = \pi x\left(x^2-1/2\right) \tag{2.109}$$

$$PV\int \xi^3 \sqrt{(1-\xi^2)}\,d\xi/(x-\xi) = \pi\left(x^4-{\textstyle\frac{1}{2}}x^2-1/8\right) \tag{2.110}$$

Consequently,

$$\begin{aligned}-I_6/\pi = &- (C_2 + {\textstyle\frac{1}{2}}C_4) + (C_1 - 3/2\,C_3)x \\ &+ 2(C_2 - C_4)x^2 + 3C_3\,x^3 + 4C_4\,x^4\end{aligned} \tag{2.111}$$

These results assist in evaluating the more complicated integrals. Let us consider, for instance, the expression for the total volume flow rate Q in the case of constant density liquids in anisotropic media. From Eq. (2.47), we find

$$\begin{aligned}f(x) = &\left[PV\int_{-1}^{+1} \{p_f^{m+1}(\xi)/(\xi-x)\}\sqrt{(1-\xi^2)}\,d\xi\right.\\ &\left.-(1/\log_e 2)\int_{-1}^{+1} p_f^{m+1}(\xi)/\sqrt{(1-\xi^2)}\,d\xi\right]/\left\{\pi^2\sqrt{(1-x^2)}\right\}\\ &+ H/\left\{\pi\log_e 2\sqrt{(1-x^2)}\right\}\\ = &\left\{-1/\left(\pi\sqrt{(1-x^2)}\right)\right\}[\{(C_0 - H) + {\textstyle\frac{1}{2}}C_2 + 3/8\,C_4\}\\ &/\log 2 - (C_2 + {\textstyle\frac{1}{2}}C_4) + (C_1 - 3/2\,C_3)x + 2(C_2 - C_4)x^2 + 3C_3x^3 + 4C_4x^4]\end{aligned} \tag{2.112}$$

which can be integrated to yield

$$\int_{-1}^{+1} f(x)\,dx = -\{(C_0 - H) + ½ C_2 + 3/8\, C_4\}/\log 2 \tag{2.113}$$

Then, using Eq. (2.65) with m = 0,

$$\begin{aligned} Q &= -2P_{ref} D\sqrt{(k_x k_y)}\,\pi/\{(m+1)\mu)\} \int_{x=-1}^{x=+1} f(x)/p_f{}^m(x)\,dx, \\ Q &= 2P_{ref}\, D\sqrt{(k_x k_y)}\,\pi\{(C_0 - H) + ½ C_2 + 3/8\, C_4\}/\{\mu \log 2)\} \end{aligned} \tag{2.114}$$

This provides the required formula. We emphasize that Eq. (2.112), which assumes $pf(\xi) = C_0\xi^0 + C_1\xi^1 + C_2\xi^2 + C_3\xi^3 + C_4\xi^4$, clearly displays the square root velocity singularity discussed earlier.

EXAMPLE 2.8. VELOCITY CONDITIONS FOR GAS FLOWS

Here we extend the discussion of Example 2.5 to gases with nonzero values of m. We start with Eqs. (2.44), (2.63), that is,

$$p^{m+1}(x, y) = \int_{-1}^{+1} f(\xi)\ \log \sqrt{\{(x-\xi)^2 + y^2\}}\, d\xi + H \tag{2.44}$$

and

$$\partial p(x,0+)/\partial y = \pi f(x)/\{(m+1)p_f{}^m(x)\} \tag{2.63}$$

Substitution of Eq. (2.63) in (2.44), and evaluation along y = 0, leads to

$$p_f{}^{m+1}(x) = \{(m+1)/\pi\} \int_{-1}^{+1} p_f{}^m(\xi)\partial p(\xi,0+)/\partial y \log \sqrt{\{(x-\xi)^2 + y^2\}}\, d\xi + H \tag{2.115}$$

The variable function $\partial p(\xi,0+)/\partial y$ is presumed to be known, since normal velocity is prescribed. Then, Eq. (2.115) provides the *nonlinear* integral equation for the fracture pressure $p_f(x)$. If m = 0, Eq. (2.115) gives p_f explicitly, and no integral equation needs to be solved. If nonzero, iterative solution methods can be used. The exact details are left as exercises for interested readers.

EXAMPLE 2.9. DETERMINING VELOCITY FIELDS

Finally, we conclude this chapter with notes on calculating velocity fields from integral expressions for pressure. To keep the discussion general, we consider the arbitrary fluid of Example 2.2 with nonzero m, taking

$$p^{m+1}(x,y) = \int_{-1}^{+1} f(\xi) \log \sqrt{\{(x-\xi)^2 + y^2\}}\, d\xi + H \tag{2.44}$$

The strength f(x) and the constant of integration H are assumed to be known. The *horizontal* velocity parallel to the fracture is obtained by first differentiating Eq. (2.44) with respect to x. This leads to

$$\partial p^{m+1}(x,y)/\partial x = \partial/\partial x \left\{ \int_{-1}^{+1} f(\xi) \log \sqrt{\{(x-\xi)^2 + y^2\}}\, d\xi + H \right\} \tag{2.116}$$

or, upon differentiation,

$$\partial p(x,y)/\partial x = \{1/((m+1)p^m)\} \int_{-1}^{+1} f(\xi)\,(x-\xi)/\left\{(x-\xi)^2 + y^2\right\} d\xi \tag{2.117}$$

where p^m, appearing on the right side of Eq. (2.117), is obtained using Eq. (2.44). Along the fracture y = 0,

$$dp_f(x)/dx = \{1/[(m+1)p_f(x)^m]\} PV \int_{-1}^{+1} f(\xi)/(x-\xi)\, d\xi \tag{2.118}$$

The integral in Eq. (2.118) is a Cauchy principal value integral and can be evaluated as discussed in Example 2.5. If the fracture pressure p_f is constant, then $dp_f/dx = 0$ and there is no flow along the fracture. This is consistent with the integral obtained by substituting Eq. (2.99) into Eq. (2.118),

$$PV \int_{-1}^{+1} 1/\{\sqrt{(1-\xi^2)}\,(x-\xi)\}\, d\xi = 0 \tag{2.119}$$

following van Dyke (1956). To obtain the *vertical* Darcy velocity, we consider

$$\partial p(x,y)/\partial y = \{1/((m+1))p^m\} \int_{-1}^{+1} f(\xi)\, y/\{(x-\xi)^2 + y^2\}\, d\xi \tag{2.120}$$

Using Eqs. (2.24), (2.27), we find that

$$\partial p(x,0+)/\partial y = \pi f(x)/\{(m+1)p_f^{\,m}(x)\} \tag{2.63}$$

This derivative is used to calculate the normal Darcy velocity at the slit. The preceding examples demonstrate the power and elegance of integral equation methods. In Chapter 3, similar methods are used to analyze flows about shales. Chapter 4 introduces modern issues in streamline tracing and the fundamentals of complex variables; this background is helpful to understanding Chapter 5, where more complicated shapes are considered.

PROBLEMS AND EXERCISES

1. Review Cauchy principal value and improper integrals in any calculus reference. Evaluate the PV integrals in Eqs. (2.95), (2.107)–(2.110) "by hand," or by using algebraic manipulation software, e.g., MathCad, Maple, or Mathematica (some programs may produce errors!).
2. Write and validate a general numerical program to evaluate Cauchy principal value integrals. What kinds of gridding problems arise? Compare results with those previously obtained analytically. Use this subroutine as the basis for a general fracture flow simulator you design.
3. Solve for the flow from a straight fracture using any reservoir flow simulator. How singular is the velocity obtained at the tips? How does volume flow rate compare with analytical results? What is the effect of tip error on total flow rate? Repeat your calculations with different mesh distributions for different assumed parameters.
4. Formulate the pressure boundary value problem for two parallel fractures separated by a fixed distance. Repeat this exercise for an array of parallel fractures. How would you use periodic boundary conditions to simplify the formulation when large numbers of fractures are involved? Explain how you would use such a formulation to evaluate the productivity of a naturally fractured reservoir or an array of repeating fractures created along a horizontal wellbore.
5. The properties of line distributions of logarithmic singularities, such as the integral and its normal derivative, were considered in this chapter and shown to useful in modeling fracture flows. Explore the properties of line distributions of arc tan or θ solutions, as introduced in Chapter 1. How can these properties be exploited to model flows past shales?
6. In Eq. (2.10), we introduced continuous line source distributions, that is, taking $p(x, y) = \int f(\xi) \log \sqrt{\{(x\text{-}\xi)^2 + y^2\}}\, d\xi + H$ over the integration limits (-1, +1), and later obtained closed form solutions. Suppose that we instead model fractures using distributions of discrete point sources, that is, $p(x, y) = H + \sum f_n(\xi_n)\, \Delta\xi_n \log \sqrt{\{(x\text{-}\xi_n)^2 + y^2\}}$, where ξ_n describes the source coordinate and the summation is taken over N singularities. This avoids the use of singular integral equations and cumbersome Cauchy principal values, but introduces other difficulties. Write a computer program in

which you (i) select the number N and the locations ξ_n of your source points and the positions of your observation points (x,y), and then (ii) solve for your unknown source strengths $f_n \Delta \xi_n$ by inverting a suitable system of coupled linear equations. How would you define the coefficient matrix of this linear system? What is the structure of this matrix? What types of problems do you encounter when prescribing pressure along the fracture $y = 0$? [Hint: the logarithm of 0 is singular.] What numerical fixes would you suggest? Run your computer program and compare pressure solutions with the exact formulas given in this chapter. How does your velocity solution behave at nodes between source points? Select a numerical reservoir flow simulator that is accessible to you, and design a similar model for fracture flow using discrete sources and sinks. How does that solution compare with yours? What are the relative advantages and disadvantages of continuous and discrete source methods in fracture flow analysis? How would you modify the previous source formula to handle curved fractures?

Chapter 3

Flows Past Shaly Bodies

In Chapter 2, we started with the well-known log r point source pressure solution, extended the idea to continuously distributed line sources, and showed how singular integral equations can be used to solve complicated problems in a straightforward way. We also saw how the normal derivative of pressure "jumps," showing discontinuous behavior through the fracture. This is no surprise: since the local velocities (proportional to $\partial p/\partial y$ via Darcy's law) on either side into the fracture are equal and opposite, it is expected. At the same time, the pressure itself is continuous and single-valued through the slit. The logarithmic solution is one of many elementary singularities of Laplace's equation; others commonly used include doublets, vortexes, source rings and horseshoe vortexes (Thwaites, 1960). Chapter 2 demonstrates how practical solutions are constructed from distributions or superpositions of logarithms. Another useful singularity is the arc tangent (Yih, 1969). In this chapter, we shall explore its usefulness in modeling flows about solid shaly bodies. Once the basic physical and mathematical notions behind logarithmic and arc tangent singularities are understood, we shall, in Chapters 4 and 5, approach simulation more generally. There, we shall consider abstractions that allow us to develop powerful mathematical models of complicated physical flows. As in Chapter 2, we develop the ideas step-by-step using simple examples.

EXAMPLE 3.1. STRAIGHT-LINE SHALE SEGMENT IN UNIFORM FLOW

Petroleum engineers in reservoir simulation do not traditionally model flows using mathematical singularities. Usually, numerical simulators with fully heterogeneous permeabilities and porosities are used; shales, for example, might be modeled by taking "k" or "ϕ" very small locally. Producing fractures, on the other hand, might be simulated by using rows of discrete wells or point sources. These approaches are sometimes acceptable, but they do not afford the physical insight that precise mathematical modeling provides. One example is the existence of square root velocity singularities at fracture tips; certainly, an awareness of its ramifications only leads to an improved understanding of the flow and to more accurate numerical models. It turns out, interestingly, that flows past shaly bodies also possess analogous singularities, which can be uncovered and studied by similar rigorous analysis.

Quantitative Methods in Reservoir Engineering. http://dx.doi.org/10.1016/B978-0-12-810518-4.00003-7

Qualitative Problem Formulation

Consider the incompressible flow of a liquid about a single straight-line shale segment located along $-c \leq X \leq +c$, and inclined at an angle $-\alpha$ relative to the uniform stream in Fig. 3.1. This shale segment is centered in an infinite reservoir. We assume that the shale is impermeable to flow; this restriction is relaxed later. The pressure P(X,Y) in a uniform isotropic medium again satisfies Laplace's equation,

$$\partial^2 P/\partial X^2 + \partial^2 P/\partial Y^2 = 0 \tag{3.1}$$

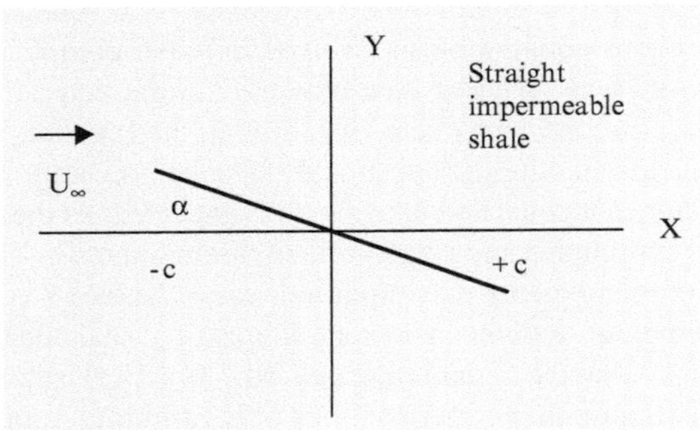

FIG. 3.1 Straight impermeable shale.

The Arc Tan Solution

While Eq. (3.1) is correct, it is not the best way to develop our ideas mathematically. To bring out the basic ideas naturally, we transform Eq. (3.1) into radial polar coordinates, first setting

$$X = r \cos \theta \tag{3.2}$$

$$Y = r \sin \theta \tag{3.3}$$

Next we determine the equation satisfied by

$$P(X, Y) = \mathit{P}(r, \theta) \tag{3.4}$$

where the italicized P indicates that the transformed function will in general be different from the nonitalicized one. Now apply the chain rule for differentiation (Hildebrand, 1948). Thus, the first and second "X" partial derivatives, for example, transform according to

$$P_X = \mathit{P}_r r_X + \mathit{P}_\theta \theta_X \tag{3.5}$$

$$\begin{aligned} P_{XX} = {} & \mathit{P}_r r_{XX} + r_X(\mathit{P}_{rr} r_X + \mathit{P}_{r\theta}\theta_X) \\ & + \mathit{P}_\theta \theta_{XX} + \theta_X(\mathit{P}_{\theta r} r_X + \mathit{P}_{\theta\theta}\theta_X) \end{aligned} \tag{3.6}$$

and similarly with "Y" derivatives. In Eqs. (3.5), (3.6), subscripts are used to indicate partial differentiation. This leads to

$$\partial^2 P/\partial X^2 + \partial^2 P/\partial Y^2 = (r_{XX} + r_{YY}) P_r + (\theta_{XX} + \theta_{YY}) P_\theta + 2(r_X\theta_X + r_Y\theta_Y) P_{\theta r} + \left(r_X{}^2 + r_Y{}^2\right) P_{rr} + \left(\theta_X{}^2 + \theta_Y{}^2\right) P_{\theta\theta} \tag{3.7}$$

Now, Eqs. (3.2), (3.3) can be explicitly inverted to give

$$r = \sqrt{(X^2 + Y^2)} \tag{3.8}$$

$$\theta = \text{arc tan } Y/X \tag{3.9}$$

If we substitute Eqs. (3.8), (3.9) in Eq. (3.7), and simplify the result using Eq. (3.1), we obtain Laplace's equation in cylindrical coordinates,

$$P_{rr} + (1/r) P_r + \left(1/r^2\right) P_{\theta\theta} = 0 \tag{3.10}$$

The Elementary Vortex Solution

The basic elementary singularities used in this book can be developed from Eq. (3.10). To show how the logarithmic solution arises, one might argue that the pressure about a circular well concentrically situated in a circular reservoir should not depend on θ. This being the case, we set the θ derivative term in Eq. (3.10) to zero, to obtain $P_{rr} + (1/r) P_r = 0$, whose fundamental solution takes the well-known log r form. On the other hand, one might ask, "What is the flow corresponding to vanishing r derivatives?" This limit leads to $P_{\theta\theta} = 0$, which has the solution

$$P(r, \theta) = \theta = \text{arc tan } Y/X \tag{3.11}$$

This solution is not considered in petroleum engineering textbooks, but it has important physical significance. In the same way that $\partial P/\partial X$ and $\partial P/\partial Y$ are proportional to Darcy velocities in the X and Y directions, respectively, we understand from vector calculus (Hildebrand, 1948) that $\partial P/\partial r$ and $(1/r)$ $\partial P/\partial \theta$ are proportional to velocities in the radial r and "azimuthal" θ directions. Thus, Eq. (3.11) is associated with vanishing radial velocities and a nonzero circumferential velocity that decays at infinity: for this reason, it is known as the point vortex solution in aerodynamics (Ashley and Landahl, 1965).

The significance of Eq. (3.11) is often understated since $P(r,\theta) = \theta$ is, in the language of mathematics, a double-valued function of space. That is, at any physical point (X,Y) in space, the pressure can be represented by some value θ, *and* also by $\theta + 2\pi$. Since the pressure at any point in a *continuous* sand possesses but a single uniquely defined value, many petroleum engineers believe that the solution given by Eq. (3.11) is nonphysical. The key to shale flow modeling lies in the word *continuous*. In a continuous sand, the pressure is certainly a "single-valued" function. However, if we introduce a solid obstacle (e.g., the impermeable shale segment assumed in this example), it is clear that this flat plate will support a pressure difference between its upper and lower sides. At any point along this infinitesimally thin shale, the mathematical representation for pressure possesses *two* values: one value just slightly above, and a completely different value just slightly below. Hence, the double-valuedness. Thus, Eq. (3.11) is as relevant to modeling flows past solid line segments (a.k.a. shales), as our log r is to modeling fracture flows. Of course, the point shale described by Eq. (3.11) is meaningless: it is a source of local swirl with no physical application. Only *continuous distributions* of such singularities have practical application in reservoir simulation.

Mathematical Formulation

Together with Eq. (3.1), we must supply boundary conditions. In the farfield, we assume a uniform horizontal Darcy velocity U_∞ as shown in Fig. 3.1. This velocity field might, for example, be the result of a distant aquifer or a row of injector wells. At the shale surface, the ratio of local vertical to horizontal velocities must be kinematically equal to the slope -tan α, if our impermeability requirement is to be met. In mathematics, we have $\{-(k/\mu)\partial P/\partial Y\}/\{-(k/\mu)\partial P/\partial X\} = -\tan\alpha$. For small inclinations, the denominator on the left side of this equation can be approximated by U_∞, while the tan α roughly equals α; in this small disturbance limit, our tangency condition is enforced along the $Y=0$ axis.

$$\partial P/\partial Y\,(X, \pm 0) = +(\mu/k)\,U_\infty\alpha,\ -1 \le X/c \le +1 \tag{3.12}$$

$$\partial P/\partial X = -(\mu/k)U_\infty,\ X^2 + Y^2 = \infty \tag{3.13}$$

This, for the time being, completes the thin airfoil mathematical formulation. The small inclination limit assumed here is taken to explore the role of edge singularities and the nonuniqueness of an integral equation solution; more general solutions not bearing this limitation are given in Chapter 5. At this point, we introduce nondimensional variables x, y and p(x,y) defined by

$$x = X/c \tag{3.14}$$

$$y = Y/c \tag{3.15}$$

$$P(X, Y) = -(\mu c U_\infty/k)\,p(x, y) \tag{3.16}$$

Then, the boundary value problem for the nondimensional p(x,y) is defined by Eqs. (3.17)–(3.19), as follows:

$$\partial^2 p/\partial x^2 + \partial^2 p/\partial y^2 = 0 \tag{3.17}$$

$$\partial p(x, \pm 0)/\partial y = -\alpha,\ -1 \le x \le +1 \tag{3.18}$$

$$\partial p(x, y)/\partial x = 1,\ x^2 + y^2 = \infty \tag{3.19}$$

Recourse to numerical methods is understandable, given the irregular "slit shape" of the reservoir nearfield. Fortunately, this is not necessary.

Singular Integral Equation Solution

An analytical solution for this problem can be obtained. To do so, note that the distributed log r solutions in Chapter 2 do *not* apply. This is seen in two ways. First, we showed in Eqs. (2.27), (2.28) that the normal derivatives of logarithmic distributions at $y = 0$ are discontinuous from one side to the next. This would violate the kinematic tangency condition at the shale surface; in fact, Eq. (3.18) insists that $\partial p/\partial y$ be single-valued through $y = 0$. Second, the use of logarithmic distributions requires that pressure going from one side to the next be continuous. In this problem, we expect pressure discontinuities since solid shale surfaces support nonzero pressure differentials. Thus, we are motivated to look for distributions based on singularities other than logarithms. To do this, examine the possibility of Eq. (3.11). Following Chapter 2, we consider this point vortex singularity, but displace it from $(x = 0, y = 0)$ to $(x = \xi, y = 0)$,

$$p(x, y) = \text{arc tan } y/(x-\xi) \tag{3.20}$$

This solution satisfies Laplace's equation (see Eq. 3.17), as does the distributed continuous line vortex

$$p(x, y) = \int_{-1}^{+1} g(\xi) \tan^{-1} y/(x-\xi)\, d\xi + x + H \tag{3.21}$$

which we anticipate models continuous shales. Here, the x handles the boundary condition at infinity (see Eq. 3.19), $g(\xi)$ is the so-called vortex strength, and H is a constant different from the one in Chapter 2. One can also verify that Eq. (3.21), which involves a linear superposition over the position coordinate ξ, satisfies Eq. (3.17) directly.

Eq. (3.21), we shall show, is the assumption that solves the boundary value problem defined by Eqs. (3.17)–(3.19), with $g(\xi)$ being the unknown function. To see that this is so, we attempt next to satisfy Eq. (3.18). Differentiation with respect to y, using standard formulas, yields

$$\partial p(x, y)/\partial y = \int_{-1}^{+1} g(\xi) \left\{ (x-\xi)/\left\{ (x-\xi)^2 + y^2 \right\} \right\} d\xi \tag{3.22}$$

If we evaluate Eq. (3.22) at $y = 0$ and substitute Eq. (3.18), we find that

$$PV \int_{-1}^{+1} g(\xi)/(x-\xi)\, d\xi = -\alpha \tag{3.23}$$

This singular integral equation, with the Cauchy kernel $1/(x-\xi)$, governs the vortex strength $g(\xi)$. The PV indicates that the integral is improper and must be evaluated using the limit process in Example 2.5.

Integral Equation Solution

Fortunately, we do not need to understand the mechanics required to solve integral equations like Eq. (3.23). Indeed, the general solution to the equation

$$\mathrm{PV}\int g(\xi)/(x-\xi)\,d\xi = -h(x) \tag{3.24}$$

is

$$g(x) = -(1/\pi^2)\sqrt{\{(1-x)/(1+x)\}}\,\mathrm{PV}\int \left\{h(\xi)\sqrt{(1+\xi)}\right\}/\left\{(\xi-x)\sqrt{(1-\xi)}\right\}d\xi$$
$$+\gamma/\sqrt{(1-x^2)} \tag{3.25}$$

where we have omitted the integration limits (-1, +1) for clarity. This solution is derived and discussed in classical math references (Mikhlin, 1965; Muskhelishvili, 1953; Carrier et al., 1966). Note, specifically, the second line of Eq. (3.25). The $\gamma/\sqrt{(1-x^2)}$ term represents the nonuniqueness associated with solutions to Eq. (3.24), with the arbitrary constant γ known in the aerodynamics literature as the circulation of a flow. We shall discuss various possibilities for circulation later.

Applying the Results

Simple results are possible for constant values of the angle $h = \alpha$. If we observe (van Dyke, 1956) that the resulting singular integral in Eq. (3.25) takes on the form

$$\mathrm{PV}\int \left\{\sqrt{(1+\xi)}\right\}/\left\{(\xi-x)\sqrt{(1-\xi)}\right\}d\xi = \pi \tag{3.26}$$

and does *not* depend on x, we obtain the vortex strength

$$g(x) = -(\alpha/\pi)\sqrt{\{(1-x)/(1+x)\}} + \gamma/\sqrt{(1-x^2)} \tag{3.27}$$

The circulation γ is still undetermined at this stage, and additional physical arguments must be considered. For this purpose, we refer to Eq. (3.21). While pressure, obviously, can take on double-valued behavior in the immediate vicinity of a shale segment, it is clear that it must remain single-valued from the viewpoint of a distant observer. This condition is necessary so that the singularities generating the local flow do not impart velocity swirl to the farfield; that is, the farfield does not see a point vortex. This requirement is guaranteed provided the *net* vortex strength vanishes,

$$\int_{-1}^{+1} g(x)dx = 0 \tag{3.28}$$

We now substitute Eq. (3.27) into Eq. (3.28) and evaluate the resulting improper integral. Thus, the circulation takes on the uniquely defined value

$$\gamma = \alpha/\pi \tag{3.29}$$

The corresponding vortex strength

$$g(x) = (\alpha/\pi)\, x/\sqrt{(1\text{-}x^2)} \tag{3.30}$$

solves Eq. (3.21). The constant H in Eq. (3.21) may be taken as zero without loss of generality since the boundary conditions in Eqs. (3.18), (3.19) involve derivatives of pressure only.

Physical Significance of Vortex Strength

Let us take the derivative of Eq. (3.21) with respect to x. Then, it follows that

$$\partial p/\partial x = -\int_{-1}^{+1} g(\xi)\, y/\left\{(x\text{-}\xi)^2 + y^2\right\}d\xi \tag{3.31}$$

The above integral was studied in Eqs. (2.24)–(2.28). In the limit $y=0$, we have, applying earlier results,

$$\partial p(x,0+)/\partial x = -\pi\, g(x) \tag{3.32}$$

$$\partial p(x,0\text{-})/\partial x = +\pi\, g(x) \tag{3.33}$$

Since the Darcy velocity parallel to the shale is proportional to $\partial p/\partial x$, the straight solid shale segment is responsible for a discontinuity in the tangential velocity, of a magnitude proportional to g(x). Eqs. (3.32), (3.33) show that the net jump in the tangential derivative (i.e., velocity slip) satisfies

$$\partial p(x,0+)/\partial x\text{-}\partial p(x,0\text{-})/\partial x = -2\pi\, g(x) \tag{3.34}$$

It is also important to observe from Eqs. (3.30)–(3.33) that the tangential velocity at the shale tips is singular, that is, infinite, behaving like $1/\sqrt{(1\text{-}x^2)}$. This velocity singularity indicates that existing numerical models are likely to yield inaccurate results. That is, any scheme that properly reproduces this infinity is likely to diverge; mesh distributions must be sufficiently coarse in order to achieve numerical stability, placing limits on accuracy. Eq. (3.30) also shows how the results for g(x) and horizontal velocity are antisymmetric, that is, $g(+x) = -g(-x)$. The pressure in Eq. (3.21), again, is double-valued through the slit, while the normal derivative in Eq. (3.18) is single-valued.

EXAMPLE 3.2. CURVED SHALE SEGMENT IN UNIFORM FLOW

We can generalize the above results to curved shale segments if we replace α by $\alpha(x)$, following strictly kinematic considerations. If we compare Eqs. (3.23), (3.24), we have

$$h(x) = \alpha(x) \tag{3.35}$$

Then, direct substitution in Eq. (3.25) yields

$$g(x) = -(1/\pi^2)\sqrt{\{(1-x)/(1+x)\}}\,\mathrm{PV}\int\left\{\alpha(\xi)\sqrt{(1+\xi)}\right\}/\left\{(\xi-x)\sqrt{(1-\xi)}\right\}d\xi + \gamma/\sqrt{(1-x^2)} \tag{3.36}$$

Again, the requirement in Eq. (3.28) applies. This leads to

$$\gamma = (1/\pi^3)\int\sqrt{\{(1-x)/(1+x)\}}\,\mathrm{PV}\int\left\{\alpha(\xi)\sqrt{(1+\xi)}\right\}/\left\{(\xi-x)\sqrt{(1-\xi)}\right\}d\xi\,dx \tag{3.37}$$

Thus $g(x)$, and hence $p(x,y)$, are completely determined.

Role of Circulation in Other Problems

The role played by the circulation γ varies from application to application (see Example 5.9 in Chapter 5), and indeed, from industry to industry. In classical aerodynamics, the circulation (used in the "direct" velocity potential description of lifting flows past airfoils) is selected to move all aft streamlines to the trailing edge, as discussed in Chapter 1. This models the global effects of viscosity and unsteady "starting vortexes," which can only be treated by a full analysis of the Navier-Stokes equations; in essence, this choice of γ *mimics* the effects of a more complete physical model. The inverse problem of aerodynamics determines the airfoil geometry that induces a prescribed surface pressure. In a series of papers, this author (Chin, 1979, 1981, 1984; Chin and Rizzetta, 1979) showed how a streamfunction formulation transforms the inverse formulation into one that is mathematically direct. The resulting circulation in this application controls trailing edge thickness; that is, numerous airfoils can exist having the same surface pressures, though with different trailing edge thicknesses and shapes. The amount of mass ejected from the trailing edge is used to model the effects of viscous wakes and flow separation.

EXAMPLE 3.3. MINERALIZED FAULTS, ANISOTROPY, AND GAS FLOW

So far, we have considered fractures, which have continuous pressures and tangential velocities but discontinuous normal velocities. Shales, by contrast, have discontinuous pressures and tangential velocities but continuous normal velocities. Special flow anomalies can be constructed by superposing linear combinations of the two. Sometimes a third class of flow arises, namely, rapid flows through mineralized faults and streaks which are not open to production. These are responsible for velocities tangent to the fracture but are different from one side of the fracture to the other. They can be visualized by imagining an otherwise homogeneous, isotropic continuous sand of permeability k with, however, a narrow isolated streak whose permeability is $k_{streak} \gg k$. Again assume a near-horizontal line segment. If this fault is *not* open to production, then equal and opposite normal velocities will *not* be present; logarithmic singularity distributions are ruled out, and normal velocities are continuous. But fluid will flow along the streak; its macroscopic effect will be a discontinuity in tangential velocity. The effect of the streak can be modeled by prescribing the distribution of g(x) in Eq. (3.34) directly, provided that the complete distribution satisfies Eq. (3.28). The exact behavior of g(x) will depend upon the results of laboratory experiments. Finally, we offer comments on anisotropic problems where the constant permeabilities k_x and k_y in the x and y directions are different, with the formation containing liquids or general gases. As shown in Chapter 2, a simple rescaling of both length dimensions leads to Laplace's equation for pressure (to the (m+1)th power) in a homogeneous medium; then, the methods of Examples 3.1 and 3.2 apply with minor modification. Before considering nontrivial flows past complicated fractures and shales, we shall review some modern issues in streamline tracing and introduce complex variables methods in Chapter 4.

PROBLEMS AND EXERCISES

1. Derive the integral solution for incompressible liquid flows past straight shales in an anisotropic medium using the rescaling methods introduced in Chapter 2. Repeat this exercise for gas flows with a general "m" exponent.
2. A "streamline" is one whose local slope is tangent to the velocity vector; that is, $dy/dx \approx -\{(k/\mu)\ \partial P(X,Y)/\partial Y\}/U_\infty$ for small inclinations. At any point in space, an equation $dy/dx = f$ can be integrated as $y_{new} \approx y_{old} + f_{old}\ \Delta x$ where Δx is a given mesh width. For the straight shale in Example 3.1, plot typical streamlines emerging from the leading and trailing edges using this integration scheme. Explain their geometric properties. Where do you expect the greatest numerical error? What is wrong with this streamline tracing approach? The starting point (x_{start}, y_{start}) in such approaches can make or break the accuracy of the solution. Show that starting points far from the shale which integrate inward lead to more accurate solutions than those integrating outward from the singularity distribution (the latter contain large initial errors that destroy the integrity of later integrations). How would you design a general streamline tracing utility? Write and program such an algorithm for use with the pressure outputs given in Chapter 7.

3. Numerical solutions depend strongly on the choice of grid. Using any reservoir simulator, compute the flowfields about stand-alone fractures and shales, say, centered in a large computational box. Do your results satisfy the symmetries and antisymmetries derived here for pressure and normal velocity? A single straight shale in an infinite medium has a perfectly antisymmetric disturbance pressure field. Show that this antisymmetry is broken by using an unsymmetric grid, therefore leading to incorrect flow results. Suppose your domain of flow contains both fractures and shales. What grid generation usage would you recommend?
4. Formulate the boundary value problem for two parallel shales separated by a fixed distance and inclined relative to the flow. Repeat this exercise for an array of parallel shales. How would you use periodic boundary conditions to simplify the formulation when large numbers of shales are involved? Can arrays of shales be modeled by θ solutions alone?
5. In this chapter, closed-form analytical solutions were obtained using exact results from singular integral equations, which were in turn developed in the 1940s. Earlier on, aerodynamicists solved the integral equation using intuitively designed trigonometric series, principally due to Glauert (Thwaites, 1960). These resembled Fourier expansions but accounted for edge singularities. Refer to basic aerodynamics textbooks, and develop a trigonometric series-based solution for simple shale flows.

Chapter 4

Streamline Tracing and Complex Variables

The streamfunction is usually discussed within the limited framework of two-dimensional, steady, constant density flow of liquids in homogeneous, isotropic media in petroleum engineering books, for example, Muskat (1937), Collins (1961), or Bear (1972). Typically, formal analysis stops with a simple proof showing how the streamfunction is constant along a streamline, and with an example displaying radial streamlines penetrating an isolated well in infinite media. The streamfunction is a step-sister to pressure and finds few published applications; even Muskat limits discussion on this crucial function, one that is potentially just as important as fluid pressure. In this chapter, we systematically develop and generalize the idea of the streamfunction. We demonstrate how, when properly posed, streamfunction formulations can be powerful weapons in the reservoir engineer's arsenal of simulation tools. The end objective is ambitious: extend streamfunction theory and streamline tracing methods to handle arbitrary heterogeneities, anisotropies, and liquids or compressible gases, in real flows with or without wells. We motivate special mathematical methods by examining different physical problems. This chapter explains the elegance offered by complex variables, but at the same time, shows its limited utility in analyzing flows with heterogeneities. We also show how pressure and streamfunction offer complementary views for flow interpretation, and we derive certain dualities between the two functions without using high-level mathematics. A knowledge of complex variables is *not* required; however, the results developed here apply to very broad classes of problems.

DISCUSSION 4.1. THE CLASSICAL STREAMFUNCTION

In this discussion, we introduce the notion of the streamfunction and explain its physical properties. Usually, fluid flow textbooks develop the basic ideas assuming familiarity with complex variables. A complex flow potential is typically introduced, whose real and imaginary parts are identified at the outset with pressure and streamfunction. Then, Cauchy-Riemann conditions are defined and used to show how each function is harmonic. The mathematical

Quantitative Methods in Reservoir Engineering. http://dx.doi.org/10.1016/B978-0-12-810518-4.00004-9

theory is established elegantly, although in a manner intimidating and lacking of physical feeling to undergraduate students.

The reader is referred to Churchill (1960), Hildebrand (1948), and Carrier et al. (1966) for advanced discussions on complex variables. This book is self-contained, however, and enough of the basic mathematics is presented as is necessary for problem solving. We deemphasize complex analysis because it cannot be used to model real-world flows. First, it cannot be extended to handle general heterogeneities: the familiar Cauchy-Riemann conditions assumed in homogeneous isotropic flows do not apply. Second, analyses are typically restricted to incompressible liquids; what of compressible gases in heterogeneous media? Third, the very important problem of streamline tracing in reservoirs containing wells cannot be treated. For these reasons, we alternatively develop the streamfunction without using complex variables at all. In our exposition, we assume only an understanding of elementary calculus and some familiarity with partial derivatives. This is not to say that our end results will be limited; quite to the contrary, the results are rigorous and powerful.

Properties of the "Simple" Streamfunction

Let us consider first the two-dimensional, steady, constant density flow of a liquid in homogeneous, isotropic media. The Darcy flow, in this case, satisfies Laplace's equation

$$\partial^2 P/\partial X^2 + \partial^2 P/\partial Y^2 = 0 \tag{4.1}$$

where P(X,Y) is pressure, X and Y are Cartesian coordinates, and P, X and Y are dimensional. The Eulerian velocities U and V in the X and Y directions are

$$U = -(k/\mu)\partial P/\partial X \tag{4.2}$$

$$V = -(k/\mu)\partial P/\partial Y \tag{4.3}$$

where k and μ are constant formation permeability and liquid viscosity, respectively. Now rewrite Eq. (4.1) in what appears to be a more cumbersome form,

$$\partial(\partial P/\partial X)/\partial X + \partial(\partial P/\partial Y)/\partial Y = 0 \tag{4.4}$$

Eq. (4.4) suggests that we can define a function $\Psi(X,Y)$ such that

$$\partial P/\partial X = \partial \Psi/\partial Y \tag{4.5}$$

$$\partial P/\partial Y = -\partial \Psi/\partial X \tag{4.6}$$

Eqs. (4.5), (4.6), after all, are simply relationships that introduce no additional assumptions; for example, substitution in Eq. (4.4) yields a trivial $0=0$. However, the function Ψ, or streamfunction, possesses interesting properties.

First, let us consider the kinematic definition of a streamline. A streamline is a flow trajectory across which fluid motion is absent; fluid moves tangentially to it. Thus, its local slope must be equal to the ratio of the vertical to the horizontal velocities,

$$dY/dX = V/U \tag{4.7}$$

If we now substitute, initially Eqs. (4.2), (4.3), and then Eqs. (4.5), (4.6) into Eq. (4.7), we obtain successively

$$dY/dX = \{\partial P/\partial Y\}/\{\partial P/\partial X\} = \{-\partial \Psi/\partial X\}/\{\partial \Psi/\partial Y\} \tag{4.8}$$

that is,

$$\{\partial \Psi/\partial X\}\, dX + \{\partial \Psi/\partial Y\}\, dY = 0 \tag{4.9}$$

From calculus, the total differential for any two-variable function is

$$d\Psi = \{\partial \Psi/\partial X\}\, dX + \{\partial \Psi/\partial Y\}\, dY \tag{4.10}$$

Comparing Eqs. (4.9), (4.10), we find that

$$d\Psi = \{\partial \Psi/\partial X\}\, dX + \{\partial \Psi/\partial Y\}\, dY = 0 \tag{4.11}$$

Hence, the streamfunction $\Psi(X,Y)$ is constant along a streamline. Since the streamfunction must take on different constant values along different streamlines, it is clear that the difference between $\Psi(X,Y)$'s measured at any two points in space should provide some measure of the nonzero volume flow rate streaming between the two points. We shall discuss this relationship later.

Now, what partial differential equation does the streamfunction satisfy? This is determined by differentiating Eq. (4.5) with respect to Y, Eq. (4.6) with respect to X, and subtracting the two. This leads to

$$\partial^2 \Psi/\partial X^2 + \partial^2 \Psi/\partial Y^2 = 0 \tag{4.12}$$

This simple proof shows that the streamfunction, like the pressure $P(X,Y)$ in Eq. (4.1), is also "harmonic"; that is, it also satisfies Laplace's equation. By postulating Eqs. (4.5), (4.6), we have shown that whenever Eq. (4.1) can be written, then so can Eq. (4.12). In complex variables, this is shown formally; P and Ψ are conjugate harmonic functions because two-dimensional harmonic functions always occur in conjugate pairs. Eqs. (4.5), (4.6) connecting the two are referred to as Cauchy-Riemann conditions. We have rederived classical streamfunction theory *without* using complex variables, and significantly, it turns out that our manner of proof is generalizable to heterogeneous flows! For now, we point out one more important property: streamlines are orthogonal, that is, perpendicular, to lines of constant pressure. This is easily demonstrated. From vector analysis (e.g., Hildebrand, 1948), the unit normal to any surface $F(X,Y) = \text{constant}$ is proportional to the gradient function $\partial F/\partial X\ \mathbf{i} + \partial F/\partial Y\ \mathbf{j}$, where **i** and **j** are unit vectors in the X and Y directions. Thus, the unit normals to lines having constant Ψ and P are, respectively, proportional to $\partial \Psi/\partial X\ \mathbf{i} + \partial \Psi/\partial Y\ \mathbf{j}$ and $\partial P/\partial X\ \mathbf{i} + \partial P/\partial Y\ \mathbf{j}$. Now, let us form the dot or scalar product between these two normals and simplify the result using Eqs. (4.5), (4.6). We obtain

$$\begin{aligned}(\partial\Psi/\partial X\,\mathbf{i} + \partial\Psi/\partial Y\,\mathbf{j}) \bullet (\partial P/\partial X\,\mathbf{i} + \partial P/\partial Y\,\mathbf{j}) \\ = (\partial\Psi/\partial X)(\partial P/\partial X) + (\partial\Psi/\partial Y)(\partial P/\partial Y) = 0\end{aligned} \tag{4.13}$$

Since the scalar product vanishes, the two normals are perpendicular; if the normals are perpendicular, then so must be their respective level surfaces. It is important to emphasize that we have obtained this result using Eq. (4.1), which is valid only for constant density liquids in isotropic, homogeneous media. In this limit, the commonly used approximation that flow from (or into) a constant pressure fracture is locally perpendicular is, in fact, exact. The flow from a producing fracture with variable lengthwise pressure, as seen from this proof, will not be orthogonal locally (this is clear because flow parallel to the fracture exists). These results do not apply to transient compressible liquids, even in isotropic, homogeneous media because the total differential in Eq. (4.10) would have to include the timewise term $(\partial\Psi/\partial T)dT$. When this is allowed, it is not possible to obtain analogous results because the total derivative $d\Psi$ does not vanish. Unsteady streamlines are known as "streaklines."

DISCUSSION 4.2. STREAMFUNCTION FOR GENERAL FLUIDS IN HETEROGENEOUS AND ANISOTROPIC FORMATIONS

We now repeat the above analysis and extend the notion of the streamfunction to liquids and compressible gases for steady flow through heterogeneous, anisotropic formations. We emphasize that, considering the various possibilities for flow permitted in this book, the physical dimensions assumed for Ψ may vary from one application to another; that is, different definitions are employed throughout that may involve changes by a constant scale factor. Consistency, however, is maintained within like applications. Our starting point is the pressure equation, again in "conservation form," that is,

$$\partial\left(k_x \partial P^{m+1}/\partial X\right)/\partial X + \partial\left(k_y \partial P^{m+1}/\partial Y\right)/\partial Y = 0 \tag{4.14}$$

where $k_x(X,Y)$ and $k_y(X,Y)$ are, respectively, different spatially dependent permeability functions in the X and Y directions. Again, the constant m is zero for liquids and nonzero for gases. If μ represents the viscosity of the fluid, the Eulerian velocities U and V can be obtained from Darcy's law as

$$U = -(k_x/\mu)\,\partial P/\partial X \tag{4.15}$$

$$V = -\left(k_y/\mu\right)\partial P/\partial Y \tag{4.16}$$

In a manner similar to Discussion 4.1, Eq. (4.14) suggests that we can define a function $\Psi(X,Y)$ such that

$$k_x \partial P^{m+1}/\partial X = \partial\Psi/\partial Y \tag{4.17}$$

$$k_y \partial P^{m+1}/\partial Y = -\partial \Psi/\partial X \quad (4.18)$$

Substitution of Eqs. (4.17), (4.18) into Eq. (4.14) yields a 0=0 identity, showing that no additional assumptions have been invoked. First, let us determine how Ψ varies, if at all, along a streamline. Eq. (4.7), which provides the kinematic description for a streamline, still holds. Thus,

$$dY/dX = V/U = \{k_y \partial P/\partial Y\}/\{k_x \partial P/\partial X\} \quad (4.19)$$

Now, if we carry out the differentiations suggested in Eqs. (4.17), (4.18), the resulting expressions for $\partial P/\partial X$ and $\partial P/\partial Y$ are

$$\partial P/\partial X = \{\partial \Psi/\partial Y\}/\{k_x(m+1)P^m\} \quad (4.20)$$

$$\partial P/\partial Y = -\{\partial \Psi/\partial X\}/\{k_y(m+1)P^m\} \quad (4.21)$$

Substitution in Eq. (4.19) leads to

$$\begin{aligned} dY/dX &= -\{\partial \Psi/\partial X\}/\{(m+1)P^m\}/\{\partial \Psi/\partial Y\}/\{(m+1)P^m\} \\ &= -\{\partial \Psi/\partial X\}/\{\partial \Psi/\partial Y\} \end{aligned} \quad (4.22)$$

that is, as before,

$$\{\partial \Psi/\partial X\}dX + \{\partial \Psi/\partial Y\}dY = 0 \quad (4.23)$$

From calculus, the total differential for any two-variable function is

$$d\Psi = \{\partial \Psi/\partial X\}\, dX + \{\partial \Psi/\partial Y\}\, dY \quad (4.24)$$

Comparing Eqs. (4.23), (4.24), we find that

$$d\Psi = \{\partial \Psi/\partial X\}\, dX + \{\partial \Psi/\partial Y\}\, dY = 0 \quad (4.25)$$

along a streamline; hence, the streamfunction $\Psi(X,Y)$ defined by Eqs. (4.17), (4.18) is constant along a streamline in heterogeneous, anisotropic flow, for liquids and real gases.

Are such streamlines perpendicular to lines of constant pressure? The answer is, "No." To see this, note that the unit normals to lines having constant Ψ's and P's are proportional to $\partial \Psi/\partial X\, \mathbf{i} + \partial \Psi/\partial Y\, \mathbf{j}$ and $\partial P/\partial X\, \mathbf{i} + \partial P/\partial Y\, \mathbf{j}$, respectively. The inner or scalar dot product formed by these normals is

$$\begin{aligned} &(\partial \Psi/\partial X)(\partial P/\partial X) + (\partial \Psi/\partial Y)(\partial P/\partial Y) \\ &= (\partial \Psi/\partial X)\{\partial \Psi/\partial Y\}/\{k_x(m+1)P^m\} - (\partial \Psi/\partial Y)\{\partial \Psi/\partial X\}/\{k_y(m+1)P^m\} \\ &= (\partial \Psi/\partial X)(\partial \Psi/\partial Y)(1/k_x - 1/k_y)/\{(m+1)\, P^m\} \end{aligned} \quad (4.26)$$

where we have used Eqs. (4.20), (4.21). Now, orthogonality is only possible when the dot product in Eq. (4.26) vanishes. This will occur at any point (X,Y) where k_x and k_y are locally equal. When k_x and k_y are equal everywhere, the product in Eq. (4.26) vanishes identically. Thus, streamlines and pressure level curves are perpendicular even in a heterogeneous (but isotropic) flow having spatially varying permeabilities. As shown in Discussion 4.1, it is not possible to obtain similar results for transient compressible flows. Next, we determine the partial differential equation satisfied by the streamfunction so defined. This is accomplished by dividing Eq. (4.17) by k_x, and similarly, Eq. (4.18) by k_y. This leads to

$$\partial P^{m+1}/\partial X = \{\partial \Psi/\partial Y\}/k_x \tag{4.27}$$

$$\partial P^{m+1}/\partial Y = -\{\partial \Psi/\partial X\}/k_y \tag{4.28}$$

If we now differentiate Eq. (4.27) with respect to Y and Eq. (4.28) with respect to X, and subtract the resulting equations, we obtain

$$\partial\{(\partial \Psi/\partial X)/k_y\}/\partial X + \partial\{(\partial \Psi/\partial Y)/k_x\}/\partial Y = 0 \tag{4.29}$$

as the required differential equation for heterogeneous, anisotropic flow. In the limit of homogeneous, isotropic flow, we recover the results of Discussion 4.1. In general, however, Eq. (4.29) applies; it is *not*, we must emphasize, Eq. (4.14) with P^{m+1} replaced by Ψ. And, further, this is *not* an equation for Ψ^{m+1}, although we might, alternatively, have proceeded along a route leading to such a result. The choice is a matter of personal preference. We do emphasize that both Eqs. (4.14), (4.29) are elliptic, so that the same elliptic algorithms can be used to solve for pressures and streamfunctions, bearing in mind several important subtleties to be discussed next.

DISCUSSION 4.3. SUBTLE DIFFERENCES BETWEEN PRESSURE AND STREAMFUNCTION FORMULATIONS

In Discussion 4.1, we showed that the pressure P(X,Y) and the streamfunction Ψ(X,Y) for a constant density liquid in a homogeneous, isotropic medium both satisfy Laplace's equation, that is,

$$\partial^2 P/\partial X^2 + \partial^2 P/\partial Y^2 = 0 \tag{4.30}$$

$$\partial^2 \Psi/\partial X^2 + \partial^2 \Psi/\partial Y^2 = 0 \tag{4.31}$$

Similar results were obtained in Discussion 4.2. Whereas Eq. (4.30) provides a meaningful way to view the effects of pressure gradients, streamline methods allow ready visualization of the flowfield. The two offer complementary and equivalent ways to describe the physics. One might conclude that once P(X,Y) has been obtained numerically on some computational box, then the streamfunction can be straightforwardly obtained by solving Eq. (4.31) subject to normal derivative boundary conditions written in terms of $\partial\Psi/\partial X$ and $\partial\Psi/\partial Y$ at the edges of the box—that is, by using the Cauchy-Riemann conditions in Eqs. (4.5), (4.6), since P(X,Y) is assumed to be known. As this Neumann formulation involves derivatives only, the exact level of Ψ would be pinned down by arbitrarily assigning it the value of zero, say, at the origin. This simple procedure turns out to be valid only when the possibility of injection or production wells is excluded: in the petroleum industry, this rules out most commercially significant flows. Thus, while this recipe may be of interest in groundwater hydrology problems (Cherry and Freeze, 1979), it is not useful in oilfield applications. But one can resurrect the basic procedure, provided certain nontrivial modifications are made in the basic formulation. To understand exactly what the problem is, we return to fundamentals: the end result will be a powerful result useful in streamline tracing. The similarities underlying Eqs. (4.30), (4.31) are superficial; there are large differences between P and Ψ that we must consider first. We explore these subtleties by reexamining the singularities in Chapters 2 and 3.

More Streamfunction Properties

To bring out the subtleties, it is convenient to transform Eqs. (4.30), (4.31) into equivalent forms in radial polar coordinates. That is, set

$$X = r\cos\theta \tag{4.32}$$

$$Y = r\sin\theta \tag{4.33}$$

in the usual manner, and determine the equations satisfied by

$$\mathrm{P}(X, Y) = P(r, \theta) \tag{4.34}$$

$$\Psi(X, Y) = \mathit{\Psi}(r, \theta) \tag{4.35}$$

where the italics indicate that the transformed functions will in general be different from the non-italicized ones. We apply the chain rule for differentiation. Thus, the first and second X derivatives, for example, transform according to

$$\mathrm{P_X} = P_r r_X + P_\theta \theta_X \tag{4.36}$$

$$\begin{aligned} \mathrm{P_{XX}} = & P_r r_{XX} + r_X(P_{rr} r_X + P_{r\theta}\theta_X) \\ & + P_\theta \theta_{XX} + \theta_X(P_{\theta r} r_X + P_{\theta\theta}\theta_X) \end{aligned} \tag{4.37}$$

where subscripts, for convenience, are used to indicate partial differentiation. Carrying through, we obtain

$$\begin{aligned} \partial^2 \mathrm{P}/\partial X^2 + \partial^2 \mathrm{P}/\partial Y^2 = & (r_{XX} + r_{YY})P_r + (\theta_{XX} + \theta_{YY})P_\theta \\ & + 2(r_X\theta_X + r_Y\theta_Y)P_{\theta r} + (r_X^2 + r_Y^2)P_{rr} \\ & + (\theta_X^2 + \theta_Y^2)P_{\theta\theta} \end{aligned} \tag{4.38}$$

If we solve Eqs. (4.32), (4.33) for r and θ explicitly in terms of X and Y, and substitute the results into Eq. (4.38), we have, on setting the Laplacian equal to zero,

$$\mathrm{P_{XX}} + \mathrm{P_{YY}} = P_{rr} + (1/r)P_r + (1/r^2)P_{\theta\theta} = 0 \tag{4.39}$$

In radial coordinates, the pressure thus satisfies Laplace's equation in the form

$$P_{rr} + (1/r)P_r + (1/r^2)P_{\theta\theta} = 0 \quad (4.40)$$

Similarly, the streamfunction satisfies

$$\Psi_{rr} + (1/r)\Psi_r + (1/r^2)\Psi_{\theta\theta} = 0 \quad (4.41)$$

As we pointed out in Chapter 3, in many elementary treatments, the dependence of P on θ is not considered. But we showed that while the log r solution is useful in modeling flows in continuous sands where pressure does not jump, the θ is ideal for modeling the effects of solid obstacles placed in the midst of larger scale flows that do support pressure differences.

The Classic Streamline Tracing Problem

As if our new concepts were not complicated or confusing enough, the streamfunction Ψ also has log r and θ solutions. What do these mean physically, and how do these relate to their pressure counterparts? Since our ultimate objective is to determine the streamfunction distribution once the pressure field is available numerically, we would like to determine the streamfunction expression that is complementary, first, to the source (or sink) solution

$$P_{source} = \log \sqrt{\{X^2 + Y^2\}} \quad (4.42)$$

To do this, we return to the Cauchy-Riemann conditions given by Eqs. (4.5), (4.6). Simple integration shows that the corresponding streamfunction is given by the expression

$$\Psi_{source} = \theta = \text{arc tan } Y/X \quad (4.43)$$

Thus, the functions log r and θ provide equivalent descriptions of the point source, depending upon whether the pressure or streamfunction formulation is employed. But there is one crucial difference. The log r is a *single-valued* function of space; that is, given some point (X,Y), the pressure has a single unique value. The streamfunction Ψ, however, is *double-valued*. What, really, does this mean? Earlier we showed that the difference between streamfunction values at any two points in space must reflect the nonzero net flow between those points. Consider a circular well situated concentrically in a circular reservoir, so that Eq. (4.43) applies. Now, envision two infinitesimally close radial streamlines with 0 degree included angle, so that the flow between the two is negligible. The alternative view, going 360 degrees about the source point, captures the entire nonzero flow. The double-valuedness in Eq. (4.43) is mathematics' way to describe the increasing capture of produced fluid as θ increases, with the rays finally returning to touching position. Point sources such as the one in Eq. (4.42) are *always* associated with double-valued streamfunctions because they create mass. In the language of math, "if one traverses about a closed circuit enclosing the origin, the streamfunction (in Eq. 4.43) increases from, say,

0 to 2π." (In a region without wells, the change in streamfunction about a closed circuit is zero.) This double-valued behavior, necessary to describe the mass-conserving nature of source flows, also contributes to obvious problems in the solution of the streamfunction equation. Most numerical simulators are developed assuming that the dependent variable (typically pressure) is uniquely defined and single-valued at any point. Thus, the streamfunction *cannot* be solved by such straightforward algorithms or standard numerics. The exception, of course, are flows without wells, which are common in the study of groundwater hydrology. This is the streamline tracing problem that has made streamfunction analyses of producing oil reservoirs impractical, but simple changes can be made to overcome this difficulty. We shall tackle this problem in Discussion 4.4, after we discuss the streamfunction corresponding to the vortex singularities introduced in Chapter 3.

The Vortex Solution

Petroleum engineers have traditionally neglected the θ solution for pressure, but we have seen how they are important in modeling flows past shale distributions. Following the lead of our aerospace colleagues, we consider the elementary vortex solution

$$P_{vortex} = \theta = \arc\tan Y/X \tag{4.44}$$

Again, use of the Cauchy-Riemann conditions given by Eqs. (4.5), (4.6) shows that the complementary streamfunction solution is

$$\Psi_{vortex} = \log \sqrt{\{X^2 + Y^2\}} \tag{4.45}$$

Whereas pressure is double-valued through a solid shale, the corresponding streamfunction, interestingly, is single-valued: the shale segment is a streamline of the flow. This is also obvious because it is not a source (or sink) that creates (or destroys) fluid. Thus, numerical methods developed for pressure fields in continuous sands, or for steady-state temperature fields in heat transfer problems without embedded insulators, can be "naively" and directly used for shale flow modeling. For such problems, there is no streamline tracing problem. In summary, Eqs. (4.42), (4.43) are the elementary solutions used to model source and sinks, that is, production or injection wells. By contrast, distributed versions of the singularities in Eqs. (4.44), (4.45) are used to model flows about shaly barriers. We also reiterate the problem with streamline tracing using Eq. (4.43). Since wells by necessity are described by double-valued streamfunctions, reservoir flows containing production and injection wells cannot be solved by using Eq. (4.31) directly. That is, any attempt to discretize the governing equation, either by finite differences or finite elements, is bound to fail: unless special double-valued operators are employed, which are beyond the scope of this book, such methods implicitly assume that the discretized function possesses at most a single value of streamfunction per node.

Solution methods for multivalued streamfunctions in aerodynamics have been developed by the author (Chin, 1979, 1981, 1984; Chin and Rizzetta, 1979), when the airfoil shape is sought that induces a prescribed surface pressure. Here the jump in streamfunction is related to the shape (or degree of closure) of the trailing edge desired. Similar ideas are presented by the author for three-dimensional electromagnetic modeling in layered media, where double-valued finite difference operators are derived and tabulated (Chin, 2000, 2014a). Finite difference schemes handling double-valued properties *are* available, of course, in the aerospace industry; for example, recall from Chapter 1 that the velocity potential must be multivalued in order to model lift. However, we shall not need to develop analogous methods for practical reasons. For one, very good pressure solvers do proliferate in the petroleum industry that handle wells adequately, so it makes sense to continue with pressure formulations. We wish only to solve Eq. (4.31) insofar that it is used as a streamline post-processor, which generates visually meaningful information augmenting the numerical pressure field. Of course, the trick is to post-process correctly an equation that accounts for double-valuedness. This is discussed next, where a complementary formulation for a single-valued Ψ is given that results from subtracting out the two-valued problem via linear superposition. This formulation is solved using standard difference or finite element methods.

DISCUSSION 4.4. STREAMLINE TRACING IN THE PRESENCE OF MULTIPLE WELLS

We demonstrated in Discussion 4.3 that, in the presence of flowing wells in a reservoir, the *known* pressure solution, whether it is numerical or analytical, implicitly contains local source solutions of the form given in Eq. (4.42). Effects representing interactions between wells will also be present. To each of these sources, one must associate a complementary double-valued streamfunction of the form given by Eq. (4.43). Thus, we cannot prescribe normal boundary conditions on computational box edges (via $\partial\Psi/\partial X$ and $\partial\Psi/\partial Y$, as obtained from a known pressure through the Cauchy-Riemann conditions of Eqs. 4.5, 4.6) *and* blindly solve Eq. (4.31) for Ψ, say, using the single-valued numerical methods discussed in Chapter 7. The actual procedure requires some modification. The key to streamline tracing in the presence of wells, as a preliminary step, requires the elimination of this troublesome multivaluedness. Before explaining the details, let us consider a steady radial liquid flow and examine first the isotropic pressure solution studied in petroleum engineering. Assume that a pressure P_W is maintained at the well radius $r = r_W$ and that a pressure of P_R is maintained at the reservoir outer radius $r = r_R$. The solution to Eq. (4.40), without the θ dependence, is

$$P = P_W + (P_R\text{-}P_W)\{\log(r/r_W)\}/\log(r_R/r_W) \tag{4.46}$$

It turns out to be more convenient to eliminate the boundary pressure P_R and rewrite Eq. (4.46) in terms of the net volume flow rate Q issuing from the well. If D is the depth into the page, the flow rate Q is determined by

$$Q = 2\pi rD(-k/\mu)\partial P/\partial r = -2\pi(kD/\mu)(P_R - P_W)/\log(r_R/r_W) \tag{4.47}$$

so that Eq. (4.46) becomes

$$P = P_W - (Q\mu/2\pi kD)\{1/2\log(X^2 + Y^2)\} - \log r_W\} \tag{4.48}$$

The corresponding expression for the streamfunction, satisfying the Cauchy-Riemann conditions in Eqs. (4.5), (4.6), is

$$\Psi = -(Q\mu/2\pi kD)\ \text{arc}\tan Y/X \tag{4.49}$$

A Simple Scheme

Now suppose that a numerical solution for the pressure field is available, for example, the finite difference solutions presented later in Chapter 7. The solution, for instance, may contain the effects of arbitrary aquifer and solid wall no-flow boundary conditions; we also suppose that this pressure solution contains the effects of multiple production and injection wells. How do we pose the streamline tracing problem using Ψ *without* dealing with multivalued functions? The solution is obvious: subtract out multivalued effects and treat the remaining single-valued formulation using standard methods. Let us assume that there exist N wells located at the coordinates (X_n, Y_n), having well volume flow rates Q_n known from the pressure solution. Here the index n varies from 1 to N. Then, it is clear from Eq. (4.49) that the *complete* streamfunction $\Psi_{total}(X,Y)$ will contain the troublesome part $\Psi_{multivalued}(X,Y)$ that specifically contains well effects, plus a single-valued trouble-free part that handles the effects of well interference and farfield interactions. Following the hint in Eq. (4.49), we can write the former as

$$\Psi_{multivalued}(X, Y) = -(\mu/2\pi kD)\sum_{n=1}^{N} Q_n\ \text{arc}\tan(Y\text{-}Yn)/(X\text{-}Xn) \tag{4.50}$$

where the (X_n, Y_n), Q_n and N are known. Thus, there remains the single-valued trouble-free part

$$\Psi_{single\text{-}valued}(X, Y) = \Psi_{total}(X, Y)\text{-}\Psi_{multivalued}(X, Y) \tag{4.51}$$

which also satisfies Eq. (4.31) owing to linearity, that is,

$$\partial^2\Psi_{single\text{-}valued}/\partial X^2 + \partial^2\Psi_{single\text{-}valued}/\partial Y^2 = 0 \tag{4.52}$$

The relevant boundary conditions on $\Psi_{\text{single-valued}}$ are likewise found by superposition. From the Cauchy-Riemann conditions in Eqs. (4.5), (4.6), we rewrite

$$\partial\Psi/\partial Y = \partial P/\partial X \tag{4.53}$$

$$\partial\Psi/\partial X = -\partial P/\partial Y \tag{4.54}$$

where P(X,Y), which again contains *all* flow effects, is known numerically. Thus, substituting

$$\Psi_{\text{total}} = \Psi_{\text{single-valued}}(X, Y) + \Psi_{\text{multivalued}}(X, Y) \tag{4.55}$$

we obtain

$$\partial\Psi_{\text{single-valued}}/\partial Y = \partial P/\partial X - \partial\Psi_{\text{multivalued}}(X, Y)/\partial Y \tag{4.56}$$

$$\partial\Psi_{\text{single-valued}}/\partial X = -\partial P/\partial Y - \partial\Psi_{\text{multivalued}}(X, Y)/\partial X \tag{4.57}$$

to be used as appropriate Neumann conditions at the horizontal and vertical edges of the computational box, respectively. As indicated earlier, since the boundary value problem involves derivatives of Ψ only, we arbitrarily fix Ψ at some point, say 0 at the origin. This has no effect on the velocity; this choice simply renders the solution unique. Ψ_{total} is then obtained from Eq. (4.55).

The Final Streamline Tracing Recipe

In summary, the problem consists of finding the streamfunction $\Psi_{\text{total}}(X,Y)$ if the pressure P(X,Y) corresponding to N wells and given farfield boundary conditions is known. Streamlines are obtained, again, by connecting lines having like values of Ψ_{total}. The solution method is as follows. (i) Construct the *analytical* expression given by Eq. (4.50), where the indexed quantities are known (for example, determine Q_n by post-processing the results for pressure). (ii) Since P(X,Y) is known, obtain $\partial\Psi_{\text{single-valued}}/\partial Y$ numerically using Eq. (4.56), and use this as the normal derivative (Neumann) boundary condition along the top and bottom sides of the box. (iii) Similarly, obtain the derivative $\partial\Psi_{\text{single-valued}}/\partial X$ numerically using Eq. (4.57), and use this as the normal derivative (Neumann) boundary condition along the left and right sides of the box. Since the problem is of the Neumann type, (iv) fix the level of $\Psi_{\text{single-valued}}$ arbitrarily at some point. Now solve this boundary value problem for the single-valued function $\Psi_{\text{single-valued}}(X,Y)$ using a standard finite difference scheme (e.g., see Chapter 7). (v) Once the converged $\Psi(X,Y)_{\text{single-valued}}$ is available, add back the multivalued part (that is, the expression obtained in (i)) as required by Eq. (4.55) to obtain the total streamfunction $\Psi_{\text{total}}(X,Y)$. (vi) Finally, draw lines through constant values of the complete streamfunction $\Psi_{\text{total}}(X,Y)$ to obtain the flow streamlines. The streamlines so obtained will appear from and disappear into wells as required.

This last step can be performed using commercial contour plotting software. In constructing this algorithm, care must be taken *not* to differentiate $\partial P/\partial X$ and $\partial P/\partial Y$ across well points, since the first derivatives of pressure are physically discontinuous through the points; appropriate one-sided difference formulas should be used. Also, it is important in the final printout of the streamfunction solution *not* to list the value of Ψ_{total} at injection and production well positions. Again, it is multivalued, and any single printed number would likely be subject to misinterpretation. Instead, at well points, some alphanumeric name might be more appropriate.

DISCUSSION 4.5. STREAMFUNCTION EXPRESSIONS FOR DISTRIBUTED LINE SOURCES AND VORTEXES

In Chapters 2 and 3, we introduced distributions of sources and vortexes to model fractures and shales. There, we obtained explicit solutions for the pressure field and its Darcy velocities. Here we shall determine in a general manner the streamfunctions associated with those pressure solutions. We shall develop, for simplicity, the formulas for constant density liquids.

Source-Like Flows

Again, we return to first principles. Let us consider the general form for the pressure corresponding to distributed line sources having the source strength density f(x), namely,

$$p(x,y) = \int_{-1}^{+1} f(\xi) \log \sqrt{\left\{(x-\xi)^2 + y^2\right\}}\, d\xi + H \tag{4.58}$$

in the dimensionless coordinates (x,y) given in Eq. (2.10). Direct differentiation shows that

$$\partial p(x,y)/\partial x = \int_{-1}^{+1} f(\xi)\,(x-\xi)/\left\{(x-\xi)^2 + y^2\right\} d\xi \tag{4.59}$$

$$\partial p(x,y)/\partial y = \int_{-1}^{+1} f(\xi)\, y/\left\{(x-\xi)^2 + y^2\right\} d\xi \tag{4.60}$$

The conjugate pair defined by Eqs. (4.42), (4.43) suggests that the streamfunction $\Psi(x,y)$ complementary to the source solution in Eq. (4.58) should take the form

$$\Psi(x, y) = \int_{-1}^{+1} f(\xi) \arctan y/(x-\xi)\, d\xi \tag{4.61}$$

To show that this is the correct solution, we need only differentiate it, and check that it satisfies the Cauchy-Riemann conditions in Eqs. (4.4), (4.5). Straightforward differentiation shows that

$$\partial\Psi(x, y)/\partial y = \int_{-1}^{+1} f(\xi)\,(x-\xi)/\left\{(x-\xi)^2 + y^2\right\} d\xi \tag{4.62}$$

$$\partial\Psi(x, y)/\partial x = -\int_{-1}^{+1} f(\xi)\, y/\left\{(x-\xi)^2 + y^2\right\} d\xi \tag{4.63}$$

Thus, from Eqs. (4.59), (4.62), we have $\partial p(x,y)/\partial x = \partial\Psi(x,y)/\partial y$, and from Eqs. (4.60), (4.63), we obtain $\partial p(x,y)/\partial y = -\partial\Psi(x,y)/\partial x$, in complete agreement with Eqs. (4.5), (4.6). This establishes the expression in Eq. (4.61) as the streamfunction for source distributions of strength f(x).

Vortex Distributions

Next, we turn to vortex distributions, which are line singularities used to model flows past shales that support pressure differentials. Let us consider the pressure field associated with a line vortex of strength g(x); from Eq. (3.21), the expression for pressure is

$$p(x, y) = \int_{-1}^{+1} g(\xi) \arctan y/(x-\xi)\, d\xi \tag{4.64}$$

Direct differentiation yields

$$\partial p(x, y)/\partial y = \int_{-1}^{+1} g(\xi)\,(x-\xi)/\left\{(x-\xi)^2 + y^2\right\} d\xi \tag{4.65}$$

$$\partial p(x, y)/\partial x = -\int_{-1}^{+1} g(\xi)\, y/\left\{(x-\xi)^2 + y^2\right\} d\xi \tag{4.66}$$

Now, we speculate from Eqs. (4.4), (4.5) that

$$\Psi(x, y) = -\int_{-1}^{+1} g(\xi) \log \sqrt{\left\{(x-\xi)^2 + y^2\right\}}\, d\xi \tag{4.67}$$

and compute the derivatives

$$\partial\Psi(x,y)/\partial x = -\int_{-1}^{+1} g(\xi)\,(x-\xi)/\left\{(x-\xi)^2 + y^2\right\} d\xi \quad (4.68)$$

$$\partial\Psi(x,y)/\partial y = -\int_{-1}^{+1} g(\xi)\,y/\left\{(x-\xi)^2 + y^2\right\} d\xi \quad (4.69)$$

Comparison of Eqs. (4.66), (4.69) shows that $\partial p(x,y)/\partial x = \partial\Psi(x,y)/\partial y$, while Eqs. (4.65), (4.68) show that $\partial p(x,y)/\partial y = -\partial\Psi(x,y)/\partial x$. Hence, the Cauchy-Riemann conditions are identically satisfied, so that the streamfunction in Eq. (4.67) is complementary to the vortex pressure in Eq. (4.64). These results provide expressions for streamline tracing in the presence of distributed singularities. In general, as we shall show in Chapter 5, the flow past (or from) an arbitrary entity can be represented by superpositions of line sources and vortexes, respectively, having strengths f(x) and g(x). The net pressure field is obtained by summing Eqs. (4.58), (4.64), while the net streamfunction is obtained by summing Eqs. (4.61), (4.67), plus effects due to any farfield flows present (e.g., the x term in Eq. 3.21).

DISCUSSION 4.6. STREAMFUNCTION SOLUTION USING COMPLEX VARIABLES TECHNIQUES

Having discussed the properties of both pressure and streamfunction, we now provide a mathematically elegant rederivation of basic ideas. Again, the new methodology does not apply to flows in heterogeneous media, but it does open up new avenues for easily constructing solutions that in turn can be extended further, when conformal mapping is developed in Chapter 5. Toward this end, let us introduce the complex variable

$$z = x + i\,y \quad (4.70)$$

where i denotes the imaginary number $\sqrt{-1}$. Also, let w(z) be a complex "analytic" function of z, e.g., sin z, log z, or exp(z). In general, w(z) will contain real and imaginary parts which we denote by p(x,y) and Ψ(x,y),

$$w(z) = p(x,y) + i\,\Psi(x,y) \quad (4.71)$$

We emphasize that p(x,y) and Ψ(x,y) are real, not imaginary functions. Now let us take partial derivatives of Eq. (4.71) with respect to x. Using the chain rule, we obtain

$$w'(z)\partial z/\partial x = \partial p/\partial x + i\,\partial\Psi/\partial x \quad (4.72a)$$

or, since $\partial z/\partial x = 1$ from Eq. (4.70),

$$w'(z) = \partial p/\partial x + i\,\partial \Psi/\partial x \qquad (4.72b)$$

If we similarly take y derivatives, we obtain

$$w'(z)\partial z/\partial y = \partial p/\partial y + i\,\partial \Psi/\partial y \qquad (4.73a)$$

Since $\partial z/\partial y = i$ from Eq. (4.70), Eq. (4.73a) becomes, on division by i,

$$w'(z) = \partial \Psi/\partial y - i\,\partial p/\partial y \qquad (4.73b)$$

Because Eqs. (4.72b), (4.73b) describe the same quantity w'(z), equating real and imaginary parts leads to

$$\partial p/\partial x = \partial \Psi/\partial y \qquad (4.74)$$

$$\partial p/\partial y = -\partial \Psi/\partial x \qquad (4.75)$$

Observe that Eqs. (4.74), (4.75) are identical to Eqs. (4.5), (4.6), the Cauchy-Riemann conditions! And again, from Discussion 4.1, we have shown that each of the functions p(x,y) and Ψ(x,y)—namely, pressure and streamfunction—must be harmonic, satisfying Laplace's equation. What do these results show in practice? They show that for any given complex function w(z), we can take real and imaginary parts p and Ψ as suggested in Eq. (4.71). Each of these real functions will automatically satisfy Laplace's equation. Thus, if w(z) can be appropriately chosen so that the streamfunction admits streamlines of physical interest, then the functions p(x,y) and Ψ(x,y) obtained in the process comprise the complete required solution! Of course, the difficulty arises in finding the functions w(z) that do generate physically realistic flows in the (x,y) plane. Fortunately, over the past century, researchers in aerodynamics, heat transfer, and electrostatics have developed and catalogued numerous solutions. These are reinterpreted in the petroleum context in Chapter 5.

Obtaining Velocity Components

Finally, we indicate how the rectangular components of Darcy velocity are related to the complex derivative w'(z). From Eqs. (4.72b), (4.75), it is clear that

$$dw(z)/dz = \partial p/\partial x + i\partial \Psi/\partial x = \partial p/\partial x - i\partial p/\partial y \qquad (4.76)$$

$$\propto u - iv \qquad (4.77)$$

where the $\propto$ in Eq. (4.77) denotes proportionality. The relevant constant is the mobility k/μ, where k represents permeability and μ the fluid viscosity, for example, $u(x,y) = -k/\mu\ \partial p/\partial x$ where x and y would be dimensional.

Flows of Gases in Isotropic Media

Although we have considered liquids (with m = 0) exclusively, complex variables ideas readily extend to steady-state compressible gases. Consider the flow of a gas in homogeneous, isotropic media, following Discussion 4.2. Eqs. (4.17), (4.18) suggest that

$$w(z) = p^{m+1} + i\Psi \tag{4.78}$$

There is no typographical error in Eq. (4.78); the m+1 applies to pressure only. Using an approach similar to that of Eqs. (4.72a), (4.72b), (4.73a), (4.73b), we obtain Cauchy-Riemann conditions consistent with Discussion 4.2, that is,

$$\partial p^{m+1}/\partial x = \partial \Psi/\partial y \tag{4.79}$$

$$\partial p^{m+1}/\partial y = -\partial \Psi/\partial x \tag{4.80}$$

Flows in anisotropic media can be similarly treated; that is, first renormalize x and y so that the resulting equations take on an isotropic, homogeneous form. Then, the results of this section apply directly. The foregoing development for pressure and streamfunction provides the first of two powerful uses of complex variables. The second, introduced in Chapter 5, called conformal mapping, potentially transforms simple, trivial flows into exact flow solutions past complicated shapes. Before focusing on these applications, we present a powerful tool known as the Circle Theorem, used by aerodynamicists to transform seemingly artificial flows past circles into real flows past airfoils.

DISCUSSION 4.7. CIRCLE THEOREM: EXACT SOLUTIONS TO LAPLACE'S EQUATION

The flow of a nonuniform stream past a circle bears, in itself, little interest to reservoir simulation. But the flow of the same stream past a straight or curved shale or fracture segment, or an array of such segments, does: the effects of fractures and flow barriers are often paramount. It turns out that this circle flow can be conformally mapped into any of the latter, thus yielding exact closed-form solutions (see Chapter 5). For this reason, the complex potential for the flow obtained when a given nonuniform stream is altered by the introduction of a circle at the origin is of interest. The Circle Theorem, due to Milne-Thomson (1940, 1958, 1968), assumes that the flowfield is irrotational; that is, the velocity components can be found as derivatives of a potential. This is certainly so with Milne-Thomson's applications, namely, the inviscid airfoil flows briefly described in Chapter 1. But the theorem also applies to Darcy flows because the pressure field is a true potential; for example, velocities are actually computed as derivatives of the pressure. Also assumed is the absence of rigid boundaries. In practice, this is achieved by restricting one's analysis to a smaller subset of space that does not "see" the effects of farfield flow barriers.

Let f(z) denote the complex potential of the original flow without the circle. If the circle $|z| = c$ (that is, $x^2 + y^2 = c^2$) is introduced into the flow, then provided there are no other singularities within the radius c, Milne-Thomson shows that the complex potential becomes

$$w(z) = f(z) + F(c^2/z) \quad (4.81)$$

where F denotes the function obtained by replacing any explicitly appearing i's by -i's, and z by c^2/z. The proof is simple: along the circle $|z| = c$, the complex potential so defined in Eq. (4.81) is purely real. Thus, the streamfunction vanishes, showing that $|z| = c$ is in fact a streamline of the flow; this result for solid, impermeable bodies is useful in thick shale flow modeling. The reader is encouraged to demonstrate this property with various simple choices of complex potentials. A similar result known as Butler's Theorem applies to spheres in axisymmetric, three-dimensional flows (Milne-Thomson, 1968; Yih, 1969). It is not based on complex variables; instead, the transformations are given in terms of the undisturbed streamfunction. Butler's theorem, while used in airship design, is not relevant to petroleum applications.

Pressure Corollary

This author (Chin, 1978d) provided a simple extension of Milne-Thomson's result, which is applicable to the reservoir flow modeling of fractures and wells. Suppose it is desired to introduce, in an existing flow, the same circle but now having zero (that is, constant) boundary pressure; our objective is the construction of the complex potential describing the new combined flow. To do this, define the augmented complex potential

$$g(z) = -if(z) = \Psi - i\,p \quad (4.82)$$

Cauchy-Riemann conditions consistent with Eq. (4.82) are $\partial p/\partial x = \partial \Psi/\partial y$ and $\partial p/\partial y = -\partial \Psi/\partial x$, exactly the same as earlier ones given in Eqs. (4.74), (4.75). By an argument similar to that used above, consider the function

$$w(z) = g(z) + G(c^2/z) \quad (4.83)$$

The imaginary part of w(z) in Eq. (4.83) vanishes on the boundary $|z| = c$, showing that pressure satisfies $p = 0$, a constant on the circular boundary. Thus, this pressure describes a flowing well, and w(z) is the sought complex potential. This result is useful in infill drilling. In simulating aquifer-driven reservoirs, pressures are prescribed along rectangular boundaries and *fixed*, while various wells are drilled. This may be true for large aquifers. In reality, new wells will affect preexisting pressures, even those normally held constant along box boundaries. The above pressure result allows pressure changes to take place. When the initial complex potential for some nontrivial flow is available, the complex potential for the modified flow with a new well is immediately found using the construction in Eq. (4.83) without any computational effort! This process can be repeated indefinitely, with one well added after another, and continual changes in interference pressure are automatically obtained. Examples using Eqs. (4.81), (4.82) are given later.

DISCUSSION 4.8. GENERALIZED STREAMLINE TRACING AND VOLUME FLOW RATE COMPUTATIONS

We discuss streamline tracing and volume flow rate computations in planar flow; liquids and gases, however, are both allowed, and a heterogeneous, anisotropic formation is permitted. Let us consider the Darcy velocity

$$-\mathbf{q} = (k_x/\mu)\,\partial p/\partial x\,\mathbf{i} + (k_y/\mu)\,\partial p/\partial y\,\mathbf{j} \tag{4.84}$$

where **i** and **j** are unit vectors in the x and y directions, respectively. From Eqs. (4.27), (4.28), we have

$$k_x \partial p/\partial x = 1/\{(m+1)p^m\}\,\partial\Psi/\partial y \tag{4.85}$$

$$k_y\,\partial p/\partial y = -1/\{(m+1)p^m\}\partial\Psi/\partial x \tag{4.86}$$

Substitution of Eqs. (4.85), (4.86) in Eq. (4.84) leads to

$$-\mathbf{q} = 1/\{(m+1)p^m\mu\}\{\partial\Psi/\partial y\,\mathbf{i} - \partial\Psi/\partial x\,\mathbf{j}\} \tag{4.87}$$

Now the total volume flow rate for a depth D into the page is given by

$$\begin{aligned} Q_W &= -D\int_C \mathbf{q}\bullet\mathbf{n}\,dS \\ &= D/\{(m+1){p_W}^m\mu\}\int_C \{\partial\Psi/\partial y\,\mathbf{i} - \partial\Psi/\partial x\,\mathbf{j}\}\bullet\mathbf{n}\,dS \end{aligned} \tag{4.88}$$

where $p = p_W$ is assumed to be constant along the well contour C (**n** is the unit vector normal to an incremental length dS along C). First, what is the meaning of the contour integral in Eq. (4.88) which we denote by I?

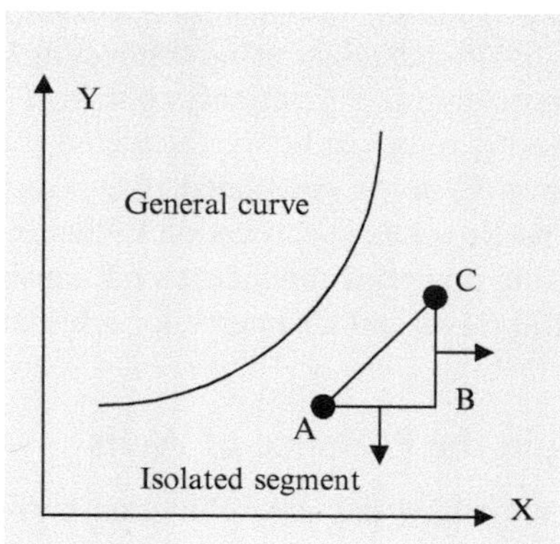

FIG. 4.1 General well contour.

$$I = \int_C \{\partial\Psi/\partial y\,\mathbf{i} - \partial\Psi/\partial x\,\mathbf{j}\}\bullet\mathbf{n}\,dS \tag{4.89}$$

To understand the meaning of I, we recognize that the general curve in Fig. 4.1 can be broken into a sequence of composite curves each of the form given by AC. The unit normals along segments AB and BC are shown in Fig. 4.1. If we evaluate the open line integral in this drawing, we obtain

$$I_{AC} = \int_{AB} \{\partial\Psi/\partial y\,\mathbf{i} - \partial\Psi/\partial x\,\mathbf{j}\} \bullet (-\mathbf{j})\,dx + \int_{BC} \{\partial\Psi/\partial y\,\mathbf{i} - \partial\Psi/\partial x\,\mathbf{j}\} \bullet \mathbf{i}\,dy$$

$$= \int_{AB} \partial\Psi/\partial x\,dx + \int_{BC} \partial\Psi/\partial y\,dy$$

$$= \Psi_B - \Psi_A + \Psi_C - \Psi_B = \Psi_C - \Psi_A = [\Psi]_{AC} \tag{4.90}$$

where the square brackets denote the jump in the streamfunction between A and C. Thus, the closed contour integral I would, after a closed circuit about a well, increase by 2π if Ψ, assuming a homogeneous isotropic flow, is solved by a simple θ. Eq. (4.88), using Eq. (4.90), simplifies to

$$Q_W = D[\Psi]/\{(m+1)\,p_W{}^m\mu\} \tag{4.91}$$

If the contour C were open and not closed, I would measure the proportional flow between A and C. For liquids with $m=0$, Eq. (4.91) becomes

$$Q_W = D[\Psi]/\mu \tag{4.92}$$

Boundary Value Problem Types

In a steady-state flow, the total volume flow rate is a constant of the problem. Sometimes, rather than pressure-pressure boundary conditions at well and far-field boundaries, pressure and flow rate specifications are desired. Observe that Eq. (4.91) represents a functional relationship connecting Q_W, P_W *and* P_R, since the pressure-pressure problem for Ψ invariably contains P_R. Now, the availability of Eq. (4.91) allows P_W to be explicitly calculated if Q_W and P_R are specified; similarly, it allows P_R to be calculated if Q_W and P_W are given. This reduces the pressure and flow rate specification to the pressure-pressure problem already solved. This powerful procedure and capability is explained in Chapter 6 for concentric flows and Chapter 9 for arbitrary planar flows.

Streamline Tracing in the Presence of Wells

We saw in Discussion 4.4 how the streamfunction is θ-like, that is, double-valued for producing and injecting wells, making standard finite difference modeling difficult. The solution to the problem was accomplished by introducing a doubled-valued function used to isolate troublesome well effects and then, to solve for the remaining single-valued disturbance streamfunction using a straightforward numerical scheme. The motivating arguments were given previously. We now extend the previous recipe to liquids and gases in anisotropic media. For our present purposes, observe that the elementary pressure function for a well

$$p^{m+1}(x, y) = A \log \sqrt{\{x^2/k_x + y^2/k_y\}} \tag{4.93}$$

satisfying $k_x \partial^2 p^{m+1}/\partial x^2 + k_y \partial^2 p^{m+1}/\partial y^2 = 0$, by virtue of the Cauchy-Riemann conditions in Eqs. (4.27), (4.28), possesses the streamfunction

$$\Psi(x, y) = A\sqrt{(k_x k_y)} \text{ arc tan}\left\{\sqrt{(k_x/k_y)}\, y/x\right\} \tag{4.94}$$

From Eq. (4.91), we have

$$[\Psi] = (m+1)\, p_W{}^m \mu Q_W/D \tag{4.95}$$

Thus, the streamfunction for a production or injection well characterized by the parameters P_W, Q_W, and gas exponent m (0, for liquids) is

$$\Psi(x, y) = \{(m+1)\, p_W{}^m \mu Q_W/(2\pi D)\} \text{ arc tan}\left\{\sqrt{(k_x/k_y)}\, y/x\right\} \tag{4.96}$$

When N wells exist, a suitable expression for the troublesome double-valued part of the total streamfunction is

$$\Psi_{d.v.}(x, y) = \{(m+1)\mu/(2\pi D)\} \times \sum_{n=1}^{N} p_{W,n}{}^m Q_{W,n} \text{ arc tan}\left\{\sqrt{(k_{x,n}/k_{y,n})}\,(y-y_n)/(x-x_n)\right\} \tag{4.97}$$

The total streamfunction $\Psi_{total}(x,y)$ satisfies

$$\Psi_{total}(x, y) = \Psi_{s.v.}(x, y) + \Psi_{d.v.}(x, y) \tag{4.98}$$

where s.v. refers to the single-valued part amenable to straightforward numerical solution. Since total streamfunction satisfies

$$\partial\{1/k_y\, \partial\Psi_{total}/\partial x\}/\partial x + \partial\{1/k_x\, \partial Y_{total}/\partial y\}/\partial y = 0 \tag{4.99}$$

it is clear that the single-valued disturbance streamfunction satisfies

$$\begin{aligned} &\partial\{1/k_y\, \partial\Psi_{s.v.}/\partial x\}/\partial x + \partial\{1/k_x\, \partial\Psi_{s.v.}/\partial y\}/\partial y \\ &= -\partial\{1/k_y\, \partial\Psi_{d.v.}/\partial x\}/\partial x - \partial\{1/k_x\, \partial\Psi_{d.v.}/\partial y\}/\partial y \end{aligned} \tag{4.100}$$

where the right side of Eq. (4.100) is evaluated using the series in Eq. (4.97). Then, the Cauchy-Riemann conditions given in Eqs. (4.27), (4.28) are used to define Neumann boundary conditions on the top and bottom, and left and right, of the computational box using the available solution for p(x,y), that is,

$$\partial\Psi_{s.v.}/\partial y = k_x\, \partial p^{m+1}/\partial x - \partial\Psi_{d.v.}/\partial y \tag{4.101}$$

$$\partial\Psi_{s.v.}/\partial x = -\,k_y\, \partial p^{m+1}/\partial y - \partial\Psi_{d.v.}/\partial x \tag{4.102}$$

Since the boundary value problem is of the Neumann type, the solution will not be unique; without loss of generality, we fix the value of Ψ_{total} to zero, say, at any point not containing a well (refer to Discussion 4.4 for details).

DISCUSSION 4.9. STREAMLINE TRACING IN 3D FLOWS

So far, we have introduced different types of streamfunctions and explained their applications to streamline tracing. Unfortunately, these methods cannot be generalized to three-dimensional problems. But for such flows, another type of 3D streamfunction or vector potential *does* exist, at least formally. This follows from the vector identity div curl $\mathbf{\Psi} = 0$, for *any* vector $\mathbf{\Psi}$. Thus, if the mass conservation requirement div $\mathbf{q} = 0$ holds, and it does for incompressible flows where $\mathbf{q}$ is the Darcy velocity, the identity guarantees that we can represent $\mathbf{q} =$ curl $\mathbf{\Psi}$, which reduces to our planar Ψ in that limit. The use of a *three-dimensional* $\mathbf{\Psi}$, however, does not render path tracing any easier than dealing with *three* velocity functions (in *two*-dimensional problems, a *single* streamfunction suffices). Thus, we shall not pursue any further discussion. But the idea of streamlines or pathlines as tangents locally parallel to the velocity vector is still attractive, and kinematically, we would expect a definition along the lines of $dx/dt \sim u$, $dy/dt \sim v$, and $dz/dt \sim w$, where $\sim$ denotes proportionality. Let us consider an interface located anywhere within a flow, that is, any surface marked by red dye, and describe it by the locus of points

$$f(x, y, z, t) = 0 \tag{4.103}$$

The interface is defined by the property that fluid does not cross it. Hence, the velocity of the fluid normal to the interface must be equal to the velocity of the interface normal to itself. From vector calculus, the normal velocity of a surface equals $-f_t/\sqrt{(f_{x^2} + f_{y^2} + f_{z^2})}$, while the normal velocity of the fluid is given by $(uf_x + vf_y + wf_z)/\{\phi\sqrt{(f_{x^2} + f_{y^2} + f_{z^2})}\}$. The condition that these be equal is

$$\partial f/\partial t + u/\phi\, \partial f/\partial x + v/\phi\, \partial f/\partial y + w/\phi\, \partial f/\partial z = 0 \tag{4.104}$$

But the left side of Eq. (4.104) is just the substantive derivative, also known as the convective or material derivative following a particle. It is usually denoted by the df/dt operator, that is,

$$df/dt = \partial f/\partial t + u/\phi\, \partial f/\partial x + v/\phi\, \partial f/\partial y + w/\phi\, \partial f/\partial z = 0 \tag{4.105}$$

Thus, we have proved that particles on a surface remain on it. To integrate Eq. (4.105), we observe that the total derivative, from calculus, satisfies

$$df = \partial f/\partial t\, dt + \partial f/\partial x\, dx + \partial f/\partial y\, dy + \partial f/\partial z\, dz \tag{4.106}$$

Division by dt leads to

$$df/dt = \partial f/\partial t + dx/dt\, \partial f/\partial x + dy/dt\, \partial f/\partial y + dz/dt\, \partial f/\partial z \tag{4.107}$$

Comparison of Eqs. (4.105), (4.107) shows that

$$df/dt = 0 \tag{4.108}$$

that is, f is constant, provided we follow the trajectories

$$dx/dt = u(x, y, z, t)/\phi \tag{4.109a}$$

$$dy/dt = v(x, y, z, t)/\phi \tag{4.109b}$$

$$dz/dt = w(x, y, z, t)/\phi \tag{4.109c}$$

Hence, when any starting coordinate (x_0,y_0,z_0) is given, its subsequent trajectory in space can be obtained by timewise integration of Eqs. (4.109a)–(4.109c) with time steps made arbitrarily small. This process is performed repeatedly with different initial starting positions to generate pathlines, also known as streaklines. These reduce to streamlines in steady flow. We also emphasize that Eqs. (4.109a)–(4.109c) are general and apply to incompressible flows (satisfying elliptic equations for pressure) as well as compressible flows (satisfying general time-dependent heat equations). While the above recipe is "straightforward," its direct use may lead to large numerical error. For example, when streamline tracing is initiated near a well, high local velocities may propel the fluid particle *out* of the computational box in a single time step! Even when this is not so, truncation errors will accumulate rapidly, destroying the integrity of the final solution. This problem can be rectified by fine-scale interpolation. Let us assume that the velocity field for a medium coarse mesh is available from a converged pressure solution (e.g., refer to Chapter 7), and consider, for example, the velocity component u(x,y,z). We denote cell block center values by asterisks. Now, the partial derivatives u_x^*, u_y^*, u_z^*, u_{xx}^*, u_{xy}^*, u_{xz}^*, u_{yx}^*, u_{yy}^*, u_{yz}^*, u_{zx}^*, u_{zy}^*, and u_{zz}^* are available from standard difference formulas. If Δ represents a typical grid size, and if also Δ_x, Δ_y, $\Delta_z \ll \Delta$, then second-order Taylor expansions give

$$\begin{aligned} u = u^* &+ \left(\Delta_x \partial/\partial x + \Delta_y \partial/\partial y + \Delta_z \partial/\partial z\right)u^* \\ &+ 1/2!\left(\Delta_x \partial/\partial x + \Delta_y \partial/\partial y + \Delta_z \partial/\partial z\right)^2 u^* + \ldots \end{aligned} \tag{4.110}$$

in operator form, where all derivatives are known. Thus, the value of u internal to any grid block is easily computed. If the streamline tracing algorithm above is applied to this subscale dimension, increased accuracy is obtained. The interpolation in Eq. (4.110) can be used independently of streamline tracing and applies also to the pressure p(x,y,z,t) if higher resolution is desired. A second approach offering improved accuracy requires initial positions (x_0, y_0,z_0) taken in low-pressure gradient locations, thus deferring the influence of large cumulative errors. However, there is no guarantee, for example, that a particular starting point will lead to a desired well if multiple wells exist in the reservoir.

Relationship to Streamfunction

The pathlines computed in Eqs. (4.109a)–(4.109c) and the streamlines (trajectories with constant streamfunction) described earlier are closely related: in steady flows, they are identical. Direct path trajectory calculations, of course, give substantially more information, since they mark individual fluid particles in time as they move. Thus, such calculations are useful for tracer, contaminant, and particle tracking analyses. Streamlines, on the other hand, yield only shape information, but they allow rapid calculations for mass flux between adjacent streamlines. The value is proportional to the differences between the streamfunction values themselves; also, narrow areas between streamlines imply high speeds. In all of the above tracing models, the effects of molecular diffusion have been ignored; this physical limitation must be considered in applications. When diffusive effects are included in any physical model, simple algorithms such as those suggested by Eqs. (4.109a)–(4.109c) cannot be used. Differential equation methods with real diffusion must be employed, which unfortunately raise the possibility of additional errors due to artificial viscosity effects that arise numerically. These subjects will be addressed in Chapter 21 in the context of molecular diffusion in fluid flows.

DISCUSSION 4.10. TRACER MOVEMENT IN 3D RESERVOIRS

Reservoir connectivity is important to sweep efficiency in all phases of production. How efficiently a formation's pore spaces are connected is determined through tracer analysis, where chemical or radioactive tracers are introduced at injection and monitored at production wells. The idea is simple: the more tracers obtained at a producer, the better the connectivity between the injectors and it. In reservoir simulation, the oilfield's permeability and porosity distributions are determined, often by trial and error, and more than likely nonuniquely, by history matching with production and well test data. In single-phase flow reservoirs, steady-state production profiles are completely determined by the pressure equation and Darcy's law, neither of which depends on porosity. In well testing, pressure buildup and drawdown depend on porosity and compressibility, factors that do not directly enter in steady-state production. Empirical tracer tests provide further information: porosity, inferred from tracer travel times, enters in steady flows where compressibility is unimportant. These three flow tests therefore provide good independent check points that are essential to good reservoir description.

As we have shown, any fluid tag in space can be tracked by the trajectory equations $dx/dt = u/\phi$, $dy/dt = v/\phi$, and $dz/dt = w/\phi$. These equations, valid for both steady and transient compressible flows, whether they contain liquids or gases, provide direct travel-time estimates for tracer breakthrough and tracer history matching. While the significance of tracer testing and analysis is appreciated operationally, the modeling of particle trajectories and time histories is plagued with unneeded numerical confusion. Very often, investigators infer streamlines and particle paths from computed two-phase saturation results, correlating local saturation changes with particle behavior in time. However, many such Eulerian-based schemes are contaminated by unnecessary truncation error and diffusion. Actually, the problem is simpler than many realize. If the Eulerian velocities u, v, and w are known for any constant density or compressible flow, for any liquid or gas phase, we recognize that the particle interface described by the surface $f(x,y,z,t)=0$ satisfies the first-order equation $\partial f/\partial t + (u/\phi)\, \partial f/\partial x + (v/\phi)\, \partial f/\partial y + (w/\phi)\, \partial f/\partial z = 0$ where ϕ is the porosity. This equation, obtained by combining Eqs. (4.107), (4.108) and derived for nonporous flows by Lord Kelvin over a century ago (e.g., see Lamb, 1945), is *exact*, and its Lagrangian solution contains the complete kinematics of the flow. Unfortunately, Kelvin's equation is used in every industry but ours. One commercial group does, however, "solve" this equation, though incorrectly labeling it as a simplified saturation equation without capillary pressure for unit mobility flows. The company uses explicit IMPES difference schemes, where pressure is solved implicitly and saturation is solved explicitly. In particular, it solves our Lagrangian f function using differencing techniques not unlike

$$
\begin{aligned}
\left(f_{i,j,k,n} - f_{i,j,k,n-1}\right)/\Delta t = & \left(u_{i,j,k,n-1}/\phi_{i,j,k}\right)\left(f_{i+1,j,k,n-1} - f_{i-1,j,k,n-1}\right)/(2\Delta x) \\
& + \left(v_{i,j,k,n-1}/\phi_{i,j,k}\right)\left(f_{i,j+1,k,n-1} - f_{i,j-1,k,n-1}\right)/(2\Delta y) \\
& + \left(w_{i,j,k,n-1}/\phi_{i,j,k}\right)\left(f_{i,j,k+1,n-1} - f_{i,j,k-1,n-1}\right)/(2\Delta z)
\end{aligned}
\tag{4.111}
$$

Since this representation is highly unstable, proprietary damping terms are introduced to offset numerical errors. The result is a scheme beset with high levels of computational diffusion. In subsequent 3D visualization, saturation fronts introduced as tracer elements, initially consisting of a single color, evolve into continuously changing multicolor displays as the saturations change along trajectories (in clear violation of $df(x,y,z,t)/dt=0$), thus giving the illusion of multiphase flow even in single-phase applications.

Of course, correct solutions to Kelvin's equation *never* produce such results. Since its trajectory equations require that f move with the particle and remain unchanged, it is clear that red water must remain red water and blue water will always be blue water. Precise methods are available to solve Kelvin's equation. For example, conservation laws of the form $\mathbf{W}_t + \{\mathbf{F}(\mathbf{W})\}_x = 0$ where $\mathbf{W}$ is a vector function of x and t are amenable to solution by high-order accurate

Lax-Wendroff schemes and their extensions (e.g., see Ames (1977)). However, unless the physical application for **W** actually requires values for individual nodes at all instances in time, the following exact, nondiffusive algorithm developed by this author can be used. To construct a simple, exact scheme, it is sufficient to observe that along each trajectory defined by dx/dt = u(x,y,z,t)/ϕ(x,y,z), dy/dt = v(x,y,z,t)/ϕ(x,y,z), and dz/dt = w(x,y,z,t)/ϕ(x,y,z), the function f(x,y,z,t) must remain unchanged by virtue of df/dt = 0. This implies, as we have suggested, that red water remains red water. We take advantage of this property by allowing the trajectory equations to update the path coordinates x(t), y(t), and z(t). We initialize f(x,y,z,t) to zero for display purposes, but once a tracer element enters a particular grid block, its f is forever marked by the *same* color and it is henceforth left alone. This introduces no diffusion beyond the simple truncation error implied by the resolution of the mesh. Some Fortran features offer useful advantages for this scheme. In Fortran, the on-off *only* nature of the function f can be coded as a logical variable, although in Fig. 4.2, we have chosen instead to use the integer array MARK(I,J,K), whose elements take on either 0 or 1 values. The entire flowfield is initially marked by 0 s, at least until individual grid blocks are penetrated by particles, at which point a Fortran switch permanently changes the particular element in MARK(I,J,K) to 1. New time-dependent indexes are defined by Fortran integer statements such as I = X/DX + 1, which track the particle to the nearest grid block. Travel-times at any point in the particle tracking are stored in the value T, which can be rewritten as an array if desired. The numerical engine behind this exact algorithm is shown in Fig. 4.2.

The Fortran engine we have described was recoded as a subroutine for a discrete input set of tracer particles. Consecutive subroutine calls repeatedly mark the array MARK(I,J,K) by 1's wherever any tracer activity is detected, leaving as 0's those grid blocks that remain unaffected. The complete particle path description is consequently embodied in the simple integer matrix MARK(I, J,K), which can *then* be plotted using off-the-shelf software.

Consider a simple but rigorous test. The right-side cubes of Fig. 4.3 display one-eighth of the spherically symmetric pressure field due to an isolated point source, providing two rotated views, for the side and front. The left cubes show purely radial trajectories, obtained for an array of tracer elements initially positioned on a side plane of the cube, as shown in the upper left plot. As is obvious, there is no numerical diffusion; particles are accurately tracked in seconds on PCs, and not hours on workstations, without concern for numerical diffusion or instability. Now, the effects of *real* diffusion can be important in practice, for example, environmental problems where contaminants convect and diffuse with the flow. A paper showing how Darcy flows couple with concentration diffusion, entitled "Modeling of Subsurface Biobarrier Formation," surveys modern numerical methods and their limitations (Chen-Charpentier and Kojouharov, 2001).

```
      .
      .
C     Define maximum dimensions (imax,jmax,kmax) of grid, and block
C     sizes dx, dy, dz.  Also, define Eulerian velocities u(i,j,k),
C     v(i,j,k), and w(i,j,k) from analytic solutions, or calculated
C     single or multiphase results. Then provide the initial tracer
C     particle coordinates, Xstart,Ystart, and Zstart.
.
C     Mark each 3D node by "0", indicating that it has not yet seen
C     tracer activity, using MARK(i,j,k) integer array.
      DO 100  I = 1,IMAX
      DO 100  J = 1,JMAX
      DO 100  K = 1,KMAX
      MARK(I,J,K) = 0
 100  CONTINUE
      .
C     Initialize position vector (x,y,z) and time.
      X = XSTART
      Y = YSTART
      Z = ZSTART
      T = 0.
      .
C     Start marching in time, for NMAX time steps.
      DO 400  N = 1,NMAX
      .
C     Define new initial (i,j,k) indexes.
      I = X/DX +1
      J = Y/DY +1
      K = Z/DZ +1
      .
C     Select time step, e.g., using
      TOP = MIN(DX,DY,DZ)
      BOT = MAX(U(I,J,K),V(I,J,K),W(I,J,K))
      DT  = 0.1 * ABS(TOP/BOT)
      .
C     If particle moves, then (i,j,k) changes.  Mark change at the
C     new coordinate with "1" (if there is no change, marking same
C     (i,j,k) repeatedly with "1s" is harmless.
      MARK(I,J,K) = 1
      .
C     Calculate new position coordinates, and update time.
      X = X + U(I,J,K)*DT
      Y = Y + V(I,J,K)*DT
      Z = Z + W(I,J,K)*DT
      T = T + DT
      .
 400  CONTINUE
      .
C     Store array of "1's" traced by particles  in "MARK.DAT" file.
C     In 3D graphics cube, "color" if "1", but do not color if "0".
C     Include header information for plotting.
```

FIG. 4.2 Rapid and exact streamline tracer algorithm.

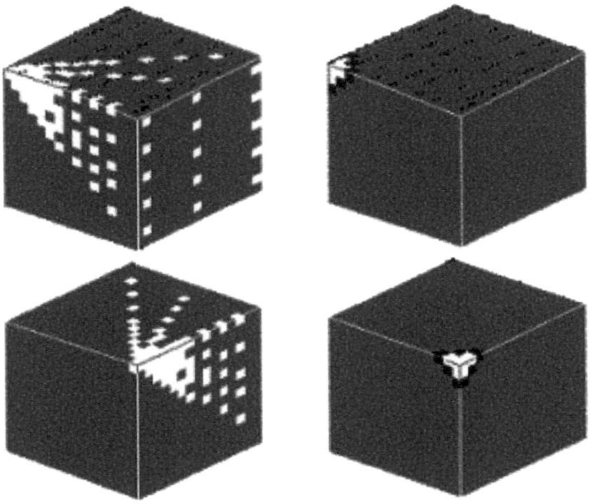

FIG. 4.3 Tracer trajectories from a spherical source.

FLUID FLOW INSTABILITIES

In deriving analytical results and preparing streamline patterns, mathematicians all-too-frequently search for steady solutions—*and find them*. Whether or not these exist in reality, however, is not obvious: the basic questions that arise form the subject that fluid-dynamicists call stability theory. Consider, for example, the flow of air past a flag that is aligned with the wind direction. The steady flow solution is $U_\infty = \text{constant}$, where U_∞ is the constant wind velocity, assuming that the flag remains perfectly still and aligned with the streamlines of the straight flow. The corresponding (aerodynamic) streamfunction is just $\Psi(x,y) = U_\infty y$. In practice, of course, flags in wind always flap: hence, the well-known "flapping flag instability." Flows in petroleum reservoirs may also be unstable, for example, the viscous fingering that is possible in enhanced oil recovery when one viscous fluid displaces another. When instability occurs, the streamlines traced on a steady basis do not exist. In the simplest cases based on linear stability theory, pathlines and interfaces oscillate sinusoidally in space and time. At critical wavelengths, neutral stability may give way to exponential growth, turbulent transition or other forms of laminar flow breakdown. To model the linear stability and existence of a steady flow, the small disturbances theory is almost always used.

The method is easily illustrated by considering the single-phase plug flow displacement model of Chapter 17, dealing with the flow of two liquids through a lineal core without mudcake (e.g., see Fig. 17.6). There, we obtain Eq. (4.112) as the equation describing the invasion front $x_{f,1}(t)$,

$$(\mu_1/\mu_2 - 1)x_{f,1} + L = +\left\{\{(\mu_1/\mu_2 - 1)x_{f,o} + L\}^2 + \{2k(P_l - P_r)/(\phi\mu_2)\}(\mu_1/\mu_2 - 1)\,t\right\}^{1/2} \tag{4.112}$$

where we have appended a subscript 1 to $x_f(t)$ in Eq. (17.13) to facilitate this discussion. Thus, depending on the relative values of μ_1 and μ_2, the displacement front may accelerate or decelerate. To address stability, consider its predecessor equation

$$dx_f/dt = -(k/(\phi\mu_1))\,(\mu_1/\mu_2)(P_r - P_l)/\{L + x_f(\mu_1/\mu_2 - 1)\} \qquad (4.113)$$

not yet restricted to solutions satisfying the initial condition $x_{f,1}(0) = x_{f,o}$. Using this less restrictive equation, let us consider the broader class of solutions $x_f = x_{f,1} + x_{f,2}$ where $|x_{f,2}| \ll |x_{f,1}|$ and $x_{f,1}$ satisfies both Eqs. (4.112), (4.113). That is, we study the disturbance $x_{f,2}$ to the baseline flow defined by Eq. (4.112). If we substitute $x_f = x_{f,1} + x_{f,2}$ into Eq. (4.113) and subtract the mean $x_{f,1}$ terms, we obtain the disturbance differential equation

$$dx_{f,2}/dt \approx (k/(\phi\mu_2))\left[(P_r - P_l)/\{L + x_{f,1}\,(\mu_1/\mu_2 - 1)\}^2\right] x_{f,1}\,(\mu_1/\mu_2 - 1) \quad (4.114)$$

Since $P_r - P_l$ is negative, and $x_{f,1}$ is positive, it follows that $dx_{f,2}/dt < 0$ when we have $\mu_1 > \mu_2$ and $dx_{f,2}/dt > 0$ when $\mu_1 < \mu_2$. This shows that if the displacing fluid is more mobile than the displaced fluid, small perturbations to the front give rise to irregularities that rapidly grow. On the other hand, if the displacing fluid is less mobile, disturbances to the front tend to decay with time. In reality, small-scale viscous fingers arise at unstable flow interfaces. We have given a simple example of hydrodynamic stability analysis. In more complicated problems, the approach is essentially the same: obtain the simple mean solution, substitute it *plus* a general disturbance back into the complete governing system, equations that will vary from problem to problem, to obtain a differential equation describing the disturbance flow, and finally, solve for the disturbance properties and determine the stability envelope. Different mathematical formulations are possible. For example, the stability of laminar, Newtonian pipe flow leads to a classic eigenvalue problem with a critical Reynolds number of 2000 that separates stable from unstable flows. Marle (1981) surveys the early literature on two-phase Darcy flows, while the books by Chin (1994, 2014b) discuss general kinematic wave aspects of hydrodynamic stability theory.

PROBLEMS AND EXERCISES

1. What advantages are gained in streamline tracing using $\Psi(x,y)$, rather than by direct numerical integration of $dy/dx = v/u$? And of $dy/dt = v/\phi$ and $dx/dt = u/\phi$? In the latter approaches, where are errors greatest? Least? Which starting flow domains for integration are best or worst for accuracy, that is, near or away from the well?
2. Verify by direct differentiation that Eqs. (4.42), (4.43), and (4.44), (4.45), satisfy the Cauchy-Riemann conditions. Repeat this exercise for Eqs. (4.58), (4.61), and (4.64), (4.67).

3. Modify the finite difference solver in Chapter 7 to allow uniform flow past three wells and compute the corresponding pressure field. Follow the procedure given in Discussion 4.4 to construct the streamlines for this flow.
4. Define the complex potential for general gas flows with arbitrary m in anisotropic media, and verify that your Cauchy-Riemann conditions satisfy Eqs. (4.17), (4.18).
5. Consider two wells in an infinite, isotropic, homogeneous reservoir, with both having identical volume flow rates. Sketch the streamline pattern when (i) both are producers, and (ii) one is a producer while the other is an injector. Repeat this exercise when the volume flow rates are different. Identify all symmetries that you observe. How would you use this information to simplify the numerical modeling of petroleum reservoirs?
6. Consider four wells located at the vertices of a square in an infinite, isotropic, homogeneous reservoir, all having identical flow rates. Sketch the streamline pattern when (i) all four are producers, and (ii) two diametrically opposite wells are producers and the other two are injectors. Identify all symmetries that you observe.

Chapter 5

Flows in Complicated Geometries

In Chapters 2 and 3, we introduced the use of integral equations in modeling flows past single fractures and shales. For these simple geometries, we were successful at developing closed-form solutions for very general boundary conditions. However, to obtain similar results for more complicated shapes, conformal mapping methods must be used. Conformal mapping is usually studied in complex variables, a graduate-level mathematics course. The transformation generates more complicated solutions once simple ones to Laplace's equation, such as those obtained previously, are available. Most elementary textbooks (Churchill, 1960; Hildebrand, 1948) offer the standard mappings—for example, finite strips to infinite domains, or angular sectors and polygons to geometrically simpler half planes. But it is in theoretical aerodynamics that complex variables and conformal mapping find their greatest application and accomplishment. Over the years, airplane and missile designers have invented numerous ingenious transforms that map simple flows into complicated ones; these mappings, unfortunately, remain unknown outside the aerospace engineering community. In this chapter, we shall apply these methods to reservoir flows. Fortunately, it is not necessary to understand the theory of complex variables, or related methods in contour integration and residue calculus, in order to obtain useful results. We shall first explain the rudimentary elements of conformal mapping using no more than simple undergraduate mathematics, expanding on the discussion in Chapter 4. Second, we shall apply the new transformation methods to Darcy flows past complicated geometries and provide useful formulas for practical use. And third, we shall give general, powerful, and simple-to-understand recipes for constructing still more general analytical solutions, and we shall also provide more solutions based on integral equation methods.

WHAT IS CONFORMAL MAPPING?

The theory behind conformal transformations is elegant and simple and it turns out that the basic notions can be introduced using no more than undergraduate math. To keep the ideas elementary, we deal with a simple fluid

Quantitative Methods in Reservoir Engineering. http://dx.doi.org/10.1016/B978-0-12-810518-4.00005-0

model first. Consider the two-dimensional, planar, steady flow of a constant density liquid in an isotropic, homogeneous medium, satisfying Laplace's equation

$$\partial^2 P/\partial x^2 + \partial^2 P/\partial y^2 = 0 \tag{5.1}$$

where P(x,y) is the pressure, and x and y are Cartesian coordinates. The Eulerian velocities u and v in the x and y directions, respectively, are

$$u = -(k/\mu)\, \partial P/\partial x \tag{5.2}$$

$$v = -(k/\mu)\, \partial P/\partial y \tag{5.3}$$

where k and μ are permeability and viscosity. As noted in Chapters 2 and 3, (x,y) coordinates are useful when flow entities of interest, for example, fractures or shale lenses, lie roughly along coordinate lines. This is not the case for most practical problems. Thus, we seek general coordinates in the hope for usable results.

A General Transformation

For this reason, we consider the transformation pair

$$\xi = \xi(x, y) \tag{5.4}$$

$$\eta = \eta(x, y) \tag{5.5}$$

and re-express Eq. (5.1) in a different but exactly equivalent way. Using the chain rule (Thomas, 1960), we differentiate with respect to x to find

$$P(x, y) = p(\xi, \eta) \tag{5.6}$$

$$\partial P/\partial x = (\partial p/\partial \xi)(\partial \xi/\partial x) + (\partial p/\partial \eta)(\partial \eta/\partial x) \tag{5.7}$$

$$\begin{aligned} \partial^2 P/\partial x^2 = {} & (\partial p/\partial \xi)\left(\partial^2 \xi/\partial x^2\right) + (\partial p/\partial \eta)\left(\partial^2 \eta/\partial x^2\right) \\ & + (\partial \xi/\partial x)\left\{\partial^2 p/\partial \xi^2 \partial \xi/\partial x + \partial^2 p/\partial \xi \partial \eta\, \partial \eta/\partial x\right\} \\ & + (\partial \eta/\partial x)\left\{\partial^2 p/\partial \eta \partial \xi\, \partial \xi/\partial x + \partial^2 p/\partial \eta^2 \partial \eta/\partial x\right\} \end{aligned} \tag{5.8}$$

Similarly, differentiation with respect to y yields

$$\partial P/\partial y = (\partial p/\partial \xi)(\partial \xi/\partial y) + (\partial p/\partial \eta)(\partial \eta/\partial y) \tag{5.9}$$

$$\begin{aligned} \partial^2 P/\partial y^2 = {} & (\partial p/\partial \xi)\left(\partial^2 \xi/\partial y^2\right) + (\partial p/\partial \eta)\left(\partial^2 \eta/\partial y^2\right) \\ & + (\partial \xi/\partial y)\left\{\partial^2 p/\partial \xi^2 \partial \xi/\partial y + \partial^2 p/\partial \xi \partial \eta\, \partial \eta/\partial y\right\} \\ & + (\partial \eta/\partial y)\left\{\partial^2 p/\partial \eta \partial \xi\, \partial \xi/\partial y + \partial^2 p/\partial \eta^2 \partial \eta/\partial y\right\} \end{aligned} \tag{5.10}$$

Therefore, it follows upon adding Eqs. (5.8), (5.10) that

$$\begin{aligned}\partial^2 P/\partial x^2 + \partial^2 P/\partial y^2 = &(\partial p/\partial \xi)\left(\partial^2 \xi/\partial x^2 + \partial^2 \xi/\partial y^2\right)\\ &+ (\partial p/\partial \eta)\left(\partial^2 \eta/\partial x^2 + \partial^2 \eta/\partial y^2\right)\\ &+ 2\left(\partial^2 p/\partial \eta \partial \xi\right)(\partial \xi/\partial x\, \partial \eta/\partial x + \partial \xi/\partial y\, \partial \eta/\partial y)\\ &+ \left(\partial^2 p/\partial \xi^2\right)\left\{(\partial \xi/\partial x)^2 + (\partial \xi/\partial y)^2\right\}\\ &+ \left(\partial^2 p/\partial \eta^2\right)\left\{(\partial \eta/\partial x)^2 + (\partial \eta/\partial y)^2\right\} \quad (5.11)\end{aligned}$$

The lower case p indicates that its functional dependence on (ξ,η) generally differs from that of P on (x,y). So far, Eqs. (5.4), (5.5) are quite general, as demonstrated by the complicated right side of Eq. (5.11).

Conformal Mapping

It is possible to make an important simplification to Eq. (5.11) that is also physically useful. Let us impose the constraint that

$$\partial \xi/\partial x = \partial \eta/\partial y \quad (5.12)$$

$$\partial \eta/\partial x = -\partial \xi/\partial y \quad (5.13)$$

If we differentiate Eq. (5.12) with respect to x and Eq. (5.13) with respect to y, and eliminate the resulting $\partial^2\eta/\partial x\partial y$ term, we obtain

$$\partial^2 \xi/\partial x^2 + \partial^2 \xi/\partial y^2 = 0 \quad (5.14)$$

Likewise, if we differentiate Eq. (5.12) with respect to y and Eq. (5.13) with respect to x, and eliminate the resulting $\partial^2\xi/\partial x\partial y$ term, we obtain

$$\partial^2 \eta/\partial x^2 + \partial^2 \eta/\partial y^2 = 0 \quad (5.15)$$

Thus, the first two lines of Eq. (5.11) vanish identically. Next, the third line of Eq. (5.11) also vanishes; this is apparent from Eqs. (5.12), (5.13), since direct substitution shows

$$\begin{aligned}&\partial \xi/\partial x\, \partial \eta/\partial x + \partial \xi/\partial y\, \partial \eta/\partial y\\ &= (\partial \eta/\partial y)(-\partial \xi/\partial y) + \partial \xi/\partial y\, \xi \partial \eta/\partial y = 0\end{aligned} \quad (5.16)$$

Now, the quantity j(x,y), known as the Jacobian of the transformation, is *defined* by the expression

$$j(x, y) = (\partial \xi/\partial x)(\partial \eta/\partial y) - (\partial \xi/\partial y)(\partial \eta/\partial x) \quad (5.17)$$

Using Eqs. (5.12), (5.13), we have the equivalent forms

$$j(x, y) = (\partial \xi/\partial x)^2 + (\partial \xi/\partial y)^2 > 0 \quad (5.18)$$

$$= (\partial \eta/\partial x)^2 + (\partial \eta/\partial y)^2 > 0 \quad (5.19)$$

That is,

$$(\partial\xi/\partial x)^2 + (\partial\xi/\partial y)^2 = (\partial\eta/\partial y)^2 + (\partial\eta/\partial x)^2 = j(x, y) \qquad (5.20)$$

Thus, the result of assuming Eqs. (5.12), (5.13) in Eq. 5.11 is the remarkable fact that

$$\partial^2 P/\partial x^2 + \partial^2 P/\partial y^2 = j(x, y)\{\partial^2 p/\partial\xi^2 + \partial^2 p/\partial\eta^2\} \qquad (5.21a)$$

Since the left side of Eq. (5.21) vanishes by virtue of Eq. (5.1), it therefore follows, since $j > 0$, that

$$\partial^2 p/\partial\xi^2 + \partial^2 p/\partial\eta^2 = 0 \qquad (5.21b)$$

which has neither first-order nor second-derivative cross-terms. Thus, when the "conformal" constraints in Eqs. (5.12), (5.13) are invoked, any harmonic function P(x,y) that originally satisfies Laplace's equation in (x,y) coordinates now satisfies the same equation in (ξ,η) coordinates.

In the language of mathematics, Laplace's equation remains invariant under conformal transformation, which is equivalent to saying that "harmonic functions stay harmonic." There are also geometric interpretations associated with Eqs. (5.12), (5.13) discussed elsewhere, for example, Churchill (1960). All of this is elegant, powerful and practical: when a closed form or numerical solution for P(x,y) is available, another free solution $p(\xi,\eta)$ is easily generated *if* we can find transforms $\xi = \xi(x,y)$ and $\eta = \eta(x,y)$ that satisfy Eqs. (5.12), (5.13). (When $p(\xi,\eta)$ is available, a mirror P(x,y) can be obtained.)

Note that both (ξ,η) and (x,y) are rectangular Cartesian coordinates and that in any application, either system can represent the physical plane. In any event, the idea is to have *additionally* chosen $\xi = \xi(x,y)$ and $\eta = \eta(x,y)$ so that, say, lines of constant ξ (or η) *happen* to fall along geometrical boundaries of interest to the physical problem at hand (e.g., a curved fracture or the irregular outer boundary of a reservoir). If so, the solution of a difficult problem in (ξ,η) coordinates is transformed into one that is purely topological and simpler (again, the converse is true). Note that the *constant* value of the harmonic function along a curve carries over to the transformed curve, and similarly, the vanishing normal derivative of the function. These two boundary conditions promise direct applications to wells, fractures or boundaries held at fixed pressures and to impermeable shales subject to no flow constraints. When a function is variable along a curve, the transformed function will take on variable values; however, the exact variation in the mapped plane will depend on the details of the mapping. For variable values of the function, conformal mapping is less direct; integral equation methods such as those described in Chapters 2 and 5 are preferable if the problem permits geometrical simplifications like those used.

These conditions, taken collectively, demand much of the engineer. It turns out that two powerful methods may be used to solve the topological problem introduced. The first uses complex variables, and is discussed next; the second, newly developed by this author, uses results based on numerical grid generators, and is discussed in Chapters 9 and 10. Finally, we discuss second-derivative cross-terms, for example, $\partial^2 p/\partial\xi\partial\eta$, mentioned earlier. These do not appear in Eq. (5.21b) because Eqs. (5.12), (5.13) introduce restrictions, fortunately useful ones, as will be shown in this chapter. More general nonconformal mappings, however, generally introduce cross-terms into transformed partial differential equations, which can complicate the solution of the new equation. Such general transformations, in the petroleum industry, are known as corner point methods, but their benefits are lost because cross-terms in the transformed equation are ignored by many simulators due to numerical stability problems. However, stable schemes can be constructed using central difference representations for $\partial^2/\partial\xi\partial\eta$ in finite difference relaxation methods. For example, the author's recent books (Chin, 2000, 2012a,b) on computational rheology make use of such algorithms to calculate (nonlinear) non-Newtonian flowfields in highly clogged noncircular pipes and eccentric annuli.

USING "SIMPLE" COMPLEX VARIABLES

The basic tools needed for conformal mapping can be developed without advanced background material related to contour integration, residue analysis, branch cuts, and so on. It turns out that the key ideas follow from simple manipulations using no more than elementary calculus. Let us employ the usual notation $i = \sqrt{-1}$ for "imaginary numbers," and introduce the complex coordinate

$$z = x + i y \tag{5.22}$$

Also, consider any function w(z) of the single variable z, for example, sin z, cosh z, or exp(z). In general, it is obvious that w(z) must be complex. Thus, it should be possible to resolve it into real and imaginary parts $\xi(x,y)$ and $\eta(x,y)$,

$$w(z) = \xi(x, y) + i\,\eta(x, y) \tag{5.23}$$

Now let us differentiate Eq. (5.23) with respect to x. Using the chain rule,

$$w'(z)\partial z/\partial x = \partial\xi(x, y)/\partial x + i\,\partial\eta(x, y)/\partial x \tag{5.24}$$

where the prime indicates an ordinary derivative taken with respect to z. But since $\partial z/\partial x = 1$ from Eq. (5.22), Eq. (5.24) becomes

$$w'(z) = \partial\xi(x, y)/\partial x + i\,\partial\eta(x, y)/\partial x \tag{5.25}$$

Similarly, if we differentiate Eq. (5.23) with respect to y and make use of the fact that $\partial z/\partial y = i$ from Eq. (5.22), we have

$$i\,w'(z) = \partial\xi(x, y)/\partial y + i\,\partial\eta(x, y)/\partial y \tag{5.26}$$

If next we multiply Eq. (5.26) through by -i, we obtain

$$w'(z) = \partial\eta(x, y)/\partial y - i\,\partial\xi(x, y)/\partial y \tag{5.27}$$

Observe that Eqs. (5.25), (5.27) describe the same derivative w'(z). Setting the two expressions equal and equating real and imaginary parts lead to

$$\partial\xi/\partial x = \partial\eta/\partial y \tag{5.28}$$

$$\partial\eta/\partial x = -\partial\xi/\partial y \tag{5.29}$$

But Eqs. (5.28), (5.29) are identical to Eqs. (5.12), (5.13)! Thus, we have uncovered a simple, yet powerful, method for generating mappings that satisfy Eqs. (5.12), (5.13): simply write down *any* analytic function w(z) of $z = x + iy$, and take its real and imaginary parts (see Eq. 5.23).

The topological problem stated at the end of the previous section is now reduced to a simpler search for appropriate functions of z, that is, those functions w(z) whose level curves $\xi(x,y) = \text{constant}$ and $\eta(x,y) = \text{constant}$ happen to coincide with geometric boundaries of interest. A classic theorem due to Riemann (e.g., see Carrier et al., 1966) guarantees the *existence* of most conformal transformations. That is, the transform taking a simply connected domain into a circle, and those taking a doubly connected ("donut") annulus into a concentric annular ring are assured. Unfortunately, how this is achieved is *not* addressed. The Schwarz-Christoffel transformation, which maps the interior of a general N-sided polygon to a half plane, though, can be used in many problems. This requires the approximation of smooth curves by sequences of straight-line segments; the process, which can lead to cumbersome integrals and unwieldy algebra when the segments exceed a few, may produce unphysical results at nonexistent corners. Finding useful transformations becomes a matter of experience and insight; fortunately, mappings are available in the literature outside of the petroleum industry that we can take advantage of. Standard ones are available in Hildebrand (1948), Kober (1957), and Spiegel (1964); less conventional ones appear in Ashley and Landahl (1965), Thwaites (1960), Woods (1961), and Carrier et al. (1966).

EXAMPLE 5.1. THE CLASSIC RADIAL FLOW SOLUTION

In this first example, we shall reproduce the logarithmic pressure solution for radial flow starting with a trivial uniform flow in an auxiliary plane. To do this, let us consider the logarithmic mapping

$$w(z) = \log z \tag{5.30}$$

where log denotes the natural logarithm. From geometry, we have $x = r\cos\theta$ and $y = r\sin\theta$, where (r,θ) are cylindrical coordinates. Thus, we can write Eq. (5.22) in the polar form $z = x + i\,y = re^{i\theta}$ so that $\log z = \log r + i\,\theta$. Applying Eq. (5.23), we have

$$w(z) = \xi(x, y) + i\,\eta(x, y) = \log r + i\theta \tag{5.31}$$

so that

$$\xi(x, y) = \log r = \log \sqrt{(x^2 + y^2)} \tag{5.32}$$

$$\eta(x, y) = \theta = \tan^{-1} y/x \tag{5.33}$$

We now determine the manner in which boundary contours transform under the mapping given above. From Eq. (5.32), the straight-line $\xi = \log R_{well} = \text{constant}$ maps into the circle $x^2 + y^2 = R_{well}^2 = \text{constant}$; the line $\xi = \log R_{ff} = \text{constant}$ maps into $x^2 + y^2 = R_{ff}^2 = \text{constant}$, where the well and ff subscripts refer to radii at the borehole and at a prescribed farfield location. Eq. (5.33) shows that lines of constant η map into radial lines having constant θ or y/x values. These results are summarized in Fig. 5.1, which shows the correspondence between the x-y plane and the ξ-y plane.

To demonstrate how nontrivial flow results can be obtained from simple ones, let us consider the pressure field

$$p(\xi, \eta) = p_{well} + (p_{ff} - p_{well})\,(\xi - \log R_{well})/(\log R_{ff} - \log R_{well}) \tag{5.34}$$

satisfying $p(\xi,\eta) = p_{ff}$ when $\xi = \log R_{ff}$, and $p(\xi,\eta) = p_{well}$ when $\xi = \log R_{well}$. In the ξ,η plane, Eq. (5.34) is associated with a trivial (and uninteresting) uniform flow having velocity components that are proportional to the pressure gradients

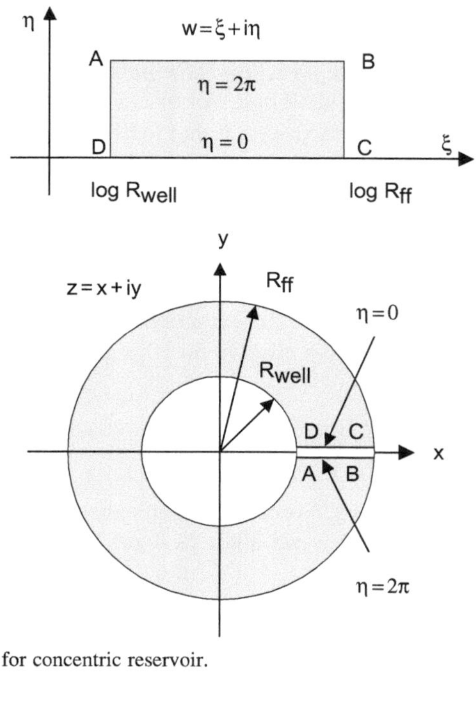

FIG. 5.1 Mapping for concentric reservoir.

$\partial p/\partial \xi = (p_{ff} - p_{well})/(\log R_{ff} - \log R_{well})$ and $\partial p/\partial \eta = 0$. Note that Eq. (5.34) is a simple function that depends on ξ only. Now, according to the results of the previous section, a harmonic pressure function satisfying Eq. (5.1) is constructed by substituting Eq. (5.32) in Eq. (5.34), leading to

$$\begin{aligned} P(x,y) &= p_{well} \\ &\quad + (p_{ff} - p_{well})\left(\log\sqrt{(x^2+y^2)} - \log R_{well}\right)/(\log R_{ff} - \log R_{well}) \\ &= p_{well} \\ &\quad + \{(p_{ff} - p_{well})/(\log R_{ff}/R_{well})\}\left\{\log\sqrt{(x^2+y^2)}/R_{well}\right\} \end{aligned} \tag{5.35}$$

which is the well-known logarithmic solution, that is, the solution for radial flow usually expressed in the more familiar log r form,

$$p(r) = p_{well} + \{(p_{ff} - p_{well})/(\log R_{ff}/R_{well})\}(\log r/R_{well}) \tag{5.36}$$

This example demonstrates the elegance and power behind simple-looking transformations such as Eq. (5.30). It is important to emphasize that we transformed a pressure function solving $p_{\xi\xi} + p_{\eta\eta} = 0$ to one solving the rectangular form $P_{xx} + P_{yy} = 0$, and *not* $p_{rr} + (1/r)\,p_r + (1/r^2)\,p_{\theta\theta} = 0$ directly.

EXAMPLE 5.2. CIRCULAR BOREHOLE WITH TWO SYMMETRIC RADIAL FRACTURES

Let us consider the steady-state flow of a liquid or a gas issuing from (or into) the straight-line fracture $-c \leq X \leq +c$, $Y = 0$, centered in an isotropic, homogeneous circular reservoir of radius $R \gg c$. We also assume that a *constant* pressure P_f is enforced along the fracture, and that a different constant pressure P_R holds at the farfield boundary. The analytical solution to this problem appears as a special limit of the $P_{Ref}^{m+1} p^{m+1}(x,y)$ formula developed in Example 2.2 of Chapter 2. Here, as in that chapter, all upper case variables will be dimensional, and all lower case variables will be dimensionless.

We shall, for convenience, denote the limit solution of Example 2.2 by the function $P_{Ref}^{m+1} p^{m+1}(X/c, Y/c)$, where the parametric dependence on the fracture half-length c is emphasized. Also observe that this function is harmonic, satisfying Laplace's equation

$$\partial^2 P_{Ref}^{m+1} p^{m+1}(X/c, Y/c)/\partial X^2 + \partial^2 P_{Ref}^{m+1} p^{m+1}(X/c, Y/c)\partial Y^2 = 0 \tag{5.37}$$

in rectangular (X,Y) coordinates, thus rendering conformal mapping applicable. The scale factor P_{Ref}^{m+1} is constant and cancels out, but we intentionally retain it to emphasize that we shall be working in dimensional (X,Y) coordinates.

The problem is simply stated: can we transform this limit solution for a straight line, available analytically, into one describing the flow from a circular borehole with two symmetric radial fractures? The answer is, "Yes." To do this, let us first define a new function

$$G^*(X, Y) = P_{Ref}^{m+1} p^{m+1}\left(X/\{s + R_w^2/s\}, Y/\{s + R_w^2/s\}\right) \tag{5.38}$$

obtained by replacing the half-length c in $P_{Ref}^{m+1} p^{m+1}(X/c, Y/c)$ with the quantity $s + R_w^2/s$, where R_w is to be the radius of the circular well (the motivation for this will be clear later). We now show that this solution, with the fracture length so defined, can be related to a wellbore with intersecting, symmetric, radial fractures of equal length. This is done by considering the Joukowski mapping from aerodynamics (Milne-Thomson, 1958),

$$Z = \zeta + R_w^2/\zeta \tag{5.39}$$

where

$$Z = X + i\,Y \tag{5.40}$$

$$\zeta = \xi(X, Y) + i\,\eta(X, Y) \tag{5.41}$$

If we next substitute Eqs. (5.40), (5.41) into Eq. (5.39), and equate real and imaginary parts, we have

$$X = \xi + R_W^2 \xi/\left(\xi^2 + \eta^2\right) \tag{5.42}$$

$$Y = \eta - R_W^2 \eta/\left(\xi^2 + \eta^2\right) \tag{5.43}$$

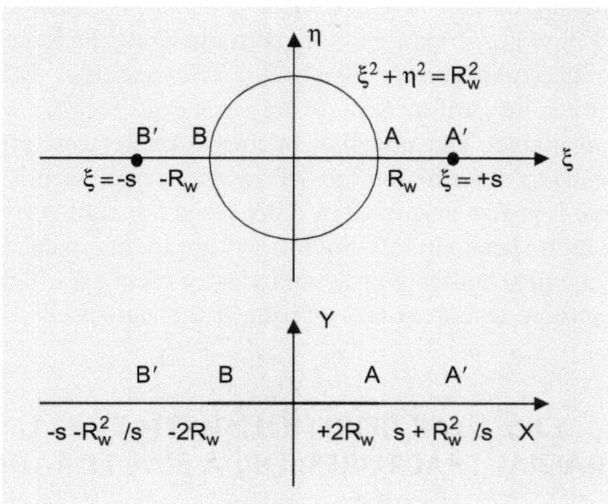

FIG. 5.2 Circular hole with symmetric radial fractures.

Let us now observe the following transformation results: (i) the circular well $\xi^2 + \eta^2 = R_w^2$ maps into the straight line (or slit) $X = 2\xi$, $Y = 0$; (ii) the line $\eta = 0$ maps into the slit $Y = 0$ of Chapter 2; and (iii) the tips ($\xi = -s$, $\eta = 0$) and ($\xi = +s$, $\eta = 0$) transform, respectively, to the points ($X = -s - R_w^2/s$, $Y = 0$) and ($X = +s + R_w^2/s$, $Y = 0$). Note that Eq. (5.39) leaves events at infinity unchanged, since $Z = \zeta$ in the farfield. These results appear in Fig. 5.2. Thus, the solution G*(X,Y) in Eq. (5.38), applying to a straight-line fracture in the (X,Y) plane spanning ($X=-s-R_w^2/s$, $Y=0$) and ($X=+s+R_w^2/s$, $Y=0$), can be used to generate the solution for a circular well of radius R_w having symmetric radial fractures in (ξ,η) coordinates with a tip-to-tip distance of 2s. The solution $P^{m+1}(\xi,\eta)$ is obtained by substitution of Eqs. (5.42), (5.43) into Eq. (5.38),

$$P^{m+1}(\xi,\eta) = G^*\{\xi + R_W^2\xi/(\xi^2+\eta^2), \eta - R_W^2\eta/(\xi^2+\eta^2)\} \qquad (5.44)$$

where the italicized P indicates that it is different functionally from p(x,y). It is important to emphasize that $P^{m+1}(\xi,\eta)$ satisfies the Cartesian form of Laplace's equation in (ξ,η) coordinates.

Finally, let $\Psi(X/c,Y/c)$ be the streamfunction associated with $p^{m+1}(X/c, Y/c)$; methods to obtain streamfunctions from pressure solutions are given in Chapter 4. We can similarly introduce a function

$$H^*(X,Y) = \Psi(X/\{s+R_W^2/s\}, Y/\{s+R_W^2/s\}) \qquad (5.45)$$

which has the corresponding italicized streamfunction

$$\Psi(\xi,\eta) = H^*(\xi + R_W^2\xi/(\xi^2+\eta^2), \eta - R_W^2\eta/(\xi^2+\eta^2)) \qquad (5.46)$$

Under conformal transformations, constant values of a function (say, pressure) transform into the same constant values in the new plane, as indicated previously; thus, the application to the function $p^{m+1}(x,y)$ is conventional. But the application to $\Psi(x,y)$ is *not*, since Ψ is *not* constant along the fracture: each portion of the fracture contributes to the net flow. Nevertheless, Eq. (5.46) is the correct transformed streamfunction, if only because $\Psi(X/c,Y/c)$ is assumed to be already available. The mapping in this example was originally given by Spreiter (1950) to describe the cross-flow about circular missile fuselages with double fins. It is also discussed in Ashley and Landahl (1965) for slender body theory in high-speed aircraft. Such flows are routinely calculated by formula in aerodynamics, but the petroleum industry often uses high-speed computers requiring more resources and offering less accuracy.

EXAMPLE 5.3. CIRCULAR BOREHOLE WITH TWO UNEVEN, OPPOSITE, RADIAL FRACTURES; OR, A SINGLE RADIAL FRACTURE

Now we consider the same circular borehole but with two uneven, opposite radial fractures. Following Example 5.2, we again consider the function $P_{Ref}^{m+1}p^{m+1}(X/c,Y/c)$ of Example 2.2, and recognize that the solution applies to the line segment $-1 \le X/c \le +1$ or $-c \le X \le +c$. But instead of the function in Eq. (5.38), we consider the X-shifted function

$$G^*(X,Y) = P_{Ref}{}^{m+1}p^{m+1}((X-\delta)/\{s + R_W^2/s\}, Y/\{s+R_W^2/s\}) \qquad (5.47)$$

which still satisfies Laplace's equation in (X,Y) coordinates. The straight-line fracture for Eq. (5.47) is located along $-1 \le (X-\delta)/\{s+R_w^2/s\} \le +1$, that is, along $-s-R_w^2/s+\delta \le X \le +s+R_w^2/s+\delta$.

Again, we consider the conformal mapping defined by Eqs. (5.39)–(5.43). As before, we have the following transformation results: (i) the circular well $\xi^2+\eta^2=R_w^2$ maps into the straight line (or slit) $X=2\xi$, $Y=0$, and also, (ii) the line $\eta=0$ maps into the slit $Y=0$ of Chapter 2. But now the fracture tips located at points ($X=-s-R_w^2/s+\delta$, $Y=0$) and ($X=+s+R_w^2/s+\delta$, $Y=0$) in the slit only plane do not map into -s and +s. To determine the corresponding tip positions in the (ξ,η) plane, we note from Eq. (5.43) that $\eta=0$ still transforms into $Y=0$. Eq. (5.42) evaluated along $\eta=0$ gives the result that $X=\xi+R_w^2/\xi$ or $\xi^2-X\xi+R_w^2=0$, so that two ξ values, obtained from the quadratic formula, correspond to any specified X.

When $X=-s-R_w^2/s+\delta$, the solution to $\xi^2+(s+R_w^2/s-\delta)\xi+R_w^2=0$ chosen is

$$\xi = \left[(-s-R_W^2/s+\delta) - \sqrt{\left\{(s+R_W^2/s-\delta)^2 - 4R_W^2\right\}}\right]/2 \tag{5.48}$$

And when $X=+s+R_w^2/s+\delta$, the solution to $\xi^2+(-s-R_w^2/s-\delta)\xi+R_w^2=0$ selected is

$$\xi = \left[(s+R_W^2/s+\delta) + \sqrt{\left\{(s+R_W^2/s+\delta)^2 - 4R_W^2\right\}}\right]/2 \tag{5.49}$$

The signs were chosen so that in the limit $R_w=0$ when the tip-to-tip fracture length greatly exceeds the wellbore radius, Eqs. (5.48), (5.49) reduce to

$$\xi = \left[(-s+\delta) - \sqrt{\{(s-\delta)^2\}}\right]/2 = -s+\delta \tag{5.50}$$

$$\xi = \left[(s+\delta) + \sqrt{\{(s+\delta)^2\}}\right]/2 = +s+\delta \tag{5.51}$$

The final schematic is given in Fig. 5.3. In the additional limit $\delta=0$, we recover single fracture only results: the foregoing abscissa values now become $\xi=-s$ and $+s$, leading to "centered fracture" results. Also, the X-shift δ can be chosen so that a single radial fracture appears in place of two.

EXAMPLE 5.4. CIRCULAR BOREHOLE WITH MULTIPLE RADIAL FRACTURES

In Examples 5.2 and 5.3, we considered a circular wellbore with double or single radial fractures. It is also possible to consider a circular well having *any number* of radial fractures of equal length. Such multiple fractures, sometimes observed in hydraulic fracturing, are often encountered in geological studies of conjugate fracture sets. Consider the fractured borehole shown in Fig. 5.4. The circle is defined by $x^2+y^2=1$, or $|z|=1$. It is drawn with N radial projections

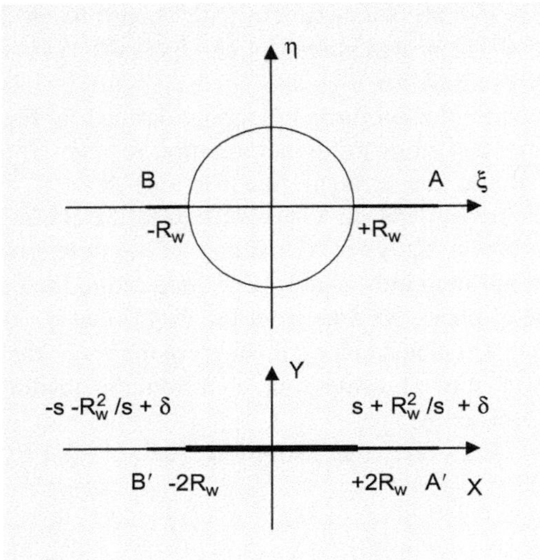

FIG. 5.3 Circular hole with nonsymmetric radial fractures.

defined by $\arg z = \arg re^{i\theta} = \theta = (2k+1)\pi/N$, $1 < |z| < a$, $k = 0, 1, \ldots, N-1$, with the parameter $a > 1$. This representation describes the fracture orientation of interest.

We emphasize that such boreholes have been observed from time to time. However, the fracture configuration of Fig. 5.4 might also represent the flow from conjugate sets of intersecting fractures, say defined by $N = 4$ and $a \gg 1$. Let us define an auxiliary quantity

$$b = \left(a^N + a^{-N}\right)/2 \tag{5.52}$$

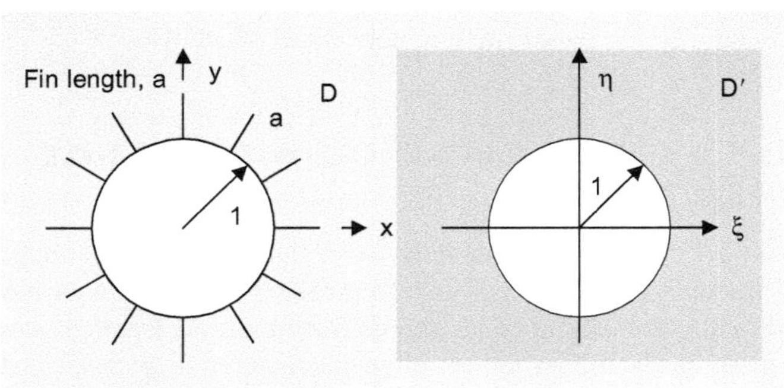

FIG. 5.4 Hole with multiple radial fractures.

Then the transformation

$$\zeta = (b+1)^{-N}\left\{z^N + z^{-N} + b - 1 + \left[\left(z^N + z^{-N} + 2b\right)\left(z^N + z^{-N} - 2\right)\right]^{1/2}\right\}^{1/N} \quad (5.53)$$

with $0 < \arg z < 2\pi$ maps the domain D in the z-plane into the exterior circular domain D' given by $|\zeta| > 1$ in the ζ plane.

This mapping is used to study the aerodynamic characteristics of multiple stabilizer fins on rocket cross-sections; for example, see Ashley and Landahl (1965). Next, a host pressure solution in the transformed coordinates is required. We start with the classical logarithmic solution for radial flow, however, expressed in Cartesian (ξ,η) coordinates. This solution is already available from Eq. (5.35); we use it with the constant $R_{well} = 1$ to be consistent with the foregoing transformation. Thus, the host pressure formula is

$$\begin{aligned} p(\xi, \eta) &= p_{well} \\ &\quad + \{(p_{ff} - p_{well})/(\log R_{ff})\} \log \sqrt{(\xi^2 + \eta^2)} \end{aligned} \quad (5.54)$$

If we now set $\zeta = \xi + i\,\eta$ and $z = x + i\,y$ in Eq. (5.53), equate real and imaginary parts to give expressions for $\xi = \xi(x,y)$ and $\eta = \eta(x,y)$, and substitute the functions into Eq. (5.54) to yield the (italicized) pressure P in (x,y) coordinates, we obtain

$$\begin{aligned} P(x, y) &= p_{well} \\ &\quad + \{(p_{ff} - p_{well})/(\log R_{ff})\} \log \sqrt{\left\{\xi(x, y)^2 + \eta(x, y)^2\right\}} \end{aligned} \quad (5.55)$$

The complexity of Eq. (5.53), of course, means that the foregoing steps will in general be performed numerically.

EXAMPLE 5.5. STRAIGHT SHALE SEGMENT AT ARBITRARY ANGLE

We turn our attention to impermeable shales, that is, to solid line objects that are impermeable to flow. In Chapter 3, we developed an analytical approach limited to small flow inclinations. The solution here applies more generally to large angles of attack.

We shall again show how conformal mapping can be used to generate flowfields past complicated structures. For reasons that will be apparent, we first obtain the flow past a circular cylinder at an arbitrary angle of attack. This initial step can be accomplished two different ways, and we demonstrate both methods.

Method 1

The traditional undergraduate approach utilizes separation of variables. We start with the pressure equation in cylindrical coordinates

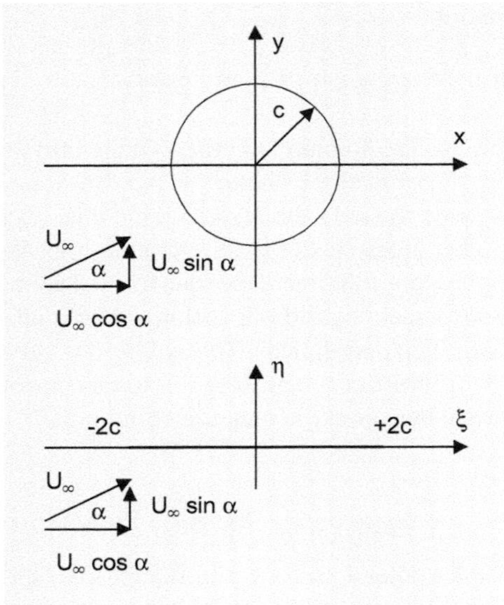

FIG. 5.5 Flow past straight shale.

$$p_{rr} + (1/r)\, p_r + \left(1/r^2\right) p_{\theta\theta} = 0 \tag{5.56}$$

and search for separable solutions of the form

$$p(r,\theta) = G(r)\, H(\theta) \tag{5.57}$$

Substitution in Eq. (5.56) shows that

$$r^2G''/G + rG'/G = -H''/H = +\sigma^2 > 0 \tag{5.58}$$

where σ^2 is a positive constant. Eq. (5.58) implies two ordinary differential equations, namely, $r^2G''+rG'-\sigma^2G=0$ and $H''+\sigma^2H=0$. For $\sigma=0$, the solutions are not meaningful physically, since H, being directly proportional to θ, implies a double-valued pressure field in the continuous sand far away from the circle. Thus, we consider a nonvanishing σ^2, for which the solutions are $G_\sigma(r)=A_\sigma r^\sigma+B_\sigma r^{-\sigma}$ and $H_\sigma(\theta)=C_\sigma \sin \sigma\theta+D_\sigma \cos \sigma\theta$. Hence, solutions take the form $p_\sigma(r,\theta)=G_\sigma(r)H_\sigma(\theta)=(A_\sigma r^\sigma+B_\sigma r^{-\sigma})(C_\sigma \sin \sigma\theta+D_\sigma \cos \sigma\theta)$.

The $\sigma=1$ solution solves the problem. In this simple limit, we find that $p_1(r,\theta)=(Ar+Br^{-1})(C \sin \theta+D \cos \theta)$, say. Now choose constants A, B, C and D so that

$$p(r,\theta) = -(U_\infty\mu/k)\left(r+c^2/r\right)(\cos\alpha\cos\theta+\sin\alpha\sin\theta) \tag{5.59a}$$

It is clear that $\partial p/\partial r$ is proportional to the factor $(1-c^2/r^2)$; hence, the radial Darcy velocity vanishes on the circle $r=c$. Next, at distances far away where r approaches infinity, Eq. (5.59a) behaves like $p(r,\theta)\approx -(U_\infty\mu/k)\, r\,(\cos\alpha\cos\theta+\sin\alpha\sin\theta)$; that is, in Cartesian (x,y) coordinates, $p(x,y)\approx -(U_\infty\mu/k)\,(x\cos\alpha+y\sin\alpha)$. Thus, the vertical Darcy velocity satisfies - $(k/\mu)\, \partial p/\partial y = U_\infty \sin \alpha$, while the horizontal velocity satisfies - $(k/\mu)\, \partial p/\partial x = U_\infty \cos \alpha$. Hence, Eq. (5.59a) gives the pressure field that will produce a flow past the circle, but inclined with angle α at infinity. This flow is shown in Fig. 5.5.

Method 2

The second method makes elegant use of the Circle Theorem outlined in Discussion 4.6. Following the recipe given there, we first write the complex potential for the uniform flow alone; that is, we consider the simple function $w(z) = -(U_\infty \mu/k)\, e^{-i\alpha} z$, where $z = x + iy$. To see that is true, direct multiplication gives $w(z) = -(U_\infty \mu/k)\{(x \cos \alpha + y \sin \alpha) + i\,(y \cos \alpha - x \sin \alpha)\}$. Since the real part of w(z) is just the pressure, we find that it satisfies the expression $p(x,y) = -(U_\infty \mu/k)\{(x \cos \alpha + y \sin \alpha)$, exactly as in Method 1.

Next, following the second part of the recipe, we form the augmented expression obtained by replacing i by -i and z by c^2/z, and add the result to the original complex potential to give

$$w(z) = -(U_\infty \mu/k)\left(e^{-i\alpha} z + e^{i\alpha} c^2/z\right) \tag{5.59b}$$

This gives the complex potential for the uniform flow *with* the circle. To see that this is true, substitute $z = r\, e^{i\theta}$ in Eq. (5.59b) and expand the result using angle summation formulas. This process leads to Eq. (5.59a). A more elegant formula can be obtained by leaving Eq. (5.59b) in complex form. Since $z = r\, e^{i\theta}$, it follows that we can write Eq. (5.59b) in the form

$$w(z) = -(U_\infty \mu/k)\left(r\, e^{i(\theta-\alpha)} + c^2/r\, e^{i(\alpha-\theta)}\right) \tag{5.59c}$$

Its real part, the pressure, is exactly the same as Eq. (5.59a) obtained, however, by brute force separation of variables. That is,

$$\begin{aligned} p(x, y) &= -(U_\infty \mu/k)\left(r + c^2/r\right)(\cos \alpha \cos \theta + \sin \alpha \sin \theta) \\ &= -(U_\infty \mu/k)\left\{1 + c^2/\left(x^2 + y^2\right)\right\}(x \cos \alpha + y \sin \alpha) \end{aligned} \tag{5.60}$$

Flow Past Straight Shale Segment

To obtain the flow past a straight shale segment, let us consider the Joukowski mapping

$$\zeta = z + c^2/z \tag{5.61}$$

where

$$z = x + i y \tag{5.62}$$

$$\zeta = \xi(x, y) + i\eta(x, y) \tag{5.63}$$

If we next substitute Eqs. (5.62), (5.63) into Eq. (5.61), and equate real and imaginary parts, we have

$$\xi = x + c^2 x/(x^2 + y^2) \tag{5.64}$$

$$\eta = y - c^2 y/(x^2 + y^2) \tag{5.65}$$

Now note the following transformation results: (i) the circular well $x^2 + y^2 = c^2$ maps into the straight line (or slit) $\xi = 2x$, $\eta = 0$; (ii) at infinity, $\zeta = z$, and both farfield flows are identical. Before progressing further, we write Eq. (5.61) in the form $z^2 - \zeta\, z + c^2 = 0$, and solve for z using the quadratic equation. Taking the root that preserves $\zeta = z$ at infinity, we have

$$z = \left\{\zeta + \sqrt{(\zeta^2 - 4c^2}\right\}/2 \tag{5.66}$$

or

$$z = \left\{\xi + i\eta + \sqrt{(\xi^2 - \eta^2 - 4c^2 + i2\xi\eta}\right\}/2 \tag{5.67}$$

This yields

$$2x = \xi + \left[(\xi^2 - \eta^2 - 4c^2)^2 + 4\xi^2\eta^2\right]^{1/4} \cos\left\{1/2 \tan^{-1} 2\xi\eta/(\xi^2 - \eta^2 - 4c^2)\right\} \tag{5.68}$$

$$2y = \eta + \left[(\xi^2 - \eta^2 - 4c^2)^2 + 4\xi^2\eta^2\right]^{1/4} \sin\left\{1/2 \tan^{-1} 2\xi\eta/(\xi^2 - \eta^2 - 4c^2)\right\} \tag{5.69}$$

Substitution of Eqs. (5.68), (5.69) into Eq. (5.60) yields the exact solution for the flow past a straight shale segment in (ξ,η) coordinates.

More General Shale Geometries

The well-known Joukowski mapping was used to generate a solution for flow past a flat plate shale. More general shapes are possible, and flows past shapes that resemble fins, rudders and struts are available in aerodynamics. For example, the Karman-Trefftz transform can be used to generate circular arc shales having nonzero thickness, with both circle centers on one side or centers taking opposite sides. The von Mises transform yields S-shaped lines that model rolling and wavy stratigraphies, and the Carofoli mapping, instead of cusps and sharpened trailing edges, yields rounded edges (Milne-Thomson, 1958, 1968).

EXAMPLE 5.6. INFINITE ARRAY OF STRAIGHT-LINE SHALES

An analytical solution for the flow past an infinite array (or cascade) of stacked periodic shales is obtained next. For this purpose, we refer to the nomenclature defined in Fig. 5.6. In this book, we give the recipe for the mapping only; for details, refer to the turbomachinery reference of Oates (1978). First, the airfoil chord c, the spacing s, and the stagger angle β are selected; next, the solidity c/s is computed. Then, the value of the constant ψ (not to be confused with the streamfunction Ψ) is obtained by iteration from the nonlinear transcendental equation

$$\begin{aligned}\pi C/(2S) &= \cos\beta \log\left[\left\{\sqrt{\left(\sinh^2\psi + \cos^2\beta\right)} + \cos\beta\right\}/\sinh\psi\right] \\ &\quad + \sin\beta \tan^{-1}\left\{(\sin\beta)/\sqrt{\left(\sinh^2\psi + \cos^2\beta\right)}\right\}\end{aligned} \tag{5.70}$$

Then, the transformation

$$z = (S/2\pi)\left\{e^{-i\beta}\log\left[(e^{\Psi}+\zeta)/(e^{\Psi}-\zeta)\right] + e^{i\beta}\log\left[(e^{\Psi}+1/\zeta)/(e^{\Psi}-1/\zeta)\right]\right\} \tag{5.71}$$

applies. This function takes a unit circle in the ζ plane and maps it into the straight-line cascade in the z-plane shown in Fig. 5.6. The points $\zeta = +e^{\Psi}$ and $-e^{\Psi}$ on the real axis map into +∞ and -∞ of the x-axis in the z-plane.

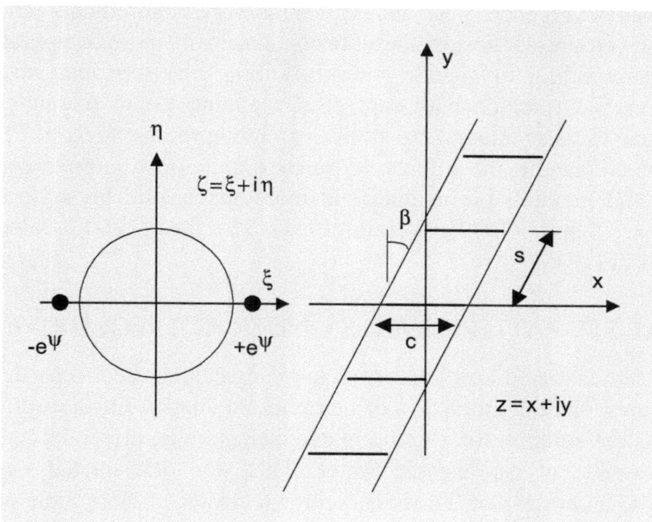

FIG. 5.6 Flow past shale array.

Using the Circle Theorem

First, we consider a uniform flow alone. In the z-plane, we would like to have the complex uniform velocity (see Eq. 4.77) at angle of attack α,

$$dw(z)/dz = U_{\infty}e^{-i\alpha} \tag{5.72}$$

To model the flow at upstream and downstream infinity in this plane, a source is placed at $-e^{\Psi}$, while a sink is placed at $+e^{\Psi}$ in the ζ plane. Then, the complex potential is obtained by superposing this flow with the image defined by the Circle Theorem of Discussion 4.7. The end result is

$$w(\zeta) = (SU_\infty/2\pi)\,\{e^{-i(\beta+\alpha)}\log[(e^\Psi+\zeta)/(e^\Psi-\zeta)] + e^{i(\beta+\alpha)}\log[(e^\Psi+1/\zeta)/(e^\Psi-1/\zeta)]\} \tag{5.73}$$

Again, the Darcy velocities u and v are obtained using

$$dw/dz = (dw/d\zeta)(d\zeta/dz) = u - iv \tag{5.74}$$

the first factor obtaining from Eq. (5.73), the second from Eq. (5.71).

More General Shale Arrays and Oncoming Fluids

The foregoing result, useful in its own right, also finds application in effective properties calculations, for instance, determining the single heterogeneity that would replicate the effects of an entire shale array in the farfield. Such applications are important because computational grids must be kept small in order to optimize convergence speed and reduce memory requirements. The aircraft turbine and compressor design literature contains many more complicated mappings corresponding to airfoils with thickness, curvature, and so on. The solution methods include both conformal mapping (as in Example 5.6), as well as dual singular integral equations, extending the methods of Chapters 2 and 3, which handle the effects of aerodynamic flow interference. These solutions and methods are available in more specialized books and papers, for example, Scholz (1977), Weinig (1964), Thwaites (1960), Oates (1978), and Lighthill (1945).

EXAMPLE 5.7. PATTERN WELLS UNDER AQUIFER DRIVE

We have demonstrated how conformal mapping can be used to produce flows past fractures, shales, and arrays of complicated shapes. Of interest to "infill drilling" is the solution for a group of producing wells, driven by an injector, in the presence of encroaching water. This was first treated by Muskat (1937), who assumed that the wells formed a cluster in the middle of a *large* reservoir; this is justified only in the initial stages of production. Here, we allow an *arbitrarily close* circular aquifer; thus, the solution applies to the later stages of production. Let us consider the well configuration in Fig. 5.7, where the circular domain $|z| < R$ contains a source of strength Q located at $z = 0$, and n sources of strength q located at the points $z_k = ae^{i(2k+1)\pi/n}$, with $k = 0, 1, \ldots, n-1$ and $0 < a < R$. The complex potential $w(z) = p + i\,\Psi$ satisfying $p = 0$ on the farfield boundary $|z| = R$ is desired. We give this solution without proof:

$$w(z) = Q \log z/R + q \log\{R^n(z^n + a^n)/(R^{2n} + z^n a^n)\} \tag{5.75}$$

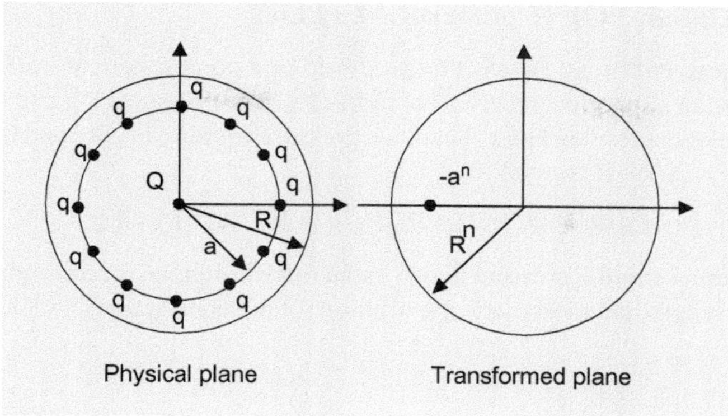

FIG. 5.7 Pattern wells.

Note that in the immediate neighborhoods of the points $z = 0$ and $z = z_k$, the lines of level pressure are approximately circles centered on these points. The complex potential, therefore, also provides an approximate solution to the pressure problem for the flow domain defined by $|z| < R$, $|z| > \delta$, $|z - z_k| > \varepsilon$, with $k = 0, 1, \ldots, n-1$ where $\delta \ll a$ and $\varepsilon \ll a$. Observe that the constants Q and q in Eq. (5.75) can be chosen to satisfy (approximately) the conditions $p = 0$ on $|z| = R$, $p = p_1$ on $|z| = \delta$, and $p = p_2$ on $|z - z_k| = \varepsilon$, where p_1 and p_2 are real. The preceding transform is used to study heat transfer in rocket nozzle applications.

THREE-DIMENSIONAL FLOWS

So far we have treated two-dimensional planar flows. However, many three-dimensional problems are also amenable to analytical solution. To proceed, we introduce the notion of the point spherical source. Actually, the concept is best taught through global mass conservation considerations. We consider two-dimensional flows first. First, the radial Darcy velocity is proportional to dp/dr. This, times the area $2\pi r$ in planar problems, must be constant; hence, in such flows, dp/dr goes like 1/r, which on integration leads to the expected logarithmic pressure. In three dimensions, $dp/dr \times 4\pi r^2$ must remain constant; thus, dp/dr goes like $1/r^2$, so that p(r) varies like 1/r. This describes the point spherical source. We could also have started more formally with the spherically symmetric form of Laplace's equation,

$$d^2p/dR^2 + (2/R)\, dp/dR = 0 \tag{5.76}$$

to obtain

$$p(r) = 1/R \tag{5.77}$$

directly as the relevant elementary solution in three-dimensional flows, where R refers to the (dimensional) radial coordinate.

EXAMPLE 5.8. POINT SPHERICAL FLOW

The simplest, nontrivial flow is that produced by a point spherical well of radius R_W located in a spherical reservoir of radius R_R. The pressures P_W and P_R apply, respectively, at these surfaces. Then, the pressure solution is obtained by assuming first $P = A/R + B$, to yield

$$P(r) = R_R P_R / R + (R_R P_R - R_W P_W)(1 - R_R/R)/(R_R - R_W) \tag{5.78a}$$

This represents the pressure due to a spherical drainhole, for example, the end of a finite length line source; it is the solution for invasion at the bit while drilling.

EXAMPLE 5.9. FINITE LINE SOURCE WITH PRESCRIBED PRESSURE

Finite line sources are of interest in horizontal well simulation; the latter are used both as injectors and producers. Moreover, their lineal shape renders them amenable to a distributed source solution, as shown in Chapter 2 for planar fractures. Let us consider a line source occupying $-c \le X \le +c$, $Y = Z = 0$, where X, Y, and Z are dimensional. Furthermore, let the pressure at the radius $R \gg c$ be P_R, while the pressure along the source is held at a constant P_W. Now, following Chapter 2, introduce the dimensionless coordinates $x = X/c$, $y = Y/c$, and $z = Z/c$, and the nondimensional pressures p_R and $p(x,0,0) = p_W$. In the present problem, we recognize that the elementary solution in Eqs. (5.77), (5.78a) can be rewritten in the equivalent form

$$p(x, y, z) = 1/\sqrt{(x^2 + y^2 + z^2)} \tag{5.78b}$$

Satisfying

$$\partial^2 p/\partial x^2 + \partial^2 p/\partial y^2 + \partial^2 p/\partial z^2 = 0 \tag{5.79}$$

The pressure distribution due to a line distribution of sources, that is,

$$p(x, y, z) = \int m(\xi)\, d\xi / \sqrt{\left\{(x-\xi)^2 + y^2 + z^2\right\}} + H \tag{5.80}$$

will also satisfy Eq. (5.79) since it is linear. As in Chapter 2, the ξ represents the source coordinate for superposition, and H is a constant. The integration limits (-1, +1) are omitted for clarity. Now, if we assume

$$p(x, 0, 0) = p_W, -1 \le x \le +1 \tag{5.81}$$

Eq. (5.80) becomes

$$PV \int m(\xi)/(\xi - x)\, d\xi = p_W - H \tag{5.82}$$

Eq. (5.82) is the governing singular integral equation of interest, satisfied by the source strength m(x).

Integral Equation Solution

This integral equation was encountered in Chapter 3, where we discussed shale flow modeling. The general solution to

$$\mathrm{PV}\int g(\xi)/(x-\xi)d\xi = -h(x) \tag{3.24}$$

is, in fact,

$$\begin{aligned} g(x) &= -(1/\pi^2)\sqrt{\{(1-x)/(1+x)\}} \\ &\times \mathrm{PV}\int \left\{h(\xi)\sqrt{(1+\xi)}\right\}/\left\{(\xi-x)\sqrt{(1-\xi)}\right\}d\xi + \gamma/\sqrt{(1-x^2)} \end{aligned} \tag{3.25}$$

where we have again omitted the integration limits (-1, +1) for clarity. This solution is derived and discussed in several classical mathematical references (Carrier et al., 1966; Mikhlin, 1965; Muskhelishvili, 1953). Eqs. (5.82), (3.25) above imply that

$$\begin{aligned} m(x) &= -(1/\pi^2)\sqrt{\{(1-x)/(1+x)\}}\,(p_W - H) \\ &\times \mathrm{PV}\int \left\{\sqrt{(1+\xi)}\right\}/\left\{(\xi-x)\sqrt{(1-\xi)}\right\}d\xi + \gamma/\sqrt{(1-x^2)} \end{aligned} \tag{5.83}$$

Note, specifically, the second lines of Eqs. (3.25), (5.83). The term $\gamma/\sqrt{(1-x^2)}$ represents the nonuniqueness associated with solutions to Eq. (3.24), with the arbitrary constant γ known in the aerodynamics literature as the circulation characterizing a flow. How is the circulation chosen here?

We proceed by evaluating Eq. (5.83). The value of the integral on the first line of Eq. (5.83) is given in van Dyke (1956) as π, a value completely independent of x. Thus, we have

$$m(x) = [\gamma - (p_W - H)/\pi + (p_W - H)x/\pi]/\sqrt{(1-x^2)} \tag{5.84}$$

Physically, we expect the source strength m(x) to be symmetrically distributed about $x=0$; that is, we anticipate an even function satisfying

$$m(x) = m(-x) \tag{5.85}$$

Consider the third term of Eq. (5.84). This is possible only if $p_W - H = 0$, implying that $H = p_W$. Hence, Eqs. (5.80), (5.83) lead to

$$p(x,y,z) = \int \gamma/\sqrt{(1-\xi^2)}\,d\xi/\sqrt{\left\{(x-\xi)^2 + y^2 + z^2\right\}} + p_W \tag{5.86}$$

If we next evaluate Eq. (5.86) at large distances from the line source, we obtain

$$\{\gamma/(R/c)\}\int d\xi/\sqrt{(1-\xi^2)} + p_W = p_R \tag{5.87}$$

Now, the definite integral in Eq. (5.87) is exactly π, so that the circulation

$$\gamma = (p_R - p_W)(R/c)/\pi \tag{5.88}$$

solves the problem. The end result, combining Eqs. (5.86), (5.88), is

$$p(x, y, z) = \int \{(p_R - p_W)(R/c)\}/\left\{\pi\sqrt{(1-\xi^2)}\right\}d\xi/\sqrt{\left\{(x-\xi)^2 + y^2 + z^2\right\}} + p_W \tag{5.89}$$

When $y = z = 0$, it is possible to show that the Cauchy principal value integral that results from Eq. (5.89) vanishes, and is independent of x (van Dyke, 1956). To determine the net inflow into (or out of) the line source, it is preferable to integrate the derivative $\partial p/\partial r$ over a spherical volume faraway rather than deal with the complicated details of the nearfield flow. This solution, again, applies to a finite line source centered in a spherical radius of length R/c. As such, it applies directly to horizontal well analysis. Of course, it can also be used for vertical wells. Since the problem is symmetric about the plane $x = 0$, the solution also represents the flow from a hemisphere into a partially penetrating well. This latter problem is considered in Muskat (1937), with slightly different boundary conditions using the method of images.

EXAMPLE 5.10. FINITE LINE SOURCE WITH PRESCRIBED FLOW RATE

Next, consider the same problem, except that we impose flow rate boundary conditions along the length of the source. As before, Eq. (5.80) applies. For convenience, we introduce the variable $\sigma^2 = z^2 + y^2$; then, for example, the radial velocity perpendicular to the line source is proportional to $\partial p/\partial \sigma$. By the same token, Eq. (5.80) can be written in the equivalent form

$$p(x, \sigma) = \int m(\xi)\, d\xi/\sqrt{\left\{(x-\xi)^2 + \sigma^2\right\}} + H \tag{5.90}$$

where the limits (-1, +1) are omitted for clarity. Now, let us form the derivative $\partial p/\partial \sigma$, and introduce the change of coordinates $\eta = (\xi - x)/\sigma$. If we evaluate Eq. (5.90) along $\sigma = 0$, and use the limit process in Example 2.1, we obtain

$$\partial p(x, 0)/\partial \sigma = -m(x)/\sigma \int_{-\infty}^{+\infty} d\eta/\left(\eta^2 + 1\right)^{3/2} = -2m(x)/\sigma \tag{5.91}$$

Thus, we can rewrite Eq. (5.90) in the alternative form

$$p(x, \sigma) = -\sigma/2 \int_{-1}^{+1} \partial p(\xi, 0)/\partial \sigma \, d\xi / \sqrt{\left\{(x-\xi)^2 + \sigma^2\right\}} + H \tag{5.92}$$

At large distances $R \gg c$ from the line source, with x-ξ held fixed, Eq. (5.92) can be approximated by

$$p_R = -\tfrac{1}{2} \int_{-1}^{+1} \partial p(\xi, 0)/\partial \sigma \, d\xi + H \tag{5.93}$$

where the integral is known, since the Darcy velocity is prescribed along the x-axis. Thus, Eq. (5.93) fixes H, so that Eq. (5.92) becomes

$$\begin{aligned} p(x, \sigma) = &-\sigma/2 \int_{-1}^{+1} \partial p(\xi, 0)/\partial \sigma \, d\xi / \sqrt{\left\{(x-\xi)^2 + \sigma^2\right\}} \\ &+ \tfrac{1}{2} \int_{-1}^{+1} \partial p(\xi, 0)/\partial \sigma \, d\xi + p_R \end{aligned} \tag{5.94}$$

EXAMPLE 5.11. FINITE CONDUCTIVITY PRODUCING FRACTURE HAVING LIMITED AREAL EXTENT

In Chapter 2, we considered the two-dimensional planar flow for a producing line fracture that is infinite in the direction into the page. It is also possible to repeat the analysis for a producing planar fracture of limited areal extent, say, located on the $z = 0$ plane and defined by the region S(x,y). In this limit, areal (as opposed to line) sources of strength m(x,y) can be distributed over S(x,y) on $z = 0$ to generate the flow, say, corresponding to an imposed Darcy velocity w(x,y) normal to the fracture plane. The required extension to Eq. (5.80), taking superpositions of 1/r spherical sources, takes the form

$$p(x, y, z) = \iint_S m(\xi, \eta) \, d\xi d\eta / \sqrt{\left\{(x-\xi)^2 + (y-\eta)^2 + z^2\right\}} + H \tag{5.95}$$

Because Eq. (5.95) assumes sources only, we have ruled out tangential velocity discontinuities that may result from mineralization effects.

Eq. (5.95) provides the multidimensional integral equation for the source strength m(x,y). Such equations are used in lifting surface theory by aerodynamicists to model wing flows, and are discussed in Mikhlin's (1965) classic book. Using a limit process similar to that in Example 2.1, for example, refer to Bisplinghoff et al. (1955), it is possible to show that

$$\partial p(x,y,0+)/\partial z = +\, m(x,y)/2 \tag{5.96}$$

$$\partial p(x,y,0-)/\partial z = -m(x,y)/2 \tag{5.97}$$

Thus, the normal derivatives, as in the two-dimensional planar limit, are antisymmetric, being equal and opposite. Hence, Eq. (5.95) becomes

$$p(x,y,z) = 2\iint_S \partial p(\xi,\eta,0+)/\partial z\, d\xi d\eta/\sqrt{\left\{(x-\xi)^2 + (y-\eta)^2 + z^2\right\}} + H \tag{5.98}$$

At large distances R compared to some diametric length characteristic of S, Eq. (5.98) satisfies the constant pressure constraint $p = p_R$. Thus,

$$p_R = (2/R)\iint_S \partial p(\xi,\eta,0+)/\partial z\, d\xi d\eta + H \tag{5.99}$$

and the constant H, for use in Eq. (5.95), is completely determined.

EXAMPLE 5.12. FINITE CONDUCTIVITY NONPRODUCING FRACTURE HAVING LIMITED AREAL EXTENT

In Examples 2.1 and 5.11, the assumption of a *producing* fracture motivated the use of logarithms as singularities; these are responsible for equal and opposite Darcy velocities normal to the fracture plane and simulate production. In practice, flow also moves parallel to the fracture, toward the penetrating well that taps the fluid. For this motion to be possible, a pressure gradient must exist along the fracture, and the variable $p_f(x)$ considered in Chapter 2 applies. Often the fracture contains solids and debris, and the parallel velocity on one side of the fracture will not be the same as that on the other. For this flow, we specify along $z = 0$ the discontinuous velocity

$$\partial p(x,y,0+)/\partial x - \partial p(x,y,0-)/\partial x = \lambda(x,y) \tag{5.100}$$

noting that x is the coordinate tangent to the $z = 0$ plane, and $\lambda(x,y)$ must be chosen subject to zero rotationality, that is,

$$\iint_S \lambda(\xi,\eta)\, d\xi\, d\eta = 0 \tag{5.101}$$

in the distant farfield. The solution assuming a unit speed in the farfield is

$$p(x,y,z) = (1/4\pi)\iint_S z\lambda(\xi,\eta)/\left[(y-\eta)^2 + z^2\right] \times \left[1 + (x-\xi)/\sqrt{\left\{(x-\xi)^2 + (y-\eta)^2 + z^2\right\}}\right] d\xi\, d\eta \tag{5.102}$$

For further details, the reader is referred to Bisplinghoff et al. (1955), Ashley and Landahl (1965), and Thwaites (1960). Eq. (5.102) arises from the "lifting surface theory" well known in aerodynamics, where the nonzero value of the integral shown in Eq. (5.101) is proportional to lift.

BOREHOLE INTERACTIONS

Muskat (1937) gave an exact solution for an infinite line source ("the line drive") into a well; his line source was held at fixed uniform pressure, and the solution was obtained using the method of images. He also gave a solution for a finite line source using infinite sets of images, again with pressure uniform over the distributed source. Muskat noted that the integral equation approach (being more difficult) is avoided; he gave a cumbersome infinite series whose correctness, in light of comments made in Example 2.1, is in doubt. So far, we have not considered radial flows in detail, but only simple log r solutions for constant density liquids. Radial flows are studied in Chapter 6, where new results and numerical schemes are developed. We now draw upon the results reiterated in Example 6.1 for steady-state (m = 0) *liquids*, in order to discuss borehole interactions with complicated fracture and shale flows. The extension to gases with nonzero m is straightforward, following the pattern of generalizations used in earlier examples, and will not be pursued further.

EXAMPLE 5.13. PRODUCING FRACTURE NEAR MULTIPLE WELLS UNDER AQUIFER DRIVE

We again consider the problem in Example 2.1 where the fracture pressure $p_f(x)$ specified along a slit but now allow an arbitrary number of injection and production wells in the neighborhood of the fracture. These wells are described by normalized volume flow rates λ_n, which we assume to be prescribed positive or negative quantities. Combining Eq. (2.10) with superpositions of singularities such as those given in Eq. (6.9) yields the extension

$$\begin{aligned} p(x,y) = &\int_{-1}^{+1} f(\xi)\,\log\sqrt{\left\{(x-\xi)^2+y^2\right\}}\,d\xi + H \\ &+ \sum \lambda_n \log\sqrt{\left\{(x-x_n)^2+(y-y_n)^2\right\}} \end{aligned} \tag{5.103}$$

where the index n refers to the well number, (x_n, y_n) represents well locations, and the summation is taken over all possible wells. The constant H, as in Chapter 2, handles farfield boundary conditions in addition to well, fracture, and farfield interference effects. This integral expression also satisfies Eq. (2.7) for pressure, since each individual contribution is harmonic and the governing equation is linear. Now evaluate Eq. (5.103) at y = 0 and apply fracture pressure conditions. This leads to Eq. (2.11), but modified as follows:

$$\int_{-1}^{+1} f(\xi) \log|x-\xi| d\xi = p_f(x) - \sum \lambda_n \log \sqrt{\left\{(x-x_n)^2 + y_n^2\right\}} - H \quad (5.104)$$

Again, the solution to the equation

$$\int_{-1}^{+1} f(\xi) \log |x-\xi| \, d\xi = g(x) \quad (2.12)$$

takes the general form

$$f(x) = \left[PV \int_{-1}^{+1} \{g'(\xi)/(\xi - x)\} \sqrt{(1-\xi^2)} \, d\xi \right.$$

$$\left. -(1/\log_e 2) \int_{-1}^{+1} g(\xi)/\sqrt{(1-\xi^2)} \, d\xi \right] / \left\{\pi^2 \sqrt{(1-x^2)}\right\} \quad (2.13)$$

where g'(ξ) is the derivative of g(ξ) with respect to ξ. The substitutions

$$g(x) = p_f(x) - \sum \lambda_n \log \sqrt{\left\{(x-x_n)^2 + y_n^2\right\}} - H \quad (5.105)$$

$$g'(x) = p_f'(x) - \sum \lambda_n (x - x_n) / \left\{(x-x_n)^2 + y_n^2\right\} \quad (5.106)$$

solve the problem. The constant H is determined by expanding Eq. (5.103) for large distances away from the fracture and applying farfield pressure boundary conditions. The final result for pressure is

$$p(x, y) = \int_{-1}^{+1} f(\xi) \log \sqrt{\left\{(x-\xi)^2 + y^2\right\}} \, d\xi + \sum \lambda_n \log \sqrt{\left\{(x-x_n)^2 + (y-y_n)^2\right\}}$$

$$- \left\{ \int f(\xi) \, d\xi + \sum \lambda_n \right\} \log R/c + P_R/P_{ref} \quad (5.107)$$

where the expressions for g(x) and g'(x) in Eqs. (5.105), (5.106) are to be used in determining the source strength f(x). An analogous extension applies to the related streamfunction; the results of Chapter 4 for fractures can be combined with superpositions of solutions such as those in Eq. (6.22).

EXAMPLE 5.14. PRODUCING FRACTURES NEAR MULTIPLE WELLS IN SOLID WALL RESERVOIRS

Consider the preceding problem with different farfield conditions. Instead of prescribing pressure, which allows flow across the boundary, assume a solid wall. In the absence of wells, the fracture would produce no flow. When wells are present, the net nonzero injected flow from all the wells appears as produced flow at the fracture. We show this analytically. The assumption underlying Eq. (5.103) applies, and we expand all terms for large radial distances. Following the procedure in Example 2.1,

$$\begin{aligned} p(x, y) &= \int_{-1}^{+1} f(\xi) \log \sqrt{\left\{(x-\xi)^2 + y^2\right\}} \, d\xi + H \\ &\quad + \sum \lambda_n \log \sqrt{\left\{(x-x_n)^2 + (y-y_n)^2\right\}} \\ &\approx \left\{ \int_{-1}^{+1} f(\xi) \, d\xi + \sum \lambda_n \right\} \log r + H \end{aligned} \tag{5.108}$$

where r is the radial coordinate. Now form the normal derivative $\partial p/\partial r$ at the farfield boundary and set it to zero so the Darcy velocity vanishes. This requires

$$\left\{ \int_{-1}^{+1} f(\xi) \, d\xi + \sum \lambda_n \right\} = 0 \tag{5.109}$$

This is consistent with the mass conservation noted earlier. Now, in the nearfield, the fracture pressure $p_f(x)$ and the well flow rates λ_n are prescribed arbitrarily. These inputs are substituted in Eq. (5.105) to yield g(x), which leads to an expression for the distributed source strength f(x) via Eq. (2.13) (found just above Eq. 5.105). But H in Eq. (5.105) is still free; it must be chosen so that Eq. (5.109), a requirement for mass conservation, holds. This choice solves the problem and uniquely determines all pressures.

EXAMPLE 5.15. STRAIGHT-LINE SHALE SEGMENT NEAR MULTIPLE WELLS IN UNIFORM FLOW

In this problem, we reconsider the flowfield of Example 3.1 and place an arbitrary number of production and injection wells in the neighborhood of an impermeable shale. However, to satisfy the assumption that the same farfield flow exists at upstream and downstream infinity, we require that the net volume flow issuing from all wells sums to zero. This implies that

$$\sum \lambda_n = 0 \tag{5.110}$$

Following Eq. (3.21) in Example 3.1, assume a line distribution of arc tan singularities for the shale, but add to it an x to model a distant uniform flow, plus a discrete number of sources, giving,

$$p(x, y) = \int_{-1}^{+1} g(\xi) \arctan y/(x-\xi)\, d\xi + H + x + \sum \lambda_n \log \sqrt{\left\{(x-x_n)^2 + (y-y_n)^2\right\}} \tag{5.111}$$

Now consider the nearfield tangency condition and form the vertical derivative

$$\partial p/\partial y = \int_{-1}^{+1} g(\xi)\,(x-\xi)/\left\{(x-\xi)^2 + y^2\right\} d\xi + \sum \lambda_n (y-y_n)/\left\{(x-x_n)^2 + (y-y_n)^2\right\} \tag{5.112}$$

If we evaluate Eq. (5.112) at $y=0$ and apply tangency conditions (see Eq. 3.18), we obtain, instead of Eq. (3.23), the equation

$$\int_{-1}^{+1} g(\xi)/(x-\xi)\, d\xi - \sum \lambda_n y_n/\left\{(x-x_n)^2 + y_n{}^2\right\} = -\alpha \tag{5.113}$$

Thus, the singular integral equation for the vortex strength g(x) becomes

$$\int_{-1}^{+1} g(\xi)/(x-\xi)\, d\xi = -\alpha + \sum \lambda_n y_n/\left\{(x-x_n)^2 + y_n{}^2\right\} \tag{5.114}$$

so that the function h(x) in Eq. (3.24) takes the form

$$h(x) = \alpha - \sum \lambda_n y_n/\left\{(x-x_n)^2 + y_n{}^2\right\} \tag{5.115}$$

Once this expression is substituted in Eq. (3.25) for the vortex strength, the final expression for g(x) is obtained by choosing the circulation γ consistently with Eq. (3.28), calling for vanishing velocity swirl in the farfield.

EXAMPLES 5.16 AND 5.17. NONPRODUCING FAULTS IN SOLID WALL AND AQUIFER-DRIVEN RESERVOIRS

In Examples 5.12 and 5.13, we considered pure fractures that did not support pressure discontinuities but that yielded normal velocity jumps that produced flow. By contrast, in Example 5.15, we considered an impermeable shale that supported pressure differences between upper and lower surfaces, but that gave continuous normal derivatives of pressure and hence no production. As we had discussed in Example 3.3, a nonuniform streak (responsible for changes in the tangent velocity from one side of the fracture to the other) located in an otherwise homogeneous, isotropic reservoir can be modeled by prescribing a nonzero vortex strength g(x) along $y=0$; that is, by specifying a discontinuity in the tangential velocity as might be observed from far away. In the case of an impermeable shale, the magnitude of g(x) is determined by the geometry of the barrier; for streaks, g(x) may be prescribed from laboratory results, provided the net vorticity (see later in this section) is zero. Again, Eq. (5.111) applies without the x, since there is no uniform flow at infinity. Now, it is of interest to expand it for large distances away from the fault. Far away, arc tan y/(x-ξ) can be approximated by arc tan y/x or θ, the azimuthal angle from radial polar coordinates. Thus,

$$p(x, y) = \int_{-1}^{+1} g(\xi) \arc\tan y/(x-\xi)\, d\xi + H + \sum \lambda_n \log \sqrt{\left\{(x-x_n)^2 + (y-y_n)^2\right\}}$$

$$\approx \theta \int_{-1}^{+1} g(\xi)\, d\xi + \left(\sum \lambda_n\right) \log r + H \tag{5.116}$$

where r is the radial coordinate. The distribution of tangential velocity discontinuity must be chosen so that the integral in Eq. (5.116) vanishes, following Eq. (3.28), if velocity swirl (proportional to $1/r\ \partial p/\partial\theta$) is to vanish in the farfield; this is a physical constraint that must be enforced.

Solid Walled Aquifer

If we require solid no flow walls in the farfield, the normal radial derivative $\partial p/\partial r$ there must vanish. Thus, on differentiating Eq. (5.116), an additional constraint calls for $\sum \lambda_n = 0$; that is, the net production and injection rates from all discrete wells must sum to zero. This is obvious from physical requirements. Since the boundary value problem involves derivatives of p only, the pressure solution is nonunique to within a constant value; its value may be arbitrarily fixed at any one point, without loss of generality, for the purposes of calculation.

Aquifer Drive

In the limit when the farfield boundary is opened to an aquifer, the (normalized) reservoir pressure P_R/P_{ref} is specified at the nondimensional radial coordinate R/c. As before, the integral in Eq. (5.116) vanishes; since the farfield boundary is opened to an aquifer, the (normalized) reservoir pressure is $P_R/P_{ref} = (\sum \lambda_n) \log R/c + H$, so that the integration constant H satisfies $H = P_R/P_{ref} - (\sum \lambda_n) \log R/c$. This value completely determines the pressure distribution in Eq. (5.116).

EXAMPLE 5.18. HIGHLY CURVED FRACTURES AND SHALES

In Chapter 2, we considered straight fractures, whereas in Chapter 3, we considered shales that were both straight and mildly curved. These assumptions were invoked to keep the physical and mathematical ideas simple and to avoid cumbersome notation. It turns out that fractures and shales with general curvature are just as easily handled; instead of distributing singularities along the approximating line y = 0, as suggested by thin airfoil theory, we simply consider the line y = h(x), say. Thus, the natural extension to Eq. (2.10) for a line distribution of sources of strength f(x) is

$$p(x, y) = \int_{-1}^{+1} f(\xi) \log \sqrt{\left\{(x - \xi)^2 + (y - h(\xi))^2\right\}} \, d\xi + H \tag{5.117}$$

whereas the expression generalizing the integral in Eq. (3.21) for a line distribution of vortexes is

$$p(x, y) = \int_{-1}^{+1} g(\xi) \arctan (y - h(\xi))/(x - \xi) \, d\xi + H \tag{5.118}$$

In Eqs. (5.117), (5.118), the respective singularities are assumed to lie along the locus of points (ξ, h(ξ)). The jump properties of the above distributions remain identical to the ones previously developed. For curved line sources, differentiation of Eq. (5.117) leads to

$$\partial p/\partial x = \int_{-1}^{+1} f(\xi) \, (x - \xi)/(x - \xi)^2 + (y - h(\xi))^2\} \, d\xi \tag{5.119}$$

$$\partial p/\partial y = \int_{-1}^{+1} f(\xi) \, (y - h(\xi))/\left\{(x - \xi)^2 + (y - h(\xi))^2\right\} d\xi \tag{5.120}$$

Along y = h(x), Eq. (5.119) becomes

$$\partial p/\partial x = \int_{-1}^{+1} f(\xi)/(x - \xi) \, d\xi \tag{5.121}$$

whereas Eq. (5.120) takes the form given in Eq. (2.24), that is,

$$\partial p/\partial y = \int_{-1}^{+1} f(\xi) \, (y - h(\xi))/\left\{(x - \xi)^2 + (y - h(\xi))^2\right\} d\xi \tag{5.122}$$

$$= y^* \int_{-1}^{+1} f(\xi)/\left\{(x - \xi)^2 + y^{*2}\right\} d\xi \tag{5.123}$$

where $y^* = y - h(\xi)$ tends to zero. Eq. (5.123) should be compared with Eq. (2.24). It is clear that the same integral equations, and the same jump properties, as obtained in Chapter 2 also apply here, and that analogous end results are achieved. Similar comments apply to Eq. (5.118) for shale flow modeling, since integrals such as those found in Eqs. (5.121), (5.122) are obtained by direct differentiation of Eq. (5.118).

PROBLEMS AND EXERCISES

1. Our conformal mapping examples focused on incompressible liquids in isotropic formations for simplicity. Using the formalism of Chapters 1–3, extend these mapping results generally to include the effects of anisotropic media and of gases with constants m. *Hint:* recall that simple scale transformations map the anisotropic equation into isotropic form, while gas effects are modeled by considering p^{m+1} as in Chapter 4.
2. Derive the streamfunction for the pressure given in Eq. (5.107). Write a computer program to calculate pressures and streamfunctions for general (x,y). Integrate the output of this program with commercial contour plotting software to produce capabilities useful in enhanced oil recovery.

Chapter 6

Radial Flow Analysis

In many books, radial flow theory is studied superficially and dismissed after cursory derivation of the "log r" pressure solution. Here we shall consider single-phase radial flow in detail. We shall examine analytical formulations that are possible in various physical limits, for different types of liquids and gases, and develop efficient models for time and cost-effective solutions. Steady-state flows of constant density liquids and compressible gases can be solved analytically, and these are considered first. In Examples 6.1–6.3, different formulations are presented, solved, and discussed; the results are useful in formation evaluation and drilling applications. Then, we introduce finite difference methods for steady and transient flows in a natural, informal, hands-on way, and combine the resulting algorithms with analytical results to provide the foundation for a powerful "write it yourself" radial flow simulator. Concepts such as explicit versus implicit schemes, von Neumann stability, and truncation error are discussed in a self-contained exposition.

EXAMPLE 6.1. STEADY LIQUIDS IN HOMOGENEOUS MEDIA

Radial flow problems are generally solved with pressure-pressure boundary conditions at the well and farfield boundaries. In steady flow, however, the total flow rate Q_W—*holding pressure constant along the wellbore contour*—from (or into) a well is a uniquely defined constant of the problem. Thus, it is of interest to have this parameter appear as a potential boundary condition. The resulting formulations are useful in formation evaluation. In Chapter 9, the three boundary value problems addressed here are considered for arbitrarily shaped reservoirs, where the general extension of log r is given.

Pressure-Pressure Formulations

The most common starting point, at least for elementary analysis, is the pressure equation

$$d^2P/dr^2 + (1/r)\,dP/dr = 0 \qquad (6.1)$$

for P(r), where r is the radial coordinate, which governs Darcy flows of incompressible liquids in homogeneous, isotropic media. The usual boundary

Quantitative Methods in Reservoir Engineering. http://dx.doi.org/10.1016/B978-0-12-810518-4.00006-2

conditions assume pressures specified at the wellbore and at some distance away from the hole. We have

$$P(r_W) = P_W \tag{6.2}$$

$$P(r_R) = P_R \tag{6.3}$$

where $r = r_W$ and $r = r_R$ refer to well and farfield radii, and P_W and P_R are the assumed pressures. This formulation has the solution (see Eq. 4.46)

$$P(r) = \{(P_R - P_W)/(\log r_R/r_W)\} \log r/r_W + P_W \tag{6.4}$$

Now, the radial velocity q(r) is given by Darcy's law, requiring that

$$q(r) = -(k/\mu)\, dP(r)/dr = -(k/\mu)\{(P_R - P_W)/(\log r_R/r_W)\}\, 1/r \tag{6.5}$$

where k is the formation permeability and μ is the fluid viscosity. Hence, the total volume flow rate Q_W, assuming a reservoir depth D into the page, is

$$Q_W = -D \int_0^{2\pi} q(r)\, r_W\, d\theta,\ r = r_W \tag{6.6}$$

$$Q_W = -2\pi r_W D q(r_W) \tag{6.7}$$

leading to

$$Q_W = -(2\pi kD/\mu)\,(P_R - P_W)/(\log r_R/r_W) \tag{6.8}$$

P_W - Q_W Formulations

In a steady flow, the volume flux through any closed curve surrounding the well is the same. Thus, another formulation with a uniquely defined solution is the problem where Eq. (6.2) is prescribed at the well, along with the total volume flow rate Q_W. If the term P_R is eliminated between Eqs. (6.4), (6.8), we have

$$P(r) = -\{Q_W\mu/2\pi kD)\log r/r_W + P_W \tag{6.9}$$

for the complete solution, and

$$P_R = P_W - \{(\mu Q_W/2\pi kD)\log r_R/r_W\} \tag{6.10}$$

for the pressure at the farfield boundary.

P_R - Q_W Formulations

Still another formulation is the problem where Eq. (6.3) is prescribed in the farfield, along with the total volume flow rate Q_W. If the term P_W is eliminated between Eqs. (6.4), (6.8), we have

$$P(r) = \{Q_W\mu/2\pi kD) \log r_R/r + P_R \tag{6.11}$$

For example, once P_R and Q_W are given, the well pressure is just

$$P_W = \{Q_W\mu/2\pi kD) \log r_R/r_W + P_R \tag{6.12}$$

EXAMPLE 6.2. SIMPLE FRONT TRACKING FOR LIQUIDS IN HOMOGENEOUS, ISOTROPIC MEDIA

Very often, the front of the injecting fluid penetrating the formation is desired as a function of time. This application is important in correcting resistivity readings obtained during drilling, when fresh and saline ("red" versus "blue") waters displace each other but otherwise do not mix in a miscible or diffusive sense. Here, we consider incompressible liquids in homogeneous, isotropic flow. The sign convention of Eq. (6.8) assumes that $Q_W > 0$ when $P_W > P_R$. Combination of Eqs. (6.5), (6.8) leads to

$$q(r) = +Q_W/(2\pi Dr) \tag{6.13}$$

Let us denote the porosity by ϕ. Then, the rate of invasion into the reservoir is

$$dr/dt = q/\phi = +Q_W/(2\pi\phi Dr) \tag{6.14}$$

and, finally, integration gives

$$r(t) = \sqrt{\{R_W^2 + Q_W\, t/\pi\phi D\}} \tag{6.15}$$

for the invasion front. Here, Q_W and ϕ are constant, and the initial front position is assumed to be $r = r_W$ at $t = 0$. By the same token, we could have considered any other ring of injected or produced fluid, and obtained

$$r(t) = \sqrt{\{R_{other}^2 + Q_W t/\pi\phi D\}} \tag{6.16}$$

if $r = R_{other} > R_W$ at $t = 0$. Let us return to Eq. (6.14). For large times,

$$r(t) \approx \sqrt{\{Q_W t/\pi\phi D\}} \tag{6.17}$$

Thus, in a steady cylindrical radial flow without mudcake effects, $\sqrt{t}$ front displacement is obtained because of geometric divergence (Eq. 6.14) assumes a constant volume flow-rate with constant porosity together with $r = r_W$ at $t = 0$). We emphasize that this $\sqrt{t}$ behavior differs from the $\sqrt{t}$ obtained when linear flow invasion is controlled by highly impermeable mudcake growing in time. This subject is treated rigorously starting with Chapter 16.

Incompressible Transient Effects

We continue with our simple cake-free model. Very often in drilling, the mud circulation rate varies with time. For constant density flows, the fluid feels these changes instantaneously; the front responds immediately to changes in the pump rate $Q_W(t)$ at the surface. In addition, an initial spurt loss into the rock, which depends on formation texture and mud rheology flow interaction, may result in our having the initial radius set as $r = R_{spurt} > r_W$ at $t = 0$. The generalization to Eq. (6.14) is

$$r(t) = \sqrt{\{R^2_{spurt} + \int_0^t Q_W(\tau)\, d\tau/\pi\phi D\}} \tag{6.18}$$

for constant porosity. Now, Eqs. (6.13), (6.14) hold generally in the assumed limit; thus, when $\phi(r)$ is variable, we must integrate

$$2\pi D\phi(r)\, r\, dr = Q_W(t)\, dt \tag{6.19}$$

where $Q_W(t)$ is considered a known (fluid loss) input quantity specified by the driller. The corresponding pressure distribution is calculated by integrating

$$d\{r\, k(r)\, dP/dr\}/dr = 0 \tag{6.20}$$

This yields

$$P(r) = -\{Q_W(t)\mu/2\pi D\} \int_{r_w}^{r} dr/r\, k(r) + P_W(t) \tag{6.21}$$

One application of Eq. (6.21) lies in the area of drilling safety; the computed pressure can be used to determine if formation fracture and consequent fluid loss are possible. Again, we have not considered mudcake buildup, which would reduce permeabilities at the sandface, effects that are discussed later.

Discontinuous Properties

The preceding formulas assume continuously varying formation properties. Problems for discontinuous, but piece-wise constant properties, are also easily solved. For such problems, Laplace's equation (see Eq. 6.1) applies in each separate flow annulus; it is convenient to append a subscript i or j corresponding to each pressure function. Each solution then takes the form $P_n = A_n + B_n \log r$. The integration constants are obtained from the conditions $P = P_W$ at $r = r_W$, $P = P_R$ at $r = r_R$, plus the matching conditions $P_i = P_j$ and $k_i\, dP_i/dr = k_j\, dP_j/dr$ at all adjacent interfaces. Example calculations are given in Muskat (1937) and Collins (1961).

Radial Flow Streamfunction

For completeness, we indicate that the streamfunction Ψ (see Eqs. 4.46–4.49) corresponding to Eq. (6.4) is

$$\Psi = -(Q_W\mu/2\pi kD) \arctan y/x = -(Q_W\mu/2\pi kD)\,\theta \tag{6.22}$$

Lines of constant y/x or θ form radial flow streamlines, with θ being the usual angle variable in polar coordinates. The streamfunction is not too important a variable in radial flow, as it is in general planar flows, since the total volume flow rate is adequately described by Q_W. But Eq. (6.22) is important to readers new to the double-valued functions in Chapter 4. Note how $\theta = 0$, say, initially at some point. After one complete circuit around the origin, θ returns with the increased value 2π; the streamfunction, at the same time, varies from $\Psi = 0$ to $\Psi = -(Q_W \mu/kD)$, taking into account the net nonzero outflow from the well. Thus, ignoring the scale factor $-\mu/kD$ for now, the streamfunction tracks increases in volume flow rate ranging from 0 to Q_W. Incremental increases in θ, for example, lead to proportional incremental increases in flow rate. This exact proportionality is not, in general, true of single or multi-well flows in general heterogeneous, anisotropic media.

EXAMPLE 6.3. STEADY-STATE GAS FLOWS IN HOMOGENEOUS, ISOTROPIC MEDIA

As in Example 6.1, we consider three boundary value problems for steady gas flows in homogeneous, isotropic media. Two are easily posed and solved, but the third requires nonlinear iteration.

Pressure-Pressure Formulations

In general, when arbitrary liquids and gases are modeled, Eq. (6.1) is replaced by

$$d^2P^{m+1}/dr^2 + (1/r)\, dP^{m+1}/dr = 0 \tag{6.23}$$

for $P^{m+1}(r)$, where "r" is the radial coordinate. Here, pressure is specified at the wellbore $r = r_W$ and some distance away at $r = r_R$. But the pressure-pressure boundary value problem is *not* nonlinear, since we can write

$$P^{m+1}(r_W) = P_W^{m+1} \tag{6.24}$$

$$P^{m+1}(r_R) = P_R^{m+1} \tag{6.25}$$

for the function $P^{m+1}(r)$, where P_W and P_R are the respective assumed pressures. This formulation has the simple solution

$$P^{m+1}(r) = P_W^{m+1} + \left\{ \left(P_W^{m+1} - P_R^{m+1}\right) / \left(\log r_W/r_R\right) \right\} \log r/r_W \tag{6.26a}$$

or

$$P(r) = \{P_W^{m+1} + \{\left(P_W^{m+1} - P_R^{m+1}\right) / \left(\log r_W/r_R\right)\} \log r/r_W\}^{1/(m+1)} \tag{6.26b}$$

Note that taking the first derivative of Eq. (6.26b) leads to

$$dP(r)/dr = \{(P_W^{m+1} - P_R^{m+1})/((m+1)P^m \log r_W/r_R)\}1/r \tag{6.27}$$

Thus, the formula for the radial velocity q(r) corresponding to Eq. (6.5) is

$$\begin{aligned} q(r) &= -(k/\mu)\, dP(r)/dr \\ &= -(k/\mu)\, \{(P_W^{m+1} - P_R^{m+1})/((m+1)P^m \log r_W/r_R)\}1/r \end{aligned} \tag{6.28}$$

where k is formation permeability and μ is fluid viscosity. Using Eq. (6.6), the corresponding total volume flow rate is

$$Q_W = (2\pi kD/\mu)\, \{(P_W^{m+1} - P_R^{m+1})/((m+1)P_W^m \log r_R/r_W)\} \tag{6.29}$$

assuming a reservoir depth D into the page.

P_W-Q_W Formulations

In a steady flow, the total volume flow rate through any closed curve surrounding the well is the same. Thus, another formulation with a uniquely defined solution is the problem where the volume flow rate Q_W is prescribed at the well. If P_W and Q_W are given, we can use Eq. (6.29) to rewrite the flow rate boundary condition in terms of

$$P_R^{m+1} = P_W^{m+1} - \{Q_W\mu(m+1)P_W^m/(2\pi kD)\} \log r_R/r_W \tag{6.30}$$

The resulting pressure-pressure boundary value problem solution in Eqs. (6.26a), (6.26b) can then be used.

P_R-Q_W Formulations

Still another useful formulation is the problem where pressure is prescribed in the farfield, along with some value for the total volume flow rate Q_W. Again, the approach is to eliminate the explicit appearance of the latter in favor of an equivalent pressure-pressure formulation. Thus, we rewrite Eq. (6.30) in the form

$$P_W^{m+1} - \{Q_W\mu(m+1)P_W^m/(2\pi kD)\} \log r_R/r_W - P_R^{m+1} = 0 \tag{6.31}$$

If $m = 0$, the resulting expression is consistent with Eq. (6.12). If $m = 1$, for isothermal gases, we obtain a quadratic equation for the well pressure. The appropriate choice of sign in the quadratic formula is the one that leads to a the correct solution of the $m = 0$ problem.

In general, for arbitrary values of m, Eq. (6.31) is a nonlinear algebraic equation for P_W, which must be solved by iterative numerical methods. Once the solution is obtained, the boundary value problem is completely defined by the pressure-pressure formulas. A fast and stable solution is possible using Newton-Raphson iteration (Carnahan et al., 1969; Dahlquist and Bjorck, 1974) if we write Eq. (6.31) in the form

$$P_W - P_R^{m+1}/P_W^m - \{Q_W \mu (m+1) \log r_R/r_W\}/(2\pi kD) = 0 \tag{6.32}$$

and define

$$f(P_W) = P_W - P_R^{m+1}/P_W^m - \{Q_W \mu (m+1) \log r_R/r_W\}/(2\pi kD) \tag{6.33}$$

The variable part of Eq. (6.33) is plotted in Fig. 6.1. Then, we can easily form the derivative

$$f'(P_W) = 1 + m(P_R/P_W)^{m+1} \tag{6.34}$$

Successive improvements to an initial guess for $P_{W(n)}$ are given by the formula

$$P_{W(n+1)} = P_{W(n)} - f\left(P_{W(n)}\right)/f'\left(P_{W(n)}\right) \tag{6.35}$$

A part of the function $f(P_W)$ defined by Eq. (6.33) is plotted in Fig. 6.1. It is clear from the monotonic nature of the curve that the choice $P_{W(1)} = P_R$ always leads to rapidly convergent solutions.

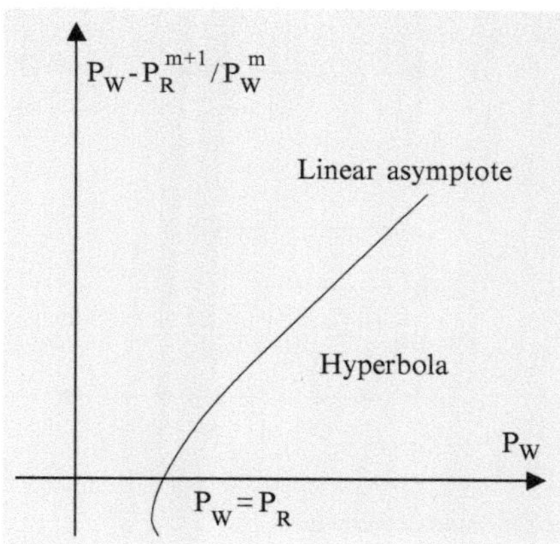

FIG. 6.1 Function for Newton-Raphson iteration.

Transient Compressible Flows

Here we shall introduce the numerical modeling of transient compressible flows. We shall develop all simulation concepts from first principles. The roles of boundary and initial conditions will be discussed. Also, the differences between explicit and implicit schemes are given, and the concept of numerical stability is developed, as are basic ideas—the advantages and disadvantages—behind variable meshes. We shall also explore intelligent ways to incorporate the analytical results derived earlier. This book does not explore the intricacies of well test interpretation. We do, however, emphasize that the great majority of real-world problems (e.g., liquids and gases in irregular reservoir domains) do *not* yield to analytically based models. These are largely limited to $m = 0$ fluids (or, liquids) which, owing to their linearity, allow convenient linear superposition; published superposition models for nonlinear gas flows are inherently incorrect. Even for linear flows, the effects of arbitrary reservoir geometry cannot be handled by analytical models, which call for drastic simplification. Ultimately, numerical well test simulations such as those described in Chapter 10 provide the best hope for accurate reservoir characterization. For the fundamentals behind well test analysis, the reader is referred to Collins (1961), or Richardson's exposition in Streeter (1961). Streltsova (1988) provides a state-of-the-art summary on closed-form solutions with idealized heterogeneities and complicated reservoir geometries.

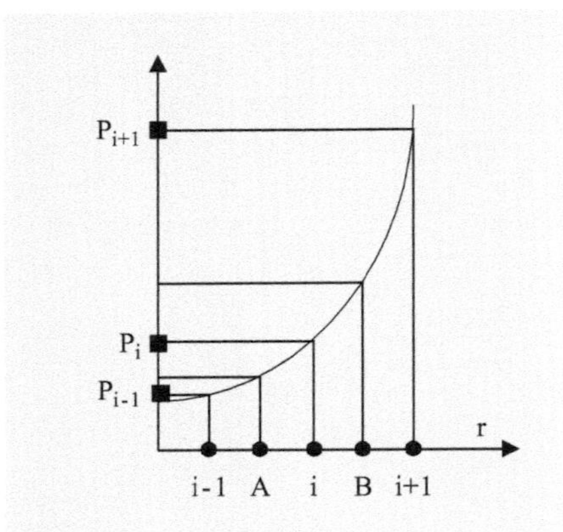

FIG. 6.2 Finite difference formula development.

EXAMPLE 6.4. NUMERICAL SOLUTION FOR STEADY FLOW

We motivate the ideas behind numerical modeling by considering steady-state flows. The resulting schemes, insofar as grid density and mesh variability are concerned, can be tested against the analytical solutions in Examples 6.1–6.3. The reader is encouraged to code, compile, and run the algorithms in this section, and to become familiar with simulator development.

Finite Difference Formulation

Suppose that analytical solutions were not possible, and that recourse to computational methods was necessary. Then, Eq. (6.1), for example, would have to be approximated by writing discretized algebraic equations from node to node, and solving the coupled equations using a matrix inversion technique. A simple way to introduce finite differences follows from Fig. 6.2. First consider constant mesh widths Δr, and examine the Point A lying at the midpoint between successive indices i-1 and i. It is clear that the *first* derivative $dP(A)/dr = (P_i - P_{i-1})/\Delta r$. Similarly, Point B requires $dP(B)/dr = (P_{i+1} - P_i)/\Delta r$. Then, the *second* derivative at i, situated midway between A and B, is, by definition

$$d^2P/dr^2\,(r_i) = (dP(B)/dr - dP(A)/dr)/\Delta r \tag{6.36}$$

$$= (P_{i+1} - 2P_i + P_{i-1})/(\Delta r)^2 \tag{6.37}$$

The complementary expression for the first derivative at i is

$$dP/dr\,(r_i) = (P_{i+1} - P_{i-1})/(2\Delta r) \tag{6.38}$$

Because our derivatives are taken using values of P from both left and right of i, Eqs. (6.37), (6.38) are called central difference approximations. Backward and forward one-sided derivatives are also possible, though less accurate for the same number of meshes. Eqs. (6.37), (6.38) are second-order accurate, with errors on the order $O(\Delta r^2)$.

Now, we substitute Eqs. (6.38), (6.39a), (6.39b) into Eq. (6.1), and we rearrange terms to form the difference equation

$$(1-\Delta r/2r_i)\,P_{i-1} - 2P_i + (1 + \Delta r/2r_i)\,P_{i+1} = 0 \tag{6.39a}$$

where the radial variable satisfies

$$r_i = r_W + (i - 1)\Delta r \tag{6.39b}$$

Suppose the index i varies from 1 to i_{max}, so that $i_{max} - 1$ meshes exist. The procedure is to write Eq. (6.39a) for $i = 2$ to $i = i_{max} - 1$ inclusively, and then, to supplement these equations with the boundary conditions $P_1 = P_W$ at the well and $P_{imax} = P_R$ at the reservoir farfield. This leads to a system of tridiagonal linear algebraic equations containing, at most, three unknowns per equation. The resulting system for $i_{max} = 4$, for example, takes the form

$$\begin{aligned} &P_1 = P_W \\ &(1-\Delta r/2r_2)\,P_1 - 2\,P_2 + (1 + \Delta r/2r_2)\,P_3 = 0 \\ &(1-\Delta r/2r_3)\,P_2 - 2\,P_3 + (1 + \Delta r/2r_3)\,P_4 = 0 \\ &P_4 = P_R \end{aligned} \tag{6.40}$$

which is a special instance of the general "tridiagonal matrix"

$$\begin{vmatrix} b_1 & c_1 & & & & & & \\ a_2 & b_2 & c_2 & & & & & \\ & a_3 & b_3 & c_3 & & & & \\ & & a_4 & b_4 & c_4 & & & \\ & & & a_5 & b_5 & c_5 & & \\ & & & . & & & & \\ & & & . & & & & \\ & & & & & a_{n-1} & b_{n-1} & c_{n-1} \\ & & & & & & a_n & b_n \end{vmatrix} \begin{vmatrix} v_1 \\ v_2 \\ v_3 \\ v_4 \\ v_5 \\ . \\ . \\ v_{n-1} \\ v_n \end{vmatrix} = \begin{vmatrix} w_1 \\ w_2 \\ w_3 \\ w_4 \\ w_5 \\ . \\ . \\ w_{n-1} \\ w_n \end{vmatrix} \tag{6.41}$$

In Eq. (6.41), the indexed quantities a_i, b_i, c_i, v_i, and w_i are called column vectors of dimension n; sometimes, they are vectors denoted by the bold-faced symbols **a**, **b**, **c**, **v**, and **w**. The square matrix at the left reveals its obvious "tri-diagonal" structure; it is a special case of a diagonal banded matrix. We shall not deal with matrix inversion in this book. Suffice it to say that the last (or the first) equation, which involves two unknowns only, is usually used to reduce the number of unknowns along each row, right up (or down) the matrix, thus resulting in a two or bidiagonal system. Repeating the process in the opposite direction yields the solution vector **v**.

When all the tridiagonal matrix coefficients in Eq. (6.41) are defined, the solution v_i can be obtained by calling standard tridiagonal solvers found in most numerical analysis books. If the programming language used is Fortran, as is assumed in this book, the subroutine listed in Fig. 6.3 can be used. The statement CALL TRIDI with the appropriate A, B, C, W, and N defined will return the desired pressure solution in the vector V. The linear boundary value problem for the function p^{m+1} considered previously is similarly solved. Darcy velocities are obtained by post-processing computed pressures, using the difference

```
      SUBROUTINE TRIDI(A,B,C,V,W,N)
      DIMENSION A(5), B(5), C(5), V(5), W(5)
      A(N) = A(N)/B(N)
      W(N) = W(N)/B(N)
      DO 100  I = 2,N
      II = -I+N+2
      BN = 1./(B(II-1)-A(II)*C(II-1))
      A(II-1) = A(II-1)*BN
      W(II-1) =  (W(II-1)-C(II-1)*W(II))*BN
100   CONTINUE
      V(1) = W(1)
      DO 200  I = 2,N
      V(I) = W(I)-A(I)*V(I-1)
200   CONTINUE
      RETURN
      END
```

FIG. 6.3 Tridiagonal matrix solver.

formula for dP/dr in Eq. (6.38) for internal nodes, the forward difference formula dP/dr $(r_1) = (P_2 - P_1)/\Delta r$ for $i = 1$, and the backward difference formula dP/dr $(r_{imax}) = (P_{imax} - P_{imax-1})/\Delta r$ for $i = i_{max}$.

EXAMPLE 6.5. EXPLICIT AND IMPLICIT SCHEMES FOR TRANSIENT COMPRESSIBLE LIQUIDS

When transient effects are present, as in well test pressure buildup or drawdown, the effects of compressibility and porosity enter. The governing equation for constant properties takes the form

$$\partial^2 P/\partial r^2 + 1/r\, \partial P/\partial r = (\phi \mu c/k)\, \partial P/\partial t \tag{6.42}$$

Now, Eqs. (6.36)–(6.38) apply to the left side of Eq. (6.42). The right side $\partial P/\partial t$ suggests that two time steps, at the very minimum, must be considered, say t_n and t_{n+1}. We can certainly write $\partial P/\partial t$ at $t = t_{n+1}$ in the form $(P_{i,n+1} - P_{i,n})/\Delta t$, where Δt is the time step. However, the question arises as to which time value the spatial derivatives are to be evaluated at.

Explicit Schemes

The choice t_n, of course, allows us to solve the resulting equation for the new pressure $P_{i,n+1}$ in terms of previously available values $P_{i-1,n}$, $P_{i,n}$ and $P_{i+1,n}$. This simple approach means that a pocket calculator is all that is required; there are no matrixes to invert, and given initial values are used to start the calculations at $n = 1$. This differencing scheme is known as an explicit scheme. Such schemes, while convenient, require very small time steps for both accuracy and numerical stability. Unstable schemes are those that lead to unrealistic infinite values after a finite number of time step calculations; thus, they may not suitable for physical modeling.

Numerical Stability

The stability of a scheme can be determined by a relatively simple von Neumann stability test (Carnahan et al., 1969; Richtmyer and Morton, 1957). This test, we emphasize, is only qualitatively accurate, since it does not account for the detailed effects of boundary and initial conditions, and for the role of heterogeneities when variable coefficients are present in the given equation. Note that a stable finite difference scheme does not necessarily converge to solutions of the PDE even if Δr, Δt are vanishingly small. This subject is treated in advanced courses.

Implicit Schemes

If we had instead evaluated the spatial derivatives at the t_{n+1} time level, we would have obtained a difference equation that takes the more complicated form

$$(P_{i-1,n+1} - 2P_{i,n+1} + P_{i+1,n+1})/(\Delta r)^2$$
$$+\{1/[r_W + (i-1)\Delta r]\}\,(P_{i+1,n+1} - P_{i-1,n+1})/(2\Delta r)$$
$$= (\phi\mu c/k)\,(P_{i,n+1} - P_{i,n})/\Delta t \quad (6.43)$$

or

$$\{1 - \Delta r/[2r_W + 2(i-1)\Delta r]\}P_{i-1,n+1} - \left\{2 + \phi\mu c(\Delta r)^2/(k\Delta t)\right\}P_{i,n+1}$$
$$+\{1 + \Delta r/[2r_W + 2(i-1)\Delta r]\}P_{i+1,n+1} = -\left\{\phi\mu c(\Delta r)^2/(k\Delta t)\right\}P_{i,n} \quad (6.44)$$

If right-side starting values $P_{i,n}$ for all values of i are known, then Eq. (6.44) can be solved for the subsequent time step using exactly the same tridiagonal matrix procedure described in Example 6.4. This scheme is known as an implicit method; it is inherently more stable and more accurate than the explicit scheme. Larger time steps may be taken, but matrix inversion is required, implying additional programming and computer requirements.

For radial and linear flows, the index i alone appears, but in higher dimensions, the pressure might be indexed by two space indexes. For example, one might have Pi,j,n, where i and j represent x and y, and the resulting difference equation is integrated one line (e.g., of constant i, with j fixed) at a time. This procedure is described in detail in Chapter 10. The process of taking $i=1$ to imax and $j=1$ to jmax in a Fortran do-loop implies that the flow domain is rectangular; rectangular shapes, in fact, are easily analyzed by implicit schemes because, for example, lines of constant i are parallel to a boundary and permit ready implementation of boundary conditions. When the boundaries are irregular, it is difficult to set up systems of lines each having the same numbers of grids. This makes implicit approaches difficult to implement, and explicit schemes, which allow point-by-point calculation, are more popular. The use of curvilinear grids, introduced later, overcomes this objection.

Variable Grids

So far, we have considered constant mesh spacings Δr for simplicity. Clearly, given rapid changes in pressure at the well, this does not provide sufficient near-field resolution. What is desired is a grid that is fine near the well and coarse far away. This is possible using variable grids, which are easily understood. Let us examine Point A in Fig. 6.2 lying at the midpoint between indices i-1 and i. The first derivative satisfies

$$dP(A)/dr = (P_i - P_{i-1})/(r_i - r_{i-1}) \quad (6.45a)$$

Similarly, Point B requires

$$dP(B)/dr = (P_{i+1} - P_i)/(r_{i+1} - r_i) \quad (6.45b)$$

The second derivative at i, situated midway between A and B, is

$$d^2P/dr^2(r_i) = (dP(B)/dr - dP(A)/dr)/\{1/2\,(r_{i+1} - r_{i-1})\} \tag{6.46}$$

or, after some algebra,

$$\begin{aligned} d^2P/dr^2(r_i) = {} & 2P_{i+1}/\{(r_{i+1} - r_{i-1})(r_{i+1} - r_i)\} \\ & + 2P_{i-1}/\{(r_{i+1} - r_{i-1})(r_i - r_{i-1})\} \\ & - 2P_i\,[1/\{(r_{i+1} - r_{i-1})(r_{i+1} - r_i)\} + 1/\{(r_{i+1} - r_{i-1})(r_i - r_{i-1})\}] \end{aligned} \tag{6.47}$$

Then, the transient model analogous to Eq. (6.44) becomes

$$\begin{aligned} & \{1 - (r_i - r_{i-1})/2r_i\}\,P_{i-1,n+1} \\ & -[\{\phi\mu c/(2k\Delta t)\}(r_{i+1} - r_{i-1})(r_i - r_{i-1}) + (r_i - r_{i-1})/(r_{i+1} - r_i) + 1]\,P_{i,n+1} \\ & + \{(r_i - r_{i-1})/(r_{i+1} - r_i) + (r_i - r_{i-1})/(2r_i)\}\,P_{i+1,n+1} \\ = {} & -[\{\phi\mu c/(2k\Delta t)\}(r_{i+1} - r_{i-1})(r_i - r_{i-1})]\,P_{i,n} \end{aligned} \tag{6.48}$$

Like Eq. (6.44), this formula, provided size increases from mesh to mesh are slow, is second-order accurate in space and first-order accurate in time. Note that multi-level time schemes and even higher-order accurate space discretizations are possible, and are discussed in advanced courses.

EXAMPLE 6.6. TRANSIENT COMPRESSIBLE GAS FLOWS

Here we consider compressible gases, as well as special topics on variable meshes, superposition, and flow initialization. This example completes our treatment of radial flows and sets the stage for general discussions on planar flows. The transient behavior characteristic of radial flows is described in petroleum textbooks and we direct interested readers to these references. Our primary concern is the transient modeling of irregular reservoirs in Chapter 10. Now, transient compressible liquids satisfy Eq. (6.42), which is linear. On the other hand, gases satisfy Eq. (6.49), where c is replaced by m/p (see Chapter 1). From an analytical viewpoint, Eqs. (6.42), (6.49) are vastly different: linear superposition methods apply to the former but not the latter.

$$\partial^2 p^{m+1}/\partial r^2 + 1/r\,\partial p^{m+1}/\partial r = (\phi\mu m/kP)\,\partial p^{m+1}/\partial t \tag{6.49}$$

Linearity Versus Nonlinearity

The meaning of linearity and superposition can be demonstrated by writing Eq. (6.42) first for pressure P_1 and then for P_2. The sum $P_1 + P_2$, by direct substitution, also satisfies Eq. (6.42). This is not so with Eq. (6.49) because the presence of P in $\phi\mu m/kP$ causes $P_1 + P_2$ to satisfy an equation other than Eq. (6.49). Thus, superposition does not hold; for nonlinear systems like Eq. (6.49), classic superposition methods for liquids do not apply to gases. On the other hand, Eq. (6.49) takes a form nearly identical to that of Eq. (6.42). For the purposes of numerical simulation, Eq. (6.49) can be treated identically as for linear flows, provided we regard the m/P as a fictitious compressibility that is updated using the latest available values at the *previous* time step. This allows us to use the linear solver TRIDI at each time step, and avoids time consuming Newton-Raphson methods. This solution is numerically stable.

Nonlinear Superposition

Very often in pressure transient testing, the pressure (or flow rate) is changed in time; for liquids, flow rate (or pressure) response is obtained by linear superposition of elementary solutions. For gases, superposition is not possible because nonlinear solutions are not linearly additive. How does one calculate the response when pressure or flow rate at the well vary, say stepwise, in time? Fortunately, the governing equations can be numerically integrated with respect to t. It remains for us to represent stepwise changes in any particular variable using convenient mathematical devices (Fig. 6.4).

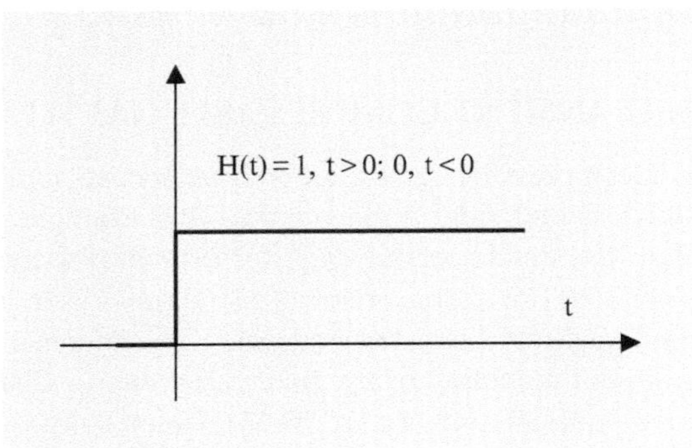

FIG. 6.4 Heaviside step function.

One such device is the Heaviside step function, defined as $H(t)=1, t \geq 0; 0, t<0$. Then, a "hump" of constant amplitude A in $0<t_{left}<t<t_{right}$, that is zero otherwise, can be represented by $A\{H(t-t_{left})-H(t-t_{right})\}$. A sequence of imposed pressure or flow rate changes at the well is just a linear superposition of such functions, for example, $p_{well}=\sum A_m\{H(t-t_{left,m})-H(t-t_{right,m})\}$ where m is a summation index and A_m, $t_{left,m}$, and $t_{right,m}$ are input arrays. This representation is easily stored in a function statement for ready access. The computed pressure response based on Eq. (6.49) contains the required nonlinear interactions. The term nonlinear superposition is misleading, but it does hint at the programming method suggested.

Choosing Variable Meshes

Modern ideas in mesh generation are treated later in this book. For now, it suffices to show how successively magnified meshes can be generated using simple formulas developed in geometrical series analyses. Let us multiply the identity

$$1 + a + a^2 + a^3 + \cdots + a^{N-1} = \left(1 - a^N\right)/(1 - a) \tag{6.50}$$

by the initial spacing Δ adjacent to the well to give

$$\Delta + a\Delta + a^2\Delta + a^3\Delta + \cdots + a^{N-1}\Delta = \left\{\left(1 - a^N\right)/(1 - a)\right\}\Delta \tag{6.51}$$

Let the mesh index i vary from $i = 1$ to $i = i_{max}$, so that the number of grids N is equal to $N = i_{max} - 1$. If the total radial distance L from the edge of the well to the outer boundary is given, and the mesh amplification rate a is specified, setting the right-hand side of Eq. (6.51) equal to L leads to the initial mesh length

$$\Delta = (1 - a)L/\left(1 - a^N\right)) \tag{6.52}$$

On the other hand, if a, L, and Δ are specified, the required final i index is given by rounding (to the nearest integer) the right-hand side of

$$i_{max} = 1 + \log\{1 + L(a - 1)/\Delta\}/\log a \tag{6.53}$$

Generally, $a > 1$, with the choice $a = 1$ defaulting to uniform constant grids.

Initialization Procedures

Well test field procedures fall into two varieties, namely, pressure drawdown and pressure buildup. Very often, an initially quiescent, pressurized reservoir of uniform pressure P_{init} (typically equal to the farfield pressure P_R in an aquifer-driven flow) is opened to a lower well pressure P_W and allowed to produce. The pressure in the neighborhood of the well declines rapidly, resulting in a drawdown; the large time behavior might be the steady-state solution obtained in Examples 6.1–6.3. On the other hand, a reservoir already producing at steady state may be shut in, resulting in a pressure buildup. For such problems, the analytical results of Examples 6.1–6.3 can be used to initialize the flow. Our earlier analytical results should also be used to calibrate the mesh; numerical solutions always contain some degree of mesh dependence, and calibration ensures that the numerical model is correct in at least some limit. There are, of course, no guarantees in simulation.

Flow Rate Boundary Conditions

In constant density flow, a producing well always results in flow across the farfield boundary; a nonproducing well always implies that the farfield flow is stagnant. This is not the case when the fluid is compressible. Even when a well is shut-in, fluid may continue to migrate across farfield boundaries because of expansion effects. Both pressure and flow rate boundary conditions may be used at the well or at the farfield radius. The differencing procedure is straightforward. For example, if an input volume flow rate Q(t) is assumed at the well, then,

$$q = (-k/\mu)\, dp/dr = Q(t)/(2\pi D r_w) \tag{6.54}$$

leads to

$$dp/dr = (P_{2,n} - P_{1,n})/(r_2 - r_1) = (-\mu/k)\, Q(t_n)/(2\pi D r_w) \tag{6.55}$$

Thus, this boundary condition affects only one line of the tridiagonal matrix. Again, the volume flux at the farfield need *not* equal Q(t), except at steady state. We shall build upon the notions and ideas introduced here in the subsequent chapters, step by step, in order to introduce state-of-the-art ideas in modeling. In Chapter 18, exact solutions for spherical, transient, compressible flow with skin, storage, and anisotropy are developed for formation tester use.

PROBLEMS AND EXERCISES

1. Write an explicit time-marching program to compute pressure transients in liquids when the initial and farfield pressures are identical, on a programmable calculator. Perform computations for two cases, first when the well pressure is high and second when it is low, relative to the initial value. Can you identify buildup and drawdown behavior?
2. Write a general "implicit" program for transient compressible liquids and gases, taking constant spatial meshes for simplicity. For both flows, assume an initially hydrostatic reservoir, with the sandface suddenly exposed to a prescribed pressure level different from hydrostatic. Run the simulations to steady-state and monitor the flow rate history at the well. Show that the asymptotic results agree with the three complementary steady flow formulations and solutions given in this chapter.
3. Study the finite difference literature on parabolic equations and transient modeling, and summarize the von Neumann stability criterion for explicit and implicit schemes. What are its strengths and limitations? What new stability tests are available to study nonlinearities and heterogeneities? Comment on group velocity and wave-based stability analysis.
4. Consider a reservoir radius of 1000 ft and a well that is 6 in. in diameter. What Δr is appropriate in the nearfield? How many such constant meshes are required to model the entire reservoir? How many variable meshes are required if a grid amplification rate of 10% is assumed? To obtain accurate solutions, do your answers depend on the physical properties of the formation and fluid, for example, ϕ, μ, c, and k? If so, on what combinations of these and mesh variables?
5. Generalize the implicit program written in (2) to accept variable meshes, quiescent and flowing initial conditions, and arbitrary step changes in pressure or volume flow rate at the well. *Hint:* Program the Heaviside step

function in a separate function statement, and write arbitrary step changes in pressure or rate in terms of this function. After you validate your program, write a user-friendly interface using Visual Basic, in order to understand user-based issues in software design.

6. Steady-state solutions for circular wells in isotropic reservoirs were given at the beginning of the chapter. Now consider anisotropic flows of liquids and general gases. Starting with the steady anisotropic pressure equation written in x, y coordinates, show that the simplest well that can be modeled analytically has an *elliptic* wellbore shape. What is the significance of this result? How would you model a *circular* well in anisotropic media?

Chapter 7

Finite Difference Methods for Planar Flows

In this chapter, we shall introduce the finite difference method and its application to solving partial differential equations. Although this subject is usually offered only in advanced numerical analysis courses, there is no reason to impose artificial requirements or prerequisites. Both basic and sophisticated ideas can be developed from simple notions in elementary calculus. The intuitive "how-to" approach taken is down-to-earth, comprehensive, and, importantly, rigorous. But we shall discuss only those ideas necessary to accomplish our objective, that is, solving Laplace's equation $\partial^2 P/\partial x^2 + \partial^2 P/\partial y^2 = 0 \left(\text{or}, \partial^2 P^{m+1}/\partial x^2 + \partial^2 P^{m+1}/\partial y^2 = 0\right)$ for steady, planar reservoir flows. We shall develop the terminology and ideas naturally, and avoid excessive mathematical formalism. We shall augment the discussions with Fortran examples and source code to make the ideas clear and the methodologies widely accessible. This presentation is no substitute for a truly rigorous and formal study of numerical methods. However, like the rest of this book, it is written to be self-contained so that the development of computational methods in petroleum engineering can proceed without undue interruption.

FINITE DIFFERENCES: BASIC CONCEPTS

Before we introduce numerical relaxation schemes and their applications to partial differential equations, we need to develop the basic ideas and working vocabulary underlying finite difference discretization methods.

Finite Difference Approximations

Let us consider the function F(x) as shown in Fig. 7.1 and examine several representations for its derivatives. Because F(x) will be approximated at a discrete set of points, we introduce a grid or mesh of i_{max} points $x_1, x_2, x_3, \ldots, x_{imax}$. In fact, we shall denote any three consecutive points by x_{i-1}, x_i and x_{i+1}, where i is an index ordered so that it increases as x increases. When the distance between successive points in this discretization process is the same, the grid is constant.

Quantitative Methods in Reservoir Engineering. http://dx.doi.org/10.1016/B978-0-12-810518-4.00007-4

On the other hand, the grid is variable if the distances vary spatially; it is adaptive in time if it adapts locally in order to track key physical events like rapid saturation changes.

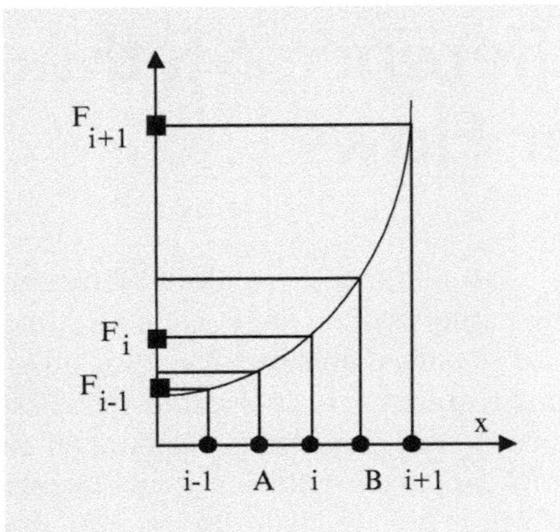

FIG. 7.1 Finite difference formula development.

For simplicity, let us assume constant (or slowly varying meshes) first, and refer to the points A and B in Fig. 7.1. The first derivative of F(x) at $x = x_A$ is approximated by evaluating its slope using points to its left and right,

$$\partial F(x_A)/\partial x = (F_i - F_{i-1})/(x_i - x_{i-1}) \tag{7.1}$$

At $x = x_B$, we likewise have

$$\partial F(x_B)/\partial x = (F_{i+1} - F_i)/(x_{i+1} - x_i) \tag{7.2}$$

Thus, the second derivative at $x = x_i$, or simply i, takes the form

$$\begin{aligned} \partial^2 F/\partial x^2 &= \{\partial F(x_B)/\partial x - \partial F(x_A)/\partial x\}/\{1/2(x_{i+1} - x_{i-1})\} \\ &= \{(F_{i+1} - F_i)/(x_{i+1} - x_i) - (F_i - F_{i-1})/(x_i - x_{i-1})\} \\ &\quad /\{1/2\,(x_{i+1} - x_{i-1})\} \end{aligned} \tag{7.3}$$

where the length ½ $(x_{i+1} - x_{i-1})$ in Eq. (7.3) applies if the meshes expand or contract slowly. The corresponding formula for the first derivative at i is

$$\partial F(x_i)/\partial x = (F_{i+1} - F_{i-1})/(x_{i+1} - x_{i-1}) \tag{7.4}$$

Eqs. (7.3), (7.4) are finite difference representations for $\partial^2 F/\partial x^2$ and $\partial F/\partial x$. Our use of left and right values to define geometric slopes (for both first and second derivatives) is called central differencing. Backward and forward one-sided differencing are also possible, though less accurate for the same number of points.

A Simple Differential Equation

We discuss a simple application for finite differences. In particular, let us consider the boundary value problem defined by the second-order linear differential equation and the boundary conditions

$$d^2y/dx^2 = 0 \tag{7.5}$$

$$y(0) = 0 \tag{7.6a}$$

$$y(2) = 2 \tag{7.6b}$$

This equation set has the simple straight-line solution

$$y(x) = x \tag{7.7}$$

The idea, of course, is to replicate this function numerically. For now, we specialize the analysis to constant meshes to illustrate the basic procedure. This choice of mesh system reduces Eq. (7.3) to the simple form

$$d^2y/dx^2 = (y_{i-1} - 2\,y_i + y_{i+1})/(\Delta x)^2 \tag{7.8}$$

where Δx is an assumed mesh length, so that Eq. (7.5) becomes

$$y_{i-1} - 2y_i + y_{i+1} = 0 \tag{7.9}$$

Eq. (7.9) is the finite difference model relating different y values at different positions x_i. It shows that these y's are coupled and must be determined simultaneously. To find the equations that must be solved, write Eq. (7.9) for each of the *internal* nodes $i = 2, 3, \ldots, (i_{max}-1)$. This leads to

$$i = 2: \quad y_1 - 2y_2 + y_3 = 0 \tag{7.10b}$$

$$i = 3: \quad y_2 - 2\,y_3 + y_4 = 0 \tag{7.10c}$$

$$i = 4: \quad y_3 - 2y_4 + y_5 = 0 \tag{7.10d}$$

$$i = 5: \quad y_4 - 2y_5 + y_6 = 0 \tag{7.10e}$$

$$i = i_{max} - 1: \quad y_{imax-2} - 2y_{imax-1} + y_{imax} = 0 \tag{7.10f}$$

Observe that there are two more unknowns than there are equations. The additional required equations are obtained from Eqs. (7.6a), (7.6b), that is,

$$y_1 = 0 \tag{7.10a}$$

$$y_{imax} = 2 \tag{7.10g}$$

which we might introduce at the top and bottom, respectively, of the equation block. Note that Eqs. (7.10a)–(7.10g) so written assume a tridiagonal structure; the exact form will be important to the iterative schemes we consider later. For now, in our direct single-pass solution to Eqs. (7.5), (7.6a), (7.6b), we can rewrite the foregoing equations in the matrix form

$$
\begin{vmatrix}
1 & 0 & & & & & & \\
1 & -2 & 1 & & & & & \\
 & 1 & -2 & 1 & & & & \\
 & & 1 & -2 & 1 & & & \\
 & & & 1 & -2 & 1 & & \\
 & & & . & & & & \\
 & & & . & & & & \\
 & & & & & 1 & -2 & 1 \\
 & & & & & & 0 & 1
\end{vmatrix}
\begin{vmatrix}
y_1 \\ y_2 \\ y_3 \\ y_4 \\ y_5 \\ . \\ . \\ y_{imax-1} \\ y_{imax}
\end{vmatrix}
=
\begin{vmatrix}
0 \\ 0 \\ 0 \\ 0 \\ 0 \\ 0 \\ 0 \\ 0 \\ 2
\end{vmatrix}
\qquad (7.11)
$$

This is a special instance of the more general tridiagonal matrix

$$
\begin{vmatrix}
b_1 & c_1 & & & & & & \\
a_2 & b_2 & c_2 & & & & & \\
 & a_3 & b_3 & c_3 & & & & \\
 & & a_4 & b_4 & c_4 & & & \\
 & & & a_5 & b_5 & c_5 & & \\
 & & & . & & & & \\
 & & & . & & & & \\
 & & & & & a_{n-1} & b_{n-1} & c_{n-1} \\
 & & & & & & a_n & b_n
\end{vmatrix}
\begin{vmatrix}
v_1 \\ v_2 \\ v_3 \\ v_4 \\ v_5 \\ . \\ . \\ v_{n-1} \\ v_n
\end{vmatrix}
=
\begin{vmatrix}
w_1 \\ w_2 \\ w_3 \\ w_4 \\ w_5 \\ . \\ . \\ w_{n-1} \\ w_n
\end{vmatrix}
\qquad (7.12)
$$

The indexed quantities a_i, b_i, c_i, v_i, and w_i are called column vectors of dimension n, although sometimes, they are simply vectors denoted by the bold-faced symbols **a**, **b**, **c**, **v**, and **w**. The matrix at the left is a tridiagonal matrix, a special case of a diagonal banded matrix. We shall not deal with matrix inversion in this book. Suffice it to say that the last (or the first) row, which involves two unknowns only, is usually used to reduce the number of unknowns along each row, right up (or down) the matrix, thus resulting in a bidiagonal system. Then, repeating the process in the opposite direction yields the solution vector **v**.

When all the coefficients in Eq. (7.12) are defined in Eq. (7.11), say, the solution vector v_i is obtained by calling standard tridiagonal solvers found in numerical analysis books. If the programming language used is Fortran, as will be assumed in this book, the subroutine in Fig. 7.2 can be used. The routine as coded destroys all original input coefficients upon inversion; if it is called successively, as the solution of partial differential equations requires, the relevant coefficients must be redefined prior to each call of TRIDI. Also, we emphasize that A(1) and C(i_{max}) should be defined and set to dummy values, "99" in our examples, even though they do not play a role in the solution. Unless this is done, certain computers will initialize their registers improperly and produce incorrect solutions. The reader should verify that the solution of Eq. (7.11) does agree with the exact solution in Eq. (7.7).

```
      SUBROUTINE TRIDI(A,B,C,V,W,N)
      DIMENSION A(11), B(11), C(11), V(11), W(11)
      A(N) = A(N)/B(N)
      W(N) = W(N)/B(N)
      DO 100  I = 2,N
      II = -I+N+2
      BN = 1./(B(II-1)-A(II)*C(II-1))
      A(II-1) = A(II-1)*BN
      W(II-1) = (W(II-1)-C(II-1)*W(II))*BN
100   CONTINUE
      V(1) = W(1)
      DO 200  I = 2,N
      V(I) = W(I)-A(I)*V(I-1)
200   CONTINUE
      RETURN
      END
```

FIG. 7.2 Tridiagonal matrix solver.

Variable Coefficients and Grids

Ordinary differential equations often contain variable coefficients, for example,

$$d^2y/dx^2 + f_1(x)dy/dx + f_2(x)y = f_3(x) \tag{7.13}$$

The f_1, f_2, and f_3 might describe spatially dependent properties in the physical problem being modeled. If they vary rapidly, the use of constant meshes may be inappropriate. If so, Eqs. (7.3), (7.4) must be used, and the discretized values of the arrays x_i, $f_1(x_i)$, $f_2(x_i)$ and $f_3(x_i)$ must be additionally stored in computer memory. Needless to say, the matrix coefficients in Eq. (7.12) now become much more complicated. We warn against blindly using the method given. For example, if the f's are singular or discontinuous, special treatment is required, and an understanding of fundamental mathematical theory is needed. For well-behaved coefficients, grid selection is straightforward and follows several rules of thumb. If a f coefficient varies rapidly in some region of space, it is reasonable to increase local mesh density to improve physical resolution. However, there is always the danger that, since the value of y_i affects each and all of its neighbors, spurious effects can contaminate the complete solution. Otherwise, constant grids may suffice. In any event, the typical mesh size should be small compared to the length scale of the problem, and mesh-to-mesh expansion rates should not exceed 10%. Detailed testing of the solution for mesh dependence should accompany program development.

FORMULATING STEADY FLOW PROBLEMS

In this section, we discuss numerical solutions to Laplace's equation for the pressure P(x,y), with and without wells and fractures, using both aquifer boundary conditions specifying pressure, and solid wall conditions assuming zero normal flow. We consider, for purposes of exposition, the Cartesian form

$$\partial^2 P/\partial x^2 + \partial^2 P/\partial y^2 = 0 \tag{7.14}$$

to be solved on a rectangular grid, and defer to Chapters 8–10 the subject of curvilinear coordinates and grid generation. Again, PDEs involve partial derivatives of the unknown function, and fall into three fundamental classifications. Eq. (7.14), for example, is elliptic (transient compressible flows satisfy parabolic equations like $\partial^2 u/\partial x^2 + \partial^2 u/\partial y^2 = \partial u/\partial t$, while seismic waves satisfy hyberbolic equations like $\partial^2 u/\partial x^2 + \partial^2 u/\partial y^2 = \partial^2 u/\partial t^2$). In this chapter, we shall discuss elliptic equations only.

Eq. (7.14) assumes the two-dimensional, constant density flow of a liquid in an isotropic homogeneous medium. First consider for simplicity a singly connected region, for example, a simple pie, square, or triangle. It is known that the solution is completely defined whenever pressure is prescribed over the entire boundary. For such Dirichlet problems, the solutions, in mathematical lingo, exist and are uniquely defined. Now suppose that the flux of mass or the velocity is given, that is, that the derivative of pressure in a direction normal to the boundary contour is prescribed. What can we expect for the solution? Since the normal derivative $\partial p/\partial n$ is proportional to the Darcy velocity, we would expect that it cannot be arbitrarily specified. It must be given in such a way that just enough flow leaves as enters the flow domain. Moreover, the value of the p's obtained will be indeterminate to within a constant, since we have prescribed only derivatives. This additive level of pressure will not affect flow rates, since it differentiates to zero; the exact pressure level is unimportant and can be conveniently set to any value at a given point.

Boundary value problems where the normal derivative $\partial p/\partial n$ is specified at the boundaries are known as Neumann problems. Their solutions are not unique, but only to the extent just described. If the flow rate, which is proportional to $\partial p/\partial n$, is prescribed over part(s) of the boundary, and pressure itself is given over the remainder, the solution is again completely determined and unique. The reason is simple: we have not unreasonably created mass. The required mass conservation will manifest itself at the boundaries where pressure was prescribed, and a net outflow or inflow will be obtained, that is, physically sound. Problems where both $\partial p/\partial n$ and p are specified are referred to as mixed Dirichlet-Neumann problems or mixed problems.

So far, we have restricted our discussion to singly connected domains, that is, uninteresting reservoir floods *without* wells. The presence of a well effectively punctures a hole in the circular or rectangular region of flow, creating a donut-like reservoir; such shapes are said to be doubly connected. Again, common-sense ideas related to mass conservation apply. If the velocity (via $\partial p/\partial n$) is prescribed over the complete outer reservoir boundary, then one cannot arbitrarily assign $\partial p/\partial n$ at the well; however, specifying the pressure level itself is completely legitimate and will lead to unique and reasonable solutions. Similar considerations apply to reservoirs with multiple wells; the corresponding domains of flow are said to be multiply connected. Of course, different considerations apply when compressible transient flow is allowed, for example, flow can be produced at a single well even when there is no flow through farfield boundaries. An understanding of pressure behavior is essential to good simulator development. It turns out that the insight gained in petroleum flow simulation is also useful in developing robust grid generation algorithms.

STEADY FLOW PROBLEMS

In this section, we obtain numerical solutions to Eq. (7.14) for several different flow problems. For simplicity, we consider constant meshes, with lengths Δx and Δy in the x and y directions. The resolution achieved with such meshes systems near wells is limited. The programs are given for illustrative purposes, but the question of resolution will be addressed when we deal with curvilinear meshes. Now, the second derivative in Eq. (7.3) applies to a function F(x) at $x = x_i$, but P(x,y) depends on an additional y, say indexed by j. At any point (i,j), use of Eq. (7.3) in both x and y directions with Eq. (7.14) leads to the simple model

$$\begin{aligned}&\left(P_{i-1,j} - 2P_{i,j} + P_{i+1,j}\right)/(\Delta x)^2 \\ &\qquad + \left(P_{i,j-1} - 2P_{i,j} + P_{i,j+1}\right)/(\Delta y)^2 = 0\end{aligned} \tag{7.15}$$

Note that this seemingly straightforward use of Eq. (7.3) is actually subtle. For our constant density fluid, the pressure at (i,j) *must* depend on its neighbors at (i-1,j), (i+1,j), (i,j-1), and (i,j+1). That is, the flow at any point is influenced by every other point, and each point affects all other points. The situation is different for hyperbolic problems; for example, disturbances created by a supersonic aircraft cannot propagate ahead of the plane, so that a difference approximation that violates domains of influence and dependence cannot be used. Similarly, in unsteady wave propagation, computations cannot depend on future time. Hence, there are areas in physics where use of central differencing throughout is inappropriate, and one-sided models must be used. However, for Laplace's equation, the approximation in Eq. (7.15) is perfectly valid.

We now consider the rectangular reservoir domain defined by the index ranges $1 \le i \le 11$ and $1 \le j \le 11$, and specifically, a Dirichlet formulation where pressures of 10, 20, 30, and 40 are specified in clockwise fashion along the four edges of the box. This no-well formulation, as discussed earlier, is associated with a unique solution. If Eq. (7.15) is written for each and every node (i,j) internal to the computational box, and the assigned boundary values are included into the set of linear equations, we obtain 11×11, or 121 unknowns that are fully determined by 121 linearly independent equations. Over 100 coupled equations are required for this very coarse mesh!

Direct Versus Iterative Solutions

The mechanics of setting up the necessary system for direct solvers, that is, for algorithms that obtain pressures in a single pass using a full matrix solver, have been discussed by Peaceman (1977), Aziz and Settarri (1979), and Thomas (1982a, 1982b). Even for the coarse mesh considered, the resulting 121×121 matrix is large and requires monumental inversion efforts. Usually, the unknowns are cleverly ordered, and cleaner inversion algorithms are used; other methods take advantage of the sparseness (i.e., the large number of 0's) in the system. Many matrix solvers, unfortunately, are company-proprietary. Instead, we shall consider iterative solvers that require minimal memory resources. These algorithms work well in both two and three dimensions; they are robust, stable, and fast.

Iterative Methods

Since an objective of this book is the development of portable tools, we shall not discuss direct solvers. Suffice it to say that such solvers, the most notorious being Gaussian elimination, are well documented in the literature (e.g., see Carnahan et al., 1969). We shall, by contrast, emphasize iterative techniques, since these require minimal computer resources and allow the greatest flexibility. As we shall show, they are also very useful in designing smart and robust algorithms. For reasons that will become obvious, let us rewrite Eq. (7.15) in the form

$$P_{i,j-1} - 2\left\{1 + (\Delta y/\Delta x)^2\right\}P_{i,j} + P_{i,j+1} = -(\Delta y/\Delta x)^2\left(P_{i-1,j} + P_{i+1,j}\right) \quad (7.16)$$

Eq. (7.16) contains the tridiagonal form given in our ordinary differential equations example. On the left, the index i stands alone. When i is fixed and "j" is incremented, a sequence of tridiagonal equations is generated.

Let us assume that some suitable first guess for the pressure field P(i,j) is available. If so, the idea is to first freeze i at $i = 2$, write Eq. (7.16) for each of the *internal* nodes $j = 2, 3, \ldots, (j_{max} - 1)$, apply boundary conditions at both $j = 1$ and $j = j_{max}$, and solve for updated values of P(2, j) along the column $i = 2$. Then, the same process is repeated for $i = 3$, $i = 4$, and so on, until the last column $i = (i_{max} - 1)$ is completed: one sweep of the box is said to have taken place.

This sweeping, called column relaxation, is repeated for multiple sweeps until satisfactory convergence is achieved. The columns located at $i = 1$ and i_{max} are not solved because pressures have been specified along them.

Relaxation is the mathematical name synonymous with the method of successive approximations. Line relaxation may proceed by columns, as we have demonstrated; or, it may proceed by rows, that is, through row relaxation by freezing j and incrementing i's. Special schemes employing combined row and columnar operations are referred to as alternating direction implicit or ADI schemes. In all cases, the basic idea is to disseminate boundary conditions rapidly and to approach convergence as quickly as possible. All of these are improvements on point relaxation, developed by earlier workers; we shall give examples for comparison later. In Fig. 7.3A, we list the Fortran source code required to implement those iterations, assuming a rectangular box with 10, 20, 30, 40 boundary conditions, without any wells, a formulation corresponding to an aquifer alone flow. Fig. 7.3B gives computed pressures at various stages in the sweeping process. Note from Fig. 7.3A that the initial guess for pressure was taken as "zero" throughout, an arbitrary choice since we knew nothing about the solution. In fact, the initial guess might have been *anything*; by contrast, the results in Fig. 7.4A and B assume an initial $P_{i,j} = i^2 + j^2$ devoid of physics, a guess having *nothing* to do with the solution or reality. Both calculations converge quickly to the same pressures, requiring much less than a second on modern personal computers.

Convergence Acceleration

The implication here is that an initial guess close to the final solution will converge more rapidly than one that is not. This can be used beneficially when performing a sequence of flow simulations where single (or multiple) parameter(s), such as well position, rate constraint, or pressure level, or reservoir heterogeneity size or shape, vary only incrementally from one run to the next. The results of each run can be used to intelligently initialize the next, with each run using close physical information that accelerates convergence. Whereas direct methods will solve N problems using N calls of a (complicated) matrix solver, iterative methods applied in the foregoing sense solve subsequent problems much more rapidly and make minimal use of computer memory. In code development or project work, it is also conceivable to have libraries of close solutions stored on disk to initialize solutions. Such a philosophy should prove productive in infill drilling and production planning. That our calculations converge to the same answer regardless of starting guess is more than fortuitous. This may surprise beginning students in numerical analysis, who are forever seeking (unstable) roots to nonlinear equations. Unlike the iterative root solvers used for such problems, where the initial closeness to different multiple roots will cause problems, the convergence of steady-state flow problems to unique solutions is assured for several reasons. For one, mathematical theory guarantees that solutions to Dirichlet and

```
C       LAPLACE EQUATION SOLVER, CASE_1.
        PROGRAM MAIN
        DIMENSION P(11,11), A(11), B(11), C(11), V(11), W(11)
        OPEN(UNIT=4,FILE='CASE_1.DAT',STATUS='NEW')
C       DEFINE GRID PARAMETERS
        DX = 1.
        DY = 1.
        RATIO2 = (DY/DX)**2
C       INITIALIZE P(I,J) TO ZERO EVERYWHERE
        DO 100  I=1,11
        DO 100  J=1,11
        P(I,J) = 0.
  100   CONTINUE
C       SET "10-20-30-40" BOUNDARY CONDITIONS
        DO 150  I=1,10
        P(I,1) =  10.
  150   CONTINUE
        DO 151  J=1,10
        P(11,J) = 20.
  151   CONTINUE
        DO 152  I=2,11
        P(I,11) = 30.
  152   CONTINUE
        DO 153  J=2,11
        P(1,J) = 40.
  153   CONTINUE
C       LINE RELAXATION BEGINS
        DO 400  NSWEEP=1,200
        IF(MOD(NSWEEP,10).NE.0) GO TO 170
C       PRINT OUT "X-Y" RESULTS
        WRITE(*,154)
        WRITE(4,154)
        WRITE(*,155) NSWEEP
        WRITE(4,155) NSWEEP
  154   FORMAT(' ')
  155   FORMAT(' P(I,J) SOLUTION FOR NSWEEP = ',I3)
        DO 160  J=1,11
        WRITE(*,157) (P(I,J),I=1,11)
        WRITE(4,157) (P(I,J),I=1,11)
  157   FORMAT(1X,11F6.1)
  160   CONTINUE
C       ITERATE COLUMN BY COLUMN WITHIN BOX
  170   DO 300  I=2,10
C       DEFINE MATRIX COEFS FOR INTERNAL POINTS
        DO 200  J=2,10
        A(J) = 1.
        B(J) = -2.*(1.+RATIO2)
        C(J) = 1.
        W(J) = -RATIO2*(P(I-1,J)+P(I+1,J))
  200   CONTINUE
```

FIG. 7.3 (A and B) Aquifer-alone, solved with 0 guess.

```
C       RESTATE UPPER/LOWER BOUNDARY CONDITIONS
C       NOTE "99" DUMMY VALUES
        A(1) = 99.
        B(1) = 1.
        C(1) = 0.
        W(1) = P(I,1)
        A(11) = 0.
        B(11) = 1.
        C(11) = 99.
        W(11) = P(I,11)
C       INVOKE TRIDIAGONAL MATRIX SOLVER
        CALL TRIDI(A,B,C,V,W,11)
C       UPDATE AND STORE COLUMN SOLUTION
        DO 250  J=2,10
        P(I,J) = V(J)
  250   CONTINUE
  300   CONTINUE
  400   CONTINUE
        CLOSE(4,STATUS='KEEP')
        STOP
        END
```

(A)

```
                P(I,J) SOLUTION FOR NSWEEP =  10
10.0  10.0  10.0  10.0  10.0  10.0  10.0  10.0  10.0  10.0  20.0
40.0  24.1  17.6  14.4  12.7  11.9  11.8  12.1  13.0  15.0  20.0
40.0  29.3  22.1  17.6  15.0  13.7  13.3  13.8  15.0  17.0  20.0
40.0  31.4  24.6  19.8  16.8  15.2  14.7  15.1  16.3  18.0  20.0
40.0  32.4  26.1  21.4  18.3  16.6  16.0  16.4  17.3  18.6  20.0
40.0  33.1  27.2  22.8  19.7  18.0  17.4  17.7  18.3  19.2  20.0
40.0  33.6  28.2  24.1  21.3  19.7  19.1  19.2  19.5  19.9  20.0
40.0  34.1  29.2  25.6  23.2  21.8  21.2  21.1  21.0  20.8  20.0
40.0  34.3  30.1  27.2  25.3  24.3  23.8  23.5  23.1  22.1  20.0
40.0  33.6  30.5  28.7  27.6  27.0  26.7  26.5  26.0  24.5  20.0
40.0  30.0  30.0  30.0  30.0  30.0  30.0  30.0  30.0  30.0  30.0

                P(I,J) SOLUTION FOR NSWEEP =  20
10.0  10.0  10.0  10.0  10.0  10.0  10.0  10.0  10.0  10.0  20.0
40.0  24.9  18.8  16.0  14.5  13.7  13.3  13.3  13.8  15.4  20.0
40.0  30.7  24.5  20.8  18.5  17.1  16.3  16.1  16.5  17.7  20.0
40.0  33.3  28.0  24.2  21.6  19.9  18.9  18.4  18.4  19.0  20.0
40.0  34.7  30.2  26.6  24.0  22.2  21.0  20.2  19.9  19.8  20.0
40.0  35.5  31.6  28.3  25.9  24.0  22.7  21.7  21.1  20.5  20.0
40.0  36.0  32.4  29.5  27.3  25.5  24.2  23.1  22.1  21.1  20.0
40.0  36.1  32.9  30.3  28.3  26.8  25.6  24.5  23.3  21.8  20.0
40.0  35.8  32.8  30.6  29.1  27.9  27.0  26.0  24.7  22.9  20.0
40.0  34.4  31.9  30.5  29.6  29.0  28.4  27.8  26.9  24.9  20.0
40.0  30.0  30.0  30.0  30.0  30.0  30.0  30.0  30.0  30.0  30.0
```

FIG. 7.3, CONT'D

(Continued)

```
                 P(I,J) SOLUTION FOR NSWEEP =  50
10.0  10.0  10.0  10.0  10.0  10.0  10.0  10.0  10.0  10.0  20.0
40.0  25.0  19.1  16.3  14.9  14.0  13.6  13.5  13.9  15.4  20.0
40.0  30.9  25.0  21.3  19.1  17.7  16.8  16.5  16.8  17.8  20.0
40.0  33.7  28.6  25.0  22.4  20.7  19.5  18.9  18.8  19.2  20.0
40.0  35.1  30.9  27.6  25.0  23.1  21.8  20.8  20.3  20.0  20.0
40.0  36.0  32.3  29.3  26.9  25.0  23.5  22.4  21.5  20.7  20.0
40.0  36.4  33.2  30.4  28.2  26.5  25.0  23.7  22.5  21.3  20.0
40.0  36.5  33.5  31.1  29.2  27.6  26.3  25.0  23.6  22.0  20.0
40.0  36.1  33.2  31.2  29.7  28.5  27.5  26.4  25.0  23.0  20.0
40.0  34.6  32.2  30.8  30.0  29.3  28.7  28.0  27.0  25.0  20.0
40.0  30.0  30.0  30.0  30.0  30.0  30.0  30.0  30.0  30.0  30.0

                 P(I,J) SOLUTION FOR NSWEEP = 150
10.0  10.0  10.0  10.0  10.0  10.0  10.0  10.0  10.0  10.0  20.0
40.0  25.0  19.1  16.3  14.9  14.0  13.6  13.5  13.9  15.4  20.0
40.0  30.9  25.0  21.3  19.1  17.7  16.8  16.5  16.8  17.8  20.0
40.0  33.7  28.7  25.0  22.4  20.7  19.5  18.9  18.8  19.2  20.0
40.0  35.1  30.9  27.6  25.0  23.1  21.8  20.8  20.3  20.0  20.0
40.0  36.0  32.3  29.3  26.9  25.0  23.5  22.4  21.5  20.7  20.0
40.0  36.4  33.2  30.5  28.2  26.5  25.0  23.7  22.5  21.3  20.0
40.0  36.5  33.5  31.1  29.2  27.6  26.3  25.0  23.6  22.0  20.0
40.0  36.1  33.2  31.2  29.7  28.5  27.5  26.4  25.0  23.0  20.0
40.0  34.6  32.2  30.8  30.0  29.3  28.7  28.0  27.0  25.0  20.0
40.0  30.0  30.0  30.0  30.0  30.0  30.0  30.0  30.0  30.0  30.0

                 P(I,J) SOLUTION FOR NSWEEP = 200
10.0  10.0  10.0  10.0  10.0  10.0  10.0  10.0  10.0  10.0  20.0
40.0  25.0  19.1  16.3  14.9  14.0  13.6  13.5  13.9  15.4  20.0
40.0  30.9  25.0  21.3  19.1  17.7  16.8  16.5  16.8  17.8  20.0
40.0  33.7  28.7  25.0  22.4  20.7  19.5  18.9  18.8  19.2  20.0
40.0  35.1  30.9  27.6  25.0  23.1  21.8  20.8  20.3  20.0  20.0
40.0  36.0  32.3  29.3  26.9  25.0  23.5  22.4  21.5  20.7  20.0
40.0  36.4  33.2  30.5  28.2  26.5  25.0  23.7  22.5  21.3  20.0
40.0  36.5  33.5  31.1  29.2  27.6  26.3  25.0  23.6  22.0  20.0
40.0  36.1  33.2  31.2  29.7  28.5  27.5  26.4  25.0  23.0  20.0
40.0  34.6  32.2  30.8  30.0  29.3  28.7  28.0  27.0  25.0  20.0
40.0  30.0  30.0  30.0  30.0  30.0  30.0  30.0  30.0  30.0  30.0
```

(B)

FIG. 7.3, CONT'D

```
C       LAPLACE EQUATION SOLVER, CASE_2.
        PROGRAM MAIN
        DIMENSION P(11,11), A(11), B(11), C(11), V(11), W(11)
        OPEN(UNIT=4,FILE='CASE_2.DAT',STATUS='NEW')
C       DEFINE GRID PARAMETERS
        DX = 1.
        DY = 1.
        RATIO2 = (DY/DX)**2
C       INITIALIZE P(I,J) TO SOMETHING ABSURD EVERYWHERE
        DO 100  I=1,11
        DO 100  J=1,11
```

FIG. 7.4 (A) Aquifer-alone, with (absurd) $P_{i,j} = i^2 + j^2$ guess.

```
      P(I,J) = I**2 + J**2
 100  CONTINUE
C     SET "10-20-30-40" BOUNDARY CONDITIONS
      DO 150  I=1,10
      P(I,1) =  10.
 150  CONTINUE
      DO 151  J=1,10
      P(11,J) = 20.
 151  CONTINUE
      DO 152  I=2,11
      P(I,11) = 30.
 152  CONTINUE
      DO 153  J=2,11
      P(1,J) = 40.
 153  CONTINUE
C     LINE RELAXATION BEGINS
      DO 400  NSWEEP=1,200
      IF(MOD(NSWEEP,10).NE.0) GO TO 170
C     PRINT OUT "X-Y" RESULTS
      WRITE(*,154)
      WRITE(4,154)
      WRITE(*,155) NSWEEP
      WRITE(4,155) NSWEEP
 154  FORMAT(' ')
 155  FORMAT(' P(I,J) SOLUTION FOR NSWEEP = ',I3)
      DO 160  J=1,11
      WRITE(*,157) (P(I,J),I=1,11)
      WRITE(4,157) (P(I,J),I=1,11)
 157  FORMAT(1X,11F6.1)
 160  CONTINUE
C     ITERATE COLUMN BY COLUMN WITHIN BOX
 170  DO 300  I=2,10
C     DEFINE MATRIX COEFS FOR INTERNAL POINTS
      DO 200  J=2,10
      A(J) = 1.
      B(J) = -2.*(1.+RATIO2)
      C(J) = 1.
      W(J) = -RATIO2*(P(I-1,J)+P(I+1,J))
 200  CONTINUE
C     RESTATE UPPER/LOWER BOUNDARY CONDITIONS
C     NOTE "99" DUMMY VALUES
      A(1) = 99.
      B(1) = 1.
      C(1) = 0.
      W(1) = P(I,1)
      A(11) = 0.
      B(11) = 1.
      C(11) = 99.
      W(11) = P(I,11)
C     INVOKE TRIDIAGONAL MATRIX SOLVER
      CALL TRIDI(A,B,C,V,W,11)
```

FIG. 7.4, CONT'D

(Continued)

```
C       UPDATE AND STORE COLUMN SOLUTION
        DO 250  J=2,10
        P(I,J) = V(J)
 250    CONTINUE
 300    CONTINUE
 400    CONTINUE
        CLOSE(4,STATUS='KEEP')
        STOP
        END
```

(A)

```
                P(I,J) SOLUTION FOR NSWEEP =  10
10.0  10.0  10.0  10.0  10.0  10.0  10.0  10.0  10.0  10.0  20.0
40.0  27.1  23.0  21.7  21.0  20.3  19.2  18.0  16.9  16.9  20.0
40.0  34.9  32.5  31.6  30.9  29.7  27.7  25.1  22.5  20.6  20.0
40.0  39.3  39.2  39.4  39.0  37.5  34.7  30.9  26.8  23.0  20.0
40.0  41.9  43.7  45.0  45.0  43.3  39.9  35.2  29.9  24.6  20.0
40.0  43.4  46.3  48.3  48.5  46.8  43.0  37.8  31.7  25.6  20.0
40.0  43.9  47.1  49.1  49.5  47.7  44.0  38.7  32.5  26.0  20.0
40.0  43.2  45.8  47.6  47.8  46.2  42.8  38.0  32.2  26.1  20.0
40.0  41.1  42.5  43.6  43.6  42.3  39.7  35.9  31.3  26.0  20.0
40.0  37.3  37.2  37.5  37.4  36.7  35.2  33.1  30.4  26.6  20.0
40.0  30.0  30.0  30.0  30.0  30.0  30.0  30.0  30.0  30.0  30.0
                P(I,J) SOLUTION FOR NSWEEP =  20
10.0  10.0  10.0  10.0  10.0  10.0  10.0  10.0  10.0  10.0  20.0
40.0  25.4  19.8  17.2  15.8  15.0  14.4  14.1  14.3  15.6  20.0
40.0  31.7  26.4  23.1  21.0  19.5  18.4  17.7  17.6  18.2  20.0
40.0  34.8  30.6  27.4  25.0  23.2  21.7  20.6  19.9  19.7  20.0
40.0  36.5  33.2  30.4  28.1  26.0  24.3  22.8  21.6  20.7  20.0
40.0  37.4  34.8  32.3  30.1  28.1  26.2  24.5  22.8  21.4  20.0
40.0  37.7  35.5  33.3  31.3  29.4  27.5  25.7  23.8  21.9  20.0
40.0  37.6  35.5  33.5  31.8  30.1  28.4  26.7  24.7  22.5  20.0
40.0  36.9  34.7  33.0  31.6  30.3  29.0  27.6  25.8  23.4  20.0
40.0  35.0  32.9  31.8  31.0  30.3  29.5  28.7  27.4  25.2  20.0
40.0  30.0  30.0  30.0  30.0  30.0  30.0  30.0  30.0  30.0  30.0
                P(I,J) SOLUTION FOR NSWEEP =  50
10.0  10.0  10.0  10.0  10.0  10.0  10.0  10.0  10.0  10.0  20.0
40.0  25.0  19.1  16.3  14.9  14.0  13.6  13.5  13.9  15.4  20.0
40.0  30.9  25.0  21.4  19.1  17.7  16.8  16.5  16.8  17.8  20.0
40.0  33.7  28.7  25.0  22.4  20.7  19.6  18.9  18.8  19.2  20.0
40.0  35.1  30.9  27.6  25.0  23.1  21.8  20.8  20.3  20.0  20.0
40.0  36.0  32.4  29.3  26.9  25.0  23.5  22.4  21.5  20.7  20.0
40.0  36.4  33.2  30.5  28.2  26.5  25.0  23.7  22.5  21.3  20.0
40.0  36.5  33.5  31.1  29.2  27.6  26.3  25.0  23.6  22.0  20.0
40.0  36.1  33.2  31.2  29.7  28.5  27.5  26.4  25.0  23.0  20.0
40.0  34.6  32.2  30.8  30.0  29.3  28.7  28.0  27.0  25.0  20.0
40.0  30.0  30.0  30.0  30.0  30.0  30.0  30.0  30.0  30.0  30.0
```

FIG. 7.4, CONT'D (B) Aquifer-alone, with $P_{i,j} = i^2 + j^2$ guess.

P(I,J) SOLUTION FOR NSWEEP = 150

10.0	10.0	10.0	10.0	10.0	10.0	10.0	10.0	10.0	10.0	20.0
40.0	25.0	19.1	16.3	14.9	14.0	13.6	13.5	13.9	15.4	20.0
40.0	30.9	25.0	21.3	19.1	17.7	16.8	16.5	16.8	17.8	20.0
40.0	33.7	28.7	25.0	22.4	20.7	19.5	18.9	18.8	19.2	20.0
40.0	35.1	30.9	27.6	25.0	23.1	21.8	20.8	20.3	20.0	20.0
40.0	36.0	32.3	29.3	26.9	25.0	23.5	22.4	21.5	20.7	20.0
40.0	36.4	33.2	30.5	28.2	26.5	25.0	23.7	22.5	21.3	20.0
40.0	36.5	33.5	31.1	29.2	27.6	26.3	25.0	23.6	22.0	20.0
40.0	36.1	33.2	31.2	29.7	28.5	27.5	26.4	25.0	23.0	20.0
40.0	34.6	32.2	30.8	30.0	29.3	28.7	28.0	27.0	25.0	20.0
40.0	30.0	30.0	30.0	30.0	30.0	30.0	30.0	30.0	30.0	30.0

P(I,J) SOLUTION FOR NSWEEP = 200

10.0	10.0	10.0	10.0	10.0	10.0	10.0	10.0	10.0	10.0	20.0
40.0	25.0	19.1	16.3	14.9	14.0	13.6	13.5	13.9	15.4	20.0
40.0	30.9	25.0	21.3	19.1	17.7	16.8	16.5	16.8	17.8	20.0
40.0	33.7	28.7	25.0	22.4	20.7	19.5	18.9	18.8	19.2	20.0
40.0	35.1	30.9	27.6	25.0	23.1	21.8	20.8	20.3	20.0	20.0
40.0	36.0	32.3	29.3	26.9	**25.0**	23.5	22.4	21.5	20.7	20.0
40.0	36.4	33.2	30.5	28.2	26.5	25.0	23.7	22.5	21.3	20.0
40.0	36.5	33.5	31.1	29.2	27.6	26.3	25.0	23.6	22.0	20.0
40.0	36.1	33.2	31.2	29.7	28.5	27.5	26.4	25.0	23.0	20.0
40.0	34.6	32.2	30.8	30.0	29.3	28.7	28.0	27.0	25.0	20.0
40.0	30.0	30.0	30.0	30.0	30.0	30.0	30.0	30.0	30.0	30.0

(B)

FIG. 7.4, CONT'D

mixed flow problems—when proper boundary conditions are used—exist and are unique. And, as we shall show later, the iterative process mimics the search for steady solutions to the transient heat equation. Any homemaker will explain that the equilibrium steady-state (room) temperature that a loaf of bread seeks is independent of its origins from the oven or the refrigerator! Here again, convergence to a unique solution is independent of the guess.

WELLS AND INTERNAL BOUNDARIES

Cases 1 and 2 deal with uninteresting pressure distributions, corresponding to flowing reservoirs without wells. Here we consider uniform boundary pressures (say, 100 psi) specified at the edges of the computational box, plus the effect of constant pressure (say, 1 psi) prescribed at the center of the domain of flow. This example illustrates doubly connected well effects crudely.

Peaceman Well Corrections

Reference to the converged pressure in Fig. 7.5B shows that results are symmetric about the x and y axes passing through the box center, where the well pressure is unity. The solutions are also symmetric with respect to 45 degree diagonals passing through this origin. Note that pressure changes near the well, while rapid, are not quite logarithmic; the use of Cartesian meshes, in this sense, does not provide enough flow resolution near producers and injectors. There are ad hoc numerical procedures used to repair such solutions after-the-fact, (see, e.g., Peaceman, 1978, 1983; Williamson and Chappelear, 1981; Chappelear and Williamson, 1981), but these Peaceman corrections are not rigorous and are approximate. They are based on radial flow solutions for liquids in steady flow,

```
C       LAPLACE EQUATION SOLVER, CASE_3.
        PROGRAM MAIN
        DIMENSION P(11,11), A(11), B(11), C(11), V(11), W(11)
        OPEN(UNIT=4,FILE='CASE_3.DAT',STATUS='NEW')
C       DEFINE GRID PARAMETERS
        DX = 1.
        DY = 1.
        RATIO2 = (DY/DX)**2
        ONE = 1.
C       INITIALIZE P(I,J) TO ZERO EVERYWHERE
        DO 100  I=1,11
        DO 100  J=1,11
        P(I,J) = 0.
  100   CONTINUE
C       SET "100" BOUNDARY CONDITION AT BOX EDGES
        DO 150  I=1,10
        P(I,1) =  100.
  150   CONTINUE
        DO 151  J=1,10 P(11,J) = 100.
  151   CONTINUE
        DO 152  I=2,11
        P(I,11) = 100.
  152   CONTINUE
        DO 153  J=2,11
        P(1,J) = 100.
  153   CONTINUE
C       LINE RELAXATION BEGINS
        DO 400  NSWEEP=1,200
        IF(MOD(NSWEEP,10).NE.0) GO TO 170
C       PRINT OUT "X-Y" RESULTS
        WRITE(*,154)
        WRITE(4,154)
        WRITE(*,155) NSWEEP
        WRITE(4,155) NSWEEP
  154   FORMAT(' ')
  155   FORMAT(' P(I,J) SOLUTION FOR NSWEEP = ',I3)
```

FIG. 7.5 (A and B) Centered well in aquifer-driven reservoir.

```
      DO 160  J=1,11
      WRITE(*,157)  (P(I,J),I=1,11)
      WRITE(4,157)  (P(I,J),I=1,11)
 157  FORMAT(1X,11F6.1)
 160  CONTINUE
C     ITERATE COLUMN BY COLUMN WITHIN BOX
 170  DO 300  I=2,10
C     DEFINE MATRIX COEFS FOR INTERNAL POINTS
      DO 200  J=2,10
      A(J) = 1.
      B(J) = -2.*(1.+RATIO2)
      C(J) = 1.
      W(J) = -RATIO2*(P(I-1,J)+P(I+1,J))
C     SET INTERNAL BOUNDARY CONDITION
      IF(I.EQ.6.AND.J.EQ.6) A(J) = 0.
      IF(I.EQ.6.AND.J.EQ.6) B(J) = 1.
      IF(I.EQ.6.AND.J.EQ.6) C(J) = 0.
      IF(I.EQ.6.AND.J.EQ.6) W(J) = ONE
 200  CONTINUE
C     RESTATE UPPER/LOWER BOUNDARY CONDITIONS
C     NOTE "99" DUMMY VALUES
      A(1) = 99.
      B(1) = 1.
      C(1) = 0.
      W(1) = P(I,1)
      A(11) = 0.
      B(11) = 1.
      C(11) = 99.
      W(11) = P(I,11)
C     INVOKE TRIDIAGONAL MATRIX SOLVER
      CALL TRIDI(A,B,C,V,W,11)
C     UPDATE AND STORE COLUMN SOLUTION
      DO 250  J=2,10
      P(I,J) = V(J)
 250  CONTINUE
 300  CONTINUE
 400  CONTINUE
      CLOSE(4,STATUS='KEEP')
      STOP
      END
```

(A)

FIG. 7.5, CONT'D

(Continued)

require effective radius and productivity index inputs, have been used to model multiphase, multiwell, compressible flows—and do not apply to deviated wells. It is known that blind use of well models can yield flow rates from 50% to 200% in error. These problems disappear (and the Peaceman approach is unnecessary) when fine enough grids are used. This is impractical with Cartesian meshes but possible using the boundary-conforming meshes in Chapters 8–10.

P(I,J) SOLUTION FOR NSWEEP = 10

99.9	99.9	99.9	99.9	99.9	99.9	99.9	99.9	99.9	99.9	99.9
99.9	95.9	92.5	89.8	88.0	87.7	88.9	91.2	94.1	97.1	99.9
99.9	92.2	85.5	80.1	76.2	75.0	77.7	82.6	88.5	94.4	99.9
99.9	89.2	79.7	71.3	64.4	60.9	66.0	74.3	83.3	92.0	99.9
99.9	87.2	75.5	64.1	52.2	41.2	53.5	67.1	79.3	90.2	99.9
99.9	86.5	73.8	60.3	41.4	**1.0**	42.1	63.0	77.6	89.5	99.9
99.9	87.2	75.5	64.1	52.2	41.2	53.5	67.1	79.3	90.2	99.9
99.9	89.2	79.7	71.3	64.4	60.9	66.0	74.3	83.3	92.0	99.9
99.9	92.2	85.5	80.1	76.2	75.0	77.7	82.6	88.5	94.4	99.9
99.9	95.9	92.5	89.8	88.0	87.7	88.9	91.2	94.1	97.1	99.9
99.9	99.9	99.9	99.9	99.9	99.9	99.9	99.9	99.9	99.9	99.9

P(I,J) SOLUTION FOR NSWEEP = 20

99.9	99.9	99.9	99.9	99.9	99.9	99.9	99.9	99.9	99.9	99.9
99.9	97.9	95.8	93.8	92.3	91.6	92.3	93.9	95.9	97.9	99.9
99.9	95.7	91.5	87.3	83.7	82.1	83.8	87.5	91.7	95.9	99.9
99.9	93.7	87.2	80.2	73.4	69.1	73.5	80.5	87.5	93.9	99.9
99.9	92.2	83.6	73.3	60.5	47.8	60.7	73.6	83.9	92.4	99.9
99.9	91.5	81.9	69.0	47.8	**1.0**	47.9	69.3	82.2	91.7	99.9
99.9	92.2	83.6	73.3	60.5	47.8	60.7	73.6	83.9	92.4	99.9
99.9	93.7	87.2	80.2	73.4	69.1	73.5	80.5	87.5	93.9	99.9
99.9	95.7	91.5	87.3	83.7	82.1	83.8	87.5	91.7	95.9	99.9
99.9	97.9	95.8	93.8	92.3	91.6	92.3	93.9	95.9	97.9	99.9
99.9	99.9	99.9	99.9	99.9	99.9	99.9	99.9	99.9	99.9	99.9

P(I,J) SOLUTION FOR NSWEEP = 100

99.9	99.9	99.9	99.9	99.9	99.9	99.9	99.9	99.9	99.9	99.9
99.9	98.0	96.0	94.0	92.5	91.9	92.5	94.0	96.0	98.0	99.9
99.9	96.0	91.8	87.7	84.1	82.4	84.1	87.7	91.8	96.0	99.9
99.9	94.0	87.7	80.8	73.9	69.6	73.9	80.8	87.7	94.0	99.9
99.9	92.5	84.1	73.9	61.0	48.2	61.0	73.9	84.1	92.5	99.9
99.9	91.9	82.4	69.6	48.2	**1.0**	48.2	69.6	82.4	91.9	99.9
99.9	92.5	84.1	73.9	61.0	48.2	61.0	73.9	84.1	92.5	99.9
99.9	94.0	87.7	80.8	73.9	69.6	73.9	80.8	87.7	94.0	99.9
99.9	96.0	91.8	87.7	84.1	82.4	84.1	87.7	91.8	96.0	99.9
99.9	98.0	96.0	94.0	92.5	91.9	92.5	94.0	96.0	98.0	99.9
99.9	99.9	99.9	99.9	99.9	99.9	99.9	99.9	99.9	99.9	99.9

P(I,J) SOLUTION FOR NSWEEP = 200

99.9	99.9	99.9	99.9	99.9	99.9	99.9	99.9	99.9	99.9	99.9
99.9	98.0	96.0	94.0	92.5	91.9	92.5	94.0	96.0	98.0	99.9
99.9	96.0	91.8	87.7	84.1	82.4	84.1	87.7	91.8	96.0	99.9
99.9	94.0	87.7	80.8	73.9	69.6	73.9	80.8	87.7	94.0	99.9
99.9	92.5	84.1	73.9	61.0	48.2	61.0	73.9	84.1	92.5	99.9
99.9	91.9	82.4	69.6	48.2	**1.0**	48.2	69.6	82.4	91.9	99.9
99.9	92.5	84.1	73.9	61.0	48.2	61.0	73.9	84.1	92.5	99.9
99.9	94.0	87.7	80.8	73.9	69.6	73.9	80.8	87.7	94.0	99.9
99.9	96.0	91.8	87.7	84.1	82.4	84.1	87.7	91.8	96.0	99.9
99.9	98.0	96.0	94.0	92.5	91.9	92.5	94.0	96.0	98.0	99.9
99.9	99.9	99.9	99.9	99.9	99.9	99.9	99.9	99.9	99.9	99.9

(B)

FIG. 7.5, CONT'D

Derivative Discontinuities

Another problem with the computed solution in Fig. 7.5B seems to be the behavior of the derivatives at the well where $P = 1$. Just to the left, the pressure decreases; that is, $\partial P/\partial x$ is a nonzero negative number. And just to the right, $\partial P/\partial x$ is a nonzero positive number having the same magnitude. The same behavior is seen above and below the well; $\partial P/\partial y$ takes on equal and opposite nonzero values. Of course, there is nothing wrong with these results. Since the Darcy velocities are proportional to the first derivatives of pressure, the change in sign only indicates that fluid is always flowing from opposite directions into the well. We have seen this before: recall from Chapter 2 that source points are always associated with discontinuities of the first derivative. This is also clear from radial flow analysis, which shows that fluids must flow toward a well; however, the discontinuous change or jump in the first derivative of P(x,y) may be disturbing at first.

This oddity is a consequence of the (x,y) coordinates used. The derivative discontinuities do not appear using radial coordinates because the point r is always greater than zero and there is nothing to the left or right. We have uncovered one truism: elliptic equations with *internal* Dirichlet conditions always yield jumps in the first derivative (this property is used in grid generation later). Thus, never blindly differentiate across such a discontinuity because the derivative does not exist! This type of jump was demonstrated in the analytical solution for fracture flow in Chapter 2, where it was shown that $\partial P/\partial y$ reverses sign across the horizontal slit $y = 0$. We can replicate that qualitatively by modifying the above Fortran; we change the single well logic, now allowing unit pressures to extend horizontally for several gridblocks. The results are shown in Fig. 7.6A and B (the 100 boundary pressures assumed in the source code, for the next several examples, were changed to 99.9 *after* the calculations, for typesetting and formatting reasons). Note how, in Fig. 7.6B, the vertical derivative $\partial P/\partial y$ at the fracture slit are equal and opposite; also observe that the fracture singularity in Chapter 2 is captured poorly in the numerics. As in the earlier calculations, our solutions were stably and rapidly obtained, requiring only minor changes to add or delete wells, or to change wells to line fractures.

POINT RELAXATION METHODS

So far, we have discussed column and row implementations of line relaxation. These methods require matrix inversion, but inverting tridiagonal matrices is a relatively straightforward task. But this was not so when computing machines were not widely available. Historically, point relaxation methods requiring simple hand calculations only (and no matrix inversion) were the first ones developed. This solution is useful for several reasons: (1) it is easily programmed, (2) it is easily implemented in irregular domains where rows and columns of constant length are difficult to define, and (3) large-scale calculations may be divided among different machines in parallel processing.

```
C      LAPLACE EQUATION SOLVER, CASE_4.
       PROGRAM MAIN
       DIMENSION P(11,11), A(11), B(11), C(11), V(11), W(11)
       OPEN(UNIT=4,FILE='CASE_4.DAT',STATUS='NEW')
C      DEFINE GRID PARAMETERS
       DX = 1.
       DY = 1.
       RATIO2 = (DY/DX)**2
       ONE = 1.
C      INITIALIZE P(I,J) TO ZERO EVERYWHERE
       DO 100  I=1,11
       DO 100  J=1,11
       P(I,J) = 0.
 100   CONTINUE
C      SET "100" BOUNDARY CONDITION AT BOX EDGES
       DO 150  I=1,10
       P(I,1) =  100.
 150   CONTINUE
       DO 151  J=1,10
       P(11,J) = 100.
 151   CONTINUE
       DO 152  I=2,11
       P(I,11) = 100.
 152   CONTINUE
       DO 153  J=2,11
       P(1,J) = 100.
 153   CONTINUE
C      LINE RELAXATION BEGINS
       DO 400  NSWEEP=1,200
       IF(MOD(NSWEEP,10).NE.0) GO TO 170
C      PRINT OUT "X-Y" RESULTS
       WRITE(*,154)
       WRITE(4,154)
       WRITE(*,155) NSWEEP
       WRITE(4,155) NSWEEP
 154   FORMAT(' ')
 155   FORMAT(' P(I,J) SOLUTION FOR NSWEEP = ',I3)
       DO 160  J=1,11
```

FIG. 7.6 (A and B) Centered fracture in aquifer-driven reservoir.

```
      WRITE(*,157) (P(I,J),I=1,11)
      WRITE(4,157) (P(I,J),I=1,11)
 157  FORMAT(1X,11F6.1)
 160  CONTINUE
C     ITERATE COLUMN BY COLUMN WITHIN BOX
 170  DO 300  I=2,10
C     DEFINE MATRIX COEFS FOR INTERNAL POINTS
      DO 200  J=2,10
      A(J) = 1.
      B(J) = -2.*(1.+RATIO2)
      C(J) = 1.
      W(J) = -RATIO2*(P(I-1,J)+P(I+1,J))
C     SET INTERNAL BOUNDARY CONDITION
      MODE = 0
      IF(I.GE.4.AND.I.LE.8) MODE = 1
      IF(MODE.EQ.1.AND.J.EQ.6) A(J) = 0.
      IF(MODE.EQ.1.AND.J.EQ.6) B(J) = 1.
      IF(MODE.EQ.1.AND.J.EQ.6) C(J) = 0.
      IF(MODE.EQ.1.AND.J.EQ.6) W(J) = ONE
 200  CONTINUE
C     RESTATE UPPER/LOWER BOUNDARY CONDITIONS
C     NOTE "99" DUMMY VALUES
      A(1) = 99.
      B(1) = 1.
      C(1) = 0.
      W(1) = P(I,1)
      A(11) = 0.
      B(11) = 1.
      C(11) = 99.
      W(11) = P(I,11)
C     INVOKE TRIDIAGONAL MATRIX SOLVER
      CALL TRIDI(A,B,C,V,W,11)
C     UPDATE AND STORE COLUMN SOLUTION
      DO 250  J=2,10
      P(I,J) = V(J)
 250  CONTINUE
 300  CONTINUE
 400  CONTINUE
      CLOSE(4,STATUS='KEEP')
      STOP
      END
```

(A)

FIG. 7.6, CONT'D

(Continued)

P(I,J) SOLUTION FOR NSWEEP = 10

99.9	99.9	99.9	99.9	99.9	99.9	99.9	99.9	99.9	99.9	99.9
99.9	94.7	90.0	86.4	84.3	83.7	84.8	87.3	91.0	95.4	99.9
99.9	89.4	79.8	72.2	67.8	66.7	68.6	73.6	81.5	90.6	99.9
99.9	84.2	69.2	56.2	49.5	47.9	50.4	57.7	70.9	85.5	99.9
99.9	79.5	57.9	35.5	27.8	26.0	28.3	36.4	59.2	80.6	99.9
99.9	76.7	48.4	1.0	1.0	1.0	1.0	1.0	49.2	77.6	99.9
99.9	79.5	57.9	35.5	27.8	26.0	28.3	36.4	59.2	80.6	99.9
99.9	84.2	69.2	56.2	49.5	47.9	50.4	57.7	70.9	85.5	99.9
99.9	89.4	79.8	72.2	67.8	66.7	68.6	73.6	81.5	90.6	99.9
99.9	94.7	90.0	86.4	84.3	83.7	84.8	87.3	91.0	95.4	99.9
99.9	99.9	99.9	99.9	99.9	99.9	99.9	99.9	99.9	99.9	99.9

P(I,J) SOLUTION FOR NSWEEP = 20

99.9	99.9	99.9	99.9	99.9	99.9	99.9	99.9	99.9	99.9	99.9
99.9	95.6	91.6	88.2	86.1	85.4	86.1	88.2	91.6	95.6	99.9
99.9	91.0	82.4	75.2	70.8	69.4	70.8	75.2	82.4	91.0	99.9
99.9	86.0	71.9	59.3	52.6	50.6	52.6	59.3	71.9	86.0	99.9
99.9	81.0	60.0	37.5	29.7	27.8	29.7	37.5	60.1	81.0	99.9
99.9	77.9	49.7	1.0	1.0	1.0	1.0	1.0	49.8	77.9	99.9
99.9	81.0	60.0	37.5	29.7	27.8	29.7	37.5	60.1	81.0	99.9
99.9	86.0	71.9	59.3	52.6	50.6	52.6	59.3	71.9	86.0	99.9
99.9	91.0	82.4	75.2	70.8	69.4	70.8	75.2	82.4	91.0	99.9
99.9	95.6	91.6	88.2	86.1	85.4	86.1	88.2	91.6	95.6	99.9
99.9	99.9	99.9	99.9	99.9	99.9	99.9	99.9	99.9	99.9	99.9

P(I,J) SOLUTION FOR NSWEEP = 100

99.9	99.9	99.9	99.9	99.9	99.9	99.9	99.9	99.9	99.9	99.9
99.9	95.7	91.6	88.2	86.1	85.4	86.1	88.2	91.6	95.7	99.9
99.9	91.0	82.4	75.2	70.9	69.4	70.9	75.2	82.4	91.0	99.9
99.9	86.0	72.0	59.3	52.6	50.6	52.6	59.3	72.0	86.0	99.9
99.9	81.0	60.1	37.5	29.7	27.8	29.7	37.5	60.1	81.0	99.9
99.9	77.9	49.8	1.0	1.0	1.0	1.0	1.0	49.8	77.9	99.9
99.9	81.0	60.1	37.5	29.7	27.8	29.7	37.5	60.1	81.0	99.9
99.9	86.0	72.0	59.3	52.6	50.6	52.6	59.3	72.0	86.0	99.9
99.9	91.0	82.4	75.2	70.9	69.4	70.9	75.2	82.4	91.0	99.9
99.9	95.7	91.6	88.2	86.1	85.4	86.1	88.2	91.6	95.7	99.9
99.9	99.9	99.9	99.9	99.9	99.9	99.9	99.9	99.9	99.9	99.9

P(I,J) SOLUTION FOR NSWEEP = 150

99.9	99.9	99.9	99.9	99.9	99.9	99.9	99.9	99.9	99.9	99.9
99.9	95.7	91.6	88.2	86.1	85.4	86.1	88.2	91.6	95.7	99.9
99.9	91.0	82.4	75.2	70.9	69.4	70.9	75.2	82.4	91.0	99.9
99.9	86.0	72.0	59.3	52.6	50.6	52.6	59.3	72.0	86.0	99.9
99.9	81.0	60.1	37.5	29.7	27.8	29.7	37.5	60.1	81.0	99.9
99.9	77.9	49.8	1.0	1.0	1.0	1.0	1.0	49.8	77.9	99.9
99.9	81.0	60.1	37.5	29.7	27.8	29.7	37.5	60.1	81.0	99.9
99.9	86.0	72.0	59.3	52.6	50.6	52.6	59.3	72.0	86.0	99.9
99.9	91.0	82.4	75.2	70.9	69.4	70.9	75.2	82.4	91.0	99.9
99.9	95.7	91.6	88.2	86.1	85.4	86.1	88.2	91.6	95.7	99.9
99.9	99.9	99.9	99.9	99.9	99.9	99.9	99.9	99.9	99.9	99.9

FIG. 7.6, CONT'D

P(I,J) SOLUTION FOR NSWEEP = **200**

99.9	99.9	99.9	99.9	99.9	99.9	99.9	99.9	99.9	99.9	99.9
99.9	95.7	91.6	88.2	86.1	85.4	86.1	88.2	91.6	95.7	99.9
99.9	91.0	82.4	75.2	70.9	69.4	70.9	75.2	82.4	91.0	99.9
99.9	86.0	72.0	59.3	52.6	50.6	52.6	59.3	72.0	86.0	99.9
99.9	81.0	60.1	37.5	29.7	27.8	29.7	37.5	60.1	81.0	99.9
99.9	77.9	49.8	**1.0**	**1.0**	**1.0**	**1.0**	**1.0**	49.8	77.9	99.9
99.9	81.0	60.1	37.5	29.7	27.8	29.7	37.5	60.1	81.0	99.9
99.9	86.0	72.0	59.3	52.6	50.6	52.6	59.3	72.0	86.0	99.9
99.9	91.0	82.4	75.2	70.9	69.4	70.9	75.2	82.4	91.0	99.9
99.9	95.7	91.6	88.2	86.1	85.4	86.1	88.2	91.6	95.7	99.9
99.9	99.9	99.9	99.9	99.9	99.9	99.9	99.9	99.9	99.9	99.9

(B)

FIG. 7.6, CONT'D

Point relaxation is best explained by assuming equal constant mesh widths $\Delta x = \Delta y$. Then, Eq. (7.16) can be rewritten in the form

$$P_{i,j} = \left(P_{i-1,j} + P_{i+1,j} + P_{i,j-1} + P_{i,j+1}\right)/4 \tag{7.17}$$

This states that center values equal the arithmetic average of their neighbors to the left, right, top, and bottom, when the mesh widths are equal. This remarkable property holds everywhere in the flowfield; that is, it holds in the small locally. And from Figs. 7.3B and 7.4B, we find that it holds in the large also, the center value of 25 being the arithmetic average of the four boundary values 10, 20, 30, and 40. This explains why elliptic operators are used to smooth numerical fields in image processing. In Fig. 7.7A and B, we revisit the 10, 20, 30, 40, no well problem in Fig. 7.3A. However, we solve it using a simple scheme, taking Eq. (7.17) as the recursion formula, again assuming $P_{i,j} = 0$ as the initial guess. Similarly, we reconsider the fracture flow problem in Fig. 7.6A and B and solve it with point relaxation. The results are shown in Fig. 7.8A and B. In both cases, pressures are identical to earlier ones.

For our last example, we treat the implementation of no-flow solid wall boundary conditions. We have chosen to rework Case 3 (see Fig. 7.5A and B), and add no-flow conditions along the vertical line $i = 1$ as well as the horizontal line $j = 1$. Now, Darcy's law guarantees zero flow in any direction provided two consecutive pressures along the tangent vector are identical. This condition is enforced along $j = 1$ by choosing B(1) = 1, C(1) = -1 and W(1) = 0. In other words, P(I,1) - P(I,2) = 0; P(I,1) and P(I,2) are solved simultaneously along with other columnar unknowns. Along $i = 1$, which falls outside the I = 2,10 range of the sweeping process, the simple update procedure P(1,J) = P(2,J) suffices. The required Fortran is shown in Fig. 7.9A, while the corresponding results are shown in Fig. 7.9B. Note how the top two rows and the left two columns, respectively, satisfy vanishing values of $\partial P/\partial y$ and $\partial P/\partial x$.

```
C       LAPLACE EQUATION SOLVER, CASE_5.
        PROGRAM MAIN
        DIMENSION P(11,11)
        OPEN(UNIT=4,FILE='CASE_5.DAT',STATUS='NEW')
C       DEFINE GRID PARAMETERS
        DX = 1.
        DY = 1.
C       INITIALIZE P(I,J) TO ZERO EVERYWHERE
        DO 100  I=1,11
        DO 100  J=1,11
        P(I,J) = 0.
 100    CONTINUE
C       SET "10-20-30-40" BOUNDARY CONDITION AT BOX EDGES
        DO 150  I=1,10
        P(I,1) =  10.
 150    CONTINUE
        DO 151  J=1,10
        P(11,J) = 20.
 151    CONTINUE
        DO 152  I=2,11
        P(I,11) = 30.
 152    CONTINUE
        DO 153  J=2,11
        P(1,J) = 40.
 153    CONTINUE
C       POINT RELAXATION BEGINS
        DO 400  NSWEEP=1,200
        IF(MOD(NSWEEP,10).NE.0) GO TO 170
C       PRINT OUT "X-Y" RESULTS
        WRITE(*,154)
        WRITE(4,154)
        WRITE(*,155) NSWEEP
        WRITE(4,155) NSWEEP
 154    FORMAT(' ')
 155    FORMAT(' P(I,J) SOLUTION FOR NSWEEP = ',I3)
        DO 160  J=1,11
        WRITE(*,157) (P(I,J),I=1,11)
        WRITE(4,157) (P(I,J),I=1,11)
 157    FORMAT(1X,11F6.1)
 160    CONTINUE
C       ITERATE POINT BY POINT WITHIN BOX
 170    DO 300  I=2,10
        DO 300  J=2,10
        P(I,J) = (P(I-1,J) +P(I+1,J) +P(I,J-1) +P(I,J+1))/4.
 300    CONTINUE
 400    CONTINUE
        CLOSE(4,STATUS='KEEP')
        STOP
        END
```

(A)

FIG. 7.7 (A and B) Aquifer alone, point relaxation.

P(I,J) SOLUTION FOR NSWEEP = 10

10.0	10.0	10.0	10.0	10.0	10.0	10.0	10.0	10.0	10.0	20.0
40.0	23.0	15.6	11.9	9.9	9.0	8.9	9.5	11.1	14.0	20.0
40.0	27.3	18.6	13.1	9.9	8.2	8.0	9.1	11.5	15.2	20.0
40.0	28.8	20.0	13.8	9.8	7.8	7.5	8.8	11.6	15.5	20.0
40.0	29.4	20.7	14.2	10.0	7.7	7.4	8.9	11.8	15.7	20.0
40.0	29.9	21.4	15.0	10.7	8.5	8.2	9.6	12.5	16.2	20.0
40.0	30.5	22.5	16.6	12.6	10.5	10.2	11.5	13.9	17.0	20.0
40.0	31.4	24.3	19.1	15.7	13.9	13.6	14.6	16.4	18.4	20.0
40.0	32.4	26.6	22.6	20.0	18.7	18.5	19.0	19.9	20.4	20.0
40.0	32.7	28.8	26.4	25.0	24.3	24.1	24.3	24.4	23.7	20.0
40.0	30.0	30.0	30.0	30.0	30.0	30.0	30.0	30.0	30.0	30.0

P(I,J) SOLUTION FOR NSWEEP = 20

10.0	10.0	10.0	10.0	10.0	10.0	10.0	10.0	10.0	10.0	20.0
40.0	24.2	17.6	14.4	12.7	11.8	11.6	11.9	12.8	14.9	20.0
40.0	29.4	22.3	17.8	15.1	13.6	13.1	13.5	14.7	16.8	20.0
40.0	31.7	25.1	20.3	17.1	15.3	14.6	14.9	16.0	17.8	20.0
40.0	32.9	26.8	22.1	18.9	17.0	16.2	16.3	17.1	18.5	20.0
40.0	33.7	28.2	23.8	20.7	18.8	17.9	17.8	18.3	19.1	20.0
40.0	34.3	29.3	25.4	22.6	20.8	19.8	19.5	19.6	19.9	20.0
40.0	34.8	30.3	26.9	24.5	22.9	22.0	21.6	21.3	20.8	20.0
40.0	34.9	31.0	28.3	26.5	25.3	24.5	24.0	23.4	22.2	20.0
40.0	33.9	31.1	29.4	28.3	27.7	27.2	26.8	26.2	24.6	20.0
40.0	30.0	30.0	30.0	30.0	30.0	30.0	30.0	30.0	30.0	30.0

P(I,J) SOLUTION FOR NSWEEP = 150

10.0	10.0	10.0	10.0	10.0	10.0	10.0	10.0	10.0	10.0	20.0
40.0	25.0	19.1	16.3	14.9	14.0	13.6	13.5	13.9	15.4	20.0
40.0	30.9	25.0	21.3	19.1	17.7	16.8	16.5	16.8	17.8	20.0
40.0	33.7	28.7	25.0	22.4	20.7	19.5	18.9	18.8	19.2	20.0
40.0	35.1	30.9	27.6	25.0	23.1	21.8	20.8	20.3	20.0	20.0
40.0	36.0	32.3	29.3	26.9	25.0	23.5	22.4	21.5	20.7	20.0
40.0	36.4	33.2	30.5	28.2	26.5	25.0	23.7	22.5	21.3	20.0
40.0	36.5	33.5	31.1	29.2	27.6	26.3	25.0	23.6	22.0	20.0
40.0	36.1	33.2	31.2	29.7	28.5	27.5	26.4	25.0	23.0	20.0
40.0	34.6	32.2	30.8	30.0	29.3	28.7	28.0	27.0	25.0	20.0
40.0	30.0	30.0	30.0	30.0	30.0	30.0	30.0	30.0	30.0	30.0

P(I,J) SOLUTION FOR NSWEEP = 200

10.0	10.0	10.0	10.0	10.0	10.0	10.0	10.0	10.0	10.0	20.0
40.0	25.0	19.1	16.3	14.9	14.0	13.6	13.5	13.9	15.4	20.0
40.0	30.9	25.0	21.3	19.1	17.7	16.8	16.5	16.8	17.8	20.0
40.0	33.7	28.7	25.0	22.4	20.7	19.5	18.9	18.8	19.2	20.0
40.0	35.1	30.9	27.6	25.0	23.1	21.8	20.8	20.3	20.0	20.0
40.0	36.0	32.3	29.3	26.9	**25.0**	23.5	22.4	21.5	20.7	20.0
40.0	36.4	33.2	30.5	28.2	26.5	25.0	23.7	22.5	21.3	20.0
40.0	36.5	33.5	31.1	29.2	27.6	26.3	25.0	23.6	22.0	20.0
40.0	36.1	33.2	31.2	29.7	28.5	27.5	26.4	25.0	23.0	20.0
40.0	34.6	32.2	30.8	30.0	29.3	28.7	28.0	27.0	25.0	20.0
40.0	30.0	30.0	30.0	30.0	30.0	30.0	30.0	30.0	30.0	30.0

(B)

FIG. 7.7, CONT'D

```
C       LAPLACE EQUATION SOLVER, CASE_6.
        DIMENSION P(11,11)
        OPEN(UNIT=4,FILE='CASE_6.DAT',STATUS='NEW')
C       DEFINE GRID PARAMETERS AND INITIALIZE P(I,J) TO ZERO
        DX = 1.
        DY = 1.
        ONE = 1.
        DO 100  I=1,11
        DO 100  J=1,11
        P(I,J) = 0.
 100    CONTINUE
C       SET "100" BOUNDARY CONDITION AT BOX EDGES
        DO 150  I=1,10
        P(I,1) =  100.
 150    CONTINUE
        DO 151  J=1,10
        P(11,J) = 100.
 151    CONTINUE
        DO 152  I=2,11
        P(I,11) = 100.
 152    CONTINUE
        DO 153  J=2,11
        P(1,J) = 100.
 153    CONTINUE
C       POINT RELAXATION BEGINS
        DO 400  NSWEEP=1,200
        IF(MOD(NSWEEP,10).NE.0) GO TO 170
        WRITE(*,155) NSWEEP
        WRITE(4,155) NSWEEP
 155    FORMAT(' P(I,J) SOLUTION FOR NSWEEP = ',I3)
        DO 160  J=1,11
        WRITE(*,157)  (P(I,J),I=1,11)
        WRITE(4,157)  (P(I,J),I=1,11)
 157    FORMAT(1X,11F6.1)
 160    CONTINUE
 170    DO 300  I=2,10
        DO 300  J=2,10
        MODE = 0
        IF(I.GE.4.AND.I.LE.8) MODE = 1
        IF(MODE.EQ.1.AND.J.EQ.6) MODE = 2
        P(I,J) = (P(I-1,J) + P(I+1,J) + P(I,J-1) + P(I,J+1))/4.
        IF(MODE.EQ.2) P(I,J) = ONE
 300    CONTINUE
 400    CONTINUE
        CLOSE(4,STATUS='KEEP')
        STOP
        END
```

(A)

FIG. 7.8 (A and B) Fracture flow, point relaxation.

P(I,J) SOLUTION FOR NSWEEP = 10

99.9	99.9	99.9	99.9	99.9	99.9	99.9	99.9	99.9	99.9	99.9
99.9	92.1	85.6	80.9	78.1	77.1	78.2	81.4	86.6	93.2	99.9
99.9	85.5	73.4	64.3	58.9	57.2	59.1	65.0	74.9	87.1	99.9
99.9	80.1	62.7	48.9	41.6	39.4	41.8	49.6	64.1	81.8	99.9
99.9	75.6	52.6	31.1	23.4	21.4	23.5	31.4	53.5	77.1	99.9
99.9	72.8	43.8	1.0	1.0	1.0	1.0	1.0	44.4	74.2	99.9
99.9	74.9	50.9	28.3	20.0	17.8	20.2	29.0	52.5	76.8	99.9
99.9	79.5	61.3	46.6	38.7	36.4	39.3	48.0	63.7	81.8	99.9
99.9	85.7	73.5	64.0	58.5	56.9	59.3	65.6	75.8	87.7	99.9
99.9	92.8	86.7	82.1	79.3	78.6	79.9	83.1	88.1	94.0	99.9
99.9	99.9	99.9	99.9	99.9	99.9	99.9	99.9	99.9	99.9	99.9

P(I,J) SOLUTION FOR NSWEEP = 20

99.9	99.9	99.9	99.9	99.9	99.9	99.9	99.9	99.9	99.9	99.9
99.9	95.2	90.9	87.4	85.2	84.6	85.4	87.6	91.2	95.4	99.9
99.9	90.4	81.4	73.9	69.5	68.2	69.7	74.3	81.8	90.7	99.9
99.9	85.3	70.9	58.1	51.4	49.5	51.6	58.5	71.3	85.7	99.9
99.9	80.4	59.2	36.8	29.0	27.1	29.2	37.0	59.5	80.7	99.9
99.9	77.4	49.1	1.0	1.0	1.0	1.0	1.0	49.4	77.7	99.9
99.9	80.3	59.0	36.6	28.8	26.9	28.9	36.9	59.5	80.7	99.9
99.9	85.3	70.8	58.0	51.3	49.3	51.5	58.4	71.3	85.7	99.9
99.9	90.4	81.5	74.1	69.7	68.3	69.9	74.5	81.9	90.8	99.9
99.9	95.3	91.1	87.6	85.5	84.8	85.6	87.8	91.3	95.5	99.9
99.9	99.9	99.9	99.9	99.9	99.9	99.9	99.9	99.9	99.9	99.9

P(I,J) SOLUTION FOR NSWEEP = 150

99.9	99.9	99.9	99.9	99.9	99.9	99.9	99.9	99.9	99.9	99.9
99.9	95.7	91.6	88.2	86.1	85.4	86.1	88.2	91.6	95.7	99.9
99.9	91.0	82.4	75.2	70.9	69.4	70.9	75.2	82.4	91.0	99.9
99.9	86.0	72.0	59.3	52.6	50.6	52.6	59.3	72.0	86.0	99.9
99.9	81.0	60.1	37.5	29.7	27.8	29.7	37.5	60.1	81.0	99.9
99.9	77.9	49.8	1.0	1.0	1.0	1.0	1.0	49.8	77.9	99.9
99.9	81.0	60.1	37.5	29.7	27.8	29.7	37.5	60.1	81.0	99.9
99.9	86.0	72.0	59.3	52.6	50.6	52.6	59.3	72.0	86.0	99.9
99.9	91.0	82.4	75.2	70.9	69.4	70.9	75.2	82.4	91.0	99.9
99.9	95.7	91.6	88.2	86.1	85.4	86.1	88.2	91.6	95.7	99.9
99.9	99.9	99.9	99.9	99.9	99.9	99.9	99.9	99.9	99.9	99.9

P(I,J) SOLUTION FOR NSWEEP = 200

99.9	99.9	99.9	99.9	99.9	99.9	99.9	99.9	99.9	99.9	99.9
99.9	95.7	91.6	88.2	86.1	85.4	86.1	88.2	91.6	95.7	99.9
99.9	91.0	82.4	75.2	70.9	69.4	70.9	75.2	82.4	91.0	99.9
99.9	86.0	72.0	59.3	52.6	50.6	52.6	59.3	72.0	86.0	99.9
99.9	81.0	60.1	37.5	29.7	27.8	29.7	37.5	60.1	81.0	99.9
99.9	77.9	49.8	**1.0**	**1.0**	**1.0**	**1.0**	**1.0**	49.8	77.9	99.9
99.9	81.0	60.1	37.5	29.7	27.8	29.7	37.5	60.1	81.0	99.9
99.9	86.0	72.0	59.3	52.6	50.6	52.6	59.3	72.0	86.0	99.9
99.9	91.0	82.4	75.2	70.9	69.4	70.9	75.2	82.4	91.0	99.9
99.9	95.7	91.6	88.2	86.1	85.4	86.1	88.2	91.6	95.7	99.9
99.9	99.9	99.9	99.9	99.9	99.9	99.9	99.9	99.9	99.9	99.9

(B)

FIG. 7.8, CONT'D

```
C      LAPLACE EQUATION SOLVER, CASE_7.
       PROGRAM MAIN
       DIMENSION P(11,11), A(11), B(11), C(11), V(11), W(11)
       OPEN(UNIT=4,FILE='CASE_7.DAT',STATUS='NEW')
C      DEFINE GRID PARAMETERS
       DX = 1.
       DY = 1.
       RATIO2 = (DY/DX)**2
       ONE = 1.
C      INITIALIZE P(I,J) TO ZERO EVERYWHERE
       DO 100  I=1,11
       DO 100  J=1,11
       P(I,J) = 0.
 100   CONTINUE
C      SET "100" BOUNDARY CONDITION AT BOX EDGES
       DO 151  J=1,10
       P(11,J) = 100.
 151   CONTINUE
       DO 152  I=2,11
       P(I,11) = 100.
 152   CONTINUE
C      LINE RELAXATION BEGINS
       DO 400  NSWEEP=1,200
       IF(MOD(NSWEEP,10).NE.0) GO TO 170
C      PRINT OUT "X-Y" RESULTS
       WRITE(*,154)
       WRITE(4,154)
       WRITE(*,155) NSWEEP
       WRITE(4,155) NSWEEP
 154   FORMAT(' ')
 155   FORMAT(' P(I,J) SOLUTION FOR NSWEEP = ',I3)
       DO 160  J=1,11
       WRITE(*,157) (P(I,J),I=1,11)
       WRITE(4,157) (P(I,J),I=1,11)
 157   FORMAT(1X,11F6.1)
 160   CONTINUE
C      ITERATE COLUMN BY COLUMN WITHIN BOX
 170   DO 300  I=2,10
C      DEFINE MATRIX COEFS FOR INTERNAL POINTS
       DO 200  J=2,10
       A(J) = 1.
       B(J) = -2.*(1.+RATIO2)
       C(J) = 1.
       W(J) = -RATIO2*(P(I-1,J)+P(I+1,J))
C      SET INTERNAL BOUNDARY CONDITION
       IF(I.EQ.6.AND.J.EQ.6) A(J) = 0.
       IF(I.EQ.6.AND.J.EQ.6) B(J) = 1.
       IF(I.EQ.6.AND.J.EQ.6) C(J) = 0.
       IF(I.EQ.6.AND.J.EQ.6) W(J) = ONE
 200   CONTINUE
C      RESTATE UPPER/LOWER BOUNDARY CONDITIONS
C      NOTE "99" DUMMY VALUES
       A(1) = 99.
       B(1) = 1.
       C(1) = -1.
       W(1) = 0.
       A(11) = 0.
       B(11) = 1.
       C(11) = 99.
       W(11) = P(I,11)
C      INVOKE TRIDIAGONAL MATRIX SOLVER
       CALL TRIDI(A,B,C,V,W,11)
C      UPDATE AND STORE COLUMN SOLUTION
```

FIG. 7.9 (A and B) Implementing no-flow boundary conditions.

```
      DO 250  J=1,11
      P(I,J) = V(J)
 250  CONTINUE
 300  CONTINUE
C     SET NO-FLOW CONDITION
      DO 350  J=1,11
      P(1,J) =  P(2,J)
 350  CONTINUE
 400  CONTINUE
      CLOSE(4,STATUS='KEEP')
      STOP
      END
```

(A)

P(I,J) SOLUTION FOR NSWEEP = 10

2.5	2.5	3.9	6.7	11.7	20.0	31.9	47.0	64.1	82.1	99.9
2.5	2.5	3.9	6.7	11.7	20.0	31.9	47.0	64.1	82.1	99.9
3.5	3.5	4.9	7.4	12.1	19.9	31.9	47.2	64.4	82.3	99.9
5.6	5.6	7.1	9.2	12.7	19.1	31.7	47.6	64.9	82.6	99.9
9.5	9.5	11.0	12.3	13.4	15.5	31.1	48.5	66.1	83.3	99.9
15.8	15.8	17.5	18.1	15.5	1.0	31.2	51.2	68.5	84.7	99.9
25.4	25.4	27.6	28.9	29.4	30.2	43.8	58.5	73.0	86.8	99.9
38.9	38.9	41.3	43.2	45.4	49.1	57.5	67.8	78.8	89.6	99.9
56.4	56.4	58.4	60.3	62.5	65.9	71.3	78.1	85.5	92.9	99.9
77.2	77.2	78.4	79.5	80.8	82.7	85.5	88.9	92.6	96.4	99.9
99.9	99.9	99.9	99.9	99.9	99.9	99.9	99.9	99.9	99.9	99.9

FIG. 7.9, CONT'D

(Continued)

OBSERVATIONS ON RELAXATION METHODS

In this section, we summarize important observations and facts about relaxation methods. These comments are based on the author's experience in developing aerodynamics and reservoir simulation models over two decades.

Easy to Program and Maintain

By modifying the source code in Fig. 7.3A to handle problems of increasing difficulty in a sequence of examples, we have shown how a finite difference model can be easily understood and extended to describe wells, fractures, aquifers, and solid walls. Multiple wells and fractures, and general combinations of aquifer and no-flow boundary conditions, of course, are just as easily treated: the basic engine driving the models requires but twenty lines of Fortran. Importantly, this powerful methodology requires little programming or numerical analysis experience.

P(I,J) SOLUTION FOR NSWEEP = 20

22.1	22.1	24.9	28.7	33.8	40.9	50.4	61.7	74.1	87.0	99.9
22.1	22.1	24.9	28.7	33.8	40.9	50.4	61.7	74.1	87.0	99.9
23.2	23.2	25.7	28.8	33.2	39.6	49.3	61.1	73.8	86.9	99.9
25.7	25.7	27.5	29.3	31.6	36.1	46.9	60.0	73.5	86.9	99.9
29.9	29.9	30.9	30.6	28.8	27.0	42.8	58.9	73.6	87.1	99.9
36.6	36.6	36.8	34.6	26.6	1.0	38.9	59.7	74.9	87.9	99.9
46.0	46.0	46.3	45.0	41.9	38.5	52.3	66.3	78.5	89.6	99.9
57.7	57.7	58.2	58.1	57.8	59.0	66.0	74.8	83.5	91.9	99.9
71.0	71.0	71.5	71.9	72.5	74.2	78.2	83.4	89.0	94.5	99.9
85.3	85.3	85.6	85.9	86.4	87.5	89.3	91.8	94.5	97.3	99.9
99.9	99.9	99.9	99.9	99.9	99.9	99.9	99.9	99.9	99.9	99.9

P(I,J) SOLUTION FOR NSWEEP = 150

52.6	52.6	52.6	53.0	54.3	57.6	63.3	71.2	80.2	90.0	99.9
52.6	52.6	52.6	53.0	54.3	57.6	63.3	71.2	80.2	90.0	99.9
52.6	52.6	52.3	52.0	52.4	55.1	61.3	69.9	79.6	89.7	99.9
53.0	53.0	52.0	50.2	48.4	48.9	56.9	67.5	78.5	89.3	99.9
54.3	54.3	52.4	48.4	41.9	35.4	49.9	64.6	77.5	89.1	99.9
57.6	57.6	55.1	48.9	35.4	1.0	42.7	63.7	77.9	89.4	99.9
63.3	63.3	61.3	56.9	49.9	42.7	56.1	69.5	80.9	90.8	99.9
71.2	71.2	69.9	67.5	64.6	63.7	69.5	77.4	85.4	92.8	99.9
80.2	80.2	79.6	78.5	77.5	77.9	80.9	85.4	90.3	95.2	99.9
90.0	90.0	89.7	89.3	89.1	89.4	90.8	92.8	95.2	97.6	99.9
99.9	99.9	99.9	99.9	99.9	99.9	99.9	99.9	99.9	99.9	99.9

P(I,J) SOLUTION FOR NSWEEP = 200

52.6	**52.6**	**52.6**	**53.0**	**54.3**	**57.6**	**63.3**	**71.2**	**80.2**	**90.0**	**99.9**
52.6	**52.6**	**52.6**	**53.0**	**54.3**	**57.6**	**63.3**	**71.2**	**80.2**	**90.0**	**99.9**
52.6	**52.6**	52.3	52.0	52.4	55.1	61.3	69.9	79.6	89.7	99.9
53.0	**53.0**	52.0	50.2	48.4	48.9	56.9	67.5	78.5	89.3	99.9
54.3	**54.3**	52.4	48.4	41.9	35.4	49.9	64.6	77.5	89.1	99.9
57.6	**57.6**	55.1	48.9	35.4	1.0	42.7	63.7	77.9	89.4	99.9
63.3	**63.3**	61.3	56.9	49.9	42.7	56.1	69.5	80.9	90.8	99.9
71.2	**71.2**	69.9	67.5	64.6	63.7	69.5	77.4	85.4	92.8	99.9
80.2	**80.2**	79.6	78.5	77.5	77.9	80.9	85.4	90.3	95.2	99.9
90.0	**90.0**	89.7	89.3	89.1	89.4	90.8	92.8	95.2	97.6	99.9
99.9	**99.9**	99.9	99.9	99.9	99.9	99.9	99.9	99.9	99.9	99.9

(B)

FIG. 7.9, CONT'D

Minimal Computing Resources

A rectangular grid with IMAX and JMAX meshes in x and y will have IMAX × JMAX unknowns. An unoptimized direct matrix solver that does not account for sparseness and bandedness will require numerous computations for inversion. The worst case possibility is Gaussian elimination, which requires

(IMAX × JMAX)^3 multiply and divide operations. The problem is compounded in three dimensions. In our scheme, only a single tridiagonal matrix solver is needed; inverting a JMAX line solution requires 3 × JMAX operations, although this is repeated NSWEEP × IMAX times. This still represents a significant improvement over direct matrix inversion methods.

Good Numerical Stability

Our programs are extremely stable numerically; that is, they do *not* lead to 10^9 pressures often. The procedures are conditionally stable on a linear von Neumann stability basis. This is so because the coefficient matrixes are diagonally dominant, becoming even more so when 3D problems are solved in a columnar fashion as in our examples. Often, a planar problem that does not converge on a 2D basis can be successfully and quickly solved as the limit of the 3D problem. An unstable 2D problem can be artificially embedded in a suitable 3D problem to facilitate convergence.

Fast Convergence

Relaxation schemes are known to converge rapidly, at least, initially; then, the rate of convergence slows somewhat, although computing times are still tolerable. Various methods are used to accelerate convergence—for example, over-relaxation (Jameson, 1975), Shanks extrapolation (van Dyke, 1964), or multigrid methods (Wesseling, 1992) which use alternating sequences of fine and coarse meshes to host the relaxation. Perhaps the most important advantage of relaxation methods is the ability to initialize the solution to an approximate one that is already available, ideally a solution with nearly the same heterogeneities or well configuration. This is important to the study of reservoir description. Suppose a number of geological realizations are created, for example, using a geostatistical model, with each successive model being slightly different from the preceding. Then, each pressure solution should require only minimal incremental effort, when compared to a direct solution technique that assumes nothing at the outset. On this basis, relaxation methods exceed direct solvers in speed. As we have seen, the method *always* seems to converge to the same answer, regardless of the initial guess. The proper initialization, of course, reduces computation times significantly. This advantage is important in the design of software that offers instantaneous user response while requiring minimal hardware resources.

Why Relaxation Methods Converge

We conclude this section by offering some quantitative insight showing why convergence to a unique solution, regardless of the initial guess, is expected, assuming of course a stable solution. Let us multiply Eq. (7.17) by 4 and rewrite it with superscripts n and n-1 to describe the recursion relation used in the iteration below.

$$4\,P^{n}_{i,j} = P^{n-1}_{i-1,j} + P^{n-1}_{i+1,j} + P^{n-1}_{i,j-1} + P^{n-1}_{i,j+1} \tag{7.18}$$

We subtract 4 $P^{n-1}_{i,j}$ from each side of Eq. (7.18) to obtain

$$P^{n-1}_{i-1,j} - 2\,P^{n-1}_{i,j} + P^{n-1}_{i+1,j} + P^{n-1}_{i,j-1} - 2\,P^{n-1}_{i,j} + P^{n-1}_{i,j+1} = 4\left(P^{n}_{i,j} - P^{n-1}_{i,j}\right) \tag{7.19}$$

If we now divide Eq. (7.19) throughout by $(\Delta x)^2$, we have

$$\begin{aligned} &\left(P^{n-1}_{i-1,j} - 2\,P^{n-1}_{i,j} + P^{n-1}_{i+1,j}\right)/(\Delta x)^2 \\ &+\left(P^{n-1}_{i,j-1} - 2P^{n-1}_{i,j} + P^{n-1}_{i,j+1}\right)/(\Delta x)^2 \\ &= \left(P^{n}_{i,j} - P^{n-1}_{i,j}\right)/\Delta t \end{aligned} \tag{7.20}$$

where $\Delta t = (\Delta x)^2/4$. We recognize Eq. (7.20) as the explicitly differenced form of the dimensionless heat equation

$$\partial^2 P/\partial x^2 + \partial^2 P/\partial y^2 = \partial P/\partial t \tag{7.21}$$

which governs heat propagation in solids when P(x,y,t) is the transient temperature. As is well known (Carslaw and Jaeger, 1946, 1959), the final steady-state solution (satisfying $\partial^2 P/\partial x^2 + \partial^2 P/\partial y^2 = 0$) is independent of initial conditions.

Thus, it is not surprising that our solutions for steady-state pressure can be obtained independently of the initial guess. This comment applies to Eq. (7.14), for a liquid, assuming pressure or flow rate boundary conditions; but it is also valid for gases, which satisfy a *linear* equation for P^{m+1}. The usual analogy comparing relaxation with polynomial root solvers is not strictly correct, since elliptic problems, at least the ones considered here, have unique solutions. In several commercial publications, the claim is made that modern direct matrix solvers help pressure fields converge much faster than older relaxation approaches. This may be true in blind comparisons where nothing is known about the solution; but as we have seen, iterative models can be quite flexible when used cleverly. When direct solvers are used, the selection of proper matrix conditioning parameters is crucial, which requires some knowledge of the structure of the coefficient matrix. This often takes longer than the pressure solution itself. The resulting parameters, in fact, may depend on the physical characteristics of the oilfield, and will vary from problem to problem and change as oil and water saturations evolve with time. But relaxation methods, very often "dummy proof," also allow users to initialize their solutions with analytical solutions, such as those derived in Chapters 1–5.

Over-Relaxation

Researchers have made great strides in accelerating the convergence of relaxation methods. Note from Eq. (7.20) that the point relaxation scheme of Eq. (7.17) is associated with a hard-coded value of heat conductivity, namely unity, in Eq. (7.21). There is nothing sacred about this value; a higher conductivity will decrease the convergence time required to achieve steady-state results, while a lower one will increase it. One way to increase conductivity is to over-relax. Previously, we updated the Fortran solution using P(I,J) = V(J), where V(J) is the latest solution obtained from the columnar matrix inversion. Instead, let us update the pressure field using P(I,J) = RELAX* V(J) + (1-RELAX)*P(I,J). The choice RELAX = 1 reduces to doing nothing. However, convergence can be accelerated by over-relaxing with RELAX > 1. At other times when numerical stability is a problem, under-relaxing with RELAX < 1 may stabilize the calculations. Other authors "embed" their Laplace operators within unsteady systems that are more rapidly convergent than that of Eq. (7.21). After all, the transient phases of iterative processes are unimportant if only steady results are desired; any fast artificial time variable will do. For a discussion on modern relaxation methods, the reader is referred to the pioneering work of Jameson (1975).

Line and Point Relaxation

Line relaxation is used for several reasons. First, the algorithm is simple to construct and maintain. Second, the tridiagonal solver requires only 3N multiplies and divides to invert an $N \times N$ system. Of course, it is called dozens of times until convergence; still, the cumulative effort needed to solve a problem is small by comparison to, say, direct solutions via Gaussian elimination. If sufficiently close solutions are available for initialization, large decreases in convergence time can be achieved. Importantly, line relaxation handles two-point boundary conditions easily. Pressure data from upper and lower boundaries are communicated instantly along columns, and left and right boundary conditions quickly propagate along rows. By contrast, point relaxation methods are sluggish; they require longer computation times to converge. However, they are easily adaptable to irregular geometries, where lines having constant program dimension or vector length are difficult to define. (The curvilinear grid methods in Chapters 8–10 are an exception.) If irregular geometries must be simulated on rectangular meshes, point relaxation is recommended because it is easily programmed, with the logic in Eq. (7.17) performed only for points inside the flow domain. This simplifies development since constant mesh number lines need not be defined.

Finally, there is the issue of vectorization, also referred to as scalar versus parallel computing. Serial computers execute instructions sequentially, in specific order; parallel machines execute multiple instructions simultaneously. Often, different flow domains are apportioned to different machines, and message passing interfaces must be designed so that these domains communicate with each other in an optimal way that minimizes computation time. Point relaxation gave way to line methods when serial computers were predominant because they were slower in converging. However, they are now used on vector machines because many points can be iterated upon in parallel. On parallel machines, it is argued that the implicit schemes associated with line methods require step-by-step matrix inversion, sequential operations that do not take advantage of computer architecture. On the other hand, researchers have vectorized line methods so that large bundles of lines are solved simultaneously. Whether the reader prefers direct or iterative methods, he is cautioned against quick and simplifying recommendations. In either approach, the issues are not as straightforward as they seem, and there is always room for ingenuity.

In Chapters 8–10, grid generation, generalized elementary solutions, transient compressible flows, alternating-direction-implicit (or ADI) methods, and 3D steady and unsteady flow analysis will be studied, building on the ideas developed here. Special curvilinear grid methods eliminate the disadvantages associated with rectangular meshes, providing fine resolution near wells where detailed solutions are required. Whatever the meshing model, grid block size and attendant flow properties are often chosen indiscriminately; for example, permeabilities obtained from small cores are applied to grid blocks hundreds of feet across, and the simple averaging methods often used lead to unforeseen consequences. In the next section, we demonstrate (using the planar finite difference methods discussed) how the upscaling process for cross-bedded sands can be subtle indeed and not-at-all straightforward. This application shows how simple numerical models can be used to analyze interesting physical concepts.

ISOTROPY AND ANISOTROPY: FLUID INVASION IN CROSS-BEDDED SANDS

Much has been written about grid block-averaging, designed to minimize memory usage and computing times, among them, arithmetic, geometric, and harmonic averaging methods, and the newer ones, such as geostatistically based models and pseudos that reservoir engineers use. Unfortunately, the averaging process itself is often the sole focus of research, with minimal concern for physical consequences. Chapter 11 introduces simple problems with exact solutions that highlight proper usage of effective properties and that point to their dangers. Dangers arise because there is always a loss of information whenever averages are taken; for example, those chosen to match production rates will yield incorrect tracer travel-times. Also, reduced grid block structures optimized for one scenario may be less applicable when a well constraint is changed or another well is added or removed. In this section, we demonstrate how upscaling can introduce new effects that are not apparent from the original small-scale description. In continuum mechanics, isotropy and anisotropy are often the results of observations on contrasting scales. For example, different small wood particles in trees may behave isotropically, but in the aggregate, they may behave anisotropically as mechanical strength varies differently across the grain than along. Here we address anisotropy from computational and well logging perspectives. Taken simply, anisotropy occurs whenever k_h and k_v differ. It is not difficult to imagine a homogeneous rock sample satisfying this requirement, and indeed, many do. However, it turns out that sedimentary layers that are isotropic individually can also behave anisotropically, when they are stacked in alternating sequences and oriented at dip. Thus, a log analyst might falsely conclude that a formation is anisotropic when it is isotropic. On the other hand, in simulation studies, isotropic laminated sequences that are not modeled anisotropically when they should be will yield incorrect large-scale results. For the problems considered, when a critical parameter is reached, streamlines that are nominally straight abruptly turn and change direction, mimicking the behavior of rocks that are anisotropic over larger scales. This is shown for steady Darcy flow through linear cores.

To understand the issues, consider the fluid motion in Fig. 7.10A and B. For flow incident upon parallel layers, the emerging flow rate is nonuniform vertically, depending on individual layer resistances; for flow incident upon perpendicular layers, this rate is uniform vertically. In both examples, incident and emerging streamlines do not change direction, and both remain straight. The usual parallel and series circuit averaging formulas (e.g., for DC electrical resistance) apply, and indeed suffice for describing the aggregate flow in isotropic terms. Anisotropic behavior never arises in Fig. 7.10A and B.

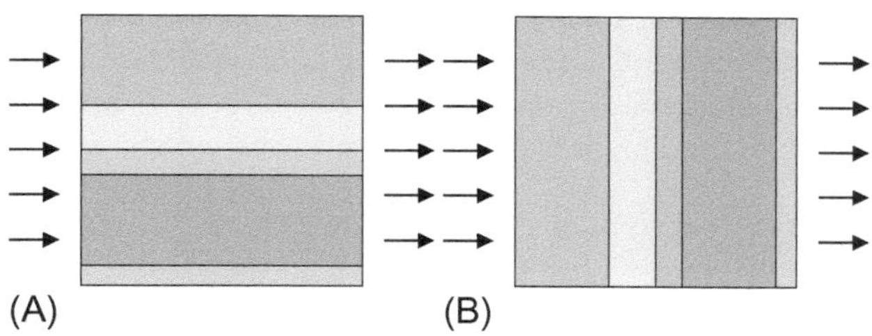

FIG. 7.10 (A) Flow parallel to layers. (B) Flow perpendicular to layers.

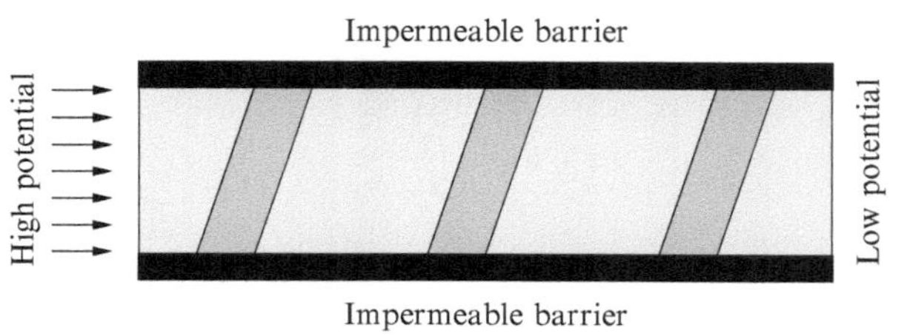

FIG. 7.11 Flow through dipping layers.

Now consider the core sample in Fig. 7.11. It is clear that the entering uniform flow must become nonuniform at the right; the emerging streamlines will turn from a horizontal direction, even when the laminates in the core are purely isotropic. An observer without knowledge of the test setup would interpret, from the orientation of the emerging streamlines, that the test sample is anisotropic over the scale of the fixture. This anisotropy follows as a consequence of a purely isotropic formulation, that is, a finite difference solution of $\partial\ (k(x,y)\ \partial P/\partial x)/\partial x + \partial\ (k(x,y)\ \partial P/\partial y)/\partial y = 0$ for nonuniform but isotropic media, and not necessarily $\partial\,(k_h\,\partial P/\partial x)/\partial x+\partial\,(k_v\,\partial P/\partial y)/\partial y=0$ for anisotropic media. This observation also suggests that an anisotropic reading from a logging instrument can be further interrogated to determine if a more detailed isotropic fine-scale structure exists.

Numerical Results

Several parameters describe our core: its overall size, the dip angle, layer thicknesses, and properties. For the simulations, a 45 degree dip was taken, and a background permeability of 1 md was assumed for light gray rocks, which are three times wider than dark ones. The differential equation $\partial\,(k(x,y)\,\partial P/\partial x)/\partial x+\partial\,(k(x,y)\partial P/\partial y)/\partial y=0$ for isotropic but variable permeabilities k(x,y) was solved for P(x,y), which provided boundary conditions for the Ψ streamfunction formulation (see Chapter 4) solved. Streamlines were accurately traced by drawing level contours of Ψ. (Streamfunction methods suppress cumulative errors that normally arise from direct velocity vector integration.) For the system described, oblique permeability streaks taking the form in Fig. 7.12 were considered in turn, with 1-1, 1-2, 1-3, 1-4, and 1-5 periodicity. Note that rapid permeability variations lead to grainy pressure fields, and thus, to slightly wiggly streamlines, which are expected. Exact P and Ψ magnitudes are not given since our emphasis is on streamline patterns.

```
////////////////////////////////////////////////////////////
1  1  5  1  1  1  5  1  1  1  5  1  1  1  5  1  1  1  5  1
1  5  1  1  1  5  1  1  1  5  1  1  1  5  1  1  1  5  1  1
5  1  1  1  5  1  1  1  5  1  1  1  5  1  1  1  5  1  1  1
1  1  1  5  1  1  1  5  1  1  1  5  1  1  1  5  1  1  1  5
1  1  5  1  1  1  5  1  1  1  5  1  1  1  5  1  1  1  5  1
1  5  1  1  1  5  1  1  1  5  1  1  1  5  1  1  1  5  1  1
5  1  1  1  5  1  1  1  5  1  1  1  5  1  1  1  5  1  1  1
1  1  1  5  1  1  1  5  1  1  1  5  1  1  1  5  1  1  1  5
1  1  5  1  1  1  5  1  1  1  5  1  1  1  5  1  1  1  5  1
1  5  1  1  1  5  1  1  1  5  1  1  1  5  1  1  1  5  1  1
5  1  1  1  5  1  1  1  5  1  1  1  5  1  1  1  5  1  1  1
////////////////////////////////////////////////////////////
```

FIG. 7.12 Typical permeability setup.

Typical second-order accurate results are shown in Fig. 7.13A–F. For small permeability variations, computed streamlines are more or less straight. But at critical values, 1-5 for the present core setup, strong streamline divergence is seen as fluid moves to the right. An observer stationed at the outlet in Fig. 7.13F would infer the presence of strong anisotropy in the core sample. Whether or not this interpretation is correct is unimportant: fluid *does* move in the general direction shown. However, to determine the structure of the rock on smaller scales, finer measurement instruments would be needed. For all 1-1, 1-2, 1-3, 1-4, and 1-5 streaks, total volume flow rate was monitored. With our 1-1 uniform core, a normalized 0.56 value was obtained. Since subsequent average permeabilities increase, total flow rates must increase; the respective numbers are 0.56, 0.69, 0.80, 0.88, and 0.87, this last exception to the trend arising from finite end effects. Fig. 7.13F shows how initially parallel and equidistant streamlines all migrate to the upper right, leaving a dead stagnation zone at the lower right. This implies poor areal sweep, here obtained as a consequence of heterogeneities. The streamline pattern in Fig. 7.13F suggests that additional production is possible from a well placed at the bottom right.

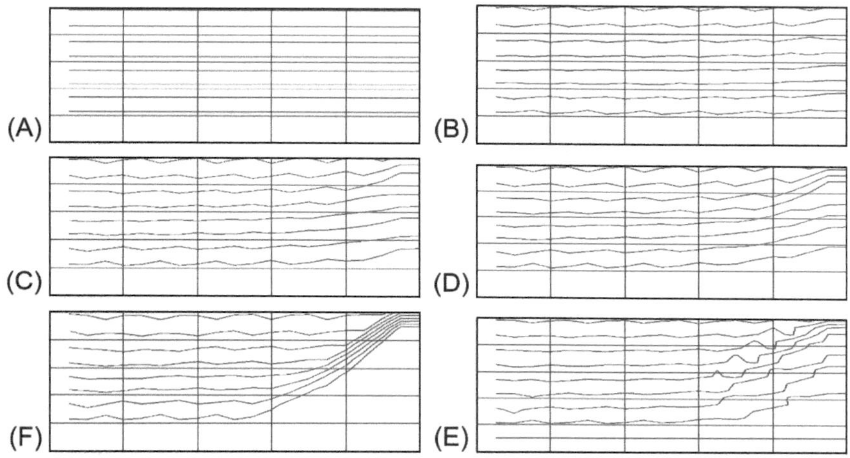

FIG. 7.13 (A) Uniform 1-1 core, straight streamlines. (B) 1-3 core, slight streamline deviation. (C) 1-4 core with minor turning. (D) 1-4.5, initial turning seen. (E) 1-4.75, strong turning. (F) 1-5, very strong streamline deflection.

As a second example, calculations were repeated with separations between dark streaks doubled. Because the overall averaged permeabilities are decreased relative to the values just considered, lower volume flow rates should result; we consistently obtained, 0.62 as opposed to 0.69, 0.67 versus 0.80, 0.70 versus 0.88, and 0.67 versus 0.87, but streamline convergence was again observed. Again, the cumulative effects of locally weak reservoir heterogeneities and nonuniformities turned out to be important in changing streamline orientation.

The baseline 1-1 uniform medium gave a flow rate of 0.56. As a check, this distribution was increased to 5-5, and the expected flow rate of 2.8 (five times greater than 0.56) was obtained. Finally, instead of Fig. 7.12, the low and high permeability streaks were altered every-other-diagonal in a simple checkerboard fashion consisting of 1s and 5s only, to create a purely random permeability distribution with no angular or directional bias, whose arithmetic average is (1+5)/2 or 3. The calculated flow rate of 1.67 is *exactly* three times the 0.56 obtained for unit permeabilities; this *three* is identical to the *arithmetic average* of (1+5)/2, obtained on a naïve basis. Thus, arithmetic averaging appears to be useful for more or less random distributions of permeability, whereas sands with crossbeds and formations with bedding planes clearly require more thoughtful anisotropic or directional consideration.

Electrical Analogy

Again, ∂ (k(x,y) ∂P/∂x)/∂x + ∂ (k(x,y) ∂P/∂y)/∂y = 0 was solved for isotropic but variable k(x,y), and results were obtained that behaved anisotropically, as if ∂ (k_h ∂P/∂x)/∂x + ∂ (k_v ∂P/∂y)/∂y = 0 had been solved. This was the case for high permeability contrasts and dip angles. Now, electric currents satisfy "∂ (σ(x,y) ∂V/∂x)/∂x + ∂ (σ(x,y) ∂V/∂y)/∂y = 0 in the low-frequency limit, where V and σ represent potential and electrical conductivity in isotropic media. Because the analogy with fluid flow is exact, our conclusions on streamline pattern apply to electric pathlines. Thus, dipping laminates that are isotropic microscopically can behave anisotropically macroscopically electrically. Because fluid and electrical models appear hand-in-hand, fluid flow anisotropy implies electrical anisotropy, and vice-versa.

PROBLEMS AND EXERCISES

1. Consider steady-state liquid flow in a homogeneous isotropic medium satisfying $\partial^2P/\partial x^2 + \partial^2P/\partial y^2 + \partial^2P/\partial z^2 = 0$. Finite difference this equation assuming that $\Delta x = \Delta y = \Delta z = \Delta$, and show that the value of $P_{i,j,k}$ at any node (x_i, y_j, z_k) is given by the arithmetic average of its neighboring values, that is, $P_{i,j,k} = 1/6\ (P_{i-1,j,k} + P_{i+1,j,k} + P_{i,j-1,k} + P_{i,j+1,k} + P_{i,j,k-1} + P_{i,j,k+1})$. Earlier, the formula $P_{i,j} = 1/4\ (P_{i-1,j} + P_{i+1,j} + P_{i,j-1} + P_{i,j+1})$ was used to design a point relaxation scheme for a well centered in a plane (e.g., Fig. 7.7B). For the present 3D problem, set P at all six faces of the computational box to 0, and place a point well at its center (ctr) having the pressure 100. Solve for the three-dimensional pressure field with a point relaxation method. For example, your Fortran may look something like,

```
C     Initialize and set "0" boundary conditions on six faces
      do 100 i = 1, imax
      do 100 j = 1, jmax
      do 100 k = 1, kmax
      p(i,j,k) = 0.
  100 continue
C     Iterate by point relaxation
      do 300 n = 1, nmax
      do 200 i = 2, imax-1
      do 200 j = 2, jmax-1
      do 200 k = 2, kmax-1
      p(i,j,k) = (p(i-1,j,k) + p(i+1,j,k) + . . . )/6.
      if(i.eq.ictr.and.j.eq.jctr.and.k.eq.kctr)
     1 then p(i,j,k) = 100.
  200 continue
      .
  300 continue
```

In particular, (i) introduce a suitable convergence criterion, (ii) run the simulator until convergence for several box sizes, and (iii) determine how pressure decays with distance. Does this agree with the steady solution for a spherical source? Next, (iv) surround the source point using different closed control surfaces and calculate the total mass flux through the surfaces. Are the mass flows identical? Should they be identical? If not, why not? (v) Repeat this flux calculation for a closed volume that does not enclose the source point. What should the result be? Now, (vi) instead of specifying $P = 100$ at the well, modify the program so that total mass flow rate is prescribed. Finally, (vii) extend this solver to arbitrary grid block sizes by removing the restriction that $\Delta x = \Delta y = \Delta z = \Delta$. Discuss possible numerical strategies needed that will ensure mass conservation.

2. Write a point relaxation program to solve the planar isotropic equation $\partial\,(k(x,y)\ \partial P/\partial x)/\partial x + \partial\,(k(x,y)\partial P/\partial y)/\partial y = 0$. Assume solid walls at top and bottom, and high and low constant pressures at the left and right. Initialize the code with $P = 0$, and other functions; show that convergence to a unique solution is independent of the function. Run this program for uniform permeabilities and show that pressures decrease linearly with x.

3. Using the computer code in (2), determine P(x,y) assuming the cross-bedded permeabilities on the following page and plot the streamlines as follows. (i) Derive the streamfunction equation, and solve it by a similar relaxation method. (ii) Required normal derivative boundary conditions in $\partial\Psi/\partial Y$ at top and bottom may be related to (available) solutions in $\partial P/\partial x$ using Cauchy-Riemann conditions. (iii) An equidistant distribution of vertical Ψ's may be assumed at the left for a starting condition, and the simplification $\partial\Psi/\partial x = 0$ may be taken at the right. Streamlines are represented by lines of constant streamfunction; the difference between any two Ψ values represents the mass flow through the two points. Show that the streamfunction distribution takes on the values shown on the following page, and sketch the streamline field. Import these pressure and streamfunction solutions into contour plotting software to generate curves that can be interpreted meaningfully, as in Fig. 7.13A–F. Repeat these calculations using 1-4 and 1-5 permeabilities.

PERMEABILITY

//

1	1	**3**	1	1	1	**3**	1	1	1	**3**	1	1	1	**3**	1	1	1	**3**	1
1	**3**	1	1	1	**3**	1	1	1	**3**	1	1	1	**3**	1	1	1	**3**	1	1
3	1	1	1	**3**	1	1	1	**3**	1	1	1	**3**	1	1	1	**3**	1	1	1
1	1	1	**3**	1	1	1	**3**	1	1	1	**3**	1	1	1	**3**	1	1	1	**3**
1	1	**3**	1	1	1	**3**	1	1	1	**3**	1	1	1	**3**	1	1	1	**3**	1
1	**3**	1	1	1	**3**	1	1	1	**3**	1	1	1	**3**	1	1	1	**3**	1	1
3	1	1	1	**3**	1	1	1	**3**	1	1	1	**3**	1	1	1	**3**	1	1	1
1	1	1	**3**	1	1	1	**3**	1	1	1	**3**	1	1	1	**3**	1	1	1	**3**
1	1	**3**	1	1	1	**3**	1	1	1	**3**	1	1	1	**3**	1	1	1	**3**	1
1	**3**	1	1	1	**3**	1	1	1	**3**	1	1	1	**3**	1	1	1	**3**	1	1
3	1	1	1	**3**	1	1	1	**3**	1	1	1	**3**	1	1	1	**3**	1	1	1

//

STREAMFUNCTION

//

100	100	100	100	100	100	100	100	100	100	100	100	100	100	100	100	100	100	100	100
90	88	87	91	92	89	87	90	92	89	87	90	92	89	87	90	92	87	82	82
80	77	80	83	80	77	80	83	80	77	80	82	79	76	79	82	78	73	74	74
70	71	73	70	67	70	72	70	67	70	72	69	66	69	71	68	64	66	68	68
60	62	60	57	60	62	60	57	59	62	59	56	59	61	58	55	57	59	56	56
50	49	47	50	52	50	47	49	52	49	47	49	51	48	45	47	49	45	41	41
40	37	40	42	40	37	39	42	39	37	39	41	39	36	38	40	37	33	34	34
30	30	32	30	27	29	32	29	27	29	31	29	26	28	31	28	25	26	28	28
20	22	20	17	19	22	19	16	19	22	19	16	19	21	18	15	18	20	18	18
10	10	7	9	12	10	7	8	11	10	7	8	11	10	6	8	11	9	5	5
0	0	0	0	0	0	0	0	0	0	0	0	0	0	0	0	0	0	0	0

//

Chapter 8

Curvilinear Coordinates and Numerical Grid Generation

OVERVIEW

Few exact solutions are ever available in reservoir flow analysis, simply because irregular boundaries and heterogeneities render the mathematics extremely difficult. Thus, researchers and practitioners concentrate their efforts on numerical models. Great strides have been achieved in petroleum reservoir simulation using such approximate techniques. However, most of the computational literature deals with rectangular and circular grid systems, where finite difference equations take on particularly simple forms.

Problems With Idealized Grids

Cartesian and radial grids produce notoriously inaccurate results insofar as resolution is concerned. For example, farfield boundaries are difficult to describe unless they take rectangular or circular form. When ideal geometric conditions are not met, the resulting matrices contain large numbers of inactive grid blocks that degrade computer performance, while active blocks are used ineffectively. Interpolative techniques are often employed, but again, at the expense of increased computations and numerical noise. In general, the problem of fitting stratigraphic boundaries to computer models is an important one to reservoir simulation, well test interpretation, and history matching. And there are other problems. Typically, an areal grid block may span hundreds of feet across, and contain multiple wells, each bearing 6-in.-diameter-length scales. Questions arise as to how computed grid block pressures relate to flowing pressures at the wells. Well-known correlations are available which address this problem; see van Poollen et al. (1968), Peaceman (1977, 1978, 1983), Williamson and Chappelear (1981), and Chappelear and Williamson (1981). These *ad hoc* approaches apply to vertical wells, but none has been developed for horizontal wells in layered media. Even accepted well models yield 50–200% of actual flow rates. Local mesh refinement and embedded grids provide two alternatives, but depending on how the additional equations are ordered in the matrix solution, severe performance degradation can result, leading to requirements for faster and more powerful computers. Also, for the

Quantitative Methods in Reservoir Engineering. http://dx.doi.org/10.1016/B978-0-12-810518-4.00008-6

Cartesian meshes used, increased nearfield grid density forces portions of the farfield to be likewise (but inconsistently) gridded.

Alternative Coordinate Systems

We have seen how coordinate systems can be cleverly exploited to advantage. For example, consider the elementary log r and θ solutions for point sources and vortexes obtained in cylindrical polar coordinates. In Chapters 2 and 3, they were rewritten in (x,y) coordinates in order to develop solutions for line fractures and shales. Or consider the conformal mappings introduced in Chapter 5; there, the simple solutions in Chapters 2 and 3 were extended to flows in complicated geometries. A newer, more powerful approach involves the use of boundary-conforming grid systems that wrap around wells and fractures in the nearfield and at the same time conform to the external boundaries of the farfield. The simplest example is provided by cylindrical coordinates, used to model circular wells concentrically located in circular reservoirs. Another is furnished by elliptical coordinates, used to model flows into straight, finite-length fractures in infinite systems.

The literature on boundary-conforming grid generation is not new. While the models are shrouded in mystery and specialized jargon, at least in reservoir simulation, the techniques are in fact highly developed. In the aerospace industry, they are routinely used to model complicated interference effects, for example, wing and fuselage juncture flows, engine flow blockage near wings, and so on. This author, in fact, has used this method extensively to study the highly eccentric annular borehole flow typically encountered in drilling and cementing applications, for instance, see Chin (1992a,b, 2001a,b, 2012). Grid generation techniques will be discussed in this chapter. We shall introduce the fundamentals of grid generation using simple, readable, algebraic derivations that avoid the complications of complex variables, differential geometry, and topology. Direct arguments and proofs are presented at a level comprehensible to undergraduate engineers, without delving into profound aspects of an inherently mathematical subject. The requisite derivations may not be as elegant as the mathematician may like; however, they are equally rigorous, providing the fundamentals in an important new area, and fall in the realm of must reading.

We emphasize that the grid generation discussions in this book are *not* exhaustive. But needless to say, the Thompson-based methods are by far the most popular, since finite difference flow models are readily implemented on the resulting structured grids. In this sense, software and algorithm development are straightforward. This is not the case with triangular, hexagonal and similar unstructured grids; these coordinate systems usually host finite element models which require (thoughtful) variational formulations, and of course, finite volume models, which require some manipulation of the integral conservation laws. The interested reader is referred to the research literature for the state of the art; the method discussed in this book is, in a sense, mature technology that can be discussed with confidence. The grid generation technique is a flexible one, and importantly, it is used to host a variety of steady and transient flow simulators designed in the later chapters.

GENERAL COORDINATE TRANSFORMATIONS

Suppose we wish to express a function f(x,y) in terms of convenient independent variables ξ and η. If the transformations $x = x(\xi,\eta)$ and $y = y(\xi,\eta)$ are available, direct substitution allows us to rewrite f(x,y) in the form

$$f(x, y) = F(\xi, \eta) \tag{8.1}$$

In Eq. (8.1), the functional relation $F(\xi,\eta)$ between ξ and η is generally different from the relation f(x,y) connecting x and y. Derivatives of f(x,y) with respect to x and y are easily related to derivatives of $F(\xi,\eta)$ taken with respect to ξ and η. By applying the chain rule (Hildebrand, 1948), we have

$$F_\xi = f_x x_\xi + f_y y_\xi \tag{8.2}$$

$$F_\eta = f_x x_\eta + f_y y_\eta \tag{8.3}$$

where subscripts, along with ∂'s, will be used to indicate partial derivatives. Eqs. (8.2), (8.3) can be algebraically solved for f_x and f_y to yield

$$f_x = \left(y_\eta F_\xi - y_\xi F_\eta\right)/J \tag{8.4}$$

$$f_y = \left(x_\xi F_\eta - x_\eta F_\xi\right)/J \tag{8.5}$$

where

$$J(\xi, \eta) = x_\xi y_\eta - x_\eta y_\xi \tag{8.6}$$

is known as the Jacobian of the transformation. For reasons that will be apparent later, we shall refer to this Jacobian as "big jay."

Most boundary value problems occurring in mathematical physics involve second-order differential equations (Tychonov and Samarski, 1964). To express such equations in (ξ,η) coordinates, transformations similar to those in Eqs. (8.4), (8.5) are therefore needed for f_{xx}, f_{xy}, and f_{yy}. Throughout this book, f and F are considered to be sufficiently smooth, so that it is possible to interchange the order of differentiation between any two independent variables. By smooth, we mean that sudden discontinuities are not expected in physical solutions. Application of the chain rule to Eqs. (8.2), (8.3) leads to

$$\begin{aligned} F_{\xi\xi} &= f_x x_{\xi\xi} + x_\xi\left(f_{xx}x_\xi + f_{xy}y_\xi\right) + f_y y_{\xi\xi} + y_\xi\left(f_{yx}x_\xi + f_{yy}y_\xi\right) \\ &= x_{\xi\xi}f_x + y_{\xi\xi}f_y + x_\xi^2 f_{xx} + y_\xi^2 f_{yy} + 2x_\xi y_\xi f_{xy} \end{aligned} \tag{8.7}$$

Similarly,

$$F_{\eta\eta} = x_{\eta\eta}f_x + y_{\eta\eta}f_y + x_\eta^2 f_{xx} + y_\eta^2 f_{yy} + 2x_\eta y_\eta f_{xy} \tag{8.8}$$

and

$$F_{\eta\xi} = x_{\eta\xi}f_x + y_{\eta\xi}f_y + x_\eta x_\xi f_{xx} + y_\eta y_\xi f_{yy} + \left(x_\eta y_\xi + x_\xi y_\eta\right) f_{xy} \tag{8.9}$$

Now, let us rewrite Eqs. (8.7)–(8.9), treating the functions f_{xx}, f_{xy}, and f_{yy} as algebraic unknowns on the left-hand side of a three by three system. That is, we write the foregoing equations in the usual format,

$$x_\xi{}^2 f_{xx} + 2x_\xi y_\xi f_{xy} + y_\xi{}^2 f_{yy} = F_{\xi\xi} - x_{\xi\xi}f_x - y_{\xi\xi}f_y \tag{8.10}$$

$$x_\eta{}^2 f_{xx} + 2x_\eta y_\eta f_{xy} + y_\eta{}^2 f_{yy} = F_{\eta\eta} - x_{\eta\eta} f_x - y_{\eta\eta}f_y \tag{8.11}$$

$$x_\eta x_\xi f_{xx} + \left(x_\eta y_\xi + x_\xi y_\eta\right) f_{xy} + y_\eta y_\xi f_{yy} = F_{\eta\xi} - x_{\eta\xi} f_x - y_{\eta\xi}f_y \tag{8.12}$$

In this form, the solutions for f_{xx}, f_{xy}, and f_{yy} can be easily obtained using determinants. However, we need not write down individual solutions, since we have no use for them in this book. We shall, however, make use of the Laplace operator $f_{xx}+f_{yy}$ which takes the form

$$\begin{aligned} f_{xx} + f_{yy} = {} & \left(\alpha F_{\xi\xi} - 2\beta F_{\xi\eta} + \gamma F_{\eta\eta}\right)/J^2 \\ & + \left[\left(\alpha x_{\xi\xi} - 2\beta x_{\xi\eta} + \gamma x_{\eta\eta}\right)\left(y_\xi F_\eta - y_\eta F_\xi\right)\right. \\ & \left. + \left(\alpha y_{\xi\xi} - 2\beta y_{\xi\eta} + \gamma y_{\eta\eta}\right)\left(x_\eta F_\xi - x_\xi F_h\right)\right]/J^3 \end{aligned} \tag{8.13}$$

where the Greek letter coefficients represent the nonlinear functions

$$\alpha = x_\eta{}^2 + y_\eta{}^2 \tag{8.14}$$

$$\beta = x_\xi x_\eta + y_\xi y_\eta \tag{8.15}$$

$$\gamma = x_\xi{}^2 + y_\xi{}^2 \tag{8.16}$$

THOMPSON'S MAPPING

So far, we have not imposed any constraints on the functions $x = x(\xi,\eta)$ and $y = y(\xi,\eta)$, or their inverses $\xi = \xi(x,y)$ and $\eta = \eta(x,y)$. One well-known transformation is Thompson's mapping, discussed in Thompson (1978, 1984), Thompson et al. (1985), White (1982), and Tamamidis and Assanis (1991). It was originally developed to solve the Navier-Stokes equations for viscous flows past planar airfoils, and later extended to three dimensions to study wing-fuselage effects in aerospace applications (Thomas, 1982b). This method was also used in Chin (1992a,b, 2001a,b, 2012b) to study non-Newtonian flows in eccentric annuli and noncircular pipes. In Thompson's approach, $\xi(x,y)$ and $\eta(x,y)$ are *defined* as solutions to the elliptic equations

$$\xi_{xx} + \xi_{yy} = P^*(\xi, \eta) \tag{8.17}$$

$$\eta_{xx} + \eta_{yy} = Q^*(\xi, \eta) \tag{8.18}$$

where P* and Q* are functions chosen (by very ingenious guess work) to control local grid density. We shall explain the exact motivation behind Thompson's choice later. For now, we ask the more immediate question, "What are the governing equations for $x=x(\xi,\eta)$ and $y=y(\xi,\eta)$, given Eqs. (8.17), (8.18)?"

At this point, it is helpful to understand that Eq. (8.13) holds for *any* function f. That is, for any prescribed set of transformations, Eq. (8.13) can be viewed as a source of useful identities. Let us take $f(x,y)=\xi(x,y)$, in which case $F(\xi,\eta)=\xi$; then, $F_\eta=0$, and all second derivatives of F with respect to ξ and η vanish. Substitution in Eq. (8.13) and replacement of the resulting Laplacian of ξ with respect to x and y using Eq. (8.17) lead to

$$-y_\eta\left(\alpha x_{\xi\xi} - 2\beta x_{\xi\eta} + \gamma x_{\eta\eta}\right) + x_\eta\left(\alpha y_{\xi\xi} - 2\beta y_{\xi\eta} + \gamma y_{\eta\eta}\right) = P^*J^3 \tag{8.19}$$

Similarly, consider $f(x,y)=\eta(x,y)$, so that $F(\xi,\eta)=\eta$. It follows that $F_\xi=0$, and that all second derivatives of F with respect to ξ and η vanish. Substitution in Eq. (8.13) and replacement of the Laplacian of η with respect to x and y using Eq. (8.18) lead to

$$+y_\xi\left(\alpha x_{\xi\xi} - 2\beta x_{\xi\eta} + \gamma x_{\eta\eta}\right) - x_\xi\left(\alpha y_{\xi\xi} - 2\beta y_{\xi\eta} + \gamma y_{\eta\eta}\right) = Q^*J^3 \tag{8.20}$$

If we now regard $(\alpha\, x_{\xi\xi} - 2\beta\, x_{\xi\eta} + \gamma\, x_{\eta\eta})$ and $(\alpha\, y_{\xi\xi} - 2\beta\, y_{\xi\eta} + \gamma\, y_{\eta\eta})$ as algebraic unknowns in a simple two by two system, Eqs. (8.19), (8.20) can be solved, thus yielding Thompson's well-known elliptic equations

$$\alpha x_{\xi\xi} - 2\beta x_{\xi\eta} + \gamma x_{\eta\eta} + J^2\left(P^* x_\xi + Q^* x_\eta\right) = 0 \tag{8.21}$$

$$\alpha y_{\xi\xi} - 2\beta y_{\xi\eta} + \gamma y_{\eta\eta} + J^2\left(P^* y_\xi + Q^* y_\eta\right) = 0 \tag{8.22}$$

Eqs. (8.21), (8.22) are nonlinearly coupled because the coefficients α, β, and γ in Eqs. (8.14)–(8.16) depend on both $x(\xi,\eta)$ and $y(\xi,\eta)$. Their complementary geometric boundary conditions are derived in Chapter 9.

SOME RECIPROCITY RELATIONS

For practical reasons, we need to convert results between physical and computational planes. Thus, reciprocity relationships are needed. Let us return to general considerations and for now refrain from invoking Thompson's assumptions. In particular, we examine the general transformations

$$x = x(\xi, \eta) \tag{8.23}$$

$$y = y(\xi, \eta) \tag{8.24}$$

From calculus, the total differentials dx and dy are given by

$$x_\eta d\eta + x_\xi d\xi = dx \tag{8.25}$$

$$y_\eta d\eta + y_\xi d\xi = dy \tag{8.26}$$

Eqs. (8.25), (8.26) can be solved in terms of $d\xi$ and $d\eta$, thus leading to

$$d\eta = - y_\xi dx/J + x_\xi dy/J \tag{8.27}$$

$$d\xi = + y_\eta dx/J - x_\eta dy/J \tag{8.28}$$

where the "big jay" Jacobian is given by Eq. (8.6). Now, we can similarly consider the inverse transformation. If we write

$$\eta = \eta(x, y) \tag{8.29}$$

$$\xi = \xi(x, y) \tag{8.30}$$

it follows that

$$d\eta = \eta_x dx + \eta_y dy \tag{8.31}$$

$$d\xi = \xi_x dx + \xi_y dy \tag{8.32}$$

Comparison of Eq. (8.27) with Eq. (8.31), and Eq. (8.28) with Eq. (8.32), leads to

$$\eta_x = - y_\xi / J \tag{8.33}$$

$$\eta_y = x_\xi / J \tag{8.34}$$

$$\xi_x = y_\eta / J \tag{8.35}$$

$$\xi_y = - x_\eta / J \tag{8.36}$$

On the other hand, we might have proceeded from the definitions for the total differentials $d\xi$ and $d\eta$, and reconsidered Eqs. (8.31), (8.32) in the form

$$\eta_x dx + \eta_y dy = d\eta \tag{8.37}$$

$$\xi_x dx + \xi_y dy = d\xi \tag{8.38}$$

Eqs. (8.37), (8.38) can be solved algebraically for dx and dy to give

$$dx = - \xi_y d\eta/j + \eta_y d\xi/j \tag{8.39}$$

$$dy = + \xi_x d\eta/ j - \eta_x d\xi /j \tag{8.40}$$

where the "little jay" Jacobian satisfies

$$j(x, y) = \xi_x \eta_y - \xi_y \eta_x \tag{8.41}$$

Comparison of Eq. (8.25) with Eq. (8.39), and Eq. (8.26) with Eq. (8.40), leads to

$$x_\eta = -\xi_y / j \tag{8.42}$$

$$x_\xi = \eta_y / j \tag{8.43}$$

$$y_\eta = \xi_x / j \tag{8.44}$$

$$y_\xi = -\eta_x / j \tag{8.45}$$

Finally, comparison of Eq. (8.33) with Eq. (8.45), Eq. (8.34) and Eq. (8.43), Eq. (8.35) and Eq. (8.44), and Eq. (8.36) and Eq. (8.42), leads to

$$J(\xi, \eta) j(x, y) = 1 \tag{8.46}$$

or

$$\left(x_\xi y_\eta - x_\eta y_\xi \right) \left(\xi_x \eta_y - \xi_y \eta_x \right) = 1 \tag{8.47}$$

It is important to understand that the equations obtained in this section are generally valid, regardless of Thompson's (or any other) set of transformations. They allow us to move conveniently between quantities expressed in the physical (x,y) and computational (ξ,η) planes. Incidentally, the derivation of the above reciprocity relations parallels that for the transonic hodograph equations in classical aerodynamics (Liepmann and Roshko, 1957).

CONFORMAL MAPPING REVISITED

In Chapter 5, conformal mapping was introduced as a vehicle that transforms simple solutions into those for flows past complicated shapes. Here, we explore its general transformation properties, and attempt to understand conformal mapping from a mathematical viewpoint. We *now* formally reintroduce the Cauchy-Riemann conditions, that is,

$$\xi_x = \eta_y \tag{8.48}$$

$$\eta_x = -\xi_y \tag{8.49}$$

Let us differentiate Eq. (8.48) with respect to x and Eq. (8.49) with respect to y; elimination of the cross-derivative term between the two results leads to Eq. (8.50). A similar procedure yields Eq. (8.51).

$$\xi_{xx} + \xi_{yy} = 0 \tag{8.50}$$

$$\eta_{xx} + \eta_{yy} = 0 \tag{8.51}$$

Eqs. (8.50), (8.51) are both elliptic; they are, in fact, *exactly* Thompson's equations (8.17) and (8.18), however, with $P^* = Q^* = 0$. (Eqs. 8.48–8.51 are identical to Eqs. 5.12–5.15.) Since ξ(x,y) and η(x,y) satisfy Laplace's equation, they are said to be harmonic. And because harmonic functions are obtained as real and imaginary parts of complex analytical functions, as we had demonstrated, Eqs. (8.50), (8.51) are usually derived more elegantly using complex variables methods, as was the case in Eqs. (5.22)–(5.29).

To understand the implications of Eqs. (8.48), (8.49) in transformed coordinates, it is helpful to use the reciprocity relations developed previously. If the ξ_x and η_y in Eq. (8.48) are replaced by their equivalents using Eqs. (8.43), (8.44), and if η_x and ξ_y in Eq. (8.49) are replaced by their equivalents using Eqs. (8.42), (8.45), we obtain

$$y_\eta = x_\xi \tag{8.52}$$

$$y_\xi = -x_\eta \tag{8.53}$$

which imply, using the same procedure we have described, that

$$x_{\xi\xi} + x_{\eta\eta} = 0 \tag{8.54}$$

$$y_{\xi\xi} + y_{\eta\eta} = 0 \tag{8.55}$$

Thus, $x(\xi,\eta)$ and $y(\xi,\eta)$ are likewise harmonic, but in the variables ξ and η. Eqs. (8.54), (8.55) are simpler than Eqs. (8.21), (8.22), with $P^* = Q^* = 0$. The use of our reciprocity relationships shows that there exists a duality between physical and mapped planes, and *vice versa*, for conformal transformations; that is, Eqs. (8.50), (8.51) are mirror images of Eqs. (8.54), (8.55). One might have anticipated this type of reversibility, but it is not directly evident from Eqs. (8.21), (8.22). Eqs. (8.54), (8.55) are consistent with Thompson's original equations (8.21) and (8.22). Use of the Cauchy-Riemann conditions in the transformed plane, that is, Eqs. (8.52), (8.53), in Eqs. (8.14)–(8.16), leads to $\alpha = \gamma$ and $\beta = 0$.

In this book, our discussions of grid generation include derivations for results of broad theoretical interest; but due to space constraints and the research nature of ongoing work, our applications will be restricted to $P^* = Q^* = 0$. For further information, consult Thomas and Middlecoff (1980), Thompson (1984), and Thompson et al. (1985). We emphasize that Eqs. (8.54), (8.55) are linear, unlike Eqs. (8.21), (8.22). However, they do *not* generally uncouple for true conformal mappings, as they might superficially suggest, since x and y cannot be arbitrarily specified along boundaries: to be conformal, x and y must satisfy Eqs. (8.52), (8.53) everywhere.

SOLUTION OF MESH GENERATION EQUATIONS

In this section, we show how the geometrical transformations derived above are actually used to solve boundary value problems (e.g., Laplace's equation on an irregular domain). The numerical solutions to Thompson's equations and to the steady-state pressure equation are new. This research was supported by the United States Department of Energy under Small Business Innovation Research Grant DE-FG03-99ER82895, and the algorithms are presently used in computational rheology work for non-Newtonian pipeline and annular flows (Chin, 2012b). In order to explain the problems and issues clearly, we shall avoid the formal mathematical approach used in earlier sections.

Commercial reservoir simulators calculate pressures, saturations, and other flow properties on rectangular grids. Again, their x-y coordinate lines do not conform to the irregular curves defining farfield reservoir boundaries; also, high grid densities imposed near wells require similarly high densities far away, where such resolution is unnecessary. This results in large, inefficient computing domains containing dead flow and extremely large matrices. Sometimes, coarse meshes are used everywhere, together with high-density corner point modeling to provide grid refinement close to (possibly fractured) wells. However, many companies refrain from their usage because cross-derivative terms in the transformed flow equations, which increase computing time, are ignored in the matrix inversion for numerical expediency!

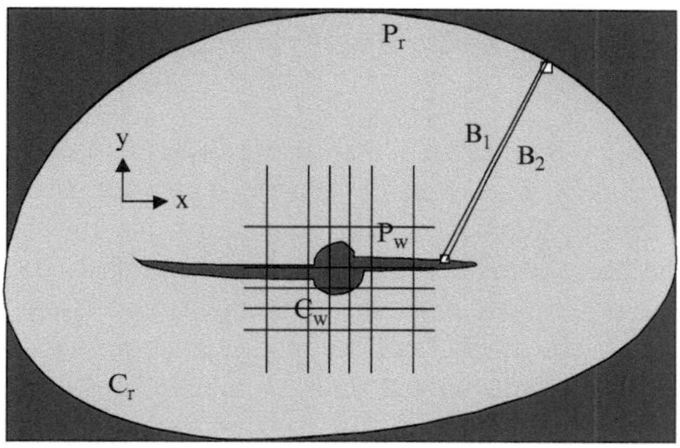

FIG. 8.1 Irregular domain with inefficient rectangular meshes.

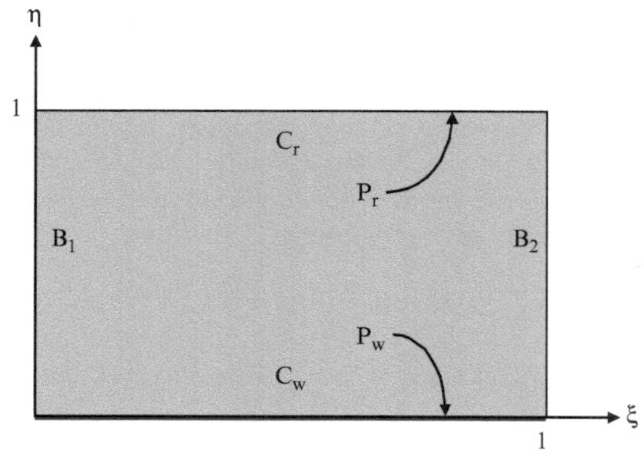

FIG. 8.2 Irregular domain mapped to rectangular ξ-η computational space.

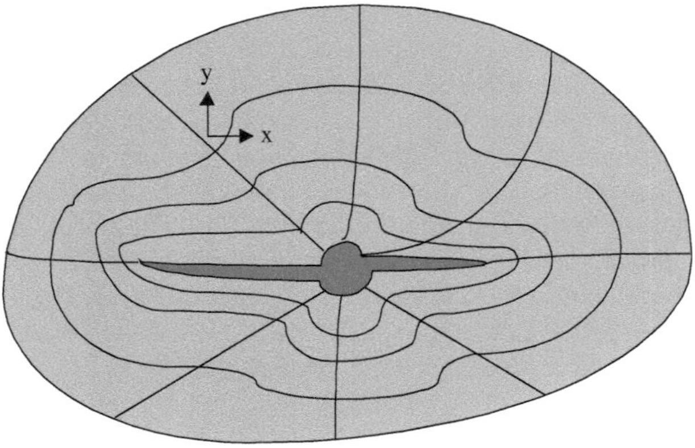

FIG. 8.3 Physical domain in boundary-conforming coordinates.

Boundary Conditions

Although the industry focuses on Cartesian meshes, more effective boundary-conforming, curvilinear grids *can* be generated, as discussed in Thompson et al. (1985), adapting to both farfield boundaries *and* individual wells. We now reiterate the basic ideas because they are essential to understanding our research contributions, but we focus on the boundary conditions needed to supplement Thompson's equations. Let us suppose that a transformation $\xi = \xi(x,y)$, $\eta = \eta(x,y)$ exists which maps the irregular domain defined by the general well and farfield reservoir boundaries of Fig. 8.1 into the rectangle of Fig. 8.2.

Physically insignificant branch cuts B_1 and B_2 have been introduced, which will be discussed. Such a mapping effectively allows calculations to be performed on more desirable high-resolution grids such as the one in Fig. 8.3. It is clear that more meaningful flow models can be formulated using "ξ, η" coordinates; improved reservoir description is possible, with fewer grids and less matrix inversion. Now, it is known from complex variables that conformal transformations satisfy linear Laplace equations in x and y, but Riemann's Lemma unfortunately does not explain *how* the mappings are obtained.

Thompson, again, developed a novel approach. Rather than dealing with $\xi = \xi(x,y)$ and $\eta = \eta(x,y)$ directly, he (equivalently) considered the inverse functions $x = x(\xi,\eta)$ and $y = y(\xi,\eta)$ satisfying *nonlinear* coupled Eqs. (8.21), (8.22), for the purposes of this chapter, considered in the form

$$\left(x_\eta{}^2 + y_\eta{}^2\right) x_{\xi\xi} - 2\left(x_\xi x_\eta + y_\xi y_\eta\right) x_{\xi\eta} + \left(x_\xi{}^2 + y_\xi{}^2\right) x_{\eta\eta} = 0 \tag{8.56}$$

$$\left(x_\eta{}^2 + y_\eta{}^2\right) y_{\xi\xi} - 2\left(x_\xi x_\eta + y_\xi y_\eta\right) y_{\xi\eta} + \left(x_\xi{}^2 + y_\xi{}^2\right) y_{\eta\eta} = 0 \tag{8.57}$$

where ξ and η are independent variables. How are these used to create mappings? Suppose that contour C_W in Fig. 8.1 is to map into $\eta = 0$ of Fig. 8.2.

The user first discretizes C_W in Fig. 8.1 by penciling along it a sequence of dots chosen to represent the curve. If these are selected in an orderly, say, clockwise fashion, they define the direction in which ξ increases. Along $\eta = 0$, values of x and y are known (e.g., from measurement on graph paper) as functions of ξ. Similarly, x and y values along C_r are known as functions of ξ on $\eta = 1$ of Fig. 8.2. These provide the boundary conditions for Eqs. (8.56), (8.57), which are augmented by single-valuedness constraints at arbitrarily chosen branch cuts B_1 and B_2.

In Thompson's and similar approaches, Eqs. (8.56), (8.57) are discretized by finite differences and solved by point or line relaxation (e.g., see Chapter 7), starting with guesses for the dependent variables x and y. The problem is linearized by approximating all nonlinear coefficients using values from earlier iterations. Typically, several updates to Eq. (8.56) are taken, followed by updates to Eq. (8.57), with this cycling process, often unstable, repeated until convergence. Variations of the approach are known, with 100×100 mesh systems in the ξ-η plane requiring minutes of computing time on typical Intel computers. Once $x = x(\xi,\eta)$ and $y = y(\xi,\eta)$ are solved and tabulated as functions of ξ and η, physical coordinates are generated. First, η is fixed; for each node ξ along this η, computed values of (x,y) pairs are successively plotted in the x-y plane to produce the required closed contour. This procedure is repeated for all values of η, until the entire family of closed curves is obtained, with limit values $\eta = 0$ and $\eta = 1$ again describing C_w and C_r. Orthogonals are constructed by repeating the procedure, with η and ξ roles reversed.

This process provides the mapping only. The equation describing the physics (e.g., the Navier-Stokes equation) must be transformed into (ξ,η) coordinates and solved. For instance, in reservoir simulation, Darcy's pressure equation must be expressed in terms of ξ,η and solved. Thompson's simplification lies not in the transformed equation, which may contain mixed derivatives and variable coefficients, but in the computational domain itself, because it takes on a rectangular form amenable to simple numerical solution.

Fast Iterative Solutions

Existing solution methods solving $x(\xi,\eta)$ and $y(\xi,\eta)$ stagger the solutions for Eqs. (8.56), (8.57). For example, crude solutions are used to initialize the coefficients of Eq. (8.56), and improvements to $x(\xi,\eta)$ are obtained. These are used to evaluate the coefficients of Eq. (8.57), in order to obtain an improved $y(\xi,\eta)$; then, attention turns to Eq. (8.56) again, and so on, until convergence is achieved. Various means are used to implement these iterations, as noted in the review paper of Thompson (1984), for example, point SOR, line SLOR, line SOR with explicit damping, alternating-direction-implicit, and multigrid, with varying degrees of success. Often these schemes diverge computationally. In any event, the staggering noted earlier introduces different artificial time levels while iterating. However, classic numerical analysis suggests that faster convergence and improved stability are possible by reducing their number.

A new approach to rapidly solve Thompson's equations was proposed by this author and based on a very simple idea. This idea has since been validated. Consider $z_{\xi\xi} + z_{\eta\eta} = 0$, for which $z_{i,j} \approx (z_{i-1,j} + z_{i+1,j} + z_{i,j-1} + z_{i,j+1})/4$ holds on constant grid systems (e.g., as derived in Chapter 7 for real functions of two variables). This well-known averaging law motivates the *recursion formula* $z_{i,j}{}^{n} = (z_{i-1,j}{}^{n-1} + z_{i+1,j}{}^{n-1} + z_{i,j-1}{}^{n-1} + z_{i,j+1}{}^{n-1})/4$ often used to illustrate and develop multilevel iterative solutions; an approximate, and even trivial solution, can be used to initialize the calculations, and nonzero solutions are always produced from nonzero boundary conditions.

But the well-known Gauss-Seidel method is fastest: as soon as a new value of $z_{i,j}$ is calculated, its previous value is discarded and overwritten by the new value. This speed is accompanied by low memory requirements, since there is no need to store both n and n-1 level solutions: only a single array, $z_{i,j}$ itself, is required in programming. Our approach to Eqs. (8.56), (8.57) was motivated by the following idea. Rather than solving for $x(\xi,\eta)$ and $y(\xi,\eta)$ in a staggered, leap-frog manner, is it possible to *simultaneously* update x and y in a similar once only manner? Are convergence rates significantly increased? What formalism permits us to solve in Gauss-Seidel fashion? What are the programming implications?

Complex variables are used in harmonic analysis problems; for example, the real and imaginary parts of an analytical function f(z), where $z = x + i\, y$, provide solutions satisfying Laplace's equation. Here we use complex analysis differently. We *define* a dependent variable z by $z(\xi,\eta) = x(\xi,\eta) + i\, y(\xi,\eta)$, and then add Eq. (8.56) plus i times Eq. (8.57), in order to obtain the result $(x_\eta{}^2 + y_\eta{}^2)\, z_{\xi\xi} - 2(x_\xi x_\eta + y_\xi y_\eta)\, z_{\xi_\eta} + (x_{\xi^2} + y_{\xi^2}) z\eta\eta = 0$. Now, the complex conjugate of z is $z^*(\xi,\eta) = x(\xi,\eta) - i\, y(\xi,\eta)$, from which we find that $x = (z + z^*)/2$ and $y = -i\,(z - z^*)/2$. Substitution produces the simple and equivalent one equation result

$$(z_\eta\, z_\eta{}^*)\, z_{\xi\xi} - (z_\xi z_\eta{}^* + z_\xi{}^* z_\eta)\, z_{\xi_\eta} + (z_\xi\, z_\xi{}^*)\, z_{\eta\eta} = 0 \qquad (8.58)$$

This form yields significant advantages. First, when z is declared as a complex variable in a Fortran program, Eq. (8.58) represents, for all practical purposes, a *single* equation in $z(\xi,\eta)$. There is no need to leap-frog between x and y solutions now, since a single formula analogous to the classical model $z_{i,j} = (z_{i-1,j} + z_{i+1,j} + z_{i,j-1} + z_{i,j+1})/4$ is easily written for the $z_{i,j}$ related to Eq. (8.58) using second-order central differences. Because both x and y are simultaneously resident in computer memory, the extra time level present in staggered schemes is completely eliminated, as in the Gauss-Seidel method. In hundreds of test simulations conducted using point and line relaxation, convergence times are shorter by factors of two to three, with convergence rates far exceeding those obtained for cyclic solutions between $x(\xi,\eta)$ and $y(\xi,\eta)$. Convergence appears to be unconditional, monotonic, and stable. Because Eq. (8.58) is nonlinear, von Neumann tests for exponential stability and traditional estimates for convergence rate do not apply, but the evidence for stability and convergence, while empirical, remains very strong and convincing.

Fast Solutions for Reservoir Pressure

The new approach implies very rapid generation of irregular, boundary-conforming, curvilinear grids that better accommodate the physics, and is, in itself, significant. The effects of complicated well/fracture contours and outer boundaries can now be modeled very accurately. But unlike rectangular systems, fewer grids not only provide better resolution: matrices are smaller, and at the same time, the above Gauss-Seidel-like scheme accelerates their inversion. Sophisticated meshes are simple to create, and refinements to them are no more complicated.

However, the best is yet to come: solutions to several classes of steady-state problems with different boundary conditions are automatic and free in a literal sense! Solutions to numerous practical problems (e.g., flows of liquids and gases, having general exponent "m") can be solved in the field, with little computational power. In the aerospace industry, the $x(\xi,\eta)$ and $y(\xi,\eta)$ define coordinates that might host solutions to the Navier-Stokes equations. In petroleum engineering, the grid in Fig. 8.3 supports calculations for properties like pressure. To keep the ideas simple, let us consider liquid Darcy flows satisfying $p_{xx}+p_{yy}=0$, which commercial simulators discretize and solve on variable x-y grids, subject to boundary conditions for pressure and rate. Conventionally, grid generation is followed by pressure analysis: *first* create the (rectangular) grid, *then* obtain pressure. But this is unnecessary. Under the assumed transformation, our $p_{xx}+p_{yy}=0$ becomes $p_{\xi\xi}+p_{\eta\eta}=0$ for $p(\xi,\eta)$. However, we do *not* need to numerically solve for $p(\xi,\eta)$, because an analytical solution is easily obtained in terms of the metrics of the mapping already available. If well and farfield reservoir pressures $p_{well}(t)$ and $p_{res}(t)$ are prescribed at $\eta=0$ and 1, the required solution is just

$$p(\xi,\eta;t) = (p_{res}\text{-}p_{well})\eta(x,y) + p_w \tag{8.59}$$

which is a linear function of $\eta(x,y)$ alone! In other words, once $x(\xi,\eta)$ and $y(\xi,\eta)$ are available and inverted (by table) to give $\xi=\xi(x,y)$ and $\eta=\eta(x,y)$, the solution to the pressure-pressure problem is available by *rescaling* using Eq. (8.59). There is *never* a need to solve for pressure since our grid generation problem is formulated to make steady-state pressure solution unnecessary! Note that the time dependence given in Eq. (8.59) does not refer to compressibility transients, but instead, to possible changes in $p_{well}(t)$ and $p_{res}(t)$ during drilling or by injection elsewhere in the reservoir.

In the next chapter, we shall show how solutions to pressure and flow rate boundary value problems, when either well pressure or farfield pressure is given, are also automatically generated. That is, *a single procedure* provides the complete range of pressure solutions whenever any two of the total flow rate

Q, and the pressures p_{res} and p_{well} are prescribed, and furthermore, for both liquids as well as gaseous fluids. In addition, we shall draw on our topological findings to extend the notion of the elementary logarithmic and arc tan solutions (for cylindrical coordinates) to arbitrary geometric domains, for example, a reservoir having the shape of Texas! Applications to transient compressible flows, in which curvilinear grids are combined with alternating direction implicit numerical integration schemes, are given in Chapter 10.

PROBLEMS AND EXERCISES

1. Use central differencing to discretize the grid generation formulation in Eq. (8.58), and develop a recursion formula (based on line or point relaxation) for the complex variable z, as described in the text. Following Chapter 7, develop a Fortran or C program to solve this complex equation iteratively, assuming the near and farfield boundary conditions discussed. As a program test, map the annular domain defined by a circular well in a concentric reservoir to a rectangular box. Sketch both computed sets of grid lines. Are they orthogonal? Should they be? What are the advantages of orthogonality, if any? Then, use Eq. (8.59) to compute steady-state pressure as a function of the radial distance r, and compare your results with the logarithmic solution derived analytically in Chapter 6.
2. Using the program developed above, consider a Texas shaped reservoir with a circular well located in "Houston," as shown in the text-based sketch below at the left. Assuming the boundary points suggested in the figure, show that a curvilinear grid can be obtained that takes the more attractive form presented at the right. Then, solve the general pressure-pressure boundary value problem by implementing Eq. (8.59) in a subroutine. Output pressure contours can be displayed in different colors using new features available with the Compaq Digital Fortran 6.5 compiler.

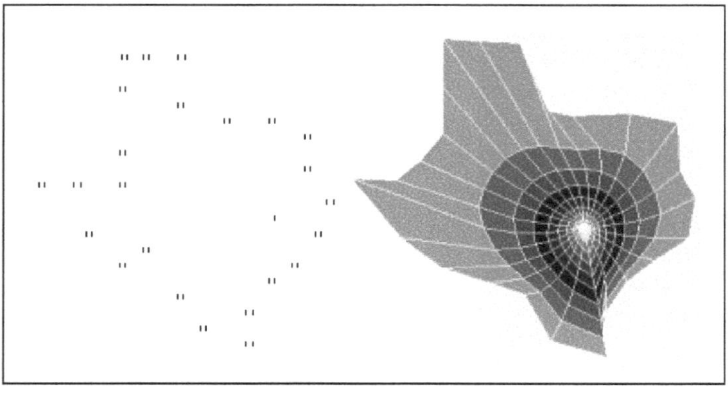

3. As a comparison, solve for the Texas pressure above with rectangular (x,y) grids, using the Fortran programs given in Chapter 7. Describe the nature of your solution in the vicinity of "Houston." Does pressure vary logarithmically near the well? How accurate is the curvilinear grid solution? How fine must the rectangular grid be in order to approach the accuracy afforded by curvilinear mesh systems?

Chapter 9

Steady-State Reservoir Applications

OVERVIEW

In this chapter, we shall transform the Darcy equations for transient compressible liquid and gas flows into curvilinear coordinates. For simplicity, the transforms are applied in the horizontal areal plane only, while the vertical coordinate is left unstretched. This enables us to keep the exposition simple and free of unnecessary algebra, at the same time, allowing us to address some important issues. Once the equations are obtained, we show how the fundamental log r and θ elementary solutions obtained for pure radial flow can be extended to arbitrary wells and fractures in reservoirs with general outer boundaries … *a major breakthrough*. Mathematical analogues to these elementary solutions can be obtained for any particular reservoir geometry, once and for all, and can then be used to solve large supersets of steady-state problems, following the ideas developed in Chapter 6 for radial flows.

Three Motivating Pressure Problems

In developing any new idea, it is useful to reexamine older, similar and simpler analogies that directly motivate the mathematics. In reservoir simulation, the fundamental governing PDE is Laplace's equation for pressure, assuming constant density, single-phase liquids in homogeneous, isotropic formations. This takes different forms depending on the coordinate system used. Consider, for example, linear Darcy flow. Here, the equation $d^2p(x)/dx^2 = 0$ holds. If $p = p_w$ at the well $x = 0$, and $p = p^*$ at the reservoir boundary $x = x^*$, the solution is the well-known linear pressure drop

$$p(x) = p_w + (x/x^*)(p^* - p_w) \tag{9.1a}$$

On the other hand, for radial flows with cylindrical symmetry, the pressure p(r) satisfies $d^2p(r)/dr^2 + 1/r\ dp/dr = 0$. If $p = p_w$ at the well $r = r_w$, and $p = p^*$ at the reservoir boundary $r = r^*$, the solution is the logarithmic pressure drop

$$p(r) = p_w + \{(\log r/r_w)/(\log r^*/r_w)\}\ (p^* - p_w) \tag{9.1b}$$

Quantitative Methods in Reservoir Engineering. http://dx.doi.org/10.1016/B978-0-12-810518-4.00009-8

For spherically symmetric flows, p(r) satisfies $d^2p(r)/dr^2 + 2/r\, dp/dr = 0$, where r is the spherical radial coordinate. If $p = p_w$ at the well $r = r_w$, and $p = p^*$ at the reservoir boundary $r = r^*$, the solution is the algebraic pressure drop

$$p(r) = p_w + \{r^*(1 - r_w/r)/(r^* - r_w)\}\,(p^* - p_w) \qquad (9.1c)$$

Reservoir Simulation as a Topology Problem

If we compare Eqs. (9.1a)–(9.1c), one might go so far as to speculate that

$$p(r) = p_w + GF\,(p^* - p_w) \qquad (9.1d)$$

can represent the general solution of a more complete Dirichlet problem. Here GF is a *simple* geometric factor, whose existence is contingent upon the availability of a *single* coordinate variable to represent distances measured from a suitable origin (this is *not* the geometric factor used in induction logging). Such a representation, if possible and practical, has the potential of rendering the most complicated geometries amenable to simple analysis. We can, in this sense, view the pressure problem as the search for an appropriate GF, and fluid flow analysis literally becomes a topological problem.

The practical benefit one incurs from this line of thinking is twofold. First, once the GF is obtained for any particular flow domain (e.g., linear, cylindrical, spherical, *or whatever*), the complete superset of flow boundary value problems (following Chapter 6) should be expressible in terms of GF alone, which is obtained just once and for all. Second, the solution for the relevant GF, as suggested in the foregoing examples, should follow from ordinary, and not difficult partial differential equations. So far, we have discussed the pressure solution only as it depends on GF, but similar considerations apply to streamline tracing. For example, the streamfunction in a linear flow is simply y, whereas that in a cylindrical radial flow is θ; these simple dependences are again the direct consequence of having available the canonical Cartesian and polar coordinates natural to the problem. We shall show that general pressure solutions depend on the function $\eta(x,y)$ alone, and also that streamfunction calculations can be considerably simplified.

A Practical Problem

To put these abstractions in down-to-earth form, consider the following hypothetical Oil Patch example. Aerial reconnaissance reveals a large, irregular petroleum reservoir in the size and shape of Texas. It is not known whether the formation contains liquid or gaseous hydrocarbons, so that the exponent m must be left as a "what if" parameter. The operating company decides to drill its exploratory well in "Houston." It would like to solve, for the purposes of production and formation evaluation, the following three boundary value problems, pressure-pressure, well pressure-flow rate, and farfield pressure-flow rate, following the discussion in Chapter 6, for all possible values of m. Normally, this would require numerous fine grid, rectangular mesh, and approximate computer simulations, repeated for large ensembles of parameters varied in nested do-loops. With the formalism developed in this chapter, the geometric factor GF is obtained once and for all, in seconds and stored in computer memory. Note that the GF is an array of real numbers, just as log r and θ also are arrays of numbers. Each of the boundary value problems, then, can be solved simply by rescaling this array of numbers, just as Eqs. (9.1a)–(9.1c) also represent simple renormalizations.

GOVERNING EQUATIONS

To fix ideas, we shall formulate the transformed problem for a single-phase, compressible flow with a constant horizontal permeability kh in the areal (x,y) coordinates, and a variable permeability kv(z) in the layered, vertical z direction. The matrix porosity φ(z) may vary with z. The fluid viscosity μ is assumed to be constant. The governing equation then takes the following form,

$$\begin{aligned} kh\left(\partial^2 p^{m+1}/\partial x^2 + \partial^2 p^{m+1}/\partial y^2\right) + \partial\left(kv\,\partial p^{m+1}/\partial z\right)/\partial z \\ = \phi\,\mu\,c^*\,\partial p^{m+1}/\partial t \end{aligned} \tag{9.2a}$$

which is a special case of Eq. (1.6). Here p(x,y,z,t) is the pressure in (x,y,z) Cartesian coordinates, and t is time. The exponent m is zero for liquids, while gases are modeled by nonzero values. For example, m = 1 for isothermal expansion, while m = Cv/Cp for adiabatic expansion. (Cv and Cp are specific heats at constant volume and pressure, respectively.) The reader should refer to Eq. (1.8) and, for example, to Saad (1966) for details. When m = 0, c* represents the net compressibility of the fluid and underlying rock; for nonzero values of m, c*(p) is variable with

$$c^*(p) = m/p \tag{9.2b}$$

This casts Eq. (9.2a) in a linear-like form for p^{m+1}, which facilitates numerical analysis, noting that full nonlinearity is assumed in this book. Now, let us consider the function $p^{m+1}(x,y,z,t)$ and its image $P^{m+1}(\xi,\eta,z,t)$ in transformed curvilinear (ξ,η,z) coordinates. Such transformed coordinates are often called corner point geometries in the reservoir simulation literature for their ability to adapt to stratigraphic boundaries. In particular, we shall examine the combination $\partial^2 p^{m+1}/\partial x^2 + \partial^2 p^{m+1}/\partial y^2$, keeping the function p^{m+1} together as a whole. That is, if

$$P^{m+1}(\xi, \eta, z, t) = p^{m+1}(x, y, z, t) \tag{9.3}$$

then the first derivative with respect to ξ transforms according to

$$P^{m+1}{}_{\xi} = p^{m+1}{}_{x} x_{\xi} + p^{m+1}{}_{y} y_{\xi} \tag{9.4}$$

and

$$P^{m+1}_{\xi\xi} = + x_{\xi\xi}p^{m+1}_x + y_{\xi\xi}p^{m+1}_y + x_\xi{}^2 p^{m+1}{}_{xx} + y_\xi{}^2 p^{m+1}{}_{yy} + 2x_\xi y_\xi p^{m+1}{}_{xy} \tag{9.5}$$

Similarly,

$$P^{m+1}{}_{\eta\eta} = + x_{\eta\eta}p^{m+1}{}_x + y_{\eta\eta}p^{m+1}{}_y + x_\eta{}^2 p^{m+1}{}_{xx} + y_\eta{}^2 p^{m+1}{}_{yy} + 2\, x_\eta y_\eta p^{m+1}{}_{xy} \tag{9.6}$$

It is important to emphasize how additional first-order and second-order mixed cross-derivatives are produced by general transformations (compare Eqs. 9.2a, 9.7). The latter terms are just as important as the unmixed second derivatives and must be accounted for in numerical analysis. This is not so with several widely used simulators. Quoting from one oil company report, "The discretization of these terms is neglected in several [commercial reservoir] simulators because their structure is not amenable to solution by existing direct matrix solvers." For this reason, "finite difference solutions may not converge to the solution of the flow equations, even with refinement of the grid." The reader should exercise caution in his use of corner point options. Fortunately, this is not the case with public domain software available in the aerospace industry; access to a wide body of literature on grid generation, including documented Fortran algorithms, is available from government research agencies for a nominal fee. Our emphasis is primarily analytical, and we return to the basic equations. The addition of Eqs. (9.5), (9.6) leads to a complicated equation, namely,

$$P^{m+1}{}_{\xi\xi} + P^{m+1}{}_{\eta\eta} = (x_{\xi\xi} + x_{\eta\eta})\, p^{m+1}{}_x + \left(y_{\xi\xi} + y_{\eta\eta}\right) p^{m+1}{}_y + \left(x_\xi{}^2 + x_\eta{}^2\right)p^{m+1}{}_{xx} + \left(y_\xi{}^2 + y_\eta{}^2\right) p^{m+1}{}_{yy} + 2\left(x_\eta y_\eta + x_\xi y_\xi\right) p^{m+1}{}_{xy} \tag{9.7}$$

At this point, we introduce Thompson's ideas, with $P^* = Q^* = 0$, but use the simplified, equivalent transformations derived in Chapter 8. Eqs. (8.54), (8.55) require that the first derivative terms disappear. The p_{xy} cross-derivative term drops out because the combination $x_\eta y_\eta + x_\xi y_\xi$ vanishes identically. Since Eqs. (8.52), (8.53) also imply that

$$J(\xi, \eta) = x_\xi y_\eta - x_\eta y_\xi = x_\eta{}^2 + x_\xi{}^2 = y_\eta{}^2 + y_\xi{}^2 > 0 \tag{9.8}$$

it follows that

$$P^{m+1}{}_{\xi\xi} + P^{m+1}{}_{\eta\eta} = J(\xi, \eta)\left(p^{m+1}{}_{xx} + p^{m+1}{}_{yy}\right) \tag{9.9}$$

Hence, we have, using Eq. (9.2a),

$$kh\left\{\left(P^{m+1}{}_{\xi\xi} + P^{m+1}{}_{\eta\eta}\right)/J(\xi, \eta)\right\} + \partial\left(kv\, \partial P^{m+1}/\partial z\right)/\partial z = \phi\mu c^*\, \partial P^{m+1}/\partial t \tag{9.10}$$

or

$$kh\left(P^{m+1}{}_{\xi\xi} + P^{m+1}{}_{\eta\eta}\right) + J(\xi,\eta)\,\partial\left(kv\,\partial P^{m+1}/\partial z\right)/\partial z = J(\xi,\eta)\phi\mu c^*\,\partial P^{m+1}/\partial t \tag{9.11}$$

STEADY AREAL FLOW: GENERALIZED "LOG R" SOLUTION

Following Chapter 6 for radial flows, we consider three problems relevant to drilling and formation evaluation: the pressure-pressure and two pressure-flow rate models. Again, in steady flows of liquids and gases, total volume flow rate is a *constant* of the flow. We consider homogeneous, isotropic systems.

Pressure-Pressure Formulations

In the steady, areal flow limit, both time and vertical (z) derivatives vanish, leaving a transformed pressure $P(\xi,\eta)$ whose (m+1)th power satisfies Laplace's equation,

$$P^{m+1}{}_{\xi\xi} + P^{m+1}{}_{\eta\eta} = 0 \tag{9.12}$$

We suppose that the reservoir in Fig. 9.1A can be mapped into the rectangle of Fig. 9.1B as shown; that is, we assume that Eqs. (8.54), (8.55) can be solved along with suitable boundary conditions to produce the mappings indicated. Once the algorithm has generated the transformations $\xi = \xi(x,y)$ and $\eta = \eta(x,y)$, these are stored in memory. At this point, we have solved a purely topological problem, without considering physical boundary conditions. Now, according to Fig. 9.1A and B, the well pressure P_W maps into the lower boundary $\eta = 0$, while the farfield pressure P_R maps into $\eta = \eta_{max}$ (η_{max} and ξ_{max} can be arbitrarily chosen). Since these pressures are constants, the boundary value problem specified by Eq. (9.12) is rendered independent of the coordinate ξ. The PDE becomes an ordinary differential equation. In fact, the quantity $P^{m+1}(\eta)$ varies linearly with η since $P^{m+1}{}_{\eta\eta} = 0$, taking the value

$$P^{m+1}(\eta) = \left(P_R{}^{m+1} - P_W{}^{m+1}\right)\eta/\eta_{max} + P_W{}^{m+1} \tag{9.13}$$

Hence, the solution $P(\eta)$ is obtained as

$$P(\eta) = \left\{\left(P_R{}^{m+1} - P_W{}^{m+1}\right)\eta/\eta_{max} + P_W{}^{m+1}\right\}^{1/(m+1)} \tag{9.14}$$

Let us summarize our results. When the radial-like function $\eta(x,y)$ is known for a given reservoir, the pressure $P(\eta)$ is available analytically as a function of P_W, P_R, m, and $\eta(x,y)$, thus solving a family of boundary value problems. Eq. (9.14) is the analogy, for general reservoirs, to the well-known x pressure variation for lineal flow or the log r solution of radial flow. The mapping functions $\xi(x,y)$ and $\eta(x,y)$ are obtained *once and for all* for a reservoir, and $\eta(x,y)$ may be viewed as a composite log radial coordinate. The solution to Eqs. (8.54), (8.55) yields $x = x(\xi,\eta)$ and $y = y(\xi,\eta)$, and not $\xi = \xi(x,y)$ and $\eta = \eta(x,y)$. But the inversion is easily accomplished as follows. Select in the (ξ,η) plane *any* point, and record its grid indexes and corresponding η and ξ values; then look up the computed (x,y) values for this point, and assign to this set of (x,y)'s the known values of ξ and η. This is repeated for all (ξ,η) points in the rectangular computational plane.

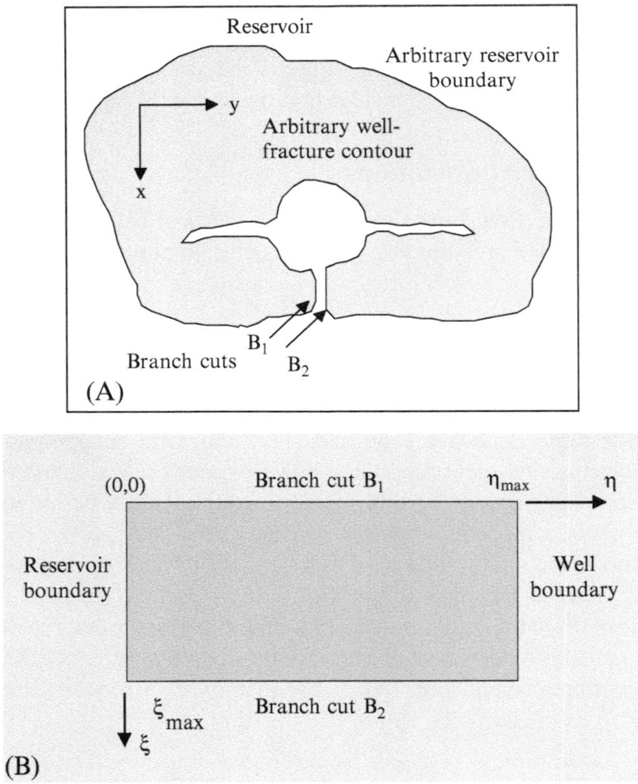

FIG. 9.1 (A) General reservoir-well in physical coordinates. (B) Thompson-mapped rectangular coordinates.

Pressure-Flow Rate Formulations

In many applications, P_W and P_R may not represent the best way to pose the practical problem. For example, it may be more convenient to specify P_W and the total volume flow rate Q_W, or perhaps, P_R and Q_W. We shall consider these formulations here. A trial-and-error iterative solution that guesses at flow rate is possible, but there is an elegant and faster alternative. Let us introduce a normalized function satisfying

$$\left(P^{m+1}\right)^* = 1 + \left(P^{m+1} - {P_R}^{m+1}\right)/\left({P_R}^{m+1} - {P_W}^{m+1}\right) \tag{9.15}$$

If we solve for P^{m+1} and substitute the result in Eq. (9.12), we find that this function is likewise harmonic, with

$$\left(P^{m+1}\right)^*_{\xi\xi} + \left(P^{m+1}\right)^*_{\eta\eta} = 0 \tag{9.16}$$

Here $(P^{m+1})^*$ equals 1 along the outer reservoir boundary, and 0 at the inner well or fracture boundary. It is important to observe that the function $(P^{m+1})^*$ so defined is independent of m, P_R, and P_W, depending *only* on the details of the prescribed reservoir geometry. We shall also require an expression for flow rate. Now, the volume flow rate Q can be computed through any closed contour C using the general line integral

$$Q = -(khD/\mu)\int_C \text{grad } P \bullet \mathbf{n}\ dl \tag{9.17}$$

where grad $P = (P_x\,\mathbf{i} + P_y\,\mathbf{j})$ assumes the usual vector notation, $\mathbf{n}$ is the outward unit normal along C, dl is an incremental length, • is the scalar product, and D is the depth of the reservoir into the page. Eq. (9.17) can be rewritten in terms of $(P^{m+1})^*$ and evaluated along the inner well or fracture boundary C_W where the pressure is a constant P_W. The result is

$$Q_W = -(khD/\mu)({P_R}^{m+1} - {P_W}^{m+1})/\{(m+1){P_W}^m\}\int_{C_W} \text{grad } \left(P^{m+1}\right)^* \bullet \mathbf{n}\ dl \tag{9.18}$$

We are ready to define a convenient algorithm. We observe that the boundary value problem for $(P^{m+1})^*$, which satisfies simple 1 and 0 pressures at the boundaries, is solved by

$$\left(P^{m+1}\right)^* = \eta/\eta_{max} \tag{9.19}$$

following Eq. (9.13), where mapping $\eta(x,y)$ is *known*. We therefore use Eq. (9.19) to evaluate the integral in Eq. (9.18). Since the argument $(P^{m+1})^*$ appears behind the gradient operation in (x,y) coordinates, the integral

$$I = \int_{C_W} \text{grad } \left(P^{m+1}\right)^* \bullet \mathbf{n}\ dl = \int_{C_W} \text{grad } \left(\eta(x,y)/\eta_{max}\right) \bullet \mathbf{n}\ dl \tag{9.20}$$

depends, like $(P^{m+1})^*$, only on geometrical details. With the value of I known, say calculated as a part of the generic mapping process, Eq. (9.18) takes on the particularly simple form

$$Q_W = -(khD/\mu)\left({P_R}^{m+1} - {P_W}^{m+1}\right)/\{(m+1){P_W}^m\}\, I \tag{9.21}$$

For steady-state areal problems where P_W and Q_W are specified, the value of the farfield reservoir pressure P_R can be immediately obtained from Eq. (9.21) since the solution for ${P_R}^{m+1}$ can be written down without difficulty. When P_R and Q_W are specified, Eq. (9.21) leads to a nonlinear algebraic equation for P_W which can be solved numerically by the fast Newton-Ralphson scheme in Chapter 6. And whether P_W or P_R are specified in conjunction with Q_W, Eq. (9.21) is always used to transform the boundary value formulation back to a pressure-pressure problem whose solution is known from Eq. (9.14). The foregoing results are powerful from a practical standpoint since they collapse several boundary condition models, for the complete range of P_W, P_R, Q_W, and m, into one simple formula whose only unknown is $\eta(x,y)$. Again, this function is obtained once and for all at the outset of any mapping. Solutions to all reservoir simulation problems are then obtained by simple rescaling. Users interested in Windows software implementing the above ideas should contact the author directly.

STREAMLINE TRACING IN CURVILINEAR COORDINATES

We have represented all possible pressure solutions through a single function $\eta(x,y)$. It is possible, it turns out, to treat a generalized streamfunction Ψ in a similar manner. To motivate the analysis, we shall write the steady, areal form of Eq. (9.2a) in the conservation form

$$\left(\partial p^{m+1}/\partial x\right)_x + \left(\partial p^{m+1}/\partial y\right)_y = 0 \tag{9.22}$$

This suggests that a function $\Psi(x,y)$ satisfying the derivative relationships

$$\Psi_y = \partial p^{m+1}/\partial x \tag{9.23}$$

$$\Psi_x = -\,\partial p^{m+1}/\partial y \tag{9.24}$$

exists, since backward substitution in Eq. (9.22) yields an exact $0 = 0$ identity. Eqs. (9.23), (9.24) are consistent with Eqs. (4.17), (4.18). Differentiation of Eq. (9.23) with respect to y, and Eq. (9.24) with respect to x, and addition, lead to

$$\Psi_{xx} + \Psi_{yy} = 0 \tag{9.25}$$

When $m = 0$, as for liquid flows satisfying the linear pressure equation, our $\Psi(x,y)$ reduces to the classical streamfunction. But the concept applies equally to steady, nonlinear gas flows, and similar properties for Ψ are obtained. To see this, let us divide Eq. (9.23) by Eq. (9.24); that is,

$$\begin{aligned} \Psi_y/\Psi_x &= -\left(\partial p^{m+1}/\partial x\right)/\left(\partial p^{m+1}/\partial y\right) \\ &= -\left(\partial p/\partial x\right)/\left(\partial p/\partial y\right) \text{ on expansion and simplification} \\ &= -\left\{(-kh/\mu\,\partial p/\partial x)/(-kh/\mu\,\partial p/\partial y)\right\} \\ &= -u/v \end{aligned} \tag{9.26}$$

where u(x,y) and v(x,y) are Darcy velocities in the x and y directions. Along a streamline, kinematical considerations require that the streamline slope dy/dx equal the velocity ratio v/u. Using Eq. (9.26), we find that

$$dy/dx = v/u = -\Psi_x/\Psi_y \tag{9.27}$$

Now, the total differential $d\Psi$ satisfies, using Eq. (9.27),

$$d\Psi = \Psi_x dx + \Psi_y dy = 0 \tag{9.28}$$

Hence, the streamfunction $\Psi(x,y)$ is constant along a streamline, with different streamlines having different values of streamfunction. Once the values of an array for $\Psi(x,y)$ are available, streamlines can be constructed by drawing lines of constant Ψ elevation (see Chapter 4).

From Eq. (9.25), the boundary value problem governing $\Psi(x,y)$ is still formulated in (x,y) Cartesian coordinates. However, the same transformation used for the pressure equation will lead to Laplace's equation

$$\Psi_{\xi\xi} + \Psi_{\eta\eta} = 0 \tag{9.29}$$

in the computational coordinates (ξ,η). In order to trace streamlines, Eq. (9.29) must be solved subject to appropriate boundary conditions. To develop these conditions, we note that the total volume flow rate per unit depth produced by the well may be written

$$\begin{aligned} Q_W/D &= -(kh/\mu) \int_{C_W} (P_x \mathbf{i} + P_y \mathbf{j}) \bullet \mathbf{n}\, dl \\ &= -(kh/\mu)\{1/((m+1)P_W{}^m)\} \int_{C_W} (\Psi_y \mathbf{i} - \Psi_x \mathbf{j}) \bullet \mathbf{n}\, dl \end{aligned} \tag{9.30}$$

where P_x and P_y were replaced using the differentiated forms of Eqs. (9.23), (9.24), and also, setting $P = P_W$. The last integral in Eq. (9.30) represents the jump or increase in streamfunction obtained by traversing once around the well or fracture contour. Physically, this measures the nonzero net volume outflow from the well.

In the mathematics literature, this jump is denoted by the bracketed quantity $[\Psi]$. Since mass is conserved, the same $[\Psi]$ must be obtained for *any* closed contour surrounding the well. We can rewrite Eq. (9.30) in a form that directly relates $[\Psi]$ to the attributes of the reservoir problem at hand, that is,

$$[\Psi] = -(m+1)\mu Q_W P_W{}^m/kh \tag{9.31}$$

where Q_W and P_W are known from the solution of the corresponding pressure problem. Unlike the solution for $P(\xi,\eta)$, which is independent of the variable ξ, the streamfunction $\Psi(\xi,\eta)$ *must* depend on ξ, since Ψ increases on traversing around the well. The nonzero difference between $\Psi(\xi,\eta)$ found at any two points (ξ,η), of course, is proportional to the flow rate passing through those points. The coordinate ξ plays an identical role to the polar coordinate θ of Chapter 4 in concentric radial flow.

We now formulate the required numerical boundary value problem for Eq. (9.29). Without loss of generality, we first arbitrarily assign a reference value of $\Psi=0$ at the origin $\xi=\eta=0$. Second, along the lower boundary $\eta=0$, we specify the *exact* distribution of volume efflux $\Psi(0,\xi)$ at the well, as obtained from a direct integration based on the known pressure solution. Third, at the end $\xi=\xi_{max}$ of each horizontal line of constant elevation $\eta>0$ surrounding the well, we additionally impose the *same* net jump $[\Psi]$ given by Eq. (9.31) to ensure mass conservation. Finally, since Eqs. (9.16), (9.29) imply that Ψ_η and P_ξ are proportional, the vanishing of P_ξ at the constant pressure farfield boundary implies the edge condition Ψ_η $(\xi,\eta_{max})=0$. Note that we have *not* assigned values for $\Psi(0,\eta)$ itself, since this assumes knowledge of the steady flow paths.

Now, the results obtained for the pressure equation indicate that all possible flows solutions can be reduced to a simple rescaling of the mapping function $\eta(x,y)$. The streamline problem, for a given reservoir configuration, is in a sense reducible to a single scalable streamfunction: in our modeling, we obtain the baseline streamfunction $\Psi^*(\xi,\eta)$ corresponding to the unit problem in Eq. (9.16), and rescale $\Psi(\xi,\eta)$ *via* Eq. (9.31) as necessary to match particular Q_W's obtained with specialized parameters. Observe that while the simple result in Eq. (9.14) applies to pressure, an equally simple closed-form analytical function is not available for the streamfunction. However, the streamfunction calculations outlined here require but minimal effort. Aside from the numerical effort needed to solve Eqs. (8.21), (8.22), or Eqs. (8.54), (8.55), the only other integrations required are those leading to the solution of Eq. (9.29). Again, $\eta(x,y)$ and $\Psi^*(\xi,\eta)$ are each obtained once and for all, for a given reservoir configuration. Thereafter, all pressure and streamfunction solutions are obtained by renormalizing stored arrays. Because the mesh systems used are boundary conforming and variable, a high degree of resolution is possible near the well or the fracture, even with coarse (curvilinear) grids. Accuracy is ensured by using second-order accurate central differenced equations in any numerical scheme. Because the mesh can be coarse, computing times required for convergence are small. Also, the two stored arrays require only minimal computer memory.

CALCULATED STEADY FLOW EXAMPLES

In this section, we describe results obtained with the new methods. Although Windows software is available to compute mappings, pressures, and streamfunctions in seconds, while displaying results in color, a text-based display will be used in this chapter to facilitate the presentation of quantitative information since color print is unavailable. The four gray-scale plots in the following Examples illustrate the capabilities of the algorithms. In Fig. 9.2A–D, the gray bands represent constant-pressure contours corresponding to producing wells and fractures located in a Texas-shaped reservoir. The interactive program first displays the internal well-fracture boundary and then prints the shape of the outer reservoir boundary. Finally, it shows the well/fracture superposed on the reservoir and asks the user to verify that the internal boundary is entirely contained within the exterior one. Once reservoir geometry is validated, pressure and streamfunction analysis proceed. The simple program is portable and does not require any investment in graphics hardware or software. As with any numerical scheme, computed solutions show some mesh dependence. However, this dependence appears to be less than that obtained on rectangular grids.

EXAMPLE 9.1. WELL IN HOUSTON

To illustrate the new approach, consider a Texas-shaped reservoir about $1400' \times 1400'$ in lateral extent as in Fig. 9.3A. A 6-in. diameter well is located in Houston. The 1 indicates the location of the first $\eta = 1$ grid line conforming to the circular well (see Fig. 9.2A for details). The 11's indicate that 11 radial grids are assumed, with the last $\eta = 11$ line coinciding with the far boundary. Twenty-five ξ grids in the circumferential ξ direction are taken. Although this 25×11 mesh system is coarse by conventional Cartesian (x,y) standards, it is fully adequate here since the grids are variable and aligned with all relevant flow boundaries. Furthermore, the matrix equations are derived using second-order accurate, central difference formulas.

Once the grid functions $x = x(\xi,\eta)$ and $y = y(\xi,\eta)$ have been computed, Eqs. (9.13), (9.14) indicate that the inverse function $\eta = \eta(x,y)$ can be rescaled to provide the solution to *any* steady-state pressure-pressure problem in the parameters P_W, P_R, Q_W, and m. This simple renormalization, representing the *only* required calculation, implies almost instantaneous pressure calculation and display. In our Texas examples, grid transformations were obtained in less than one second on Intel Core i5 computers. Data for the calculated η's are shown in Fig. 9.3B. By connecting points of constant η elevation, η grid lines are obtained. Again, the $\eta = 1$ line conforms to the circular well, while the $\eta = 11$ line adheres to the Texas boundary. The intermediate lines $\eta = 2, 3, \ldots, 10$ provide the detailed mesh, and the remaining family of intersecting grid lines is obtained by similarly tracing lines of constant ξ.

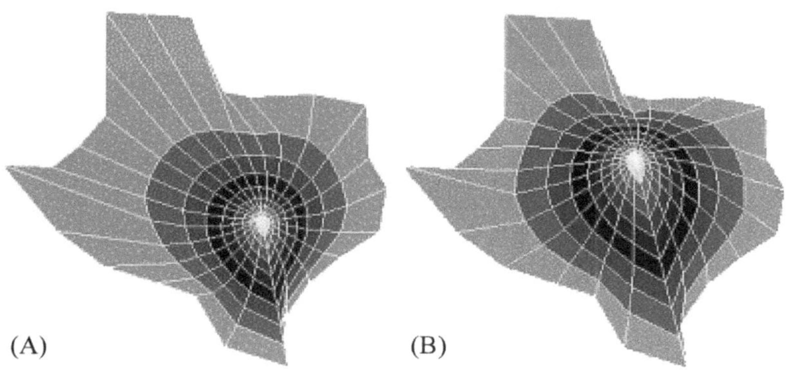

(A) (B)

FIG. 9.2 (A) Well in Houston. (B) Well in Dallas.

(Continued)

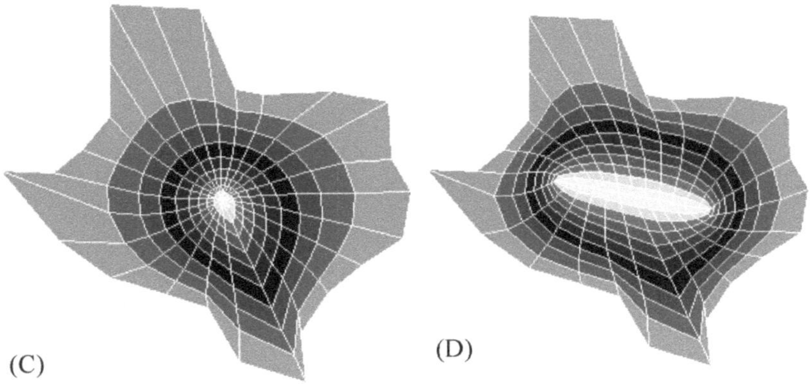

FIG. 9.2, CONT'D (C) Well in Austin. (D) Fracture across Texas.

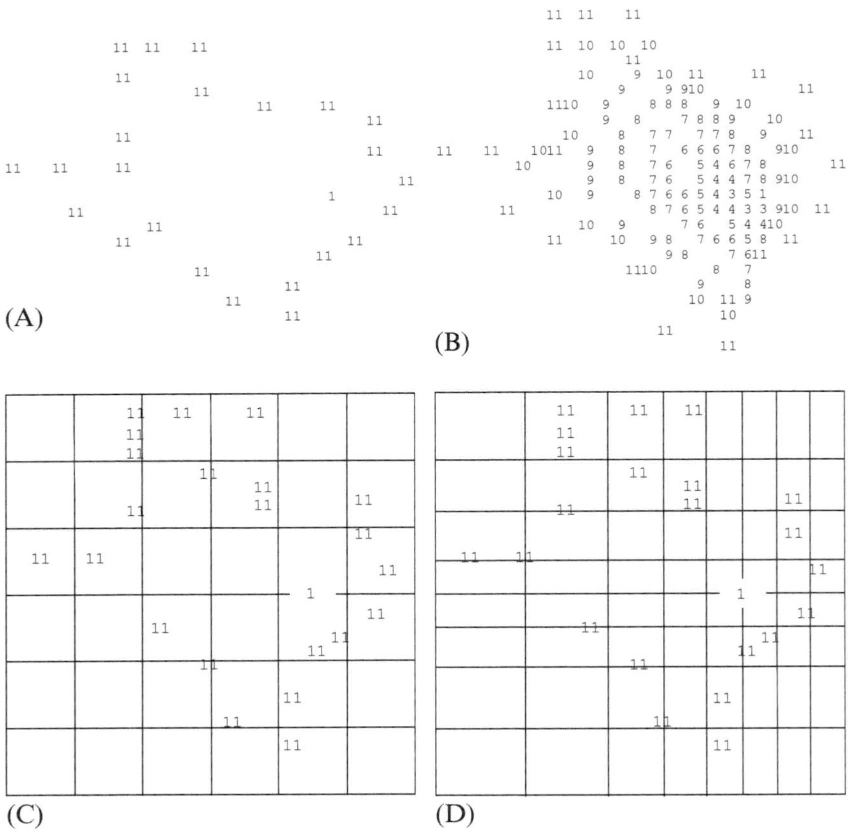

FIG. 9.3 (A) Houston well in Texas. (B) Boundary-conforming Houston-Texas grid. (C) Constant rectangular Houston-Texas grid. (D) Variable (refined) rectangular Houston-Texas grid.

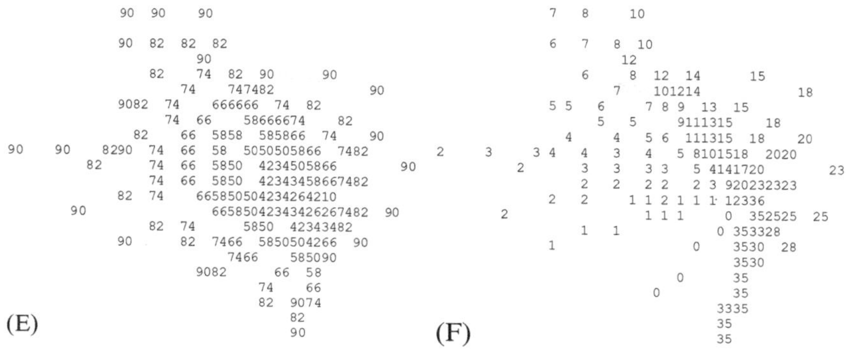

FIG. 9.3, CONT'D (E) Houston-Texas pressure (also see Fig. 9.2A). (F) Houston-Texas streamfunction.

It is important that high grid density and good physical resolution are obtained near the well at 1, while lower densities are obtained further away. The mesh *automatically* coarsens up at faraway reservoir boundaries. This contrasts with conventional grids, typified by Fig. 9.3C and D. The grid in Fig. 9.3C, for example, provides little resolution near the well; attempts to refine the mesh near 1, as shown in Fig. 9.3D, lead to unnecessary and inconsistent refinement in the farfield. In both rectangular meshes, the corners of the rectangular mesh system contain inactive gridblocks that imply inefficient computational overhead. These problems do not appear in the boundary-conforming grids of Fig. 9.2. Also note how the grids in Fig. 9.2 are polar-coordinate-like, in the sense that they can be topologically deformed into cylindrical coordinate systems. This behavior is essential to the geometrical factor objectives stated in the introduction. The grid generation method proposed by Sharpe and Anderson (1991), in contrast, produces meshed square structures not unlike that in Fig. 9.3D. The farfield mesh is rectangular, with lines evenly spaced, while in the nearfield at 1, the grid lines are locally clustered using an attraction algorithm. The exact differences between the two approaches will be discussed later.

To obtain representative numbers, we assumed a steady flow with pressures given at both well and outer boundary, being 100 and 900 psi, respectively. The permeability was 1 Darcy, the viscosity was 1 centipoise, and a liquid with $m = 0$ was assumed. The computed volume flow rate per unit depth is 360.4 ft^3/h. No attempt was made to calibrate the mesh with exact known solutions, since the results are presented for illustrative purposes only. The display program previously described will also overlay the computed pressures at their exact physical locations within the reservoir, showing the first two significant digits of the solution, in psi units. Pressure results are given in Fig. 9.3E.

The streamfunction Ψ(x,y), like the pressure, is obtained by renormalization only and is again instantaneously available. The difference between Ψ at any two points yields the volume flow rate per unit depth in cubic feet/hour, flowing between the points under consideration. Fig. 9.3F gives the computed streamfunction, plotted using the display program. We emphasize that, alternatively, when $P_W = 100$ psi and $Q_W = 360.4$ ft^3/h. are specified, the program *immediately* yields $P_R = 900$ psi. Similarly, when we specify $P_R = 900$ psi and $Q_W = 360.4$ ft^3/h, the program *immediately* yields $P_W = 100$ psi. Fig. 9.3E and F is also *immediately* provided by mathematically rigorous rescaling. This computational consistency behind three different boundary value problems is desirable because of its high accuracy. Because their solutions are almost instantaneous, the total time from problem setup to comprehensive solution with color display is less than one second.

EXAMPLE 9.2. WELL IN DALLAS

We repeat our calculations with the same permeability, viscosity, m, grid parameters, and P_W and P_R values, but instead locate our well in Dallas. Fig. 9.4A and B displays the reservoir geometry and the computed mesh system. The volume flow rate increases to 448.8 ft^3/h. The corresponding pressure and streamfunction solutions are given in Fig. 9.4C and D.

EXAMPLE 9.3. WELL IN CENTER OF TEXAS

We repeat our calculations with the same parameters but locate our well in the center of Texas. Fig. 9.5A and B displays the reservoir and the mesh. Volume flow rate decreases to a low 1.28 ft^3/h. This is well known: for pressure-pressure problems, optimal flow rates are obtained, with wells positioned near outer boundaries, thus maximizing net pressure gradients.

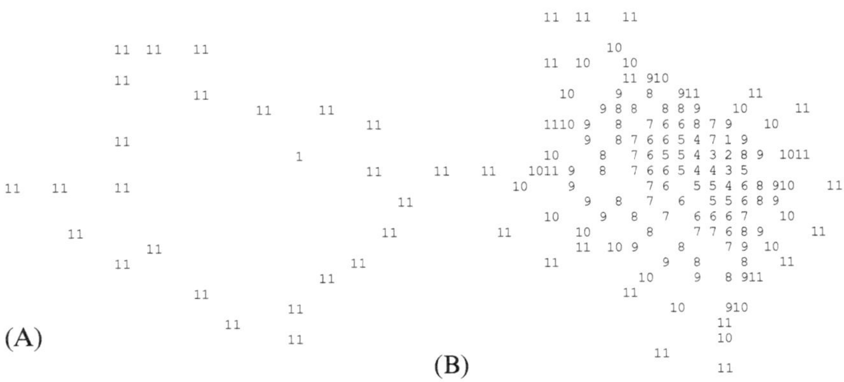

FIG. 9.4 (A) Dallas well in Texas. (B) Boundary-conforming Dallas-Texas grid.

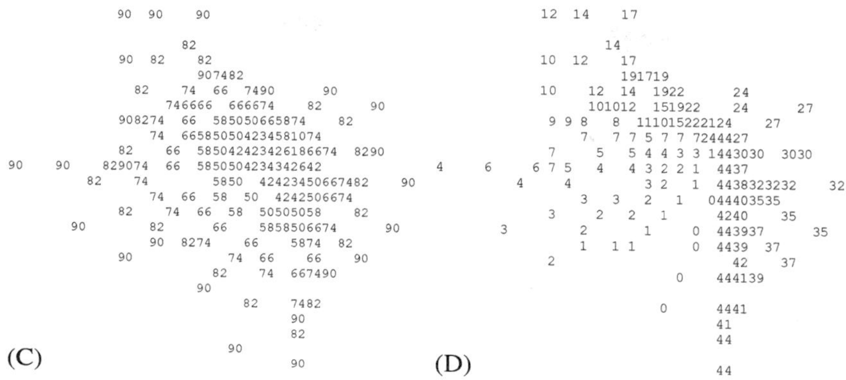

FIG. 9.4, CONT'D (C) Dallas-Texas pressure (also see Fig. 9.2B). (D) Dallas-Texas streamfunction.

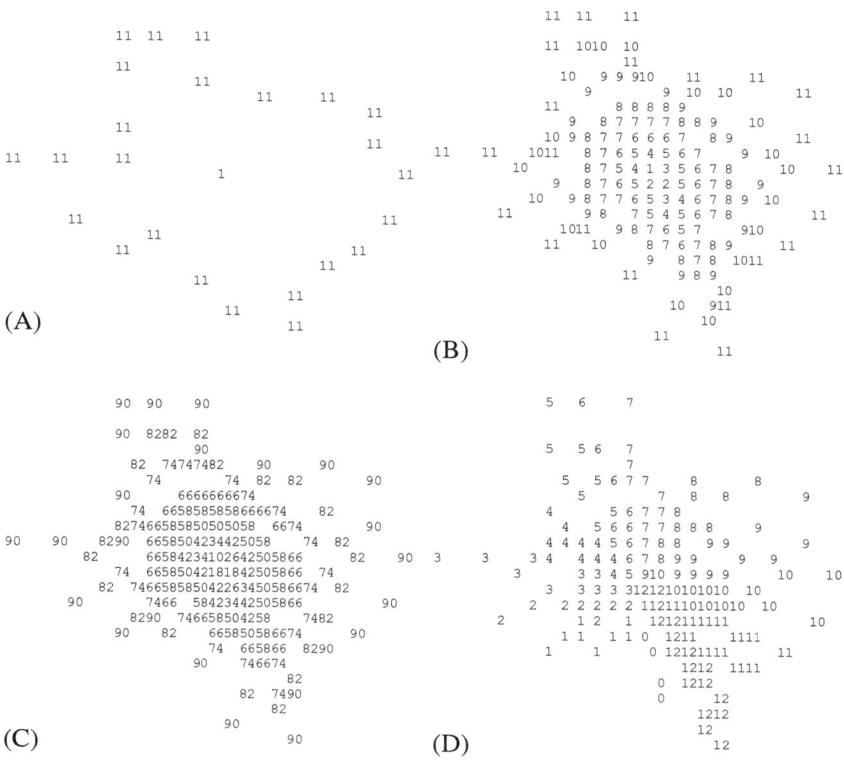

FIG. 9.5 (A) Centered well in Texas. (B) Boundary-conforming center-Texas grid. (C) Center-Texas pressure (also see Fig. 9.2C). (D) Center-Texas streamfunction.

Pressure and streamfunction solutions appear in Fig. 9.5C and D. To infill drillers and petroleum engineers, the position of the optimal well is crucial to reservoir economics. Numerical solutions based on boundary-conforming grids provide a better estimate of total production than crude rectangular grids.

EXAMPLE 9.4. FRACTURE ACROSS TEXAS

Management has decided to produce the Texas-shaped reservoir from a massive hydraulic fracture passing through Texas. We assumed a slightly curved fracture with a one-inch width, whose lateral position is given in Fig. 9.6A. Otherwise, all input parameters remain unchanged. Volume flow rate can be computed, but here we examine streamfunction differences along the fracture to find flow rate contributions to total production.

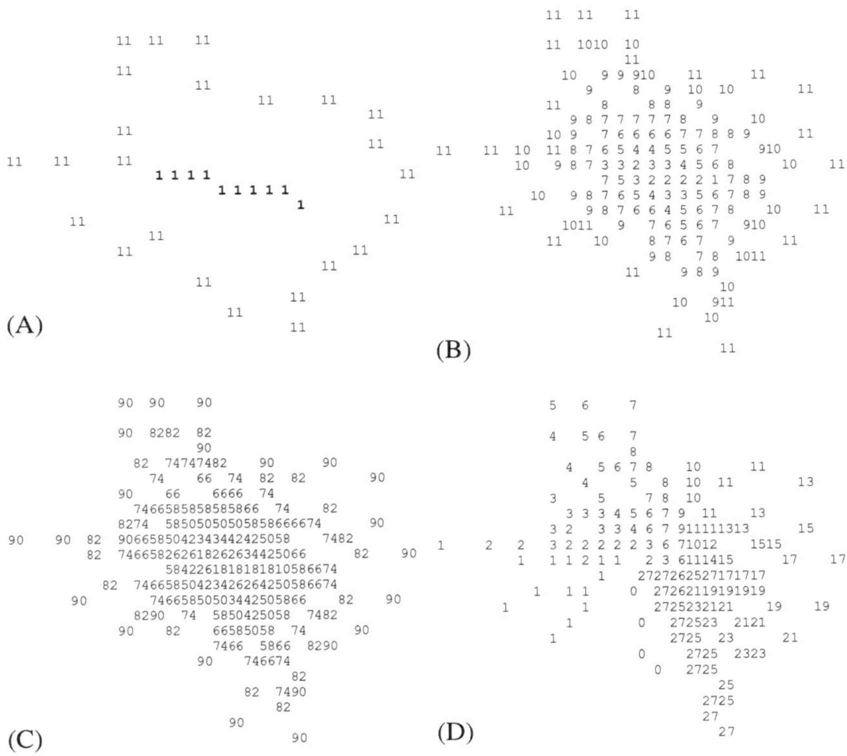

FIG. 9.6 (A) Massive fracture in Texas. (B) Boundary-conforming fracture/Texas grid. (C) Fracture/Texas pressure (also see Fig. 9.2D). (D) Fracture/Texas streamfunction.

A partial bookkeeping is given below, showing contributions between adjacent notes; grid, pressures, and streamfunction appear in Fig. 9.6B–D.

```
Computed flow contribution between fracture points:
O  No.  1 and No.  2  is   .2946E+02 cubic ft/hour
O  No.  2 and No.  3  is   .3239E+02 cubic ft/hour
O  No.  3 and No.  4  is   .3801E+02 cubic ft/hour
```

EXAMPLE 9.5. ISOTHERMAL AND ADIABATIC GAS FLOWS

We consider fluids other than liquids (for which m = 0). We choose a permeability of 0.001 Darcy, a fluid viscosity of 0.1 centipoise, a well pressure of 100 psi, and a reservoir pressure of 900 psi. In the first calculation, we assume an *isothermal* gas with m = 1. The flow rate is 0.0639 ft^3/h per unit foot depth. Pressures and streamfunctions appear in Fig. 9.7A and B.

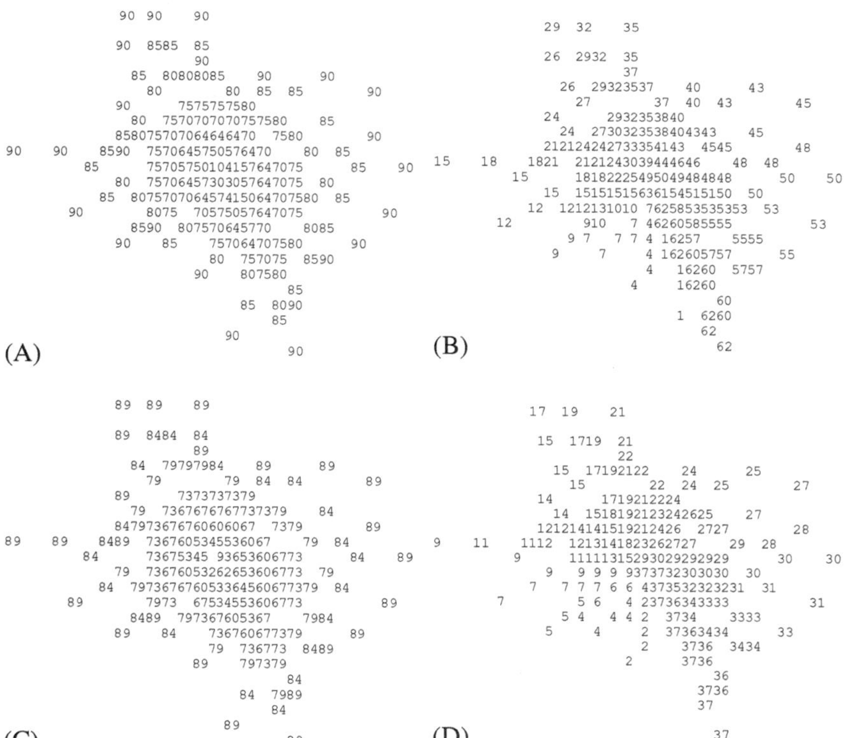

FIG. 9.7 (A) Pressure for isothermal gas. (B) Streamfunction for isothermal gas. (C) Pressure for adiabatic gas. (D) Streamfunction for adiabatic gas.

(Continued)

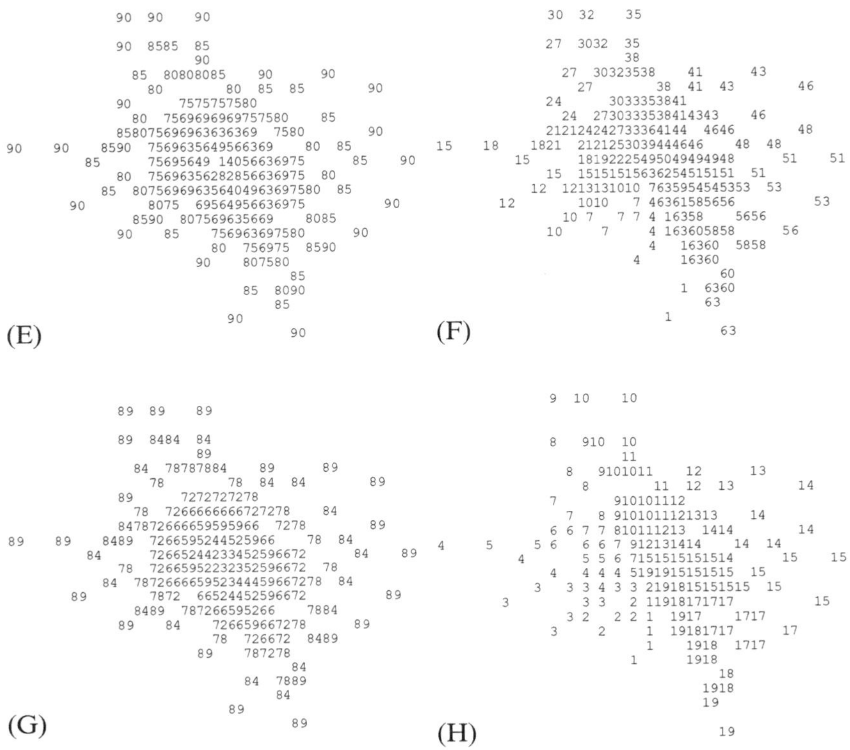

FIG. 9.7, CONT'D (E) Pressure for high-gradient run. (F) Streamfunction for high-gradient run. (G) Pressure for high-gradient run. (H) Streamfunction for high-gradient run.

Next we retain the same parameters throughout, but assume an *adiabatic* process instead, with the exponent $m = 1/1.4 = 0.7$ approximately. The flow rate is substantially different, at 0.0384 ft^3/h. The corresponding pressure and streamfunction solutions are shown in Fig. 9.7C and D.

We return to our *isothermal* gas but assume a farfield pressure of 9000 psi instead of 900. All other parameters remain unchanged. The flow rate increases to 6.47 ft^3/h from 0.0639, a $100\times$ increase for a $10\times$ pressure increase, a result of gas nonlinearities. Pressure and streamfunction solutions are given in Fig. 9.7E and F. The 90 in Fig. 9.7E refers to 9000 now, and not 900.

As a final study, we reconsider the 9000 psi problem just treated but set $m = 0.7$ to model an *adiabatic* gas. The flow rate is 1.97 ft^3/h, contrasting with the 0.0384 ft^3/h for the 900 psi adiabatic flow treated earlier. Pressure and streamfunction results are shown in Fig. 9.7G and H.

The point served by these steady flow comparisons is not so much that results can be obtained numerically. We emphasize, first, that flows on irregular boundaries *can* be modeled accurately using boundary-conforming grids.

Second, the complete gamut of boundary value problems in the parameters P_W, P_R, Q_W, and m can be solved *almost instantaneously* on the slowest computers by simple renormalization using the analytical approach developed in this chapter: the $\eta(x,y)$ function generalizes x for linear, and log r for radial flows. Third, streamline tracing and particle tracking are just as easily performed as pressure calculation. Thus, the availability of natural coordinate systems opens up more advantages than those purely numerical; it paves the way for smarter, simpler, faster analytical procedures for daily practical use.

MESH GENERATION: SEVERAL REMARKS

The "radial, polar-coordinate-like" mesh systems we have studied so far are typified by Fig. 9.2A–D, computed for a Texas-shaped reservoir. This type of mesh offers high resolution in tight spaces. Other variations are possible. Here, we discuss different options available for reservoir simulation; the ideal mesh depends on the particular application, for example, well position, boundary condition type, locations of stratigraphic boundaries, and so on.

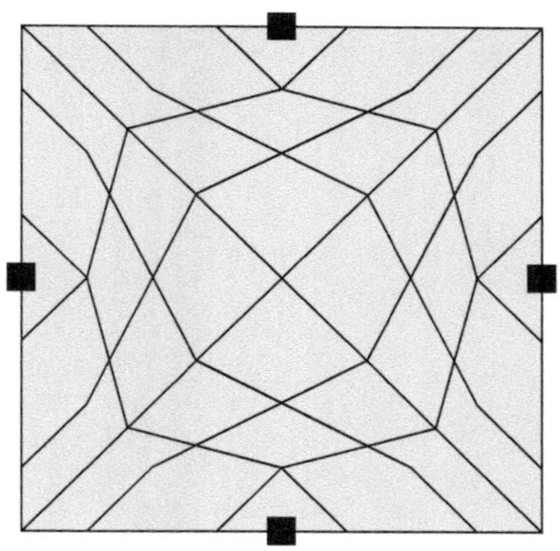

FIG. 9.8 Lopsided square grid (with dot corners).

Lopsided Square Grids

Consider a conventional square reservoir. A rectangular x-y grid might be used in routine applications (e.g., Fig. 9.3C and D), but it is not unnatural to develop a grid, *via* Thompson's equations, that produces corners in the mapped plane that are coincident with the black dots in Fig. 9.8. The resulting grid bears none of the features that we expect of square, Cartesian-based systems, but it can be useful if the distribution of heterogeneities follows the coordinate lines shown. This would permit improved reservoir modeling of nonideal flow effects and geological structures.

Square Grid for Circles

We appreciate that there is nothing sacred about the circular grids classically used to model flows in circular reservoirs. In Fig. 9.9, a rectangular grid, at least in the mapped Thompson plane, is fitted to a circular boundary and used to host numerical computations.

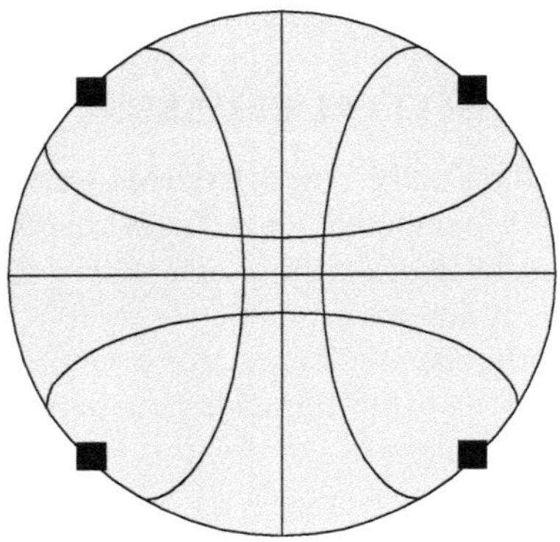

FIG. 9.9 Circle with square-like grid.

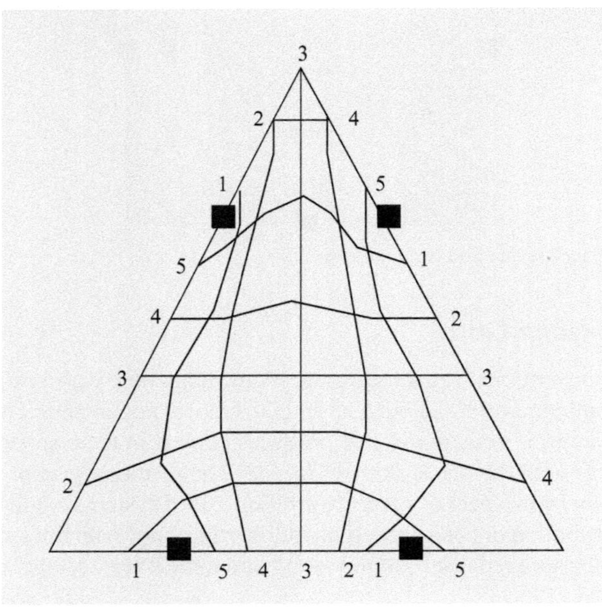

FIG. 9.10 Triangle with square-like grid.

Grids for Odd Shapes

Fig. 9.10 shows a rectangular 6×6 computational grid fitted to a triangular physical boundary; the black dots indicate the corners of the mapped rectangle. Normally, the sharp corners characteristic of the triangular shape would preclude analytical mappings based on algebraic formulas. The Thompson scheme generates smooth grid lines internal to the triangle, while slope discontinuities are preserved at boundaries.

Grids for Faulted Sections

Faulted cross-sections are easily treated using Thompson's mappings. Fig. 9.11 shows a rectangular computational mesh fitted to both upper and lower stratigraphic bed boundaries, with standard vertical ones retained at the left and right. This type of grid is useful in modeling left-to-right or almost unidirectional flow. When flows into wells and fractures are required, the grids used in Fig. 9.2A–D are preferable.

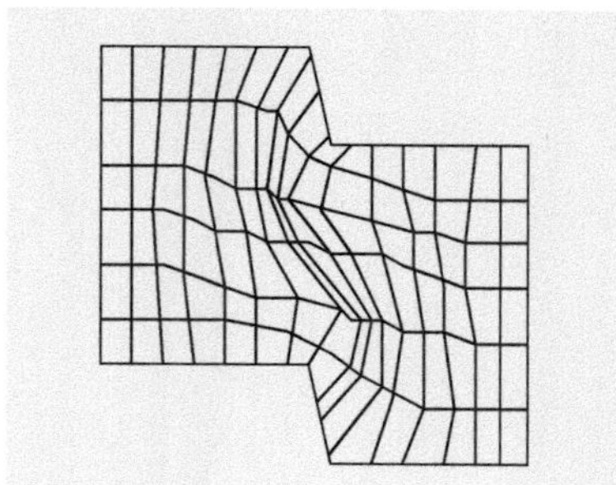

FIG. 9.11 Fault with Cartesian-like rectangular grid.

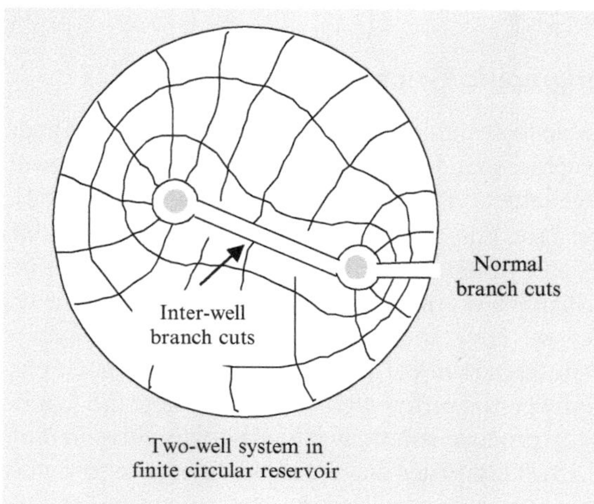

FIG. 9.12 Boundary-conforming two-well grid.

Multiple Wells

So far, we have discussed grid generation for single-well domains only. But the treatment of computational flowfields with multiple wells or holes is, again, well established (Thompson, 1978, 1984; Thompson et al., 1985). This area of grid generation is considered in detail in aerodynamics, where it is used to study flows past multielement airfoils, wing flaps, high-lift devices, and so on. Fig. 9.12 displays a grid system that might be useful for a two-well simulation. Here, the inner contour does not completely wrap around each of the wells, although in practice it does. Interwell branch cuts render the innermost contour dumb-bell-like, with the outermost contour developing into the obvious circle. Boundary conditions at interwell branch cuts preserve continuity of physical quantities such as pressure and velocity. The mapped domain is again rectangular and simple; multiple wells are similarly treated with multiple internal branch cuts.

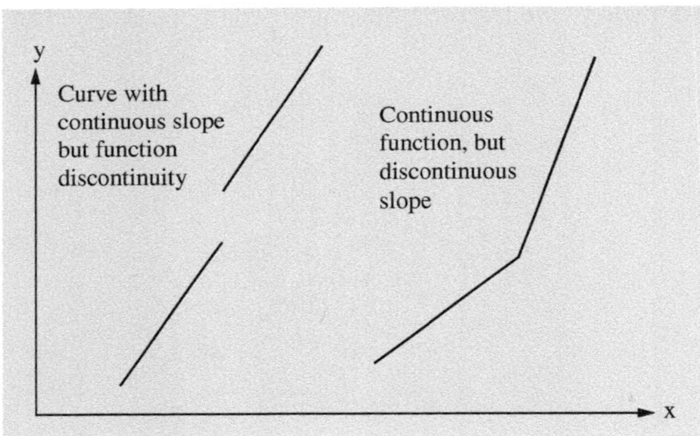

FIG. 9.13 General functional behavior.

General Stratigraphic Grids, Internal Boundaries

It is often desirable to design grid systems whose coordinate lines coincide with internal stratigraphic boundaries, or perhaps with other known boundaries of isolated heterogeneities. This amounts to solving Eqs. (8.54), (8.55), or in the more general case, Eqs. (8.21), (8.22), subject to specifications of the dependent variables x and y *internal* to the outer boundary of the flow domain. Such schemes (e.g., Sharpe and Anderson, 1991), are embody some dangers and hidden pitfalls. As we have seen from our closed-form fracture solutions of Chapter 2, and in the numerical well and fracture computations of Chapter 7, solutions to elliptic partial differential equations where the function is internally specified typically produce solutions with discontinuities in the first derivative (e.g., see Fig. 9.13). This means that the usual difference formulas for all second pressure derivatives *must* be modified in order to account for internal jumps that will vary from problem to problem. Unless this change is carried out, any flow simulations are likely to be incorrect. This subject is treated in Chin (2000, 2014a), which deals with the subject in the context of Maxwell's equations in layered media. That work presents formulas for differentiating across surfaces bearing discontinuities in the function or its derivative.

Because of the existence of first-derivative discontinuities when internal boundary conditions are specified, flows past multielement airfoils in the aerospace industry, as cases in point, are simulated using singly connected computational domains such as that shown in Fig. 9.12 which displace the source of the discontinuity to a computational boundary. Whereas the modeling by Sharpe and Anderson (1991) of wells as internal fixed points produces undesired discontinuities, aerospace methods produce meshes where all metrics and derivatives are continuous. Sharpe and Anderson also embed their elliptic operators in first-order, time-like systems. The complete process yields shocks in some instances, perhaps because the embedded system possesses nonlinear hyperbolic properties. Jameson (1975) has shown how various transient diffusive systems can be derived to host relaxation-based techniques; these methods are further optimized for computational speed.

The extremely stable scheme for Eq. (8.58), used in Chin (2001a,b) for mesh generation in highly eccentric borehole annuli and noncircular ducts, for example, can produce 50×50 grids systems in approximately one second on typical Intel machines. Figs. 9.8–9.12 show that alternatives to the radial-like grid systems typified by Fig. 9.2 are possible. While Thompson's scheme is quite flexible, and can be readily adapted to mappings with multiple branch cuts and lopsided grid structures, it is not the only host formulation available. Many other mesh generation schemes are available from public Internet websites devoted to this very interesting area.

PROBLEMS AND EXERCISES

1. The calculations performed in this chapter mapped the single-well reservoir in Fig. 9.1A to the rectangular computational domain of Fig. 9.1B using two types of boundary conditions, discussed previously. The first deals with the physical coordinates of the well/fracture and the reservoir, while the second focuses on the use of branch cuts drawn in the flow domain. Review these conditions, and explain in your own words why the latter are needed. One possible grid system for *two* wells in a reservoir is shown in Fig. 9.12. Draw the computational domain analogous to Fig. 9.1B, and label the requisite branch cut pairs required to render the domain rectangular. What functional conditions must be satisfied at these cuts?
2. How would you extend the solution of (1) to three wells? Four wells? And finally, to N number of wells? What are the minimum numbers of grids in the computational η and ξ directions needed to ensure good physical resolution as a function of N?

Chapter 10

Transient Compressible Flows: Numerical Well Test Simulation

OVERVIEW

In this chapter, we first consider two-dimensional, single-well, planar reservoir flows with boundary-conforming curvilinear meshes. The transient solver is based on the alternating-direction-implicit (ADI) method, which is introduced for simple systems. Then, other aspects of steady and unsteady flow simulation in three dimensions are discussed, and the basic algorithms are given.

Two-Dimensional Planar Flows

For planar flows, the vertical z derivatives in Eq. (9.11) vanish, leaving the simpler result

$$P^{m+1}{}_{\xi\xi} + P^{m+1}{}_{\eta\eta} = J(\xi, \eta)\, \phi\mu c^*/kh\, \partial P^{m+1}/\partial t \qquad (10.1)$$

Eq. (10.1) must be solved with appropriate boundary and initial conditions (e.g., see Fig. 9.1A and B). The well pressure P_W or the volume flow rate Q_W may be specified along the lower $\eta = 0$ boundary of the computational box. At the edge of the reservoir where $\eta = \eta_{max}$, the pressure P_R may be specified, while a zero flow boundary represents another option. At the sides $\xi = 0$ and $\xi = \xi_{max}$, the pressure may vary with η, but will be single-valued at any particular η. Finally, at the initial time $t = 0$, the pressure $P(\xi,\eta,0)$ may take on some constant value P_0. Or it may take on any of the steady-state, *flowing* pressure distributions obtained in the previous chapter.

Alternating-Direction-Implicit (ADI) Methods

Let us consider the classical, dimensionless, two-dimensional heat equation for u(x,y,t),

$$\partial u/\partial t = \partial^2 u/\partial x^2 + \partial^2 u/\partial y^2 \qquad (10.2)$$

Quantitative Methods in Reservoir Engineering. http://dx.doi.org/10.1016/B978-0-12-810518-4.00010-4

A simple finite difference method developed along the lines of Chapter 6 for radial flows leads to

$$\left(u_{i,j,n+1} - u_{i,j,n}\right)/\Delta t = \delta_x{}^2 u_{i,j,n} + \delta_y{}^2 u_{i,j,n} \tag{10.3}$$

Here, i and j are indexes in the x and y directions, δ_x^2 and δ_y^2 represent central difference operators and not mesh lengths, while n and n+1 represent successive time steps. Because the right side of Eq. (10.2) has been approximated at the old time step, Eq. (10.3) provides a simple explicit scheme that is amenable to pocket calculator solution. It is, however, very unstable and requires extremely small time steps for convergence.

On the other hand, if the right side of Eq. (10.2) is approximated at the new time step, we have the implicit scheme

$$\left(u_{i,j,n+1} - u_{i,j,n}\right)/\Delta t = \delta_x{}^2 u_{i,j,n+1} + \delta_y{}^2 u_{i,j,n+1} \tag{10.4}$$

This five-point scheme, however, leads to a complicated coefficient matrix that requires extensive calculation for inversion. In radial flow, recall that the implicit scheme was useful only because it led to tridiagonal matrices. Alternating-direction-methods are based on a clever motivating question. Is it possible to write Eq. (10.4) in a form that requires two successive passes using a tridiagonal solver only? The answer is, "Yes." And this implies much faster, more stable, easier-to-code computations. The idea is this: add special high-order terms to the base equation, so that the resulting multipoint formula can be approximately factorized into two successive tridiagonal operations. We shall omit the details, which are research oriented. But it turns out that, if $u_{i,j}{}^*$ is an intermediate (nonphysical) variable at the end of the first pass (described by Eq. 10.5), then a second pass (defined by Eq. 10.6) yielding $u_{i,j,n+1}$ completes the calculation for the net time step Δt.

$$\left(u_{i,j}{}^* - u_{i,j,n}\right)/(\Delta t/2) = \delta_x{}^2 u_{i,j}{}^* + \delta_y{}^2 u_{i,j,n} \tag{10.5}$$

$$\left(u_{i,j,n+1} - u_{i,j}{}^*\right)/(\Delta t/2) = \delta_x{}^2 u_{i,j}{}^* + \delta_y{}^2 u_{i,j,n+1} \tag{10.6}$$

Such schemes, also known as ADI and approximate factorization methods, are found in numerical analysis books, and in texts on reservoir simulation, for example, Carnahan et al. (1969) and Peaceman (1977).

Solving the Mapped Equation

Eq. (10.1), which governs transient liquid and gas flows in mapped coordinates, obviously resembles Eq. (10.2). It is similar to the pressure transient equation used in physical (x,y) coordinates, except for the multiplicative presence of the nondimensional Jacobian $J(\xi,\eta)$ given by Eq. (8.6). Thus, existing numerical techniques devised for $J=1$ (and Eq. 10.2, in particular) are applicable with minor modification. We emphasize that conventional boundary value problems in (x,y) space treat all four sides of the computational rectangle as exterior reservoir boundaries, and represent wells as internal singularities or delta functions. In the present approach, the boundaries $\eta = 0$ and $\eta = \eta_{max}$ represent, respectively, the well and the farfield; the side boundaries $\xi = 0$ and $\xi = \xi_{max}$, however, are branch cuts where the coordinate function ξ and the streamfunction $\Psi(\xi,\eta)$ both jump, while $P(\xi,\eta,t)$ is single-valued. Since there are no internal sources, the carryover from conventional methods is not only straightforward—it is, in fact, simpler.

EXAMPLE 10.1. TRANSIENT PRESSURE DRAWDOWN

We reconsider the circular well centered in Texas, but relax our steady flow requirement by modeling the transient flow of a liquid having $m = 0$, a viscosity of 20 centipoise, and a compressibility of 0.000015/psi. A 20% porous formation is taken, along with a permeability of 0.001 Darcy.

Our pressure boundary conditions are unchanged, and an initial uniform pressure of 900 psi is assumed. Initially, at $t = 0$ h, the reservoir pressure is 90 throughout, as shown in Fig. 10.1A. Successive snapshots of the pressure distribution in time appear in Fig. 10.1B–G. The wellbore flow rate increases from zero initially, to an asymptotic steady-state value.

In the above, variable time step sizes were taken, ranging from 0.1 h at the outset to 100 h near steady-state conditions, for a total of 1000 steps. At this point, the computations were terminated, with the results almost unchanging from one step to the next, the final flow rate being 0.00006487 ft^3/h.

We asked how this compares with results from a direct steady flow calculation. The same fluid and formation parameters, and boundary conditions, were assumed. The calculated flow rate was 0.00006393 ft^3/h, which is consistent with the large time solution of the unsteady problem. The steady pressure and streamfunction appear in Fig. 10.1H and I.

Finally, because the "$\eta = 1$" coordinate line is exactly coincident with the internal well (or fracture) contour, it is possible to compute quite accurately the incremental volume flow rates per unit (foot) depth entering different portions of the inner contour. This capability is important in evaluating the quality of artificially stimulated fractures. Fig. 10.1J provides (automatically computed) flow contributions at steady state for selected points, showing that, unlike purely radial problems, flow rates *will* vary circumferentially because the farfield boundary is irregular.

EXAMPLE 10.2. TRANSIENT PRESSURE BUILDUP

We shall now reconsider the reservoir in Example 9.5. There, the well was in a sense newly drilled and was allowed to attain its maximum flow rate in time, under pressure-pressure boundary conditions. Here, we shall consider the

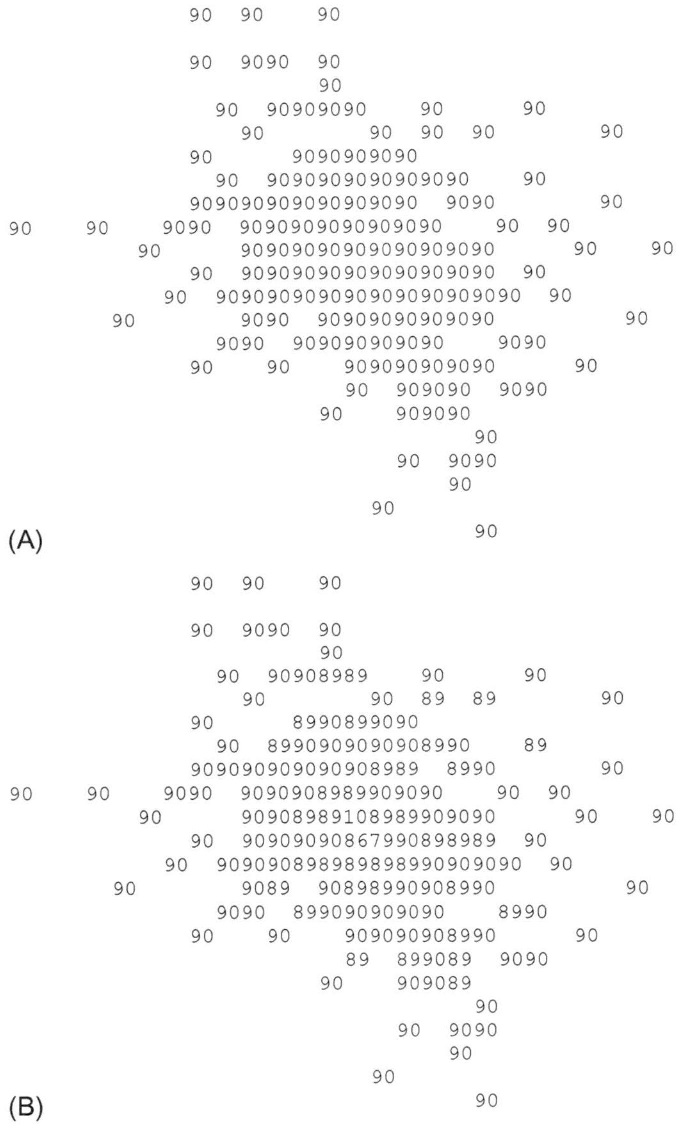

FIG. 10.1 (A) Pressure distribution (0 h). (B) Pressure distribution (10 h).

reverse situation. The reservoir is assumed to be flowing at steady state initially in accordance with Fig. 10.1H and I. Suddenly at $t=0+h$, the well is completely shut in, so that $Q_W=0\ ft^3/h$. Thus, we expect P_W to increase with time, and the reservoir to achieve everywhere the 900 psi set at the farfield. But the problems

```
              90  90    90

              90  8990  90
                        90
                89  89908990    90      90
                  89        90  89  90        90
              90      8989898990
                90  8989898989908989    90
              909090898989898989  9090        90
90    90    8990  9089898984898989    90  90
          90      90898983107189899090      90    90
              90  89898989605189898989  89
            89  899089898989858789899090  89
        90        9090  89898989899090          90
                9090  899089898990    9089
              90    90    909089899090      90
                          90  899089  8990
                        90    909089
                                    90
                              90  8990
                                  90
                            90
                                    90
```

(C)

```
              90  90    90

              90  8990  90
                        90
                89  89898989    90      90
                  89        89  89  90        90
              90      8989898989
                89  8989898989898989    90
              898989888885858689  8989        90
90    90    8990  8988857663788689    89  90
          89      89897662104979868989      90    90
              89  89898577353481878989  89
            90  898989898680617187898989  90
        90        8989  89818084888989          90
                9090  898989888789    8989
              90    89    898989898989      90
                          89  898989  9090
                        90    898989
                                    89
                              90  8990
                                  90
                            90
                                    90
```

(D)

FIG. 10.1, CONT'D (C) Pressure distribution (5 days). (D) Pressure distribution (1 month).

(Continued)

here and in Example 9.5 are not identical, with the direction of time simply reversed. Numerically, we have different truncation and cumulative error histories, and stability issues arising from contrasting initial-boundary conditions are unlike.

```
                 90  90     90

                 90  8989   89
                            90
                   89  89898989     90       90
                     89         89  89  89         90
                 90      8888888889
                   89  87838383848888889     89
                 898987838375767684  8889          90
90     90     8990  8783756452667785    89  89
            89      888364521040677885 88      89     90
                 89  888476652827697885 88  89
               89  89888485786946597986 8989  89
        90          8988  85716474818689             90
                   8990  898886827886    8989
                 90     89     898786878989      90
                               89  898989  8990
                            90    898989
                                       89
                                  89  8990
                                      89
                              90
                                       90
```

(E)

```
                 90  90     90

                 90  8888   88
                            90
                   88  84848487     90       90
                     84         84  88  88         90
                 90      7878787984
                   84  7871717171797985      88
                 878478707062626272  8085          90
90     90     8790  7871615242536372    86  88
            88      78715241103254647380       88     90
                 85  79716253222255647381  86
               88  85797273645534456574818 6  88
        90          8580  74564758677481             90
                   8890  858075685975    8688
                 90     88     817770778286      90
                               86  837983  8890
                            90    878587
                                       88
                                  89  8890
                                      89
                              90
                                       90
```

(F)

FIG. 10.1, CONT'D (E) Pressure distribution (2 months). (F) Pressure distribution (8 months).

Fig. 10.2A–E displays the reservoir pressure distribution at various stages in the shut-in process. Finally, in Fig. 10.2F, selected well shut-in pressures are tabulated. The overall time scales for both problems are comparable in order-of-magnitude, as expected.

```
              90  90    90

              90  8383  83
                        90
                83  77777783    90      90
                  77        77  83  84        90
              90      6969696977
                76  6961616161707077      84
              8376696161535362  7078          90
90    90    8390  6961534436455462    78  84
          84      69614436102745546270      84    90
              77  70625345191946546371  78
            84  777062625446283755637178  84
        90        7770  63463847556371          90
                8490  787164564764    7884
              90  84    716456647279      90
                          78  726572  8490
                        90    797379
                                    85
                              85  8090
                                  85
                            90
                                    90
```

(G)

```
              90  90    90

              90  8282  82
                        90
                82  74747482    90      90
                  74        74  82  82        90
              90      6666666674
                74  6658585858666674      82
              8274665858505050 58  6674          90
90    90    8290  6658504234425058    74  82
          82      66584234102642505866      82    90
              74  66585042181842505866  74
            82  746658585042263450586674  82
        90        7466  58423442505866          90
                8290  746658504258    7482
              90  82    665850586674      90
                          74  665866  8290
                        90    746674
                                    82
                              82  7490
                                  82
                            90
                                    90
```

(H)

FIG. 10.1, CONT'D (G) Pressure distribution (2 years). (H) Steady-state pressure distribution.

(Continued)

Steady Three-Dimensional Flow

While we have addressed two-dimensional planar flows for simplicity, three-dimensional steady flows are just as convenient to formulate, code, and solve. For example, consider

```
                 29  32    35

                 26  2932  35
                           37
                   26  29323537    40      43
                     27       37  40  43         45
                 24      2932353840
                   24  2730323538404343     45
                 21212424273335414 3  4545         48
 15    18     1821  2121243039444646    48  48
            15      1818222549504948 4848      50    50
                15  15151515636154515150  50
              12  1212131010 7625853535353  53
        12         910   7 46260585555         53
                 9 7   7 7 4 16257    5555
               9     7     4 162605757      55
                           4   16260  5757
                         4     16260
                                    60
                               1  6260
                                   62
                             1
                                    62
```

(I)

```
          Breakdown of total volume flow rate
   from well/fracture:  Contribution between points
 O  No.  1 and No.  2  is   .1863E-05 cubic ft/hour
 O  No.  2 and No.  3  is   .2244E-05 cubic ft/hour
 O  No.  3 and No.  4  is   .2860E-05 cubic ft/hour
 O  No.  4 and No.  5  is   .3455E-05 cubic ft/hour
```

(J)

FIG. 10.1, CONT'D (I) Steady-state streamfunction solution. (J) Flow contributions around the well.

$$\partial^2 p/\partial x^2 + \partial^2 p/\partial y^2 + \partial^2 p/\partial z^2 = 0 \tag{10.7}$$

This can be central differenced in all of the coordinate directions, straightforwardly, yielding the seven-point representation

$$\left(p_{i-1,j,k} - 2p_{i,j,k} + p_{i+1,j,k}\right)/(\Delta x)^2 + \left(p_{i,j-1,k} - 2p_{i,j,k} + p_{i,j+1,k}\right)/(\Delta y)^2 + \left(p_{i,j,k-1} - 2p_{i,j,k} + p_{i,j,k+1}\right)/(\Delta z)^2 = 0 \tag{10.8}$$

assuming constant mesh lengths. This finite difference molecule is, of course, extremely cumbersome; it leads to a sparse coefficient matrix requiring significant effort for direct inversion (see Peaceman, 1977). On the other hand, we can rearrange Eq. (10.8) in such a way that the left side contains an operator of the form $(\,)\,p_{i-1,j,k} + (\,)\,p_{i,j,k} + (\,)\,p_{i+1,j,k} = \ldots$, where the … represents all terms not on

```
                      90  90    90

                      90  8182  82
                                90
                        81  73737482    90      90
                          74        73  82  82        90
                      90      6666656573
                        74  6657575857666674    81
                      827466575850505058  6674        90
  90    90    8190  6658504234424958    73  82
            82      6658423423274250586 6      82    90
                    74  6558504219214250 5865  74
                  82  74655757494226345058 6674  81
              90      7466  58423442495865          90
                        8290  736658504258    7481
                      90    82    665850586674      90
                                74  655865  8190
                              90    746674
                                          82
                                    82  7390
                                        82
                                  90
(A)                                       90
```

```
                      90  90    90

                      90  8182  82
                                90
                        81  74747482    90      90
                          74        74  82  82        90
                      90      6666666674
                        74  6658585858666674    82
                      827466585850505058  6674        90
  90    90    8290  6658514541445058    74  82
            81      6658454137374450586 6      82    90
                    74  6658504431324350 5866  74
                  82  74665858504432385058 6674  82
              90      7466  58433542505866          90
                        8290  746658504258    7481
                      90    82    665850586674      90
                                74  665866  8290
                              90    746674
                                          82
                                    82  7390
                                        82
                                  90
(B)                                       90
```

FIG. 10.2 (A) Shut-in pressure (50 h). (B) Shut-in pressure (3 weeks).

(Continued)

the left. This result is reminiscent of the row relaxation method developed in Chapter 7, except that the relaxation must be extended over an extra spatial dimension—a simple procedure easily implemented by introducing an additional do-loop level (e.g., see Exercise 1 in Chapter 7). This line relaxation

FIG. 10.2, CONT'D (C). Shut-in pressure (4 months). (D) Shut-in pressure (10 months).

can be carried in any of the three (x,y,z) directions. If the heterogeneous extension of Eq. (10.7) (for variable permeabilities) contains destabilizing terms, alternative columns or lines with increased diagonal dominance can be selected. All of the advantages of the relaxation method cited in Chapter 7 apply to three dimensions.

```
                         90  90     90

                         90  8888   88
                                    90
                            88  86868688    90      90
                               86        86  88  87          90
                         90        8585858586
                            86  8584848484858586      87
                         888685848484848384  8486            90
90      90          8890  8584848383838384      85  87
                 87       85848383828283838384        87     90
                      86  85848383818182838384  85
                   87  868584838382818182838485  87
             90           8684  83828181828384               90
                       8790  858483828182       8587
                      90    87      848281828385        90
                                    85  838283  8790
                                 90     858385
                                              87
                                        87  8490
                                              87
                                     90
(E)                                           90
```

TIME SIMULATION RESULTS

	Time (h)	Flowrate (cu ft/h)	Cum-Vol (cu ft)	Pw-Avg (psi)	PR-Avg (psi
1	.1000E+01	.0000E+00	.0000E+00	.1809E+03	.9000E+03
50	.5000E+02	.0000E+00	.0000E+00	.2357E+03	.9000E+03
100	.5500E+03	.0000E+00	.0000E+00	.3697E+03	.9000E+03
150	.1550E+04	.0000E+00	.0000E+00	.4615E+03	.9000E+03
200	.2550E+04	.0000E+00	.0000E+00	.5141E+03	.9000E+03
254	.5250E+04	.0000E+00	.0000E+00	.6004E+03	.9000E+03
300	.7550E+04	.0000E+00	.0000E+00	.6483E+03	.9000E+03
340	.1555E+05	.0000E+00	.0000E+00	.7512E+03	.9000E+03
400	.2755E+05	.0000E+00	.0000E+00	.8282E+03	.9000E+03

(F)

FIG. 10.2, CONT'D (E) Shut-in pressure (3 years). (F) Shut-in pressure history.

Transient 3D Flow, ADI Methods

Again, for purposes of illustration, consider the classical, dimensionless, 3D heat equation for u(x,y,t),

$$\partial u/\partial t = \partial^2 u/\partial x^2 + \partial^2 u/\partial y^2 + \partial^2 u/\partial z^2 \tag{10.9}$$

A straightforward extension of the two-level scheme for planar 2D flows leads to the four-level, tridiagonal based scheme

$$(u^* - u_n)/(\Delta t/2) = \delta_x{}^2 u^* + \delta_y{}^2 u_n + \delta_z{}^2 u_n \tag{10.10}$$

$$(u^{**} - u_n) / (\Delta t/2) = \delta_x{}^2 u^* + \delta_y{}^2 u^{**} + \delta_z{}^2 u_n \tag{10.11}$$

$$(u^{***} - u_n) / (\Delta t/2) = \delta_x{}^2 u^* + \delta_y{}^2 u^{**} + \delta_z{}^2 u^{***} \tag{10.12}$$

$$(u_{n+1} - u_n) / \Delta t = \delta_x{}^2 u^* + \delta_y{}^2 u^{**} + \delta_z{}^2 u^{***} \tag{10.13}$$

This scheme is discussed in Carnahan et al. (1969) and Peaceman (1977), where von Neumann stability analyses are also given. As in the two-dimensional case, it is easily implemented, although it is more memory intensive. Its application to Eq. (9.2a) is direct and straightforward, and is left to the reader as an exercise. In Chapter 15, we shall show how to implement horizontal, deviated, and multilateral well logic in three-dimensional heterogeneous and anisotropic reservoirs. This chapter is not intended to "replace" well test interpretation and analysis. Rather, the methods proposed can be used to supplement existing methods; available solutions can be used to calibrate grid-generated solutions, in order to increase their reliability insofar as extrapolation to newer problems is concerned.

PROBLEMS AND EXERCISES

1. Consider two-dimensional transient heat transfer in a square, where the initial temperature T_0 is uniform throughout, and the four sides are subsequently held at constant temperatures T_1, T_2, T_3 and T_4. Assume suitable physical properties as required. (i) Approximate all spatial derivatives using central differencing, and write a finite difference program for T(x,y,t) that is explicit in time. Perform a von Neumann stability analysis for your scheme, building its requirements into your program logic. (ii) Program the ADI method given in Eqs. (10.5), (10.6). (iii) Run both codes and show that the steady-state solution in each case is independent of the initial temperature; also show that the steady temperature at the center of the square equals the arithmetic average of the given boundary values. (iv) Derive the exact analytical Fourier series solution for this problem, and determine the time required to achieve steady state. (v) Does this time agree with your computations? (vi) If steady solutions are the objective, how much more efficient is the ADI scheme than your explicit scheme? Explain and quantify your answers.
2. Repeat Exercise 1 for a cube, satisfying the three-dimensional transient heat equation. What is the steady-state value of temperature at the center of the cube? You should obtain this value by taking the large time limit of your exact Fourier series solution. For your ADI method, program Eqs. (10.10)–(10.13).
3. Now suppose that in (1) and (2), respectively, square and cubic holes are to be introduced at the centers of the computational domains, respectively. Temperatures or heat fluxes are to be prescribed internally, allowing these holes to model heat sources and sinks. How would you modify the above

programs to include the internal well suggested here? What programming difficulties and computational inaccuracies do you anticipate? Make the appropriate changes for (2) and run your program. Compare your solutions to those obtained using boundary-conforming, curvilinear grids for a square hole in a square domain. Which program runs faster? Which produces more accurate solutions?

4. Transient, compressible flows of liquids and gases in real reservoirs satisfy the three-dimensional heat equation for pressure. Suppose a multilateral well with several out-of-plane drainholes is drilled into the formation. The total volume flow rate is the sum of all Darcy contributions, subject to the constancy of an unknown wellbore pressure (to be determined as part of the solution) when friction and gravity are neglected. How do general non-neighboring grid blocks imposed by well topology adversely affect the matrix structure of your ADI method? Can you save the tridiagonal scheme? What solution strategies would you suggest?

Chapter 11

Effective Properties in Single and Multiphase Flows

In this chapter, we consider effective properties in single and two-phase flows. We shall explore the pitfalls that workers are likely to encounter and point out subtleties that are not well known. We shall not give a general theory applicable to all problems, nor is any claim made to. Our objectives are, in fact, quite limited but to the point. Consistent with the philosophy underlying this book, we shall develop exact, closed-form solutions in order to study their analytical structure, with a view toward understanding the nuances of these solutions and their potential (or lack of) for generalization. Toward this end, we begin with simple models in order to keep both the mathematics and the physics comprehensible, and thereafter, we shall consider problems of increasing difficulty.

EXAMPLE 11.1. CONSTANT DENSITY LIQUID IN STEADY LINEAR FLOW

We consider the steady flow of a constant density liquid through two linear cores placed in series, as shown in Fig. 11.1. We emphasize the stringent assumptions enforced in the next paragraphs: liquid (with an exponent of $m = 0$), steady-state, one-dimensional, single-phase flow, homogeneous cores in series.

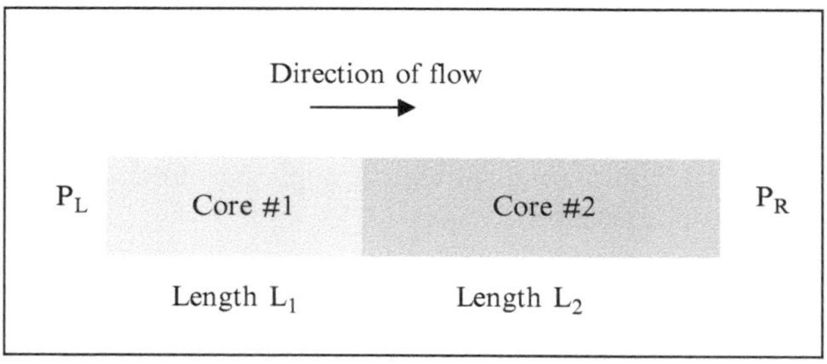

FIG. 11.1 Two linear cores in series.

Quantitative Methods in Reservoir Engineering. http://dx.doi.org/10.1016/B978-0-12-810518-4.00011-6

In this limit, the governing equations are $\partial^2 P_1/\partial x^2 = 0$ and $\partial^2 P_2/\partial x^2 = 0$, where the subscripts refer to the core sample. The solutions are linear functions of x taking the form Ax + B, where A and B are constants. Their values are found using the left and right conditions $P_1(0) = P_L$ and $P_2(L_1 + L_2) = P_R$, and also pressure and flow rate matching conditions at the core interface, $P_1(L_1) = P_2(L_1)$ and $(k_1/\mu)\partial P_1(L_1)/\partial x = (k_2/\mu)\partial P_2(L_1)/\partial x$. Here, k_1 and k_2 represent core permeabilities, and μ is the fluid viscosity of the single fluid. The required solutions are straightforwardly obtained as

$$P_1(x) = P_L + k_2(P_R - P_L)x/(k_1 L_2 + k_2 L_1) \tag{11.1}$$

$$P_2(x) = P_R + k_1(P_R - P_L)(x - L_1 - L_2)/(k_1 L_2 + k_2 L_1) \tag{11.2}$$

The two corresponding Darcy velocities are obtained from

$$q_1 = -(k_1/\mu)\,\partial P_1/\partial x = -k_1 k_2 (P_R - P_L)/\{\mu(k_1 L_2 + k_2 L_1)\} \tag{11.3}$$

$$q_2 = -(k_2/\mu)\,\partial P_2/\partial x = -k_2 k_1 (P_R - P_L)/\{\mu(k_1 L_2 + k_2 L_1)\} \tag{11.4}$$

Effective Permeability and Harmonic Averaging

Very often, it is desired to visualize these two sequential flows as the equivalent flow through a single system; that is, the steady flow through a single core of length $(L_1 + L_2)$. In this case, we observe from Eqs. (11.3), (11.4) that

$$\text{"}q\text{"} = q_1 = q_2 = -(1/\mu)\{k_l k_2(L_1 + L_2)/(k_1 L_2 + k_2 L_1)\}\{(P_R - P_L)/(L_1 + L_2)\} \tag{11.5}$$

In Eq. (11.5), which takes on the familiar Darcy form, it is possible to interpret $(P_R - P_L)/(L_1 + L_2)$ as the effective pressure gradient yielding the equivalent flow rate q, provided an effective permeability is defined with

$$k_{eff} = k_1 k_2 (L_1 + L_2)/(k_1 L_2 + k_2 L_1) \tag{11.6}$$

Eq. (11.6) contains the well-known definition for the harmonic average. Our derivation brings out clearly the very stringent set of circumstances under which the foregoing formula for effective permeability applies. Thus, one cannot blindly use Eq. (11.6) for constant density radial flows, for transient compressible linear flows, for two-phase flows, for gaseous flows having nonzero values of m, and so on. Each flow limit, in general, must be studied on its own merits. Nonetheless, several commercial simulators apply harmonic averaging outside the scope of its valid derivation.

Cores Arranged in Parallel

Only when cores are arranged in parallel, for example, does the usual arithmetic summing process apply. Consider two independent cores open to the same pressure sources; these produce at the flow rates $q_1 = -(k_1/L_1)\{(P_R - P_L)/\mu\}$

and $q_2 = -(k_2/L_2)\{(P_R - P_L)/\mu\}$, where we have rearranged Darcy's law but otherwise left it unchanged. The net flow rate from both cores is $q_{total} = q_1 + q_2 = (-1/\mu)k_{eff}\,[(P_R - P_L)/\{1/2(L_1 + L_2)\}]$ for an effective core having the average length $\frac{1}{2}(L_1 + L_2)$ and a permeability of $k_{eff} = \{(k_1L_2 + k_2L_1)/(L_1L_2)\}\{1/2(L_1 + L_2)\}$. Here k_{eff} is *different* from Eq. (11.6) for flows in series; only when both core lengths are identical do we obtain the simple recipe $k_{eff} = k_1 + k_2$.

Effective Porosity and Front Tracking

We now return to the serial flow problem in Fig. 11.1. To illustrate the dangers behind the blind use of effective permeabilities, consider the practical problem of monitoring tracer breakthrough. Dyes, salts, and radioactive tracers are conventionally used, of course, to map reservoir connectivity; the time required for the tracer to traverse the distance between two wells provides an indicator of reservoir flow resistance. Now, we suppose that the initial position of the tracer (dye) is $x = 0$ at $t = 0$. The speed of the front is a constant, taking on the value

$$dx_1/dt = q_1/\phi_1 = -k_1k_2(P_R - P_L)/\{\mu\phi_1(k_1L_2 + k_2L_1)\} \tag{11.7}$$

where ϕ_1 is the porosity in the first core. Thus, the time T_1 required for the dye to traverse from $x = 0$ to $x = L_1$, ignoring molecular diffusion, is

$$T_1 = L_1/(dx_1/dt) = \mu\phi_1L_1(k_1L_2 + k_2L_1)/\{k_1k_2(P_L - P_R)\} \tag{11.8}$$

Similarly, the time T_2 required to move from $x = L_1$ to $x = L_1 + L_2$ is

$$T_2 = \mu\phi_2L_2(k_1L_2 + k_2L_1)/\{k_1k_2(P_L - P_R)\} \tag{11.9}$$

Hence, the average speed of the tracer front is

$$\begin{aligned}(L_1 + L_2)/(T_1 + T_2) = -(1/\mu)\,&\{k_1k_2(L_1 + L_2)/(k_1L_2 + k_2L_1)\}\\ &\{(L_1 + L_2)/(\phi_1L_1 + \phi_2L_2)\}\,\{(P_R - P_L)/(L_1 + L_2)\}\end{aligned} \tag{11.10}$$

where the { } indicate three consecutive products. In Eq. (11.10), we recognize the expression k_{eff} on the first line. However, on the second line, we observe the presence of a new product $1/\phi_{eff}$, that is, an effective porosity

$$\phi_{eff} = (\phi_1L_1 + \phi_2L_2)/(L_1 + L_2) \tag{11.11}$$

The averaging process defined by Eq. (11.11) is not harmonic; it is the commonly used length-weighted average.

The Lessons Learned

There are several lessons to be learned from these exact calculations. First, the notion of k_{eff} as obtained here is very restrictive. Second, an exact calculation performed using such an effective property may be useful in replicating net flow rates. However, in extracting tracer arrival time, for example, using effective permeabilities alone without appropriately modifying the porosity will lead to error. This limitation arises because there is an inherent loss of flow information whenever *any* kind of averaging is considered. Our calculations assumed serial flow. In any real reservoir flow simulation containing numerous grid blocks and multiple wells, combinations of parallel and serial flow will be achieved. The resulting effective properties for permeability and porosity will depend on the number and orientation of the grid blocks selected. Also, they will apply only to the particular simulations, well topologies, and relative pressure drops at hand: effective properties are not absolute properties defined at the microscopic level, since they are process-dependent. Thus the use of effective properties, like the mysterious role played by pseudos in reservoir simulation, while of interest in practice, may be difficult to assess academically.

EXAMPLE 11.2. LINEAL MULTIPHASE FLOW IN TWO SERIAL CORES

The results of Example 11.1 are exact, and we might legitimately ask what their value or role might be in the study of immiscible two-phase constant density flow. Again, we shall assume the same geometric flow configuration and obtain closed-form results. The entire derivation is given for completeness to highlight the assumptions. The reader is referred to standard texts for discussions of basic equations (e.g., Collins, 1961; Peaceman, 1977).

Darcy's Laws

Let the subscripts w and nw denote wetting and nonwetting, and let 1 and 2 denote the first and second cores, respectively, having lengths L_1 and L_2. We now consider the first core in detail. Here, the relevant Darcy velocities are

$$q_{1w} = -(k_{1W}/\mu_W)\,\partial P_{1W}/\partial x \tag{11.12}$$

$$q_{1nW} = -(k_{1nW}/\mu_{nW})\,\partial P_{1nW}/\partial x \tag{11.13}$$

where μ_w and μ_{nw} are viscosities. For simplicity, we assume high flow rates, that is, zero capillary pressures p_c, so that

$$P_{1nW} - P_{1W} = P_c = 0 \tag{11.14}$$

This simplification is *not* used later in this book. For now, since $p_{1nw} = p_{1w}$, the pressure gradient terms in Eqs. (11.12), (11.13) are identical. If we divide Eq. (11.12) by Eq. (11.13), these terms cancel, and we obtain

$$q_{1nW} = (k_{1nW}\,\mu_W/k_{1W}\,\mu_{nW})\,q_{1W} \tag{11.15}$$

Mass Conservation

At this point, we invoke mass conservation, recalling the assumption of a constant density flow. Then, it follows that

$$\partial q_{1w}/\partial x = -\phi_1\, \partial S_{1w}/\partial t \tag{11.16}$$

$$\partial q_{1nw}/\partial x = -\phi_1\, \partial S_{1nw}/\partial t \tag{11.17}$$

Since the wetting and nonwetting fluid saturations S_{1w} and S_{1nw} in Eqs. (11.16), (11.17) must sum to a constant value of unity, that is,

$$S_{1w} + S_{1nw} = 1 \tag{11.18}$$

it follows from adding Eqs. (11.16), (11.17) that

$$\partial(q_{1w} + q_{1nw})/\partial x = 0 \tag{11.19}$$

Hence, a total speed q can be defined from the integral of Eq. (11.19),

$$q_{1w} + q_{1nw} = q = q(t) \tag{11.20}$$

where a dependence on time is allowed.

Fractional Flow Functions

Now, it is convenient to define the fractional flow function f_{1w} for the wetting phase by the quotient

$$f_{1w} = q_{1w}/q \tag{11.21}$$

Then, for the nonwetting phase, we obtain

$$f_{1nw} = q_{1nw}/q = (q - q_{1w})/q = 1 - f_{1w} \tag{11.22}$$

where we have used Eq. (11.20). The foregoing result can be written in the form $q_{1nw} = q\,(1 - f_{1w})$. Eq. (11.21) is similarly recast as $q_{1w} = qf_{1w}$. If these equations for q_{1w} and q_{1nw} are substituted in Eq. (11.15), the q's cancel, and we obtain

$$f_{1w} = 1/\{1 + (k_{1nw}\,\mu_w/k_{1w}\,\mu_{nw})\} \tag{11.23}$$

Saturation Equations

The function $f_{1w} = f_{1w}\,(S_{1w}, \mu_w/\mu_{nw})$ in Eq. (11.23) is a function of the viscosity ratio μ_w/μ_{nw} and the saturation S_{1w}. Thus, the left side of Eq. (11.16), for example, transforms according to

$$\partial q_{1w}/\partial x = q\,\partial f_{1w}/\partial x = q\,\{df_{1w}(S_{1w})/dS_{1w}\}\,\partial S_{1w}/\partial x \tag{11.24}$$

Eqs. (11.16), (11.24) together imply that the saturation S_{1w} satisfies the first-order nonlinear partial differential equation

$$\partial S_{1w}/\partial t + (q/\phi_1)\,\{df_{1w}(S_{1w})/dS_{1w}\}\,\partial S_{1w}/\partial x = 0 \tag{11.25}$$

A similar derivation for the second core (downstream to the first) leads to

$$\partial S_{2w}/\partial t + (q/\phi_2)\,\{df_{2w}(S_{2w})/dS_{2w}\}\,\partial S_{2w}/\partial x = 0 \tag{11.26}$$

where the q(t)'s in Eqs. (11.25), (11.26) are identical, since the fluid leaving the first core enters the second in its entirety.

Solving the Saturation Equations

We now consider a physical formulation for the two-phase flow boundary-initial value problem that is complementary to Example 11.1. We shall assume that both core samples are initially held at the same constant saturation S_{w^i} throughout, that is,

$$S_{1w}(x, 0) = S_w{}^i \tag{11.27a}$$

$$S_{2w}(x, 0) = S_w{}^i \tag{11.27b}$$

Also, we assume that at subsequent times, the left boundary $x = 0$ at the entrance to the first core is flooded with a saturation equal to

$$S_{1w}(0, t) = S_{1w}{}^L \tag{11.28}$$

The resulting boundary value formulations for $S_{1w}(x,t)$ and $S_{2w}(x,t)$ lead to the so-called Buckley-Leverett problem well known to reservoir engineers. Their solutions can contain shockwaves or steep saturation discontinuities, depending on the form of the fractional flow functions and the initial conditions. The basic issues are discussed in Collins (1961) and will not be repeated here.

Various schemes (e.g., Welge's integration) are used to render the saturation a single-valued function, which are analogous to the shock-fitting used in high-speed gasdynamics in aerospace engineering. The simplest, and by far the most elegant, is described in the classic nonlinear wave mechanics book of Whitham (1974). There, he shows that the well-known signaling problem defined by Eqs. (11.25), (11.27a), (11.28) possesses a shock speed V_{1shock} available in exact closed analytical form. It is

$$V_{1shock} = \{Q_{1w}(S_{1w}{}^L) - Q_{1w}(S_w{}^i)\}/\{S_{1w}{}^L - S_w{}^i\} \tag{11.29}$$

where we have denoted, for simplicity,

$$Q(S_{1w}) = (q/\phi_1)\{df_{1w}(S_{1w})/dS_{1w}\} \tag{11.30}$$

(refer to Eq. 11.25). Thus, assuming now that q(t) is a constant q, the time T_1 required for the influence of the flood at $x = 0$ to be felt at the end of the first core (and hence, the beginning of the second) is

$$T_1 = L_1/V_{1shock} = L_1\{S_{1W}{}^L - S_W{}^i\}/\{Q_{1W}(S_{1W}{}^L) - Q_{1W}(S_W{}^i)\} \tag{11.31a}$$

The saturation profile, incidentally, takes the form of a step function having constant values $S_{1w}{}^L$ and $S_w{}^i$ at the left and right sides of the shock. Once the shock front reaches the second core, which starts at $x = L_1$ and ends at the position $x = L_1 + L_2$, the flood saturation $S_{1w}{}^L$ acts, just as it had in the case of the first core. Hence, the time T_2 required for the shock to traverse the length L_2 can be obtained by inference from Eq. (11.31a). That is,

$$T_2 = L_2/V_{2shock} = L_2\{S_{1W}{}^L - S_W{}^i\}/\{Q_{2W}(S_{1W}{}^L) - Q_{2W}(S_W{}^i)\} \quad (11.31b)$$

The total elapsed travel time

$$\begin{aligned} T_1 + T_2 &= L_1\{S_{1W}{}^L - S_W{}^i\}/\{Q_{1W}(S_{1W}{}^L) - Q_{1W}(S_W{}^i)\} \\ &\quad + L_2\{S_{1W}{}^L - S_W{}^i\}/\{Q_{2W}(S_{1W}{}^L) - Q_{2W}(S_W{}^i)\} \\ &= (S_{1W}{}^L - S_W{}^i)[L_1/\{Q_{1W}(S_{1W}{}^L) - Q_{1W}(S_W{}^i)\} \\ &\quad + L_2/\{Q_{2W}(S_{1W}{}^L) - Q_{2W}(S_W{}^i)\}] \end{aligned} \quad (11.32)$$

Thus, the average shock speed takes the form

$$\begin{aligned} V_{avg\ shk} &= (L_1 + L_2)/(T_1 + T_2) \\ &= (L_1 + L_2)/\{(S_{1W}{}^L - S_W{}^i)[L_1/\{Q_{1W}(S_{1W}{}^L) - Q_{1W}(S_W{}^i)\} \\ &\quad + L_2/\{Q_{2W}(S_{1W}{}^L) - Q_{2W}(S_W{}^i)\}]\} \end{aligned} \quad (11.33)$$

Characteristic Speeds in Reservoir Analysis

There are several different measures by which we can characterize speed in a reservoir. In the simple case treated in Example 11.1, consisting of two lineal cores having lengths L_1 and L_2 arranged in series, we first identified a net flow velocity q which motivated effective permeability. Then, the tracer problem suggested that the concept of effective permeability was not enough to describe all aspects of the problem, and an effective porosity was identified. Finally, we considered here a simple, but physically important, two-phase flow problem that yielded to exact analysis. In this limit, we identified still another speed, the shock velocity, which is given in Eq. (11.33). This velocity is important because it leads to the time required for water breakthrough. It is evident from Eq. (11.33) that, given the nontrivial presence of the Q's, the value of $V_{avg\ shk}$ cannot be easily related to the effective permeability or the effective porosity defined for single-phase flows. These results suggest that each of the effective properties, while interesting in its own right, has little to do with the other in a strictly rigorous sense. Many commercial simulators use effective properties definitions indiscriminately. Often used recipes, for example, the use of effective permeabilities based on Example 11.1 in three-dimensional, multiphase applications with different boundary conditions, completely outside the range of validity of the original derivation, are just that: recipes that, while plausible, are in fact very incorrect. This conclusion is all the more true in real reservoirs containing multiple wells, where numerous grid blocks are used to simulate combinations of parallel and serial flow. Calculated effective properties and breakthrough velocities and times will, in general, depend on relative pressure drops, well locations, reservoir shape, boundary conditions, and so on.

The Multiphase Pressure Field

We conclude this example by demonstrating how a closed-form solution for the transient pressure distribution can be obtained. This exercise is important because, to this author's knowledge, analytical solutions to similar nontrivial problems have not been given. For this purpose, let us substitute Eqs. (11.12), (11.13) into Eq. (11.20); then, using Eq. (11.14), we obtain

$$\{(k_{1w}(S_{1w})/\mu_w) + (k_{1nw}(S_{1w})/\mu_{nw})\}\partial P_{1w}/\partial x = -q(t) \tag{11.34}$$

On rearranging, we find that the spatial pressure gradient satisfies

$$\partial P_{1w}/\partial x = -q(t)/\{(k_{1w}(S_{1w})/\mu_w) + (k_{1nw}(S_{1w})/\mu_{nw})\} \tag{11.35}$$

Since the saturation $S_{1w}(x,t)$ (following Whitham's solution to the signaling problem) is a step function in the x direction whose hump moves at the shock velocity, the pressure gradient in Eq. (11.35) takes on either of two constant values, depending on whether S_{1w} equals S^L_{1w} or S^i_w locally. Thus, on either side of the shock front, we have different but linear pressure variations when time is held fixed. This situation is shown in Fig. 11.2. At the shock front itself, the requirement that pressure be continuous is sufficient to uniquely define the time-varying pressure drop across the core sample. This completes our general analysis; details are left to the reader.

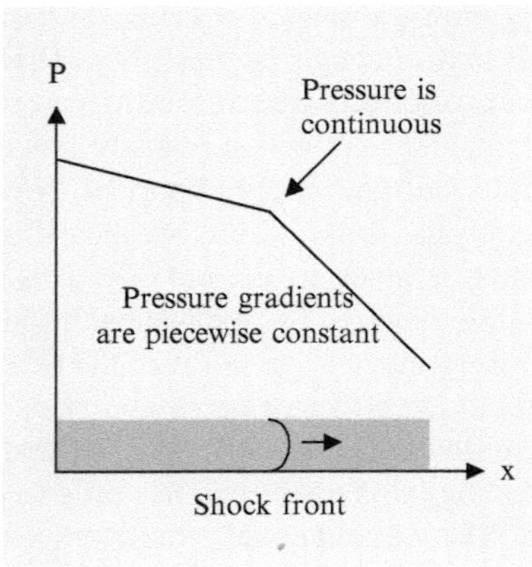

FIG. 11.2 Spatial pressures in two-phase flow.

EXAMPLE 11.3. EFFECTIVE PROPERTIES IN STEADY CYLINDRICAL FLOW

In Example 11.1, we considered two lineal cores in series, assuming steady, single-phase flow, and derived the harmonic average rule from first principles. It is straightforward, extending these results to radial flow. For example, concentric annular rings having different permeabilities can be treated with $d^2p/dr^2 + 1/r\ dp/dr = 0$ as the host model, yielding solutions $A + B \log r$, as opposed to $d^2p/dx^2 = 0$ which produces $A + Bx$ solutions. The analytical solution can be found as before, by matching pressures and velocities at the interface. A harmonic average rule with logarithm-based weighting functions can be derived by addressing the following. What is the effective radius? The effective pressure gradient? How must effective permeability be defined so that the solution looks like that for a single medium problem?

EXAMPLE 11.4. STEADY, SINGLE-PHASE, HETEROGENEOUS FLOWS

Here we generalize the harmonic rule for discrete cores having distinctly different rock properties, to single lineal cores having continuous heterogeneous properties. This is accomplished by considering

$$d\{k(x)\,dp/dx\}/dx = 0 \tag{11.36}$$

where k(x) is variable (a liquid is assumed, noting that p^{m+1} replaces p for gas flows). If the left and right pressure boundary conditions of Fig. 11.1 are assumed for a single core of length L, we obtain on integration

$$p(x) = (P_R - P_L) \int_0^x d\zeta/k(\zeta) \Big/ \int_0^L d\zeta/k(\zeta) + P_L \tag{11.37a}$$

This solution can be validated by differentiation; it may find use in reducing grid numbers in reservoir flow simulations. Also note that the Darcy flow velocity $q = -\{k(x)/\mu\}dp/dx$ is constant for steady flow. It can be written as

$$q = -(1/\mu)\left\{L \Big/ \int_0^L d\zeta/k(\zeta)\right\}[(P_R - P_L)/L] \tag{11.37b}$$

Therefore, the quantity in the curly { } brackets represents the effective permeability for heterogeneous flow, reducing to a single constant k in homogeneous media.

EXAMPLE 11.5. TIME SCALE FOR COMPRESSIBLE TRANSIENTS

So far, we have considered steady flows. In practice, nonzero compressibility means that transient effects can be important, and one may be concerned as to the applicability of steady-state results. We shall not address multicore examples, but will treat a single core of length L with constant properties. The objective is a simple expression defining the dimensionless time scale τ over which transient effects can be confidently predicted to have mostly dissipated. At that time, steady flow results hold. The boundary conditions used in Example 11.3 again apply here, and the governing equation for single-phase, liquid flows is

$$\partial^2 p/\partial x^2 = \phi\mu c/k \, \partial p/\partial t \tag{11.38}$$

In Eq. (11.38), p(x,t) is the transient pressure, and ϕ, μ, c, and k represent porosity, viscosity, compressibility, and permeability. Standard separation of variables procedures lead to a Fourier series solution of the form

$$\begin{aligned} p(x,t) &= P_L + (x/L)(P_R - P_L) \\ &+ \sum_{n=1}^{\infty} A_n \exp\left\{-n^2\pi^2 kt/\phi\mu c L^2\right\} \sin n\pi x/L \end{aligned} \tag{11.39}$$

where the coefficients A_n are determined from initial conditions assumed for p(x,0). If we set t = 0 in Eq. (11.39), we obtain

$$\sum_{n=1}^{\infty} A_n \sin n\pi x/L = p(x,0) - P_L - (x/L)(P_R - P_L) \tag{11.40}$$

Formulas for the Fourier coefficients A_n, once the right side of Eq. (11.40) is specified, are available in mathematical references (e.g., Hildebrand, 1948). The first line of Eq. (11.39) displays the steady-state response, while the second describes transient behavior. The strongest temporal response is given by n = 1, defining the relevant nondimensional time scale as

$$\tau = \pi^2 kt/\phi\mu c L^2 \tag{11.41}$$

This combination of terms, and not the rock and fluid properties parameters individually, controls the asymptotic convergence of our unsteady flow to steady state. In this chapter, we have presented several exact closed-form solutions describing single- and two-phase flow through cores in series. These solutions, not unexpectedly, demonstrate that the flowfields are completely different; hence, any attempt to apply effective properties obtained from simple problems (e.g., Example 11.1) to more complicated ones will result in error.

PROBLEMS AND EXERCISES

1. We shall derive a harmonic average rule for radial flow. Consider constant density liquid flowing into a well of radius R_{well}. The well is surrounded by a first concentric ring having permeability k_1 and ending at radial position R_1, which is, in turn, surrounded by a second concentric ring having a permeability k_2 and ending at position R_2. The viscosity of the single fluid is μ. The well pressure is P_W and the farfield reservoir pressure is P_R. Each pressure distribution satisfies $d^2p/dr^2 + 1/r \, dp/dr = 0$, which yields solutions A + B log r in each concentric domain. Obtain the complete two-layer analytical solution for this problem by matching pressures and velocities at the interface. What is the solution if the second outer layer did not exist? Rearrange the

two-layer solution into the form taken by the single-layer solution, and compare the two solutions. How would you define effective permeability for two-layer radial problems? For N-layer problems?

2. In the above, the equation $d^2p/dr^2 + 1/r\ dp/dr = 0$ was used because constant density liquids were assumed. Extend the analysis to steady-state gas flows satisfying $d^2p^{m+1}/dr^2 + 1/r\ dp^{m+1}/dr = 0$. Note that this is *not* a nonlinear problem, since the governing equation is linear in p^{m+1}.
3. In the text, we considered effective properties for liquids in linear cores with heterogeneous permeabilities satisfying $d\{k(x)\ dp/dx\}/dx = 0$. What is the governing equation for a radial flow with variable properties k(r)? Derive the expression for effective permeability. What is the expression assuming steady-state gas flows with general exponents m?
4. In Example 11.5, compressible transient flow in a single homogeneous lineal core was studied. Repeat this exercise for two lineal cores in series. Also, extend the analysis of Example 11.5 to single and dual radial cores.

Chapter 12

Modeling Stochastic Heterogeneities

The appearance of geological patterns bearing somewhat random features, or complicated, but periodic, well-defined physical structures, very often raises questions among flow modelers regarding possibilities for quantitative flow simulation. The reservoir description process—the ability to describe geological structures accurately—is certainly not useful unless the ability to simulate flows is equally well developed. Thus, a fractal description may not be meaningful unless a suitable fractal calculus is at hand or unless supercomputers are used for direct, fine-scale analysis. Likewise, a good stochastic description of a petroleum reservoir completes only part of the job: efficient flow formulations taking advantage of the particular description methodology should be diligently researched and posed. We again caution against blind use of effective properties and upscaling methods, since these methods can be quite restrictive, as noted in Chapter 11. However, despite geometric complication, analytical methods are possible, drawing on literature from other areas of continuum mechanics.

OBSERVATIONS ON EXISTING MODELS

In this section, comments on several existing models will be made. Many models attempt to appear quantitative, giving the impression of scientific rigor when, instead, crucial flow details and dynamical effects are ignored and lost.

Dual Porosity Models

The best known attempts at simple continuum models are the dual porosity approaches for naturally fractured reservoirs (Aguilera, 1980; van Golf-Racht, 1982). These are usually idealized by considering three-dimensional arrays of stacked sugar cubes. The cubes represent oil-bearing matrix rock, which release fluid to the highly permeable surrounding fractures (i.e., the surfaces between the sugar cubes) for transport to the well. But the stochastic nature of the problem is avoided; even the periodic flow simplification provided by this idealization is ignored, in favor of a less clearly defined two-continuum description. The first medium describes the rock matrix, which satisfies Darcy's

Quantitative Methods in Reservoir Engineering. http://dx.doi.org/10.1016/B978-0-12-810518-4.00012-8

equations, already familiar to engineers. However, this continuum, in turn, appears as a source-like term to a second continuum describing macroscopic fractures; the discrete nature of the matrix rock and the fractures in dual porosity approaches is lost and disappears.

The end modeling results are coupled immiscible flow equation systems, containing twice as many input parameters as the more rational single-porosity model would have: two sets of relative permeability curves, two sets of capillary pressure curves, and so on. Consequently, such hopelessly ill-defined approaches, given the dearth of real-world data, not to mention errors likely to be found in laboratory measurement, may never see complete validation. Simpler flow models for periodic shales and fractures, such as those introduced in Chapter 5, shed greater physical insight.

Geostatistical vs. Direct Modeling

Mathematical geostatisticians often develop their reservoir models by minimizing suitably defined error functions that are consistent with measured statistics, subject to auxiliary boundary constraints. These functions are typically positive definite, so that a solution to the minimization process exists. Usually, the particular function used is chosen only to simplify the mathematics, and its connection to the actual depositional process, the flow properties, or the physical appearance of the rock is not considered. In contrast, many civil engineers, hydrologists, and fluid dynamicists have designed large-scale computer models that solve sophisticated formulations that couple the flow equations together with empirical erosion laws determined in the laboratory. They have been successful in generating solutions that, qualitatively anyway, often approach reality. A case in point is the numerical simulation of meandering rivers and streams as they develop and evolve over geological time scales. Geostatistical research needs to focus on reality before it can be accepted by the general user community.

Mathematical Connections

The differential equation methods used in modeling and the optimization approaches used in geostatistics are closely related. The similarities are explored in variational calculus, a well-known mathematical specialty that relates differential equations to global minimization problems (Garabedian, 1964; Hildebrand, 1965; Stakgold, 1968). Such approaches are not new; for instance, structural engineers have employed differential equation models side by side with equivalent minimum total strain energy methods for decades. In the fluid dynamics context, the solution to the Laplace equation $\partial^2 p/\partial x^2 + \partial^2 p/\partial y^2 + \partial^2 p/\partial z^2 = 0$ subject to appropriate boundary conditions, can be *exactly* translated into a variational problem that minimizes an energy-like integral, in particular,

$$\text{Min} \int \left\{ (\partial p/\partial x)^2 + (\partial p/\partial y)^2 + (\partial p/\partial z)^2 \right\} dx\,dy\,dz + \text{constraints}$$

over the flow domain, when augmented by additional conditions. In fact, most differential equation formulations can be posed as equivalent variational problems and solved by techniques not covered in this book. Minimization algorithms, Lagrange multipliers, optimization methods, and the like, would replace the relaxation methods of Chapters 7 and 9, and it can be shown that identical solutions can be obtained using approaches that are very different.

We are not proposing solutions of the flow equations using optimization methods, which are a definite possibility. However, the fact that a one-to-one correspondence relates the two indicates that every optimization formulation designed by a geostatistician implies an equivalent differential equation model and physical process. Something about the physics is implicitly assumed that must be subject to external tests. What is it? Is it reasonable? For example, suppose we have no experience with differential equations and that someone new to the profession proposes mimimization as a means to determine pressure. Is it correct? Yes, in that we expect pressure gradients to be smooth globally. But the fact that the method *is* related to $\partial^2 p/\partial x^2 + \partial^2 p/\partial y^2 + \partial^2 p/\partial z^2 = 0$ means that the variational problem applies to isotropic media only. And the fact that this *is* Laplace's equation means that mass is conserved.

In closing, we emphasize that geostatistical methods may contain implicit assumptions about the flow that deserve clarification. Also, we note that by rewriting the differential equation formulations in variational form, additional solution methods, algorithms, and physical interpretations are available. Having established the equivalence between differential equation approaches and optimization or global minimization methods, we next show how the discrete nature of periodic heterogeneities *can* be retained and modeled by direct analysis using exact but conventional mathematical methods. The simple limits considered here are designed to motivate further research along these lines; they have the potential for providing detailed descriptions of flow past stochastic shales and fractures.

A MATHEMATICAL STRATEGY

For simplicity, let us consider a three-dimensional reservoir having the isotropic, variable permeability k(x,y,z). Assuming a constant density, single-phase, steady liquid flow, we have the partial differential equation

$$\partial(k\,\partial p/\partial x)/\partial x + \partial(k\partial p/\partial y)/\partial y + \partial(k\partial p/\partial z)/\partial z = 0 \qquad (12.1)$$

for the pressure p(x,y,z). The relaxation methods developed in Chapters 7 and 15 can be used to solve Eq. (12.1), of course, when k(x,y,z) is specified and treated as an indexed array. In this section, we demonstrate how detailed classes of k(x,y,z) can be defined to both simplify the mathematics and provide highly detailed solutions throughout the reservoir.

Permeability Modeling

Direct recourse to high-powered numerical analysis is not often necessary or even desired. Following Bear (1972), we introduce an auxiliary function g(x,y,z) defined by

$$g(x, y, z) = p(x, y, z) \sqrt{k(x, y, z)} \tag{12.2}$$

Then, it is possible to verify that

$$\partial^2 g/\partial x^2 + \partial^2 g/\partial y^2 + \partial^2 g/\partial z^2 + \alpha^2 g = 0 \tag{12.3a}$$

if

$$\partial^2 \sqrt{k}/\partial x^2 + \partial^2 \sqrt{k}/\partial y^2 + \partial^2 \sqrt{k}/\partial z^2 + \alpha^2 \sqrt{k} = 0 \tag{12.3b}$$

where α is a constant. Eqs. (12.3a), (12.3b) describe the so-called Helmholtz inhomogeneous medium. If $\alpha = 0$, direct algebraic manipulations show that

$$\partial^2 g/\partial x^2 + \partial^2 g/\partial y^2 + \partial^2 g/\partial z^2 = 0 \tag{12.4a}$$

if

$$\partial^2 \sqrt{k}/\partial x^2 + \partial^2 \sqrt{k}/\partial y^2 + \partial^2 \sqrt{k}/\partial z^2 = 0 \tag{12.4b}$$

Eqs. (12.4a), (12.4b) define the harmonically inhomogeneous medium. Bear (1972), however, does not provide flow solutions to these formation models. It turns out, though, that simple solutions are possible. We now explain how these equations can be solved using methods in heat transfer and structural vibrations.

Physical Implications

As we indicated in our introductory discussion, differential equation models are associated with equivalent energy minimization problems. For Eq. (12.3b), the appropriate model is

$$\text{Min} \int_V \left\{ \left(\partial\sqrt{k}/\partial x\right)^2 + \left(\partial\sqrt{k}/\partial y\right)^2 + \left(\partial\sqrt{k}/\partial z\right)^2 + \alpha^2\left(\sqrt{k}\right)^2 \right\} dxdydz \tag{12.3c}$$

whereas for Eq. (12.4b), the minimization

$$\text{Min} \int_V \left\{ \left(\partial\sqrt{k}/\partial x\right)^2 + \left(\partial\sqrt{k}/\partial y\right)^2 + \left(\partial\sqrt{k}/\partial z\right)^2 \right\} dxdydz \tag{12.4c}$$

applies; in Eqs. (12.3c), (12.4c), the integration volume V represents the domain of flow. Thus, while Eqs. (12.3b), (12.4b) appear as analytical conditions proposed for mathematical simplification only, their physical interpretation following Eqs. (12.3c), (12.4c) suggests that they imply a certain global smoothness in the permeability distribution—a reasonable basis for a sound depositional model given the smoothing effects of geological time (e.g., imposed natural elements like wind and erosion). Of course, the fact that Eqs. (12.3b), (12.4b) *are* differential equations means that infinite classes of solutions can be generated as test cases against actual reservoir constraints, such as tracer data, production histories, and seismic tests. More importantly, this opens up the possibility of powerful mathematical techniques developed in recent years in other disciplines.

Mathematical Approaches

In structural mechanics, researchers normally deal with wave equations of the form $\partial^2 u/\partial x^2 + \partial^2 u/\partial y^2 + \partial^2 u/\partial z^2 = \partial^2 u/\partial t^2$, where t is time. Steady state oscillations are studied by assuming sinusoidal Fourier components of the form $u(x,y,z,t) = U(x,y,z)e^{i\omega t}$, where ω is the frequency. Direct substitution leads to a governing equation for U(x,y,z), namely, $\partial^2 U/\partial x^2 + \partial^2 U/\partial y^2 + \partial^2 U/\partial z^2 + \omega^2 U = 0$. This time-independent equation is often studied from the viewpoint of eigenfunction analysis: its standing wave vibrating modes are sought. Now, this equation for U(x,y,z) is identical to Eq. (12.3b) for $\sqrt{k}$; thus, for example, the repeating patterns of crests and troughs seen on vibrating plates can be interpreted as similar patterns of periodic heterogeneities in reservoirs. It goes without saying that the wealth of vibrations solutions available in structural analysis can be profitably used for interdisciplinary work in reservoir description. In the illustrative example that follows, we adapt a solution from transient heat transfer analysis to solve the governing permeability equations for a complicated fracture geology. Interestingly, in the new approach, we need not make use of dual porosity or direct statistical methods in order to model the flow exactly.

EXAMPLE 12.1. CONTRACTIONAL FRACTURES

Let us complete this chapter with an illustrative example that describes the flow through the complicated contractional fracture system in Fig. 12.1. We shall presume that the harmonic permeability assumption applies: again, a smooth depositional model that globally minimizes the spatial rates of variation of $\sqrt{k}$ is taken, so that Eq. (12.4b) holds.

$$\partial^2 \sqrt{k}/\partial x^2 + \partial^2 \sqrt{k}/\partial y^2 + \partial^2 \sqrt{k}/\partial z^2 = 0 \qquad (12.4b)$$

Using standard separation of variables procedures, we set

$$\sqrt{k} = \sqrt{k_m} + F(x,y)\,G(z) \qquad (12.5a)$$

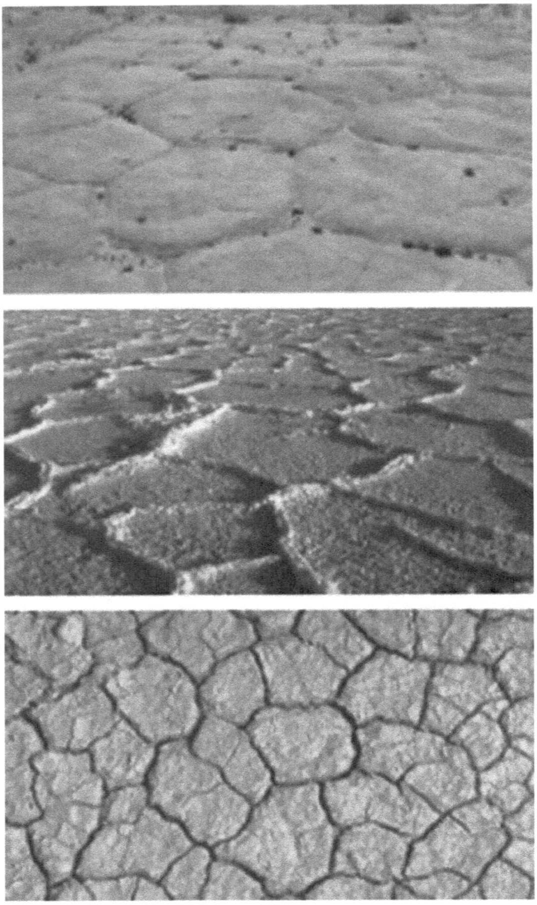

FIG. 12.1 Contractional fractures.

where

$$\sqrt{k_m} = \sigma + \alpha x + \beta y + \gamma z \tag{12.5b}$$

can be a constant or variable mean reference, or background average level assumed for the reservoir (σ, α, β, and γ are free constants). Actually, even more general second-degree polynomials can be used provided their Laplacian vanishes. This leads straightforwardly to the reduced equation

$$F_{xx}(x, y)\,G + F_{yy}(x, y)\,G + FG''(z) = 0 \tag{12.6}$$

If we next divide Eq. (12.6) through by G, we find that

$$F_{xx}(x, y) + F_{yy}(x, y) + (G''(z)/G)\,F = 0 \tag{12.7}$$

Now, let us consider rock formations for which $G''(z)/G(z)$ is a constant, say, a^2. For such problems, the ordinary differential equation

$$G''(z) - a^2 G = 0 \tag{12.8}$$

applies. The assumed vertical permeability varies exponentially, but the variation can be made as weak as desired. Then, Eqs. (12.7), (12.8) imply

$$F_{xx}(x, y) + F_{yy}(x, y) + a^2 F = 0 \tag{12.9}$$

Methods From Heat Transfer

The modeling of Benard convection cells in hydrodynamic stability is relevant to our purposes, since it leads to a modal equation identical to Eq. (12.9) (Yih, 1969). It turns out interestingly that Eq. (12.9) itself can be solved by separation of variables once more. For example, the choice

$$F(x, y) = \cos mx \cos ny, \; m^2 + n^2 = a^2 \tag{12.10}$$

leads to rectangular cells, while, for example, the selection

$$F(x, y) = f_0 \left[\cos \left\{ a \left(\sqrt{3}\, x + y \right) / 2 \right\} + \cos \left\{ a \left(\sqrt{3}\, x - y \right) / 2 \right\} + \cos ay \right] \tag{12.11}$$

yields hexagonal patterns (f_0, along with a^2, are free parameters available for geological modeling). Other patterns are similarly possible. At this juncture, we have, in closed form, the permeability function

$$\sqrt{k} = \sqrt{k_m} + f_0 G(Z) \left[\cos \left\{ a \left(\sqrt{3}\, x + y \right) / 2 \right\} + \cos \left\{ a \left(\sqrt{3}\, x - y \right) / 2 \right\} + \cos ay \right] \tag{12.12}$$

where f_0 and a can be tuned to provide the desired level of periodic variation to the heterogeneous background permeability $\sqrt{k_m}$. This capability permits both shale and fracture modeling, for the appropriate parameter values.

Pressure Solution

Next, consider the corresponding pressure field. We recall from Eqs. (12.2), (12.4a) that $g(x,y,z) = p(x,y,z)\sqrt{k(x,y,z)}$ satisfies $\partial^2 g/\partial x^2 + \partial^2 g/\partial y^2 + \partial^2 g/\partial z^2 = 0$. If we assume that both the permeabilities and pressures are known at all well positions and boundaries, it follows that $g = p\sqrt{k}$ can be prescribed as known Dirichlet boundary conditions. Then, the numerical methods devised in Chapter 7 for elliptic equations can be applied directly; on the other hand, analytical separation of variables methods can be employed for problems with idealized pressure boundary conditions. The general approach in this example is desirable for two reasons. First, the analytical constructions devised for the permeability function (see Eqs. 12.5b, 12.10, 12.11) allow us to retain full control over the details of small-scale heterogeneity. Second, the equation for the modified pressure g(x,y,z) (see Eq. 12.4a) does *not* contain variable, heterogeneity-dependent coefficients. It is, in fact, smooth; thus, it can be solved with a coarser mesh distribution than is otherwise possible.

Alternative Solutions for Permeability

We emphasize that we *could have* solved Eq. (12.4b) as a partial differential equation, subject to known values of $\sqrt{k}$ at prescribed boundaries numerically; this approach would have been purely "brute force." On the other hand, in Eqs. (12.10), (12.11), we chose analytical solutions that unmistakably bring out the pseudo-periodic structure of many geological entities found in nature. Other solution approaches are also possible. For example, consider a two-dimensional reservoir whose heterogeneities satisfy $\partial^2\sqrt{k}/\partial x^2 + \partial^2\sqrt{k}/\partial y^2 = 0$; that is, $\sqrt{k}$ is a harmonic function satisfying Laplace's equation. Following the mathematical ideas developed in complex variables in Chapters 4 and 5, it is clear that the real or imaginary part of *any* function f(z) of the complex variable $z = x + iy$ can be taken as a candidate solution for $\sqrt{k}$. For example, $f(z) = az + \sin bz$, or perhaps $f(z) = \sin(az + bz^2 + cz^3 + \ldots)$, might do; the trick is to allow enough free parameters to fit constrained values, in order to define solutions elsewhere. The solution to the differential equation for permeability then reduces to the search for appropriate level contours in f(z).

PROBLEMS AND EXERCISES

1. The equivalence between $\partial^2 p/\partial x^2 + \partial^2 p/\partial y^2 + \partial^2 p/\partial z^2 = 0$ and the formulation $\mathrm{Min}\int\left\{(\partial p/\partial x)^2 + (\partial p/\partial y)^2 + (\partial p/\partial z)^2\right\}dx\,dy\,dz$ for isotropic flow has been noted. What is the differential equation governing anisotropic problems with different permeabilities in the three coordinate directions? Write down the equivalent variational formulation. Review the basic numerical methods used to solve variational problems, and state their advantages and disadvantages over finite difference or finite element methods. What analytical methods, exact and approximate, are available?
2. Review the literature in structural mechanics, vibration theory, acoustics, heat transfer, and electrical engineering, in particular, with respect to the key words mode shape, eigenvalue problem, eigenvalue, natural modes, vibrating mode, waveguide, and related terminology. Start a pictorial catalog of one-, two-, and three-dimensional mode shapes, and compare these with geological heterogeneities having similar periodic or random patterns. What types of equations are solved? How do they compare with the pressure and permeability equations used in this chapter?
3. Geological entities with obvious periodicity include rolling sand dunes with undulating crests and troughs, and meandering streams and rivers that wind in and out. Describe their overall physical features, for example, amplitude, wavenumber, three-dimensional variations, and so on. What types of physical correlations can be developed to describe these features? Can you develop (readily solved) differential equation models whose solutions can be used to generalize these correlations?

Chapter 13

Real and Artificial Viscosity

In Chapter 11, we showed how a first-order nonlinear partial differential equation arises in modeling immiscible two-phase flows. In fact, we derived the saturation equation $\partial S_{1w}/\partial t + (q/\phi_1)\{df_{1w}(S_{1w})/dS_{1w}\}\partial S_{1w}/\partial x = 0$, which describes the well-known Buckley-Leverett problem that governs in the absence of capillary pressure. Then, we applied Whitham's (1974) closed-form results for flows containing shockwaves and developed a formula for the shock speed. When capillary pressure is accounted for, the shocks predicted on a low-order basis need not exist. Depending on the exact value and form of the small higher derivative term, the flow may, in fact, be smooth. Actual computations are complicated by artificial viscosity, incorrect numerical entropy generation, and other issues that have led to, among other matters, confusion on the purported benefits of upwind or upstream differencing. Fortunately, these problems have been addressed in gasdynamics, plasma physics, and water waves, and the basic ideas are discussed in the petroleum context.

REAL VISCOSITY AND SHOCKWAVES

Here, we shall review basic properties of low-order wave equations that admit shocks, demonstrate that correct entropy conditions follow as direct consequences of high-order derivative terms, and show how artificial viscosity and upstream differencing can lead to errors in modeling important physical quantities and also in describing shock front speed.

Low-Order Nonlinear Wave Model

Let us consider the first-order, convective, nonlinear wave equation for a function u(x,t) satisfying

$$\partial u/\partial t + u\,\partial u/\partial x = 0 \tag{13.1}$$

which should be compared with the low-order saturation equation in Chapter 11,

$$\partial S_{1w}/\partial t + (q/\phi_1)\{df_{1w}(S_{1w})/dS_{1w}\}\partial S_{1w}/\partial x = 0 \tag{11.25}$$

Quantitative Methods in Reservoir Engineering. http://dx.doi.org/10.1016/B978-0-12-810518-4.00013-X

Eq. (13.1) possesses a simple general solution for arbitrary initial conditions. Recall that the total change for *any* function u(x,t) satisfies

$$du/dt = \partial u/\partial t + dx/dt\, \partial u/\partial x \tag{13.2}$$

If we compare Eqs. (13.1), (13.2), we find that du/dt = 0, provided we identify dx/dt = u, that is, u(x,t) does not change following a ray path dx/dt = u(x,t). This is succinctly expressed by the statement

$$u(x, t) = G(x - ut) \tag{13.3}$$

which emphasizes that u(x,t) must be a function of x - ut. Here, G can be any function. If we must enforce the initial condition

$$u(x, 0) = F(x) \tag{13.4}$$

where F(x) is given, it is clear that the choice G = F solves the problem. Hence, we have the general solution

$$u(x, t) = F\{x - u(x, t)\, t\} \tag{13.5}$$

Singularities in the Low-Order Model

Eq. (13.5) may or may not lead to shock formation or singularities in the first derivatives. To see how these may arise, differentiate Eq. (13.5) with respect to x using the chain rule to obtain $\partial u/\partial x = F'\{x-u(x,t)t\}\{1 - t\, \partial u/\partial x\}$. If we solve for $\partial u/\partial x$, we find

$$\partial u/\partial x = F'\{x - u(x, t)t\}/\{1 + tF'\} \tag{13.6}$$

If the initial condition F(x) is such that F' > 0, the denominator 1 + t F' > 0 is positive and the gradient $\partial u/\partial x$ is well behaved. If F' is negative, shockwave solutions with infinite values of $\partial u/\partial x$ form in a finite amount of time. These shocks are analogous to the water breakthrough (a.k.a., saturation discontinuity) phenomena familiar in waterflooding.

Existence of the Singularity

Eq. (13.1) embodies a class of solutions containing shockwaves, but do these exist in reality? Although shocks can form as a mathematical consequence of Eq. (13.5), it is often the case that Eq. (13.1) arises as a cruder model to a more accurate formulation. Fortunately, exact solutions to two higher order equations whose low-order terms are identical to Eq. (13.1) are available for study, namely, Burger's equation and the Korteweg deVries equation. The former arises in the modeling of gasdynamic shocks in high-speed aerodynamics, and is given by

$$\partial u/\partial t + u\, \partial u/\partial x = \varepsilon\, \partial^2 u/\partial x^2 \tag{13.7}$$

where $\varepsilon > 0$, a small number, is related to the *real viscosity* of the fluid. The exact Cole-Hopf solution, first reported by Cole (1949), is discussed in Whitham's book on nonlinear wave mechanics (Whitham, 1974). It is possible to show that physical systems that satisfy Eq. (13.7) can be modeled by Eq. (13.1), which is much simpler. That is, shocked solutions of Eq. (13.1) are also obtained from the more detailed description in Eq. (13.7). Why is this *not* obvious? This issue is subtle because we have not stated what ε is small with respect to; when spatial gradients (as obtained from Eq. 13.1) become large, the $\varepsilon\, \partial^2 u/\partial x^2$ term in Eq. (13.7) may no longer be small by comparison to the left-hand side. Thus, Eq. (13.1) may or may not apply near the shock, and direct recourse to the detailed physical model must be made.

Now consider the Korteweg deVries equation, obtained in the study of long, inviscid, water waves. Instead of Eq. (13.7), we have

$$\partial u/\partial t + u\, \partial u/\partial x = \delta\, \partial^3 u/\partial x^3 \tag{13.8}$$

where $\delta > 0$ is also a small number. While Eqs. (13.7), (13.8) differ only in the order of the small right-side term, the solution for Eq. (13.8) is completely different. An exact solution to the general initial value problem, using methods from inverse scattering, is again available (Whitham, 1974). It turns out that Eq. (13.8) does not admit any solutions with shocks. Thus, even if δ *is* small, Eq. (13.1) is never relevant as a simplified model. Our point is this: the high-order terms in Eqs. (13.7), (13.8) control the solution at all scales. In two-phase reservoir flow analysis, such high-order terms are related to local details provided by the capillary pressure function. If the low-order Buckley-Leverett model without capillary pressure is used, sharp shocks not unlike those uncovered in Eq. (13.6) can form, but their extent and thickness depend on the local effects of capillary and inertial forces.

Entropy Conditions

Once the high-order model is agreed upon, for example, Eq. (13.7) or Eq. (13.8), the complete physical description of the problem is self-contained. That is, the entropy conditions one "pulls from hats" in thermodynamics can be obtained from integration by parts. Let us consider Eq. (13.7). For simplicity, we move with the shock speed, so that

$$u\, \partial u/\partial x = \varepsilon\, \partial^2 u/\partial x^2 \tag{13.9}$$

applies locally, and we rewrite Eq. (13.9) in the conservation form

$$\partial\{1/2\, u^2 - \varepsilon\, \partial u/\partial x\}/\partial x = 0 \tag{13.10}$$

If we integrate from one side of the shock to the other, where each side is represented by uniform thermodynamic conditions with vanishing $\partial u/\partial x$'s, it is clear that $\{1/2\ u^2 - \varepsilon\, \partial u/\partial x\}_{upstream} = \{1/2\ u^2 - \varepsilon\, \partial u/\partial x\}_{downstream}$ leads to the global conservation law $u^2_{upstream} = u^2_{downstream}$, that is,

$$u_-^2 = u_+^2 \tag{13.11}$$

This jump condition is analogous to the global mass conservation constraint enforced in the Buckley-Leverett problem (e.g., via "Welge's construction").

Exact conservation laws like Eq. (13.11) are just one consequence of complete models like Eq. (13.9), with the explicit form of the high-order derivative term available. Its algebraic structure controls the form of energy-like quantities that are dissipated across discontinuities. For example, multiply Eq. (13.9) by u(x) throughout, so that $u^2\, \partial u/\partial x = \varepsilon u\, \partial^2 u/\partial x^2$. This can be rewritten as $\partial(1/3\ u^3)/\partial x = \varepsilon u\, \partial^2 u/\partial x^2$. If we now integrate by parts, we have

$$\begin{aligned}&(1/3u^3)_+ - (1/3u^3)_- \\ &= \varepsilon\left[\left\{u\,\partial u/\partial x - \int (\partial u/\partial x)^2 dx\right\}_+ - \left\{u\,\partial u/\partial x - \int (\partial u/\partial x)^2 dx\right\}_-\right]\end{aligned} \tag{13.12}$$

The $\partial u/\partial x$ terms on either side of the shock vanish identically, but the positive definite integral does not. This leaves

$$(1/3\,u^3)_+ - (1/3\,u^3)_- = -\varepsilon \int (\partial u/\partial x)^2\, dx < 0 \tag{13.13}$$

and hence the entropy condition below

$$u_-^3 > u_+^3 \tag{13.14}$$

Thus, we have shown that entropy conditions need not be arrived at independently via thermodynamic considerations; they, and indeed *all* of the physics, can be obtained naturally once the structure of the high-order derivative is known. Additional entropy conditions can be generated by multiplying Eq. (13.9) by other powers or functionals of u(x), and then integrating by parts. Indeed, when the correction to $\partial u/\partial t + u\, \partial u/\partial x = 0$ is not $\varepsilon\, \partial^2 u/\partial x^2$, but a high-order term satisfying rheological models where the viscosity is a given function of u and $\partial u/\partial x$, the methods just presented are the only rigorous ones to determine the structure of shock-like flows. In reservoir flow analysis, the low-order Eq. (11.25) (i.e., $\partial S_{1w}/\partial t + (q/\phi_1)\{df_{1w}(S_{1w})/dS_{1w}\}\partial S_{1w}/\partial x = 0$) is not complete without $\partial^2 S_{1w}/\partial x^2$ second-derivative capillary pressure terms included (e.g., see Chapter 21 for a full derivation). Whether or not they are important in the final analysis depends on the relative magnitudes of capillary to inertial forces, which for oil- and water-producing reservoirs, vary with time.

Related problems are found in high-speed aerodynamics. For instance, the high-order *parabolic* viscous transonic equation derived by Cole (1949) and Sichel (1966), was solved by this author (Chin, 1977, 1978a,b,c) and shown to be equivalent to the solutions of low-order, mixed elliptic-hyperbolic equations when augmented by external jump conditions. For further analysis on the role of small high-order terms in continuum mechanics, the reader is referred to Ashley and Landahl (1965), Cole (1968), and Nayfeh (1973). Finally, we emphasize that the standard Rankine-Hugoniot jump conditions connecting equilibrium thermodynamic states across shock waves, developed over a century ago from detailed physical arguments, can be derived from the one-dimensional Navier-Stokes equations in a similar straightforward manner as given here (Courant and Friedrichs, 1948). Also, note that the combined high-order equation $\partial u/\partial t + u\,\partial u/\partial x = \varepsilon\,\partial^2 u/\partial x^2 + \delta\,\partial^3 u/\partial x^3$ is treated in Whitham (1974).

ARTIFICIAL VISCOSITY AND FICTITIOUS JUMPS

Despite the importance of the present subject in numerical modeling, the basic ideas developed here are not discussed, to this author's knowledge, in industry reservoir engineering books. This is not to say that these issues have been completely ignored. In fact, problems with multivalued saturations and steep gradients cited here were noted in the 1950s by petroleum mathematicians. Important contributions were made by Sheldon et al. (1959), Cardwell (1959), and Lee and Fayers (1959). These authors (correctly) invoked an analogy found in nonlinear acoustics, and resolved the low-order problem using the method of characteristics and shock fitting. They also speculated that the inclusion of capillary pressure in the underlying formulation would render all saturations single-valued, and they compared this to the process by which viscous diffusion smears gasdynamic shocks in compressible aerodynamics. Their ideas on the role of capillary pressure were validated by ongoing research. The numerical investigations of Douglas et al. (1958, 1959), McEwen (1959), Fayers and Sheldon (1959), Hovanessian and Fayers (1961), and others, in fact, pointed out that "small" capillary effects can and will affect both shock structure and position. In the limit of "high flow rates," computed solutions correctly gave the corresponding low-order, shock-fitted Buckley-Leverett solutions.

The existence of truncation errors in finite difference approximations to differential equations is discussed in numerical analysis texts with respect to round-off error and computational instabilities (Roache, 1972; Richtmyer and Morton, 1957), but Lantz (1971) was among the first to address the *form* of the truncation error as it related to diffusion. Lantz considered a linear, convective, parabolic equation similar to $\partial u/\partial t + U\,\partial u/\partial x = \varepsilon\,\partial^2 u/\partial x^2$ and differenced it in several ways. He showed that the effective diffusion coefficient was not ε, as one might have suggested analytically, but $\varepsilon + O(\Delta x, \Delta t)$ (so that the actual diffusion term appearing in computed solutions is the modified coefficient *times* $\partial^2 u/\partial x^2$) where the $O(\Delta x, \Delta t)$ truncation errors, being functions of $u(x,t)$, are comparable in magnitude to ε. Because this artificial diffusion necessarily differs from the actual physical model, one would expect that the entropy conditions characteristic of the computed results could likely be fictitious.

When physical capillary effects are modeled, the low-order equation model $\partial S_{1w}/\partial t + (q/\phi_1)\{df_{1w}(S_{1w})/dS_{1w}\}\partial S_{1w}/\partial x = 0$ changes, in that a second-order derivative term $\partial^2 S_{1w}/\partial x^2$ with physically defined coefficients appear (Bear, 1972; Collins, 1961; Scheidegger, 1957). Additional derivations are also found in Aziz and Settari (1979), Peaceman (1977), and Thomas (1982b). The effects modeled by the new terms will be altered as discussed when "upwind" or upstream differencing is used (e.g., see Allen and Pinder, 1982). Although such differencing schemes may be successful in particular examples, for instance, in correctly modeling mass conservation, upwind methods are numerous and must be evaluated detail in any application. So much depends on the mesh system used, for example, five vs. nine point schemes, the details of the differencing, and so on, that indiscriminate use should be avoided. Again, to properly capture saturation shocks and discontinuities, the appropriate upstream differencing must be used. There is no single all-purpose scheme, and identifying the correct one often requires careful mathematical analysis (Jameson, 1975).

This leads to the following question. When the second-derivative terms *are* available, it is not necessary to use upwind differencing methods: why not solve an accurate, high-order numerical scheme that captures *real viscosity* (or capillary pressure) effects in the leading description? This idea was successfully proposed and tested in the 1970s by the author (Chin, 1977, 1978a,b,c) in high-speed aerodynamics. The same questions were posed by Moretti and Salas (1972), who suggested the possibility of solving problems containing gas-dynamic shocks with difference schemes, consistent with the high-order Navier-Stokes model and *not* artificial viscosity. Artificial viscosity, in the historical context, was popularized by von Neumann in an age when numerical analysis was not mature and computer resources were expensive (Richtmyer and Morton, 1957). As recently as 1991, Zarnowski and Hoff (1991) warned against artificial viscosity, recommending instead a direct attack on the exact equation, thus avoiding all the problems of correct entropy production by retaining the correct physical viscosity in the main formulation.

PROBLEMS AND EXERCISES

1. Research the numerical simulation literature in reservoir engineering and state the differences between upwind, central, and downwind approximations. These were first discussed by Lantz (1971) in a one-dimensional context. What are their modern multidimensional extensions? What are their implications for mass conservation? Sharpness of the saturation front discontinuity? Accuracy in predicting water breakthrough time? Relative oil and water production? Effect on numerical stability?
2. Identify a reputable immiscible, two-phase flow simulator for use in this problem, and select a validated problem set (with available solutions) where consistent relative permeability and capillary pressure curves have been successfully tested against field data. Re-run selected data sets. How do your

solutions change as the absolute magnitude of capillary pressure change? What happens when the capillary pressure vs. saturation curve is replaced by an approximate straight-line function? What if the capillary pressure is set identically to zero? In all three scenarios, note the position of the saturation discontinuity, its steepness, and the thickness of the front. Do your solutions oscillate in time? If so, numerical instability is indicated.

3. Various schemes have been proposed to model water breakthrough accurately in petroleum reservoirs, involving the use of finite difference molecules contrasting in number and orientation. It is known that solutions are highly sensitive to the grid used and to the form of the difference molecule. Can you test their correctness without resorting to field data comparisons, for example, by developing entropy-based criteria? Also, describe how moving time-adaptive grids that follow the shock might resolve the physics better while using fewer grids.

Chapter 14

Borehole Flow Invasion, Lost Circulation and Time Lapse Logging

This chapter hints at borehole flow invasion, lost circulation, and time lapse logging, that is, drilling subjects not usually treated in reservoir engineering. They are topics important operationally, but they *do* represent real problems in Darcy flow analysis near the well, just as traditional reservoir engineering treats flows away from the well. During drilling, muds are known to invade boreholes; they may or may not contain dispersed solid particles that weight up the borehole fluid, thus providing increased density for improved well control. The pressure differential between the borehole fluid column and the formation then forces mud filtrate into the reservoir, very often damaging the potential producing zone (hence, the motivation for underbalanced drilling). When the formation is permeable compared to the mudcake remaining at the sandface, the flow rate into the formation is controlled by the buildup of mudcake, that is, the time-dependent increase in filter-cake thickness arising from the pile-up of separated solid particles left by invading liquids. In this transition chapter, we shall present crude models of near-well invasion in order to introduce the ideas physically. In the next chapter, the far-well perspective is taken in which the overall effects of large-scale multilateral and horizontal well systems on reservoir flow are considered. Then, in the remainder of this book, we formulate comprehensive models of formation invasion, noting that these are equally applicable to near and far-well reservoir flooding applications.

BOREHOLE INVASION MODELING

In this chapter, we shall present some basic ideas about borehole flow invasion and lost circulation. Once the fundamentals are clearly explained, we shall introduce a relatively new concept known as time lapse logging. We shall begin our discussion with the simplest invasion models and progress to those that bear increasing sophistication. Then, we shall indicate how these results can be used to assist in formation evaluation and reservoir description, discuss possible pitfalls, and direct the reader to more recent literature.

Quantitative Methods in Reservoir Engineering. http://dx.doi.org/10.1016/B978-0-12-810518-4.00014-1

EXAMPLE 14.1. THIN LOSSY MUDS (THAT IS, WATER)

The simplest borehole flow invasion problem can be posed using the radial flow model of Chapter 6, and, in particular, applying pressure-pressure boundary conditions at the well and farfield boundaries. Because our mud is assumed to be lossy, we can ignore the presence of cake buildup; many shallow wells are, in fact, circulated with water or brine as the drilling fluid.

Pressure-Pressure Formulation

The starting point, at least for this elementary analysis, is the pressure equation

$$d^2P/dr^2 + (1/r)dP/dr = 0 \tag{14.1}$$

for P(r), where r is the cylindrical radial coordinate, which governs Darcy flows of incompressible liquids in homogeneous, isotropic media. The usual boundary conditions assume pressures specified at the wellbore and at some distance away from the hole. We have

$$P(r_W) = P_W \tag{14.2}$$

$$P(r_R) = P_R \tag{14.3}$$

where $r = r_W$ and $r = r_R$ refer to the well and farfield radius, and P_W and P_R are the assumed pressures. This formulation has the solution (see Eq. 4.46)

$$P(r) = \{(P_R - P_W)/(\log r_R/r_W)\} \log r/r_W + P_W \tag{14.4}$$

Now, the radial velocity q(r) is given by Darcy's law, requiring that

$$q(r) = -(k/\mu)dP(r)/dr = -(k/\mu)\{(P_R - P_W)/(\log r_R/r_W\}1/r \tag{14.5}$$

where k is the formation permeability and μ is the viscosity of the fluid. Hence, the total volume flow rate Q_W, assuming a reservoir depth D into the page, is

$$Q_W = -D \int_0^{2\pi} q(r) r_W \, d\theta, r = r_W \tag{14.6}$$

or, simply

$$Q_W = -2\pi r_W D \, q(r_W) \tag{14.7}$$

leading to a constant

$$Q_W = -(2\pi k D/\mu)(P_R - P_W)/(\log r_R/r_W) \tag{14.8}$$

Simple Invasion Modeling, and $\sqrt{t}$ Behavior

The front of the drilling fluid penetrating the reservoir or formation is desired as a function of time for the constant influx rate above. In fluid mechanics jargon, this requires a Lagrangian as opposed to an Eulerian description of the flow field. The mathematics, however, is straightforward. Let us denote the (constant) porosity by ϕ. Then, the rate of invasion into the reservoir is

$$dr/dt = q/\phi = +Q_W/(2\pi\phi Dr) \tag{14.9}$$

Simple integration gives

$$r(t) = \sqrt{\{R_W^2 + Q_W t/\pi\phi D\}} \tag{14.10}$$

for the invasion front. Here, Q_W and ϕ are assumed to be constant, and the initial front position is taken to be $r = r_W$ at $t = 0$. By the same token, we could have considered any other ring of invading filtrate, and obtain

$$r(t) = \sqrt{\{R_{other}^2 + Q_W t/\pi\phi D\}} \tag{14.11}$$

if $r = R_{other} > R_W$ at $t = 0$. Let us return to Eq. (14.10). For large times, this formula reduces to

$$r(t) \approx \sqrt{\{Q_W t/\pi\phi D\}} \tag{14.12}$$

Thus, in a steady radial flow without mudcake, we obtain "$\sqrt{t}$" behavior for the invasion front. If r(t) is known at some time t, say from resistivity log analysis, then the porosity ϕ can be calculated from Eq. (14.12).

EXAMPLE 14.2. TIME-DEPENDENT PRESSURE DIFFERENTIALS

In Example 14.1, we assumed that the pressure differential between the mud column and the formation is constant. In drilling operations, this is often not the case: mud weights can be increased for improved blowout control or decreased to prevent formation fracture. In general, the quantity $(P_R - P_W)$ may be a function of time, that is, $\{P_R - P_W(t)\}$, but in this example we continue to ignore the mudcake buildup that accompanies this weighting up. The basic ideas still apply. Again, the transient effects here do not arise from fluid compressibility. To obtain quantitative results, substitute Eq. (14.8) in Eq. (14.9) and carry out the integration. Let us introduce the notation

$$\Delta p(t) = P_W(t) - P_R \tag{14.13}$$

This leads to the following formula for the invasion front,

$$r(t) = \left[R_w^2 + \{2k/(\mu\phi \log r_R/r_W)\} \int_0^t \Delta p(\tau) d\tau \right]^{1/2} \tag{14.14}$$

The corresponding invasion front velocity is obtained by differentiating Eq. (14.14). This leads to

$$dr(t)/dt = k\Delta p(t)/(\mu\phi r \log r_R/r_W) \tag{14.15}$$

where r is given by Eq. (14.14). This evaluation is straightforward once the integration in Eq. (14.14) is performed. Note that the $\sqrt{t}$ behavior in Eq. (14.12) no longer applies.

EXAMPLE 14.3. INVASION WITH MUDCAKE EFFECTS

In this example, we provide a simple, approximate approach that is useful in field application when mudcake controls the flow into the reservoir. This model does not apply to tight (i.e., low-permeability) formations. First, mudcake filtration properties must be determined. Rather than postulating phenomenological models, we apply the results of laboratory measurements (denoted by asterisks). Let h^* represent the height of filtrate collected in a time t^*, under an applied pressure drop of $(\Delta p)^*$ in a linear flow vessel. It is known that h(t) grows like $\sqrt{t}$ (general results appear in Chapter 17). Thus, we write

$$h(t) = c\Delta p\sqrt{t} \tag{14.16}$$

where Δp, the pressure differential applying outside the laboratory, is taken to be constant, and c is a parameter that describes the particular cake and mud type. Since the asterisked quantities are available from empirical or field data, the constant c is uniquely determined from Eq. (14.16) as

$$c = h^*/\left\{(\Delta p)^*\left(\sqrt{t^*}\right)\right\} \tag{14.17}$$

Thus,

$$h(t) = \left[h^*/\left\{(\Delta p)^*\left(\sqrt{t^*}\right)\right\}\right]\Delta p\sqrt{t} \tag{14.18}$$

Now we assume that the mudcake is thin, that is, that the thickness of the cake is small compared to the radius of the hole. If so, and if the relatively impermeable cake controls the flow rate into the formation, the net volume of fluid Vol deposited into the formation after a time t is approximately obtained by multiplying Eq. (14.18) by $2\pi r_W D$, the area of the borehole surface,

$$\text{Vol}(t) = \left[h^*/\left\{(\Delta p)^*\left(\sqrt{t^*}\right)\right\}\right]\Delta p\sqrt{t}\,(2\pi r_W)D \tag{14.19}$$

where D is the length of the borehole. But this must be equal to the volume in the formation that is available for fluid storage, that is,

$$\text{Available volume} = \left(\pi r_f^{\,2} - \pi r_w^{\,2}\right)\phi D \tag{14.20}$$

where we have denoted the front position radius as r_f. Setting the expressions in Eqs. (14.19), (14.20) equal leads to

$$\left(\pi r_f^2 - \pi r_W^2\right)\phi = \left[h^*/\left\{(\Delta p)^*\left(\sqrt{t^*}\right)\right\}\right]\Delta p \sqrt{t}\,(2\pi r_W) \tag{14.21a}$$

or

$$r_f^2 = r_W^2 + \left[h^*/\left\{(\Delta p)^*\left(\sqrt{t^*}\right)\right\}\right]\Delta p \sqrt{t}\,(2r_W/\phi) \tag{14.21b}$$

At early times, a binomial expansion of Eq. (14.21b) shows that $r_f(t)$ behaves like $\sqrt{t}$. We emphasize that this lineal flow, cake-dominated process, while it yields $\sqrt{t}$ behavior, is physically different from the no-cake, radial model in Example 14.1. That both possess identical algebraic behavior (but different proportionality constants) increases the possibility of interpretation error in log analysis if the details of the flow process are not well understood. Again, the $\sqrt{t}$ applicable to drilling with a clean brine differs from that obtained with solids-containing mud. In either case, the front location r_f can be determined if the porosity is known; conversely, if the front location is available, say from resistivity log analysis, the porosity can be calculated.

TIME LAPSE LOGGING

Mudcakes are important because their extremely low permeabilities control the filtration rate into the formation, which is assumed to be much more permeable. This simplifies the modeling of reservoir flow: front motion becomes a purely kinematic process governed only by mudcake mechanical properties and geometric divergence within the rock. This property can be used to advantage in formation evaluation. If the location of the front is known as a function of time, then for the simplified flow considered here, the porosity should be available from simple calculation (e.g., Eqs. 14.21a, 14.21b). Let us suppose that the front position can be accurately inferred from resistivity log analysis. This is often so if salty and fresh waters have mixed only very little, in which case the assumptions behind the piston models used in tornado chart development hold. Then, it should be possible to predict porosity as a function of time. If the formation porosity is constant, then nonconstant predicted values are the result of spurt loss errors and nonuniform initial mudcake growth. With time, the porosity should tend to the constant value of the formation. In this section, we shall discuss basic supporting experiments.

Again, we consider boreholes with significant filter cake formation due to non-lossy muds. When this is the case, we solve Eqs. (14.21a), (14.21b) for porosity,

$$\phi = \left[h^*/\left\{(\Delta p)^*\left(\sqrt{t^*}\right)\right\}\right]\Delta p \sqrt{t}\,(2r_W)/\left(r_f^2 - r_W^2\right) \tag{14.22}$$

If all the parameters on the right side are known with confidence, the porosity can be determined. This method was applied successfully in some field applications. Chin et al. (1986) gave an invasion porosity log that showed both qualitative and quantitative agreement with conventional neutron and density porosity logs. In the paper, the authors determined the asterisked properties of the mud using API filtration tests and obtained $r_f(t)$ from standard resistivity analysis.

Fig. 14.1 shows the log results obtained with the method, whereas Fig. 14.2 shows laboratory porosity predictions converging to the correct (independently measured) 23% value after approximately 1 h. It is clear that understanding the filtration properties of mudcakes and the detailed dynamics of their growth is important to obtaining accurate invasion porosity logs (e.g., as in Fig. 14.1). Chin et al. (1986) studied these transient processes experimentally using the linear flow and radial flow filtration test vessels shown in Figs. 14.3 and 14.4.

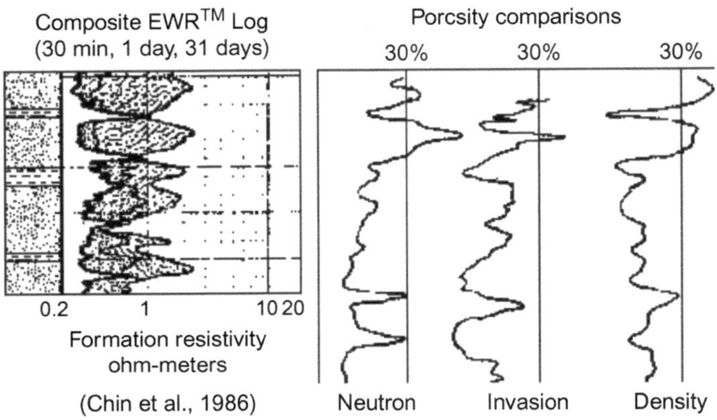

FIG. 14.1 Time lapse logging, field results.

Time (min)	Porosity (%)
1.2	10
3.9	14
9.0	17
16.1	20
25.6	21
36.1	21
49.1	22
64.1	22
81.0	23
100.0	23
121.0	23
144.0	23

FIG. 14.2 Predicted porosities, radial filtration vessel, converging to 23%.

In these laboratory fixtures, water-base muds of varying weights were allowed to flow in a direction perpendicular to the rock surface (i.e., static filtration) in order to form mudcake, but erosive shearing flow in directions tangential to the surface (i.e., dynamic filtration) were not permitted (see Chapter 17 for recent rheology-dependent models). These small vessels were placed in Catscan units, which recorded in detail the time-dependent positions of the moving cake surface and the invading filtrate within the rock. Sample linear and radial flow Catscan images, which display density contrasts, are given in Figs. 14.5 and 14.6, where mudcakes appear as very dark bands and invading fluids are lightest in color.

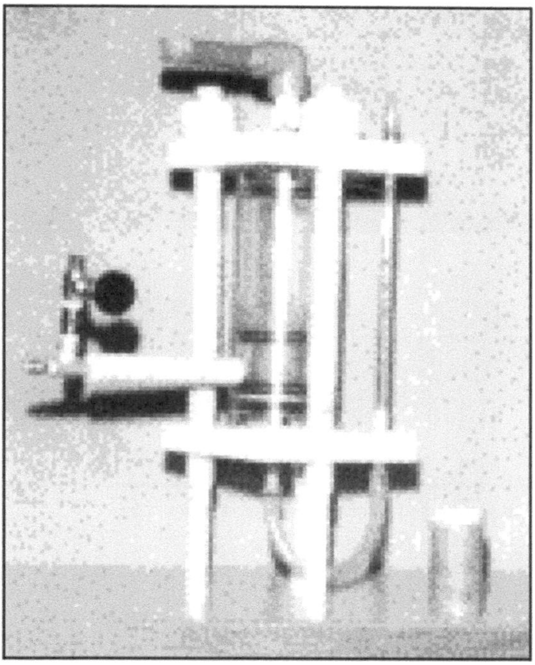

FIG. 14.3 Linear flow filtration vessel.

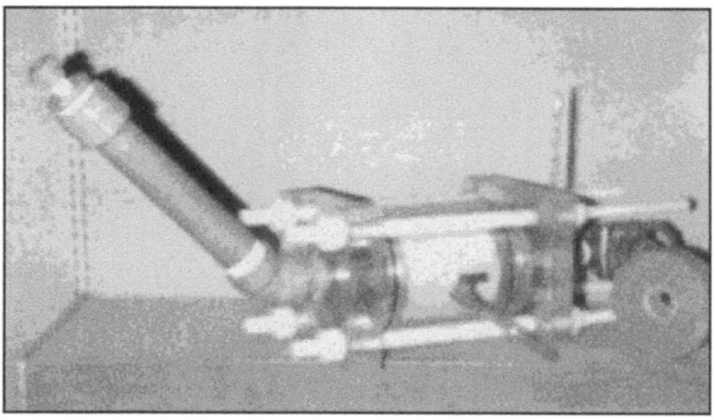

FIG. 14.4 Radial flow filtration vessel.

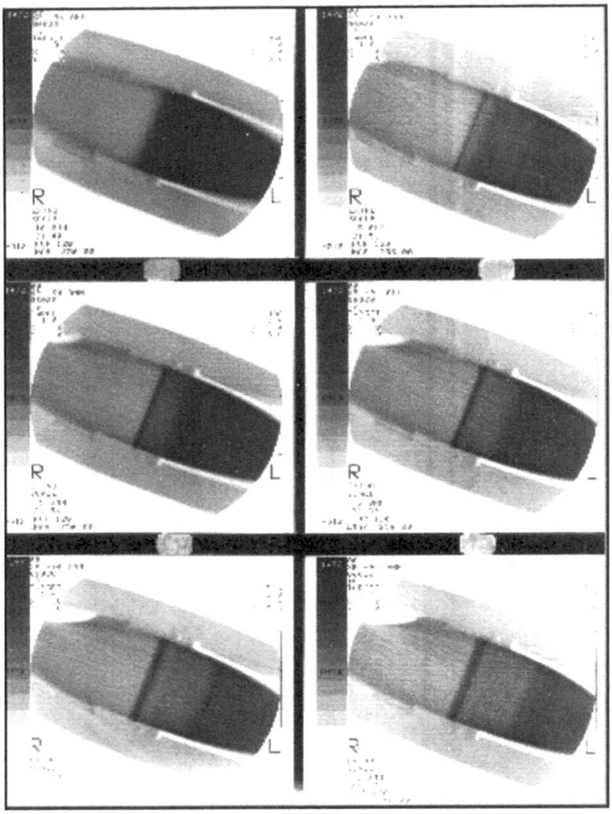

FIG. 14.5 Mudcake growth and filtration Catscan in linear core.

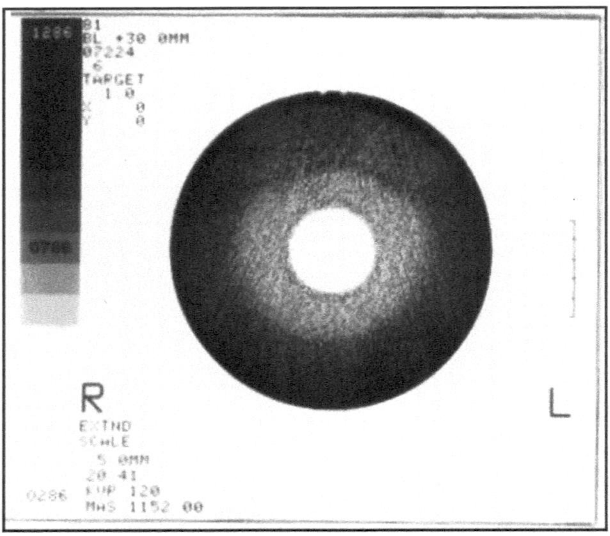

FIG. 14.6 Mudcake growth and filtration in radial core.

We emphasize that the parameter c in Eq. (14.17), which characterizes the filtration properties of the mudcake, was assumed to be constant only for simplicity. In essence, the mudcake was taken as a rigid medium that does not respond to imposed pressure. This idealization is not true. Fig. 14.7, which illustrates mudcake growth with time, also shows direct evidence that local cake density increases with time, a result of nonlinear compaction. The resulting changes in cake permeability, porosity, and particle packing, which cause c to vary, must be accounted for in improved models.

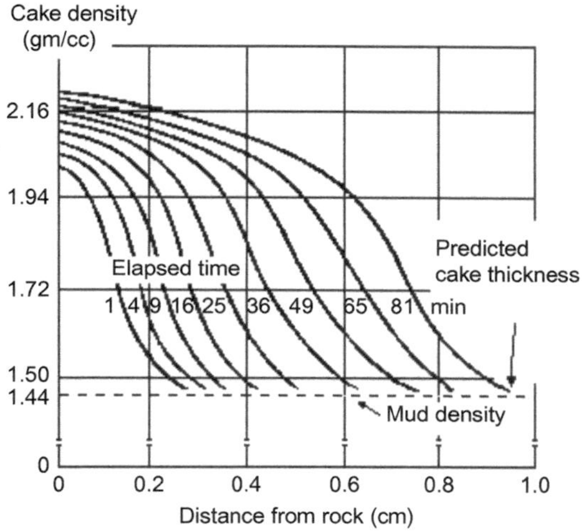

FIG. 14.7 Catscan data, mudcake compaction with time.

Although the reservoir engineering ideas behind invasion porosity may be obvious in retrospect, the work in Chin et al. (1986) focused on the potential for obtaining formation characteristics in addition to porosity. In principle, good porosity prediction can be determined from a single front position given at a single point in time. In order to obtain, for instance, permeability, viscosity, hydrocarbon mobility, and so on, the number of additional timewise measurements and the time separations between resistivity readings must be known. Plausible time scales are immediate $t=0$ values based on MWD measurements, 1-to-2 day readings while tripping, and 2-weeks to 1-month data. Data points must not be too closely spaced in time, or else, the resulting algebraic equations will be ill conditioned and error-prone.

A comprehensive fluid-dynamical solution for single-phase flow was given in the Chin et al. (1986) work. There, a flow model with coupled cake growth and filtrate invasion was constructed, comprised of a three-layer Darcy flow: the mudcake, the rock with filtrate, and the rock with reservoir fluid. Mass and pressure continuity were enforced at interfaces. At the cake surface, where it is exposed to borehole fluids, the well pressure was specified and a constitutive model for empirical cake buildup was invoked (e.g., see Collins, 1961). Cake thickness was allowed to increase with time in the moving boundary value problem formulation, thus slowing the rate of formation invasion. In the third-layer farfield, the reservoir pressure was assumed to be known. Once pressures for the different layers were available, exact solutions for invasion speed and cake growth rate were obtained. The physical ideas noted earlier regarding the controlling role played by mudcake were confirmed using the solution. Again, mudcake properties are all-important. In any practical model, mudcake compaction and erosion under dynamic conditions must be well characterized.

This chapter does not close the book on time lapse logging or invasion. Quite the opposite, many questions are raised. What time scales are typical for good repeated resistivity measurements? How do these depend on the relative mobility differences between mudcake and formation Darcy flows? How do our interpretation techniques change, going from single-phase flows (e.g., the red water displacing blue water considered here) to the two-phase water-oil or water-gas flows characteristic of real petroleum reservoirs? These reservoir flow issues are studied later, when we expand on the ideas introduced here. Some tool and resistivity related issues are addressed in early logging publications, for example, Cobern and Nuckols (1985) and Allen and Jacobsen (1987). Unfortunately, there has been minimal industry effort over the past decade to broaden the theoretical basis for time lapse logging. Although progress has been made in simulating different types of reservoir flow, complementary progress in modeling electromagnetic wave motion is lacking. As logging tools increase in sophistication, however, we expect electromagnetic modeling to mature.

LOST CIRCULATION

We close this chapter with a note on lost circulation (e.g., see Messenger, 1981 for a comprehensive discussion) in drilling applications. This important subject in rigsite safety is again founded on reservoir engineering principles. The models derived in this book for reservoir flows can be used to determine formation pressures *while drilling* in order to monitor or retard fracture initiation or propagation. For instance, in Chapter 5 we gave closed-form solutions for circular boreholes with (i) two symmetric radial fractures, (ii) a single fracture, (iii) symmetric unequal fractures, and (iv) a multiplicity of radial fractures. (These solutions were derived for use in hydraulic fracturing and horizontal well analysis.) Accurate knowledge of borehole-to-reservoir pressure loads can be used to predict equilibrium fracture configuration, say, the lateral extent of a fracture arm once other data are available. In addition, pressure transient models for the above configurations can be found in the literature, typically obtained for formation evaluation use. These open up the possibility of coupling analytical reservoir flow methods to drilling data, in particular, to real-time MWD information or rigsite information, in order to provide answers to questions related to both drilling safety and formation evaluation.

PROBLEMS AND EXERCISES

1. Problems in formation invasion involve internal moving boundaries, for example, the front formed when water displaces oil, and require new methods for boundary value problem analysis. In elementary courses, it is taught that permeability and viscosity enter only through k/μ, but k and μ effects can be quite different in real petroleum reservoirs. The example treated in Fig. 11.1 deals with a single fluid flowing through two cores having different permeabilities, noting that the core interface is fixed with time. But it is also possible to have two different fluids in a homogeneous core, one displacing the other, where now the interface separating two distinct flow regimes must move. Consider, for example, a linear core of fixed length, with water displacing oil. Does the front accelerate or decelerate? What if oil displaces water? How do your answers change in cylindrical radial flow, when the effects of geometric divergence compete with acceleration based on momentum (or viscosity contrast) considerations? We have assumed that plug displacement exists in these flows. Which flow is physically stable? Which is not? Why?
2. Assume that the flows discussed in (1) are stable, that is, they do not break down to form viscous fingers. Consider a homogeneous linear core of fixed length, containing two fluids with different viscosities. How would you formulate the general initial-boundary value problem to predict all subsequent motions? What matching conditions apply at the moving interface? Formulate and solve the problem analytically. The initial condition, which describes the proportion of water relative to oil in the core, controls the acceleration or deceleration of the moving front. Develop a formula showing how volume proportions and viscosity contrasts affect this motion. What are the relevant dimensionless parameters?

Chapter 15

Horizontal, Deviated and Modern Multilateral Well Analysis

So far we have performed detailed studies for flow over isolated bodies, for example, curved fractures, shale arrays and fractured boreholes. Here we shall focus on steady and transient, compressible, reservoir-scale flows produced by multilateral well systems. Because their topologies are not simple, we turn to computational methods. We shall highlight problems that arise in reservoir simulator development, and importantly, we shall describe a recently developed, three-dimensional algorithm that is very robust, numerically stable, exceptionally fast and extremely accurate, and now available to the user community. Engineering implementation is an objective of the work: oil companies want practical solutions that optimize operations, profits and time value of money. The model provides tools that evaluate "what if" production scenarios, infill drilling strategies and waterflood sweep efficiencies. In addition to being accurate, the solutions require minimal hardware, software and costly human resources. High-power processors and graphics accelerators are not required. The methodology, user interface and detailed validations have been works in progress for two decades and we are pleased to provide highlights on our capabilities and experiences for new and more senior reservoir engineers.

OVERVIEW

Formulation Errors

In the author's experience with many flow simulators, as many questions arose during calculations as there existed at the outset. Many offered black oil, compositional and dual porosity capabilities, yet few produced evidence that the "$p = A + B \log r$" solution for steady, single-phase, radial flow could be recovered on a rectangular mesh. Mass conservation was presumably enforced in all runs, yet frustrating time step cuts indicated that many schemes were not robust. Where intuition suggested that time scales should be measured in minutes or hours, stability considerations often kept time steps to thousandths of a second before diverging.

Quantitative Methods in Reservoir Engineering. http://dx.doi.org/10.1016/B978-0-12-810518-4.00015-3

One model applied linear superposition incorrectly: solutions from several single well, pressure-constrained runs were simply added together to generate multi-well field results, without accounting for interwell interactions. In several fracture flow models, source code analysis revealed systematic abuse of harmonic, geometric and arithmetic averaging techniques, with formulas applied to fracture and matrix continua where they were completely inapplicable. Not one simulator addressed the velocity singularities at fracture tips derived in Chapter 2; one validation cited perceived agreement with Muskat, but unfortunately an incorrect result stated that net volume flow rates were independent of fracture length.

I/O Problems

All of these problems were compounded by input/output difficulties. Numerical values for permeabilities and porosities were entered at the keyboard into eighty column worksheets, burying the geological feel of the reservoir. Well positions were defined by (i,j,k) coordinates that were not easily visualized. Checking for typographical errors entailed tedious work. Few simulators listed the default assumptions used, so that they could be available for inspection, confirmation or change. In many cases, cryptic commands replaced engineering decision making and users were forced to memorize unnatural Unix-like keywords. Flow analyses were often performed without knowing underlying assumptions or the shapes of assumed relative permeability curves, key steps ignored just to get the simulator to run by quitting time.

Computation-intensive software requires high speed machines and too many service personnel. Sometimes this gave way to unexpected problems. Oil company data centers often allocate user account memory without informing clients of arbitrarily chosen byte limits. For three months, this author was unable to resolve a simulation problem that turned out to result from new input data writing over old data, an unthinkable excuse in an age of inexpensive memory. Thus, we are led to several blunt but relevant questions, "Are there smarter, more efficient ways to simulate reservoir flows?" "Do simulators really need to be computation intensive?" "Are there good, robust algorithms that avoid the difficulties of less optimal approaches?" To address these questions, we must consider why expensive hardware, complicated software and "make work" are required in the first place. And, if need be, we must redesign the building blocks, methodically from the ground up.

Synopsis

In this chapter we update the numerical algorithms first presented Chin (1993a,b, 2002). Since the original publication, enhancements to numerical stability (with respect to fluid and rock properties, multilateral well topology and constraints, the environment itself while simulating, e.g., sidetracking, recompletions, conversions between pressure and rate constraints, shut-ins, addition of lateral extensions to main bores), key requirements for robust simulation, have been devised and applied to challenging examples.

But good stability, rapid computing speed and low hardware costs alone would not suffice. Until a simple graphical user interface could be designed which allowed users to hand sketch well systems and geological structures on the computer screen—and have these diagrams rapidly interrogated by software algorithms and automatically launched into simulations—our simulator could not become a reality. And so, for many years, the upgraded methods "sat on the back burner" awaiting the appropriate moment. As luck would have it, a consortium of operating companies would fund just such an effort, in particular, targeting novice as well as experienced users. This interesting project will be discussed later in the chapter to emphasize how, in the years since Chin (1993a,b, 2002) appeared, we have translated abstract ideas into useful planning tools.

FUNDAMENTAL ISSUES AND PROBLEMS

Many issues confront users of commercial simulators. Among these are numerical stability, convergence, matrix size and structure, computational resolution, physical modeling capabilities, graphical limitations, and, of course, hardware constraints. The prevailing opinion supports the adage that good solutions require more hardware, more grid blocks, more computer time, and more costly software and graphics. While million grid block compositional simulations modeling complex physical phenomena in heterogeneous formations should be used when they are necessary and justified, the majority of runs requiring significant computer resources are no more than the result of inadequately designed software products. For the great majority of simulation runs conducted for screening purposes, for example, determining the qualitative effects of sweep efficiency, heterogeneity assumptions, and multilateral well design and placement, there is no reason why a simple fluid model won't suffice so long as the main engineering options are built in.

The bottom line is important. There are smarter ways to simulate, and in the end, a good, robust, stable algorithm that anticipates and accommodates user needs, while introducing the fewest number of uncertainties, should provide the foundation for a simple, multipurpose flow engine that runs efficiently. It must run the first time, and every time, without "crashing." It should demand few numerical and "computerese" user inputs. It should handle complicated reservoir heterogeneities and well patterns, and it must operate with a minimum of hardware and software investment. Such algorithms, developed over the years for three-dimensional aerodynamics under government funding, are widely available and can be readily adapted to modern Darcy flow problems satisfying similar equations. We give these general algorithms later but will take the opportunity now to expand on the ideas introduced in Chapters 6–10.

Numerical Stability

Nothing strikes greater fear in simulation than instabilities. Numerical instabilities manifest themselves through unrealistic oscillations in pressure buildup or drawdown curves, wiggly spatial pressure distributions that lead to infinities and overflow. How can they be avoided? One useful tool is the von Neumann stability test, after John von Neumann, the computer pioneer who advanced finite difference methods in the 1950s. Numerical analysts employ these tests to evaluate candidate algorithms before code development begins.

Consider the heat equation $u_t = u_{xx}$ for $u = u(x,t)$. We assume that a discretized u can be represented by $v(x_i,t_n)$, or simply "$v_{i,n}$," which satisfies the explicit $(v_{i,n+1} - v_{i,n})/\Delta t = (v_{i-1,n} - 2v_{i,n} + v_{i+1,n})/(\Delta x)^2$ model, where Δt and Δx are time and spatial increments.

But how useful is this obvious difference approximation? Let us separate variables, and consider a wave component $v_{i,n} = \psi(t)\, e^{j\beta x}$, where $j = \sqrt{-1}$, leading to $\{\psi(t+\Delta t) e^{j\beta x} - \psi(t) e^{j\beta x}\}/\Delta t = \psi(t)\left[e^{j\beta(x-\Delta x)} - 2e^{j\beta x} - e^{j\beta(x+\Delta x)}\right]/(\Delta x)^2$. Thus, $\psi(t+\Delta t) = \psi(t)(1 - 4\lambda \sin^2 \beta\, \Delta x/2)$, where $\lambda = \Delta t/(\Delta x)^2$. Since $\psi(0) = 1$, we find that $\psi(t) = (1 - 4\,\lambda \sin^2 \beta\, \Delta x/2)^{t/\Delta t}$. For stability, $\psi(t)$ must remain bounded as Δt, and thus Δx, approaches zero. Thus, $|1 - 4\,\lambda \sin^2 \beta\, \Delta x/2| < 1$, thereby establishing requirements for Δx and Δt. We need not have solved for $\psi(t)$.

We could have defined an amplification factor $a = |\psi(t+\Delta t)/\psi(t)|$ and determined that $a = |1 - 4\lambda \sin^2 \beta\Delta x/2| < 1$, leading to the same requirement. Stability tests show that implicit methods are more stable than explicit ones; they allow larger time steps, reducing computer requirements. The multilevel transient ADI scheme in Chapter 10 was motivated by stability and speed.

While we have demonstrated von Neumann's test for the transient heat equation, the stability test applies equally to iterative methods for elliptic equations describing steady flows. The (artificial) time levels t and $t+\Delta t$ would refer to the approximate solutions obtained at consecutive iterations. The pressure solvers in Chapter 7 are examples of simple elliptic solvers that are stable in von Neumann's sense. Recall that the iterative method applied to single wells as it did to line fractures. Such a robust algorithm can be used to model general multilateral well drainhole trajectories where the overall topology can be arbitrarily defined by the driller or reservoir engineer.

Inadequacies of the von Neumann Test

Although von Neumann's test for an arbitrary wave component seems quite general, there are limitations. For example, it does not fully account for initial and boundary conditions; also, it does not model heterogeneities (i.e., variable coefficients). Analogies between von Neumann disturbances and propagating physical wave motions have been drawn in recent years by physicists. Actual wave motions undergo subtle changes in trajectory and wave-medium interaction as they propagate through nonuniformities, and similar effects are expected of moving numerical disturbances. Classical notions (e.g., group and phase velocity developed in wave mechanics) have recently been applied to the study of computational instabilities (such effects are not handled by older von Neumann tests).

Perhaps the greatest limitation on most tests is the restriction to linear systems. In nonlinear problems, as in transient Darcy flows of gases, a single harmonic disturbance wave component will lead to multiples of the primary frequency. This phenomenon, well known to vibrations engineers, is not accounted for in linear theory. Nonlinear models do exist, but solid, practical, fool-proof recipes are not yet available. In summary, stability on a linear von Neumann basis provides a warm level of comfort, but this is neither necessary nor sufficient for real stability. In practice, programming techniques and off-the-cuff coding decisions including stability, and some experimentation during code development is encouraged. Our requirements for stability were more demanding than von Neumann tests. As indicated, our simulator would operate stably with respect to changes in fluid and rock properties, in multilateral well topology and constraints, in the environment itself while simulating, e.g., sidetracking, recompletions, conversions between pressure and rate constraints, shut-ins, addition of lateral extensions to main bores, and so on. This objective entailed evaluation of numerous programming strategies. Finally, detailed engineering validations were performed to ensure that the simulator performed as envisioned. The software model, to include almost two dozen examples, will appear in Chin (2016a), a new book offered by John Wiley & Sons entitled *Reservoir Engineering in Modern Oilfields: Vertical, Deviated, Horizontal and Multilateral Well Systems*. This chapter gives highlights of the new publication.

Convergence

In our differencing of u(x,t), we denoted its numerical representation by $v_{i,n}$; that u may not, in fact, equal v is often a possibility. And as noted in Chapter 13, whether an equation arises as an approximation to one high-order system or another can completely seal its fate as a valid physical model. By the same token, the structure of formally small truncation errors is important in numerical analysis: without evaluating the role of higher derivatives in these terms, whose diffusive or dispersive effects always remain with the computed solution, the extent to which an "obvious" difference scheme models a differential equation cannot be ascertained. In advanced courses, examples are actually constructed showing how $\Delta x \rightarrow 0$ never yields correct solutions for certain classes of equations. Suffice it to say that nothing is straightforward about numerical analysis. From an optimistic point of view, this flexibility can be beneficial; ingenious devices can be created to accelerate the solution of elliptic equations. In Chapter 7, we demonstrated the equivalence between a relaxation scheme solving Laplace's equation and the explicit time integration of the transient heat equation. Modern researchers realize that solving elliptic problems as large time asymptotic limits of simple linear heat equations can be inefficient.

Therefore, invariant embedding techniques are developed, which embed the basic elliptic system in artificial time domains that provide rapid yet stable convergence. Consider yet another example. Thompson's grid generation method, defined by Eqs. (8.21) and (8.22), poses certain difficulties. In the form given, the coupled system of nonlinear elliptic equations in the dependent variables x and y leads, at best, to slow convergence and conditional stability (e.g., Sharpe and Anderson, 1991). But by reformulating the problem in somewhat unlikely complex conjugate coordinates $z = x + iy$ and $z^* = x - iy$, rapid convergence and absolute stability can always be guaranteed, as noted earlier. Through this nonlinear transformation, runs normally requiring minutes on workstations could be accomplished in seconds on standard computers!

Physical Resolution

Good physical resolution is the goal of reservoir analysis. Existing simulators provide high-level detail using grid refinement methods. One popular approach discretizes near-well grid blocks into even smaller blocks, effectively creating grid systems within grid systems. The resulting Cartesian formulation contains original macroscopic and new microscopic unknowns. But now, the governing difference equations are described by a completely different matrix structure, requiring new equation solvers and more research. At the very minimum, this means renaming pressure indexes and reordering equations. But by confronting the resolution issue in the formulation stage, say by using clever grid generation techniques, this needless work can be avoided and existing linear algebra techniques can be used (the large matrixes used in grid refinement imply more costly hardware and more complicated software). Consider still another problem. Chapters 8–10 discussed the ideas behind two-dimensional, areal grid generation, but gridding technology can be used in cross-sectional planes too. In Fig. 9.11, we introduced a faulted example of a boundary-conforming stratigraphic grid, whose coordinate surfaces adhered to the boundaries formed by geological layers. Simulations on such rectangular-like grids, it turns out, can be performed conveniently, and we shall give the general theory later.

Direct Solvers

In Chapter 7, we explained why direct solvers impose severe demands on computational resources, thus limiting the range of problems amenable to numerical analysis. The reasons, developed from two-dimensional examples, are even more pertinent to three-dimensional flows. Consider, for instance, $P_{xx} + P_{yy} + P_{zz} = 0$. When the lengths Δx, Δy, and Δz are constant, its finite difference representation takes the form $(P_{i-1,j,k}-2P_{i,j,k}+P_{i+1,j,k})/\Delta x^2 + (P_{i,j-1,k}-2P_{i,j,k}+P_{i,j+1,k})/\Delta y^2 + (P_{i,j,k-1}-2P_{i,j,k} \ +P_{i,j,k+1}) \ /\Delta z^2 = 0$. At each node, therefore, the difference molecule involves seven unknowns. Imagine a coarse grid simulation with *ten* grid blocks in each direction. This small model nonetheless contains $10 \times 10 \times 10$, or 1000 cells, with 1000 unknown pressures. A 1000×1000 equation set, needless to say, is undesirable.

For transient gas flows, or flows with nonlinear compaction, the intermediate use of Newton-Raphson iterations worsens these computational demands. Even if convergence is possible, truncation errors and cumulative round-offs will introduce numerous inaccuracies. Most direct solvers will solve carefully defined classes of problems efficiently. However, they do require special matrix conditioning and cumbersome fine tuning preprocessing that varies from field to field, and even within the same oil field, as changing multiphase oil production alters the coefficient structure of the governing equations with time. Whether such solvers are really more efficient than simpler, all-purpose simulators that function every time without special parameter inputs is a serious question that should be asked by all involved in reservoir modeling.

Modern Simulation Requirements

So far, we have discussed issues that apply to broad classes of problems. In petroleum engineering, however, the technological innovations of the past decade in drilling and production have brought about new requirements in computer modeling. Wells are no longer simple, fully penetrating, vertical sources or sinks that amicably coexist with rectangular grid structures. They are deviated, and even when horizontal, often take on wavy form. Most offshore wells start out vertically, but they will typically contain numerous out-of-plane horizontal or dipping drainholes whose induced flowfields interact. Fig. 15.1A illustrates two multilateral wells each producing in near-isolation, while Fig. 15.1B shows two interfering well systems that cannibalize each other's flow. But the trajectories in Fig. 15.1A, which are reasonable ones drilled using real-time logging data, may not be optimal from a more global perspective. Are they really the ones best suited to produce the most in the least amount of time? The highest total production over time? How should well topologies change as field optimization strategies change? The trajectories in Fig. 15.1B overlap and draw from the same source rock. Clearly some of these extensions can be shortened or removed. But which ones? The decision depends on drive mechanisms, and fluid and rock properties. These questions cannot be answered unless means exist to describe heterogeneities and complicated wells accurately, and numerical engines are developed to model the governing equations and specialized boundary conditions accurately.

Thus, idealized analytical solutions assuming, say, straight, centered, infinitely long horizontal wells in homogeneous formations sandwiched between impermeable layers, while elegant, may not be useful in steady-state or transient flow.

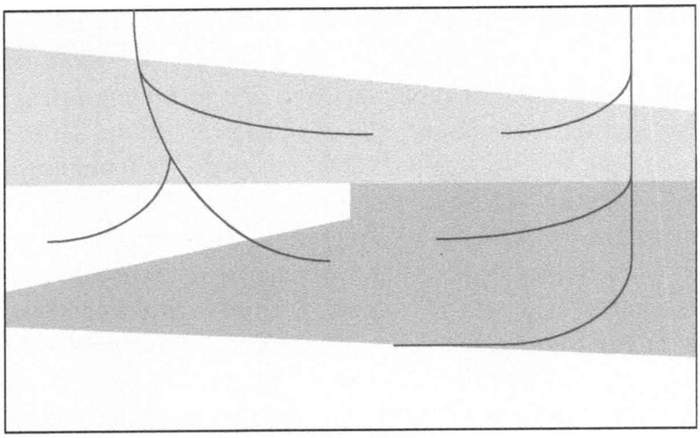

FIG. 15.1A Two multilaterals producing in near-isolation.

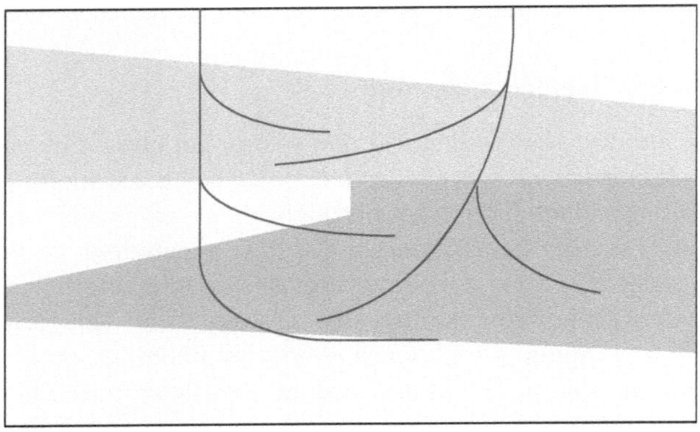

FIG. 15.1B Interfering multilaterals with cannibalizing flows.

And in reservoir description, classical well test interpretation methods related to early or late (dimensionless) time may not be relevant in highly heterogeneous rock produced by multilateral wells. Forward simulations needed to interpret well test response must be fast in order to be useful. But because solutions require lengthy Laplace transform inversions and unwieldy transcendental functions, even when crude homogeneous rock assumptions are made, they are not practical for routine use. With hardware costs declining, simulation is clearly becoming attractive. But there are mathematical issues that arise because well paths take on arbitrary form in space and time. To understand them, we must understand how boundary conditions along well paths, or simply well constraints, complicate the modern formulations.

Pressure Constraints

When a general wellbore defined along an arbitrary locus of points in three-dimensional space is pressure constrained, the equations along the well path are simple. For example, if gravity and friction are neglected, all the points satisfy $p_{iwell,jwell,kwell} = P_{well}$, where P_{well} is a specified constant. This simple boundary condition can nonetheless lead to inefficiencies and instabilities. For instance, when the sparse finite difference equation that normally applies is replaced by direct pressure specification at particular sets of (i_{well}, j_{well}, k_{well}) arbitrarily defined by the reservoir engineer, problems may or may not arise depending on the equation solver used because the structure of the coefficient matrix is determined arbitrarily by the driller.

Flow Rate Constraints

Pressure specification alone, at wells and farfield boundaries, leads to classical Dirichlet problems with completely deterministic, unique solutions. However, they will lead to internal discontinuities in the first derivatives of pressure, as discussed in Chapter 7. Specifying total wellbore volume flow rate Q, in the case of multilateral wells, leads to subtle problems that have not been discussed in the literature. In pure radial flows (e.g., see Chapter 6), any specification of Q can be equivalently re-expressed as a specification of the normal (radial) derivative dp/dr. The result is a classical Neumann problem whose solution is unique when pressure is known along some part of the reservoir, and to within an additive constant that does not affect flow rate, for a completely walled reservoir. But in prescribing the total flow rate Q for a general multilateral well system, the solution can be obtained in any number of "obvious" ways, only one of which yields the correct physics. In the absence of gravity and friction, the physically correct solution is the one reproducing Q, together with a borehole pressure that is constant along the entire completely general well path. Furthermore, this pressure level is an unknown that must be determined as part of the solution. Several flow simulators allocate or apportion Q by assigning velocity flux contributions to intersected layers based on local "permeability thickness" products. This reasonable method is *not* correct. So long as total mass is conserved, this yields a solution, but the result is incorrect because the pressure so obtained will vary along the path. So-called "kh methods," while plausible, are inherently incorrect and flawed.

For a general well path, point source contributions from all of the cell blocks making up the multilateral are needed to form the total sum Q. In other words, the flow rate formulation is not a classical Neumann problem because it involves an *integral* of pressure taken over all source points and their neighbors. To solve the problem correctly, the iterative solution of a large set of coupled finite difference equations must resolve several integrals, each summed over numerous non-neighboring connections. This destroys the idealized matrix structures (e.g., sparse, banded, or block diagonal) usually assumed in designing fast inversion routines. Moreover, this loss of ideal structure is compounded by the fact that the driller decides the topology of the multilateral well system. Failure to treat Q specifications correctly, use of flawed "kh methods," neglect of cross-derivative terms in "corner point" modeling, and so on, are routine in reservoir analysis. However, physical correctness must never be compromised for expediency and speed.

Object-Oriented Geobodies

Reservoir analysis involves entities like fault traps, channel sands, stratigraphic boundaries, dome-shaped structures, and so on. Often, the exploration geologist is able to render a reliable judgment regarding the nature of the structural geology, although the exact permeabilities, the degree of anisotropy, and the distribution of porosity remain unknowns to be refined via log analysis, seismic testing, and evaluation of production data. Why shouldn't reservoir simulators preserve the geological character of the oil field and read pictures instead? Can all of this be performed inexpensively? Once the high-level pictures are read in, the software can then interrogate the user about the values of quantities like permeability and porosity. Certainly, such an I/O approach is less prone to keyboard error, since numerical arrays are not entered; it is "fun," making simulation available more broadly and frequently. And it is certainly more intuitive than geostatistical methods, which appear to confer more accuracy than they deserve.

Plan for Remaining Sections

In the following sections, extremely stable, fast, and robust steady-state and transient compressible flow algorithms for liquids and gases in anisotropic heterogeneous media are given. Applications to deviated and horizontal wells are presented, convergence acceleration methods are demonstrated, and stratigraphic grid applications are developed. Importantly, the numerical schemes presented are user-friendly, requiring no numerical, computerese inputs; they typically lead to simulations that run the first time and every time. These algorithms were developed in aerodynamics for swept wing flows, also known as stratigraphic problems in petroleum engineering. Our discussion concludes with difficult examples of real geologies, solved by the new simulator, embodying all the features discussed next. In this chapter, five early validations from Chin (1993a,b, 2002) are replicated with the improved algorithm, which are followed by five very detailed deviated and multilateral well examples operating under a variety of field conditions. Graphical user-interface details and other nuances regarding these flows can be found in Chin (2016a).

GOVERNING EQUATIONS AND NUMERICAL FORMULATION

The equations for three-dimensional, compressible, heterogeneous, anisotropic, steady and transient, liquid and gas Darcy flows are given, as are those relating local pressures to total flow rates along arbitrary horizontal, deviated, and multilateral well paths. Stable algorithms are presented in all cases, drawing on the relaxation and ADI methods developed earlier.

Steady Flows of Liquids

The fundamental equation describing single-phase, liquid, Darcy flows in petroleum reservoirs is

$$\{(k_x/\mu)p_x\}_x + \{(k_y/\mu)p_y\}_y + \{(k_z/\mu)p_z\}_z = \phi c\, p_t + q(x, y, z, t) \qquad (15.1)$$

where $k_x(x,y,z)$, $k_y(x,y,z)$, and $k_z(x,y,z)$ denote permeabilities in the x, y and z directions, respectively, μ is the viscosity, $\phi(x,y,z)$ is the porosity, $c(x,y,z)$ is the effective compressibility characterizing the fluid and rock matrix system, and p (x,y,z,t) is the pressure field. Eq. (15.1) requires that all permeabilities vary smoothly, so that they and their corresponding pressure fields are differentiable; if there exist sudden changes in properties (e.g., as at layer interfaces), then pressure and velocity matching conditions must be used locally, as in Example 11.1, as extended to multiple dimensions.

In contrast to Chapter 1, we have explicitly introduced q(x,y,x,t), representing the local source volume flow rate per unit volume produced by any infinitesimal element of a general well. It is a three-dimensional, point singularity that applies to both injector and producer applications. For example, when q is a semi-infinite line, cylindrical radial flow is obtained over most of the source distribution, while spherical flow effects apply at the tip. In other words, partial penetration and spherical flow are modeled exactly. In this section, subscripts are used in three different contexts. First, they represent partial derivatives; for example, p_x is the partial derivative of p(x,y,z,t) with respect to the spatial coordinate x. Second, they are used as directional markers; for example, k_y (x,y,z) is the anisotropic permeability in the y direction. Finally, subscript indexes (i,j,k) in $p_{i,j,k}$ represent the centers of grid block volumes used in our finite difference discretizations. As usual, Δx, Δy, Δz, and Δt are used to denote grid sizes for the independent variables x, y, z, and t.

Difference Equation Formulation

Let us consider three-dimensional steady flows first, so that time derivatives vanish. Central differencing leads to

$$
\begin{aligned}
&[\{2\left[k_{xi,j,k}k_{xi+1,j,k}/\left(k_{xi,j,k}+k_{xi+1,j,k}\right)\right]\left(p_{i+1,j,k}-p_{i,j,k}\right)/\Delta x \\
&\quad -2\left[k_{xi-1,j,k}k_{xi,j,k}/\left(k_{xi-1,j,k}+k_{xi,j,k}\right)\right]\left(p_{i,j,k}-p_{i-1,j,k}\right)/\Delta x\}]/\mu\Delta x \\
&+[\{2[k_{yi,j,k}k_{yi,j+1,k}/\left(k_{yi,j,k}+k_{yi,j+1,k}\right)]\left(p_{i,j+1,k}-p_{i,j,k}\right)/\Delta y \\
&\quad -2\left[k_{yi,j-1,k}k_{yi,j,k}/\left(k_{yi,j-1,k}+k_{yi,j,k}\right)\right]\left(p_{i,j,k}-p_{i,j-1,k}\right)/\Delta y]/\mu\Delta y \\
&+[\{2\left[k_{zi,j,k}k_{zi,j,k+1}/\left(k_{zi,j,k}+k_{zi,j,k+1}\right)\right]\left(p_{i,j,k+1}-p_{i,j,k}\right)/\Delta z \\
&\quad -2\left[k_{zi,j,k-1}k_{zi,j,k}/\left(k_{zi,j,k-1}+k_{zi,j,k}\right)\right]\left(p_{i,j,k}-p_{i,j,k-1}\right)/\Delta z\}]/\mu\Delta z \\
&= q_{i,j,k}
\end{aligned}
\tag{15.2}
$$

where harmonic averages are used to represent permeabilities. We now multiply throughout by $\mu\Delta x\Delta y\Delta z$, where $\Delta x\Delta y\Delta z$ is the grid block volume, to obtain

$$
\begin{aligned}
&(\Delta y\Delta z/\Delta x)\,2\left[k_{xi,j,k}k_{xi+1,j,k}/\left(k_{xi,j,k}+k_{xi+1,j,k}\right)\right]\left(p_{i+1,j,k}-p_{i,j,k}\right) \\
&-(\Delta y\Delta z/\Delta x)2\left[k_{xi-1,j,k}k_{xi,j,k}/\left(k_{xi-1,j,k}+k_{xi,j,k}\right)\right]\left(p_{i,j,k}-p_{i-1,j,k}\right) \\
&+(\Delta x\Delta z/\Delta y)2\left[k_{yi,j,k}k_{yi,j+1,k}/\left(k_{yi,j,k}+k_{yi,j+1,k}\right)\right]\left(p_{i,j+1,k}-p_{i,j,k}\right) \\
&-(\Delta x\Delta z/\Delta y)2\left[k_{yi,j-1,k}k_{yi,j,k}/\left(k_{yi,j-1,k}+k_{yi,j,k}\right)\right]\left(p_{i,j,k}-p_{i,j-1,k}\right) \\
&+(\Delta x\Delta y/\Delta z)2\left[k_{zi,j,k}k_{zi,j,k+1}/\left(k_{zi,j,k}+k_{zi,j,k+1}\right)\right]\left(p_{i,j,k+1}-p_{i,j,k}\right) \\
&-(\Delta x\Delta y/\Delta z)2\left[k_{zi,j,k-1}k_{zi,j,k}/\left(k_{zi,j,k-1}+k_{zi,j,k}\right)\right]\left(p_{i,j,k}-p_{i,j,k-1}\right) \\
&= \mu q_{i,j,k}\,\Delta x\Delta y\Delta z
\end{aligned}
\tag{15.3}
$$

This suggests the following definitions for the transmissibilities TX, TY, and TZ, for convenience defined independently of the viscosity,

$$TX_{i,j,k} = (\Delta y\Delta z/\Delta x)2\left[k_{xi,j,k}k_{xi+1,j,k}/\left(k_{xi,j,k}+k_{xi+1,j,k}\right)\right] \tag{15.4a}$$

$$TX_{i-1,j,k} = (\Delta y\Delta z/\Delta x)2\left[k_{xi-1,j,k}k_{xi,j,k}/\left(k_{xi-1,j,k}+k_{xi,j,k}\right)\right] \tag{15.4b}$$

$$TY_{i,j,k} = (\Delta x\Delta z/\Delta y)2\left[k_{yi,j,k}k_{yi,j+1,k}/\left(k_{yi,j,k}+k_{yi,j+1,k}\right)\right] \tag{15.4c}$$

$$TY_{i,j-1,k} = (\Delta x\Delta z/\Delta y)2\left[k_{yi,j-1,k}k_{yi,j,k}/\left(k_{yi,j-1,k}+k_{yi,j,k}\right)\right] \tag{15.4d}$$

$$TZ_{i,j,k} = (\Delta x\Delta y/\Delta z)2\left[k_{zi,j,k}k_{zi,j,k+1}/\left(k_{zi,j,k}+k_{zi,j,k+1}\right)\right] \tag{15.4e}$$

$$TZ_{i,j,k-1} = (\Delta x\Delta y/\Delta z)2\left[k_{zi,j,k-1}k_{zi,j,k}/\left(k_{zi,j,k-1}+k_{zi,j,k}\right)\right] \tag{15.4f}$$

Then, Eq. (15.3) takes the more convenient form

$$\begin{aligned}&TX_{i,j,k}\left(p_{i+1,j,k}-p_{i,j,k}\right)-TX_{i-1,j,k}\left(p_{i,j,k}-p_{i-1,j,k}\right)\\&+TY_{i,j,k}\left(p_{i,j+1,k}-p_{i,j,k}\right)-TY_{i,j-1,k}\left(p_{i,j,k}-p_{i,j-1,k}\right)\\&+TZ_{i,j,k}\left(p_{i,j,k+1}-p_{i,j,k}\right)-TZ_{i,j,k-1}\left(p_{i,j,k}-p_{i,j,k-1}\right)\\&=\mu q_{i,j,k}\Delta x\Delta y\Delta z\end{aligned} \tag{15.5}$$

This equation, still very general, applies at all points. We consider points away from wells first. In these cases, the source term $q_{i,j,k}$ vanishes, and

$$\begin{aligned}&TX_{i,j,k}\left(p_{i+1,j,k}-p_{i,j,k}\right)-TX_{i-1,j,k}\left(p_{i,j,k}-p_{i-1,j,k}\right)\\&+TY_{i,j,k}\left(p_{i,j+1,k}-p_{i,j,k}\right)-TY_{i,j-1,k}\left(p_{i,j,k}-p_{i,j-1,k}\right)\\&+TZ_{i,j,k}\left(p_{i,j,k+1}-p_{i,j,k}\right)-TZ_{i,j,k-1}\left(p_{i,j,k}-p_{i,j,k-1}\right)=0\end{aligned} \tag{15.6}$$

which, for reasons that will become obvious, we rewrite in the form

$$\begin{aligned}&TZ_{i,j,k-1}\,p_{i,j,k-1}\\&-\left\{TZ_{i,j,k}+TZ_{i,j,k-1}+TY_{i,j,k}+TY_{i,j-1,k}+TX_{i,j,k}+TX_{i-1,j,k}\right\}p_{i,j,k}\\&+TZ_{i,j,k}\,p_{i,j,k+1}\\&=-TX_{i,j,k}\,p_{i+1,j,k}-TX_{i-1,j,k}\,p_{i-1,j,k}-TY_{i,j,k}\,p_{i,j+1,k}-TY_{i,j-1,k}\,p_{i,j-1,k}\end{aligned} \tag{15.7}$$

The Iterative Scheme

An iterative three-dimensional solution is suggested. If we fix y_j in outermost programming loop, consider a given x_i plane, write Eq. (15.7) at all internal node points z_k, and couple with upper and lower boundary conditions, the solutions of all left-hand side points can be obtained if the right side terms of Eq. (15.7) were (approximately) known. Like the planar examples in Chapter 7, Eq. (15.7) leads to tridiagonal matrices, whose inversions require only 3N multiplies or divides for O(N) systems. Eq. (15.7) not only retains its diagonal dominance, but in three dimensions, numerical stability turns out to be significantly enhanced. When grid block aspect ratios and anisotropic permeabilities reduce diagonal dominance, we simply use sister forms of Eq. (15.7) written along the alternative lines i - 1, i, i + 1 or j - 1, j, j + 1. Together with the use of over relaxation (e.g., Chapter 7), we have a new variant of SLOR or Successive Line Over Relaxation. The above lines are swept along planes, then from plane to plane, and farfield boundary conditions are used to update all end-plane lines. The computational box is treated repeatedly in this manner. Latest pressure values are used as they are available to evaluate all coefficient matrices.

The heat equation analogy in Chapter 7, justifying the convergence of this iterative method to the unique solution guaranteed by Laplace's equation, again applies here. The method is robust because it always converges and requires little in the way of matrix conditioning and parameter tuning. And the solution is, importantly, independent of the initial guess. Any guess will lead to the solution, as we have shown in Chapter 7. Of course, the closer the guess is to the actual solution, the faster the convergence; analytical solutions such as those derived in Chapters 2–6 can be used where appropriate. This property allows us to run multiple realizations of a physical problem quickly and efficiently. Thus, when the topology of a deviated horizontal well is changed, or when an existing well simply grows longer or adds drainholes, or when fluid and formation properties are modified, or when well constraints are altered, the iterations need not begin from scratch. The algorithm given here uses prior information for earlier simulations to produce fast solutions with only incremental effort. In reservoir description applications where multiple geological (or geostatistical) realizations are often evaluated, and in infill drilling problems where numerous production scenarios are often considered, this feature is important.

Modeling Well Constraints for Liquids

Now we discuss boundary conditions internal to the computational box. In reservoir simulation, well constraints provide the most important class of internal boundary conditions; other internal conditions may include symmetry and antisymmetry statements used to model fractures and shales.

Pressure constraints are the simplest to implement: at the physical location corresponding to a particular well, a simple equation explicitly enforcing a prescribed level replaces the tridiagonal equation otherwise written at that point. Modeling net volume flow rate constraints at wells, as we have already indicated, is somewhat more complicated. In many simulators, the net flow rate is allocated to the layers intercepted by the well path according to local kh product, and often disallowing interlayer flow as well. Such kh allocation is incorrect because the net production in each layer is also proportional to the difference between wellbore and grid block pressures, where both must be determined as part of the solution.

In the absence of gravity and wellbore friction, the solution process must be enforced in such a way that the pressure (under a net volume flow rate specification) is a constant along the well path. This integral constraint, obtained by integrating Darcy's velocity formula over numerous non-neighboring connections, degrades the performance of equation solvers and encourages the use of incorrect kh fixes. To be precise, we consider a locus of points L defining a general wellbore that may be vertical, horizontal, deviated, and out-of-plane, or, bifurcated with multiple clustered drainhole extensions. Let the symbol Σ

denote summations along L performed in any order. We write Eq. (15.5) for each well point along L in the form

$$
\begin{aligned}
&\{TX_{i,j,k}p_{i+1,j,k} + TX_{i-1,j,k}p_{i-1,j,k} + TY_{i,j,k}p_{i,j+1,k} \\
&+ TY_{i,j-1,k}p_{i,j-1,k} + TZ_{i,j,k}p_{i,j,k+1} + TZ_{i,j,k-1}p_{i,j,k-1}\} \\
&-p_{i,j,k}\left[TX_{i,j,k} + TX_{i-1,j,k} + TY_{i,j,k} + TY_{i,j-1,k} + TZ_{i,j,k} + TZ_{i,j,k-1}\right] \\
&= \mu q_{i,j,k}\Delta x \Delta y \Delta z
\end{aligned}
\tag{15.8}
$$

and sum the resultant set of algebraic equations over all (i,j,k)'s along L, to give

$$
\begin{aligned}
&\sum\{TX_{i,j,k}p_{i+1,j,k} + TX_{i-1,j,k}p_{i-1,j,k} + TY_{i,j,k}p_{i,j+1,k} \\
&+ TY_{i,j-1,k}p_{i,j-1,k} + TZ_{i,j,k}p_{i,j,k+1} + TZ_{i,j,k-1}p_{i,j,k-1}\} \\
&-\sum p_{i,j,k}\left[TX_{i,j,k} + TX_{i-1,j,k} + TY_{i,j,k} + TY_{i,j-1,k} + TZ_{i,j,k} + TZ_{i,j,k-1}\right] \\
&= \mu\sum q_{i,j,k}\,\Delta x \Delta y \Delta z
\end{aligned}
\tag{15.9}
$$

or, more conveniently,

$$\sum\{\,\}-\sum p_{i,j,k}[\,] = \mu\sum q_{i,j,k}\Delta x\Delta y\Delta z \tag{15.10}$$

At this point, several physical conditions can be invoked to simplify the algebra. First, because gravity and friction are neglected in the present formulation, the $p_{i,j,k}$ factor can be moved across the summation operator since the pressure at any point within the well system is a constant. This constant is prescribed when the well is pressure constrained; but when the well is volume flow rate constrained, the unknown constant pressure level, which is different from well to well, must be found as part of the solution.

Let us denote this constant pressure, whether it is known or unknown, by the symbol p_w. Now, the summation on the right-hand side of Eq. (15.10) is the volume flow rate Q_w of the producer or injector well. We denote

$$Q_w = \sum q_{i,j,k}\Delta x\Delta y\Delta z \tag{15.11}$$

so that

$$\sum\{\,\}-p_w\sum[\,] = \mu Q_w \tag{15.12}$$

Thus, it follows that

$$p_w = \left(\sum\{\,\}-\mu Q_w\right)/\sum[\,] \tag{15.13}$$

The strategy for rate-constrained wells is simple: use *this* pressure prescription as the diagonally dominant difference equation at well points. The result is a stable algorithm that looks pressure constrained, but the right side of the above (evaluated with latest values) is not really known until the iterations converge.

This procedure has the added benefit of conserving mass in the local sense since the pressure-dependent variable itself is prescribed and not its normal derivatives; numerical experiments also show that it is highly stabilizing. Once the iterations have converged globally, Eq. (15.13) is used to compute well pressures at rate-constrained wells, while the expression for Q_w from Eq. (15.12) is used to compute net flow rates at pressure-constrained wells.

Steady and Unsteady Nonlinear Gas Flows

While gas flows also satisfy Darcy's laws, the equation of state that connects density and pressure renders the governing equations somewhat intractable and less amenable to solution. Mathematically, they become nonlinear. Thus, linear superposition methods in conventional well testing, where the solutions corresponding to step-wise changing rates or pressures are directly summed, do not apply. Nonetheless, superposition is often used, assuming that mean reservoir conditions do not change much, so that nonlinear coefficients can be frozen about nearly static values. This is, in general, incorrect; with high-speed computers widely available, there is really no need to invoke such limiting assumptions. Because an unconditionally stable scheme for transient linear liquid flows turns out to be available and provided below, it makes practical sense to take advantage of it and to reformulate the general problem for nonlinear gases as closely as possible. The complete equation for mass conservation in three dimensions is

$$(\rho u)_x + (\rho v)_y + (\rho w)_z = -\phi \rho_t - q^* \tag{15.14}$$

where $\rho(x,y,z,t)$ is the mass density, and $q^*(x,y,z,t)$ is the local mass flow rate per unit volume. Now, the Cartesian velocity components u, v, and w in the x, y, and z directions are given by Darcy's laws,

$$u(x, y, z, t) = -(k_x(x, y, z)/\mu)p_x \tag{15.15a}$$

$$v(x, y, z, t) = -(k_y(x, y, z)/\mu)p_y \tag{15.15b}$$

$$w(x, y, z, t) = -(k_z(x, y, z)/\mu)p_z \tag{15.15c}$$

The pressure $p(x,y,z,t)$ and the density $\rho(x,y,z,t)$, following Muskat (1937), are assumed to be thermodynamically connected by the polytropic relationship

$$\rho = \gamma p^m \tag{15.16}$$

where m is Muskat's exponent and γ is determined from reference conditions. If we now substitute this expression for density into Eq. (15.14), we have

$$(k_x p^m p_x)_x + (k_y p^m p_y)_y + (k_z p^m p_z)_z = \phi\mu(p^m)_t + \mu q^*/\gamma \tag{15.17}$$

Thus, we are led to rewrite Eq. (15.17) in the form

$$\{k_x p^{m+1}{}_x\}_x + \{k_y p^{m+1}{}_y\}_y + \{k_z p^{m+1}{}_z\}_z = \phi \mu c^* p^{m+1}{}_t + \mu \{(m+1)/\gamma\} q^*(x, y, z, t) \quad (15.18)$$

with

$$c^* = m/p(x, y, z, t) \quad (15.19)$$

where c^* is a fictitious compressibility for the pressure-like quantity p^{m+1}. This liquid-like formulation for p^{m+1} is useful because the unconditionally stable time integration scheme developed for linear liquid transients to be given, satisfying the classical parabolic heat equation, applies with little modification. The coefficient c^* depends on the evolving pressure p(x,y,z,t); however, this nonlinear dependence turns out to be numerically stabilizing. Nowhere have we invoked linear superposition, which does not apply. We give a unified presentation applicable to both transient liquids and gases later. But before embarking on general gas flows, we consider steady problems first, in order to understand several crucial physical and mathematical formulation differences.

Steady Gas Flows

From a numerical viewpoint, the iterative solution for gases does not depart significantly from that for liquids; essentially, total mass, not volume, is conserved. Volume varies as a function of pressure, which varies with position; detailed numerical bookkeeping to track mass balances accurately is critical to error-free results. By analogy to Eq. (15.2), a similar discretization process leads to the cluster

$$\begin{aligned}
&[\{2k_{xi,j,k}k_{xi+1,j,k}/(k_{xi,j,k} + k_{xi+1,j,k})]\left(p^{m+1}{}_{i+1,j,k} - p^{m+1}{}_{i,j,k}\right)/\Delta x \\
&\quad -2\left[k_{xi-1,j,k}k_{xi,j,k}/(k_{xi-1,j,k} + k_{xi,j,k})\right]\left(p^{m+1}{}_{i,j,k} - p^{m+1}{}_{i-1,j,k}\right)/\Delta x\}]\Delta x \\
&+ [\{2k_{yi,j,k}k_{yi,j+1,k}/(k_{yi,j,k} + k_{yi,j+1,k})]\left(p^{m+1}{}_{i,j+1,k} - p^{m+1}{}_{i,j,k}\right)/\Delta y \\
&\quad -2\left[k_{yi,j-1,k}k_{yi,j,k}/(k_{yi,j-1,k} + k_{yi,j,k})\right]\left(p^{m+1}{}_{i,j,k} - p^{m+1}{}_{i,j-1,k}\right)/\Delta y\}]/\Delta y \\
&+ [\{2k_{zi,j,k}k_{zi,j,k+1}/(k_{zi,j,k} + k_{zi,j,k+1})]\left(p^{m+1}{}_{i,j,k+1} - p^{m+1}{}_{i,j,k}\right)/\Delta z \\
&\quad -2\left[k_{zi,j,k-1}k_{zi,j,k}/(k_{zi,j,k-1} + k_{zi,j,k})\right]\left(p^{m+1}{}_{i,j,k} - p^{m+1}{}_{i,j,k-1}\right)/\Delta z\}]/\Delta z \\
&= \mu\{(m+1)/\gamma\}\, q^*_{i,j,k} \qquad (15.20)
\end{aligned}$$

or

$$
\begin{aligned}
&(\Delta y \Delta z/\Delta x)2\left[k_{xi,j,k}k_{xi+1,j,k}/\left(k_{xi,j,k}+k_{xi+1,j,k}\right)\right]\left(p^{m+1}{}_{i+1,j,k}-p^{m+1}{}_{i,j,k}\right)\\
&\quad-(\Delta y \Delta z/\Delta x)\,2\left[k_{xi-1,j,k}k_{xi,j,k}/\left(k_{xi-1,j,k}+k_{xi,j,k}\right)\right]\left(p^{m+1}{}_{i,j,k}-p^{m+1}{}_{i-1,j,k}\right)\\
&\quad+(\Delta x \Delta z/\Delta y)2\left[k_{yi,j,k}k_{yi,j+1,k}/\left(k_{yi,j,k}+k_{yi,j+1,k}\right)\right]\left(p^{m+1}{}_{i,j+1,k}-p^{m+1}{}_{i,j,k}\right)\\
&\quad-(\Delta x \Delta z/\Delta y)2\left[k_{yi,j-1,k}k_{yi,j,k}/\left(k_{yi,j-1,k}+k_{yi,j,k}\right)\right]\left(p^{m+1}{}_{i,j,k}-p^{m+1}{}_{i,j-1,k}\right)\\
&\quad+(\Delta x \Delta y/\Delta z)2\left[k_{zi,j,k}k_{zi,j,k+1}/\left(k_{zi,j,k}+k_{zi,j,k+1}\right)\right]\left(p^{m+1}{}_{i,j,k+1}-p^{m+1}{}_{i,j,k}\right)\\
&\quad-(\Delta x \Delta y/\Delta z)2\left[k_{zi,j,k-1}k_{zi,j,k}/\left(k_{zi,j,k-1}+k_{zi,j,k}\right)\right]\left(p^{m+1}{}_{i,j,k}-p^{m+1}{}_{i,j,k-1}\right)\\
&\quad=\mu\{(m+1)/\gamma\}\,q^*_{i,j,k}\,\Delta x \Delta y \Delta z
\end{aligned}
\tag{15.21}
$$

Using the transmissibility definitions in Eqs. (15.4a)–(15.4f), we have

$$
\begin{aligned}
&TX_{i,j,k}\left(p^{m+1}{}_{i+1,j,k}-p^{m+1}{}_{i,j,k}\right)-TX_{i-1,j,k}\left(p^{m+1}{}_{i,j,k}-p^{m+1}{}_{i-1,j,k}\right)\\
&\quad+TY_{i,j,k}\left(p^{m+1}{}_{i,j+1,k}-p^{m+1}{}_{i,j,k}\right)-TY_{i,j-1,k}\left(p^{m+1}{}_{i,j,k}-p^{m+1}{}_{i,j-1,k}\right)\\
&\quad+TZ_{i,j,k}\left(p^{m+1}{}_{i,j,k+1}-p^{m+1}{}_{i,j,k}\right)-TZ_{i,j,k-1}\left(p^{m+1}{}_{i,j,k}-p^{m+1}{}_{i,j,k-1}\right)\\
&\quad=\mu\{(m+1)/\gamma\}\,q^*_{i,j,k}\,\Delta x \Delta y \Delta z
\end{aligned}
\tag{15.22}
$$

First we write Eq. (15.22) for points that do not contain wells, setting $q^*_{i,j,k}$ to zero. Then we cast this in tridiagonal form to facilitate the iterations, that is,

$$
\begin{aligned}
&TZ_{i,j,k-1}\,p^{m+1}{}_{i,j,k-1}\\
&\quad-\left\{TZ_{i,j,k}+TZ_{i,j,k-1}+TY_{i,j,k}+TY_{i,j-1,k}+TX_{i,j,k}+TX_{i-1,j,k}\right\}p^{m+1}{}_{i,j,k}\\
&\quad+TZ_{i,j,k}\,p^{m+1}{}_{i,j,k+1}\\
&=-TX_{i,j,k}p^{m+1}{}_{i+1,j,k}-TX_{i-1,j,k}p^{m+1}{}_{i-1,j,k}\\
&\quad-TY_{i,j,k}p^{m+1}{}_{i,j+1,k}-TY_{i,j-1,k}p^{m+1}{}_{i,j-1,k}
\end{aligned}
\tag{15.23}
$$

Then, all the comments made immediately following Eq. (15.7) apply without change, to the dependent variable p^{m+1}.

Well Constraints for Gas Flows

Consider a locus of points L defining a general well path that may be vertical, horizontal, or deviated out-of-plane and containing multiple drainholes. Let Σ denote summations performed along L. Along well paths only, in anticipation of constant pressures in the borehole, we simplify Eq. (15.22) by factoring out $p_{i,j,k}{}^{m+1}$ so that

$$\{TX_{i,j,k}\,p^{m+1}{}_{i+1,j,k} + TX_{i-1,j,k}\,p^{m+1}{}_{i-1,j,k} + TY_{i,j,k}\,p^{m+1}{}_{i,j+1,k}$$
$$+TY_{i,j-1,k}\,p^{m+1}{}_{i,j-1,k} + TZ_{i,j,k}\,p^{m+1}{}_{i,j,k+1} + TZ_{i,j,k-1}p^{m+1}{}_{i,j,k-1}\}$$
$$-p^{m+1}{}_{i,j,k}\left[TX_{i,j,k} + TX_{i-1,j,k} + TY_{i,j,k} + TY_{i,j-1,k} + TZ_{i,j,k} + TZ_{i,j,k-1}\right]$$
$$= \mu\{(m+1)/\gamma\}q^*_{i,j,k}\,\Delta x\Delta y\Delta z \tag{15.24}$$

When the foregoing equation is written for each well point along L, and the resultant equations are summed, we have

$$\sum\{TX_{i,j,k}\,p^{m+1}{}_{i+1,j,k} + TX_{i-1,j,k}\,p^{m+1}{}_{i-1,j,k} + TY_{i,j,k}\,p^{m+1}{}_{i,j+1,k}$$
$$+ TY_{i,j-1,k}\,p^{m+1}{}_{i,j-1,k} + TZ_{i,j,k}p^{m+1}{}_{i,j,k+1} + TZ_{i,j,k-1}p^{m+1}{}_{i,j,k-1}\}$$
$$-\sum p^{m+1}{}_{i,j,k}\left[TX_{i,j,k} + TX_{i-1,j,k} + TY_{i,j,k} + TY_{i,j-1,k} + TZ_{i,j,k} + TZ_{i,j,k-1}\right]$$
$$= \mu\{(m+1)/\gamma\}\sum q^*{}_{i,j,k}\Delta x\Delta y\Delta z \tag{15.25}$$

or, more conveniently,

$$\sum\{\} - \sum p^{m+1}{}_{i,j,k}\,[\,] = \mu\{(m+1)/\gamma\}\sum q^*{}_{i,j,k}\Delta x\Delta y\Delta z \tag{15.26}$$

Because gravity and wellbore friction are neglected in this formulation, the constant $p_{i,j,k}$ can be moved across the summation since the pressure at any point within the well system is a constant. This constant is prescribed when the well is pressure constrained; when it is mass-flow-rate-constrained, the constant pressure must be found as part of the solution. Let us denote this constant pressure, whether it is known or unknown, as p_w. Now, the summation on the right-hand side of Eq. (15.26) is just the total mass flow rate associated with the producer or injector well, that is,

$$M_w = \sum q^*{}_{i,j,k}\Delta x\Delta y\Delta z \tag{15.27}$$

In field practice, all measurements are reported at standard surface conditions, normally 14.7 psi and 60°F. Then, the mass flow rate satisfies

$$M_w = \rho_{sc}Q_{w,sc} \tag{15.28}$$

where $Q_{w,sc}(t)$ is the total volume flow rate at the surface, and ρ_{sc} is the surface mass density, with the subscript sc denoting standard gas conditions. Eq. (15.26) becomes, since well pressure is constant,

$$\sum\{\} - p_w{}^{m+1}\sum[\,] = \mu\{(m+1)/\gamma\}M_w \tag{15.29}$$

It follows that the wellbore pressure p_w satisfies

$$p_w{}^{m+1} = \left(\sum\{\} - \mu\,[(m+1)\rho_{sc}/\gamma]\,Q_{w,sc}\right)/\sum[\,] \tag{15.30}$$

From this point onward, the treatment of well constraints is identical to that for Darcy flows of liquids, with minor changes. It is clear that the liquid scheme is unchanged so long as we replace p by p^{m+1}, vanishing normal derivatives of p by those of p^{m+1}, and the viscosity μ by $\mu\,[(m+1)\rho_{sc}/\gamma]$. When there exists more than one multilateral well path L, that is, if there exist more than one multilateral well cluster in the reservoir, the same computational logic applies to each cluster individually. Any number of well clusters is permissible, although it is obvious that the total number of grid blocks without wells should greatly exceed the number of grid blocks used to describe wells.

Transient, Compressible Flows

Very often, oil companies produce reservoirs from a virgin static state in which the fluid is quiescent everywhere. At other times, a steady-state flow (such as that computed from our relaxation method) may be completely or partially shut-in for well testing or for economical reasons. Sometimes nonproductive intervals are sealed off, and horizontal drainholes may be drilled to enhance local production at other locations. All of these scenarios demand that any time integration scheme be especially robust, capable of withstanding sudden operational shocks to the system. The algorithm given below, like the relaxation method developed for steady flows of liquids and gases, is very stable. Without loss of generality, let us drop the source term q* from the governing equation for non-well points, understanding that we shall replace the particular difference equation with our internal constraint condition for those points affected by wells. Thus, we have

$$(k_x p^m p_x)_x + (k_y p^m p_y)_y + (k_z p^m p_z)_z = \phi\mu(p^m)_t \tag{15.31}$$

or, after some manipulation,

$$\{k_x p^{m+1}{}_x\}_x + \{k_y p^{m+1}{}_y\}_y + \{k_z p^{m+1}{}_z\}_z = \phi\mu c^* p^{m+1}{}_t \tag{15.32}$$

Eqs. (15.31) and (15.32) apply to gases and liquids (that is, m = 0 and c* = c). A differencing similar to that for steady flow can be used, provided we include time. If n and n + 1 denote times at t_n and t_{n+1}, we have the implicit scheme

$$\begin{aligned}
&[\{2k_{xi,j,k}k_{xi+1,j,k}/(k_{xi,j,k}+k_{xi+1,j,k})]\left(p^{m+1}{}_{i+1,j,k,n+1}-p^{m+1}{}_{i,j,k,n+1}\right)/\Delta x\\
&-2\left[k_{xi-1,j,k}k_{xi,j,k}/(k_{xi-1,j,k}+k_{xi,j,k})\right]\left(p^{m+1}{}_{i,j,k,n+1}-p^{m+1}{}_{i-1,j,k,n+1}\right)/\Delta x\}]/\Delta x\\
&+[\{2k_{yi,j,k}k_{yi,j+1,k}/(k_{yi,j,k}+k_{yi,j+1,k})]\left(p^{m+1}{}_{i,j+1,k,n+1}-p^{m+1}{}_{i,j,k,n+1}\right)/\Delta y\\
&-2\left[k_{yi,j-1,k}k_{yi,j,k}/(k_{yi,j-1,k}+k_{yi,j,k})\right]\left(p^{m+1}{}_{i,j,k,n+1}-p^{m+1}{}_{i,j-1,k,n+1}\right)/\Delta y\}]/\Delta y\\
&+[\{2k_{zi,j,k}k_{zi,j,k+1}/(k_{zi,j,k}+k_{zi,j,k+1})]\left(p^{m+1}{}_{i,j,k+1,n+1}-p^{m+1}{}_{i,j,k,n+1}\right)/\Delta z\\
&-2\left[k_{zi,j,k-1}k_{zi,j,k}/(kz_{i,j,k-1}+k_{zi,j,k})\right]\left(p^{m+1}{}_{i,j,k,n+1}-p^{m+1}{}_{i,j,k-1,n+1}\right)/\Delta z\}]/\Delta z\\
&=\phi_{i,j,k}\,\mu c^*{}_{i,j,k,n}\left(p^{m+1}{}_{i,j,k,n+1}-p^{m+1}{}_{i,j,k,n}\right)/\Delta t
\end{aligned} \tag{15.33}$$

Using our definitions for transmissibility, Eq. (15.33) becomes

$$\begin{aligned}
&TX_{i,j,k}\left(p^{m+1}{}_{i+1,j,k,n+1} - p^{m+1}{}_{i,j,k,n+1}\right) - TX_{i-1,j,k}\left(p^{m+1}{}_{i,j,k,n+1}\right.\\
&\left.- p^{m+1}{}_{i-1,j,k,n+1}\right) + TY_{i,j,k}\left(p^{m+1}{}_{i,j+1,k,n+1} - p^{m+1}{}_{i,j,k,n+1}\right)\\
&- TY_{i,j-1,k}\left(p^{m+1}{}_{i,j,k,n+1} - p^{m+1}{}_{i,j-1,k,n+1}\right) + TZ_{i,j,k}\left(p^{m+1}{}_{i,j,k+1,n+1}\right.\\
&\left.- p^{m+1}{}_{i,j,k,n+1}\right) - TZ_{i,j,k-1}\left(p^{m+1}{}_{i,j,k,n+1} - p^{m+1}{}_{i,j,k-1,n+1}\right)\\
&= \phi_{i,j,k}\,\mu c^*{}_{i,j,k,n}\left(p^{m+1}{}_{i,j,k,n+1} - p^{m+1}{}_{i,j,k,n}\right)\Delta x \Delta y \Delta z/\Delta t
\end{aligned} \tag{15.34}$$

If Eq. (15.34) is to be written for each (i,j,k) node and solved at the new time step (n+1), we obtain a complicated system of algebraic equations that is costly to invert computationally. When it cannot be locally linearized, the full but sparse matrix is solved using even more expensive Newton-Raphson iterations. Thus, we employ approximate factorization techniques to resolve the system into three simpler, but sequential banded ones. In this approach, especially popular in the Soviet literature, appropriate high-order terms no larger than the discretization errors implicit in the derivation of Eq. (15.33) are added to Eq. (15.34). These terms are chosen to facilitate a nested factorization of the difference operator just given. The design is structured so that the three-step process required for the integration of a typical time step is unconditionally stable on a linearized von Neumann basis. Moreover, each intermediate-time-step level employs efficient tridiagonal matrices only. The results of this factorization lead to Eqs. (15.35), (15.36) and (15.37), defining predictor Steps 1 and 2, and corrector Step 3, that is,

Step 1

$$\begin{aligned}
&TX_{i,j,k}\left(p^{m+1}{}_{i+1,j,k,n} - p^{m+1}{}_{i,j,k,n}\right) - TX_{i-1,j,k}\left(p^{m+1}{}_{i,j,k,n} - p^{m+1}{}_{i-1,j,k,n}\right)\\
&\quad + TY_{i,j,k}\left(p^{m+1}{}_{i,j+1,k,n+1/3} - p^{m+1}{}_{i,j,k,n+1/3}\right)\\
&\quad - TY_{i,j-1,k}\left(p^{m+1}{}_{i,j,k,n+1/3} - p^{m+1}{}_{i,j-1,k,n+1/3}\right)\\
&\quad + TZ_{i,j,k}\left(p^{m+1}{}_{i,j,k+1,n} - p^{m+1}{}_{i,j,k,n}\right)\\
&\quad - TZ_{i,j,k-1}\left(p^{m+1}{}_{i,j,k,n} - p^{m+1}{}_{i,j,k-1,n}\right)\\
&= \phi_{i,j,k}\,\mu c^*{}_{i,j,k,n}\left(p^{m+1}{}_{i,j,k,n+1/3} - p^{m+1}{}_{i,j,k,n}\right)\Delta x \Delta y \Delta z/\Delta t
\end{aligned} \tag{15.35}$$

Step 2

$$\begin{aligned} & TX_{i,j,k}\left(p^{m+1}{}_{i+1,j,k,n+2/3} - p^{m+1}{}_{i,j,k,n+2/3}\right) \\ & \quad - TX_{i-1,j,k}\left(p^{m+1}{}_{i,j,k,n+2/3} - p^{m+1}{}_{i-1,j,k,n+2/3}\right) \\ & = TX_{i,j,k}\left(p^{m+1}{}_{i+1,j,k,n} - p^{m+1}{}_{i,j,k,n}\right) \\ & \quad - TX_{i-1,j,k}\left(p^{m+1}{}_{i,j,k,n} - p^{m+1}{}_{i-1,j,k,n}\right) \\ & \quad + \phi_{i,j,k}\,\mu c^*{}_{i,j,k,n}\left(p^{m+1}{}_{i,j,k,n+2/3} - p^{m+1}{}_{i,j,k,n+1/3}\right)\Delta x\Delta y\Delta z/\Delta t \end{aligned} \tag{15.36}$$

Step 3

$$\begin{aligned} & TZ_{i,j,k}\left(p^{m+1}{}_{i,j,k+1,n+1} - p^{m+1}{}_{i,j,k,n+1}\right) \\ & \quad - TZ_{i,j,k-1}\left(p^{m+1}{}_{i,j,k,n+1} - p^{m+1}{}_{i,j,k-1,n+1}\right) \\ & = TZ_{i,j,k}\left(p^{m+1}{}_{i,j,k+1,n} - p^{m+1}{}_{i,j,k,n}\right) \\ & \quad - TZ_{i,j,k-1}\left(p^{m+1}{}_{i,j,k,n} - p^{m+1}{}_{i,j,k-1,n}\right) \\ & \quad + \phi_{i,j,k}\,\mu c^*{}_{i,j,k,n}\left(p^{m+1}{}_{i,j,k,n+1} - p^{m+1}{}_{i,j,k,n+2/3}\right)\Delta x\Delta y\Delta z/\Delta t \end{aligned} \tag{15.37}$$

Formal von Neumann analysis shows that this three-step process is second-order accurate in Δx, Δy, and Δz, and first-order accurate in Δt. Well constraints within each step are handled exactly as in our relaxation approach for steady-state flows. We emphasize that unconditional stability alone does not ensure convergence to physically correct solutions. Stability is necessary but not sufficient for practical solutions; (somewhat) small time steps are nonetheless required to capture the physics and provide physical resolution where needed.

Compaction, Consolidation and Subsidence

A formal approach to modeling compaction, consolidation and subsidence requires the use of well-defined constitutive equations that describe both fluid and solid phases of matter. At the same time, these would be applied to a general Lagrangian dynamical formulation written to host the deforming meshes, whose exact time histories must be determined as part of the overall solution. These nonlinear deformations are often plastic in nature, and not elastic, as in linear analyses usually employed in structural mechanics. This finite deformation approach, usually adopted in more rigorous academic researches into compressible porous media, is well known in soil mechanics and civil engineering. However, it is computationally intensive and not practical for routine use. For everyday use, a simpler empirical model is justifiable since the input parameters required of more sophisticated approaches are themselves subject to error. Simplicity is useful when order-of-magnitude effects and qualitative trends only are examined.

Despite the apparent rigor in many of the accepted mathematical models, however, most are nonetheless empirical. They typically assume a linear relationship between pore pressure and porosity; that is, they assume that instantaneous pressure affects the original $\phi(x,y,z)$ linearly. The constants appearing in the constitutive equations, moreover, can be subject to significant measurement error. In the Ekofisk reservoir where subsidence and compaction drives are important, an overall height decrease of 40 ft, compared to a 400 ft net reservoir thickness originally, has been observed; this 10% change, however, occurred over a twenty-year period. These physical scales suggest that a simpler engineering model suffices for approximate trend analysis.

In the scheme adopted here, we define $\phi(x,y,z)$ as the baseline porosity function when compaction is not important. In the numerical analysis, however, the actual porosity is $\phi(x,y,z)$ pre-multiplied by a $\{1+a\,p(x,y,z,t)\}$ factor, where "a" is a user-defined "what if" parameter; it is a negative constant (or secondary compressibility), having units of 1/psi. There are several implicit assumptions. Consider the mass balance equation $(\rho u)_x+(\rho v)_y+(\rho w)_z=-\phi\rho_t$ assumed earlier; there, ϕ was a prescribed, spatially varying function, independent of time. The right-hand side, in the more general case when temporal changes are allowed, however, would take the form $(\phi\ \rho)_t$ where ϕ now refers to $\{1+a\ p(x,y,z,t)\}$ $\phi(x,y,z)$. Thus, our procedure assumes $\phi\rho_t \gg \rho\phi_t$. The dominant effect of compaction, in this small disturbance limit, therefore arises from the porosity reduction enforced by the a p(x,y,z,t) term taken above and not from direct volume changes. This may or may not be physically valid. The correction, in this sense, plays the role of a secondary compressibility, a parameter introduced earlier. In the next correction sequence, pressure-dependent decreases to permeabilities will appear, and these would be consistently modeled on a time-varying deforming mesh.

Boundary-Conforming Grids

The grid generation technology in Chapters 8–10 should be used where possible. Consider the irregular boundaries seen by our Houston well in a Texas-shaped reservoir in Chapter 9. Whereas boundary-conforming grids will provide detailed resolution using 200 grid blocks, a Cartesian mesh would require roughly 2000 to produce equivalent results! Such meshes are capable of wrapping around multiple boreholes and fractures, of course, while conforming to irregular farfield reservoir boundaries. Thompson's grid generation technique forms the basis for the powerful normalization theory developed in Chapter 9. The theory allows solutions to supersets of problems (with different boundary condition modes and fluid types) to be expressible in terms of one set of metrics obtained once and for all. This is analogous to the approach of Chapter 6 for more obvious radial flows, which show how log r similarly solves supersets of like problems. Besides the gridding methods presented in Chapters 8–10 for areal problems, others just as powerful can be used for other reservoir applications.

We shall introduce stratigraphic meshes next, develop the general theory for slowly varying stratigraphies, and present an illustrative calculation later.

Stratigraphic Meshes for Layered Media

Most geological boundaries do not conform to the simple coordinate lines of rectangular mesh systems. Dipping stratigraphic layers with nonparallel tops and bottoms are a case in point. The use of finely gridded (x,y,z) meshes, while not incorrect, results in awkward stair-step representations of the physical boundaries, plus numerous inactive simulation grid blocks. General curvilinear coordinates provide good physical resolution, but the retention of all of the transformation terms leads to massive equations with first derivatives, second-derivative cross-terms, and numerous variable coefficients. Often, however, such a general approach is not warranted. Many layered stratigraphies are somewhat distorted or warped in a global sense, but so long as local changes in elevation are small, important simplifications can be made. Under the circumstances, stratigraphic coordinates need not be orthogonal. Thus, we retain x and y as independent variables in the areal plane, and continue to use constant values of Δx and Δy. In the vertical direction, however, z is no longer a suitable coordinate, since it does not model dip and lateral variation well. Instead, we introduce the height variable

$$Z = z\text{-}f(x, y, t) \tag{15.38}$$

and associate with it the new capital P pressure function

$$p(x, y, z, t) = P(x, y, Z, t) \tag{15.39}$$

Instead of rederiving all physical laws in x, y, and Z coordinates, we simply express Eqs. (15.1) and (15.18) in these variables via the chain rule, that is,

$$p_x = P_x + P_Z Z_x = P_x\text{-}f_x(x, y)P_Z \tag{15.40a}$$

$$p_y = P_y + P_Z Z_Y = P_y\text{-}f_y(x, y)P_Z \tag{15.40b}$$

$$p_z = P_Z Z_z = P_Z \tag{15.40c}$$

If the slopes $f_x(x,y)$ and $f_y(x,y)$ are small, Eqs. (15.40a)–(15.40c) show that $p_x = P_x$, $p_y = P_y$, and $p_z = P_z$ approximately. Thus, Eqs. (15.1) and (15.18) apply with z replaced by Z, p(x,y,z) replaced by P(x,y,Z), and (x,y,z) replaced by (x,y,Z). How does this affect our difference equations? Let us consider, for example, the representative first term of Eq. (15.3), namely,

$$(\Delta y \Delta z/\Delta x)2\left[k_{xi,j,k}k_{xi+1,j,k}/\left(k_{xi,j,k} + k_{xi+1,j,k}\right)\right]\left(p_{i+1,j,k} - p_{i,j,k}\right)\cdots \tag{15.41}$$

which earlier led to the transmissibility definition

$$TX_{i,j,k} = (\Delta y \Delta z/\Delta x)2\left[k_{xi,j,k}k_{xi+1,j,k}/\left(k_{xi,j,k} + k_{xi+1,j,k}\right)\right] \tag{15.42}$$

These two equations are now replaced, respectively, by

$$[\Delta y\{Z(i,j,k+1)\text{-}Z(i,j,k)\}/\Delta x]2\left[k_{xi,j,k}k_{xi+1,j,k}/\left(k_{xi,j,k}+k_{xi+1,j,k}\right)\right]\left(P_{i+1,j,k}\text{-}P_{i,j,k}\right)+\ldots \tag{15.43}$$

and the revised transmissibility definition

$$TX_{i,j,k}=[\Delta y\{Z(i,j,k+1)\text{-}Z(i,j,k)\}/\Delta x]2\left[k_{xi,j,k}k_{xi+1,j,k}/\left(k_{xi,j,k}+k_{xi+1,j,k}\right)\right] \tag{15.44}$$

where the index k now refers to our stratigraphic coordinates.

Thus, all the difference formulas for constant rectangular meshes with fixed Δx, Δy, and Δz apply without modification, provided we calculate our transmissibilities using $\{Z(i,j,k+1)\text{-}Z(i,j,k)\}$ and replace the volume element $\Delta x\Delta y\Delta z$ using $\{Z(i,j,k+1)\text{-}Z(i,j,k)\}\Delta x\Delta y$ in the equation for transient flow and in the flow rate summations along wellbore blocks. Eqs. (15.13) and (15.30), used to implement net flow rate constraints at wells, do not change; the sum over our incremental lengths $\{Z(i,j,k+1)\text{-}Z(i,j,k)\}$ themselves is specified and requires no additional integration. These simplifications for slowly varying stratigraphies yield large savings in computer memory and speed, while drawing upon the advantages of the highly stable rectangular schemes developed. This option, at the time this book went to press, is built into our source code implementation but not yet available in the interactive simulator.

Modeling Wellbore Storage

Wells are opened and closed at the surface, and not at the sandface downhole. When opening a well for a drawdown test (from the surface), a portion of the flow results from fluid expansion in the wellbore itself. Likewise, in a buildup test, fluid influx into the wellbore continues after shut-in of the well. Thus, total flow rate constraints cannot be applied at the sandface directly, without accounting for time delays associated with the borehole fluid compressibility C_{bh} and the wellbore storage volume V_{bh}. *We also note that storage effects are important in underbalanced drilling, where lower borehole pressures may allow free gas to exist, increasing the compressibility in the fluid column.*

How, exactly, is storage modeled? Imagine a highly pressured reservoir that is initially static. When a well is opened to production at a fixed surface volume flow rate $Q_{prod}>0$, note that the well pressure $p_w(t)$ must decrease in time. That is, $dp_w/dt<0$ because the wellbore fluid is expanding. Thus, the desired Q_{prod} is obtained as the sum of $\text{-}Vol_{bh}C_{bh}dp_w/dt$ (which is positive) and the usual reservoir flow contribution. In other words, when solving the pressure differential equation, the total flow rate (boundary condition) constraint is taken as $Q_{bc}=Q_{prod}+Vol_{bh}C_{bh}dp_w/dt$. This states that on initial production, the rate of flow Q_{bc} through the sandface is actually less than the Q_{prod} pumped at the surface. To see that this applies to an injector pumping fluid into an initially quiescent reservoir, consider $Q_{inj} < 0$ with a corresponding $dp_w/dt > 0$. Now, the initial injection acts first to compress the borehole fluid, so that the injected fluid does not entirely enter at the sandface. Thus, $Q_{bc} = Q_{inj} + Vol_{bh}C_{bh}dp_w/dt$ is again correct, this time because Q_{bc} is less negative than Q_{inj} since $Vol_{bh}C_{bh}dp_w/dt$ is positive. When a producing well (with $Q_{prod} > 0$) is shut-in with $Q_{prod} = 0$ from the surface, the compressibility of the borehole fluid allows Q_{bc} at the sandface to continue for a limited time with $Q_{bc} > 0$. Thus, the foregoing production formula leads to $Q_{bc} = 0 + Vol_{bh}C_{bh}dp_w/dt > 0$. This implies that $dp_w/dt > 0$, so that the well pressure continues to increase, as expected on physical grounds.

All of these effects can be modeled quantitatively. Because the production model $Vol_{bh}C_{bh}dp_w/dt$ is approximate anyway, with storage effects also arising from free gas bubbles, surface facilities, and so on, there is no need to attach too much significance to the exact values of C_{bh} and Vol_{bh}. For simulation purposes, we introduce the lumped storage factor F or capacity defined with $Q_{bc} = Q_{\text{desired prod or inj volume}} + F dp_w/dt$, and take F as a history-matching parameter that depends on borehole fill-up, annulus properties, and other effects that may be difficult to characterize.

GROUP 1, BASIC EXAMPLE CALCULATIONS

The steady and transient algorithms given earlier are extremely stable for general well topologies (Fig. 15.2). In numerous simulations with sudden changes to wellbore paths and constraints, wide variations to rock heterogeneity, fluid type, and gridding parameters, stability and mass conservation were always maintained. This core capability provided the foundation for a robust simulator that did not require unreasonably small time steps for well test forward analysis and primary production modeling. It allowed us to focus on another objective,

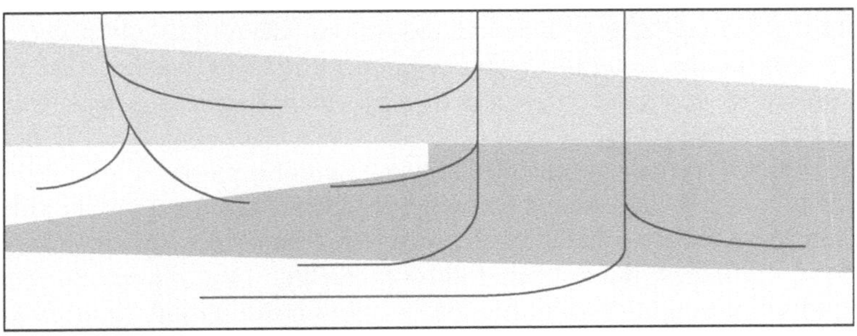

FIG. 15.2 Multilaterals in heterogeneous, layered, anisotropic reservoirs.

namely, extending ease-of-use and convenience to inexpensive personal computers.

Simulation Capabilities

Classical solutions to elliptic and parabolic equations emphasize simple boundary conditions along external box edges. But in petroleum engineering, the most significant auxiliary conditions are the internal constraints applied at injecting or producing wells. Pressures and net flow rates, the latter subject to constant pressure along well paths specified at the user's discretion, render matrix structures far from ideal. The challenges are practical. Not only are the parameters arbitrary; so are the number, position, and geometric inclination of the well systems. The numerical solution for both steady and transient flows must be stable to any heterogeneities in k_x, k_y, k_z and ϕ, to fully and partially penetrating vertical, horizontal and deviated wells, to wells with multiple side-tracked drainhole bifurcations, and to general farfield aquifer or solid wall boundary conditions.

For transient compressible flows, the new simulator described here allowed users to change well constraints in mid-stream, to drill horizontal, wiggly drainhole extensions to existing wells, to shut-in nonproducing intervals and to perforate new ones, and finally, to drill new wells having complicated wellbore paths subject to general constraints. All of these capabilities are accomplished without exacting any performance penalties from the baseline algorithm. We emphasize that these modeling options are driven by real needs, since they permit simulation of actual drilling processes as they are performed in the field. At the time of this writing, the author is not aware of any other simulator possessing these flexible runtime options.

Data Structures and Programming

The computational efficiencies demanded by our objectives were possible because the numerical algorithm was designed so that physical complications would not alter the stable tridiagonal nature of the underlying routines. The iterative nature of the steady solver required no additional three-dimensional arrays beyond the obvious ones for pressure and formation properties. Similarly, the transient algorithm employed only the minimum number of time levels dictated by stability considerations. Good memory management was critical. Information was always written to disk when possible, common blocks were used, and built-in analytical solutions and formulation checks always assured mathematical consistency.

Central to a good simulator is simple, easily modifiable reservoir and wellbore description. For example, users need not enter rows and columns of five-digit permeabilities and porosities, and tabulated (i,j,k) coordinates for multiple well paths. The entire approach should, ideally, be visually driven and easily inputted. This does not necessarily mean expensive graphics and pixel-level resolution, requirements that would compete with our algorithm for vital memory resources—an important consideration when the prototype simulator was first written in the early 1990s. A simple ASCII text "picture" file, as we shall observe, more than suffices. Five basic examples are described next, developed more than a decade ago, which support the idea that powerful simulation capabilities important to modern applications can be realized without resorting to workstations and mainframes. Results computed with early and recent versions of the simulator are identical. Later, in "Group 2, Advanced Calculations and User Interface," all features of the new software will be given together with examples for multilaterals operating under general conditions.

EXAMPLE 15.1. CONVERGENCE ACCELERATION, TWO DEVIATED HORIZONTAL GAS WELLS IN A CHANNEL SAND

We prototyped our ideas almost two decades ago by writing a personal computer program operable on 8086 machines. We focused on low-cost and ease-of-use, and preserved the geobody or object character of the reservoir by having input routines "read" layer pictures created by simple text editing. GEO and DRL layer file types describe, respectively, the geology of the field and the trajectories of any drilled wells.

For this example, consider a three-layer heterogeneous reservoir where all layers are identical to LAYER1.GEO below. The corresponding DRL files show that Wells 1 and 2, which explicitly appear in the sketches, represent vertical wells containing long horizontal drainholes drilled into a channel sand. Layer numbers increase in the downward direction, to be consistent with the drilling process. Thus, additional layers can always be added to existing data sets without reordering layer numbers.

On scanning the GEO files, the simulator automatically identifies the number of layers and the size of the areal grid; it also notes that three lithologies represented by the keyboard symbols +, &, and # are present (non-alphanumeric symbols are reserved for rock types). At this point, information related to grid block size, to the permeabilities and porosities characteristic of each of the lithology symbols, and to fluid properties, is entered interactively at the keyboard in response to on-line queries. Typed numerical arrays are never necessary; in subsequent simulations, changes in geological properties are introduced by changing the lithological symbol or by altering its assigned properties.

```
C>type layer1.geo
+ + + + + + + + + & & & & + + # # # # #
+ + + + + + + + & & & & + # # # # # + +
+ + + + + + & & & & + + # # # # # + + +
+ + + + & & & & + + # # # # + + + + + +
+ + & & & & + + + + # # # # # + + + + +
+ + & & & & + + + + + # # # # # + + + +
+ + + & & & & + + + + + # # # # # + + +
+ + + + + & & & & & + + + # # # # # + +
+ + + + + + + & & & & & + + # # # # # +
+ + + + + + + + & & & & & + + # # # # #
+ + + + + + + + & & & & + + + + # # # #
+ + + + + + + & & & & + + + + + + # # #
+ + + + + & & & & + + + + + + + + + # #
+ + + & & & & + + + + + + + + + + + + +
+ & & & & + + + + + + + + + + + + + + +
```

```
C>type layer1.drl
+ + + + + + + + + & & & & + + # # # # #
+ + + + + + + + & & & & + # # # # # + +
+ + + + + + & & & & + + # # # # # + + +
+ + + + 1 & & & + + # # # # + + + + + +
+ + & & & & + + + + # # # # # + + + + +
+ + & & & & + + + + + # # # # # + + + +
+ + + & & & & + + + + + # # # # # + + +
+ + + + + & & & & & + + + # # # # # + +
+ + + + + + + & & & & & + + # # # # # +
+ + + + + + + + & & & & & + + # # 2 # #
+ + + + + + + + & & & & + + + + # # # #
+ + + + + + + & & & & + + + + + + # # #
+ + + + + & & & & + + + + + + + + + # #
+ + + & & & & + + + + + + + + + + + + +
+ & & & & + + + + + + + + + + + + + + +
```

```
C>type layer2.drl
+ + + + + + + + + & & & & + + # # # # #
+ + + + + + + + & & & & + # # # # # + +
+ + + + + + & & & & + + # # # # # + + +
+ + + + 1 & & & + + # # # # + + + + + +
+ + & & 1 & + + + + # # # # # + + + + +
+ + & & 1 & + + + + + # # # # # + + + +
+ + + & 1 & & + + + + + # # # # # + + +
+ + + + 1 & & & & & + + + # # # # # + +
+ + + + 1 + + & & & & & + + # # # # # +
+ + 2 2 2 2 2 2 2 2 2 2 2 2 2 2 2 2 # #
+ + + + + + + + & & & & + + + + # # # #
+ + + + + + + & & & & + + + + + + # # #
+ + + + + & & & & + + + + + + + + + # #
+ + + & & & & + + + + + + + + + + + + +
+ & & & & + + + + + + + + + + + + + + +
```

```
C>type layer3.drl
+ + + + + + + + + & & & & + + # # # # #
+ + + + + + + + & & & & + # # # # # + +
+ + + + + + & & & & + + # # # # # + + +
+ + + + 1 & & & + + # # # # + + + + + +
+ + & & & & + + + + # # # # # + + + + +
+ + & & & & + + + + + # # # # # + + + +
+ + + & & & & + + + + + # # # # # + + +
+ + + + + & & & & & + + + # # # # # + +
+ + + + + + + & & & & & + + # # # # # +
+ + + + + + + + & & & & & + + # # 2 # #
+ + + + + + + + & & & & + + + + # # # #
+ + + + + + + & & & & + + + + + + # # #
+ + + + + & & & & + + + + + + + + + # #
+ + + & & & & + + + + + + + + + + + + +
+ & & & & + + + + + + + + + + + + + + +
```

The simulator next scans the DRL files automatically, identifies two wells (numbers are always reserved for well labels and are highlighted in red font in the digital version of this book), stores their coordinate information, and asks for well constraint modes and values. Note that Wells 1 and 2 are vertical wells with very long horizontal drainhole extensions, and are here oriented perpendicularly in Layer 2. However, any oblique out-of-plane orientation is permissible. Few simulations on such complicated well paths have been reported in the literature. Next, information on farfield boundary conditions is entered, and simulation commences. In this example, we shall demonstrate how accurately the steady algorithm conserves mass.

We shall also show that the solution obtained for the first run, when used to initialize a second run with different parameters, leads to much more rapid convergence. Such initializations provide quick solutions because only incremental actions are needed. The reader should examine the GEO and LAYER files displayed above, and note the likeness to typical channel sands. In the following displays, simple `Courier font` denotes screen activity, while **`bold Courier`** indicates user-entered commands.

Run 1

In this rectangular grid run, Δx, Δy, and Δz were 100, 200, and 300 ft. The + lithology was isotropic, with a permeability of 100 md and a porosity of 20%, while the # lithology, at 800 md, was 30% porous. Lithotype & was anisotropic, with kx, ky and kz equal to 500, 600 and 700 md; it was 25% porous. Well 1 was pressure constrained at 5000 psi, while Well 2 was flow rate constrained at 1,000,000 cu ft/h. Because the evaluation objective here was mass conservation, all six sides of the computational box were chosen as solid no-flow walls in order to provide a severe test. The degree to which the computed flow rate at Well 1 equaled -1,000,000 cu ft/h was to be assessed. To complicate matters, the flow of a gas was considered. This renders the formulation nonlinear and provides a good test for the algorithm. The gas viscosity was 0.018 cp, a surface density of 0.003 lbf s^2/ft^4 at 14.7 psi was selected, and a gas exponent of $m = 0.5$ was chosen. Again, our general m fluid model allows us to alter gas thermodynamics; it is not restricted to idealized isothermal solutions. In this test case, the steady flow solver was initialized to zero pressure everywhere—a worst case assumption acknowledging that nothing is known about the reservoir—and allowed to converge. Screen dumps showing iteration history and flow rate summaries are as follows.

```
Iterative solutions starting, please wait ...
Iteration     1 of maximum 99999 completed ...
Iteration     2 of maximum 99999 completed ...
.
Iteration    11, maximum 99999, .1851E+02 % error.
Iteration    12, maximum 99999, .2334E+02 % error.
Iteration    13, maximum 99999, .1178E+02 % error.
Iteration    14, maximum 99999, .2002E+02 % error.
.
Iteration    99, maximum 99999, .3337E+00 % error.
Iteration   100, maximum 99999, .3226E+00 % error.

Iteration   100, (Un)converged volume flow rates by well cluster:
Cluster 1:  P=  .5000E+04 psi, Q= -.2764E+08 cu ft/hr.
Cluster 2:  P=  .4788E+04 psi, Q=  .1745E+07 cu ft/hr.

Iteration   200, (Un)converged volume flow rates by well cluster:
Cluster 1:  P=  .5000E+04 psi, Q= -.2864E+07 cu ft/hr.
Cluster 2:  P=  .4977E+04 psi, Q=  .1057E+07 cu ft/hr.

Iteration   300, (Un)converged volume flow rates by well cluster:
Cluster 1:  P=  .5000E+04 psi, Q= -.1127E+07 cu ft/hr.
Cluster 2:  P=  .4990E+04 psi, Q=  .1003E+07 cu ft/hr.

Iteration   400, (Un)converged volume flow rates by well cluster:
Cluster 1:  P=  .5000E+04 psi, Q= -.1005E+07 cu ft/hr.
Cluster 2:  P=  .4991E+04 psi, Q=  .1001E+07 cu ft/hr.
```

Satisfactory results are not obtained until Iteration 400, at which point, Well 1's -0.1005E+07 cu ft/h and Cluster 2's 0.1001E+07 cu ft/h are close (on typical PCs, this requires just seconds). We have enforced exact mass balance for heterogeneous reservoirs with nonconventional wells, and we have calculated nonlinear pressure fields correctly, starting with a trivial zero guess.

Run 2

Next we terminate the steady-state simulation, begin another where we retain that geology (although we need not), and change the shape of Well 2 and other inputs. The redrilled, more deviated well follows.

```
C>type layer2.drl
+ + + + + + + + + & & & & + + # # # # #
+ + + + + + + + & & & & + # # # # # + +
+ + + + + + & & & & + + # # # # # + + +
+ + + + 1 & & & + + # # # # + + + + + +
+ + & & 1 & + + + + # # # # # + + + + +
+ + & & 1 & + + + + + # # # # # + + + +
+ + + & 1 & & + + + + + # # # # # + + +
+ + + + 1 & & & & & + + + # # # # # + +
+ + + + 1 + + & & & & & + + # # # # # +
+ + + + + 2 2 2 2 2 2 2 2 2 2 2 2 2 # #
+ + + + 2 + + + & & & & + + + + # # # #
+ + 2 + + + + & & & & + + + + + + # # #
+ + 2 + + & & & & + + + + + + + + + # #
+ + + & & & & + + + + + + + + + + + + +
+ & & & & + + + + + + + + + + + + + + +
```

Well 1 is still pressure constrained at 5000 psi, but we instead rate-constrain Well 2 at 1,500,000 cu ft/h. We also change our gas viscosity to 0.04 cp, the surface reference density to 0.004 lbf s^2/ft^4 at 14.7 psi, and reassign m to 0.7. These changes normally require new simulations, with detailed analysis beginning anew, but the power of our relaxation approach is seen as follows. We expect similar convergence histories, but *instead of initializing the solver to "0," we use the pressure solution available from the above run.*

```
Iterative solutions starting, please wait ...
Iteration      1 of maximum 99999 completed ...
Iteration      2 of maximum 99999 completed ...
.
Iteration     11, maximum 99999, .2650E-02 % error.
Iteration     12, maximum 99999, .7193E-03 % error.
.
Iteration     99, maximum 99999, .1469E-03 % error.
.
Iteration    100, (Un)converged volume flow rates
by well cluster:
Cluster 1:  P=  .5000E+04 psi, Q= -.1470E+07 cu ft/hr.
Cluster 2:  P=  .4989E+04 psi, Q=  .1500E+07 cu ft/hr.
```

At Iteration 11, the error measure has decreased to .2650E-02%, compared with the .1851E+02% in the first run. The results of Iteration 100 indicate that injected and produced flow rates are converging, and are much faster than before. This acceleration is possible because a close solution was used to start the iterations. It suggests that it is possible to perform successive simulations quickly, because incremental changes to fluid, well configuration, boundary condition, and geological description inputs require only incremental work.

EXAMPLE 15.2. DUAL-LATERAL HORIZONTAL COMPLETION IN A FRACTURED, DIPPING, HETEROGENEOUS, LAYERED FORMATION

In June 1992, Texaco announced the completion of a pioneering dual lateral horizontal well in a fractured formation in the Gulf of Mexico. A vertical well was drilled into the pay zone, at which point the drilling of two horizontal wells were initiated, heading in opposite directions. This example shows how the flow from such a completion is easily simulated. Fictitious input parameters are used for illustrative purposes. For brevity, the GEO files will not be shown, but the general idea about the formation can be inferred from the DRL layer pictures presented. We have a single (well) 1, headed towards the pay zone %. Once penetrated, the first horizontal well branch heads north, while the second heads south. This bifurcation is readily seen in Layers 3 and 4 below. The math model treats the one vertical and two horizontal wells as part of a single well system (collectively called Well 1) because a single well constraint applies to the entire group of three wells. Here, the = symbol represents the highly permeable and porous fracture plane.

```
C>type layer1.drl

+ + + + + + + + + +
+ + + + + + + + + +
+ + + + + + + + + +
+ + + + 1 + + + + +
+ + + + + + + + + +
+ + + + + + + + + +
+ + + + + + + + + +
+ + + + + + + + + +
```

```
C>type layer2.drl

% % % % % % % % % %
% % % % % % % % % %
+ + + + + + + + + +
+ + + + 1 + + + + +
+ + + + + + + + + +
+ + + + + + + + + +
+ + + + + + + + + +
+ + + + + + + + + +
```

```
C>type layer3.drl

% % % % % % % % % %
% % % % 1 % % % % %
% % % % 1 % % % % %
% = = = 1 = = = = %
+ + + + + + + + + +
+ + + + + + + + + +
+ + + + + + + + + +
+ + + + + + + + + +
```

```
C>type layer4.drl

% % % % % % % % % %
% % % % % % % % % %
% % % % % % % % % %
% = = = 1 = = = = %
% % % % 1 % % % % %
% % % % 1 % % % % %
+ + + + + + + + + +
+ + + + + + + + + +
```

```
C>type layer5.drl

% % % % % % % % % %
% % % % % % % % % %
% % % % % % % % % %
% % % % % % % % % %
% % % % % % % % % %
% % % % % % % % % %
% % % % 1 % % % % %
% % % % % % % % % %
```

```
C>type layer6.drl

+ + + + + + + + + +
+ + + + + + + + + +
+ + + + + + + + + +
+ + + + + + + + + +
+ + + + + + + + + +
+ + + + + + + + + +
+ + + + + + + + + +
+ + + + + + + + + +
```

Observe that storing formation attributes and well location information in three-dimensional, character-string array format provides special advantages. For example, by rearranging the print sequences in the Fortran do-loops, it is possible to print out x-z and y-z lithology cross-sections and display well trajectories within stratigraphic layers. This assists with visualization, interpretation, and error-checking. The two plots that follow show two different vertical projections of the well.

```
+---- Y/Z: I=  4  ---> Y

| + + + + 1 + + + + +
| + + + + 1 + + + + +
| % = = = 1 = = = = %
| % = = = 1 = = = = %
| % % % % % % % % % %
| + + + + + + + + + +
| Z
```

```
+---- X/Z: J=  5  ---> X

| + + + 1 + + + +
| % % + 1 + + + +
| % 1 1 1 + + + +
| % % % 1 1 1 + +
| % % % % % % 1 %
| + + + + + + + +
| Z
```

Here, we have a six-layer, 8×10 rectangular grid, with Δx, Δy and Δz equal to 300, 200, and 100 ft, respectively. All of our rock types are isotropic. Rock+ has a permeability of 50 md and a porosity of 20%; rock %, a permeability of 800 md and a porosity of 30%; while rock type =, our fracture, has a permeability of 5000 md and a porosity of 90%. Our only well, Well 1, will be pressure constrained at 1000 psi. But this constrains the entire system, both horizontal branches and the vertical, at 1000 psi. In practice, the vertical section is sealed and nonproducing. To enforce this, we use a transmissibility modification option that allows us to modify local transmissibilities everywhere, well by well, or cell by cell, if need be. For brevity, we show some of the interactive screens but modify two cell blocks only.

```
You may modify TX, TY and TZ transmissibilities for simulation
purposes WITHOUT altering values on disk .... Modify?  Y/N: y

Modify EVERYWHERE?  At WELL(S) ONLY?  E/W: w
Modify transmissibilities in Well 1?  Y/N: y
Modify "cell by cell" ?  Y/N: y

Enter cell block identification number, 1-   9: 1
Existing TX =    .359E-10, TY =    .807E-10, TZ =    .323E-09 ft^3 at
Well 1, Block   1: (i= 4, j= 5, Layer=1) ...

O  Enter cell block TX multiplier:  .01
O  Enter cell block TY multiplier:  .02
O  Enter cell block TZ multiplier:  .03

Change TX, TY and TZ in another cell block within
present Well 1?  Y/N:  y

Enter cell block identification number, 1-   9: 2
Existing TX =    .359E-10, TY =    .807E-10, TZ =    .639E-09 ft^3 at
Well 1, Block   2: (i= 4, j= 5, Layer=2) ...

O  Enter cell block TX multiplier:  .01
O  Enter cell block TY multiplier:  .01
O  Enter cell block TZ multiplier:  .01

Change TX, TY and TZ in another cell block within
present Well 1?  Y/N:  n
```

We shall also assume four aquifer side boundaries, and two solid walls at the top and bottom of the reservoir. Note the simplicity of the computer inputs and outputs; the graphics is crude but serves its purpose.

```
INPUT FARFIELD BOUNDARY CONDITION SETUP:
                          Z(k)
                           |
                           | j=1        Jmax(10)
                      i=1  |
                           +--------------+  ------  Y(j)
                          /|______________|  \
             Left       /  |___   Back ___|    \   Right
                       /   |______________|     \
                      /    +--------------+      \
      Imax( 8)      +-------/ Top, Layer 1  \-------+
                   /  |_____________________________|
                  /   |_________    Front   ________|
                / X(i)|_____________________________|
                      +-------  Bottom, Layer 6  ------+

                           COORDINATE SYSTEM

O  FRONT ... is aquifer or no flow wall?  A/W: a
O  Pressure at FRONT face (psi): 5000
O  BACK .... is aquifer or no flow wall?  A/W: a
O  Pressure at BACK face  (psi): 5000
O  LEFT .... is aquifer or no flow wall?  A/W: a
O  Pressure at LEFT face  (psi): 5000
O  RIGHT ... is aquifer or no flow wall?  A/W: a
O  Pressure at RIGHT face (psi): 5000
O  TOP ..... is aquifer or no flow wall?  A/W: w
O  BOTTOM .. is aquifer or no flow wall?  A/W: w
```

Consider a liquid with a viscosity of 10 cp and a compressibility of 0.00001/psi in a fully transient flow. The initial pressure everywhere is 5000 psi, and the assumed time step is 1 h. From the well history that follows, it is clear that the expected rate decline is obtained and computed stably.

C>**type well1.sim**

WELL #1: Step No.	Dt (Hour)	Time (Hour)	Pressure (Psi)	Flow Rate (Cu Ft/Hr)	Cum Vol (Cu Ft)
1	.100E+01	.100E+01	.100E+04	.756E+06	.756E+06
2	.100E+01	.200E+01	.100E+04	.555E+06	.131E+07
3	.100E+01	.300E+01	.100E+04	.464E+06	.178E+07
4	.100E+01	.400E+01	.100E+04	.412E+06	.219E+07
5	.100E+01	.500E+01	.100E+04	.379E+06	.257E+07
6	.100E+01	.600E+01	.100E+04	.356E+06	.292E+07
7	.100E+01	.700E+01	.100E+04	.339E+06	.326E+07
8	.100E+01	.800E+01	.100E+04	.325E+06	.359E+07
9	.100E+01	.900E+01	.100E+04	.313E+06	.390E+07
10	.100E+01	.100E+02	.100E+04	.303E+06	.420E+07
.					
20	.100E+01	.200E+02	.100E+04	.245E+06	.686E+07
50	.100E+01	.500E+02	.100E+04	.200E+06	.133E+08
100	.100E+01	.100E+03	.100E+04	.188E+06	.229E+08
150	.100E+01	.150E+03	.100E+04	.186E+06	.322E+08
199	.100E+01	.199E+03	.100E+04	.185E+06	.413E+08

The objective is threefold. First, geological object-oriented file inputs with complicated wellbore trajectories are simple to create; lithological data only enter, and then conveniently, through a lower level routine ("what" if studies do not require retyping of cumbersome numbers into rows and columns). Second, very heterogeneous transient problems can be simulated with high stability, taking relatively large time steps. And third, the complete "at the keyboard" work session, including computing time, required just minutes.

EXAMPLE 15.3. STRATIGRAPHIC GRIDS, DRILLING DOME-SHAPED STRUCTURES

Here, simulations on stratigraphic grids are performed. Such grids wrap around all relevant layers. The use of standard rectangular grids for the structure shown here would have produced numerous inactive grid blocks that decrease convergence rate. The formation is homogeneous for simplicity, but the important point here is the choice of extremely flat grids and special dome-like coordinates. These are associated with convergence problems in many elliptic solvers. (Actually, a dimensionless parameter based on grid block aspect ratio and anisotropic permeability controls convergence.) The blocks are assumed to be $100' \times 100' \times 1'$ high, residing in a four-layer, 7×7 grid system; the rock, with a permeability of 500 md, is 20% porous. Also, two wells are oppositely placed. All four DRL layers take the form of the uppermost LAYER1.DRL, with vertical Wells 1 and 2 positioned as shown.

```
C>type layer1.drl
+ + + + + + +
+ + + + + + +
+ + 1 + + + +
+ + + + + + +
+ + + + 2 + +
+ + + + + + +
+ + + + + + +
```

The elevations of the stratigraphic grid blocks themselves can be defined by using a text picture of the uppermost surface displaying the z coordinate.

```
(i=1,j=1)  85  85  85  85  85  85  85
           85  90  90  90  90  90  85
           85  90  95  95  95  90  85
           85  90  95 100  95  90  85
           85  90  95  95  95  90  85
           85  90  90  90  90  90  85
           85  85  85  85  85  85  85   (i=7,j=7)
```

This, plus the vertical thickness of the uniformly thick layers (1 ft) and the number of layers (again, four), completely defines the stratigraphy. More general topographies require detailed I/O procedures, which are avoided in this book. Next, pressure constrain Well 1 at 1000 psi, and flow rate constrain Well 2 at 50 cu ft/h. Also, model a liquid with a viscosity of 1 cp, and assume that the six sides of the computational box are solid no-flow walls in order to provide a severe test for mass conservation. The steady numerical scheme, in order to conserve mass, must determine a flow rate at Well 1 that is exactly the negative of the assumed rate at Well 2. Can this be achieved? Calculated results follow.

```
Iteration   200, (Un)converged volume flow rates
by well cluster:

Cluster 1:  P=  .1000E+04 psi, Q= -.5020E+03 cu ft/hr.
Cluster 2:  P=  .1851E+02 psi, Q=  .5008E+03 cu ft/hr.
```

At 200 iterations, requiring just seconds, the injector and producer flow rates are almost identical. From the results below for the first two layers, the computed pressures show the anticipated symmetries. The position of the prescribed 1000 psi is shown in bold type for reference; observe that there is no reason why the computed pressure at Well 2 should be -1000 psi.

```
Calculated 3D Pressures   (Intermediate Results)

Iteration   200, Pressure (psi) in Layer 1:
BACK
  .850E+03  .850E+03  .850E+03  .701E+03  .573E+03  .509E+03  .509E+03
  .850E+03  .850E+03  .850E+03  .701E+03  .573E+03  .509E+03  .509E+03
  .850E+03  .850E+03  .100E+04  .680E+03  .509E+03  .445E+03  .445E+03
  .701E+03  .701E+03  .680E+03  .509E+03  .338E+03  .317E+03  .317E+03
  .573E+03  .573E+03  .509E+03  .338E+03  .185E+02  .168E+03  .168E+03
  .509E+03  .509E+03  .445E+03  .317E+03  .168E+03  .168E+03  .168E+03
  .509E+03  .509E+03  .445E+03  .317E+03  .168E+03  .168E+03  .168E+03

FRONT

Iteration   200, Pressure (psi) in Layer 2:
BACK
  .850E+03  .850E+03  .850E+03  .701E+03  .573E+03  .509E+03  .509E+03
  .850E+03  .850E+03  .850E+03  .701E+03  .573E+03  .509E+03  .509E+03
  .850E+03  .850E+03  .100E+04  .680E+03  .509E+03  .445E+03  .445E+03
  .701E+03  .701E+03  .680E+03  .509E+03  .338E+03  .317E+03  .317E+03
  .573E+03  .573E+03  .509E+03  .338E+03  .185E+02  .168E+03  .168E+03
  .509E+03  .509E+03  .445E+03  .317E+03  .168E+03  .168E+03  .168E+03
  .509E+03  .509E+03  .445E+03  .317E+03  .168E+03  .168E+03  .168E+03
FRONT
```

With this steady flow established, we now continue with a transient compressible analysis and shut-in in both wells. A compressibility of 0.000003/psi is assumed, along with time steps of 0.005 h for 500 steps. For injector Well 1, we expect a pressure decrease with time because fluid is pulling away, while for producer Well 2, a pressure buildup is anticipated because fluid is piling up. The pressures in the first two layers display smooth, stable trends; also, the expected pressure histories at both wells are qualitatively correct.

```
LAYER RESULTS @ Step   499, Time =    .250E+01 hours:

Pressure Distribution (psi) in Layer 1:
BACK
   .837E+03  .837E+03  .837E+03  .695E+03  .573E+03  .512E+03  .512E+03
   .837E+03  .837E+03  .837E+03  .695E+03  .573E+03  .512E+03  .512E+03
   .837E+03  .837E+03  .979E+03  .674E+03  .512E+03  .451E+03  .451E+03
   .695E+03  .695E+03  .674E+03  .512E+03  .349E+03  .329E+03  .329E+03
   .573E+03  .573E+03  .512E+03  .349E+03  .447E+02  .186E+03  .186E+03
   .512E+03  .512E+03  .451E+03  .329E+03  .186E+03  .186E+03  .186E+03
   .512E+03  .512E+03  .451E+03  .329E+03  .186E+03  .186E+03  .186E+03
FRONT

Pressure Distribution (psi) in Layer 2:
BACK
   .837E+03  .837E+03  .837E+03  .695E+03  .573E+03  .512E+03  .512E+03
   .837E+03  .837E+03  .837E+03  .695E+03  .573E+03  .512E+03  .512E+03
   .837E+03  .837E+03  .979E+03  .674E+03  .512E+03  .451E+03  .451E+03
   .695E+03  .695E+03  .674E+03  .512E+03  .349E+03  .329E+03  .329E+03
   .573E+03  .573E+03  .512E+03  .349E+03  .447E+02  .186E+03  .186E+03
   .512E+03  .512E+03  .451E+03  .329E+03  .186E+03  .186E+03  .186E+03
   .512E+03  .512E+03  .451E+03  .329E+03  .186E+03  .186E+03  .186E+03
FRONT
```

```
C>type well1.sim
    WELL #1:       Dt         Time       Pressure   Flow Rate
    Step No.     (Hour)      (Hour)       (Psi)     (Cu Ft/Hr)
           0   .500E-02    .000E+00    .100E+04   -.502E+03
           1   .500E-02    .500E-02    .100E+04    .000E+00
         100   .500E-02    .500E+00    .996E+03    .000E+00
         200   .500E-02    .100E+01    .991E+03    .000E+00
         300   .500E-02    .150E+01    .987E+03    .000E+00
         400   .500E-02    .200E+01    .983E+03    .000E+00
         499   .500E-02    .250E+01    .979E+03    .000E+00 (i.e., pressure decreases)

C>type well2.sim
    WELL #2:       Dt         Time       Pressure   Flow Rate
    Step No.     (Hour)      (Hour)       (Psi)     (Cu Ft/Hr)
           0   .500E-02    .000E+00    .185E+02    .500E+02
           1   .500E-02    .500E-02    .185E+02    .000E+00
         100   .500E-02    .500E+00    .238E+02    .000E+00
         200   .500E-02    .100E+01    .291E+02    .000E+00
         300   .500E-02    .150E+01    .344E+02    .000E+00
         400   .500E-02    .200E+01    .396E+02    .000E+00
         499   .500E-02    .250E+01    .447E+02    .000E+00 (i.e., pressure increases)
```

EXAMPLE 15.4. SIMULATING-WHILE-DRILLING HORIZONTAL GAS WELLS THROUGH A DOME-SHAPED RESERVOIR

In this example, a general anisotropic matrix rock hosting a nonlinear gas flow is modeled. In particular, the transient effects of newly drilled horizontal drainholes and deviated wells just brought on stream are studied. This example shows how dome shapes can be modeled using rectangular grids. Also, well constraints will be changed while simulating, and computational stability is demonstrated. The ability to simulate while drilling implies improved formation evaluation (e.g., permeabilities can be better matched using annular pressure while drilling data). The GEO geology files are not listed for brevity; the heterogeneities can, however, be inferred from the six LAYER*.DRL 10×10 text files below, which also contain well placement information. Only Well 1 is present initially, but this is joined by a second well system that is later drilled while simulating.

```
C>type layer1.drl
+ + + + + + + + + +
+ + + + + + + + + +
+ + + + + + + + + +
+ + + 1 + + + + + +
+ + + + + + + + + +
+ + + + + + + + + +
+ + + + + + + + + +
+ + + + + + + + + +
+ + + + + + + + + +
+ + + + + + + + + +
```

```
C>type layer2.drl
+ + + + + + + + + +
+ + + + + + + + + +
+ + + + + + + + + +
+ + + 1 * * * + + +
+ + + + * * * + + +
+ + + + + + + + + +
+ + + + + + + + + +
+ + + + + + + + + +
+ + + + + + + + + +
+ + + + + + + + + +
```

```
C>type layer3.drl
+ + + + + + + + + +
+ + + + + + + + + +
+ + + # # # # + + +
+ + + 1 # # # # + +
+ + + + # # # # # +
+ + + + + # # # # #
+ + + + + + + + + +
+ + + + + + + + + +
+ + + + + + + + + +
+ + + + + + + + + +
```

```
C>type layer4.drl
+ + + + + + + + + +
+ + + + + + + + + +
+ + $ $ $ $ $ + + +
+ $ $ 1 $ $ $ $ + +
+ + $ $ $ $ $ $ $ +
+ + + $ $ $ $ $ $ $
+ + + + + $ $ $ $ +
+ + + + + + + + + +
+ + + + + + + + + +
+ + + + + + + + + +
```

```
C>type layer5.drl
+ + + + + + + + + +
+ + + + + + + + + +
+ + $ $ $ $ $ + + +
+ $ $ 1 $ $ $ $ + +
+ + $ $ $ $ $ $ $ +
+ + + $ $ $ $ $ $ $
+ + + + + $ $ $ $ +
+ + + + + + + + + +
+ + + + + + + + + +
+ + + + + + + + + +
```

```
C>type layer6.drl
+ + + + + + + + + +
+ + + + + + + + + +
+ + $ $ $ $ $ + + +
+ $ $ ! ! ! $ $ + +
+ + $ ! ! ! ! $ $ +
+ + + $ $ ! ! ! $ $
+ + + + + $ $ $ $ +
+ + + + + + + + + +
+ + + + + + + + + +
+ + + + + + + + + +
```

Our grid block sizes Δx, Δy and Δz are 100, 200, and 300 ft. The properties for the first four rock types listed are isotropic. Rock type + has a permeability of 100 md and a porosity of 20%; type * is 200 md and 30% porous; type # is 300 md and 25% porous, while type ! is 100 md and 20% porous. Next allow the $ pay sand to be complicated; it is anisotropic, with permeabilities of 700, 800, and 900 md in the x, y, and z directions, with a porosity of 25%.

Now assume that Well 1 is pressure constrained at 500 psi, that the six sides of the computational box are solid no-flow walls, and that the simulator is run in a purely transient compressible flow mode for an isothermal gas. The gas has a viscosity of 1 cp, a surface density of 0.003 lbf s^2/ft^4 at 14.7 psi, and a gas exponent of m = 1. Let us initialize our reservoir to 10,000 psi to provide a significant shock to the system, and let us study the initial history obtained at Well 1, as extracted from WELL1.SIM. Recall that Well 1 is initially pressure constrained at 500 psi. Note how the flow rate correctly decreases with time and how the cumulative volume increases in time. The computed rate behavior shows no unrealistic oscillations in time.

```
WELL #1:       Dt          Time        Pressure   Flow Rate     Cum Vol
Step No.     (Hour)       (Hour)        (Psi)     (Cu Ft/Hr)    (Cu Ft)
       1   .500E-02     .500E-02     .500E+03     .254E+10     .127E+08
       2   .500E-02     .100E-01     .500E+03     .254E+10     .254E+08
       3   .500E-02     .150E-01     .500E+03     .253E+10     .380E+08
       4   .500E-02     .200E-01     .500E+03     .252E+10     .507E+08
       5   .500E-02     .250E-01     .500E+03     .252E+10     .632E+08
       6   .500E-02     .300E-01     .500E+03     .251E+10     .758E+08
       7   .500E-02     .350E-01     .500E+03     .250E+10     .883E+08
       8   .500E-02     .400E-01     .500E+03     .250E+10     .101E+09
       9   .500E-02     .450E-01     .500E+03     .249E+10     .113E+09
      10   .500E-02     .500E-01     .500E+03     .249E+10     .126E+09
      11   .500E-02     .550E-01     .500E+03     .248E+10     .138E+09
      12   .500E-02     .600E-01     .500E+03     .247E+10     .150E+09
      13   .500E-02     .650E-01     .500E+03     .247E+10     .163E+09
      14   .500E-02     .700E-01     .500E+03     .246E+10     .175E+09
      15   .500E-02     .750E-01     .500E+03     .246E+10     .187E+09
      16   .500E-02     .800E-01     .500E+03     .245E+10     .200E+09
      17   .500E-02     .850E-01     .500E+03     .245E+10     .212E+09
      18   .500E-02     .900E-01     .500E+03     .244E+10     .224E+09
      19   .500E-02     .950E-01     .500E+03     .243E+10     .236E+09
```

Next assume that (as oilfield owners) we are dissatisfied with the flow rates. We refer back to the LAYER*.DRL pictures, and we decide to drill a horizontal drainhole starting from Layer 4 in Well 1, which cuts a four grid block path through the $ pay zone. Nine blocks define the revised path for Well 1. The simulator provides the existing coordinates of well block centers. In what follows, we also re-constrain the well at a new 55 psi, shaking it up to test numerical stability!

```
Existing Well No. 1 defined by following blocks:
Block No.    1:   i= 4, j= 4, Layer=1
Block No.    2:   i= 4, j= 4, Layer=2
Block No.    3:   i= 4, j= 4, Layer=3
Block No.    4:   i= 4, j= 4, Layer=4
Block No.    5:   i= 4, j= 4, Layer=5
```

```
Number of active gridblocks defining modified well: 9
Enter blocks in any order, they need not be contiguous -

O  Block   1, New x(i) position index, i:  4
O  Block   1, New y(j) position index, j:  4
O  Block   1, New z(k) position, Layer #:  1
O  Block   2, New x(i) position index, i:  4
O  Block   2, New y(j) position index, j:  4
O  Block   2, New z(k) position, Layer #:  2
O  Block   3, New x(i) position index, i:  4
O  Block   3, New y(j) position index, j:  4
O  Block   3, New z(k) position, Layer #:  3
O  Block   4, New x(i) position index, i:  4
O  Block   4, New y(j) position index, j:  4
O  Block   4, New z(k) position, Layer #:  4
O  Block   5, New x(i) position index, i:  4
O  Block   5, New y(j) position index, j:  4
O  Block   5, New z(k) position, Layer #:  5
O  Block   6, New x(i) position index, i:  5
O  Block   6, New y(j) position index, j:  4
O  Block   6, New z(k) position, Layer #:  4
O  Block   7, New x(i) position index, i:  6
O  Block   7, New y(j) position index, j:  4
O  Block   7, New z(k) position, Layer #:  4
O  Block   8, New x(i) position index, i:  7
O  Block   8, New y(j) position index, j:  4
O  Block   8, New z(k) position, Layer #:  4
O  Block   9, New x(i) position index, i:  8
O  Block   9, New y(j) position index, j:  4
O  Block   9, New z(k) position, Layer #:  4

Modify TX, TY or TZ in present Well 1?  Y/N:   n
New well constraint, pressure or rate?  P/R:   p
New pressure (psi):  55
```

Incidentally, to ensure that existing and new well paths do not cross each other, collision sensing background logic was added to the prototype simulator in order to enhance the drilling-while-simulating option. Let us review the results in the WELL1.SIM history file for time Steps 20–39. For reference, all of the computed Well 1 results starting from Step 1 are listed. At Step 20, the flow rate definitely increases but not as much as we had desired. However, the rate decline is not as severe as that due to the original vertical well alone. Again, the computations are completed stably.

WELL #1: Step No.	Dt (Hour)	Time (Hour)	Pressure (Psi)	Flow Rate (Cu Ft/Hr)	Cum Vol (Cu Ft)
1	.500E-02	.500E-02	.500E+03	.254E+10	.127E+08
2	.500E-02	.100E-01	.500E+03	.254E+10	.254E+08
3	.500E-02	.150E-01	.500E+03	.253E+10	.380E+08
4	.500E-02	.200E-01	.500E+03	.252E+10	.507E+08
5	.500E-02	.250E-01	.500E+03	.252E+10	.632E+08
6	.500E-02	.300E-01	.500E+03	.251E+10	.758E+08
7	.500E-02	.350E-01	.500E+03	.250E+10	.883E+08
8	.500E-02	.400E-01	.500E+03	.250E+10	.101E+09
9	.500E-02	.450E-01	.500E+03	.249E+10	.113E+09
10	.500E-02	.500E-01	.500E+03	.249E+10	.126E+09
11	.500E-02	.550E-01	.500E+03	.248E+10	.138E+09
12	.500E-02	.600E-01	.500E+03	.247E+10	.150E+09
13	.500E-02	.650E-01	.500E+03	.247E+10	.163E+09
14	.500E-02	.700E-01	.500E+03	.246E+10	.175E+09
15	.500E-02	.750E-01	.500E+03	.246E+10	.187E+09
16	.500E-02	.800E-01	.500E+03	.245E+10	.200E+09
17	.500E-02	.850E-01	.500E+03	.245E+10	.212E+09
18	.500E-02	.900E-01	.500E+03	.244E+10	.224E+09
19	.500E-02	.950E-01	.500E+03	.243E+10	.236E+09
		Drainhole drilled ...			
20	.500E-02	.100E+00	.550E+02	.269E+10	.250E+09
21	.500E-02	.105E+00	.550E+02	.268E+10	.263E+09
22	.500E-02	.110E+00	.550E+02	.268E+10	.277E+09
23	.500E-02	.115E+00	.550E+02	.267E+10	.290E+09
24	.500E-02	.120E+00	.550E+02	.267E+10	.303E+09
25	.500E-02	.125E+00	.550E+02	.266E+10	.317E+09
26	.500E-02	.130E+00	.550E+02	.266E+10	.330E+09
27	.500E-02	.135E+00	.550E+02	.265E+10	.343E+09
28	.500E-02	.140E+00	.550E+02	.265E+10	.356E+09
29	.500E-02	.145E+00	.550E+02	.264E+10	.370E+09
30	.500E-02	.150E+00	.550E+02	.264E+10	.383E+09
31	.500E-02	.155E+00	.550E+02	.263E+10	.396E+09
32	.500E-02	.160E+00	.550E+02	.263E+10	.409E+09
33	.500E-02	.165E+00	.550E+02	.262E+10	.422E+09
34	.500E-02	.170E+00	.550E+02	.262E+10	.435E+09
35	.500E-02	.175E+00	.550E+02	.261E+10	.448E+09
36	.500E-02	.180E+00	.550E+02	.261E+10	.461E+09
37	.500E-02	.185E+00	.550E+02	.260E+10	.474E+09
38	.500E-02	.190E+00	.550E+02	.260E+10	.487E+09
39	.500E-02	.195E+00	.550E+02	.260E+10	.500E+09

Now let us leave Well 1 alone and drill a completely new well (Well 2) during simulation. Observe from the following keyboard coordinate entries that the new well is highly deviated.

```
Continue  transient flow simulation modeling?  Y/N:  y
Well #1,  @ Step #   39, Time  .195E+00  hrs,
is "pressure constrained" at   .550E+02  psi.
Well status or geometry, Change or Unchanged?  C/U:  u
Drill any (more) new wells and well clusters?  Y/N:  y
```

The simulator informs you that you have brought a new well on stream and that the number of wells has increased to two. This drill new wells option always appears in the runtime menu so long as the total number of well clusters is less than the maximum allowable of nine. At this point, you will have separately determined that six blocks, say, are required to define the deviated well needed to penetrate the $ pay zone. Incidentally, the maximum number of grid blocks supported by the prototype PC simulator is $20 \times 20 \times 9$, or approximately 4000. The number of well clusters supported is 9, with a maximum number of 200 blocks defining each cluster. The total number of character based lithologies supported is 31. These numbers are easily increased by redimensioning.

```
A new well has just been brought on stream ...
Total well number has increased to 2.
Number of active cell blocks defining new well:  6

O  Block   1, New x(i) position index, i:  4
O  Block   1, New y(j) position index, j:  7
O  Block   1, New z(k) position, Layer #:  1
O  Block   2, New x(i) position index, i:  4
O  Block   2, New y(j) position index, j:  7
O  Block   2, New z(k) position, Layer #:  2
O  Block   3, New x(i) position index, i:  5
O  Block   3, New y(j) position index, j:  7
O  Block   3, New z(k) position, Layer #:  3
O  Block   4, New x(i) position index, i:  6
O  Block   4, New y(j) position index, j:  7
O  Block   4, New z(k) position, Layer #:  4
O  Block   5, New x(i) position index, i:  7
O  Block   5, New y(j) position index, j:  7
O  Block   5, New z(k) position, Layer #:  5
O  Block   6, New x(i) position index, i:  8
O  Block   6, New y(j) position index, j:  7
O  Block   6, New z(k) position, Layer #:  6

New well constraint, pressure or rate?  P/R:  p
New pressure (psi):  1000
```

There are presently two wells: Well 1, which originated from the start at Step 1, and Well 2, which was brought on stream at Step 40. The WELL1.SIM and WELL2.SIM files reflect this fact. And both wells, pressure constrained as they are in Steps 40–59, show the physically expected decline in flow rate with time.

WELL #1: Step No.	Dt (Hour)	Time (Hour)	Pressure (Psi)	Flow Rate (Cu Ft/Hr)	Cum Vol (Cu Ft)
1	.500E-02	.500E-02	.500E+03	.254E+10	.127E+08
2	.500E-02	.100E-01	.500E+03	.254E+10	.254E+08
3	.500E-02	.150E-01	.500E+03	.253E+10	.380E+08
.					
37	.500E-02	.185E+00	.550E+02	.260E+10	.474E+09
38	.500E-02	.190E+00	.550E+02	.260E+10	.487E+09
39	.500E-02	.195E+00	.550E+02	.260E+10	.500E+09
40	.500E-02	.200E+00	.550E+02	.259E+10	.513E+09
41	.500E-02	.205E+00	.550E+02	.259E+10	.526E+09
42	.500E-02	.210E+00	.550E+02	.258E+10	.539E+09
43	.500E-02	.215E+00	.550E+02	.258E+10	.552E+09
44	.500E-02	.220E+00	.550E+02	.257E+10	.565E+09
45	.500E-02	.225E+00	.550E+02	.257E+10	.578E+09
46	.500E-02	.230E+00	.550E+02	.256E+10	.591E+09
47	.500E-02	.235E+00	.550E+02	.256E+10	.603E+09
48	.500E-02	.240E+00	.550E+02	.256E+10	.616E+09
49	.500E-02	.245E+00	.550E+02	.255E+10	.629E+09
50	.500E-02	.250E+00	.550E+02	.255E+10	.642E+09
51	.500E-02	.255E+00	.550E+02	.254E+10	.654E+09
52	.500E-02	.260E+00	.550E+02	.254E+10	.667E+09
53	.500E-02	.265E+00	.550E+02	.254E+10	.680E+09
54	.500E-02	.270E+00	.550E+02	.253E+10	.692E+09
55	.500E-02	.275E+00	.550E+02	.253E+10	.705E+09
56	.500E-02	.280E+00	.550E+02	.252E+10	.718E+09
57	.500E-02	.285E+00	.550E+02	.252E+10	.730E+09
58	.500E-02	.290E+00	.550E+02	.251E+10	.743E+09
59	.500E-02	.295E+00	.550E+02	.251E+10	.755E+09

WELL #2: Step No.	Dt (Hour)	Time (Hour)	Pressure (Psi)	Flow Rate (Cu Ft/Hr)	Cum Vol (Cu Ft)
40	.500E-02	.200E+00	.100E+04	.000E+00	.000E+00
41	.500E-02	.205E+00	.100E+04	.283E+10	.142E+08
42	.500E-02	.210E+00	.100E+04	.283E+10	.283E+08
43	.500E-02	.215E+00	.100E+04	.282E+10	.424E+08
44	.500E-02	.220E+00	.100E+04	.281E+10	.565E+08
45	.500E-02	.225E+00	.100E+04	.281E+10	.705E+08
46	.500E-02	.230E+00	.100E+04	.280E+10	.845E+08
47	.500E-02	.235E+00	.100E+04	.279E+10	.985E+08
48	.500E-02	.240E+00	.100E+04	.279E+10	.112E+09
49	.500E-02	.245E+00	.100E+04	.278E+10	.126E+09
50	.500E-02	.250E+00	.100E+04	.277E+10	.140E+09
51	.500E-02	.255E+00	.100E+04	.277E+10	.154E+09
52	.500E-02	.260E+00	.100E+04	.276E+10	.168E+09
53	.500E-02	.265E+00	.100E+04	.276E+10	.182E+09
54	.500E-02	.270E+00	.100E+04	.275E+10	.195E+09
55	.500E-02	.275E+00	.100E+04	.274E+10	.209E+09
56	.500E-02	.280E+00	.100E+04	.274E+10	.223E+09
57	.500E-02	.285E+00	.100E+04	.273E+10	.236E+09
58	.500E-02	.290E+00	.100E+04	.273E+10	.250E+09
59	.500E-02	.295E+00	.100E+04	.272E+10	.264E+09

Finally, let us numerically shock our transient compressible simulation once again, this time shutting in production interactively at both wells for time Steps 60–79. From our WELL1.SIM and WELL2.SIM pressure files, we find that during Steps 60–79, we have in each case a rapid pressure buildup initially followed by a more gradual rise.

WELL #1: Step No.	Dt (Hour)	Time (Hour)	Pressure (Psi)	Flow Rate (Cu Ft/Hr)	Cum Vol (Cu Ft)
.					
.					
57	.500E-02	.285E+00	.550E+02	.252E+10	.730E+09
58	.500E-02	.290E+00	.550E+02	.251E+10	.743E+09
59	.500E-02	.295E+00	.550E+02	.251E+10	.755E+09
60	.500E-02	.300E+00	.722E+04	.000E+00	.755E+09
61	.500E-02	.305E+00	.861E+04	.000E+00	.755E+09
62	.500E-02	.310E+00	.913E+04	.000E+00	.755E+09
63	.500E-02	.315E+00	.934E+04	.000E+00	.755E+09
64	.500E-02	.320E+00	.943E+04	.000E+00	.755E+09
65	.500E-02	.325E+00	.946E+04	.000E+00	.755E+09
66	.500E-02	.330E+00	.948E+04	.000E+00	.755E+09
67	.500E-02	.335E+00	.949E+04	.000E+00	.755E+09
68	.500E-02	.340E+00	.949E+04	.000E+00	.755E+09
69	.500E-02	.345E+00	.950E+04	.000E+00	.755E+09
70	.500E-02	.350E+00	.950E+04	.000E+00	.755E+09
71	.500E-02	.355E+00	.950E+04	.000E+00	.755E+09
72	.500E-02	.360E+00	.950E+04	.000E+00	.755E+09
73	.500E-02	.365E+00	.951E+04	.000E+00	.755E+09
74	.500E-02	.370E+00	.951E+04	.000E+00	.755E+09
75	.500E-02	.375E+00	.951E+04	.000E+00	.755E+09
76	.500E-02	.380E+00	.951E+04	.000E+00	.755E+09
77	.500E-02	.385E+00	.951E+04	.000E+00	.755E+09
78	.500E-02	.390E+00	.952E+04	.000E+00	.755E+09
79	.500E-02	.395E+00	.952E+04	.000E+00	.755E+09

WELL #2: Step No.	Dt (Hour)	Time (Hour)	Pressure (Psi)	Flow Rate (Cu Ft/Hr)	Cum Vol (Cu Ft)
.					
.					
57	.500E-02	.285E+00	.100E+04	.273E+10	.236E+09
58	.500E-02	.290E+00	.100E+04	.273E+10	.250E+09
59	.500E-02	.295E+00	.100E+04	.272E+10	.264E+09
60	.500E-02	.300E+00	.973E+04	.000E+00	.264E+09
61	.500E-02	.305E+00	.977E+04	.000E+00	.264E+09
62	.500E-02	.310E+00	.977E+04	.000E+00	.264E+09
63	.500E-02	.315E+00	.978E+04	.000E+00	.264E+09
64	.500E-02	.320E+00	.978E+04	.000E+00	.264E+09
65	.500E-02	.325E+00	.978E+04	.000E+00	.264E+09
66	.500E-02	.330E+00	.978E+04	.000E+00	.264E+09
67	.500E-02	.335E+00	.978E+04	.000E+00	.264E+09
68	.500E-02	.340E+00	.978E+04	.000E+00	.264E+09
69	.500E-02	.345E+00	.978E+04	.000E+00	.264E+09
70	.500E-02	.350E+00	.978E+04	.000E+00	.264E+09
71	.500E-02	.355E+00	.979E+04	.000E+00	.264E+09
72	.500E-02	.360E+00	.979E+04	.000E+00	.264E+09
73	.500E-02	.365E+00	.979E+04	.000E+00	.264E+09
74	.500E-02	.370E+00	.979E+04	.000E+00	.264E+09
75	.500E-02	.375E+00	.979E+04	.000E+00	.264E+09
76	.500E-02	.380E+00	.979E+04	.000E+00	.264E+09
77	.500E-02	.385E+00	.979E+04	.000E+00	.264E+09
78	.500E-02	.390E+00	.979E+04	.000E+00	.264E+09
79	.500E-02	.395E+00	.979E+04	.000E+00	.264E+09

This completes our simulating while drilling example. Although we have used time steps of 0.005 h, the algorithm will simulate very stably with step sizes on the order of hours and days. The key emphasis is the robustness of the scheme when the reservoir is subjected to actual operational changes.

The simulation was designed to show how general heterogeneities and well configurations can be modeled with minimal effort. And operational changes can be implemented and studied in real time during interactive simulation, without requiring extremely small time steps for stability. Such simulations are also ideal for real-world history matching applications. Time step sizes and the number of integration cycles between runtime menu displays can be altered in accordance with operational changes, allowing users to replicate oilfield operations easily and perform what if production tests efficiently.

EXAMPLE 15.5. MODELING WELLBORE STORAGE EFFECTS AND COMPRESSIBLE BOREHOLE FLOW TRANSIENTS

Consider a two-layer homogeneous reservoir with a centered vertical well, that is, the simple $11 \times 11 \times 2$ system in LAYER1.DRL. The computational box is surrounded by aquifers on all four sides and held at 1000 psi, while the tops and bottoms are solid no-flow walls. The reservoir is initially pressurized at 1000 psi, and the well is flow rate constrained at 1000 cu ft/h once production begins. Wellbore storage effects, specifically the consequences of varying F from 0.0 ft^5/lbf (for zero storage) through a range of values, are desired.

```
C>type layer1.drl (layer2.drl is identical)
+ + + + + + + + + + +
+ + + + + + + + + + +
+ + + + + + + + + + +
+ + + + + + + + + + +
+ + + + + + + + + + +
+ + + + + 1 + + + + +
+ + + + + + + + + + +
+ + + + + + + + + + +
+ + + + + + + + + + +
+ + + + + + + + + + +
+ + + + + + + + + + +

  INPUT LITHOLOGY AND FORMATION PROPERTIES:
  O  Enter option, rectangular or stratigraphic grids?  R/S:  r
  O  Enter number of reservoir layers (# 1-9):  2
     2 layer(s)  taken in  the  "Z" direction.
     11 grid blocks  counted in "X" direction.
     11 grid blocks  counted in "Y" direction.
  O  Enter grid length DX in X direction (ft):  100
  O  Enter grid length DY in Y direction (ft):  100
  O  Enter "thickness" DZ in Z direction (ft):  100

  Reading geological files from disk, please wait ...
  Lithology definition begins:

  Is rock type + isotropic?  Y/N:  y
     Isotropic permeability (md) of lithology + is:  100

     Porosities are used in "steady state"  flows to solve
     front positions only, and are not needed for pressure
     calculations.   In compressible flow, particularly in
     well test and primary recovery, porosities are needed
     for both pressure and front computations.

     Nonzero porosity  (decimal) of lithology + is:  .20
```

Rock compressibility forms part of the input: the net compressibility of the fluid and rock system, not that of the fluid alone, affects transients. This is critical if significant variations of rock having different compressibilities exist. If these are available during analysis, the opportunity to weight both values by porosity-averaging is available, and an effective compressibility of $c_{eff} = \phi(x,y,x)c_{fluid} + (1 - \phi)\ c_{rock}$ is used, with c_{fluid} and c_{rock} denoting fluid and rock values, and $\phi(x,y,z)$ being the porosity. (At this writing, this option is available only for liquids.) Thus, a 99% porous medium would be dominated by liquid effects, while a 1% porous medium would have its flow dominated by rock effects.

```
Rock compressibility is required if averaging of rock
and liquid values is applied in transient simulation;
if not, enter dummy values (e.g., "1").

Rock compressibility (/psi) of lithology + is:  .000008

Number of lithologies identified in reservoir:   1

Lithotype + Formation Properties:
kx = .1000E+03 md, ky = .1000E+03 md, kz = .1000E+03 md,
porosity = .2000E+00, compressibility = .8000E-05 / psi.

Copying files to disk, please wait ...

Total volume of "computational box" is  .242E+09  cu ft,
total pore space volume is  .484E+08 cu ft.
```

Several simulations designed to illustrate differences between production and injection wells, and wells with and without wellbore storage, are given. The captured screens shown next are self-explanatory.

Run 1. Production Well, No Wellbore Storage Effects

```
Reading geological/drilling records, please wait -
One cluster of wells was identified in your reservoir.

Well constraint conventions:  (1)  Pressure levels
must be positive ( >0).  (2) Flow rate constraints
assume "-" for injectors,  and  "+" for producers;
for gases,  enter rates corresponding to "standard
surface conditions" (i.e.,  @ 14.7 psi, 60 deg F).
Additional properties will be required at runtime.

Units available, (1) CuFt/Hr, (2) CuFt/D, (3) B/D.
Enter option (1, 2 or 3):  1
```

```
Is "Well No. 1" pressure or rate constrained?  P/R:  r
O  Enter total cluster volume flow rate:  1000

SIMULATION SETUP, PHYSICAL FLUID MODELING, FARFIELD
AND WELLBORE RUN/TIME BOUNDARY CONDITION DEFINITION:

Reading drilling records, please wait ...

PRESENT RESERVOIR STATUS:

Reservoir  grid parameters:
To continue, type <Return>:
Imax   = 11, Dx = .1000E+03 ft, Imax   *Dx = .1100E+04 ft.
Jmax   = 11, Dy = .1000E+03 ft, Jmax   *Dy = .1100E+04 ft.
Layers =  2, Dz = .1000E+03 ft, Layers*Dz = .2000E+03 ft.

Number of initial well clusters identified: 1

To continue, type <Return>:  <Return>
Reading transmissibility files, please wait ...

You may modify TX, TY and TZ transmissibilities for simulation
purposes WITHOUT altering values on disk .... Modify?  Y/N:  n

Combining geological/drilling information, please wait ...

Well block transmissibility summary (ft^3):

To continue, type <Return>:  <Return>
Well 1, defined by   2 grid blocks, is
flow rate constrained at  .1000E+04  cu ft/hr,
that is,  .427E+04 b/d, or  .240E+05 cu ft/day.
Block   1: (I= 6, J= 6, Layer 1), Tx = .108E-09, Ty = .108E-09, Tz = .108E-09
Block   2: (I= 6, J= 6, Layer 2), Tx = .108E-09, Ty = .108E-09, Tz = .108E-09

INPUT FARFIELD BOUNDARY CONDITION SETUP:

                                Z(k)
                               |
                               | j=1        Jmax(11)
                          i=1  |
                               +--------------+  ------  Y(j)
                              /|______________| \
             Left        /     |___  Back  ___|   \   Right
                        /      |______________|    \
                       /       +--------------+     \
      Imax(11)      +-------/  Top, Layer 1  \-------+
                   /  |_______________________________|
                  /   |___________  Front ____________|
                / X(i)|_______________________________|
                      +-------  Bottom, Layer 2  ------+

                              COORDINATE SYSTEM

O  FRONT ... is aquifer or no flow wall?  A/W:  a
O  Pressure at FRONT face (psi):  1000
O  BACK .... is aquifer or no flow wall?  A/W:  a
O  Pressure at BACK face  (psi):  1000
O  LEFT .... is aquifer or no flow wall?  A/W:  a
O  Pressure at LEFT face  (psi):  1000
O  RIGHT ... is aquifer or no flow wall?  A/W:  a
O  Pressure at RIGHT face (psi):  1000
O  TOP ..... is aquifer or no flow wall?  A/W:  w
O  BOTTOM .. is aquifer or no flow wall?  A/W:  w

PHYSICAL FLUID MODEL SETUP:

O  Fluid viscosity of water and air at room temperature
   and pressure are 1 cp and 0.018 cp, respectively ...

O  Fluid viscosity (centipoise):  1

Is reservoir fluid a liquid or a gas?  L/G:  L

Analyze steady or transient compressible flow?  S/T:  t

TRANSIENT COMPRESSIBLE FLOW SIMULATION MODE SELECTED.
```

```
Transient compressible flow calculation beginning ...
Initialize solution to (A) constant pressure everywhere
or (B) variable pressure field stored in file?  A/B:  a
O  Enter initial uniform pressure (psi):  1000
WELL TEST INPUT PARAMETER SETUP:
Reading porosity array, please wait ...
O  Typical compressibilities: oil @ 0.00001/psi,
   water @ 0.000003/psi, gas @ 0.0005/psi, etc.
O  Liquid compressibility (1/psi):  .000003
```

The following command allows porosity-averaged rock-fluid compressibility, per the above discussion (again, option is available for liquid flows only),

```
Porosity-average this liquid compressibility with
matrix compressibilities  entered previously?  Y/N:  n
Time scale estimate?      Y/N:  n
O  Initial time step (hours):  .1
O  Maximum  number  of steps:  1000
Invoke "small deformation" compaction model?   Y/N:  n
Continue  transient flow simulation modeling?  Y/N:  y
Well #1,   @ Step #    0, Time  .000E+00  hrs,
is "rate constrained" at   .100E+04 cu ft/hr.
Well status or geometry, Change or Unchanged?  C/U:  u
Drill any (more) new wells and well clusters?  Y/N:  n
Time step now  .100E+00 hr, Change/Unchanged?  C/U:  u
```

Note from the following query that the simulator's default mode assumes *zero* wellbore storage. However, you may change the value of the capacity factor F periodically during the transient run and also choose F to be different for different wells. At iteration intervals you define, you are informed of all current F's and are permitted to alter them as drilling or production conditions change.

```
Well Cluster 1:
Well storage capacity, now  .000E+00 ft^5/lbf, C/U:  u
Time steps between pressure plots,  now   10,  C/U:  c
......................  Enter new time step number:  200
Time steps between well status changes,   10,  C/U:  c
......................  Enter new time step number:  200
Calculating at time step     1, please wait ...
Calculating at time step     2, please wait ...
```

On completion, the WELL#.SIM files created during simulation are stored and available for plotting. These files contain pressure and flow rate history at each single well or multilateral well cluster, for example,

```
C>type well1.sim

 WELL #1:        Dt          Time        Pressure    Flow Rate      Cum Vol
 Step No.      (Hour)       (Hour)        (Psi)     (Cu Ft/Hr)     (Cu Ft)
       1     .100E+00     .100E+00     .969E+03     .100E+04     .100E+03
       2     .100E+00     .200E+00     .954E+03     .100E+04     .200E+03
       3     .100E+00     .300E+00     .946E+03     .100E+04     .300E+03
       4     .100E+00     .400E+00     .940E+03     .100E+04     .400E+03
       5     .100E+00     .500E+00     .936E+03     .100E+04     .500E+03
       6     .100E+00     .600E+00     .932E+03     .100E+04     .600E+03
       7     .100E+00     .700E+00     .929E+03     .100E+04     .700E+03
       8     .100E+00     .800E+00     .927E+03     .100E+04     .800E+03
       9     .100E+00     .900E+00     .925E+03     .100E+04     .900E+03
      10     .100E+00     .100E+01     .923E+03     .100E+04     .100E+04
       .
       .
     199     .100E+00     .199E+02     .902E+03     .100E+04     .199E+05
```

Run 2. Production Well, With Some Wellbore Storage Effects

Observe from the preceding table that the well pressure decreases from 969 to 902 psi after 200 time steps. In Run 1, we assumed that F was identically zero, with 0.0 ft^5/lbf. Now we repeat our calculations with a slightly different capacity, assuming that $F = 0.00000001$ ft^5/lbf. All other parameters are identical to Run 1 for comparison. The effect of a nonzero (positive) capacity allows borehole fluid to expand initially. In this problem, this expansion supplies part of the produced flow, decreasing the production rate required at the sandface. Hence, the pressure decrease should in time be less rapid than in Run 1 which, again, assumed zero storage. The final results, shown here, are consistent with the physics, with a final pressure of 932 psi instead of 902 psi.

```
 WELL #1:        Dt          Time        Pressure    Flow Rate      Cum Vol
 Step No.      (Hour)       (Hour)        (Psi)     (Cu Ft/Hr)     (Cu Ft)
       1     .100E+00     .100E+00     .969E+03     .100E+04     .100E+03
       2     .100E+00     .200E+00     .969E+03     .100E+04     .200E+03
       3     .100E+00     .300E+00     .968E+03     .100E+04     .300E+03
       4     .100E+00     .400E+00     .968E+03     .100E+04     .400E+03
       5     .100E+00     .500E+00     .968E+03     .100E+04     .500E+03
       6     .100E+00     .600E+00     .968E+03     .100E+04     .600E+03
       7     .100E+00     .700E+00     .967E+03     .100E+04     .700E+03
       8     .100E+00     .800E+00     .967E+03     .100E+04     .800E+03
       9     .100E+00     .900E+00     .967E+03     .100E+04     .900E+03
      10     .100E+00     .100E+01     .967E+03     .100E+04     .100E+04
      20     .100E+00     .200E+01     .964E+03     .100E+04     .200E+04
      30     .100E+00     .300E+01     .962E+03     .100E+04     .300E+04
      40     .100E+00     .400E+01     .960E+03     .100E+04     .400E+04
      50     .100E+00     .500E+01     .957E+03     .100E+04     .500E+04
     100     .100E+00     .100E+02     .947E+03     .100E+04     .100E+05
       .
     199     .100E+00     .199E+02     .932E+03     .100E+04     .199E+05
```

Run 3. Production Well, With More Wellbore Storage Effects

In this run, we increase the F assumed in Run 2 *one-hundred-fold* to 0.000001 ft^5/lbf. This represents a case where almost all of the surface production is assumed by borehole fluid expansion; it simulates underbalanced drilling when substantial gas is released from solution into the wellbore column. This simulation was designed to test the stability limits and physical correctness of the scheme. The results of this simulation lead to a very nonproductive reservoir as expected. For example, the following plot,

```
Pressure (psi) versus time:
     Hours                          0
                                    ______________________________
      .10        .9690E+03          |                            *
      .20        .9690E+03          |                            *
      .30        .9690E+03          |                            *
      .40        .9690E+03          |                            *
      .50        .9690E+03          |                            *
      .60        .9690E+03          |                            *
      .70        .9690E+03          |                            *
      .80        .9690E+03          |                            *
      .90        .9690E+03          |                            *
     1.00        .9690E+03          |                            *
        .
        .
    19.00        .9680E+03          |                           *
        .
    19.90        .9680E+03          |                           *
```

reveals a constant level of sandface pressure is consistent with the fact that the reservoir contributes very little to production.

Run 4. Injector Well, Without Wellbore Storage Effects

We repeat Run 1 (without wellbore storage effects) but allow Well 1 to be flow rate constrained at -1000 cu ft/h, that is, assume a sign change. Thus, Well 1 is converted from a producer to an injector. The edited WELL1.SIM file that follows displays the physically expected increase in pressure with time.

```
WELL #1:       Dt          Time        Pressure   Flow Rate      Cum Vol
Step No.     (Hour)       (Hour)        (Psi)     (Cu Ft/Hr)     (Cu Ft)
       1    .100E+00     .100E+00     .103E+04   -.100E+04    -.100E+03
       2    .100E+00     .200E+00     .105E+04   -.100E+04    -.200E+03
       3    .100E+00     .300E+00     .105E+04   -.100E+04    -.300E+03
       4    .100E+00     .400E+00     .106E+04   -.100E+04    -.400E+03
       .
       .
     100    .100E+00     .100E+02     .110E+04   -.100E+04    -.100E+05
     199    .100E+00     .199E+02     .110E+04   -.100E+04    -.199E+05
```

Run 5. Injector Well, With Wellbore Storage Effects

Here, we shall repeat Run 4, except that we set F = 0.00000001 ft^5/lbf. The effect of this nonzero (positive) storage is easily envisioned. At t = 0, the injection will compress the fluid in the borehole first. Thus, the rate of timewise pressure increase obtained at the sandface should fall *below* the levels calculated in Run 4. To see that this is in fact the situation calculated, the reader should examine the WELL1.SIM file shown here. Whereas in Run 4, the pressure at 200 time steps was 1100 psi, the final value now is 1070 psi.

```
WELL #1:      Dt          Time        Pressure    Flow Rate     Cum Vol
Step No.    (Hour)       (Hour)        (Psi)     (Cu Ft/Hr)    (Cu Ft)
     1    .100E+00     .100E+00     .103E+04   -.100E+04    -.100E+03
     2    .100E+00     .200E+00     .103E+04   -.100E+04    -.200E+03
     3    .100E+00     .300E+00     .103E+04   -.100E+04    -.300E+03
     .
     .

    40    .100E+00     .400E+01     .104E+04   -.100E+04    -.400E+04
    50    .100E+00     .500E+01     .104E+04   -.100E+04    -.500E+04
   100    .100E+00     .100E+02     .105E+04   -.100E+04    -.100E+05
   199    .100E+00     .199E+02     .107E+04   -.100E+04    -.199E+05
```

As with all simulator options, the storage algorithm was very stable. Also, spatial results calculated at 200 steps, or 20 h, show the correct horizontal, vertical, and diagonal pressure symmetries about the centered well, highlighted below in bold numerals between asterisks; the pressures in Layer 1 and Layer 2 are identical, as required physically. Such simple checks are really demanding, since few algorithms claim stability without losing accuracy. We have studied a vertical well; the storage option applies to the general heterogeneities, plus arbitrary horizontal and multilateral wells.

```
Pressure Distribution (psi) in Layer 1:
BACK
.100E+04 .100E+04 .100E+04 .100E+04 .100E+04 .100E+04 .100E+04 .100E+04 .100E+04 .100E+04 .100E+04
.100E+04 .100E+04 .100E+04 .100E+04 .101E+04 .101E+04 .101E+04 .100E+04 .100E+04 .100E+04 .100E+04
.100E+04 .100E+04 .101E+04 .101E+04 .101E+04 .101E+04 .101E+04 .101E+04 .101E+04 .100E+04 .100E+04
.100E+04 .100E+04 .101E+04 .101E+04 .102E+04 .102E+04 .102E+04 .101E+04 .101E+04 .100E+04 .100E+04
.100E+04 .101E+04 .101E+04 .102E+04 .103E+04 .104E+04 .103E+04 .102E+04 .101E+04 .101E+04 .100E+04
.100E+04 .101E+04 .101E+04 .102E+04 .104E+04*.107E+04*.104E+04 .102E+04 .101E+04 .101E+04 .100E+04
.100E+04 .101E+04 .101E+04 .102E+04 .103E+04 .104E+04 .103E+04 .102E+04 .101E+04 .101E+04 .100E+04
.100E+04 .100E+04 .101E+04 .101E+04 .102E+04 .102E+04 .102E+04 .101E+04 .101E+04 .100E+04 .100E+04
.100E+04 .100E+04 .101E+04 .101E+04 .101E+04 .101E+04 .101E+04 .101E+04 .101E+04 .100E+04 .100E+04
.100E+04 .100E+04 .100E+04 .100E+04 .101E+04 .101E+04 .101E+04 .100E+04 .100E+04 .100E+04 .100E+04
.100E+04 .100E+04 .100E+04 .100E+04 .100E+04 .100E+04 .100E+04 .100E+04 .100E+04 .100E+04 .100E+04
FRONT
```

```
Pressure Distribution (psi) in Layer 2:
BACK
.100E+04 .100E+04 .100E+04 .100E+04 .100E+04 .100E+04 .100E+04 .100E+04 .100E+04 .100E+04 .100E+04
.100E+04 .100E+04 .100E+04 .100E+04 .101E+04 .101E+04 .101E+04 .100E+04 .100E+04 .100E+04 .100E+04
.100E+04 .100E+04 .101E+04 .101E+04 .101E+04 .101E+04 .101E+04 .101E+04 .101E+04 .100E+04 .100E+04
.100E+04 .100E+04 .101E+04 .101E+04 .102E+04 .102E+04 .102E+04 .101E+04 .101E+04 .100E+04 .100E+04
.100E+04 .101E+04 .101E+04 .102E+04 .103E+04 .104E+04 .103E+04 .102E+04 .101E+04 .101E+04 .100E+04
.100E+04 .101E+04 .101E+04 .102E+04 .104E+04*.107E+04*.104E+04 .102E+04 .101E+04 .101E+04 .100E+04
.100E+04 .101E+04 .101E+04 .102E+04 .103E+04 .104E+04 .103E+04 .102E+04 .101E+04 .101E+04 .100E+04
.100E+04 .100E+04 .101E+04 .101E+04 .102E+04 .102E+04 .102E+04 .101E+04 .101E+04 .100E+04 .100E+04
.100E+04 .100E+04 .101E+04 .101E+04 .101E+04 .101E+04 .101E+04 .101E+04 .101E+04 .100E+04 .100E+04
.100E+04 .100E+04 .100E+04 .100E+04 .101E+04 .101E+04 .101E+04 .100E+04 .100E+04 .100E+04 .100E+04
.100E+04 .100E+04 .100E+04 .100E+04 .100E+04 .100E+04 .100E+04 .100E+04 .100E+04 .100E+04 .100E+04
FRONT
```

Storage was modeled numerically because general multilateral well topologies and their placement in heterogeneous formations preclude analytical solution. For simpler problems in homogeneous media, closed-form solutions can be given. For example, an exact solution is derived in Chapter 18 for a non-zero radius ellipsoidal source, which includes storage, anisotropy, and skin effects, in order to demonstrate classic Laplace transform analysis methods. This model is used in formation tester pressure transient interpretation. Finally, we note that deconvolution methods are available to un-do the effects of well-bore storage, so that the formation response itself is available for analysis. These methods are strictly applicable to liquid flows without rock compaction, since they employ Duhamel's integral, a superposition method restricted to linear systems. When gases are modeled, or when liquids with compaction drives are considered, deconvolution methods cannot be used and direct simulation is required.

GROUP 2, ADVANCED CALCULATIONS AND USER INTERFACE

The author had previously described his reservoir engineering experiences (and frustrations) in simulation—these, however, importantly motivated the math models in Chin (2002), which would address the failures of then and present flow simulators to adequately model physical reality. And his day-to-day activities would shape his perspectives on workflow optimization—in particular, removing from the simulation process artificial hurdles like high hardware costs, inflexible software licenses, difficult-to-use graphics processors, and impossible Unix-like commands that turned challenging and fun reservoir analysis into boring data entry exercises. Still lacking, however, was the friendly computer user interface needed to facilitate change.

Overview

As luck would have it, the *Education for the Energy Industry* program funded by ExxonMobil, Shell, ChevronTexaco and other operating companies in the early 2000s to introduce petroleum engineering to K-12 students and teachers would provide the test bed for our new ideas on simulation. In this effort, the author was selected as lead curriculum developer and software architect, whose plans and prototype models were put to the test at Aldine Independent School District, then the second largest in Texas. That modeling should be simple, fun and productive—that it should take the boredom out of existing workflows and allow engineers time to address the real questions that mattered—took hold as our improved algorithms would see their rebirth in very user-friendly screens now piloted by true novices who would evaluate reservoir flow and integrate results with spreadsheet economic analyses.

In short, we've come a long way. Multisim™ now allows reservoir analysts to focus on petroleum engineering issues. For example, "What does the pressure field look like?" "What are the flow rates in pressure-constrained wells, and likewise, pressures in rate-constrained systems?" "How do wells interact or cannibalize each other?" "What types of events transpire across lease boundaries?" "How do changes in boundary conditions affect production?" And importantly, the focus of our modeling, "How can we optimize production and cash flow while reducing drilling and well costs?"

In developing our approach, we understand that two major problems impede simulation progress at the user level. First is an understanding of the workflow involved in setting up and solving particular problems. For example, one would normally complete a course in petroleum engineering, and then, just *might* recall that steady-state pressure field computations require only permeability and viscosity inputs—and that, only for transient flows, do porosity, compressibility and wellbore storage enter. In this sense, expertise on the part of the user is one impediment to the profession. The second is the user interface itself. Quite often, software screens are designed by programmers who do not understand reservoir engineering workflows and problem formulations—this results in the need for detailed user manuals and unnecessary input commands. When menus are not structured properly, simulation setups are not intuitive and work productivity drops. How do engineers work and in what order? Wells are first drilled. Then they are flowed assuming pressure or rate constraints. They may be shut-in, sidetracked, or recompleted. And the cycle begins all over again, perhaps with still another multilateral system. This knowledge should be built into the interface and background logic.

A good menu system must support this workflow and minimize the need for training and rote learning on the part of the user. We have developed just such a system. Because operating a simulator is just as important as developing one, we have carefully selected five examples that exercise all of our simulator options. More than a dozen examples are presented in Chin (2016a), but here we shall quickly summarize key ideas and outputs. In the discussions that follow, we integrate our discussions related to simulation approach and menu design. Results for five key examples are summarized, namely,

- Example 15.6. Multilateral and vertical wells in multilayer media.
- Example 15.7. Dual lateral with transient operations.
- Example 15.8. Producer and injector conversions.
- Example 15.9. Production with top and bottom drives.
- Example 15.10. Transient gas production from dual horizontal with wellbore storage effects.

Before proceeding, we offer a summary of key Multisim™ physical modeling and software features, so that our capabilities can be readily compared to existing reservoir flow simulators. It is important to emphasize that, aside from a quick reading of Chin (2016a), no user manuals are offered or need to be.

MULTISIM™ SOFTWARE FEATURES

Reservoir Description

- General heterogeneities, anisotropies, layering, geological structures sketched using lithological symbols like #, $, % and so on
- Fracture and flow barrier modeling
- Transmissibilities may be temporarily altered during simulation
- Incompressible and compressible liquid and gas single-phase flows
- General thermodynamic options for gas flow modeling
- Fluid and matrix rock compressibility (porosity-based) averaging
- Rigid formation versus "small deformation" compaction models available
- General drive mechanisms supported, e.g., gas, aquifer, and so on
- Stratigraphic grids built into source code (not available interactively)

Well System Modeling

- Multiple (partially or fully penetrating) vertical, deviated, horizontal and multilaterals are supported
- Arbitrary well topologies, rate or pressure constraints may be changed during simulation, while multilateral "arms" and "legs" may be altered or recompleted while computations are in progress—up to nine systems supported on general layered reservoir model
- Sidetracking, redrilling and recompletions while simulating
- Means to define local empirical "productivity indexes" offered

Additional Simulator Features

- Arbitrary specification of injectors and producers
- Steady flow solutions, fully transient modeling, or steady, then transient
- Initial pressures may be constant or variable
- Transient simulator initialization to existing pressures, e.g., a three-well solution may "start" a two-well analysis where one well is being abandoned
- Menus "activated" step-by-step guide users in data entry (internal workflow procedures automatically accessed depending on user objectives, no user manuals offered or required)
- Highly integrated three-dimensional color graphics and line plots
- Matrix inversion performed "behind the scenes" transparently to user
- Automatic equation set-up, matrix inversion and solution, "behind the scenes" computations transparent to user, three-dimensional color graphics tightly integrated with simulator
- Standard Windows computer, graphics cards, and detailed engineering expertise and user manuals not needed

EXAMPLE 15.6. MULTILATERAL AND VERTICAL WELLS IN MULTILAYER MEDIA

In this simulation, we demonstrate powerful features of the full simulator that were omitted in our "basic examples" introduction, if only to avoid initial "information overload" and possible confusion. Here we shall consider the largest and finest resolution options now possible, namely a nine-layer reservoir with 31×31 grid blocks in each layer; that is, a computing domain consisting of 8649 or nearly 10,000 grid blocks, hosting general heterogeneities and anisotropies with arbitrary well topologies. We shall show that stable and rapid convergence to meaningful results plus 3D color graphic display is possible in about ten seconds on mid-range Intel Core i5 personal computers.

In Examples 15.1–15.5, we introduced the introductory version of the simulator, a three-layer model with 15×15 grid blocks developed for the *Education for the Energy Industry* program. Subsequently, our capabilities were extended to nine layer problems with 31×31 grid blocks. This powerful version requires password access as suggested at the bottom of Fig. 15.3A. The password is machine-dependent and obtained by sending the Serial Number to Stratamagnetic Software LLC. Once "Go" is clicked, an information screen stating "License validated, all options available" is displayed.

Main workflows are listed in the main horizontal menu at the top of Fig. 15.3A, with "Geology," "Wells," "Simulate" and "3DPlots" selected in turn. The "Geology" menu is used to describe the underlying reservoir. It is first accessed to define the number of layers by selecting "Number of Layers" as shown in Fig. 15.3B. This leads to the screen in Fig. 15.3C, where all nine buttons are now active, indicating that up to nine layer are available for reservoir description—we select the maximum, namely, "9." Note, in Fig. 15.3B, that vertical menu entries beneath "Number of Layers" are "grayed" (or inactive). Only when the number of layers is defined is the next item, "Create Reservoir" in Fig. 15.3D-1 highlighted and rendered active. This menu strategy is used in our software design—it is impossible to click menus in the wrong sequence, and only those menus that must be completed (highlighted) are presented to the user.

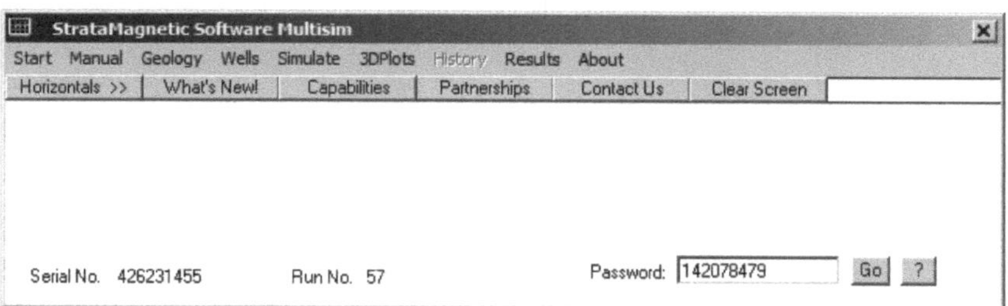

FIG. 15.3A Overview of Multisim™ menu screen.

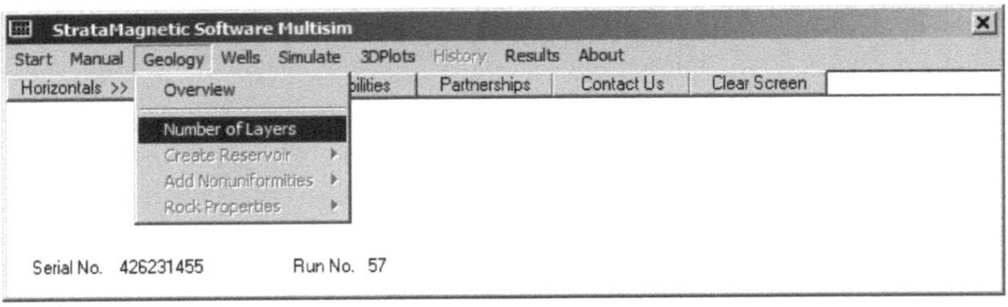

FIG. 15.3B “Number of Layers” selection.

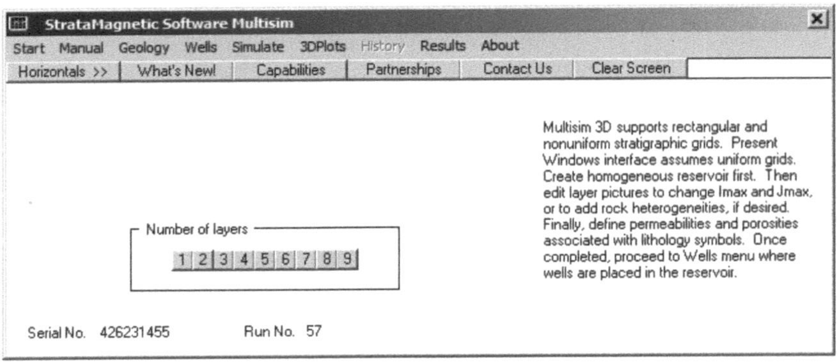

FIG. 15.3C Up to nine layers available for reservoir description.

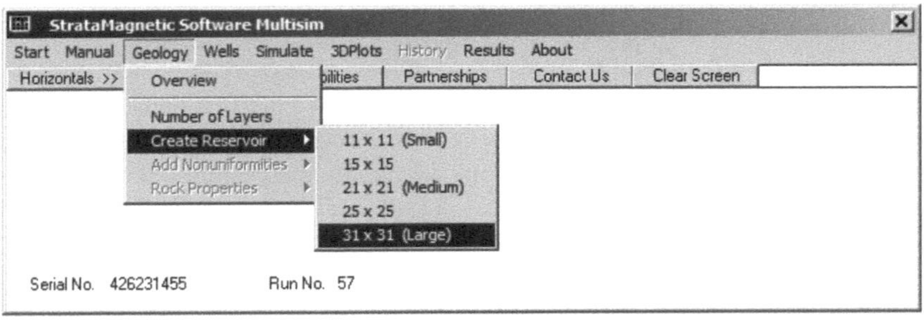

FIG. 15.3D-1 Reservoir areal grid block options.

Next, we turn to the “Create Reservoir” option and select the maximum allowable “31 × 31 (Large)” choice as shown in Fig. 15.3D-1. Fig. 15.3D-2 indicates that the overall grid structure has been defined, and on-screen notes explain that reservoir heterogeneities can be “drawn” grid by grid if desired. We have done this previously using keyboard symbols like #, $ and % in Examples 15.1–15.5. As we intend to illustrate multilateral well definition in

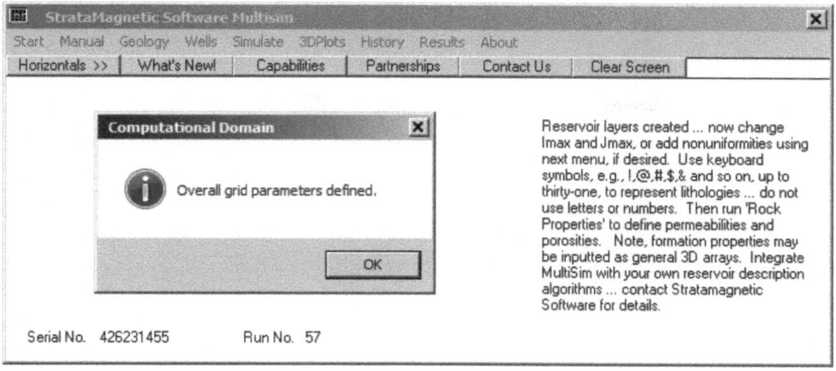

FIG. 15.3D-2 Overall grid structure defined.

the present example, we select "Uniform Medium" for now, as shown in Fig. 15.3E, which automatically creates all required LAYER*.GEO files transparently in the background with default "dots" or "periods." Fig. 15.3F confirms our selection for homogeneous media—all geology definition files LAYER1. GEO, LAYER2.GEO, … and so on, are simple "dots" or "periods" as explained and are not displayed here. The next highlighted menu for use is shown in Fig. 15.3G, indicating that rock properties are to be defined. Figs. 15.3H and 15.3I appear next and we accept their default values for the purposes of illustration.

Next we turn to the high-level horizontal "Wells" menu to define our well systems and individual structures. An "Insert Wells" selection appears, and

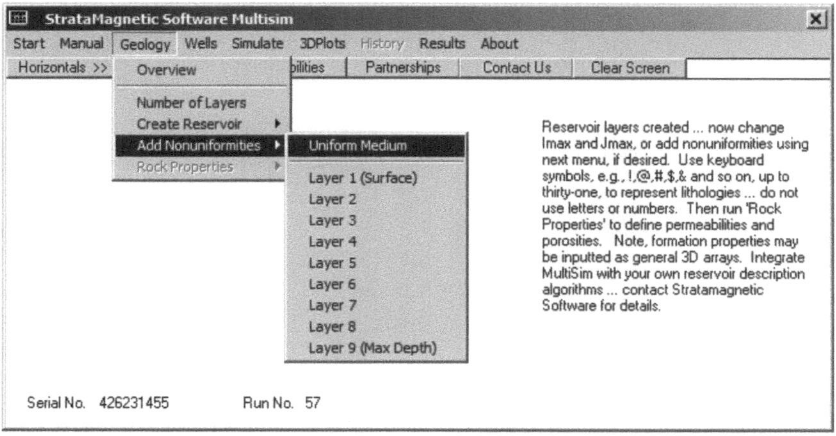

FIG. 15.3E Selecting "Uniform Medium" for now.

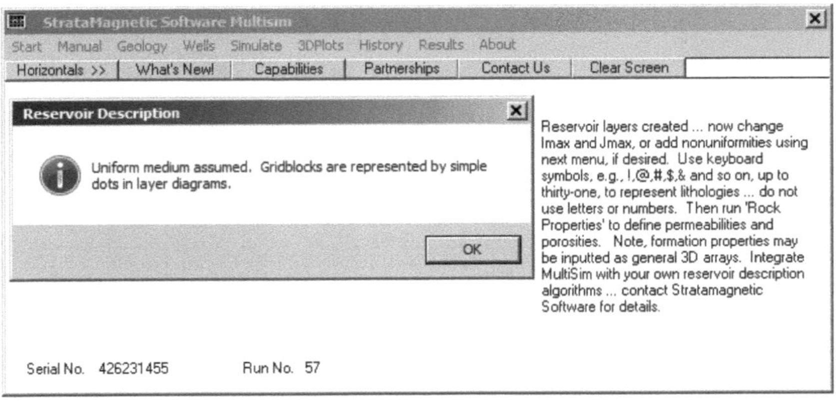

FIG. 15.3F Uniform medium selection confirmed.

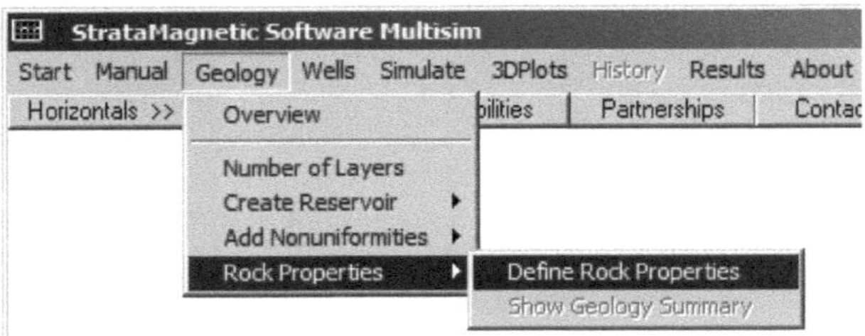

FIG. 15.3G Defining rock properties next.

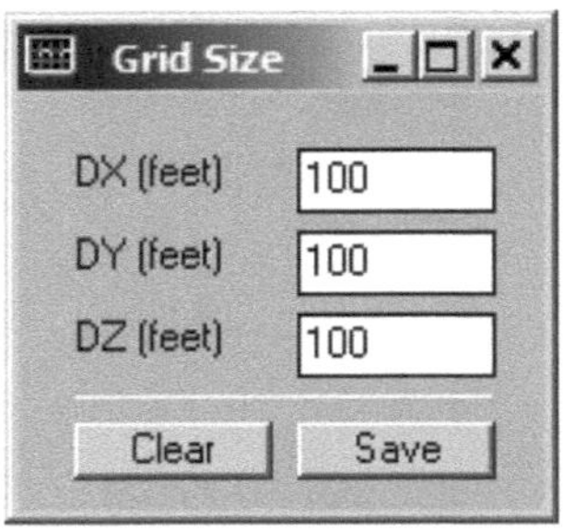

FIG. 15.3H Grid block size and shape.

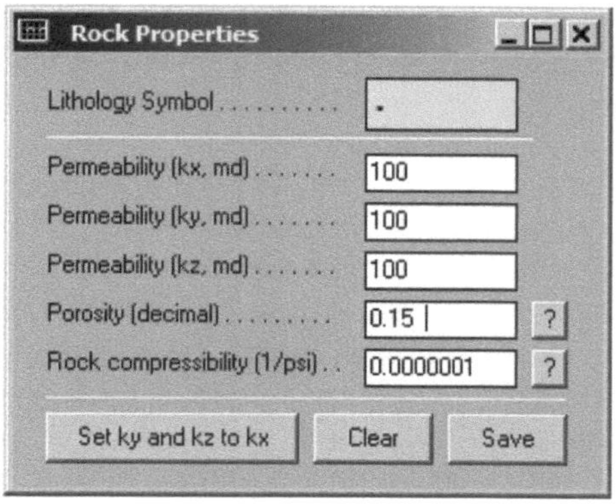

FIG. 15.3I Inputting rock properties.

since nine layers were indicated for this simulation, nine layer numbers are highlighted in Fig. 15.3J. Note that Layer 1 represents the "surface," that is, rock corresponding to the original formation drilled, while "max depth" represents the deepest portion of the reservoir. This usage is consistent with field application. For simplicity, we shall drill or "insert" only two wells, namely, Well 1 and Well 2, although up to nine well systems (the last denoted by "Well 9") could be used. Importantly, each "well" may be as simple as a rudimentary vertical well or as complicated as a three-dimensional multilateral with multiple out-of-plane appendage, each with different lengths.

Note that each well system operates under a single well constraint. For example, it is not possible to pressure constrain one leg while rate constraining another (if a nearby system requires a different constraint, simply assign it a

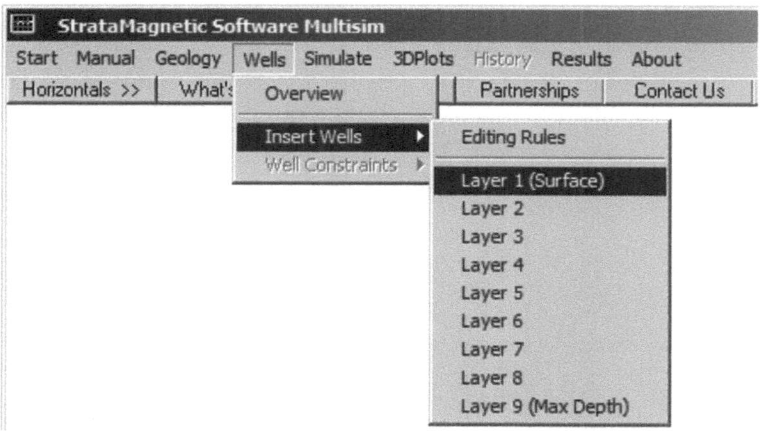

FIG. 15.3J Up to nine layers available for well placement.

different well number). In addition, the numerical value corresponding to the constraint used applies to all parts of the well system; we also note that friction and gravity are ignored in our formulation since flow rates are assumed to be sufficiently high.

We shall define Wells 1 and 2 for all nine layers, beginning with Layer 1 for convenience (actually, layers may be accessed in any order). Select "Layer 1 (Surface)." Internally, the geology for LAYER1.GEO has been copied into LAYER1.DRL. Note how "1" and "2" are inserted into LAYER1.DRL. Not inserting well numbers at all is also possible, but only physically meaningful for bottom layers that have not yet been drilled—these layers may represent hydrocarbon bearing zones that can be produced or as additional pressure sources when the bottom is under pressure drive. Now, save LAYER1.DRL. This well definition process is repeated for all nine layers in this case, with all LAYER*.DRL files to be saved by the user on completion (there are no software prompts to proceed to the next layer). Fig. 15.3K shows the placement assumed for the two wells with 31 × 31 "dots" in the background representing the assumed uniform medium. The well descriptions in the first four layers, that is, Layer 1 to Layer 4, are identical, and represent to vertical wells. These are shown in Figs. 15.3K–15.3N. Next, Fig. 15.3O (Layer 5) represents a major departure from standard reservoir simulation capabilities. We have arbitrarily "drawn" (using Windows Notepad, which is automatically "called" by our simulator) a very general multilateral system extending from Well 1, while Well 2 remains vertical. We could just as easily have introduced seven more multilateral well systems.

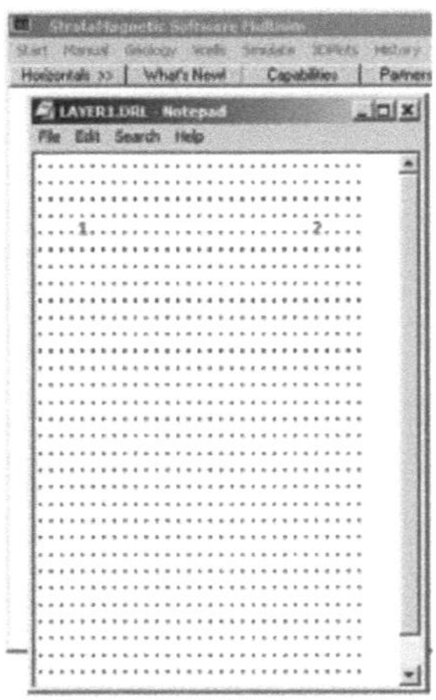

FIG. 15.3K Layer 1 well definition for Well 1 and Well 2.

Figs. 15.3P and 15.3Q for Layers 6 and 7 show that the well systems are assumed to be identical to those in Layer 1, that is, they are vertically drilled. In Fig. 15.3R, we have introduced an additional multilateral extension to Well 1

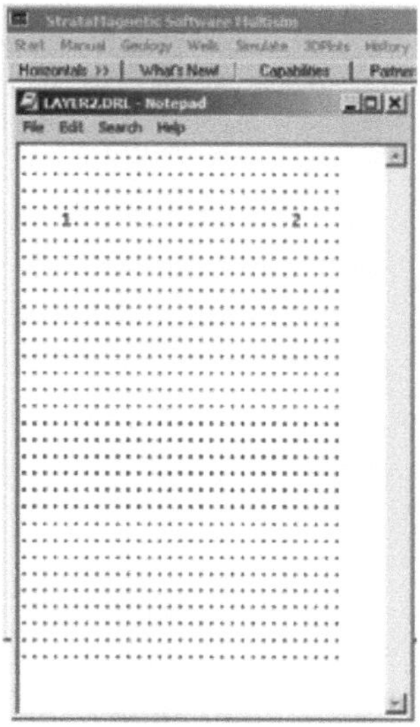

FIG. 15.3L Layer 2 well definition identical to Layer 1.

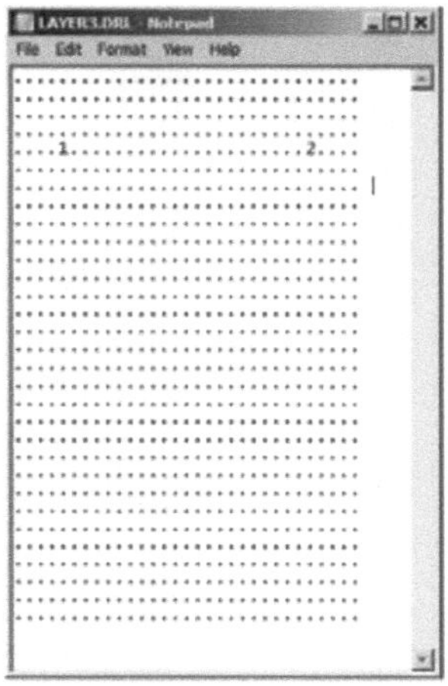

FIG. 15.3M Layer 3 well definition identical to Layer 1.

simply to illustrate the versatility of our simulator. Fig. 15.3S shows the last bottom layer in the simulation. As explained earlier, bottom layers need not be drilled—they may simply represent additional fluid bearing zones that may be present and produced, or they might host additional sources of reservoir pressure. In Fig. 15.3T, we continue drilling Well 1 vertically; however, there is no Well 2 because this well is taken as partially penetrating.

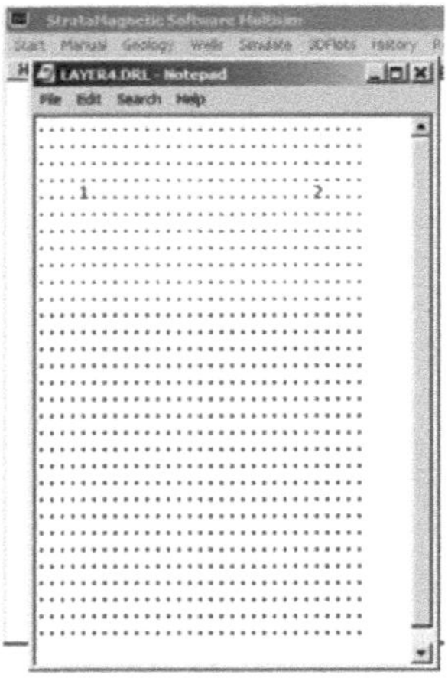

FIG. 15.3N Layer 4 well definition identical to Layer 1.

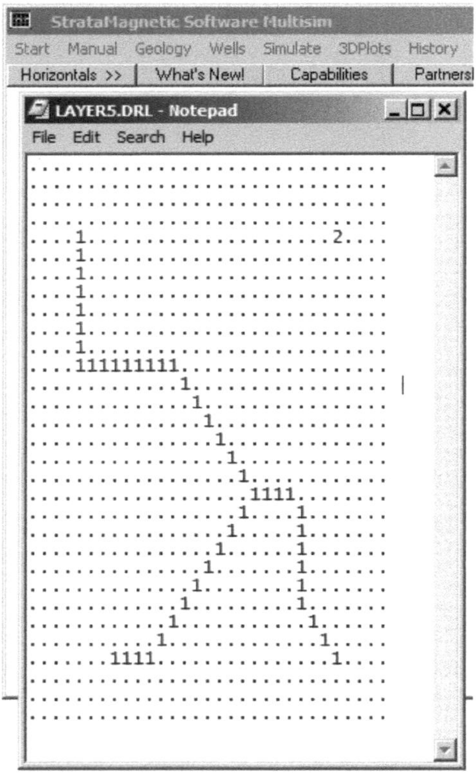

FIG. 15.3O Well 1 with very complicated multilateral structure—if desired, Well 2 may be equally complicated, and similarly for Wells 3–9 if they exist—also, wells may be slanted across multiple layers.

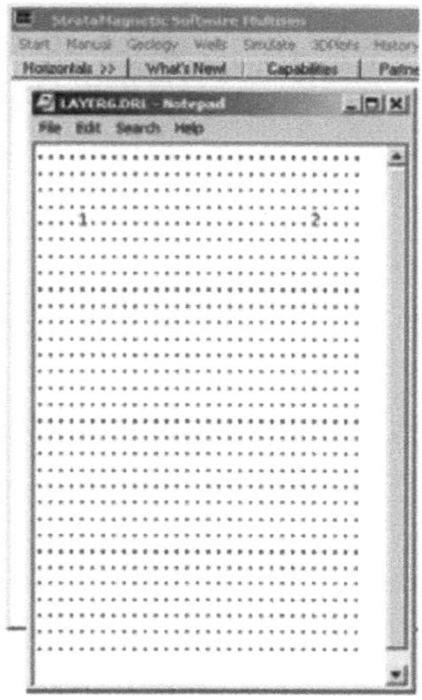

FIG. 15.3P Layer 6 well definition identical to Layer 1.

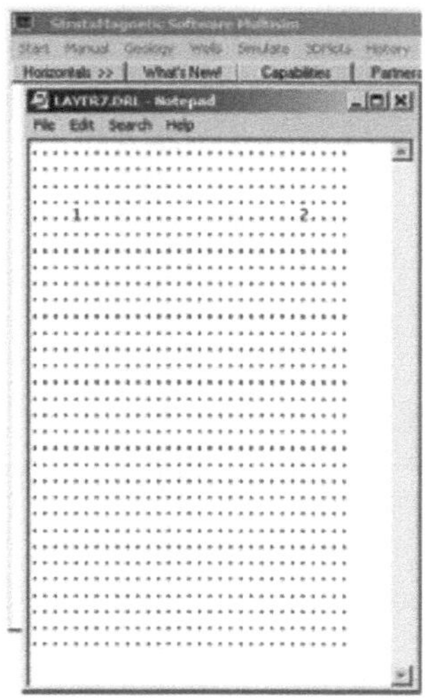

FIG. 15.3Q Layer 7 well definition identical to Layer 1.

Once wells have been inserted into different layers of the reservoir, we turn to "Well Constraints" in our "Wells" menu, and select "Production Mode" as shown in Fig. 15.3U. The screen in Fig. 15.3V shows that steady or transient modeling is permitted. We click "Yes" to perform steady-flow production

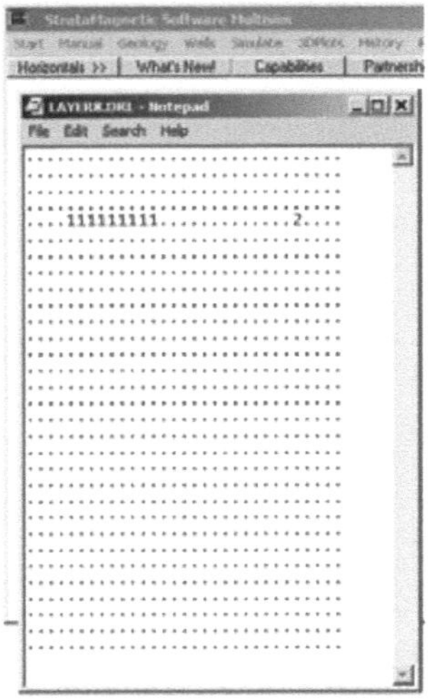

FIG. 15.3R Additional small lateral defined for Well 1.

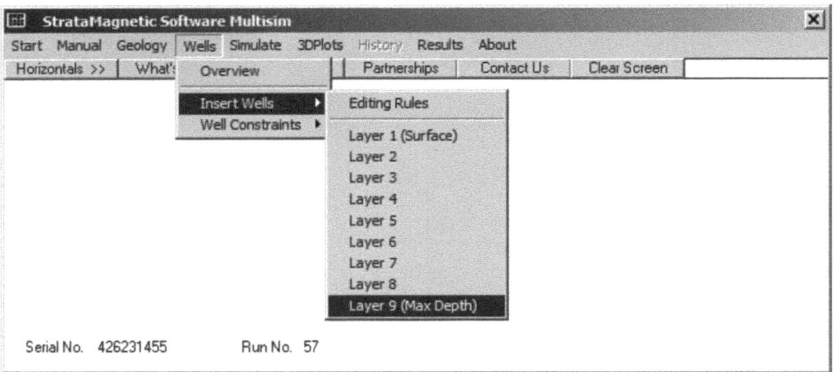

FIG. 15.3S Drilling into the final layer.

analysis for the present simulation, if only to avoid for now the lengthy menus that appear when performing transient analyses (again, our present purpose is illustrating multilateral well definition and plotted results).

The screen in Fig. 15.3W provides access to well constraint specification. Selecting the option shown leads to Fig. 15.3X which explains available options and requests flow rate units to be used (we select barrels per day, or "b/d"). In Fig. 15.3Y, we pressure constrain multilateral Well 1 at 1000 psi, and in Fig. 15.3Z-1, we also pressure constrain vertical Well 2, but at 5000 psi (from Chapter 3, Example 3.4 of Chin (2016a), we demonstrate how it is possible to pressure constrain one well while rate constraining another; in fact, when all

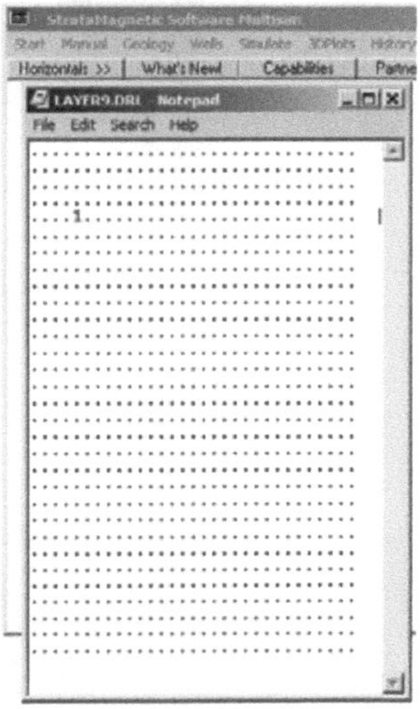

FIG. 15.3T Bottom layer (non-existent Well 2 is partially penetrating)

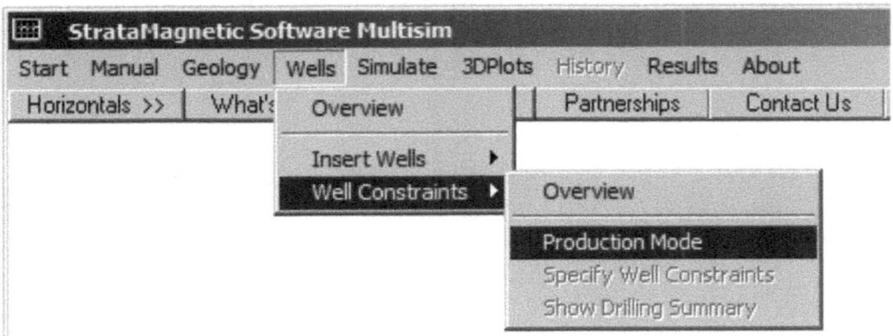

FIG. 15.3U Selecting production mode.

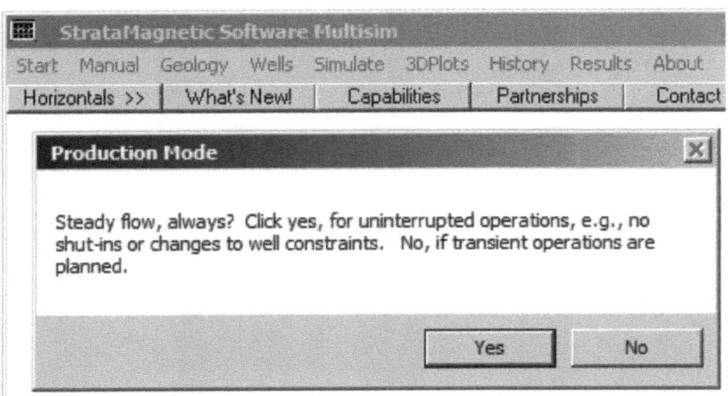

FIG. 15.3V Steady or transient operations permitted—click "Yes."

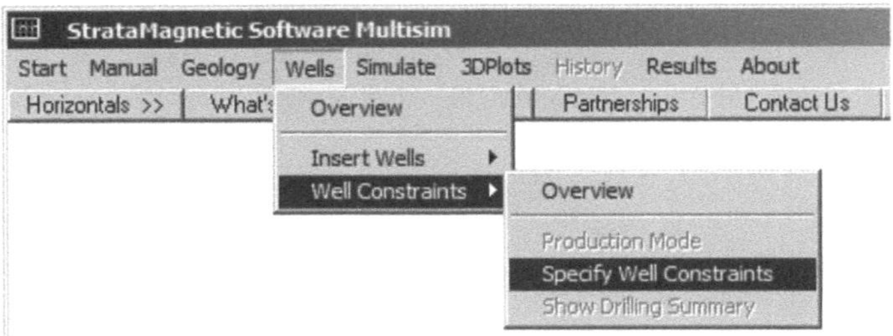

FIG. 15.3W Specifying well constraints.

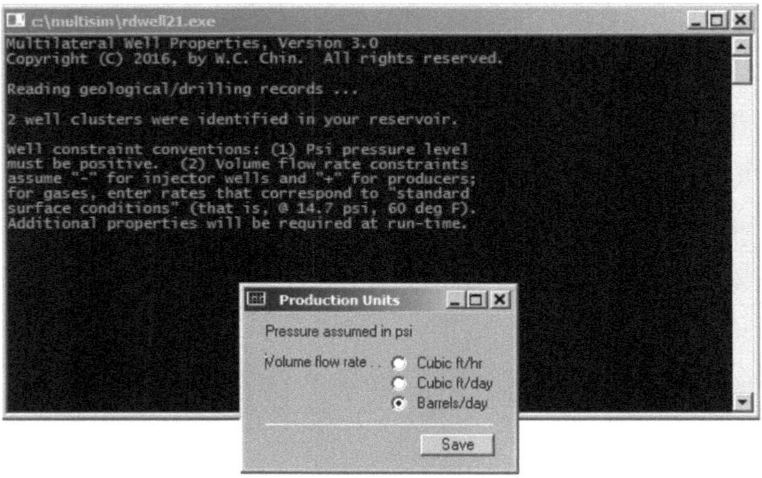

FIG. 15.3X Information screen with rate units selection menu.

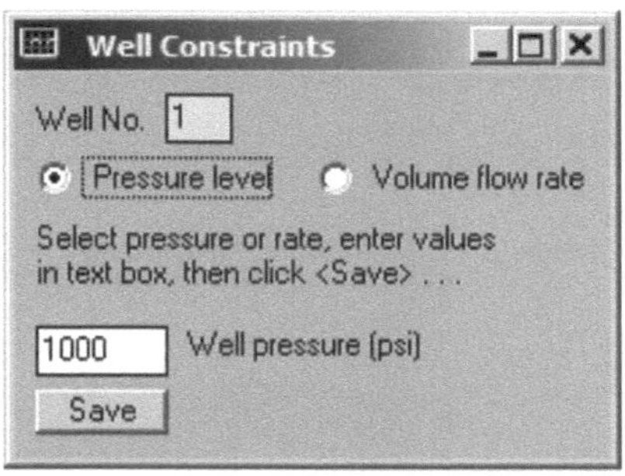

FIG. 15.3Y Pressure constraining (multilateral) Well 1 at 1000 psi.

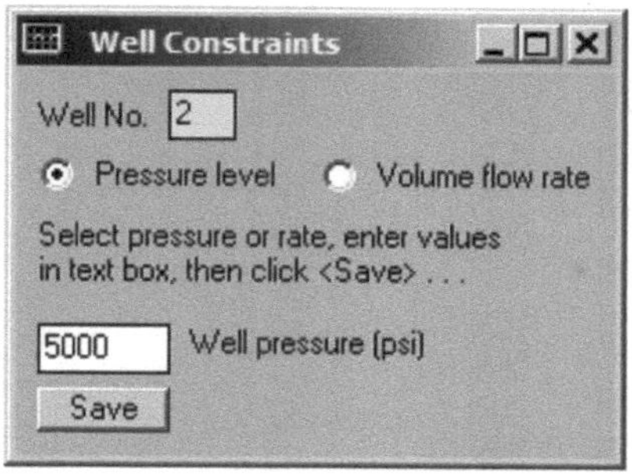

FIG. 15.3Z-1 Pressure constraining (vertical) Well 2 at 5000 psi.

nine well systems exist, pressure and rate constraints can be arbitrarily mixed). Our use of "1000" and "5000" in this simulation is simply motivated—we just want to see these numbers in our color plots where "10,000" will be the reservoir maximum.

Note that our simulator has, in the background, and transparently to the user, identified the existence of two-well systems, namely, Well 1 and Well 2. Two well constraint menus appear on-screen automatically as shown in Figs. 15.3Y and 15.3Z-1, and as explained previously, pressure constraints of 1000 and 5000 psi are assumed. Next we turn to the "Simulate" selection in the horizontal bar, and in particular, to farfield boundary condition definition as shown in Figs. 15.3Z-2 and 15.3Z-3. Note that we have assumed "no flow" conditions at the top and bottom of the reservoir, and "10,000 psi" at the four sides. Again, the numbers "1000," "5000" and "10,000" psi were taken for visualization purposes in our three-dimensional color plots.

Next, we consider the "Simulate" menu and the "Run Simulator" option to initiate iterations that solve the complete flow equations (Fig. 15.3Z-4). We emphasize that we have $31 \times 31 \times 9$ grid blocks, that is, 8649 or approximately 10,000 simultaneous equations. As we shall show, these are solved in about ten seconds on Intel Core i5 personal computers. Several intermediate screens, not shown, appear but are not discussed here for brevity—for this example, we do not modify transmissibilities stored, we choose "Liquid" fluid type and a viscosity of 1 cp, and we accept "Steady flow, for now." Computations will begin

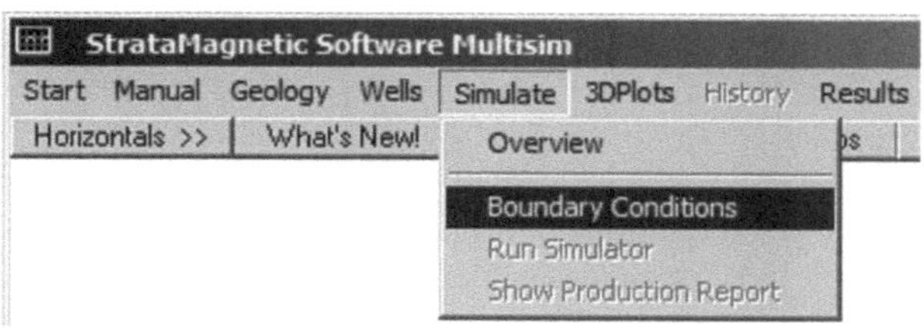

FIG. 15.3Z-2 Defining farfield boundary conditions.

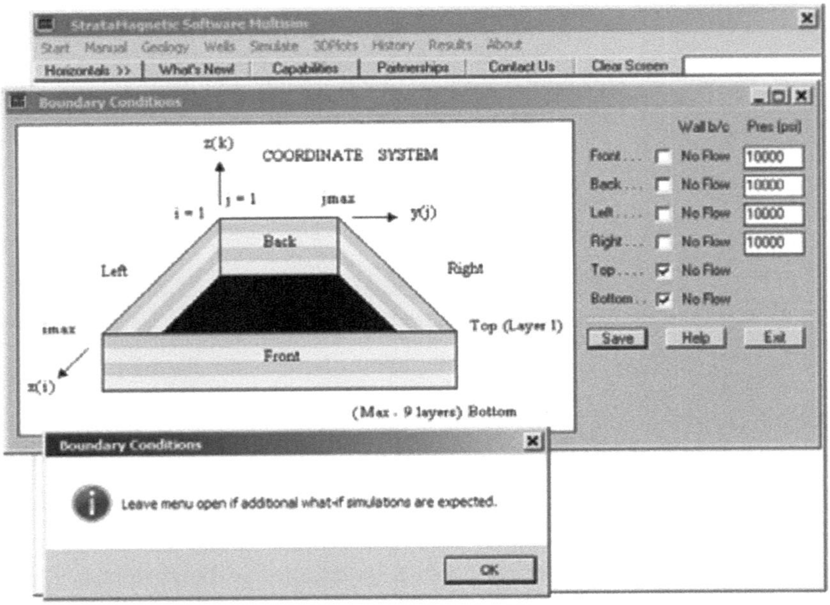

FIG. 15.3Z-3 Defining farfield boundary conditions.

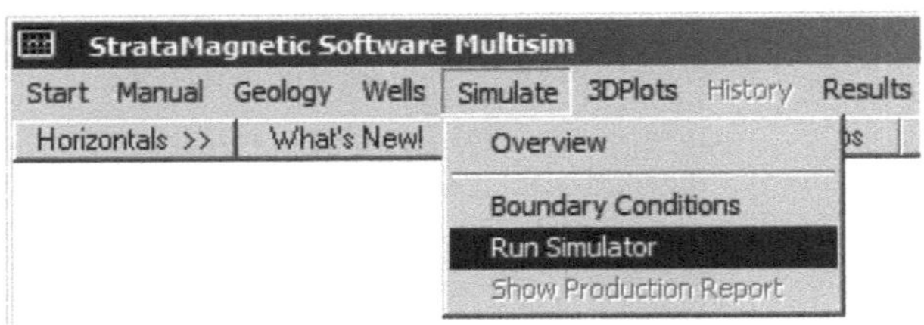

FIG. 15.3Z-4 Running the simulator.

automatically. At Iteration 100, the screen in Fig. 15.3Z-5 appears, indicating that a flow rate of "0.8169E+07" was obtained at (multilateral) Well 1, while a much smaller rate of "0.7979E+06" was obtained at (vertical) Well 2. The prompt shown asks if we wish to continue simulations, and we respond with "Yes," continuing to Iteration 500. Here every hundred iterations requires ten seconds of computing time on mid-range Intel Core i5 personal computers.

From Figs. 15.3Z-6 and 15.3Z-7, we observe that Iteration 500 flow rate results for Wells 1 and 2 are almost identical to those for Iteration 100, indicating full convergence. The label "Error" does not always tend to identically zero even if convergence is achieved; its definition depends on many properties and

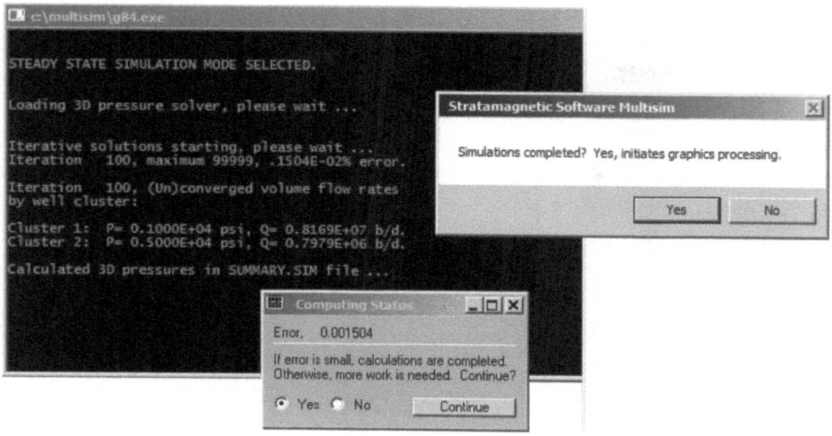

FIG. 15.3Z-5 Status screen at Iteration 100.

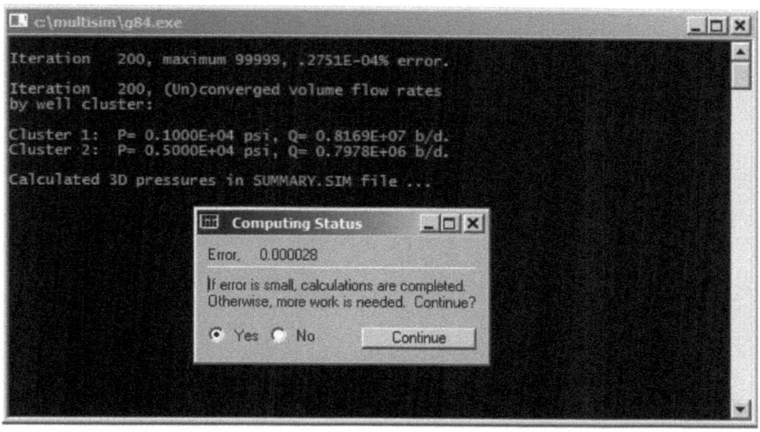

FIG. 15.3Z-6 Iteration 200 results.

can be affected by truncation error. Thus, its value should be viewed only qualitatively. We terminate calculations and select "No, Continue."

Finally, we have completed our three-dimensional steady flow calculations. We click "No" in Fig. 15.3Z-8 (because we do not intend to initiate transient operations) and "Yes" in Fig. 15.3Z-9 (to start color graphics processing). In Fig. 15.3Z-10, the "3DPlots" menu is shown, where we first select the stratum or layer to be considered. We shall plot pressures for each of the nine layers studied for this example. For example, we can select "Layer 1 (Surface)" first, although any display order is permissible—we shall, however, display results in numerical order from surface to bottom to illustrate how the strong pressures in

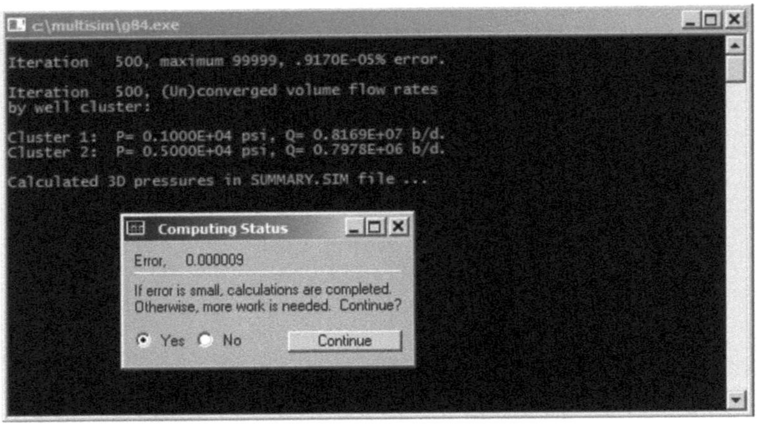

FIG. 15.3Z-7 Iteration 500 results.

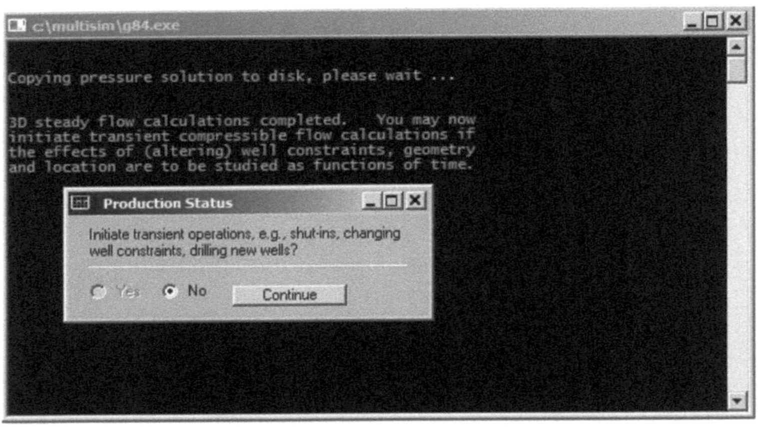

FIG. 15.3Z-8 Option to initiate transient operations.

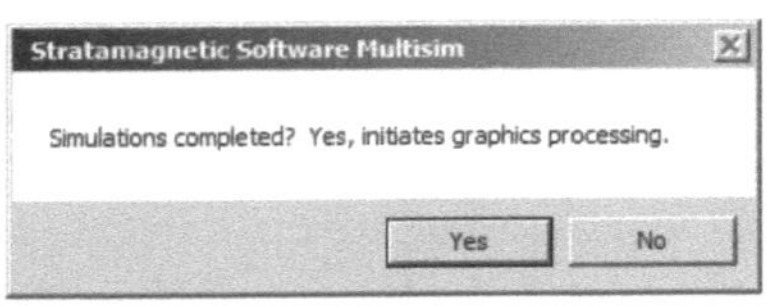

FIG. 15.3Z-9 Post-processing for graphics color displays.

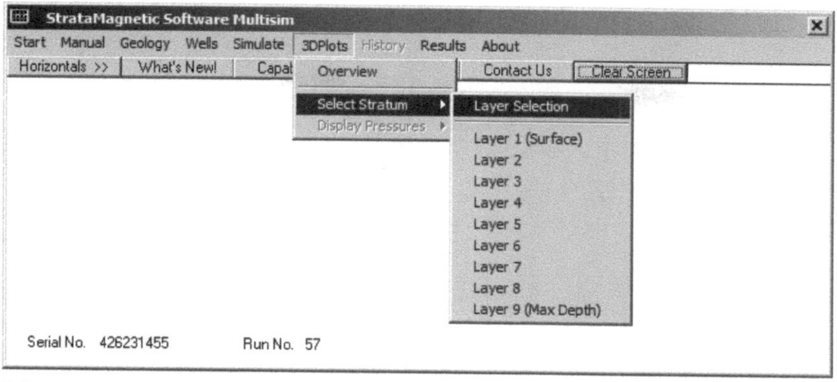

FIG. 15.3Z-10 Selecting layer for color pressure display.

Layer 5 diffuse vertically in Darcy flows. Remember, each layer must be "selected" by mouse first before the plot type can be chosen.

Static contour plots for pressure are shown for each layer in Figs. 15.3Z-11–15.3Z-23. It is instructive to start with middle Layer 5, where through Fig. 15.3O we had introduced a very complicated multilateral structure for Well 1. Color pressure plots for pressure in Layer 5 are found in

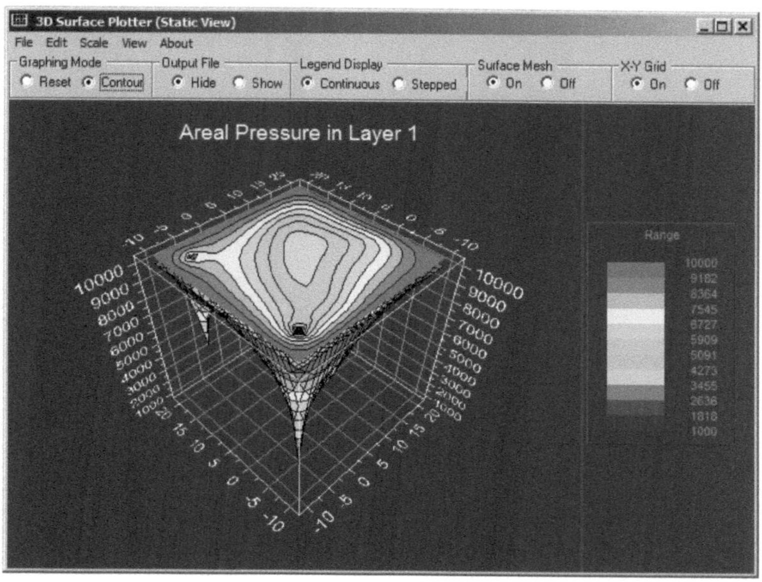

FIG. 15.3Z-11 Layer 1 pressures.

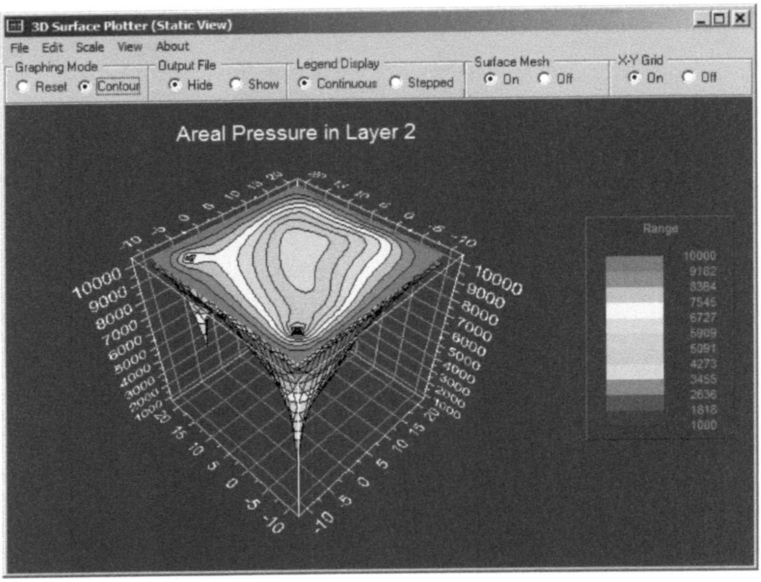

FIG. 15.3Z-12 Layer 2 pressures.

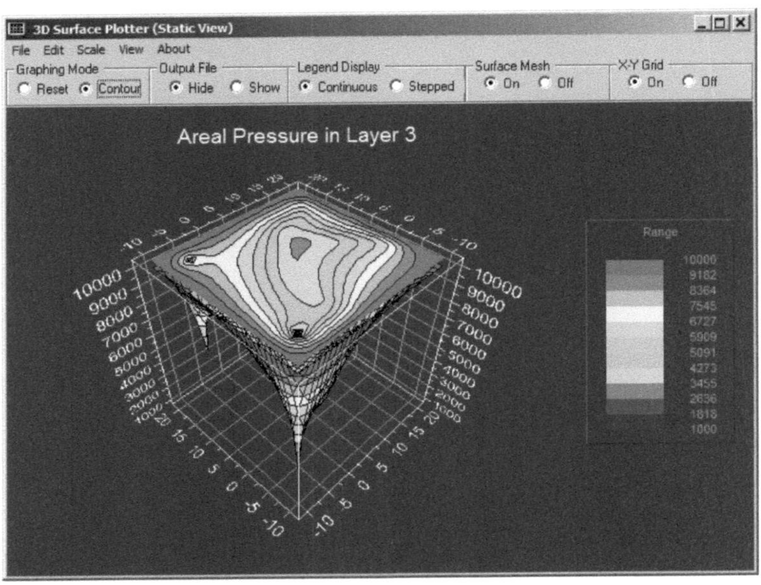

FIG. 15.3Z-13 Layer 3 pressures.

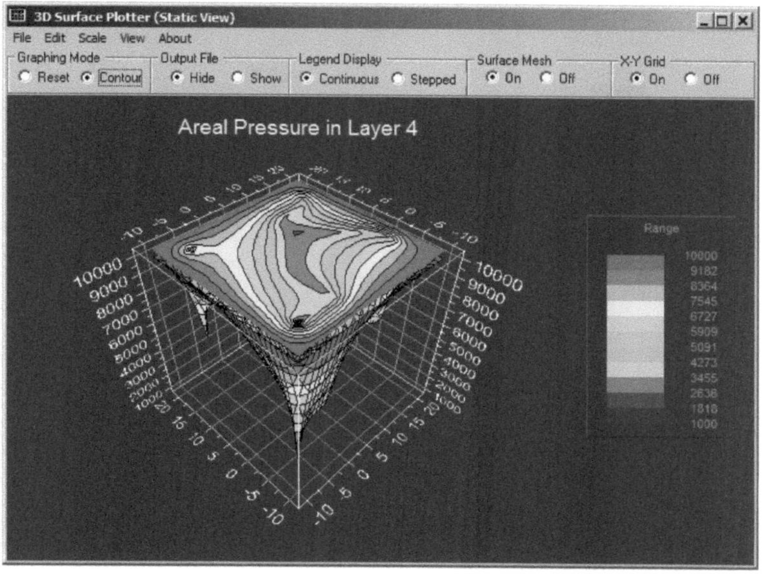

FIG. 15.3Z-14 Layer 4 pressures.

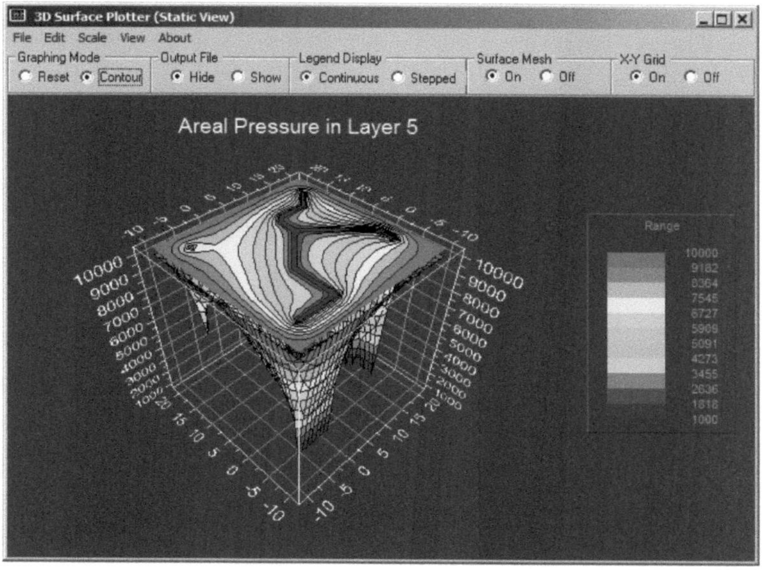

FIG. 15.3Z-15 Layer 5 pressures.

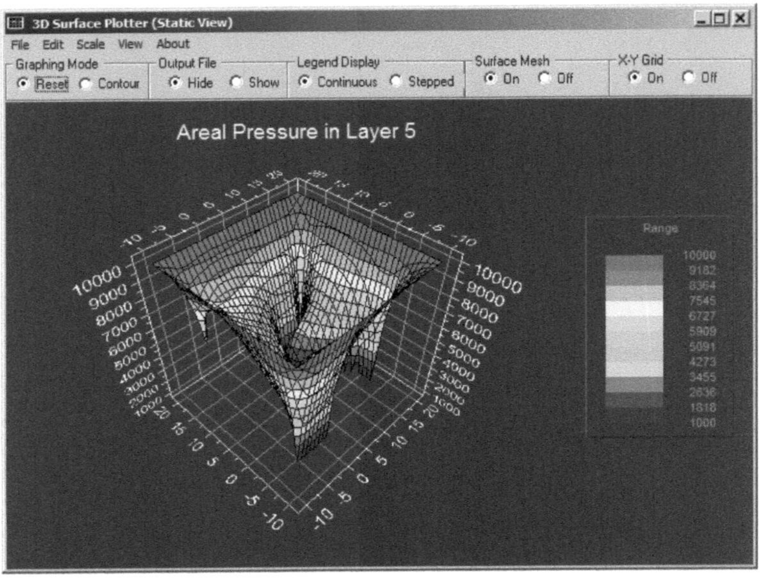

FIG. 15.3Z-16 Layer 5 pressures.

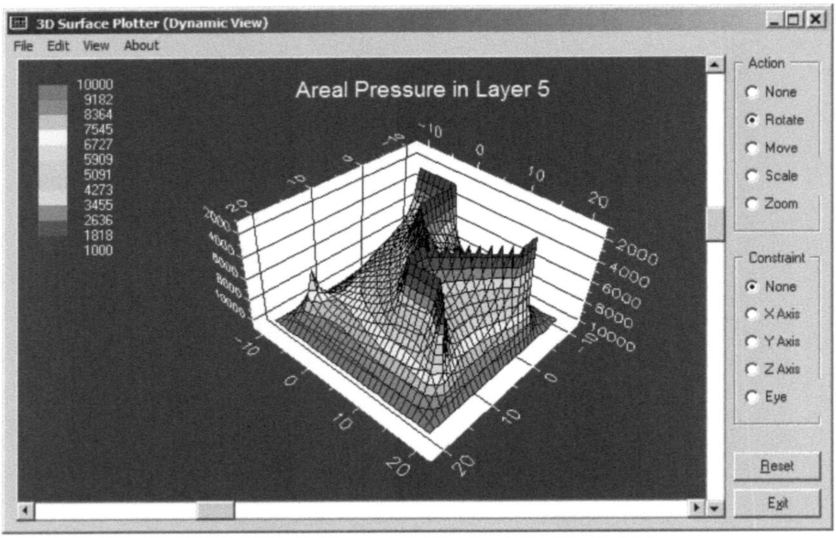

FIG. 15.3Z-17 Layer 5 pressures.

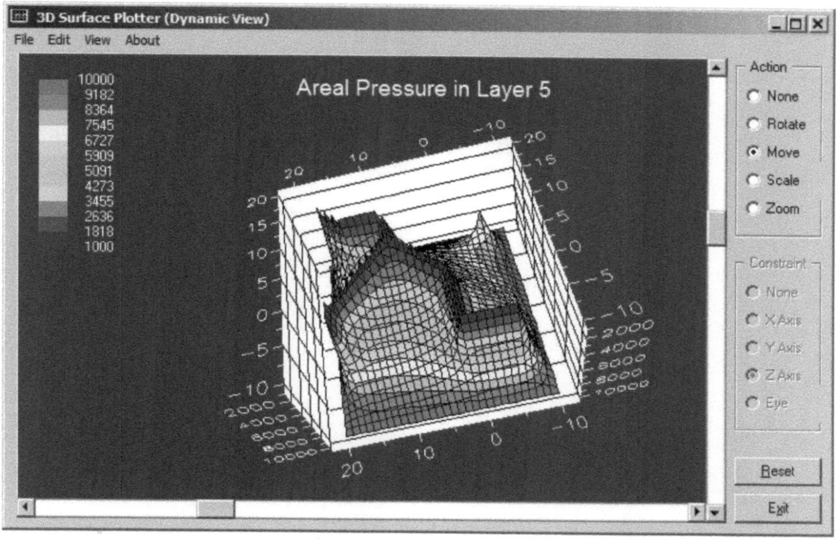

FIG. 15.3Z-18 Layer 5 pressures.

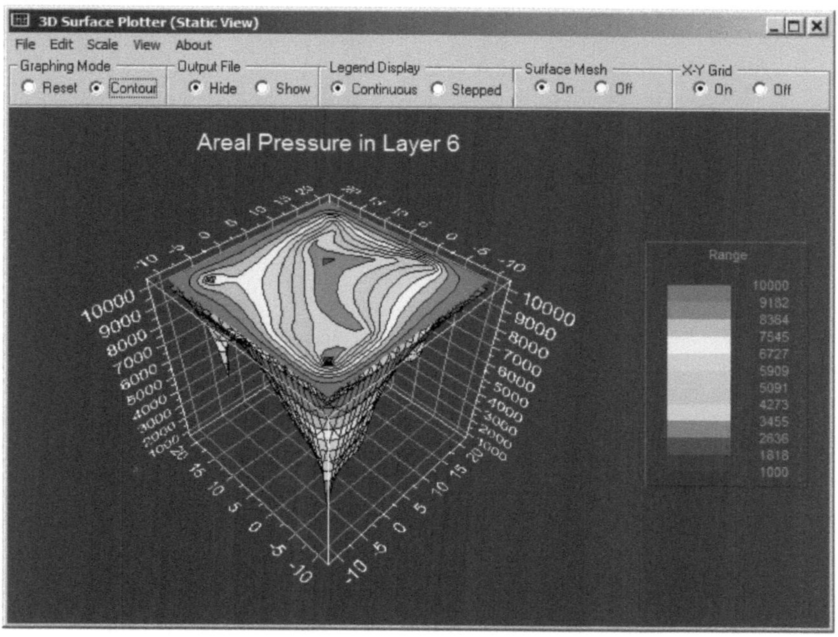

FIG. 15.3Z-19 Layer 6 pressures.

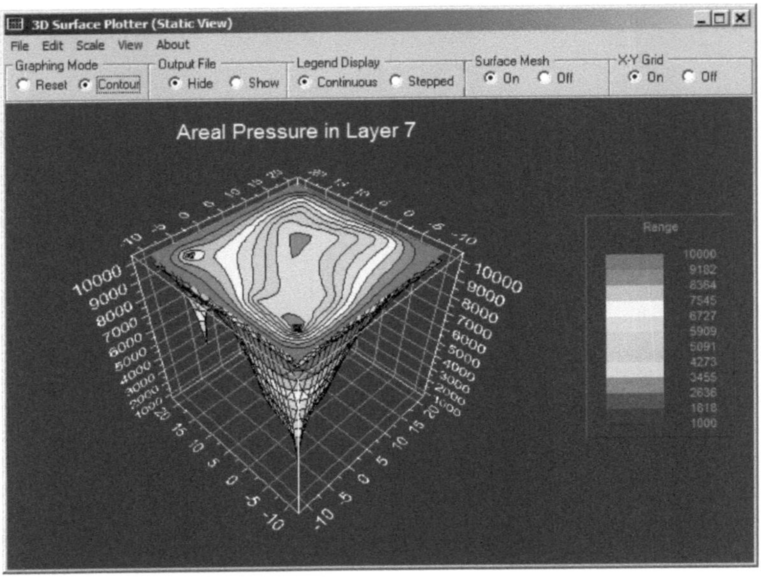

FIG. 15.3Z-20 Layer 7 pressures.

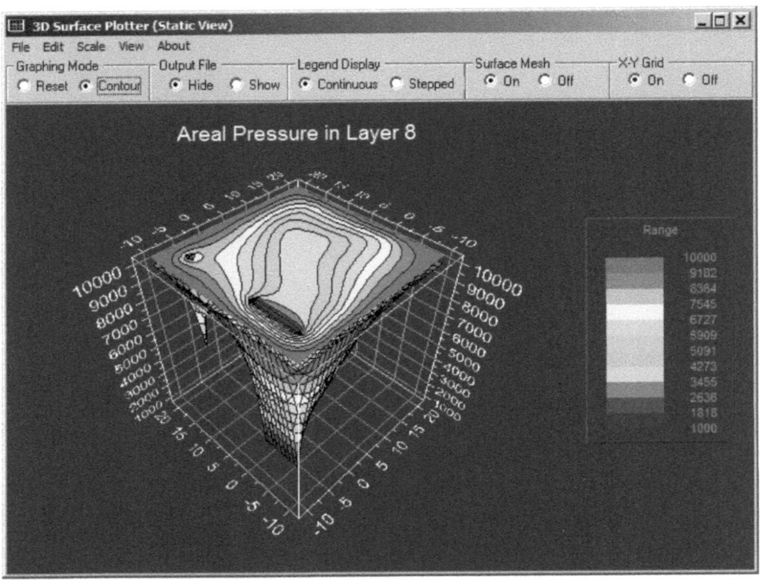

FIG. 15.3Z-21 Layer 8 pressures.

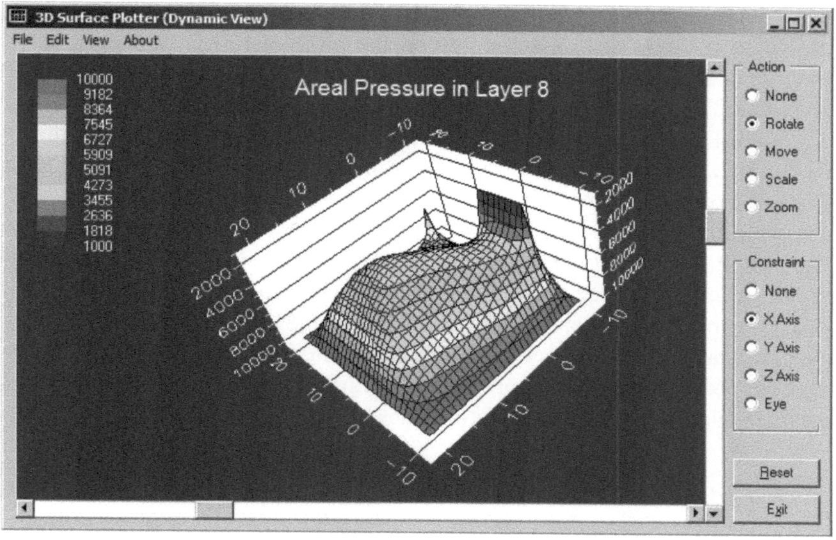

FIG. 15.3Z-22 Layer 8 pressures.

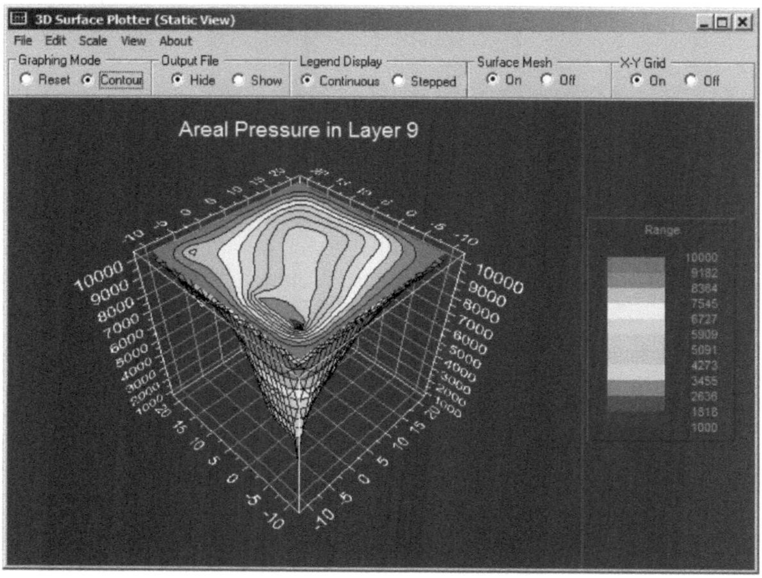

FIG. 15.3Z-23 Layer 9 pressures.

Figs. 15.3Z-15–15.3Z-18 and show the sharp blue pressure streak (indicating the low value of "1000 psi" assumed along the multilateral). This solid pattern becomes milder as we progress away from Layer 5 in both upward and downward directions. This smoothing effect arises from "pressure diffusion," not unexpectedly, since Darcy pressure satisfies the heat or diffusion equation. Recall that in Fig. 15.3R we had introduced a short horizontal lateral in Well 1 for Layer 8. Its effect is also clearly seen in Fig. 15.3Z-21 as a short blue line, with a nearby "dot" indicating the end pressure effects for partially penetrating Well 2. Finally, Fig. 15.3Z-24 shows the menu sequence "Results" > "Text Summaries" > "Reservoir Production" allowing access to geology, drilling and reservoir production reports. Portions of the file SUMMARY.SIM are produced—detailed numerical tabulations for areal pressure are omitted for brevity, although these could be ported to standard worksheets for further analysis. At the end of this summary file, we have the production rates

```
Cluster 1:  P= 0.1000E+04 psi, Q= 0.8169E+07 b/d.
Cluster 2:  P= 0.5000E+04 psi, Q= 0.7978E+06 b/d.
```

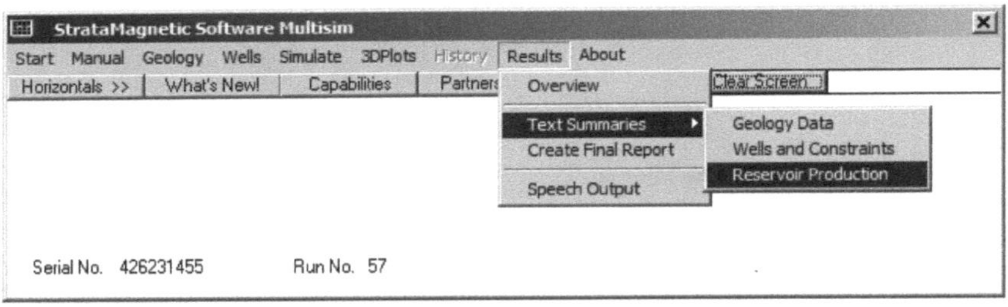

FIG. 15.3Z-24 Accessing computer generated reports.

Well 1 produces 8.169/0.7978 or about ten times as much flow as Well 2. This could be the result of three reasons, namely, well location in the reservoir, differences in pressure constraint levels, or the number of grid blocks used to create the wells—63 versus 8, as highlighted in SUMMARY.SIM, numbers that are proportional to drilling cost. The petroleum engineer could study the economic benefits, for example, of changes in pressure or well topology, with the present reservoir flow simulator.

```
SUMMARY.SIM
SIMULATION SUMMARIES, PHYSICAL FLUID MODELING, FAR FIELD
AND WELL BORE RUN TIME BOUNDARY CONDITIONS (SUMMARY.SIM):

Simulation results based on rectangular mesh system.

PRESENT RESERVOIR STATUS:

Reservoir  grid parameters:
Imax   = 31, Dx = .1000E+03 ft, Imax  *Dx = .3100E+04 ft.
Jmax   = 31, Dy = .1000E+03 ft, Jmax  *Dy = .3100E+04 ft.
Layers =  9, Dz = .1000E+03 ft, Layers*Dz = .9000E+03 ft.

Number of initial well clusters identified: 2

Combining geological/drilling information, please wait ...

Well block transmissibility (ft^3) summary:

Well 1, defined by  63 grid blocks, is
pressure constrained at 0.1000E+04 psi.
.
.
Well 2, defined by   8 grid blocks, is
pressure constrained at 0.5000E+04 psi.
.
.
INPUT FARFIELD BOUNDARY CONDITION SUMMARY:

                          Z(k)
                          |
                          | j=1        Jmax(31)
                      i=1 |
                          +--------------+  ------  Y(j)
                         / 3ooooooooooooooo3 \
            Left     /     3ooo  Back  ooo3    \   Right
                   /       3ooooooooooooooo3     \
                 /        +--------------+         \
   Imax(31)   +-------/  Top, Layer 1  \-------+
             /  3ooooooooooooooooooooooooooooooooo3
           /    3oooooooooooo  Front  oooooooooooo3
      / X(i)  3ooooooooooooooooooooooooooooooooo3
              +-------  Bottom, Layer 9  ------+

                       COORDINATE SYSTEM
```

```
o  FRONT assumed to be aquifer at pressure 10000.00 psi
o  BACK  assumed to be aquifer at pressure 10000.00 psi
o  LEFT  assumed to be aquifer at pressure 10000.00 psi
o  RIGHT assumed to be aquifer at pressure 10000.00 psi
o  TOP   assumed  to be "no flow wall"
o  BOTTOM assumed to be "no flow wall"

PHYSICAL FLUID MODEL SUMMARY:

o Fluid viscosity:   0.100E+01 centipoise

Fluid, assumed to be a liquid, satisfies linear partial
differential equation if pressure compaction  option is
not selected later.

STEADY STATE SIMULATION MODE SELECTED.

Pressure field initialized to internally
generated approximate solution.

Note:  "Jmax > 7" values cause "word wrap-around" in
screen displays and written files.  Use "File Reader"
utility to "unwrap" printouts for convenient viewing.

Iteration   100, (Un)converged volume flow rates
by well cluster:

Cluster 1:  P= 0.1000E+04 psi, Q= 0.8169E+07 b/d.
Cluster 2:  P= 0.5000E+04 psi, Q= 0.7979E+06 b/d.
.
.
.
Iteration   500, (Un)converged volume flow rates
by well cluster:

Cluster 1:  P= 0.1000E+04 psi, Q= 0.8169E+07 b/d.
Cluster 2:  P= 0.5000E+04 psi, Q= 0.7978E+06 b/d.
```

Again, smooth and fast convergence at 100 iterations (or ten seconds for a 10,000 grid block simulation containing complicated multilateral Well 1 and a partially penetrating vertical Well 2) was achieved effortlessly. The "bottlenecks" in minimizing total "desk time" are found in constructing LAYER*.DRL files like those in Figs. 15.3K–15.3T, which require some visualization experience. Denser 31 × 31 grid structures should be used to provide finer geometric description when needed. Ideally, our well definition interface would improve with "point and click" or "screen draw" capabilities, options which are presently under consideration for software upgrades.

EXAMPLE 15.7. DUAL LATERAL WITH TRANSIENT OPERATIONS

In this simulation, we demonstrate how transient operations are performed for a dual lateral whose arms are set apart by an irregular angle. For our "Geology" menu, we assume a three-layer, 15 × 15 mesh for simplicity, a fully uniform medium (described by "dots" or "periods" in the respective *.GEO files), plus default grid and rock properties as used in previous examples. In our "Wells" menu, shown in Fig. 15.4A, we shall drill a vertical well with dual laterals extending from it in the middle layer, as shown in Figs. 15.4B–15.4D. Our production mode is defined in Figs. 15.4E and 15.4F. In the latter, we click "No" since transient operations are planned. A transient flow modeling menu, asking for the number of wells to be considered in our simulation problem, not shown here, appears. While only a single well system, namely Well 1, is evident from Figs. 15.4B–15.4D, recall that our simulator allows additional wells to be drilled or created while computing. Since we do not plan to do this, we enter "1"

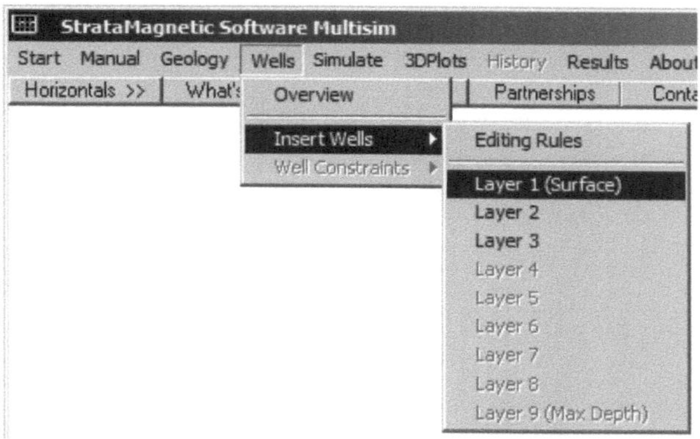

FIG. 15.4A Three-layer reservoir simulation.

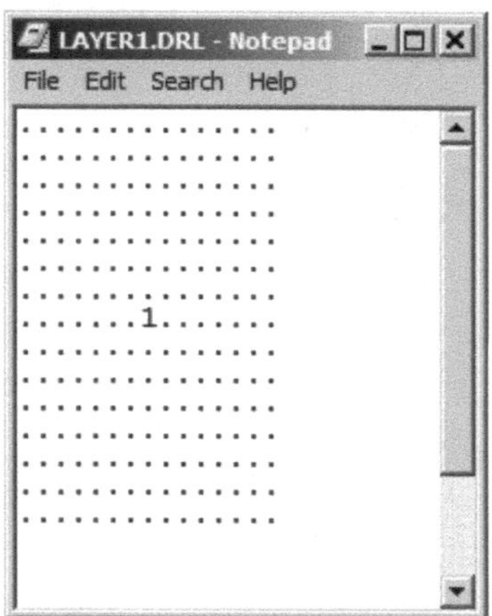

FIG. 15.4B Top layer well structure.

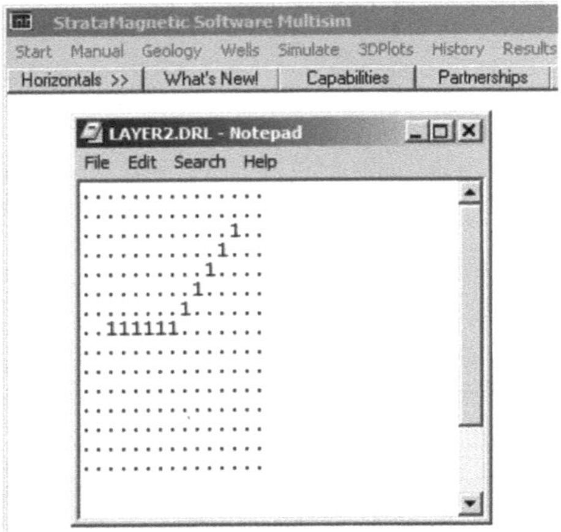

FIG. 15.4C Middle layer dual lateral well structure.

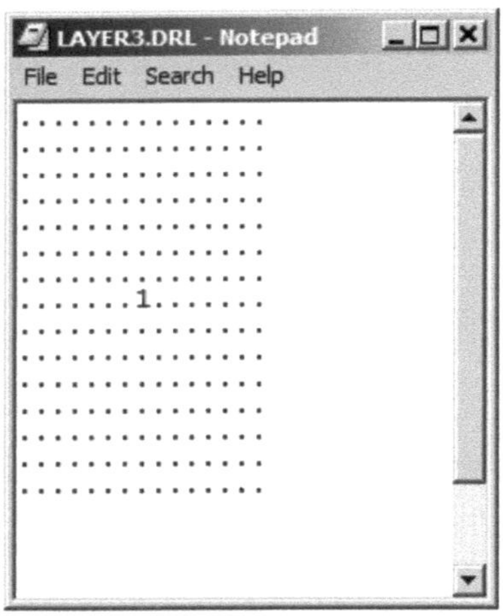

FIG. 15.4D Bottom layer well structure.

for the requested number of wells—again, this menu is not shown here. Well constraints are defined in Figs. 15.4G–15.4I. A pressure constraint of 5000 psi is assumed in Fig. 15.4I.

We turn to the "Simulate" menu and define farfield conditions at the six sides of the computational reservoir box using the "Boundary Conditions" menu in Fig. 15.4J. Numerical integrations are initiated in Fig. 15.4K. From two menus not shown, we do not change transmissibilities during drilling in the first,

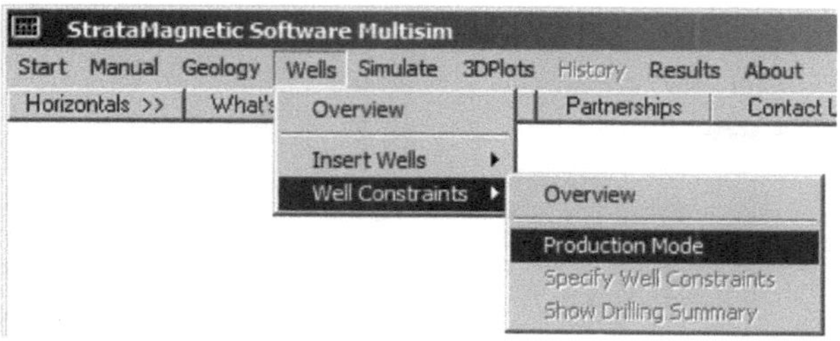

FIG. 15.4E Production mode definition.

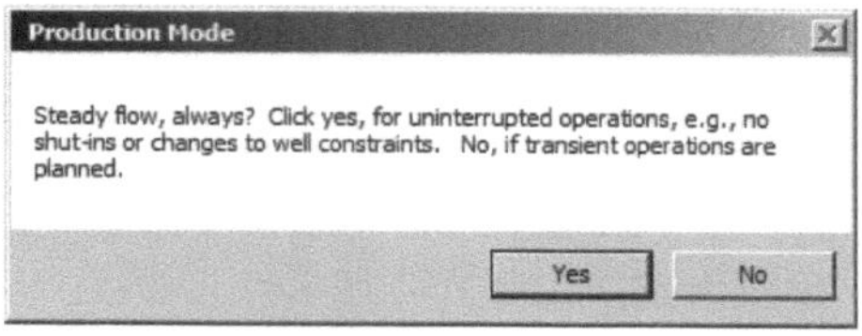

FIG. 15.4F Production mode definition (click, "No.").

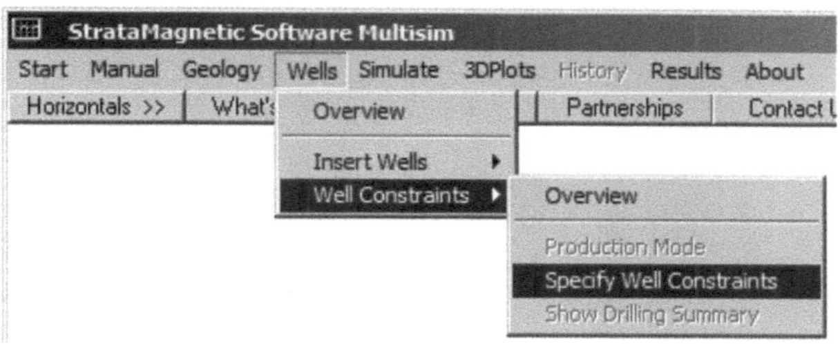

FIG. 15.4G Specifying well constraints.

while in the second, we select a liquid as the fluid type, having a viscosity of 1 cp. In a "Production Mode" menu, we select "Steady flow, for now."

Fig. 15.4L shows how steady flow calculations are in progress. In that screen, the volume flow rate at Iteration 100 is 0.1026E+07 b/d, which is very close to the result below at Iteration 200. This indicates that the iterations needed to determine the equilibrium pressure field have converged and that calculations may be safely terminated. We select "No, Continue" at this point.

```
Iteration   200, maximum 99999, .0000E+00% error.
Iteration   200, (Un)converged volume flow rate by well cluster:

Cluster 1:  P= 0.5000E+04 psi, Q= 0.1025E+07 b/d.

Calculated 3D pressures in SUMMARY.SIM file ...
.
.
.
```

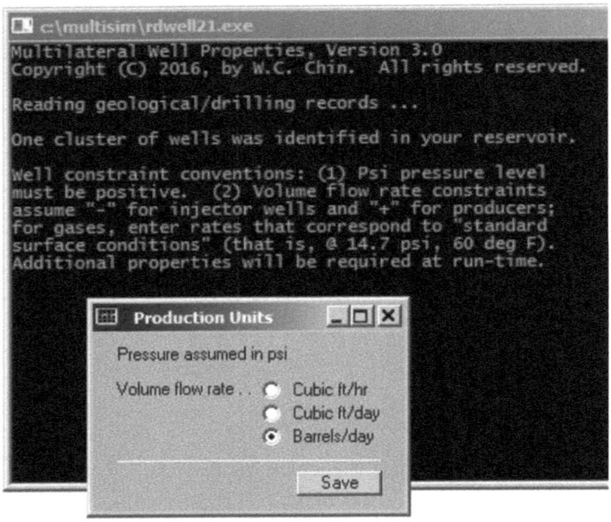

FIG. 15.4H Selecting volume flow rate units.

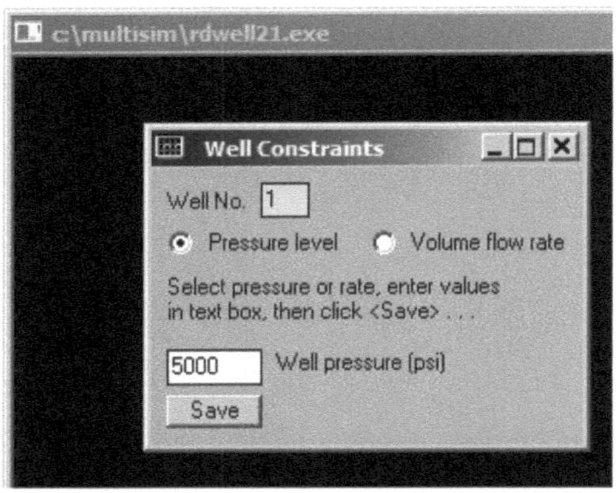

FIG. 15.4I Setting pressure constraint level.

Next, Fig. 15.4M allows us to continue with transient operations and we select this option with "Yes." From two menus not shown, we select the default oil compressibility and also do not porosity average liquid and rock compressibilities. In Fig. 15.4N for in time step recommendations, click for the suggestion, but override it with 0.1 h.

In a compaction menu not shown, we select "Rigid formation" versus "Small deformation." Time status appears in Fig. 15.4O, indicating that at the initial time t = 0, Well 1 is pressure constrained at 5000 psi. Now, Fig. 15.4M states that we have completed steady flow calculations, and from

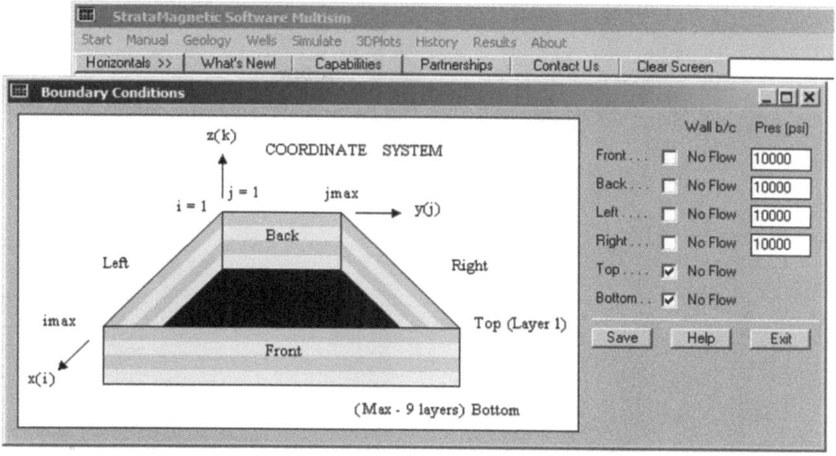

FIG. 15.4J Defining farfield boundary conditions.

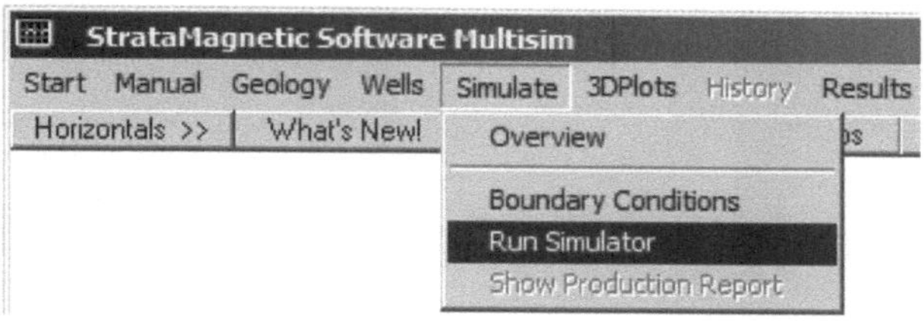

FIG. 15.4K Running the simulation.

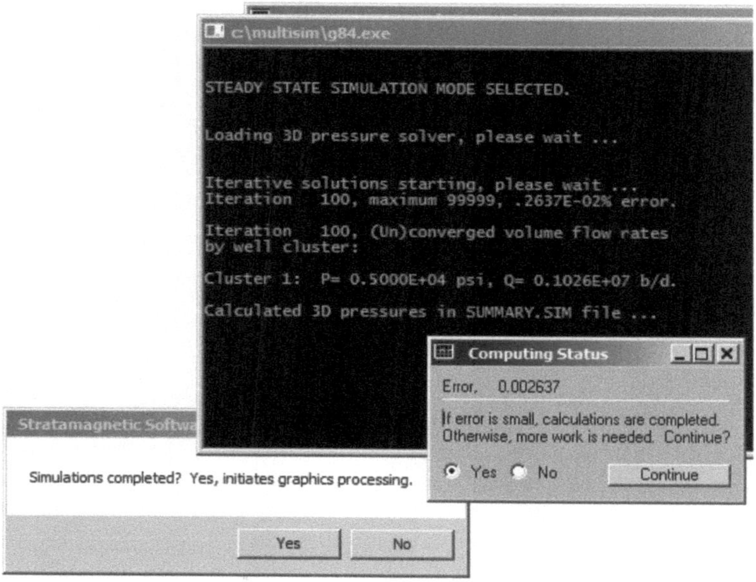

FIG. 15.4L Steady flow solutions in progress.

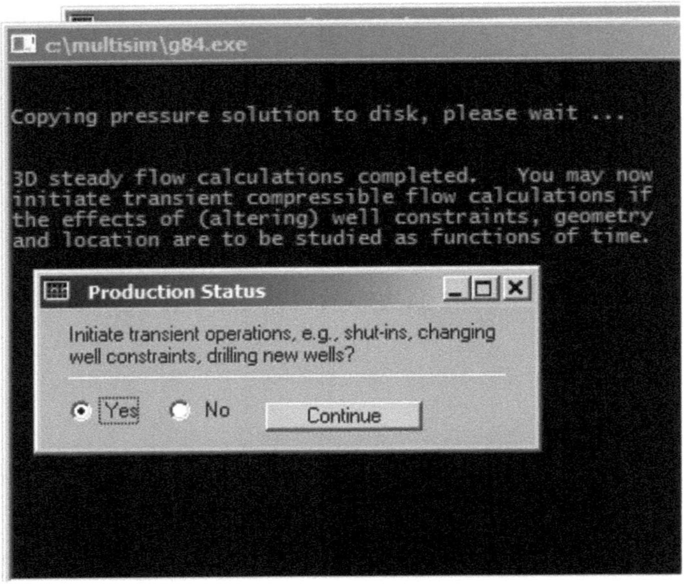

FIG. 15.4M Initiating transient operations.

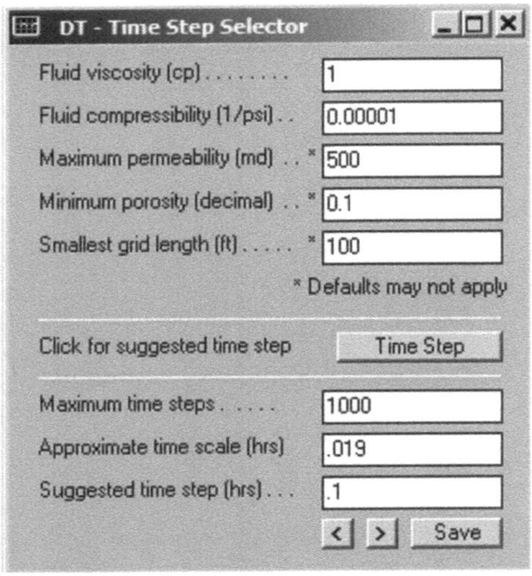

FIG. 15.4N Time step selection.

Fig. 15.4O, we click "Yes" to change constraints as indicated in Figs. 15.4P and 15.4Q. The screen sequence leading up to and including Fig. 15.4R states that we have converted the pressure constraint to one for flow rate held at a positive (production) rate of 5000 b/d. A menu asking if additional wells are to be

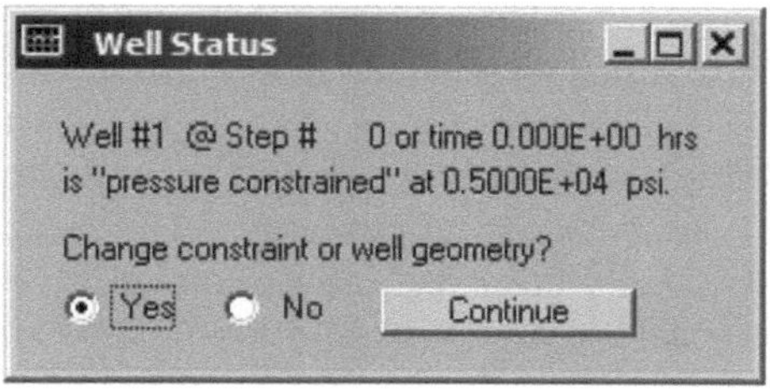

FIG. 15.4O Well status change.

FIG. 15.4P More well action options, click "No."

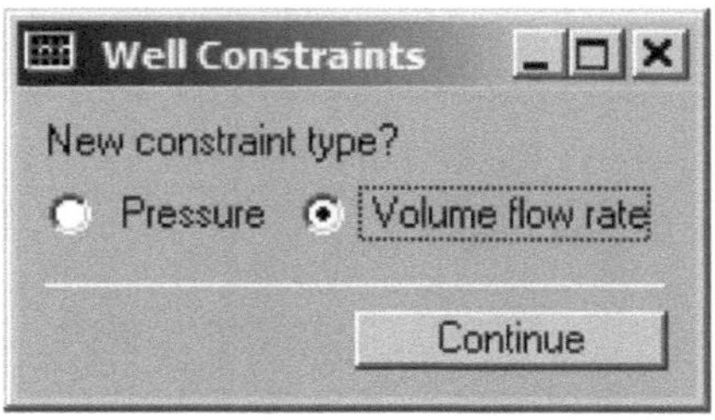

FIG. 15.4Q Volume flow rate constraint selected.

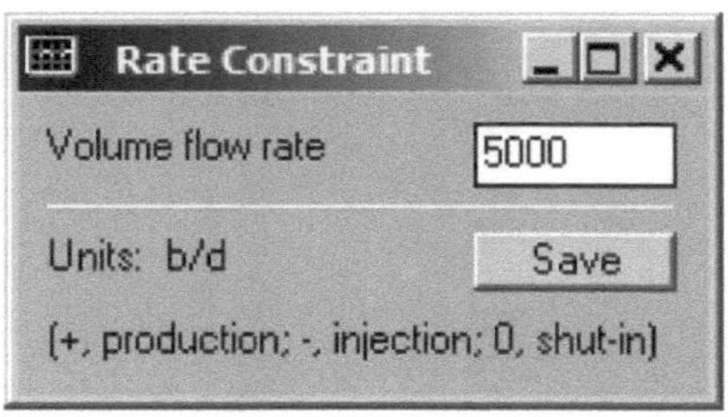

FIG. 15.4R Well 1 rate constraint defined.

drilled, not shown, appears, and we answer "No." Fig. 15.4S for display options also appears, followed by one for wellbore storage effects, which we assume are negligible.

Time-integration computations are proceeding as indicated in the status screen of Fig. 15.4T. We decide to terminate the calculations, save the pressure file to PRESSURE.OLD and click "Yes" in the menu of Fig. 15.4U to initiate plotting of three-dimensional color pressures. In Figs. 15.4V and 15.4W, we plot pressures in Layer 1 (Surface). Note that the lowest pressure in the surface layer is higher than the 5000 psi assumed (but less than the 10,000 psi farfield value) very early on, because we had switched to a flow rate constraint. We next

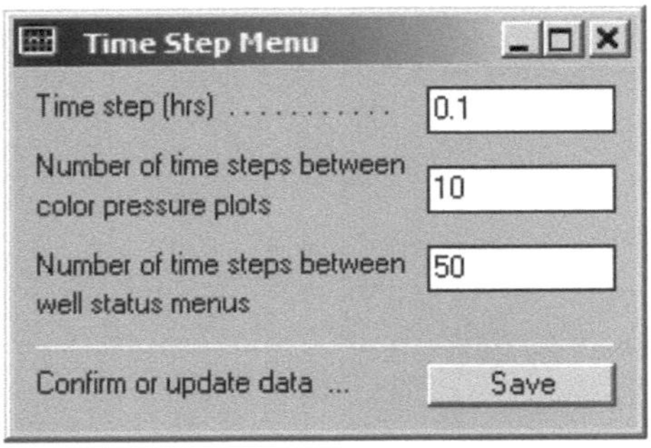

FIG. 15.4S Display options.

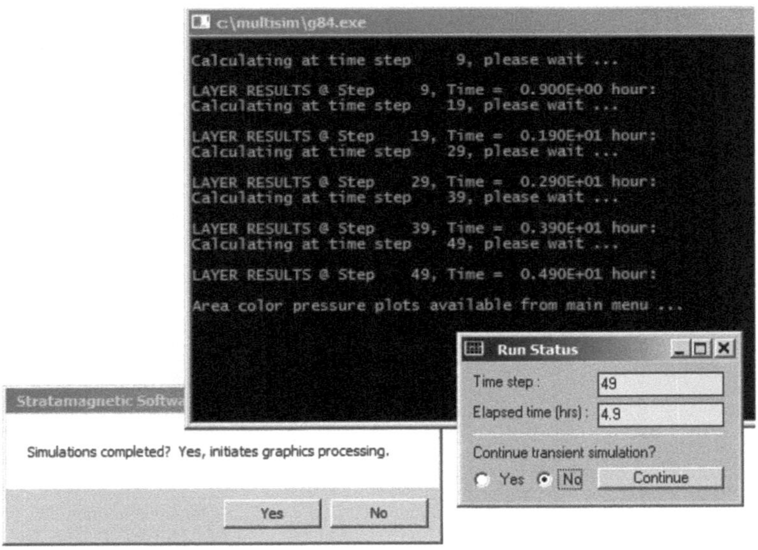

FIG. 15.4T Status screen.

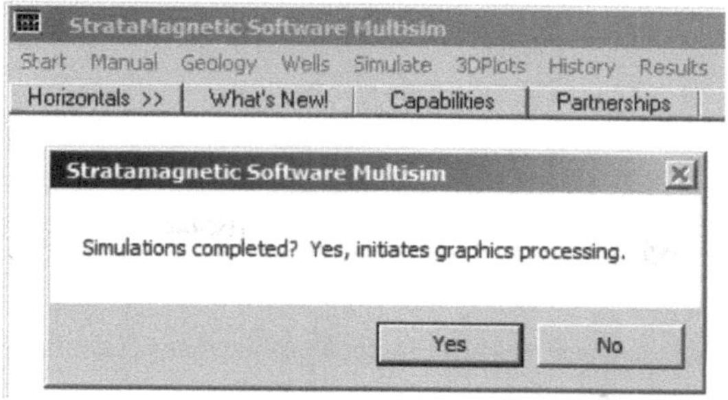

FIG. 15.4U Color processing menu, click "Yes."

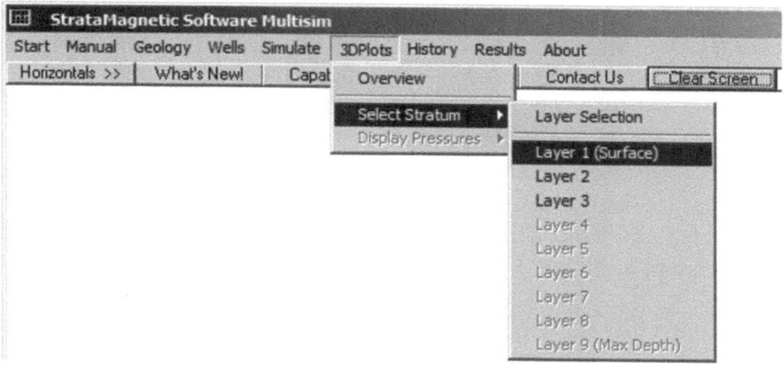

FIG. 15.4V Selecting Layer 1 (Surface) pressures.

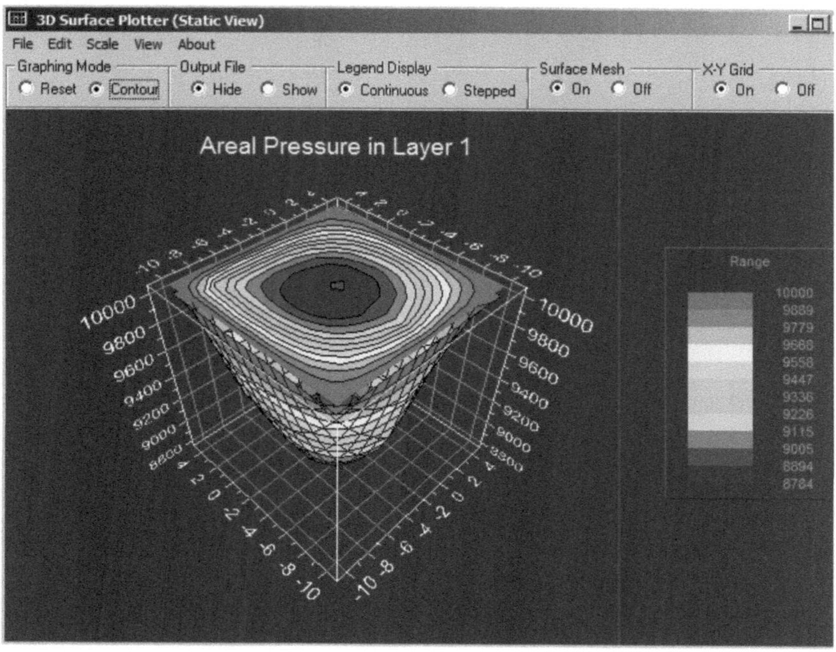

FIG. 15.4W Pressures in Layer 1 (Surface).

select "Layer 2" to show pressure fields in the layer containing the dual lateral. Figs. 15.4X and 15.4Y show the pressure field for Layer 2 where the dual lateral for Well 1 is situated. In Figs. 15.4Z-1–15.4Z-3, we access the pressure history at Well 1. Recall that the positive volume flow rate when steady flow calculations were completed was a large 0.1025E+07 b/d. We then changed the pressure constraint to a flow rate constraint producing at a much lower 5000 b/d. Physically, this will cause the pressure to increase with time, heading toward the 10,000 psi assumed for the reservoir farfield, and this is seen in Fig. 15.4Z-4. Again, the results make perfect physical sense. Tabulated results are offered in Fig. 15.4Z-5. The files SUMMARY.SIM, SUMMARY.GEO and SUMMARY.DRL, not shown due to space limitations, contain simulation, geological and drilling summaries, and are accessed from the main menu.

Finally, note that in this example, we could have easily added heterogeneities (e.g., through the use of lithological symbols like #, $, % and so on), anisotropy, multiple layers (up to nine), multiple vertical, deviated, horizontal and multilateral well systems, and a wide variety of transient well operations to include drilling while simulating. All of these would have complicated our discussion of menus, of course, but these subject areas have been discussed in carefully chosen and designed prior examples.

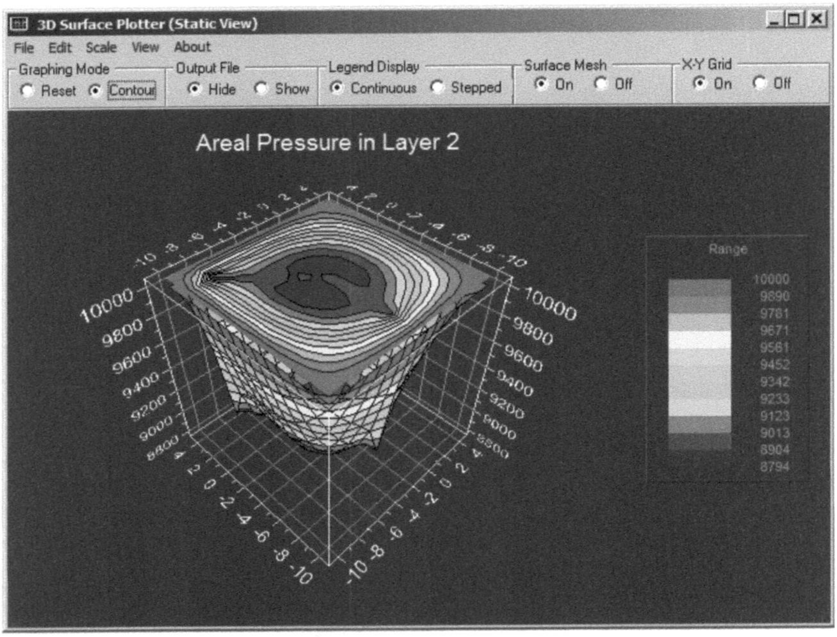

FIG. 15.4X Layer 2 dual lateral pressure field.

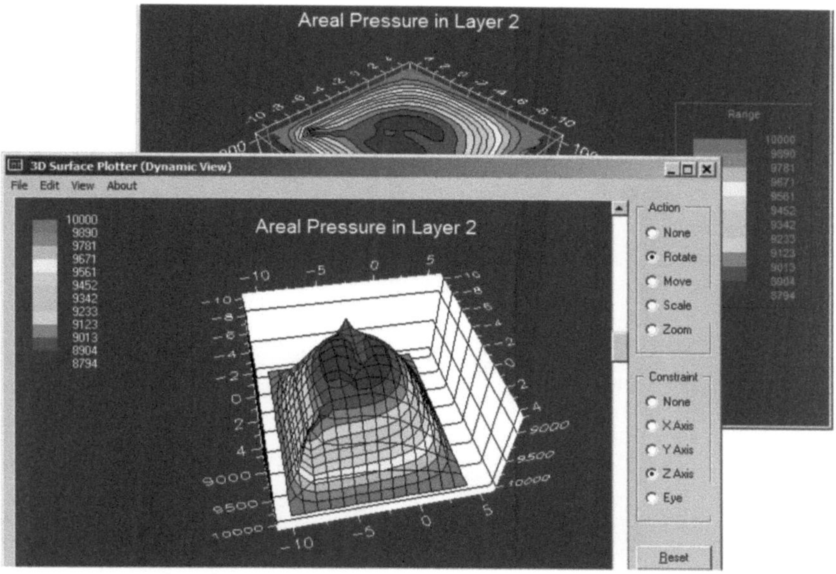

FIG. 15.4Y Layer 2 dual lateral pressure field.

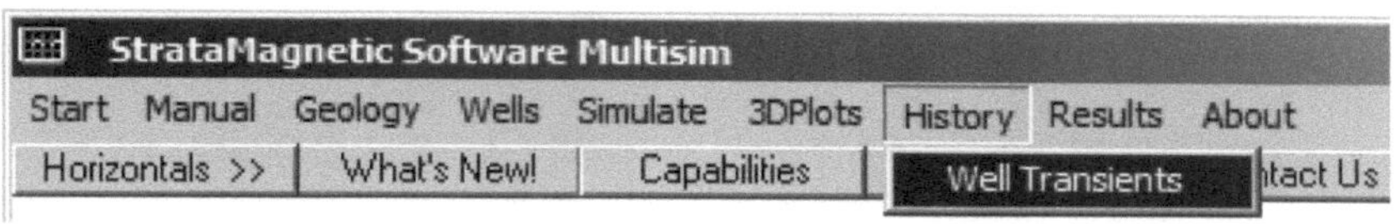

FIG. 15.4Z-1 Accessing well transient history.

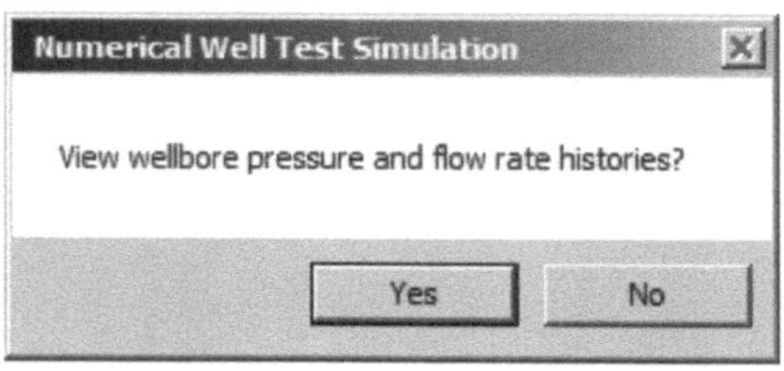

FIG. 15.4Z-2 Accessing well transient history.

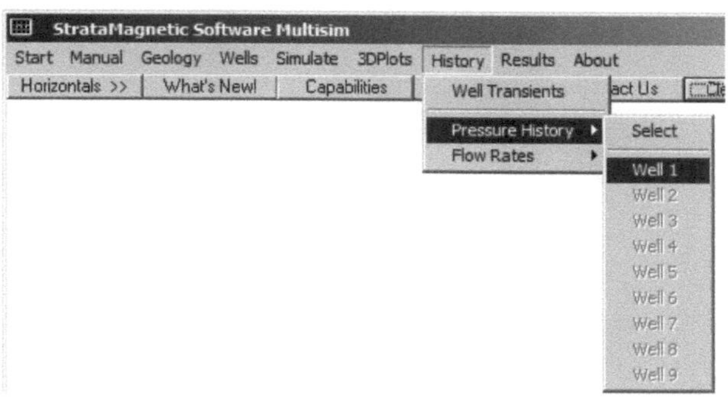

FIG. 15.4Z-3 Accessing well transient history.

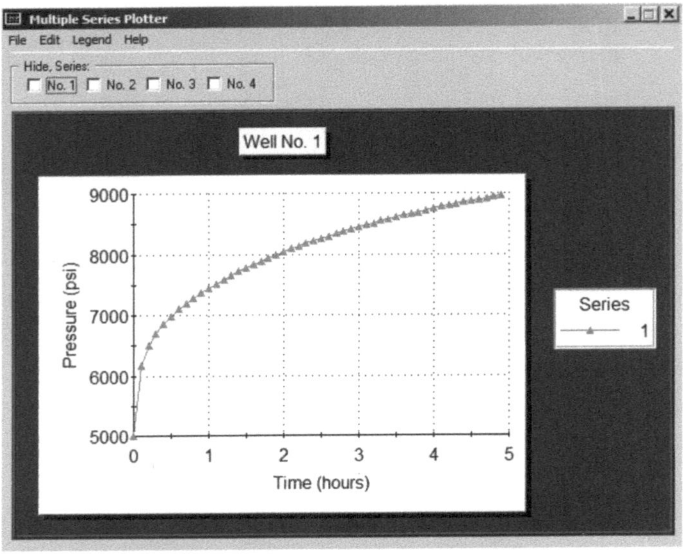

FIG. 15.4Z-4 Pressure buildup resulting from decreased production rate.

WELL1.SIM - Notepad

WELL #1: Step No.	Dt (Hour)	Time (Hour)	Pressure (Psi)	Flow Rate (Cu Ft/Hr)
0	0.100E+00	0.000E+00	0.500E+04	0.240E+06
1	0.100E+00	0.100E+00	0.616E+04	0.117E+04
2	0.100E+00	0.200E+00	0.650E+04	0.117E+04
3	0.100E+00	0.300E+00	0.670E+04	0.117E+04
4	0.100E+00	0.400E+00	0.686E+04	0.117E+04
5	0.100E+00	0.500E+00	0.699E+04	0.117E+04
6	0.100E+00	0.600E+00	0.710E+04	0.117E+04
7	0.100E+00	0.700E+00	0.720E+04	0.117E+04
8	0.100E+00	0.800E+00	0.729E+04	0.117E+04
9	0.100E+00	0.900E+00	0.737E+04	0.117E+04
10	0.100E+00	0.100E+01	0.745E+04	0.117E+04
11	0.100E+00	0.110E+01	0.752E+04	0.117E+04
12	0.100E+00	0.120E+01	0.759E+04	0.117E+04
13	0.100E+00	0.130E+01	0.766E+04	0.117E+04
14	0.100E+00	0.140E+01	0.772E+04	0.117E+04
15	0.100E+00	0.150E+01	0.778E+04	0.117E+04
16	0.100E+00	0.160E+01	0.784E+04	0.117E+04
17	0.100E+00	0.170E+01	0.789E+04	0.117E+04
18	0.100E+00	0.180E+01	0.794E+04	0.117E+04
19	0.100E+00	0.190E+01	0.799E+04	0.117E+04
20	0.100E+00	0.200E+01	0.804E+04	0.117E+04
21	0.100E+00	0.210E+01	0.809E+04	0.117E+04
22	0.100E+00	0.220E+01	0.813E+04	0.117E+04
23	0.100E+00	0.230E+01	0.818E+04	0.117E+04
24	0.100E+00	0.240E+01	0.822E+04	0.117E+04
25	0.100E+00	0.250E+01	0.826E+04	0.117E+04
26	0.100E+00	0.260E+01	0.830E+04	0.117E+04
27	0.100E+00	0.270E+01	0.834E+04	0.117E+04
28	0.100E+00	0.280E+01	0.838E+04	0.117E+04
29	0.100E+00	0.290E+01	0.841E+04	0.117E+04
30	0.100E+00	0.300E+01	0.845E+04	0.117E+04
31	0.100E+00	0.310E+01	0.848E+04	0.117E+04
32	0.100E+00	0.320E+01	0.851E+04	0.117E+04
33	0.100E+00	0.330E+01	0.855E+04	0.117E+04
34	0.100E+00	0.340E+01	0.858E+04	0.117E+04

FIG. 15.4Z-5 Tabulated well history.

EXAMPLE 15.8. PRODUCER AND INJECTOR CONVERSIONS

In this example we consider a common operational scenario where well constraints are altered during production. In our "Geology" menu, a three layer, homogeneous, isotropic medium with default properties is assumed (menus not shown). Then, in our "Wells" menu, two horizontal wells are drilled, one pressure and the other rate-constrained, and after a period of operation, their respective constraint types are reversed. The objective is to demonstrate simulator setup as well as computational stability in the presence of strong "numerical shocks" to the system. This is assessed by examining the smoothness of the end three-dimensional color pressure plot, as well as inspecting pressure and rate histories at the individual wells. For our "Geology" definition, we take a three-layer reservoir with a simple 15×15 grid. Let us now turn to the main horizontal "Wells" menu in Fig. 15.5A. Figs. 15.5B–15.5D show how vertical Wells 1 and 2 penetrate all three layers, however, horizontals are drilled that travel along Layer 2. In Fig. 15.5E, we select the production mode. In prior examples, we had created a steady flowfield first and then initiated transient operations. In the present simulator run, we are in fully transient mode from beginning to end. Thus, we click "No" in Fig. 15.5F. As in an earlier example, the menu in Fig. 15.5G allows us to drill additional wells during simulation. In this simulation, we choose not to and retain the "2" assumed initially. Well constraints are finally specified by selecting the menu in Fig. 15.5H.

In Fig. 15.5I, we select physical units for volume flow rate, in this case, barrels per day. We shall pressure constrain Well 1 in Fig. 15.5J and rate constrain Well 2 in Fig. 15.5K. Note, because Well 1 is pressure constrained at 5000 psi and the reservoir farfield pressure will be set to 10,000 psi, Well 1 will be a producer. Well 2, because the volume flow rate is assumed to be positive, will also be a producer. Farfield boundary conditions are defined using the screens in Figs. 15.5L and 15.5M. The menu in Fig. 15.5N initiates numerical time integrations. Two additional menus appear, which are not shown but discussed in

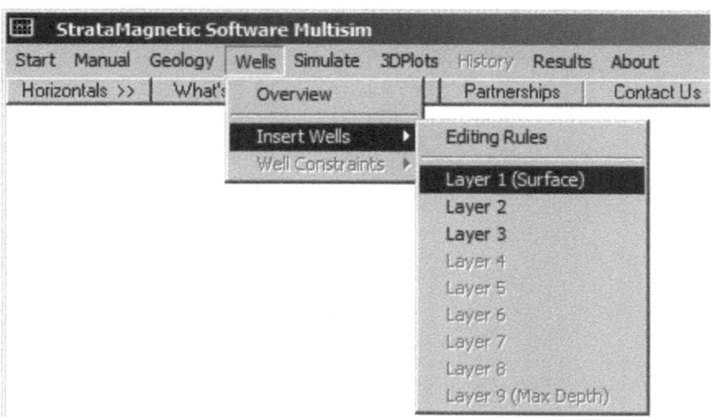

FIG. 15.5A Drilling wells into three-layer reservoir.

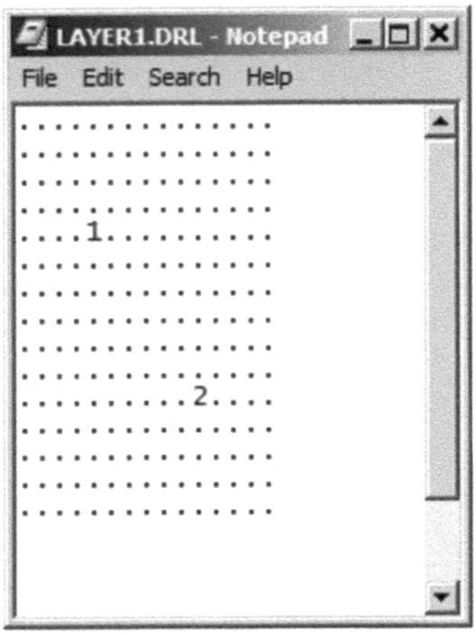

FIG. 15.5B Vertical Wells 1 and 2 with horizontals in Layer 2.

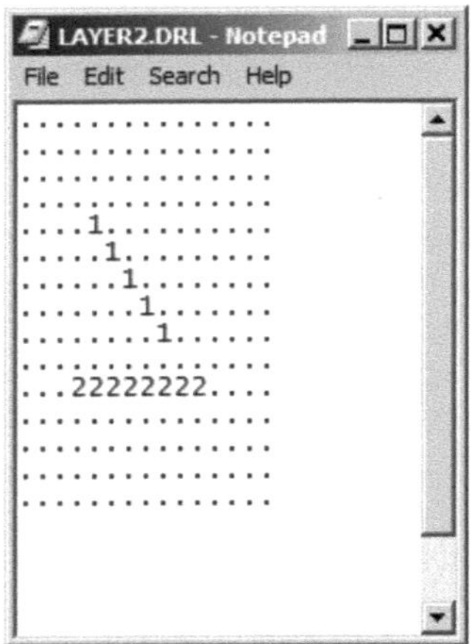

FIG. 15.5C Vertical Wells 1 and 2 with horizontals in Layer 2.

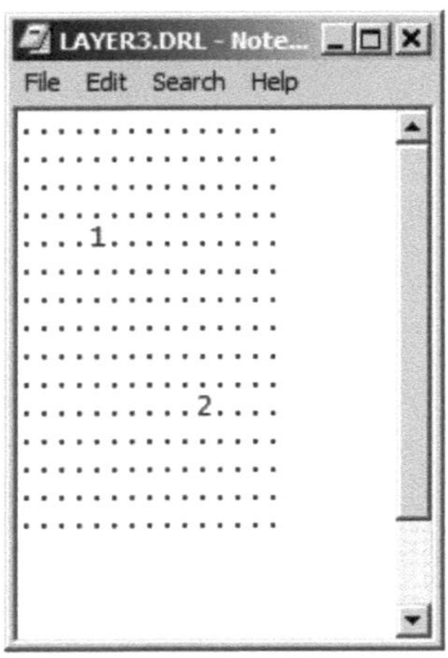

FIG. 15.5D Vertical Wells 1 and 2 with horizontals in Layer 2.

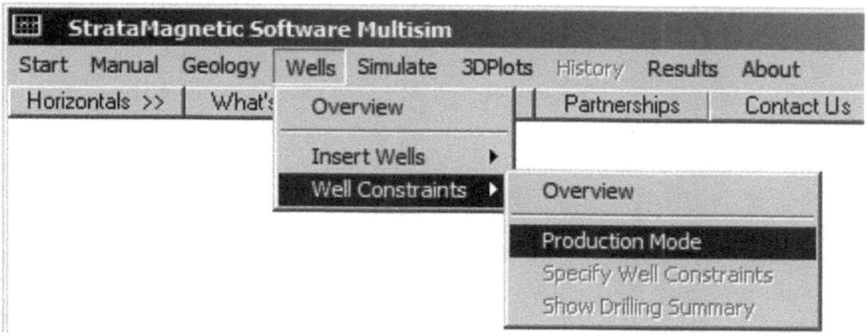

FIG. 15.5E Selecting production mode.

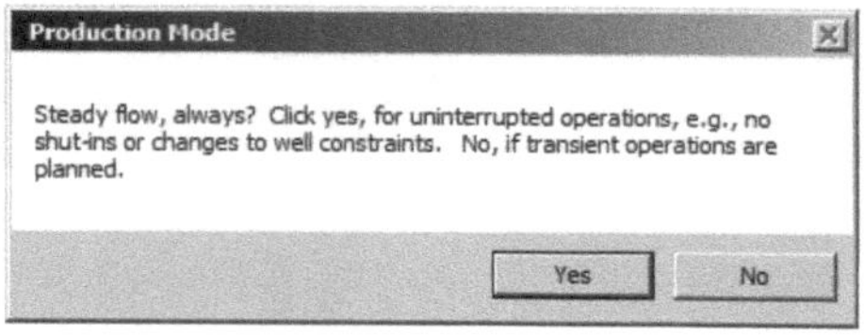

FIG. 15.5F Steady only versus transient flow (click, "No.").

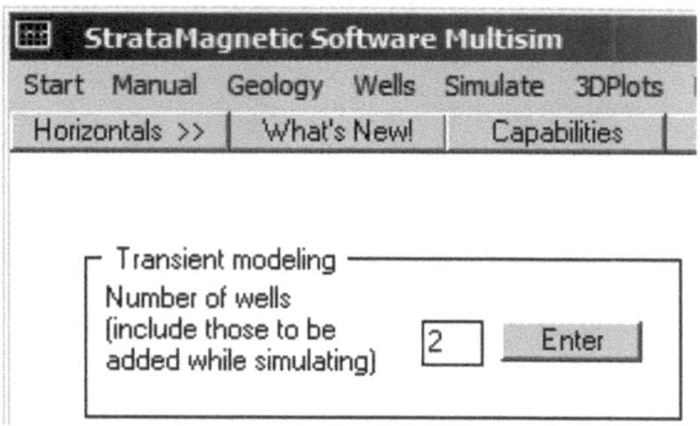

FIG. 15.5G Two wells assumed for entire duration.

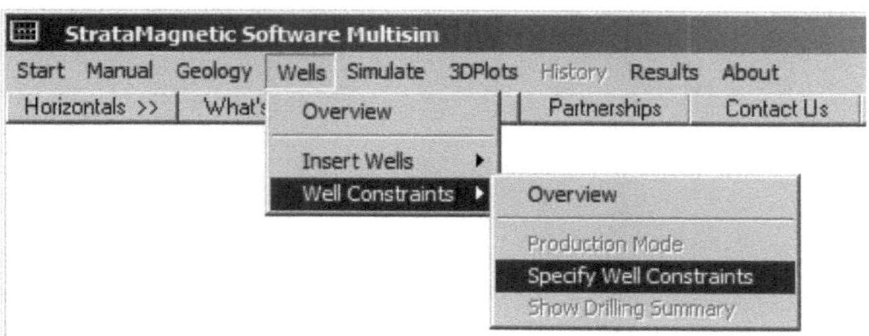

FIG. 15.5H Specifying well constraints.

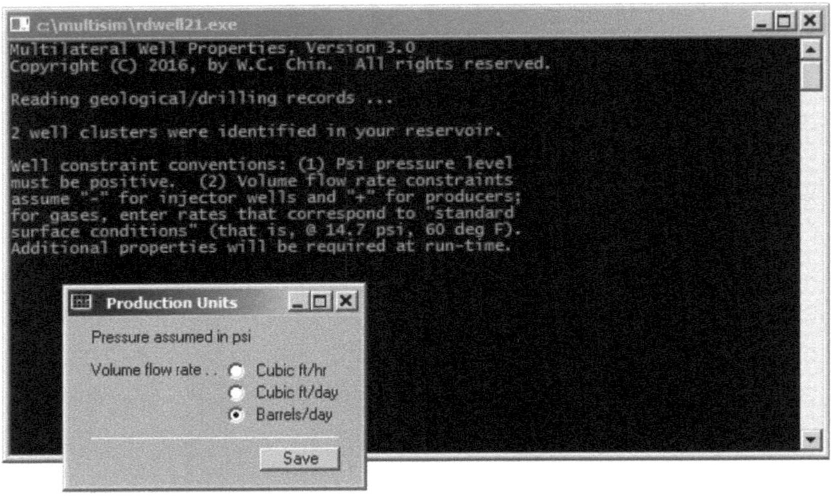

FIG. 15.5I Selecting volume flow rate units.

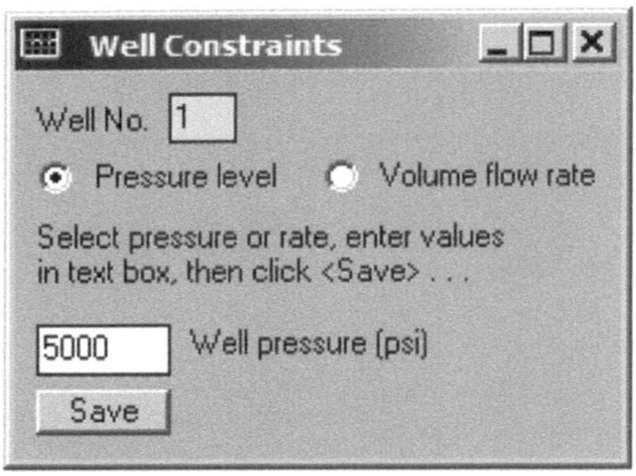

FIG. 15.5J Pressure constraining Well 1.

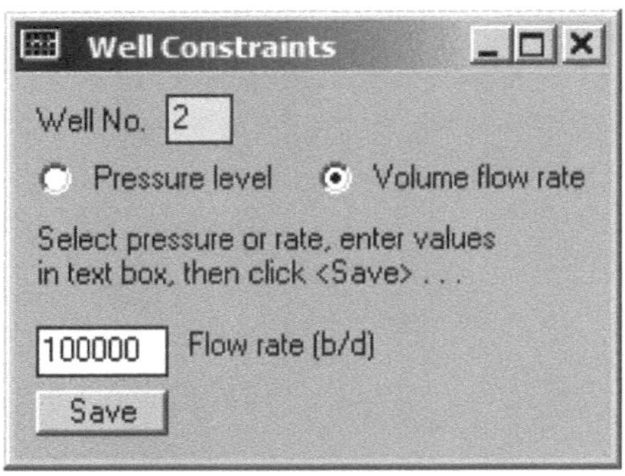

FIG. 15.5K Rate constraining Well 2.

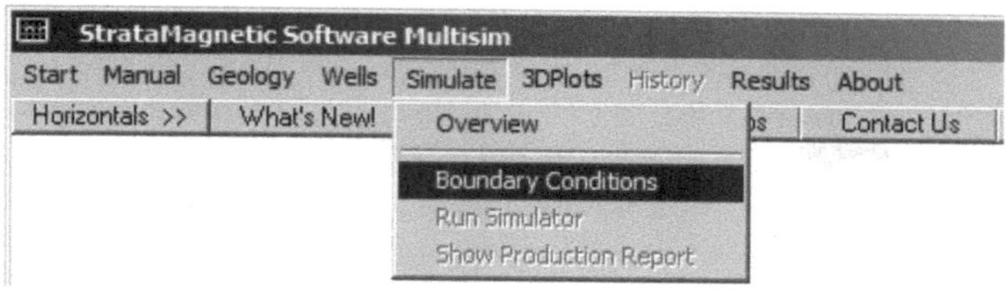

FIG. 15.5L Defining boundary conditions.

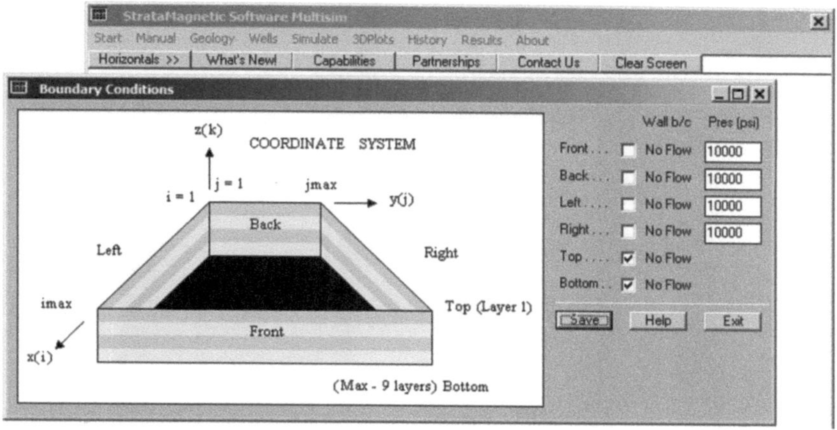

FIG. 15.5M Boundary condition definition at six box sides.

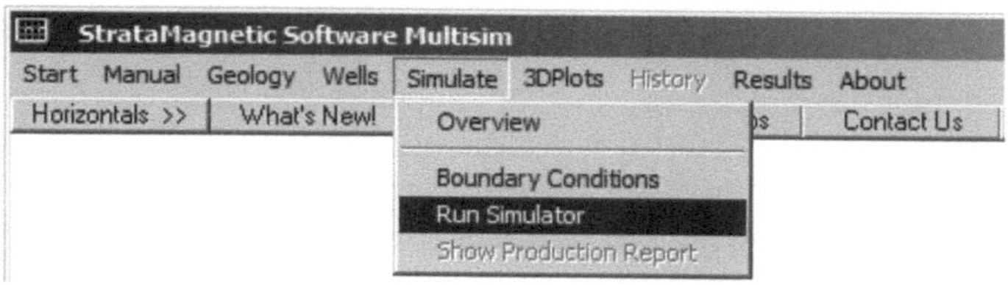

FIG. 15.5N Time-marching numerical integration process begins.

earlier examples. In the first, we do not change transmissibilities while simulating. In the second, we assume a liquid fluid type with a viscosity of 1 cp. In the following menu shown in Fig. 15.5O, we now choose "transient" flow—we had always selected "steady" for most examples. Thus initial conditions for pressure are required—we set the initial reservoir pressure equal to that assumed for the boundary (this is not physically or mathematically required). This is shown in Fig. 15.5P, which also allows initializations to variable pressure fields stored from prior runs, e.g., PRESSURE.OLD.

FIG. 15.5O Defining production mode.

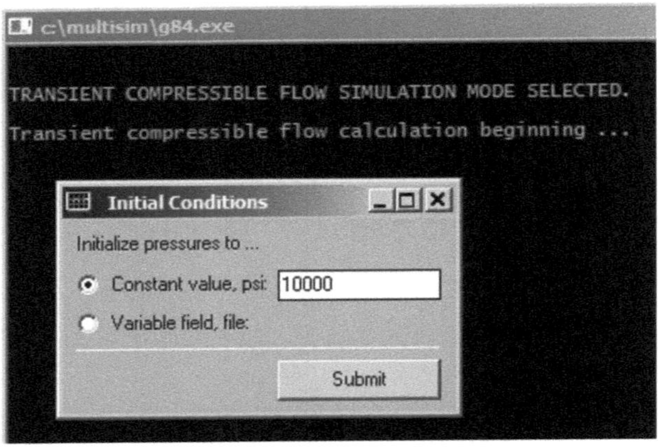

FIG. 15.5P Setting initial reservoir pressure to a constant value.

Additional menus appear, not shown here, which were discussed in prior examples. In the first, we assume default fluid compressibility values, while in the second, we choose not to porosity average liquid and rock compressibilities. In our time step recommendation menu, we click for the assumed defaults but override the selection with our usual 0.1 h time step value. We next assume a "rigid formation" as opposed to "small deformation" model for compaction analysis. The status screen in Fig. 15.5Q reminds us that at initial time $t=0$, Well 1 is pressure constrained with values assumed previously—we click "Continue." We similarly click "Continue" in Fig. 15.5R for Well 2 which is rate constrained. In a subsequent menu not shown, we choose to not drill additional wells or clusters while simulating.

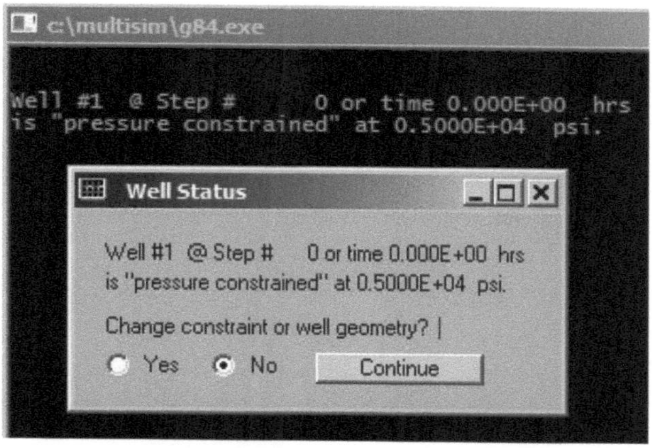

FIG. 15.5Q Pressure constrained Well 1 at t=0.

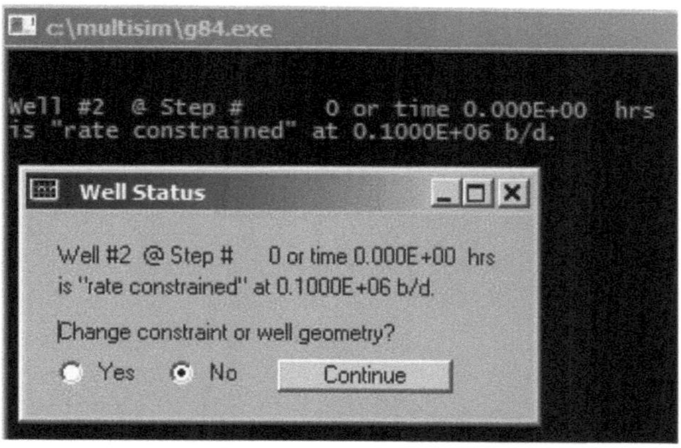

FIG. 15.5R Rate constrained Well 2 at t=0.

The menu in Fig. 15.5S allows us to control display settings. Two other menus appear, not shown, prior to actual time integration. In the first, for Well 1, wellbore storage effects are set to "0"—this is similarly done in the second menu for Well 2. In Fig. 15.5T, displayed after a number of time steps, we have the option of terminating simulations. However, we choose "Continue" in order to change well constraint types—the objective of this example.

The menus in Figs. 15.5U–15.5W show our converting the pressure constraint of Well 1 to a negative flow rate constraint characteristic of injector wells. We did not sidetrack, redrill or recomplete Well 1 when asked. Figs. 15.5X and 15.5Y show how we change constraints for Well 2.

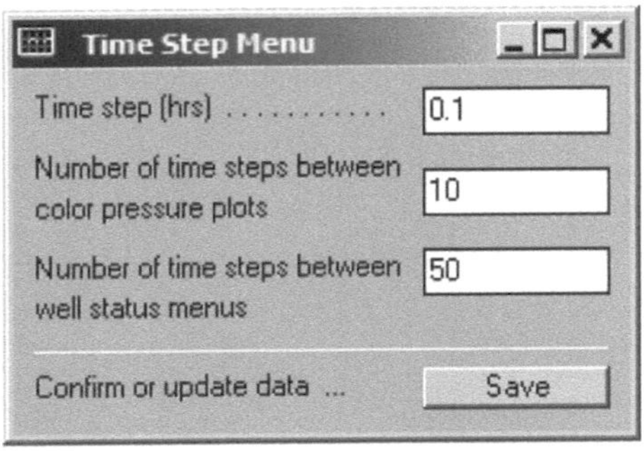

FIG. 15.5S Display menu settings.

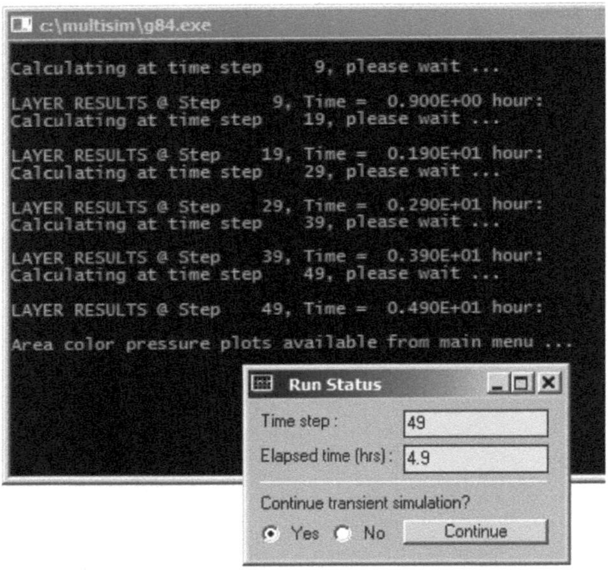

FIG. 15.5T Continue transient modeling (and change constraint types).

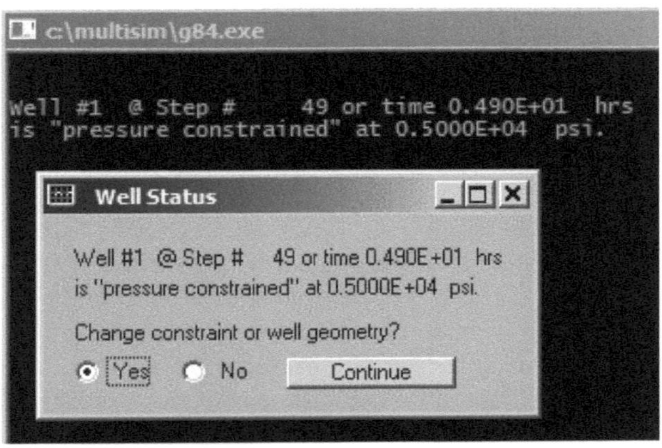

FIG. 15.5U Constraint change option.

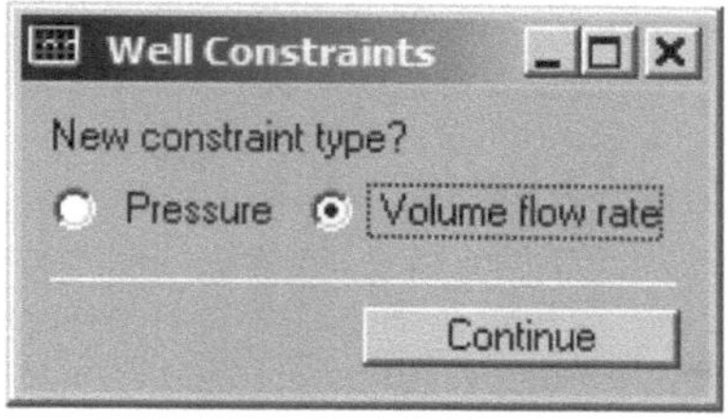

FIG. 15.5V Well constraint changed to flow rate.

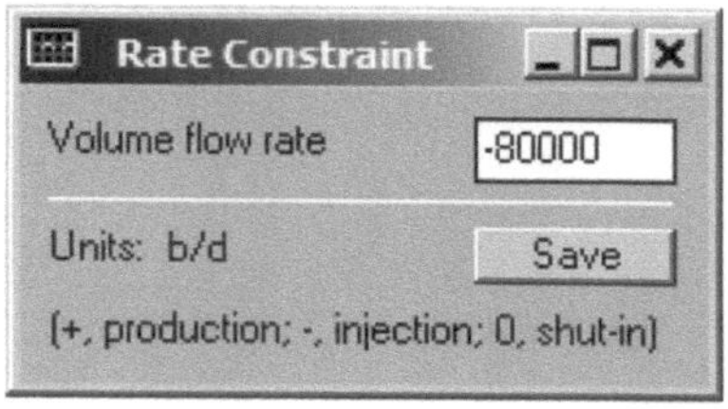

FIG. 15.5W Negative flow rate for injector status.

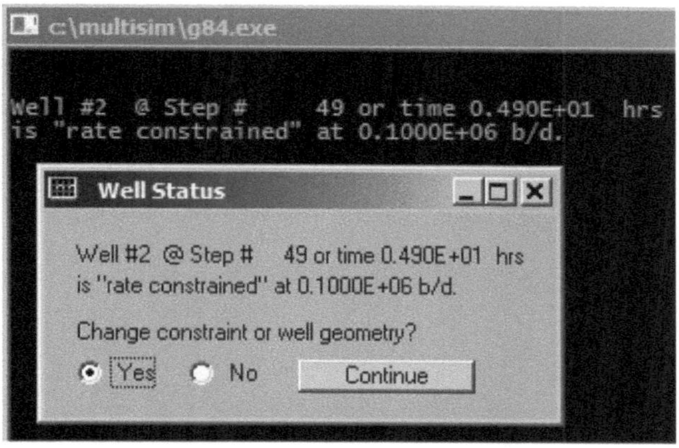

FIG. 15.5X Constraint change option.

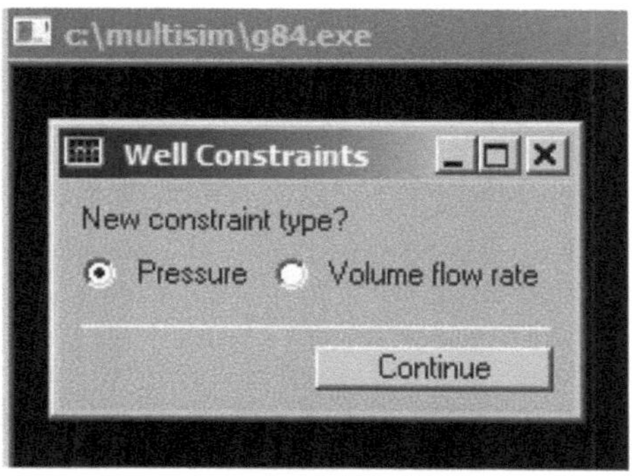

FIG. 15.5Y Rate constraint changed to one for pressure.

Note that we did not sidetrack, redrill or recomplete Well 2 when asked, in a menu not shown here. The original rate constraint is changed to one for pressure, that is, 15,000 psi as shown in Fig. 15.5Z-1. Since this is greater than the 10,000 psi assumed in the farfield, Fig. 15.5Z-1 converts Well 2 to injector status. Again, we do not drill any more new wells or clusters when asked, in a menu not shown. Then menu in Fig. 15.5Z-2 sets display parameters, and in two subsequent menus not shown, we set wellbore storage effects for Well 1 and Well 2 to zero. Calculations continue to the point shown in Fig. 15.5Z-3.

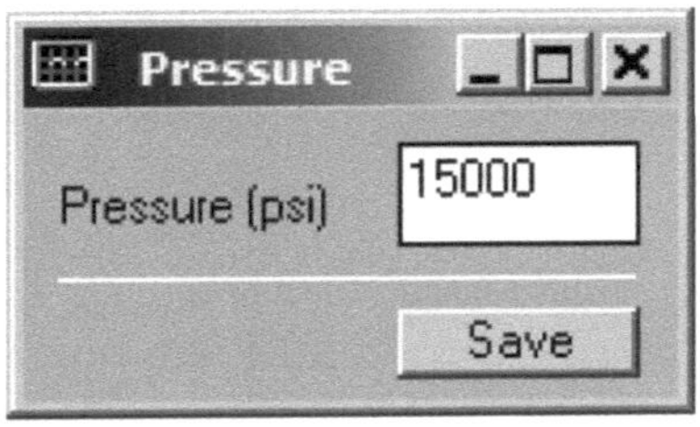

FIG. 15.5Z-1 Pressure constraint applied.

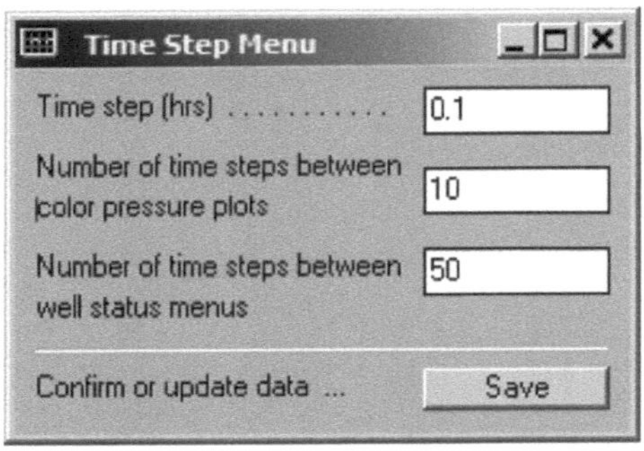

FIG. 15.5Z-2 Display menu.

At this point, the pressure field is saved in a file named as PRESSURE.OLD, and three-dimensional color graphics processing is initiated by clicking "Yes" in Fig. 15.5Z-4. Then menu in Fig. 15.5Z-5 shows that the results for middle Layer 2, where both horizontal extensions for Wells 1 and 2 are located, are selected for display. Graphics options appear in Fig. 15.5Z-6 and results follow in Figs. 15.5Z-7–15.5Z-9. The menus in Figs. 15.5Z-10 and 15.5Z-11 initiate line plotting of well transient history, and Well 1 pressure history is selected and appears in Figs. 15.5Z-12 and 15.5Z-13.

Why does Fig. 15.5Z-13 make physical sense? First, during the initial half of the simulation, we had imposed a pressure constraint of 5000 psi, and that is

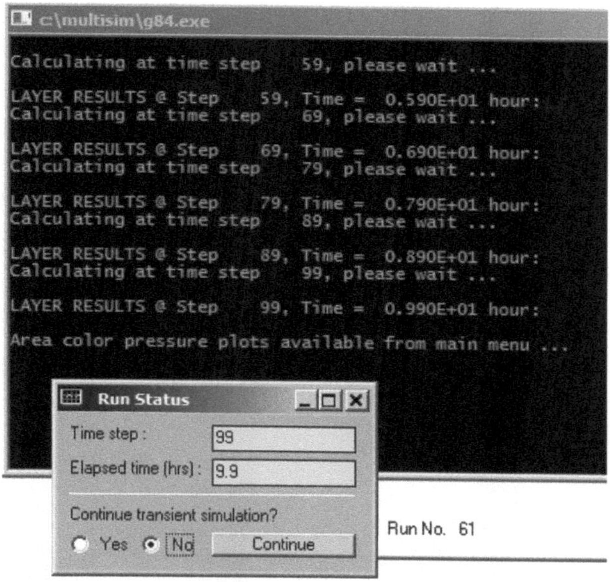

FIG. 15.5Z-3 Calculations to terminate, click "No."

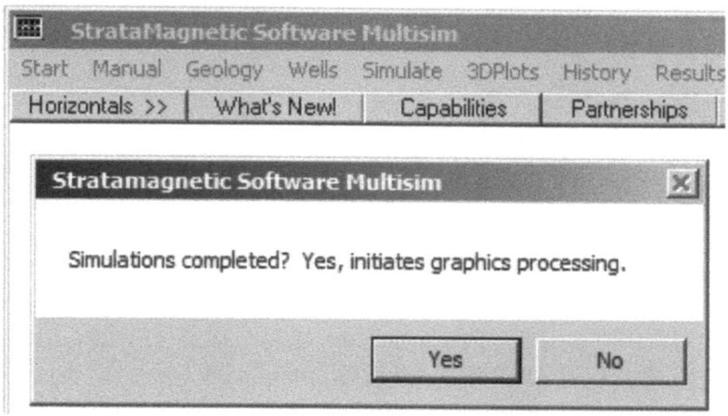

FIG. 15.5Z-4 Color graphics processing begins.

shown correctly. During the second half, we had changed Well 1's well constraint to enforce a constant negative flow rate of "-80,000" so that it would function as an injector. In order for this to happen, well pressures must increase from the start level of 5000 psi, and this increase is seen in Fig. 15.5Z-13. In this case, we see that the pressure at Well 1 eventually exceeds the pressure of 10,000 psi imposed at the farfield boundary.

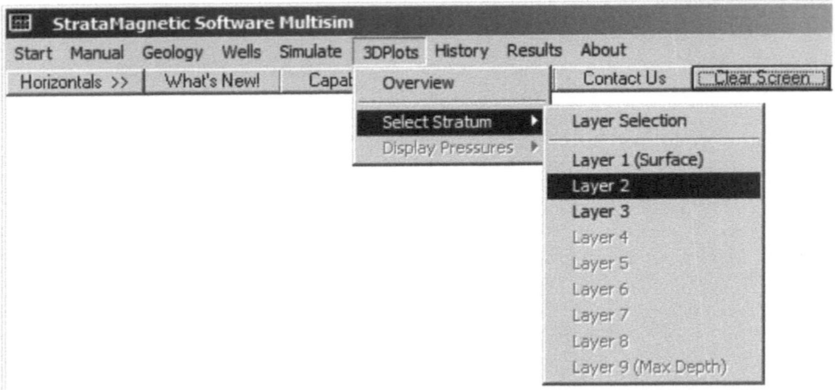

FIG. 15.5Z-5 Middle Layer 2 results selected for display.

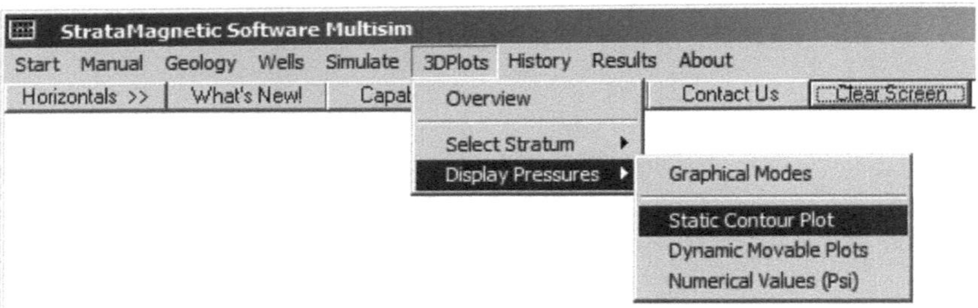

FIG. 15.5Z-6 Color graphics options.

The pressure history for Well 2 is selected for display in Fig. 15.5Z-14. Fig. 15.5Z-15 makes physical sense. Initially, the reservoir initial condition was taken as 10,000 psi, and this is seen at t = 0 h. During the first half of the simulation, Well 2 was taken as a producer with a positive constant "+100,000" flow rate. The decrease in pressure is consistent with production from the reservoir. Later, a 15,000 psi level is prescribed and shown correctly. In Fig. 15.5Z-16, we select Well 1's flow rate for display. Because its prescribed well pressure of 5000 psi is less than the specified 10,000 psi pressure at the reservoir farfield, Well 1 functions as a producer. The decrease in flow rate shown in Fig. 15.5Z-17 is physically consistent with this action. In the latter half of the simulation, a constant negative flow rate is prescribed and is properly shown.

In Fig. 15.5Z-18, we select the flow rate for Well 2 for display, and results are given in Fig. 15.5Z-19. The first half of the simulation shows a constant

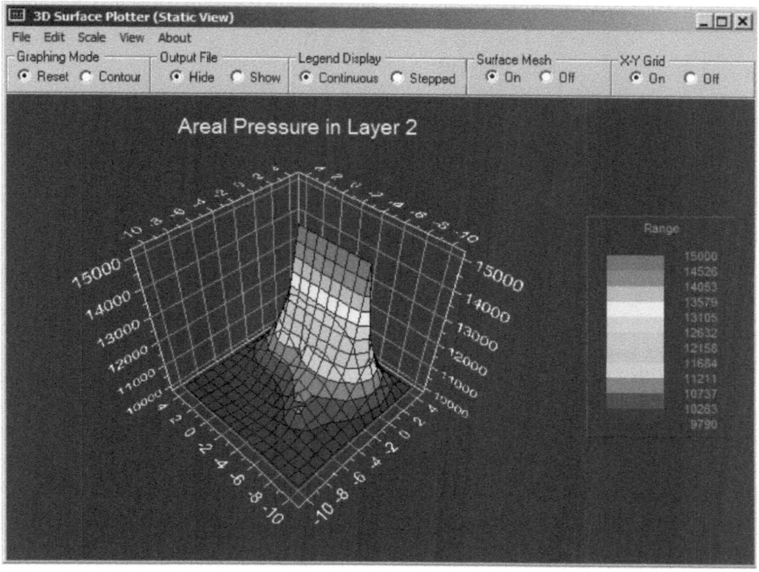

FIG. 15.5Z-7 Areal pressure in Layer 2.

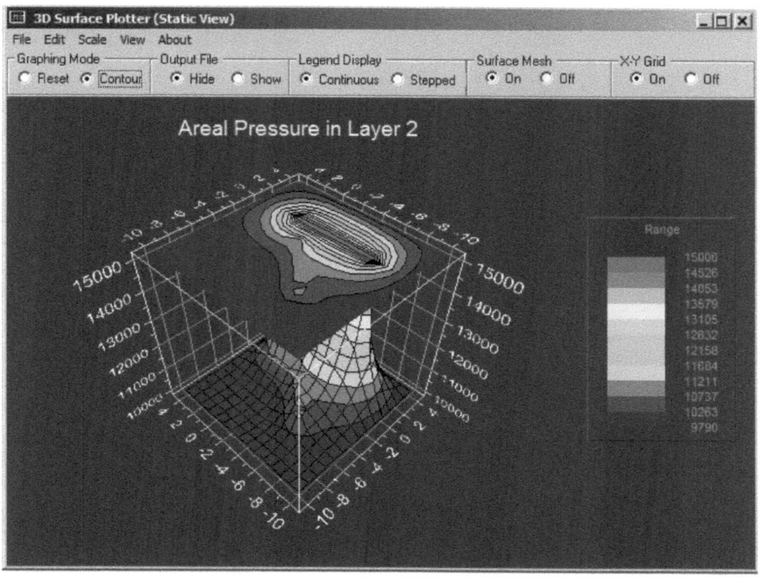

FIG. 15.5Z-8 Areal pressure in Layer 2.

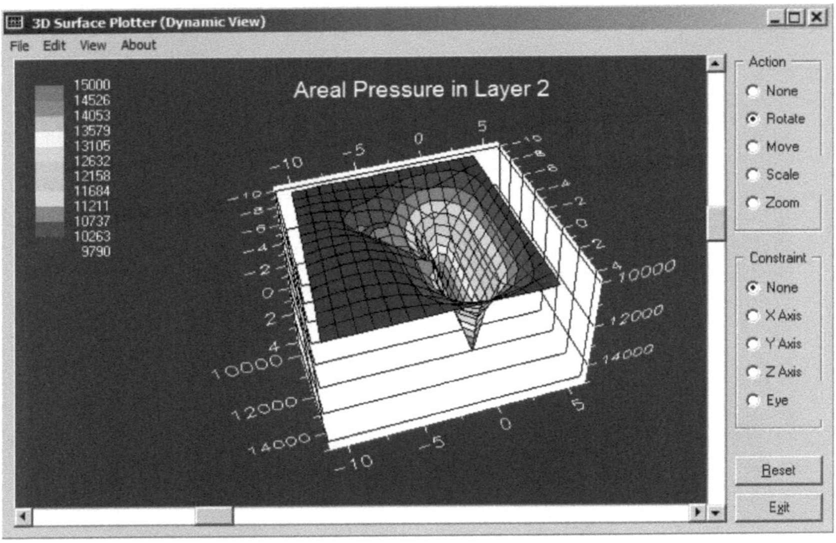

FIG. 15.5Z-9 Areal pressure in Layer 2.

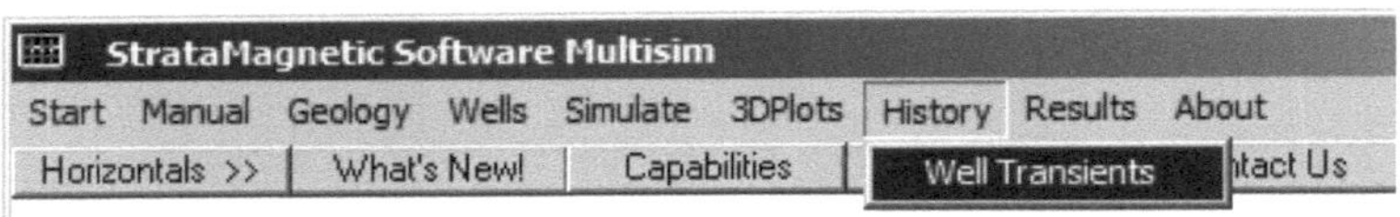

FIG. 15.5Z-10 Displaying well transient history at wells.

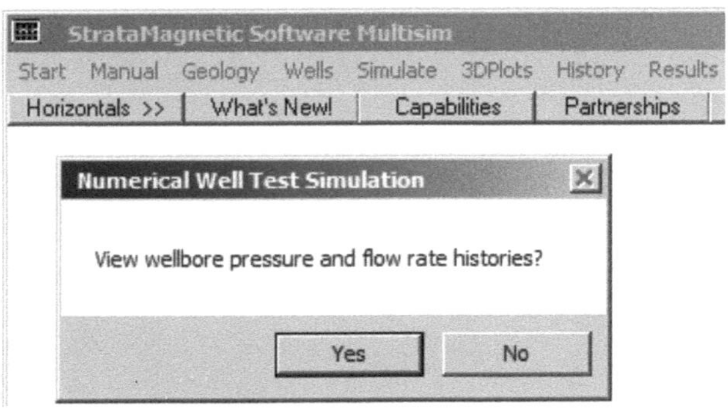

FIG. 15.5Z-11 Menu screen initiates line plotting.

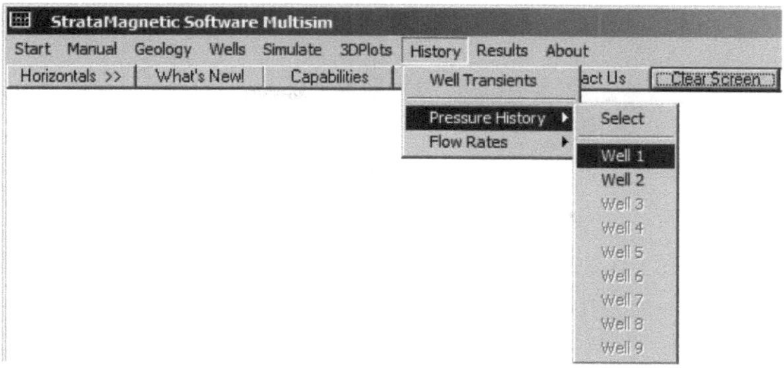

FIG. 15.5Z-12 Well 1 pressure history selected.

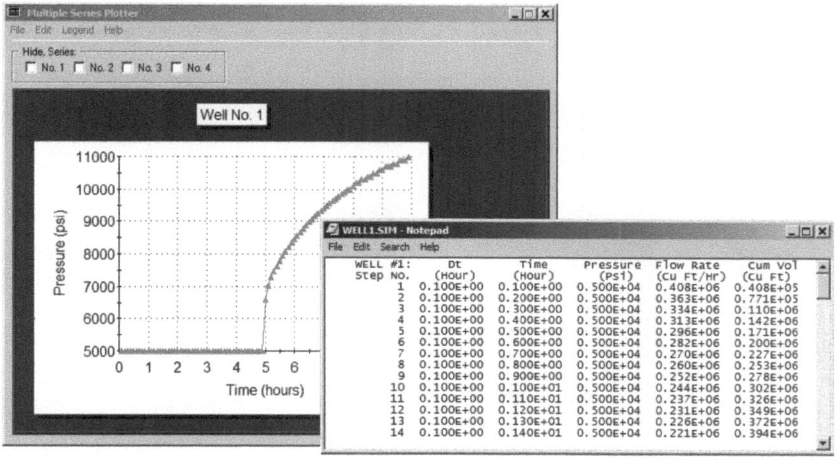

FIG. 15.5Z-13 Well 1 pressure history.

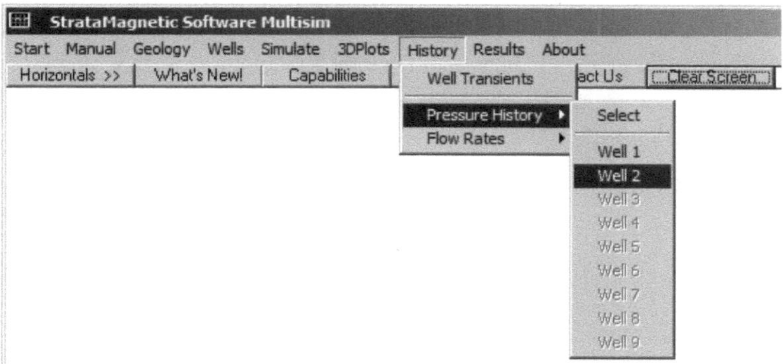

FIG. 15.5Z-14 Pressure history at Well 2 selected.

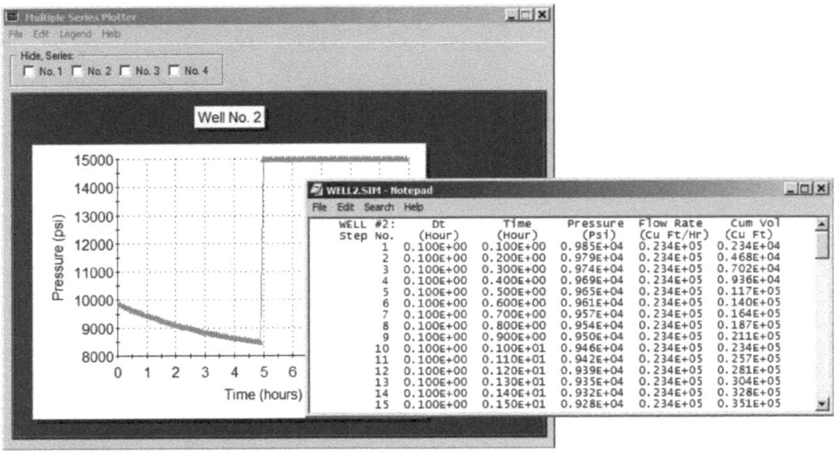

FIG. 15.5Z-15 Well 2 pressure history.

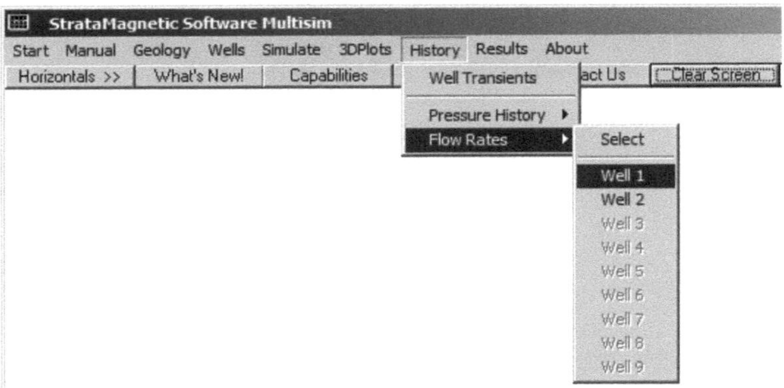

FIG. 15.5Z-16 Well 1 flow rate selected for display.

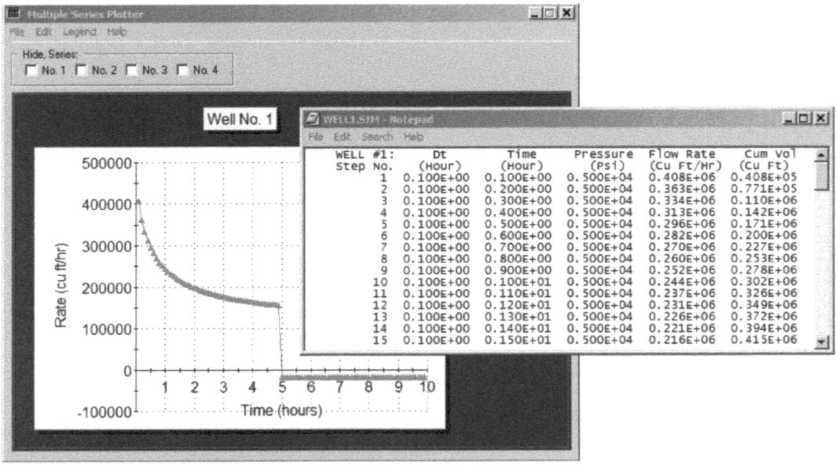

FIG. 15.5Z-17 Well 1 flow rate history.

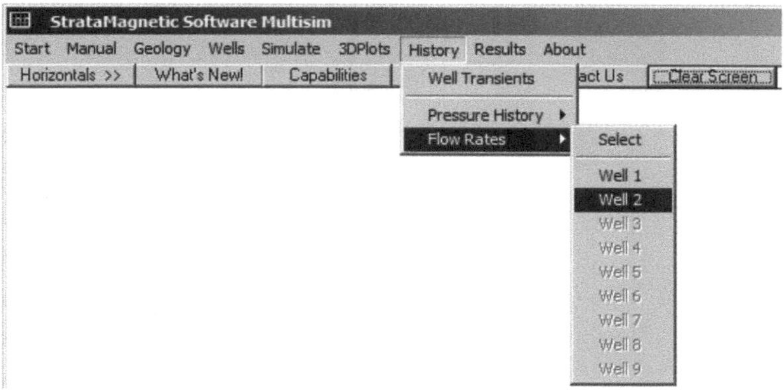

FIG. 15.5Z-18 Well 2 flow rate history selected for display.

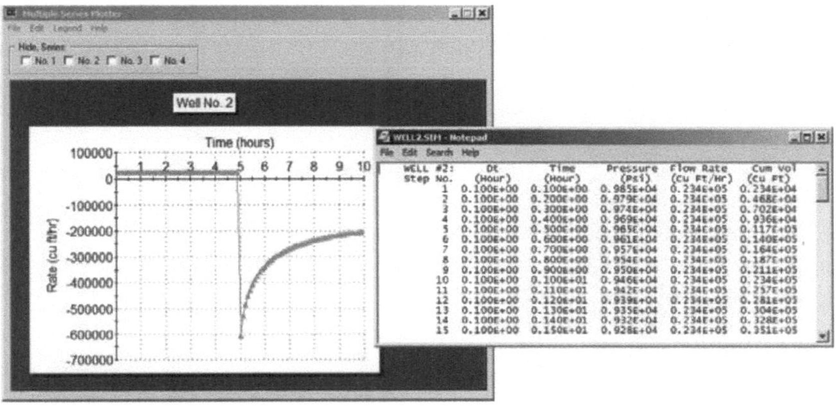

FIG. 15.5Z-19 Well 2 flow rate history.

positive flow rate constraint indicating a producer well status—note that the constraint is expressed in "b/d" whereas the volume flow rate plot appears as "cu ft/h" and this constant is properly plotted. In the second half of the simulation, a pressure constraint of 15,000 psi is applied, which is higher than the reservoir farfield pressure of 10,000 psi. Fig. 15.5Z-19 shows that suddenly the well is converted to injector status. Its rate slows as expected and all of this is properly simulated according to the physics. All computations, that is, color plots and line graphs, are smooth, indicating numerically stable calculations. As in all of our examples, the numbers themselves are not important, since we used default inputs in menus for display clarity. In Fig. 15.5Z-20, we display the menu needed to show the SUMMARY.SIM report.

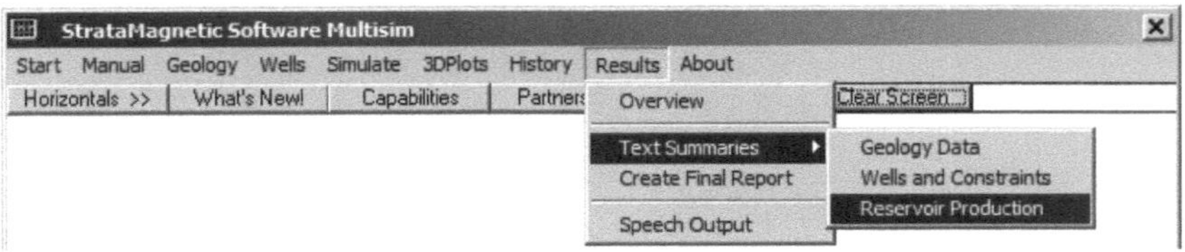

FIG. 15.5Z-20 Reservoir production report summary selected.

EXAMPLE 15.9. PRODUCTION WITH TOP AND BOTTOM DRIVES

So far our examples have considered reservoirs where the tops and bottoms are sealed "no flow" boundaries, while the four sides are held at constant pressure, that is, they are "aquifer drives." Note that there is no requirement that the four pressures be equal to each other although this was assumed for simplicity. In the present run, we create an artificial example for presentation purposes where the four sides are now "no flow" while the top and bottom are held at user-defined pressures. The former are often "gas drives" while the latter are "water drives." In our "Geology" menu, we shall take a coarse 15 × 15 areal grid system and a three-layer reservoir for ease of display. A uniform medium and a computing grid, both with default properties, will be assumed. Geology definition menus are not shown. From our "Wells" menu, we shall drill a single "Well 1" system where the main vertical bore penetrates all three layers while the middle layer contains a dual lateral—note that all of these appendages from the well nonetheless are part of a single "Well 1" system per our software architecture. We purposely draw a slanted horizontal to emphasize that it need not align with coordinate lines. The following screen captures in Figs. 15.6A–15.6I are self-explanatory (read figure captions for basic explanations).

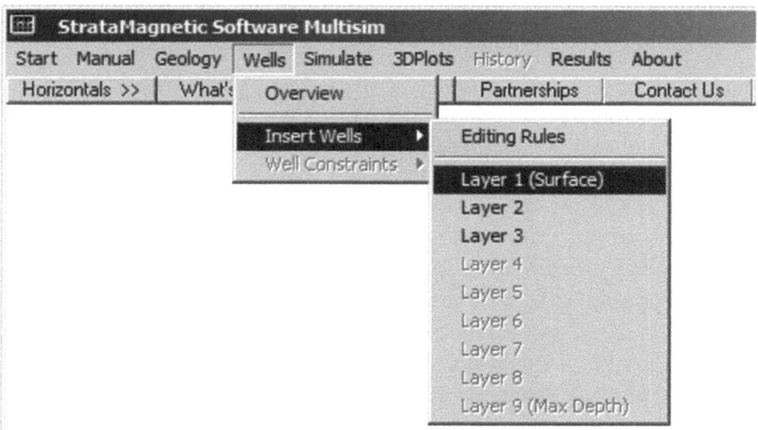

FIG. 15.6A Drilling wells into Layers 1, 2 and 3.

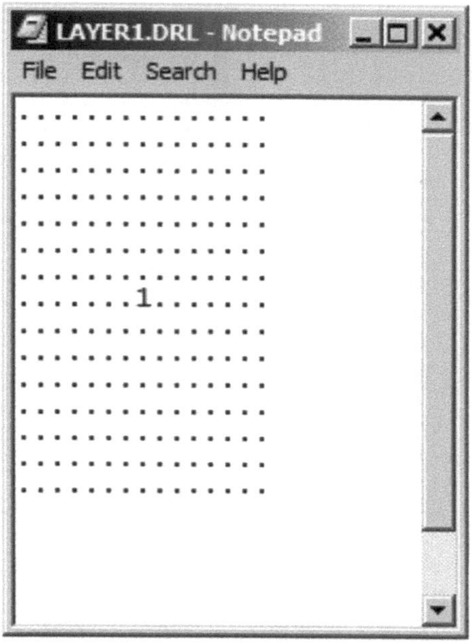

FIG. 15.6B Vertical well in Layer 1.

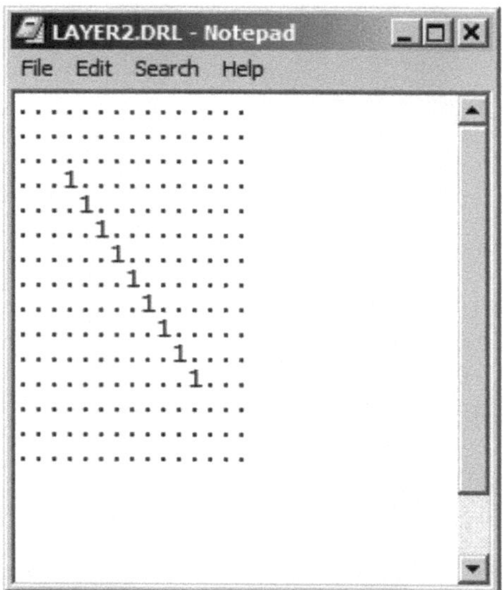

FIG. 15.6C Dual lateral in Layer 2.

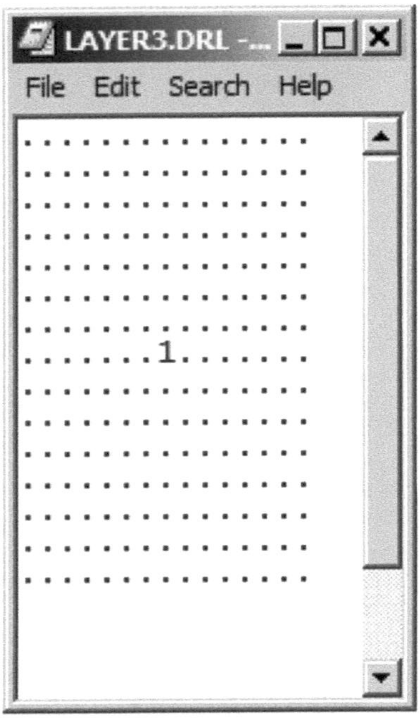

FIG. 15.6D Vertical well in Layer 3.

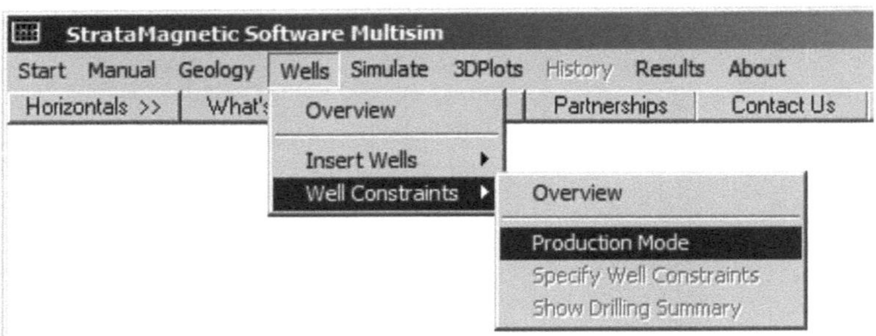

FIG. 15.6E Production mode input.

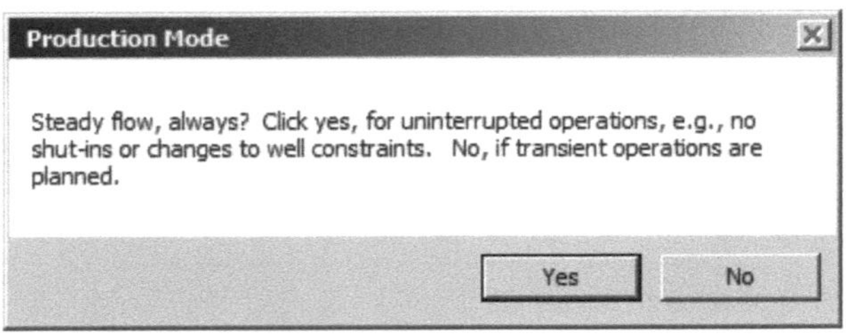

FIG. 15.6F Steady versus transient flow, click "Yes."

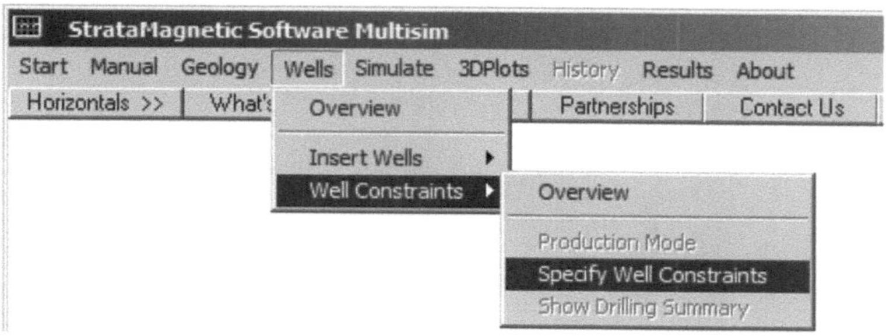

FIG. 15.6G Specifying well constraints.

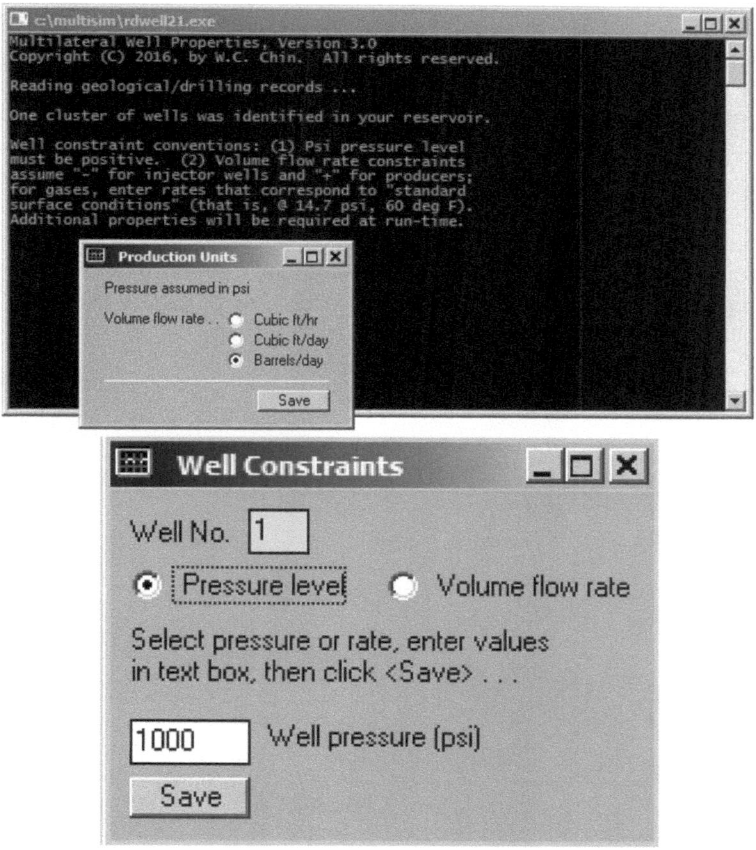

FIG. 15.6H Specifying flow rate units and pressure constraint.

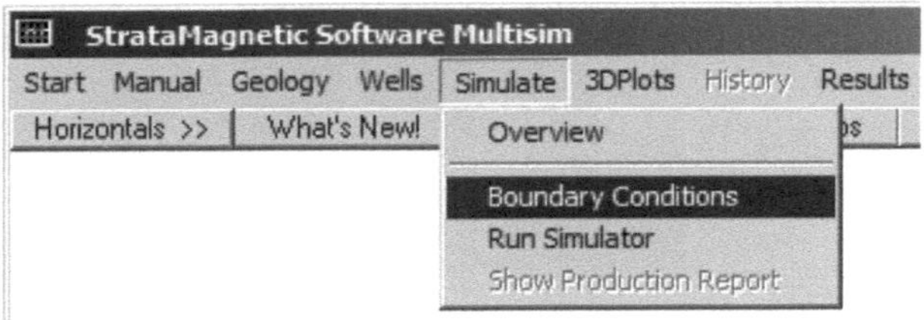

FIG. 15.6I Specifying farfield boundary conditions.

In Fig. 15.6J, we have set front, back, left and right as "no flow" boundaries, and prescribed pressures at top and bottom, opposite to our usual boundary condition definition. Next, select "Run Simulator." Several menus appear, which we shall not reproduce. In essence, we do not change transmissibilities while simulating, a liquid fluid is assumed with viscosity of 1 cp, and we consider "Steady flow, for now." Fig. 15.6K shows results at Iteration 100 while Fig. 15.6L shows results at Iteration 500. Examination of volume flow rates in the black screen shows that our computations have definitely converged, since they have not changed since Iteration 100. Thus, we terminate calculations—we do not pursue transient calculations, and initiate graphics processing. In Fig. 15.6M, we select the layer for color pressure field display, and in Fig. 15.6N, the type of graphical display.

Figs. 15.6O–15.6R show pressure traces that are consistent with the slanted horizontal assumed in Fig. 15.6C. As expected, since the drilling is identical in Layers 1 and 3, the pressure fields in Figs. 15.6O and 15.6R are likewise identical. This type of symmetry provides an additional check on correct algorithm development and software implementation.

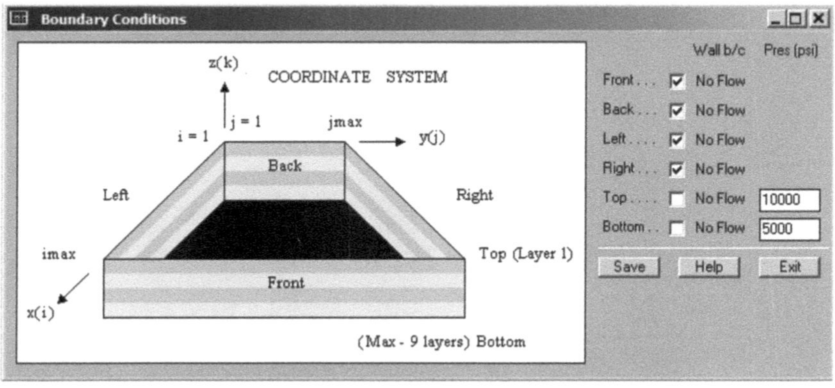

FIG. 15.6J Specifying top and bottom pressure levels.

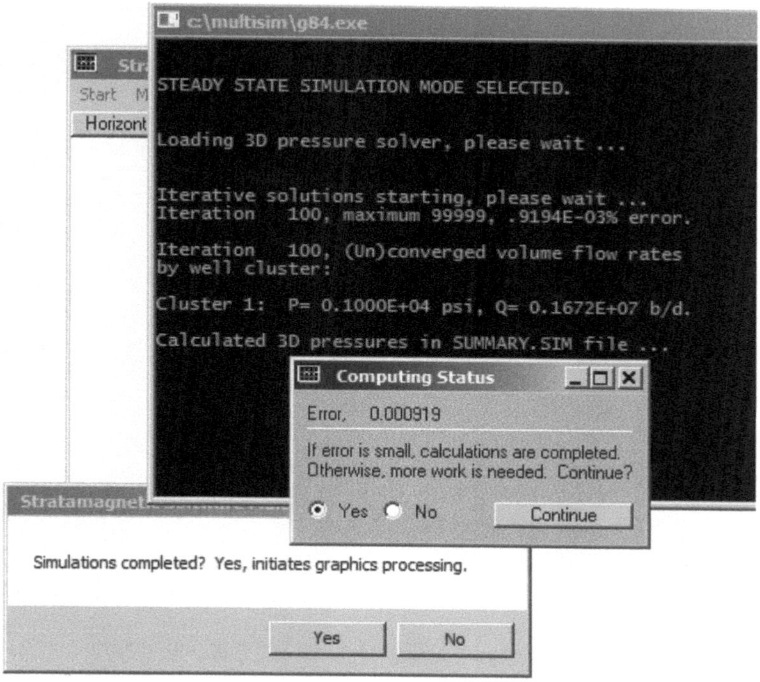

FIG. 15.6K Status screens at 100 iteration intervals.

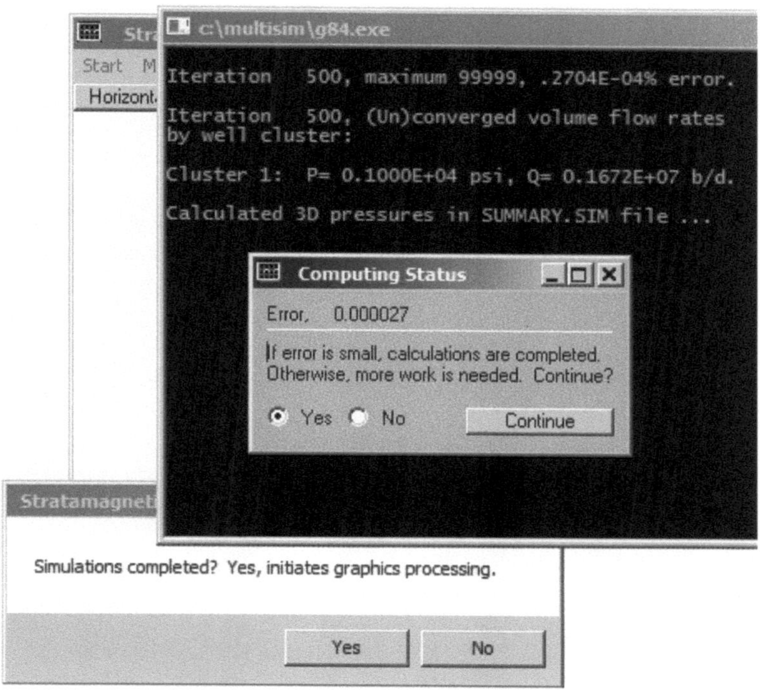

FIG. 15.6L Status screen at 500 iterations.

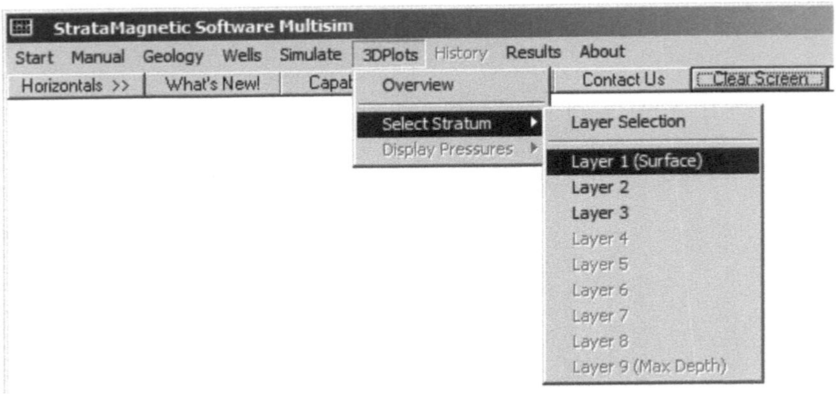

FIG. 15.6M Selecting layer for pressure field display.

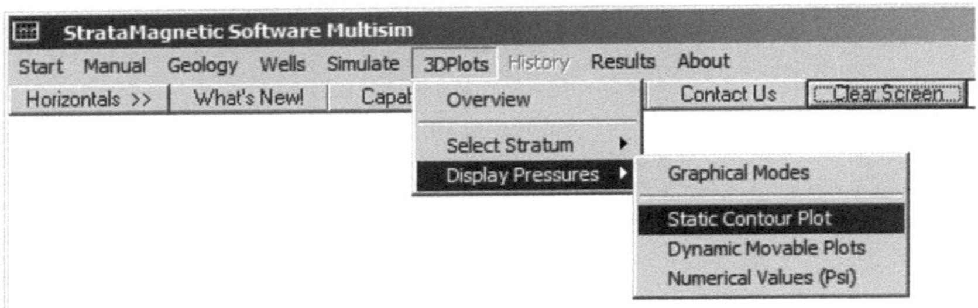

FIG. 15.6N Three display options for pressure.

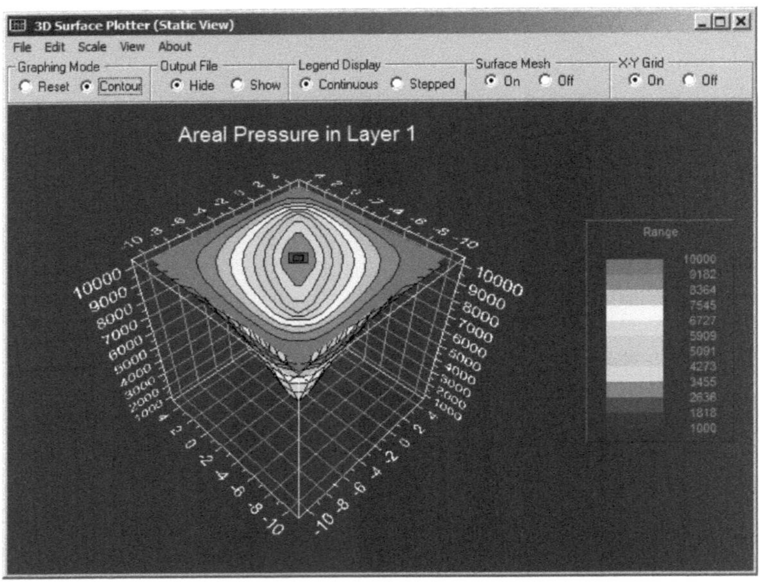

FIG. 15.6O Effect of dual lateral on top layer pressure.

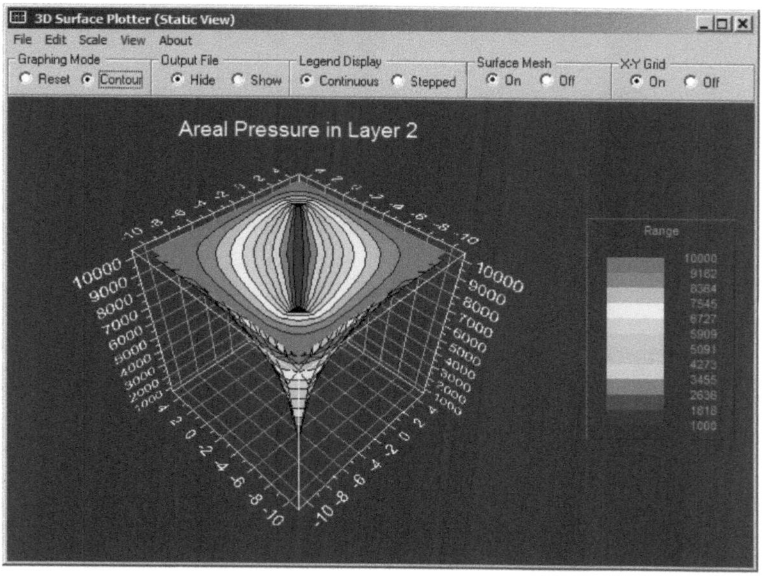

FIG. 15.6P Dual lateral effect on middle layer pressure.

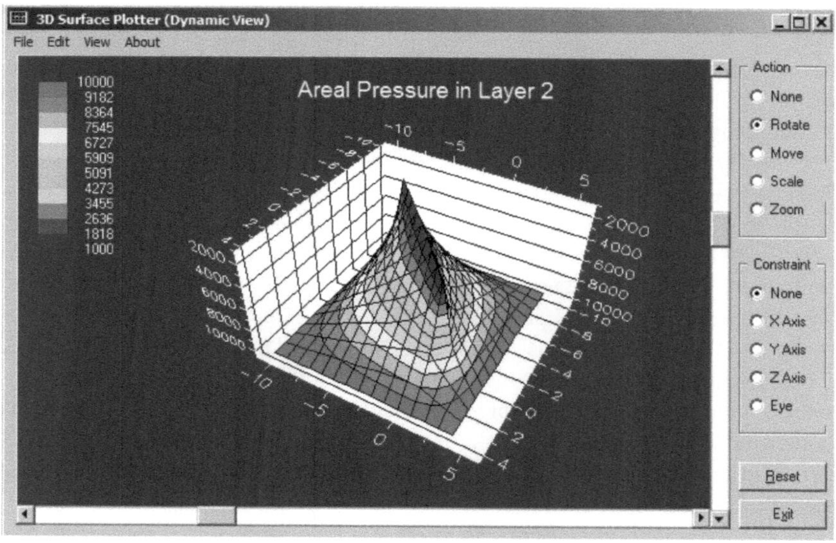

FIG. 15.6Q Dual lateral effect on middle layer pressure.

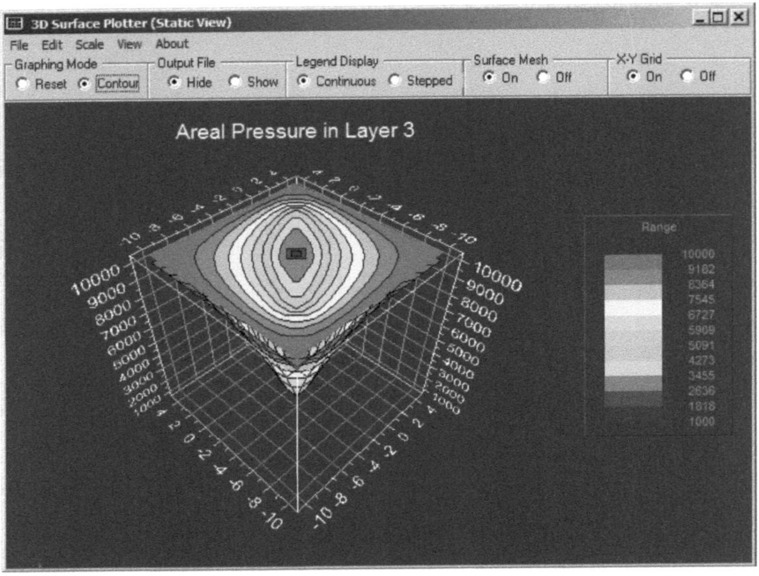

FIG. 15.6R Effect of dual lateral on bottom layer pressure.

EXAMPLE 15.10. TRANSIENT GAS PRODUCTION FROM DUAL HORIZONTAL WITH WELLBORE STORAGE EFFECTS

In this final example, we shall focus on gases as opposed to liquids. In fact, we shall consider transient gas production from a dual lateral and also demonstrate how wellbore storage effects are modeled in our simulator. In "Geology" menu, we select three-layer reservoir, a 15×15 areal grid, a uniform medium with default properties, and rock and grid parameters also assuming default values (for brevity, menus are not shown). From our "Wells" menu, we drill a single Well 1 system with a vertical well penetrating all three layers through the center of the square reservoir, but which contains a dual lateral in the middle layer. LAYER*.DRL files and other screen captures are shown (Figs. 15.7A–15.7J).

In prior examples, we selected "Liquid" in Fig. 15.7K, entered a viscosity, plus compressibility in a separate menu. Here we select "Gas" and enter a viscosity of 0.01 cp. When "Gas" is selected, the special input screen in Fig. 15.7L appears, with a help screen not shown, which explains to the user the various thermodynamic options available for modeling. In Fig. 15.7L, we assume an isothermal process. Fig. 15.7M states that a transient calculation will be initiated, and in response, Fig. 15.7N asks for initial reservoir pressure—this need not be the same as the farfield reservoir pressure as is the case here.

Elsewhere in this chapter, we mentioned that our time-step recommendation (see Fig. 15.7O) is only approximate—time scales applicable to any problem

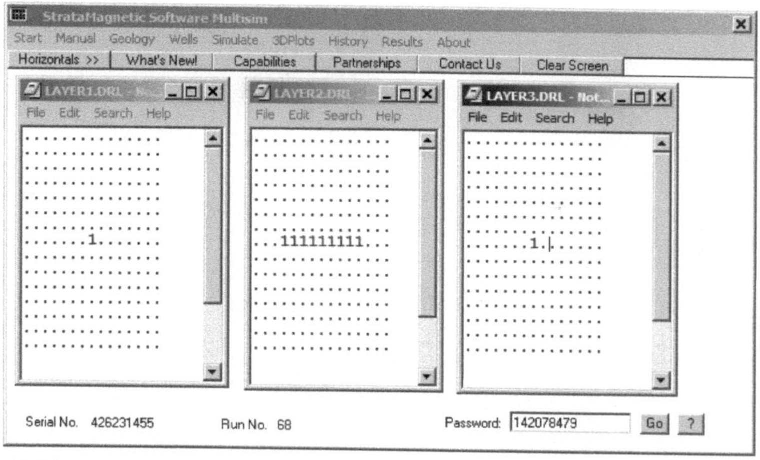

FIG. 15.7A Top, middle and bottom layer drilled well.

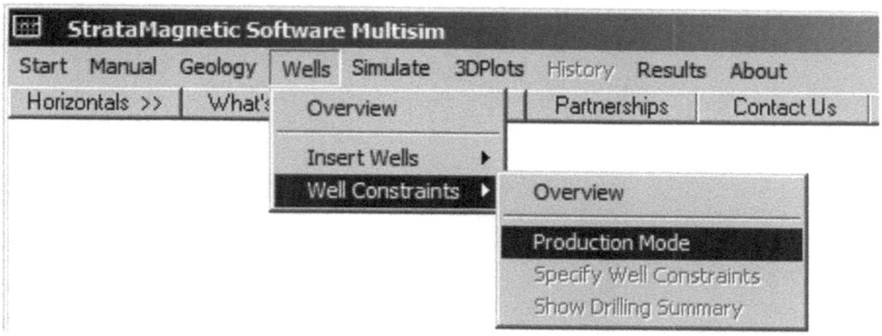

FIG. 15.7B Selecting production mode.

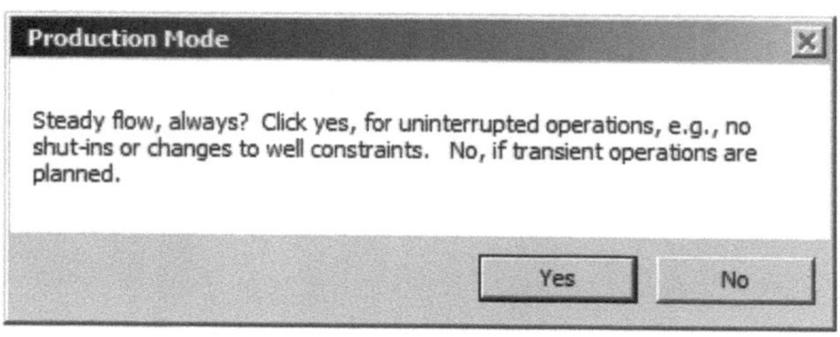

FIG. 15.7C Production mode, click "No" for fully transient analysis.

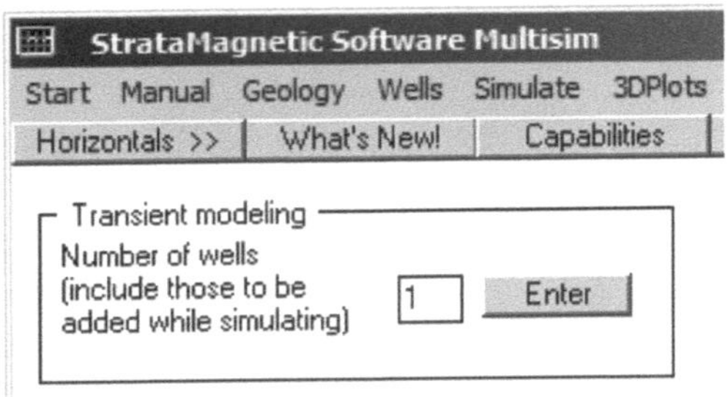

FIG. 15.7D Select "1" since no additional wells will be drilled.

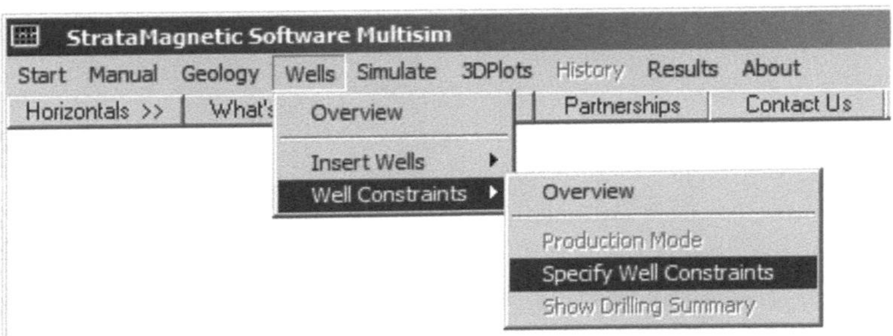

FIG. 15.7E Specifying well constraints.

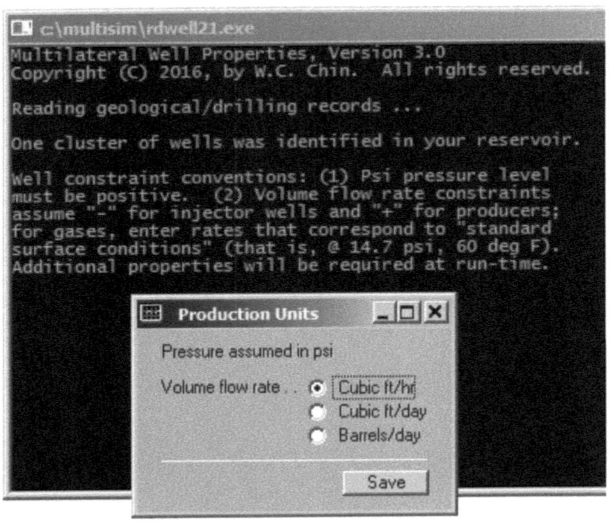

FIG. 15.7F Selecting "cubic ft/h" for gas flow.

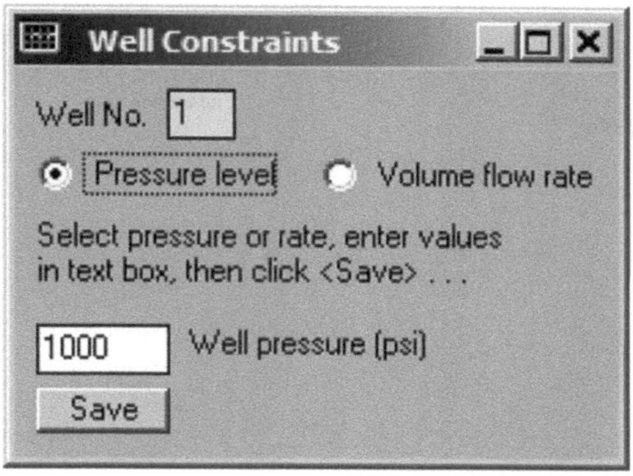

FIG. 15.7G Setting pressure constraint to 1000 psi.

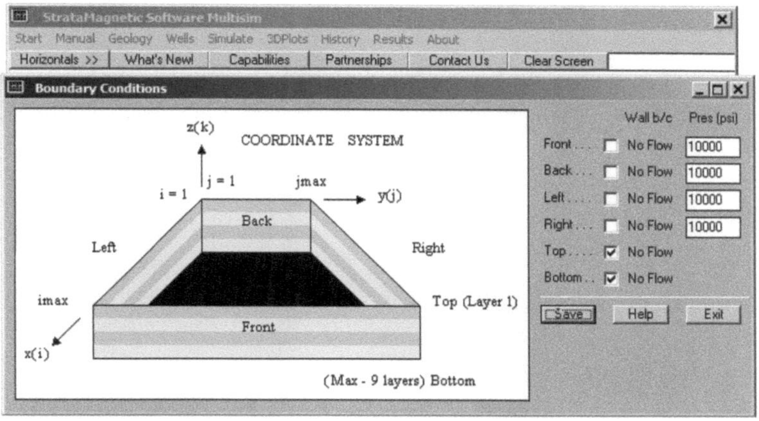

FIG. 15.7H Defining farfield aquifer drive pressures.

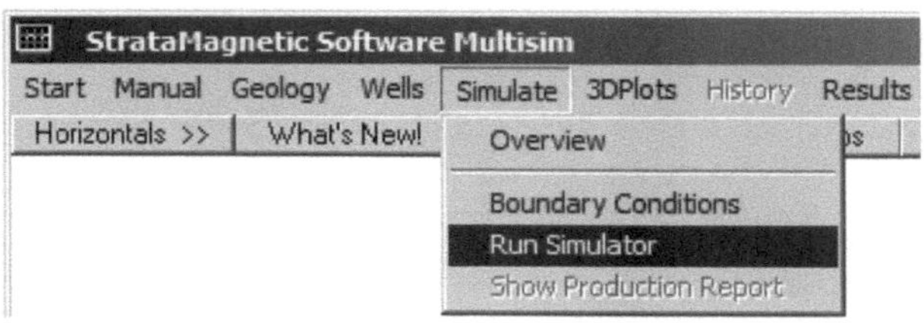

FIG. 15.7I Numerical integrations commence.

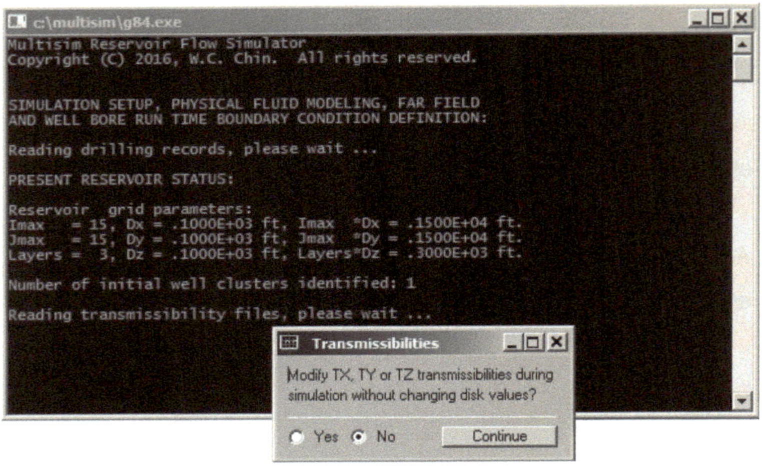

FIG. 15.7J Transmissibility modification option, select "No."

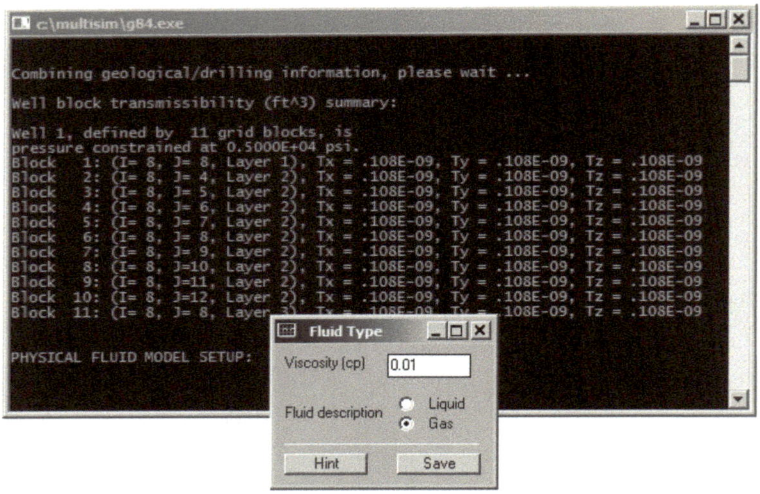

FIG. 15.7K Selecting "Gas" as the working fluid.

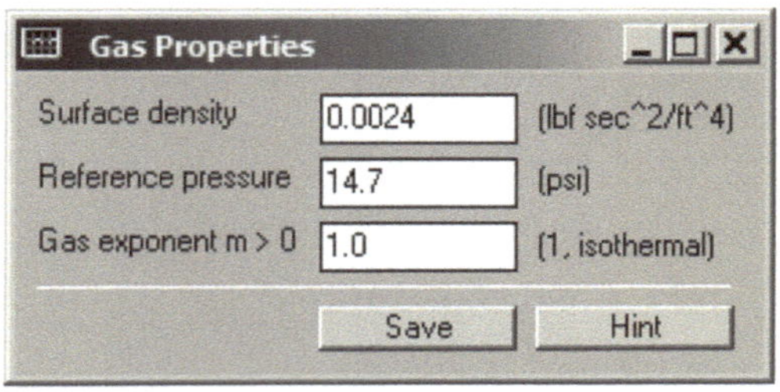

FIG. 15.7L Isothermal process assumed.

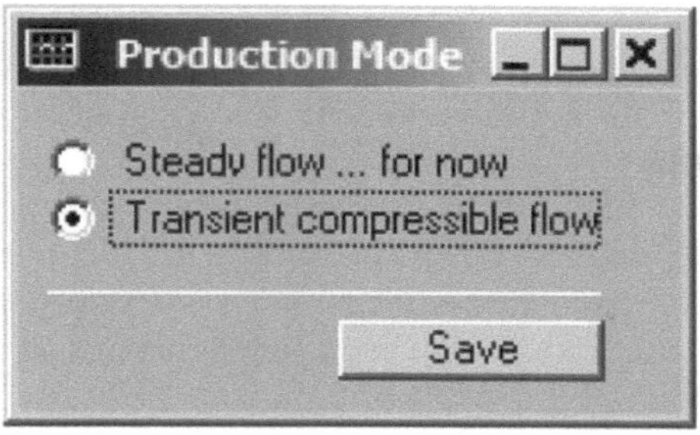

FIG. 15.7M Transient flow option selected.

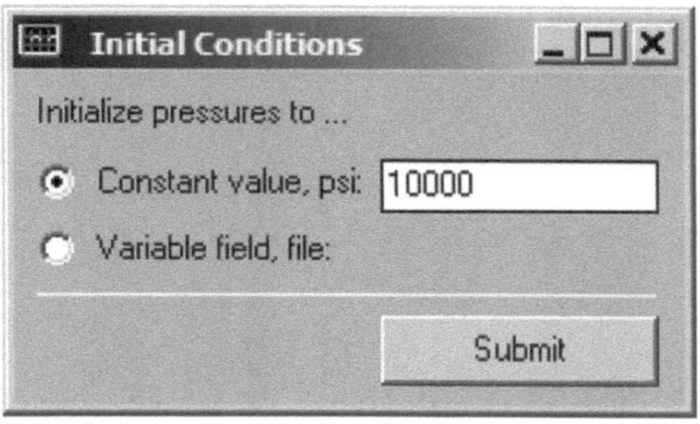

FIG. 15.7N Initial pressure level defined.

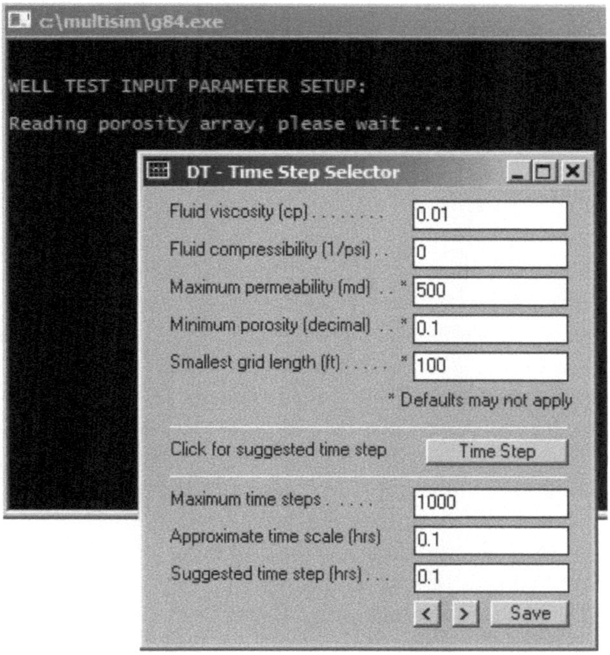

FIG. 15.7O Time step recommendations.

depend on the details of that problem. Thus, in Fig. 15.7O, this author often overrides the recommendation with 0.1 h. For gases, because viscosities are low, the bottom boxes are often left blank, and the user should type in values he deems appropriate—in the present case, we have selected 0.1 h again. "Rigid formation" versus "Small deformation" compaction options are also built into the simulator, as shown in Fig. 15.7P. However, the user is strongly advised to select only "Rigid formation" as of this writing, since the model used internally can lead to unpredictable instability in the numerical integrations. Thus, follow the suggestion offered in Fig. 15.7Q.

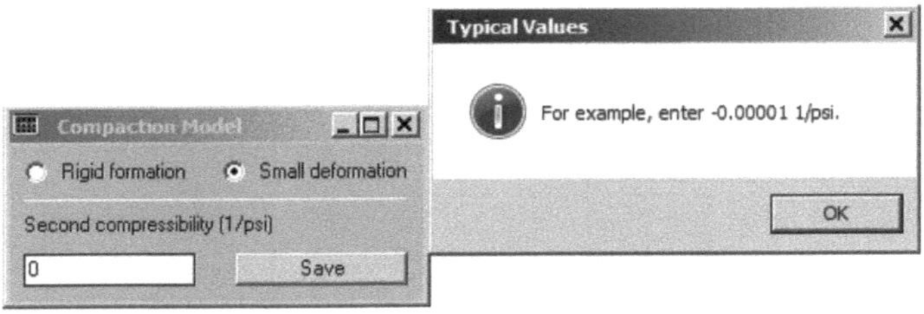

FIG. 15.7P Compaction option.

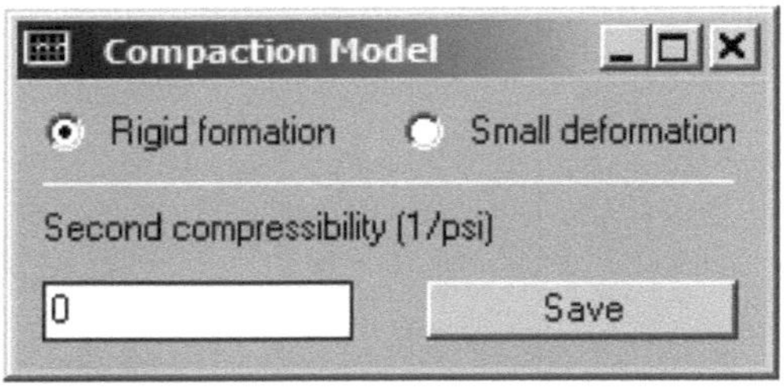

FIG. 15.7Q Rigid formation option selected.

Figs. 15.7R–15.7U display self-explanatory inputs and options discussed previously. Fig. 15.7V, however, introduces the use of "wellbore storage" effects in transient flow analysis. Our implementation, as we shall see, is numerically stable. In Fig. 15.7W, we terminate our calculations. Fig. 15.7X shows how the most recent pressure areal distribution can be stored, for example, for future plotting and analysis, or for use in pressure initialization for transient simulations as shown in Fig. 15.7N. The menus in Figs. 15.7Y, 15.7Z-1 and 15.7Z-2 illustrate the selection process needed for displays.

Figs. 15.7Z-3–15.7Z-5 provide different displays for Layer 2 pressure. In Figs. 15.7Z-6–15.7Z-8, we initiating line plotting of production flow rates for our pressure-constrained well. In Figs. 15.7Z-9 and 15.7Z-10, tabulated pressure and rate histories are shown, corresponding to our Well1.SIM

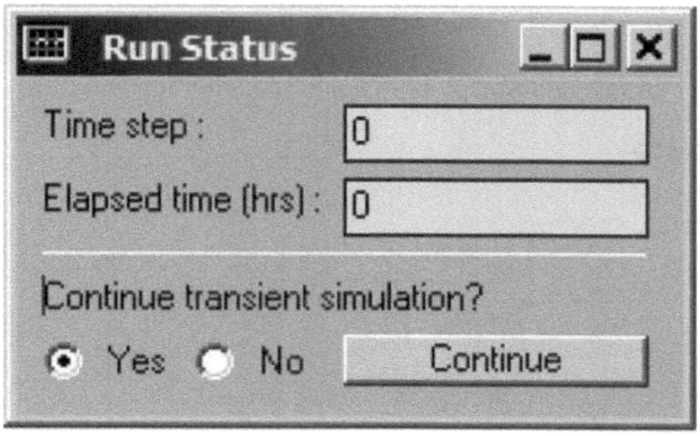

FIG. 15.7R Actual computations begin.

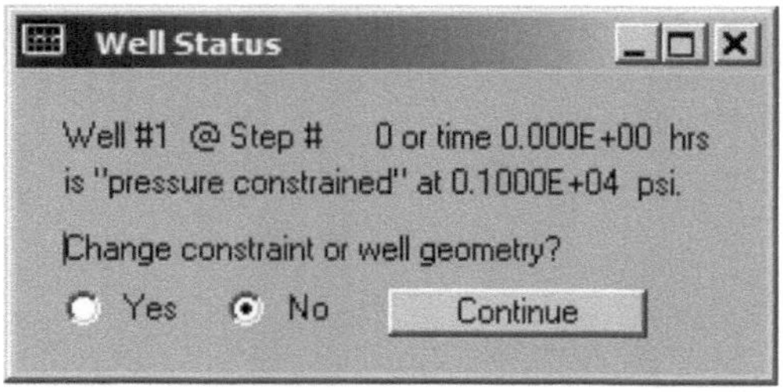

FIG. 15.7S Confirming pressure constraint.

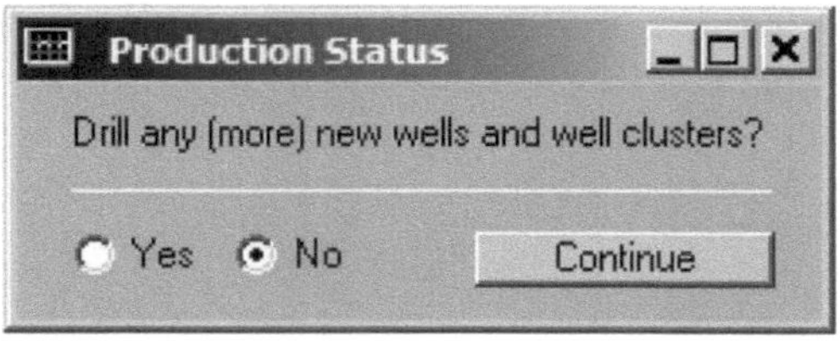

FIG. 15.7T Model existing Well 1 only.

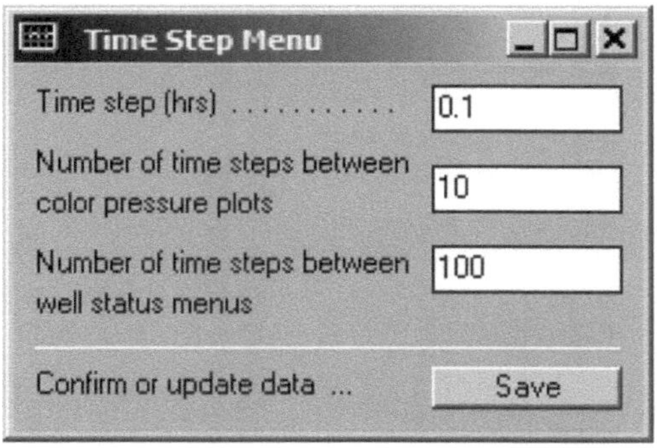

FIG. 15.7U Selecting display parameters.

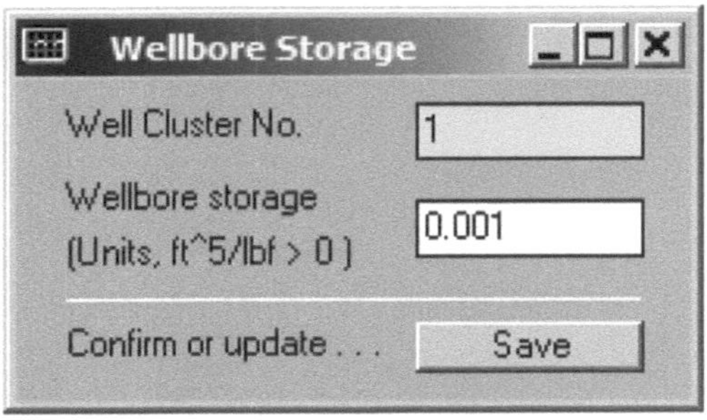

FIG. 15.7V Selecting nonzero wellbore storage factor.

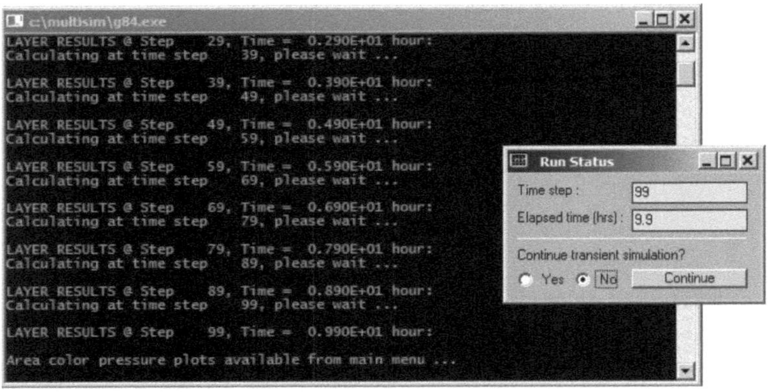

FIG. 15.7W Terminating calculations.

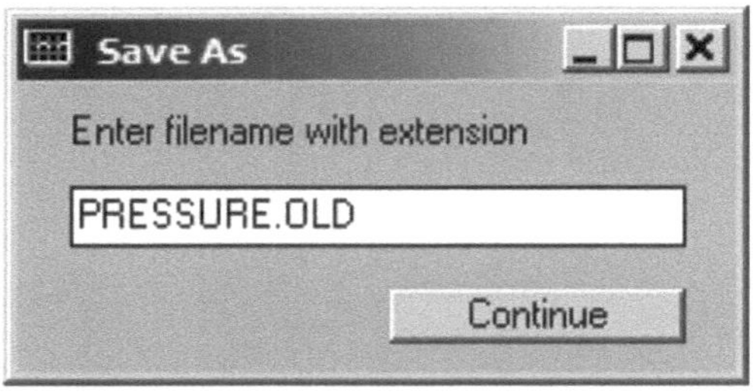

FIG. 15.7X Storing latest pressure distribution.

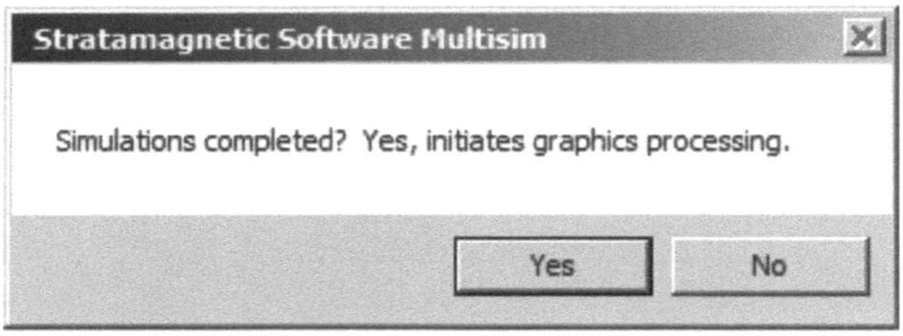

FIG. 15.7Y Initiating processing for color graphics display.

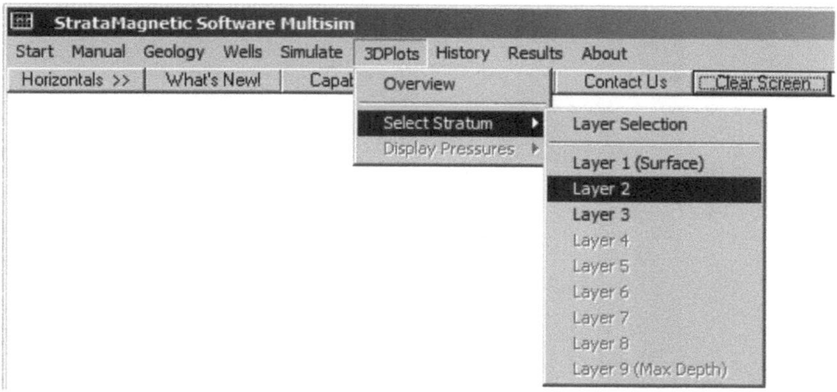

FIG. 15.7Z-1 Layer selection menu.

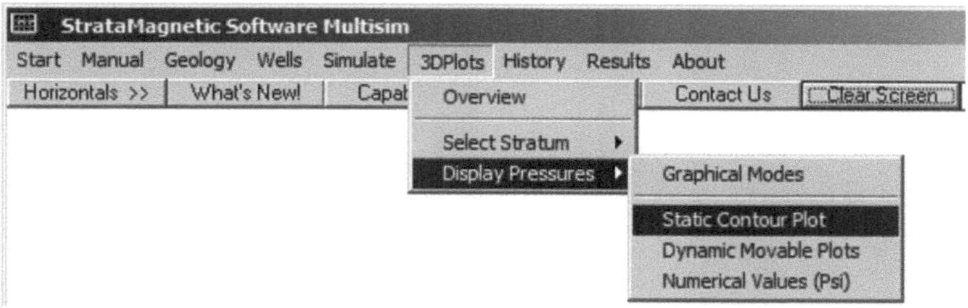

FIG. 15.7Z-2 Graphics type display menu.

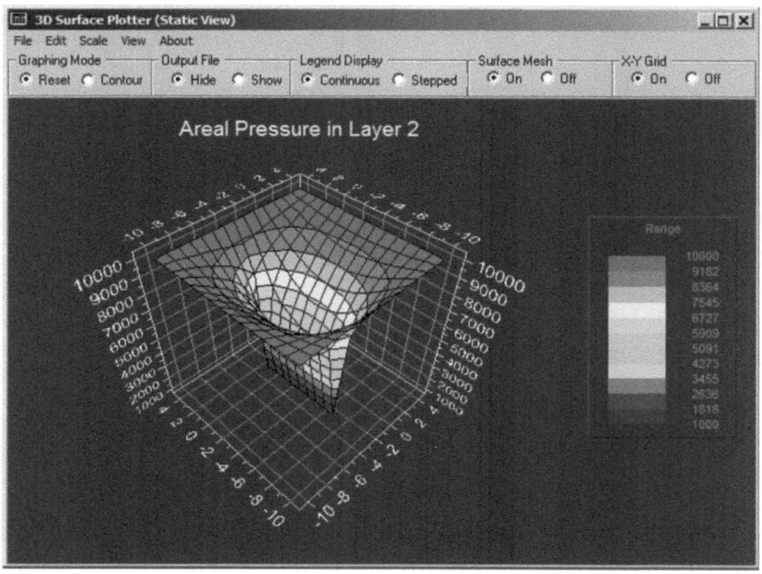

FIG. 15.7Z-3 Layer 2 pressure.

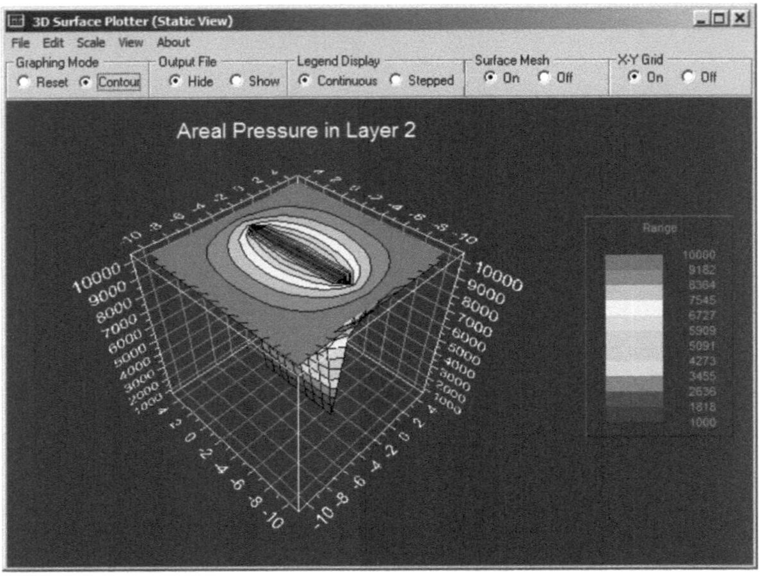

FIG. 15.7Z-4 Layer 2 pressure contour plot.

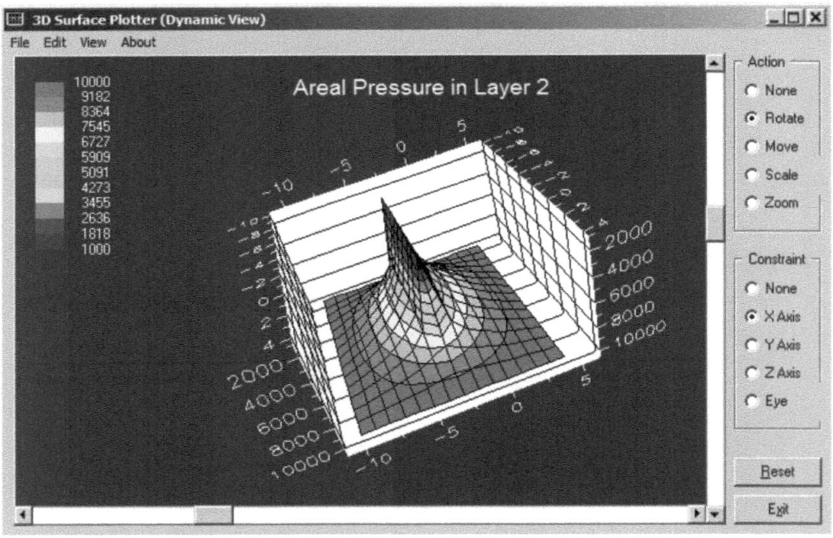

FIG. 15.7Z-5 Layer 2 dynamic pressure plot.

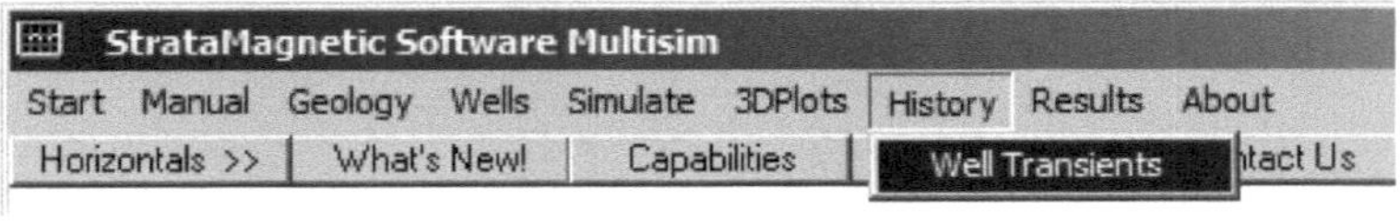

FIG. 15.7Z-6 Selecting well transient plotting.

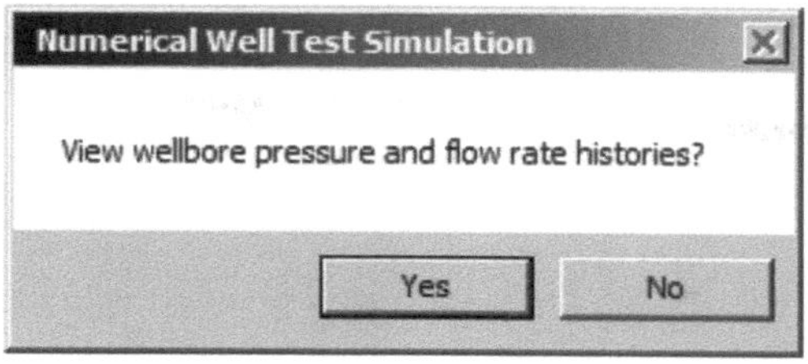

FIG. 15.7Z-7 Post-processing for well history starts.

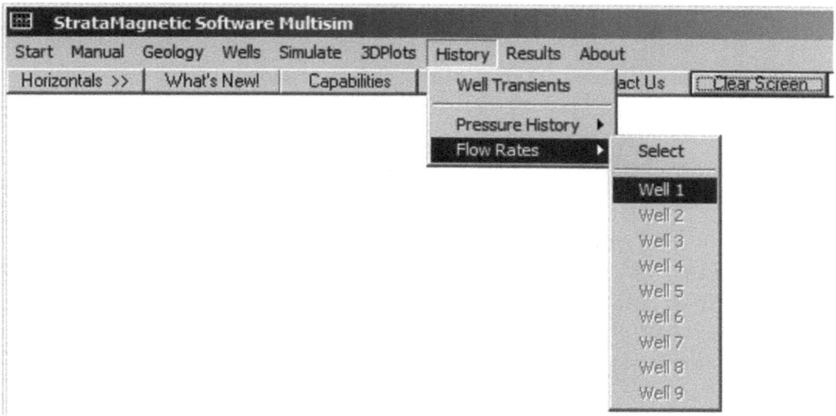

FIG. 15.7Z-8 Selecting flow rate line plots.

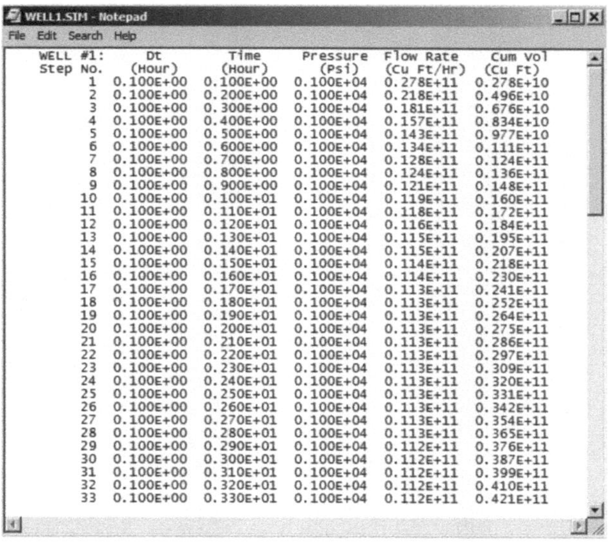

WELL #1: Step No.	Dt (Hour)	Time (Hour)	Pressure (Psi)	Flow Rate (Cu Ft/Hr)	Cum Vol (Cu Ft)
1	0.100E+00	0.100E+00	0.100E+04	0.278E+11	0.278E+10
2	0.100E+00	0.200E+00	0.100E+04	0.218E+11	0.496E+10
3	0.100E+00	0.300E+00	0.100E+04	0.181E+11	0.676E+10
4	0.100E+00	0.400E+00	0.100E+04	0.157E+11	0.834E+10
5	0.100E+00	0.500E+00	0.100E+04	0.143E+11	0.977E+10
6	0.100E+00	0.600E+00	0.100E+04	0.134E+11	0.111E+11
7	0.100E+00	0.700E+00	0.100E+04	0.128E+11	0.124E+11
8	0.100E+00	0.800E+00	0.100E+04	0.124E+11	0.136E+11
9	0.100E+00	0.900E+00	0.100E+04	0.121E+11	0.148E+11
10	0.100E+00	0.100E+01	0.100E+04	0.119E+11	0.160E+11
11	0.100E+00	0.110E+01	0.100E+04	0.118E+11	0.172E+11
12	0.100E+00	0.120E+01	0.100E+04	0.116E+11	0.184E+11
13	0.100E+00	0.130E+01	0.100E+04	0.115E+11	0.195E+11
14	0.100E+00	0.140E+01	0.100E+04	0.115E+11	0.207E+11
15	0.100E+00	0.150E+01	0.100E+04	0.114E+11	0.218E+11
16	0.100E+00	0.160E+01	0.100E+04	0.114E+11	0.230E+11
17	0.100E+00	0.170E+01	0.100E+04	0.113E+11	0.241E+11
18	0.100E+00	0.180E+01	0.100E+04	0.113E+11	0.252E+11
19	0.100E+00	0.190E+01	0.100E+04	0.113E+11	0.264E+11
20	0.100E+00	0.200E+01	0.100E+04	0.113E+11	0.275E+11
21	0.100E+00	0.210E+01	0.100E+04	0.113E+11	0.286E+11
22	0.100E+00	0.220E+01	0.100E+04	0.113E+11	0.297E+11
23	0.100E+00	0.230E+01	0.100E+04	0.113E+11	0.309E+11
24	0.100E+00	0.240E+01	0.100E+04	0.113E+11	0.320E+11
25	0.100E+00	0.250E+01	0.100E+04	0.113E+11	0.331E+11
26	0.100E+00	0.260E+01	0.100E+04	0.113E+11	0.342E+11
27	0.100E+00	0.270E+01	0.100E+04	0.113E+11	0.354E+11
28	0.100E+00	0.280E+01	0.100E+04	0.113E+11	0.365E+11
29	0.100E+00	0.290E+01	0.100E+04	0.112E+11	0.376E+11
30	0.100E+00	0.300E+01	0.100E+04	0.112E+11	0.387E+11
31	0.100E+00	0.310E+01	0.100E+04	0.112E+11	0.399E+11
32	0.100E+00	0.320E+01	0.100E+04	0.112E+11	0.410E+11
33	0.100E+00	0.330E+01	0.100E+04	0.112E+11	0.421E+11

FIG. 15.7Z-9 Steady-state flow rate achieved.

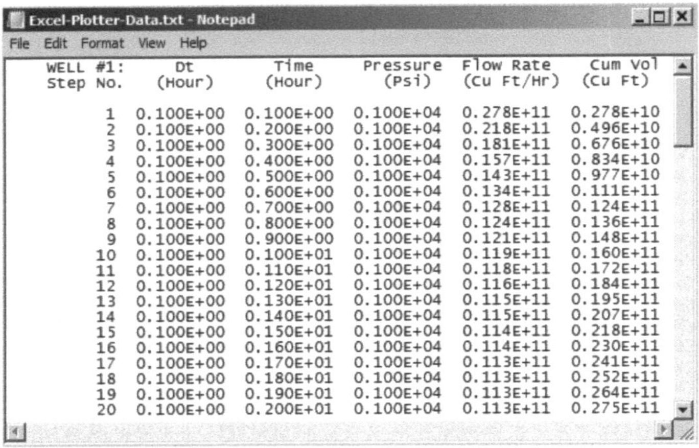
Excel-Plotter-Data.txt - Notepad

WELL #1: Step No.	Dt (Hour)	Time (Hour)	Pressure (Psi)	Flow Rate (Cu Ft/Hr)	Cum Vol (Cu Ft)
1	0.100E+00	0.100E+00	0.100E+04	0.278E+11	0.278E+10
2	0.100E+00	0.200E+00	0.100E+04	0.218E+11	0.496E+10
3	0.100E+00	0.300E+00	0.100E+04	0.181E+11	0.676E+10
4	0.100E+00	0.400E+00	0.100E+04	0.157E+11	0.834E+10
5	0.100E+00	0.500E+00	0.100E+04	0.143E+11	0.977E+10
6	0.100E+00	0.600E+00	0.100E+04	0.134E+11	0.111E+11
7	0.100E+00	0.700E+00	0.100E+04	0.128E+11	0.124E+11
8	0.100E+00	0.800E+00	0.100E+04	0.124E+11	0.136E+11
9	0.100E+00	0.900E+00	0.100E+04	0.121E+11	0.148E+11
10	0.100E+00	0.100E+01	0.100E+04	0.119E+11	0.160E+11
11	0.100E+00	0.110E+01	0.100E+04	0.118E+11	0.172E+11
12	0.100E+00	0.120E+01	0.100E+04	0.116E+11	0.184E+11
13	0.100E+00	0.130E+01	0.100E+04	0.115E+11	0.195E+11
14	0.100E+00	0.140E+01	0.100E+04	0.115E+11	0.207E+11
15	0.100E+00	0.150E+01	0.100E+04	0.114E+11	0.218E+11
16	0.100E+00	0.160E+01	0.100E+04	0.114E+11	0.230E+11
17	0.100E+00	0.170E+01	0.100E+04	0.113E+11	0.241E+11
18	0.100E+00	0.180E+01	0.100E+04	0.113E+11	0.252E+11
19	0.100E+00	0.190E+01	0.100E+04	0.113E+11	0.264E+11
20	0.100E+00	0.200E+01	0.100E+04	0.113E+11	0.275E+11

FIG. 15.7Z-10 Data selected for Microsoft Excel™ plotting.

summary file. Had we drilled other wells, similar history files would have been created for other well systems, e.g., Well9.SIM for Well 9. In Fig. 15.7Z-11, we have plotted versus time, the volume flow rate which is seen to approach steady-state together with the cumulative volume, using Microsoft Excel™. Well*. SIM files contain more data for detailed spreadsheet analysis.

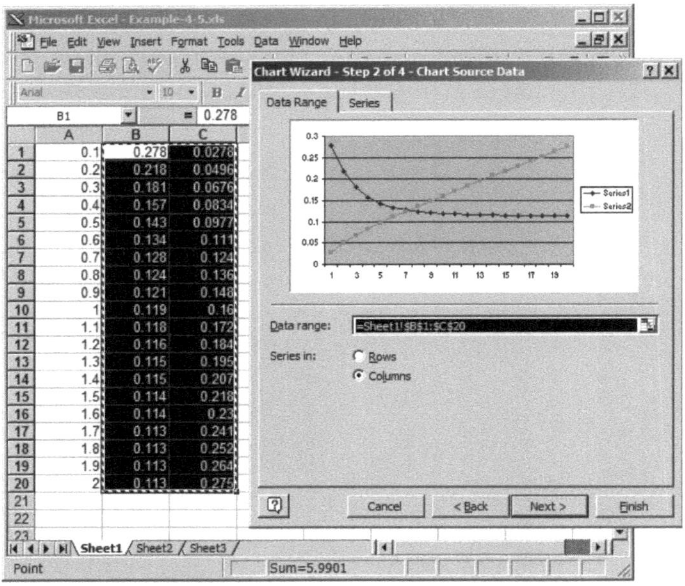

FIG. 15.7Z-11 Rate and cumulative production using Microsoft Excel™.

WELL MODELS AND PRODUCTIVITY INDEXES

Radial and three-dimensional flow modeling serve similar objectives, namely, accurate calculation of production volume and flow rate as they support economic analyses. Let R_w and R_{res} denote wellbore and farfield reservoir radii, P_w and P_{res} their corresponding pressures, k the isotropic permeability, μ the liquid viscosity and H the thickness of the circular field. When this field is produced at the center by a fully penetrating vertical well, the steady-state pressure and volume flow rate Q are $P=P_w+(P_{res}-P_w)\ (\log r/R_w)/\log(R_{res}/R_w)$ and $Q=-(2\pi kH/\mu)(P_{res}-P_w)/\log(R_{res}/R_w)$ where "r" is the radial coordinate. In transient unsteady flow, analogous time-dependent formulas are found which depend on farfield boundary conditions. These formulas are classic and accepted in the industry. For instance, the pressure equation satisfies $P=P_w$ at the well $r=R_w$ and $P=P_{res}$ at the farfield radius $r=R_{res}$. Now, from earlier Fig. 3-1-1z for Example 3-1 of Chin (2016a), it is clear that we were able to solve problems satisfying a prescribed "1000 psi" at the well and "10,000 psi" at the farfield boundary. This being so, we raised no further questions.

Radial Versus 3D Modeling—Loss of Wellbore Resolution

But, we might ask, "What happened to R_w and why doesn't it appear in our steady and transient three-dimensional formulations?" The reason is simple. To accommodate the large-scale details of the reservoir, e.g., layering, fractures, thick versus thin formations and so on, well boundary conditions were applied at nodal points of our three-dimensional grid system—thus, borehole radius would not enter on the scale of the overall reservoir model. This limitation is well known as a deficiency of finite difference modeling on rectangular mesh systems. Lee and Milliken (1993) summarize the problem concisely. "In finite difference reservoir simulation, a well is generally treated as a point source or sink. As a result, a well productivity index must be specified to relate the difference between well block pressure and wellbore pressure to the production rate." More recent related publications include Wolfsteiner et al. (2003) and Durlofsky and Aziz (2004).

Empirical fixes have been suggested by numerous researchers and all follow one common philosophy. Essentially, whereas Darcy's law for the velocity $q=-(kH/\mu)dP/dr$ applies *exactly* to radial flow, a "productivity index" PI is often used to "correct" this equation to $q=-(PI)(kH/\mu)dP/dr$ so that computed solutions in Cartesian xyz coordinates bear some semblance to reality. Different authors proposed different schemes applicable to different geologies and well geometries. None are universal and all are empirical. They are, one might call, "fudge factors" in the colloquial without any hope for redemption.

Analogies in Computational Aerodynamics

The aerospace industry dealt with similar problems in the 1970s. "Airfoils," or cross-sections of wings aligned with the direction of airflow, are usually thin and pitched at small angles to avoid massive flow separation which would destroy lift. Originally, as shown in Fig. 15.8A, rectangular computation domains were used to solve Laplace-like equations (not unlike our pressure equation) and boundary conditions related to local geometry were enforced along the solid horizontal line shown.

In reality, flows impinging at the leading edge of the airfoil are forced to turn up or down suddenly at extremely rapid speeds. These could not be captured on simple rectangular grid systems because specifications of large numbers would destabilize numerical algorithms. Thus, computational aerodynamicists "played grid games" with the so-called "leading edge problem." It was not uncommon for engineers to experiment with many grid candidates before settling on one that fortuitously matched wind tunnel results.

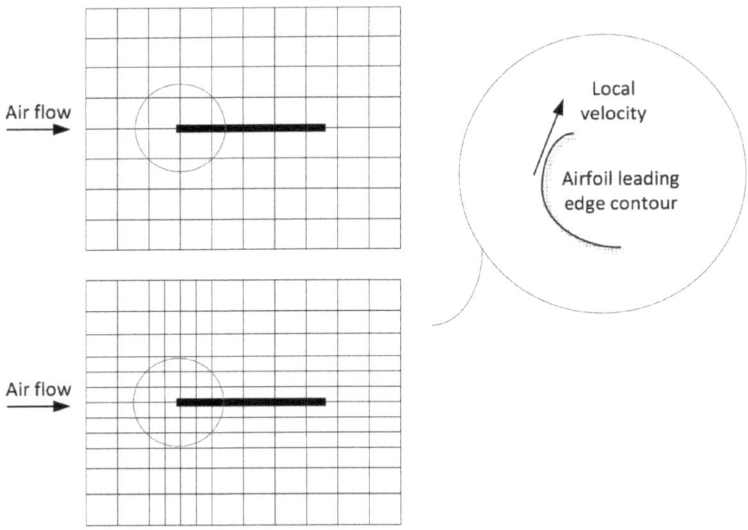

FIG. 15.8A Thin airfoil theory, leading edge details masked.

Fig. 15.8A applies to a single wing section. In general, airplane wings are built with angular "sweep" and lengths ("chords") along the spanwise direction that are not constant in length. The result is the situation Fig. 15.8B where different "fudge factors" (or, "effective leading edge" slopes) apply at different locations. The resulting models were used to match wind tunnel data. This is not unlike petroleum industry practice where different "productivity indexes" or "well indexes" apply at different stations along the wellbore and are used to match well production data. How these are obtained in any particular run vary from engineer to engineer. These methods, often masked by intimidating equations, are, in the final analysis, crude. If an "exact" solution is required, to the extent that grid dependencies are removable, one could be obtained straightforwardly using curvilinear grid systems that capture local geometric details accurately. In modern aerospace applications, grid systems like that in Fig. 15.8C *will* provide accurate solutions without using the equivalent of "productivity indexes" because they actually provide leading edge physical resolution.

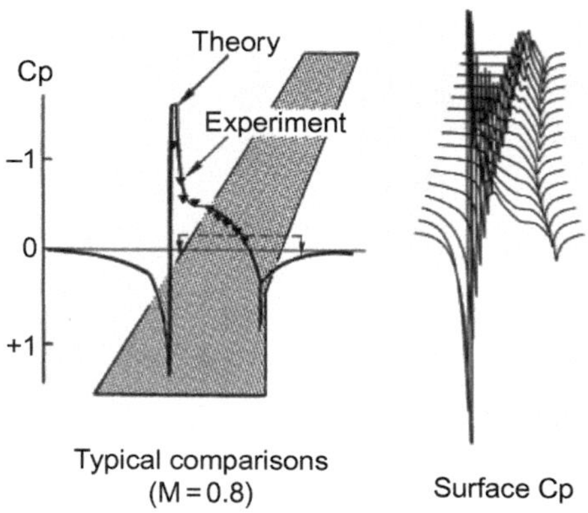

FIG. 15.8B Different correction factors applied along wing span.

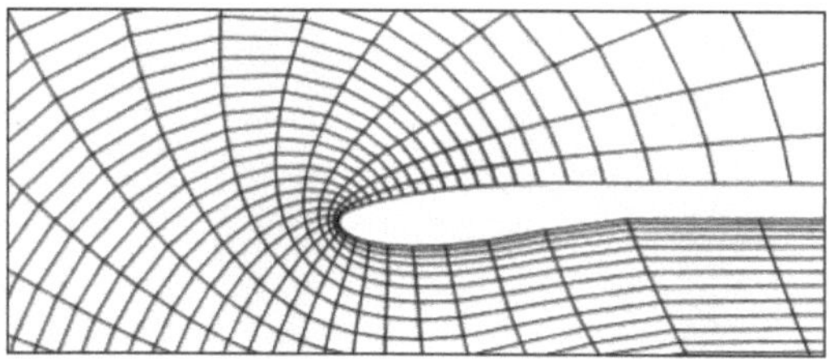

FIG. 15.8C Curvilinear mesh methods model exact details.

Curvilinear Grids in Reservoir Simulation

We have digressed from our original question, "What happened to R_w?" The foregoing discussion led to an analysis of "productivity indexes" which are commonly used in reservoir modeling—and, again, we emphasize that these are empirical and that none are generally applicable to all situations. The underlying reason lies in the choice of coordinate systems used for large-scale simulation which, by necessity, are rectangular or Cartesian because their coordinate lines coincide with stratigraphic boundaries. Grid blocks in such systems are typically hundreds of feet in length, and for practical reasons, pressures and no-flow constraints are imposed at nodes defined at coordinate line intersections. This being so, the well radius never enters; but for the time being, we are satisfied that the results of Examples 3-1 to 3-10 of Chin (2016a) are at least physically and qualitatively satisfying.

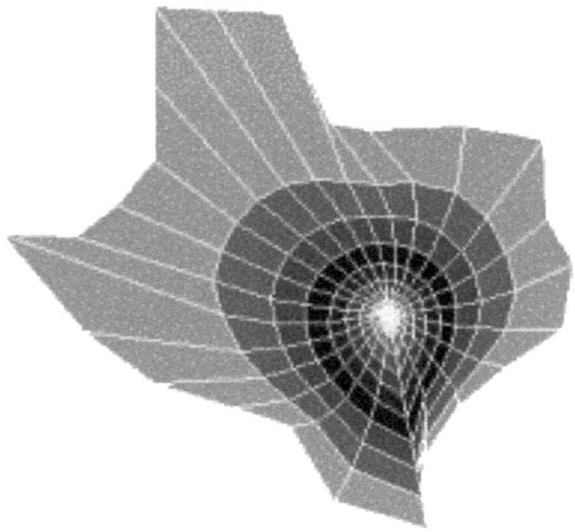

FIG. 15.8D Houston well in Texas-shaped reservoir from Chin (2002).

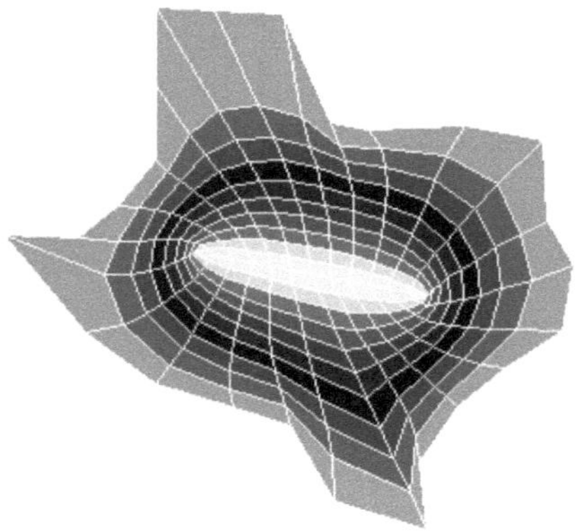

FIG. 15.8E Fracture across Texas from Chin (2002).

Had we selected a coordinate system that does accommodate a specification for borehole radius, we *would* be able to provide an explicit dependence on R_w. In fact, this author shows, in Chapter 9 of *Quantitative Methods in Reservoir Engineering* (Chin, 2002) that a general solution for pressure can be written as $P^{m+1}(\eta) = (P_R{}^{m+1} - P_W{}^{m+1})\eta/\eta_{max} + P_W{}^{m+1}$ where "m" is the Muskat exponent, taken as "0" for liquids and nonzero for gases. The function $\eta(x,y)$ is the solution of a topological (or grid generation) problem formulated and solved in that reference (which extends the logarithmic solution applicable to radial coordinates). In Chin (2002), the exact procedure for calculating $\eta(x,y)$ is explained and algorithms are given (updates to the methodology are provided in this book). The total volume flow rate Q can be similarly expressed in closed form.

Figs. 15.8D and 15.8E provide examples, the first for a "circular well in a Texas-shaped reservoir" and the second for a "fracture across Texas." Reservoirs produced by multiple wells can be analogously studied. For example, a two-well system with a boundary-conforming curvilinear grid is shown in Fig. 15.8F. The advantage behind Chin (2002) is the availability of explicit algebraic formulas written in terms of mapping functions that are produced once and for all for any given reservoir. Direct treatment of well details can, of course, be studied using finite element methods, which are numerical and do not offer as much physical insight. Such three-dimensional methods are recommended when the two-dimensional exact methods behind examples like Figs. 15.8D–15.8F cannot be easily performed.

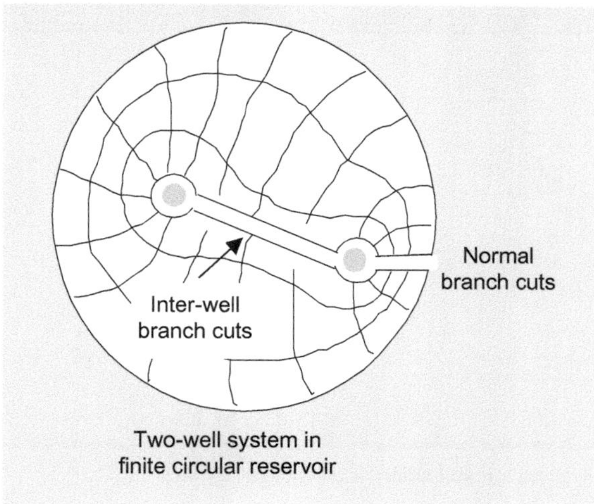

FIG. 15.8F Boundary-conforming two-well grid.

Problems also arise when vertical wells are only partially penetrating, or when reservoirs contain horizontal or multilateral wells, or both. While a fully three-dimensional finite element method (with grid structures accommodating well boundaries) will offer "exact" solutions, to the extent that grid dependencies can be eliminated, they generally do not offer the flexibility that rectangular (or near-rectangular) systems offer. As we shall find, say, from Figs. 15.3O and

15.3Z-18 for a complicated multilateral well, problem set-up and solution display can be extremely simple and offer cost and time-effective physical insight that is otherwise not possible. So we return to the question, "How do we accommodate R_w, or more generally, how do we introduce the details of wellbore topology in more complicated multilateral well systems?"

Productivity Index Modeling

The answer is straightforward. Recall that Darcy's law for the velocity q = -(kH/μ)dP/dr applies *exactly* to radial flow, and a "productivity index" PI is often used to "correct" this equation to q = -(PI) (kH/μ)dP/dr so that computed rectangular grid solutions bear some semblance to the exact solution (or, more often than not, the results of production data). A considerable amount of effort over the past several decades has addressed PI calculations—all are empirically

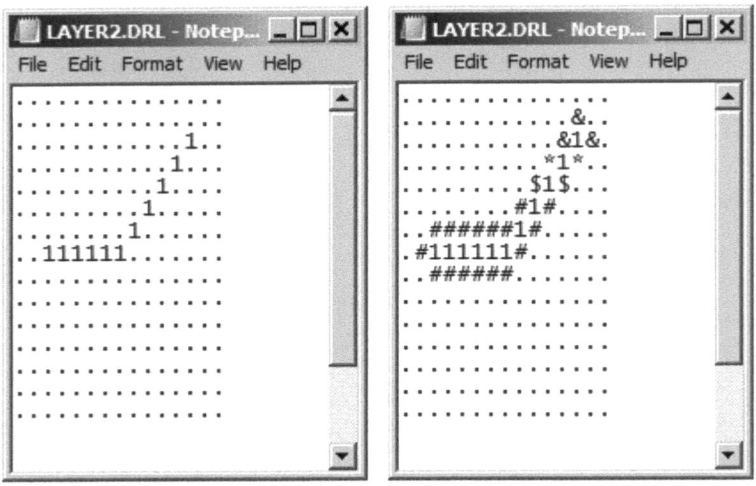

FIG. 15.8G Implementing well model or productivity indexes.

motivated, apply to very narrow problem sets, and are unlikely to be extended to modern well systems.

This author discourages readers from applying these techniques, whose esoteric published descriptions imply more rigor than is actually found. Since PI's are empirical anyway, "all methods are equal." For instance, the ideal well at the left of Fig. 15.8G could be replaced by the one at the right using our modeling approach, where local lithological changes indicated by &, #, and $ are used to represent the required effective permeability based on well test of production data (these replace the simple dots representing uniform media). The flow attributes associated with these symbols can be selected by any user-defined criteria. The procedure is simple, stable and easy to implement.

APPLICATIONS TO UNCONVENTIONAL RESOURCES

The economic significance of "unconventional resources" in recent production cannot be understated, the relatively modern term referring to tight oils and shale gases requiring stimulation via hydraulic fracturing. Fracturing is facilitated by the use of horizontal and multilateral wells, all of which are necessary to produce formations whose permeabilities normally do not exceed 0.1 md and which are functions of fluid pressures that change with production. At this writing, there are no standard petrophysical methods for formation evaluation in such applications—uncertainty abounds in extrapolating reserves and future behavior trends from production data. Suffice it to say, though, that questions related to optimum well spacing, multilateral topology, flow interference and compaction arise, but that our Multisim™ reservoir simulator is positioned to address the relevant issues. Because the methodology is standalone and well documented, "what if" questions can be answered by rapid source code prototype changes and easily visualized validations. This book will not delve into the numerous issues surrounding unconventional resources, but excellent references are available, among them the recent publications by

- Durst et al. "Unconventional Shale Play Selective Fracturing Using Multilateral Technology," SPE Paper 151989-MS, SPE/EAGE European Unconventional Resources Conference and Exhibition, Vienna, Austria, March 20–22, 2012.
- Heckman et al. "Best Practices for Reserves Estimation in Conventional Reservoirs—Present and Future Considerations," SPE Unconventional Resources Conference, The Woodlands, Texas, April 10–12, 2013.
- Ding et al. "Numerical Simulation of Low Permeability Unconventional Gas Reservoirs," SPE Paper 167711, SPE/EAGE European Unconventional Conference and Exhibition, Vienna, Austria, February 25–27, 2014.

PROBLEMS AND EXERCISES

1. Our wellbore model ignores friction and gravity for simplicity, so that p_w can be moved across the summation operator; for example, see Eqs. (15.12), (15.13) and (15.29), (15.30). Extend the finite difference model to include these effects.
2. Consider an underbalanced drilling situation where low pressures in the borehole release free gas at the shallower depths. Formulate a wellbore storage model for use in such highly compressible two-phase flows.
3. Explain the physical significance of wellbore friction on the flow rates predicted by ideal reservoir simulation models under pressure constraints. Are

they lower or higher? Why? How should computed results be used in economic forecasting?

4. The three-dimensional model in this chapter applies to single-phase flows of liquids and gases only. How should the approach be modified to handle multiphase flows where a saturation variable enters which supplements that for pressure? Refer to Chin and Proett (2005) for hints and suggestions.

Chapter 16

Fluid Mechanics of Invasion

Few subjects in petroleum engineering have elicited as much interest and practical concern among a broad spectrum of specialists, for example, workers in drilling, reservoir simulation, stimulation, and well log analysis, as the study of formation invasion. This interest has spanned decades: where hydrocarbon deposits are likely to be found, the role that invasion plays is as critical to physical understanding as it is to economics. Early on, the effects of invasion on resistivity interpretation dominated industry interest. Now, with high-data-rate Measurement-While-Drilling (or, MWD) tools on the horizon, reservoir engineers believe that formation characteristics can be reliably extracted from resistivity data, taken at different points in time, possibly using different types of electromagnetic tools, applied in conjunction with other types of logging data. This optimism has spawned a new specialty known as time lapse analysis, introduced briefly in Chapter 14. As we indicated, formation invasion is of interest in numerous applications. For example, invasion modeling is useful to squeeze cementing operations, designed to plug undesired completion perforations in well casing. It is useful in assessing formation damage due to mud contamination of near-well rock zones, thus contributing to improved production and stimulation planning. And it is vital to designing good hydraulic fracturing programs, insofar as understanding the nature and effects of proppant transport and buildup is concerned. Finally, invasion modeling is instrumental in interpreting real-time MWD measurements. How does invasion contaminate early-time data, and what corrections are possible? What types of information can be inferred from differences between early MWD and later wireline logs?

OVERVIEW

Despite wide industry interest, little in the way of scientifically rigorous flow modeling is available, and a long history of neglect and abuse has plagued invasion analysis with regard to mathematical description and understanding. Classic formulas have been systematically misused, and hundreds of papers have literally appeared, mistakenly applying a universal $\sqrt{t}$ law to areas that are entirely inappropriate. Well log analysts understand the physical role of low-to-medium permeability mudcakes in dominating reservoir invasion and have *qualitatively* described the kinds of formation information (e.g., porosity, permeability, and hydrocarbon saturation), that can be reliably inferred under

Quantitative Methods in Reservoir Engineering. http://dx.doi.org/10.1016/B978-0-12-810518-4.00016-5

different circumstances. But additional parameters affect invasion, and these must be identified and understood before *quantitative* time lapse analysis is possible. The tools needed to pursue this effort have been lacking, perhaps because the modeling ideas are specialized and not well known to the industry. Formation invasion analysis can, it turns out, be crafted on firm scientific ground. In the end, one need only realize that the invasion process represents but a Darcy flow, governed by the classic equations of reservoir engineering, although complicated by the presence of moving boundaries.

Such problems with moving boundaries, that is, the mud-to-mudcake interface and the filtrate-to-formation-fluid displacement front, or moving boundary value problems, have been addressed to some extent in the literature on interface dynamics. For example, models are available in heat transfer analysis, where "phase changes" (say, solid-to-liquid, or liquid-to-gas transitions) are prominent. Well-known solution methods have been applied to reservoir flow problems dealing with oil production from permafrost zones, where moving heat fronts are used to significantly lower fluid viscosities. And finally, reservoir engineers have modeled and monitored water breakthrough and moving saturation fronts in oilfields for decades. These examples deal with moving fronts *within* the rock. In formation invasion, this reservoir flow is further coupled to an *external* front defined by the borehole fluid and mudcake interface. Since the cake grows with time as a result of filtration at the sandface, transient mudcake growth models must be developed that are dynamically coupled to the single or multiphase flow in the rock.

Despite the complexity, it turns out that the mathematics needed to understand the complete formation invasion process is simple and easily explained. The basic ideas and principles require some calculus and only a different perspective on Darcy's equations. These ideas, all formulations, analytical solutions, and numerical algorithms with Fortran code, are presented in the remaining chapters in a readable, self-contained, down-to-earth manner. We elect to discuss invasion from the fluid-modeling point of view, because fluid flow is the principal physical mechanism: it drives and controls resistivity measurement, but is not affected by dielectrics, conductivities, or nuclear properties. Logging tool responses depend on the motion of formation fluids (e.g., resistivities move with the fluid and not vice-versa. Of course, there are disciplines where fluids and electromagnetics cannot be considered independently (e.g., plasma physics), but invasion is not one of them. Even so, invasion analysis can be intimidating because the fluid model that applies in any logging situation is never known with certainty. And since unsteady front positions are inferred from resistivity calculations that presuppose some fluid model, time lapse analysis may be a highly iterative endeavor.

This implies that time lapse logging can be subjective, drawing upon field experience. Log interpretation may direct the analyst to some of the fluid models in this book; it becomes his job to understand how his tool responds with respect to the model and to interpret his readings based on environmental variables. Log analysts who deal with resistivity interpretation and tornado charts have criticized the deficiencies behind the modeling of resistivity variations using multilayer step and even straight-line ramped profiles because they do not represent reality. Radii used in the former are arbitrarily selected, while the latter ramped profiles do not resemble actual ones with smoothed corners. The models developed later adapt single- and two-phase methods from reservoir engineering to make overly simplified models unnecessary. While the prior chapters have discussed the forward modeling of fluid flow, we shall also address the inverse problem, in which rock and fluid properties are inferred from logging tool readings. Using ideas borrowed from seismic migration, where physical models are used to image subsurface formations, we show how complicated flows can be tracked *backwards* in time to unveil a simpler flow more easily interrogated for formation properties. For example, an arbitrarily smeared, transient concentration profile can be undiffused, or migrated to an earlier time to reveal the original step discontinuity. Then, the distinct front radius so obtained can be interpreted using formulas for piston-like flows. Alternatively, saturation discontinuities can be unshocked in two-phase immiscible flow, in order to recover the original smooth flows for study.

QUALITATIVE IDEAS ON FORMATION INVASION

In this section, we qualitatively discuss the mechanics of formation invasion and describe different physical mechanisms that occur simultaneously, with a view toward quantitatively modeling these phenomena. We emphasize that all of the gross fluid motions within our formation pore spaces, whether they are miscible, immiscible, compressible, or simple, single-phase, incompressible flows, acting individually or in combination, are unaffected by electrical or nuclear properties. But the reverse is not true. Measured electrical, electromagnetic, and nuclear properties obtained by logging tools depend on the exact distribution of fluids (e.g., mud filtrate, movable and immovable connate water, oil and gas, miscible and immiscible displacements), while their time-dependent changes depend on different space and time scales that characterize each fluid component. Detailed fluid dynamics analysis takes precedence in simulation since good electrical modeling (say) is premised on the use of correct fluid distributions. True reservoir planning and economic modeling cannot take place until the uncertainty in measured parameters can be understood. Thus, reservoir flow modeling plays crucial roles at two scales: the production level which is traditionally the domain of reservoir engineers, and the logging tool level, which forces us to ask what is really being measured.

For any given resistivity logging instrument, say, tool response is complex, depending on wave frequencies, formation properties, and transmitter and receiver locations. Since the response at any point represents some type of volumetric average based on Maxwell's equations, what the exact properties are and where cannot be uniquely determined, even when all fluid motions are known. These uncertainties are in addition to those posed by fluid flow. What happens near the well? During oil well drilling, drilling fluids or muds, are circulated down the drillpipe, through the drill bit nozzles, and finally, up the borehole annulus to the surface (see Fig. 16.1). Muds serve numerous practical purposes, for example, lubricating the drill bit, bringing cuttings to the surface, delivering ingredients vital to downhole formation chemistry, and providing increased hydrostatic pressure for well control. While some wells are drilled using water and brines as the primary drilling fluid, the great majority contain solid particles that are suspended in the fluid by viscosifiers. These solids (e.g., barite) increase the weight of the fluid column and allow drillers to drill overbalanced (as opposed to underbalanced), with the excess pressure over the formation pressure preventing fluid influx and possible blowouts.

During those initial instants when the drill bit exposes the virgin formation to drilling mud, pore bridging is accomplished by mud particles that migrate into those pore spaces very close to the rock surface, forming internal mudcake. In this spurt invasion phase, whole mud enters the formation quite freely, although the overall motion still satisfies Darcy's law for low Reynolds number flows. The rate of spurt invasion and the total volume of mud deposited as spurt loss depends, of course, on the relative sizes of the mud solids and the pore throats, and the geometrical arrangement and connectivity of the latter. The amount of spurt lost into the formation is important to early-time resistivity interpretation. Since a relatively large volume of fresh filtrate can combine with saline formation waters at small radii, the interaction with logging instruments during this time period can be important.

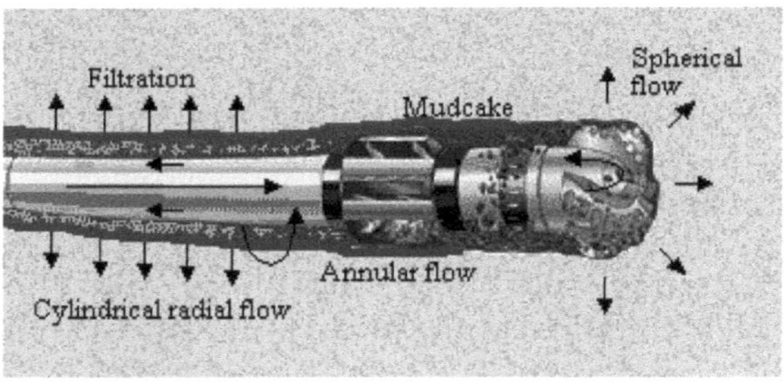

FIG. 16.1 Annular flow and filtration in the borehole.

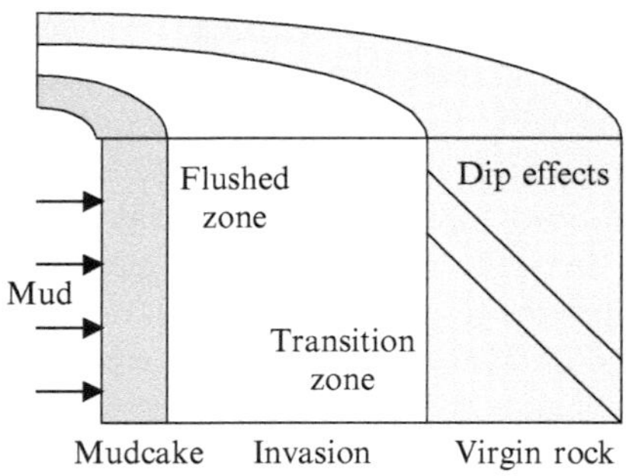

FIG. 16.2 Fluid invasion in the reservoir.

Once internal cake forms and pore bridging stabilizes, external mudcake buildup becomes more pronounced. (Internal cake can be viewed as a separate layer in a multilayer scheme, but for simplicity, we shall book-keep its effects together with those of the rock.) As drilling mud is forced into the formation under high pressure, its constituent solids are left behind outside the invaded rock. This mudcake thickens with time as filtrate continues its penetration into the rock. During transient invasion, mudcake thickness increases, providing ever-growing resistance to flow and thus causing consequent declines in filtration rate. This process is known as static filtration, noting that invasion *is* ongoing and there is nothing static about it. In dynamic filtration, the invasion process acts in the presence of mud flowing parallel to the cake surface. This flow gives rise to viscous shear stresses at the surface of the mudcake that act to erode it, or possibly convect moving solids away in such a way that cake never forms. At some critical invasion rate, dynamic equilibrium is achieved and cake thickness remains constant. In this limit only does two-layer, concentric, steady flow modeling apply. In general, the critical rate will depend on whether the annular flow is laminar or turbulent, and in either case, the degree to which it obeys power law, Bingham plastic, Herschel-Bulkley, or other rheological flow models. Fig. 16.1 shows the basic processes that occur within the borehole annulus, while Fig. 16.2 outlines invasion at various depths. Note the possibility of formation dip, bedding, and shoulder effects. Also, the invaded zone is broken down into flushed zones dominated by filtrate and transition zones where invading and displaced fluids are present in comparable quantities.

What happens beyond the borehole is actually more complicated than Fig. 16.2 suggests and is just as problematic as the flow near the borehole. Initially, an untapped hydrocarbon reservoir may contain large concentrations of oil and gas, together with smaller proportions of immovable connate water. As water-base filtrate invades the water-wet, hydrocarbon-bearing formation, a saturation front separating invading water from preexisting oil and gas propagates into the formation. If permeabilities in the formation greatly exceed those in the cake, then mudcake and formation flows decouple, and flow rates are dictated by the cake. On the other hand, in tight zones where both permeabilities are comparably small, the

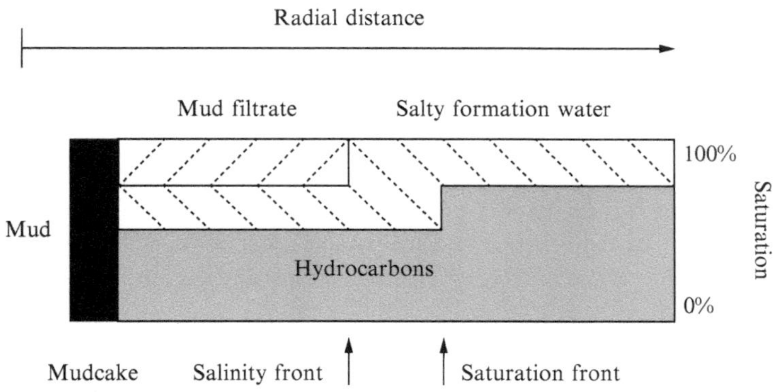

FIG. 16.3 Detailed displacement process in the reservoir.

flows are dynamically coupled: cake buildup is controlled by slow flow in the formation, which in turn decreases the rate at which cake grows in thickness.

As the oil and gas retreat further into the formation, original connate water is left behind, which mixes with newer water. If the two waters have different salinities, for example, the result of fresh mud filtrate displacing salt-saturated formation water, ionic diffusion takes place. The invading bank of water of mud filtrate salinity displaces water having formation water salinity. The two banks of water, each characterized by different electrical properties, are separated by a moving "salinity front" which typically lags the saturation front. The process described so far is diagrammed in Fig. 16.3, which shows residual oil and formation water left behind by moving filtrate.

The filtrate invasion of oil saturated rock, which displaces resident oil and water, causing displaced formation water to concentrate into a bank advancing ahead of the filtrate, results in a region sometimes called an annulus (not to be confused with the borehole annulus) comprising high formation water saturation at the boundary of an invaded oil- or gas-bearing interval. If a hydrocarbon-saturated interval is invaded by filtrate fresher than formation water, the annulus will have a lower resistivity than either the flushed or the undisturbed formation. Gravity effects and differences in vertical and horizontal rock permeability will complicate the interpretation of invaded zone well logs. A discussion on interpretation subtleties is given by Broussard (1989).

We have spoken of saturation fronts propagating into oil pools, and moving salinity fronts that follow saturation fronts. Such terminology presumes that some type of rapid motion is ongoing. If this is the case, then piston-like, slug, or plug-like descriptions of fluid motion are possible, and these form the basis of the simple models that we shall study first. Under the circumstances, the displacement of salty formation water by fresh water filtrate, *and* the displacement of oil and gas by incoming water, both shown in Fig. 16.3, are amenable to simplified plug modeling. These two types of flows are quite different. The first water-to-water flow is a single-phase flow, augmented by miscible diffusion processes, that complicates log interpretation because different salinity distributions lead to different tool responses. The second is an immiscible multiphase flow (with possible shock discontinuities) whose description is characterized by nonlinear relative permeability and capillary pressure effects. Each of the foregoing flows possesses its own time, space, and rate scales. Whether a flow is rapid, or to the contrary slow, depends on the parameters particular to a specific problem, and especially on the value of the relevant dimensionless numbers defining that problem. The physical implications arising from having two distinct phenomena simultaneously present, namely, salinity equilibration through ionic diffusion between filtrate and irreducible trapped formation waters, and capillary pressure and relative permeability dominated immiscible flows, must be understood. In either flow, idealized discontinuities may in fact be wide, and fronts may crawl and not propagate. A sharp salinity front, for example, may coexist with a wide capillary-pressure-dominated transition zone.

The most frequently cited example encompassing all of these elements has fresh mud filtrate invading a zone containing both salty formation water and movable hydrocarbons. Moving from the wellbore, resistivity at first decreases and then increases: this requires the development of both a salinity front and a saturation front, which can result only from moved hydrocarbons, creating the so-called annulus effect described earlier. Other types of resistivity distributions are possible, for example, that are due to invasion by oil-base mud filtrates into water and hydrocarbon-bearing zones. As if these possibilities were not complex enough, there are always local permeability heterogeneities that speed up or slow down fronts, effects that are to be distinguished from the viscous fingering that arises from flow instabilities. Gravity effects, which are important when fluids have contrasting densities, result in vertical flows. Since hydrocarbons are typically lighter than fresh waters, which are in turn lighter than saline waters, vertical flow movements can be induced by density variations. Fluid flows are further complicated by differences between horizontal and vertical permeabilities, which can be large. The survey article by Allen et al. (1991) provides a qualitative review of invasion. It cites interesting logging examples, showing how invasion manifests itself and how information about permeability and porosity can be inferred from resistivity tool measurements.

BACKGROUND LITERATURE

By far, the most often cited work is the early paper by Outmans (1963), which applied differential equation-based filtration methods developed in chemical engineering to static and dynamic invasion in the borehole. In this single-phase flow study, where lineal flow was assumed and the applied differential pressure was completely supported by the mudcake, Outmans derived the well-known $\sqrt{t}$ law, subject to the further proviso of cake incompressibility. (The effects of cake nonlinearity and compaction can be important over time; e.g., see Fig. 14.7.) Thus, the $\sqrt{t}$ law cannot be used when the net flow resistance offered by the formation is comparable to that of the mudcake (e.g., thin muds in permeable formations, or thick muds in very impermeable rocks). Also, the law does not apply to slimholes, where the radial geometry is important. Finally, the $\sqrt{t}$ law does not generally apply to reservoirs with two-phase, immiscible flow, or miscible flow, or both. Only under these restrictive assumptions does Outmans' correctly derived law hold.

But the $\sqrt{t}$ law has evolved over the past three decades into the industry vehicle by which well logs are interpreted, and it is commonly used in situations that are inappropriate. Literally hundreds of papers have been written. It is not the purpose of this book to survey this vast literature. To keep our objectives focused, only those experimental, analytical, and recent review papers deemed material to the development of the predictive models in this book are referenced here. Before we proceed with our analysis, let us consider some approaches used in the industry. The study of Phelps et al. (1984) dealing with invaded zone characteristics and their influence on wireline log and well test interpretation, like many investigations on two-phase flow invasion, assumes that mudcake controls the overall flow rate into the formation and, in particular, invokes the $\sqrt{t}$ law. This usage is acceptable for higher permeability reservoirs, but there is no real reason why this simplified formulation should apply generally to reservoirs containing multiple immiscible fluids or, for example, to problems dealing with flows in tight gas sands. Thus, this accepted industry practice is both unnecessary and unwarranted. In Chapter 21, the coupled boundary value problem governing cake growth and immiscible invasion is transformed into a simple reservoir flow formulation with a source-like boundary condition, which is solved using finite difference methods.

Semmelbeck and Holditch (1988) investigated mud filtrate invasion effects on induction log interpretation. In their Abstract, they state that "a rigorous model of mud filtrate invasion has been developed with a numerical fluid flow simulator." But cake permeability was simply set to 0.001 md for the first 24 h, and it was reduced to 0.00001 md thereafter. Tobola and Holditch (1989) determined *small* formation permeabilities by history matching the change in induction tool response over time, using the immiscible-diffusive finite difference fluid flow simulator just discussed. But for *small* permeabilities, the interaction between reservoir flow and cake growth cannot be ignored. Although the authors state that mud filter cake permeability profile with time must be accurately simulated to properly interpret logging data, they do not follow that suggestion. Their claims behind a successful history match were premised upon ad hoc assumptions (e.g., mudcake permeability was fixed at an arbitrary 0.001 md for the duration of the 30-day simulation). Not only was the milli-Darcy cake permeability level arbitrarily chosen, but the transient character of the invasion and mudcake growth was neglected. As noted earlier, since mudcake develops differently on low-permeability formations than it does in laboratory filter press tests, precisely the type of low perm rock the authors targeted with time lapse analysis, it is unclear from their paper how laboratory mudcake data obtained under idealized conditions is to be used successfully in any inverse application.

Holditch and Dewan (1991), again dealing with formation permeability prediction from time lapse analysis measurements during and after drilling, introduce an additional adhesion fraction deemed to be important to dynamic filtration. This parameter characterizes how resilient mudcake responds to the shear stress imparted by passing annular flow. Certainly, its relevance is clear and well accepted, but introducing terminology merely acknowledges a problem and leaves fundamental issues unresolved. Their adhesion fraction, analogous to the critical invasion rate due to Fordham et al. (1991), does not model uncertain but important erosion. Dewan and Holditch (1992), again using the limiting empirical $\sqrt{t}$ law as the basis for analysis, take their work further and calculate radial response functions for borehole logging tools on this ad hoc basis. In this book, the quantitative basis for *predictive* evaluation is developed for both static and dynamic filtration from first principles.

Lane (1993) presented a refreshing discussion on numerical filtrate invasion simulation from the viewpoint of reservoir engineering. Noting that the step invasion profile assumed in developing resistivity correction (or tornado) charts represented more the exception than the rule, he applied the two-phase flow concepts used in waterflooding to model diffusion processes and capillary effects. The discussion is enlightening from a physical viewpoint, and although a high-quality simulator was used, only qualitative results were offered. As is typical, filtrate invasion rate was specified as an input to the simulator, with mudcake properties totally controlling the fluid loss rate. The best references on invasion are the experimental papers by Doll (1955) and Gondouin and Heim (1964), pointing to the complications that the real world imposes. Numerous nonideal effects are present. The former paper, for example, notes that mud filtrate, being less saline than the interstitial water originally located in the pore spaces, with its density appreciably smaller, will tend to ascend. This gravitational effect, due to density differences, acts in concert with radial pressure gradients that arise from differences between borehole mud and reservoir pore pressure. Using radioactive tracers, the latter paper explained how extensive transition zones between flushed and uncontaminated zones existed. (The authors carefully used lineal cores cut parallel to bedding planes.) Physical mechanisms of importance included capillarity, which draws formation water from the annulus into the uncontaminated zone, and the miscible displacement of formation water by mud filtrate, together with ionic diffusion. These features hint at the inadequacy of the simplified models used to model formation resistivity distributions, namely, step, slope, and ramp profiles, and of course, the limiting $\sqrt{t}$ law. Given the profusion of recent papers on invasion, especially its effect on logging tool response and time lapse analysis, a direct response to invasion's role and growing importance to MWD, the need for rigorous mathematical modeling and clearly stated assumptions and limitations is critical now more than ever.

Dewan and Chenevert (1993) noted that no method for calculating transient mudcake buildup and the corresponding invasion rate variation in low-permeability formations was available in the literature. Actually, the first solution that *did* solve for dynamically coupled mudcake growth and filtrate displacement of a dissimilar fluid in a rock core, was given by Chin et al. (1986). In that 1986 paper, it is explained how, assuming incompressibility, three separate Laplace equations governed mudcake, flushed zone, and virgin rock, and how they should be coupled using pressure and velocity continuity matching conditions at common boundaries; exact pressure magnitudes are, of course, specified at extreme boundaries. Then, the moving boundary value problem formulation is completed by introducing differential equations that describe the Lagrangian kinematics of the interface and the constitutive buildup properties of the mudcake. These ordinary differential equations, together with the standard Darcy pressure equations, form a coupled system whose exact, closed-form, analytical solution is displayed in Fig. 16.4. While that published solution was exact, making no assumptions about the relative mobilities behind cake and rock fluid flows, its extension to radial flows was not as straightforward because of algebraic difficulties. To solve the radial problem, for example, zero spurt was assumed, rendering the unpublished solution of limited value in studying fresh-to-saline water resistivity averaging. Also, numerical integration was employed as a last resort because of overwhelming algebraic complexities.

These difficulties have now been removed. For incompressible mudcakes, filtrates, and formation fluids, the single-phase radial flow problem allowing dynamical interaction between cake and reservoir flow, including front displacement by fluids with dissimilar viscosities, plus nonzero spurt, is solved, and the exact, closed-form, analytical solution is given. This general solution uncovers the dimensionless groups that control mudcake formation and displacement front motion, and has led us to postulate different flow phenomena that are obviously possible but not as yet considered. For example, in a lineal flow where the right-side outlet boundary is fixed (e.g., see Fig. 16.4), neglecting mudcake for now, the displacement of oil by less viscous water leads to an accelerating front as oil is forced from the core sample. By the same token, the displacement of water by oil produces a decelerating front. Whether the same mobility ratios lead to acceleration or deceleration in a radial flow, of course, will depend on the severity of geometric spreading, which in turn depends on the hole radius. Geometric spreading (or lack of it) affects filtration rate and therefore cake buildup. The smaller the hole, the more important radial effects are: hence, slimholes may behave differently from normal holes.

Now, the introduction of mudcake leads to unforeseen, but interesting real events. Consider cake buildup where a water filtrate displaces a core saturated with more viscous oil. At first, when the mudcake is thin, the usual $\sqrt{t}$ law governing cake growth applies so long as the flow is geometrically linear and the formation flow is much more mobile than that in the cake. Mudcake growth will initiate rapidly and slow down, as expected. However, at some critical time, cake growth may accelerate as the more viscous oil disappears, and rapid frontal advance draws increased amounts of filtrate flow and hence increased solids deposition. Transient effects such as these certainly introduce new twists in well log interpretation and suggest that parallel problems in well test and repeat formation testing analyses are likely to be no less subtle. In this book, we shall discuss the physics of such water-oil-cake flows and extend our discussions to handle diffusion, compressibility, and immiscible flow effects.

DARCY RESERVOIR FLOW EQUATIONS

Petroleum engineers use partial differential equation models to simulate reservoir flows, to interpret well tests, to characterize formation heterogeneities, and to assist in infill drilling planning and secondary recovery. Many hierarchies for fluid flow modeling exist, ranging from simple single-phase oil alone or gas only flows to multiphase descriptions, encompassing both miscible and immiscible limits, to black oil and compositional models. In this book, we shall address all but the latter two flow models. Since invasion modeling requires a slightly different perspective than that taken in earlier chapters, it is worthwhile to recapitulate the basic governing equations.

Single-Phase Flow Pressure Equations

Fluid flows are governed by partial differential equations. For example, single-phase flows of constant density liquids in homogeneous, isotropic media satisfy Laplace's equation for pressure,

Dynamically Coupled Lineal Flow

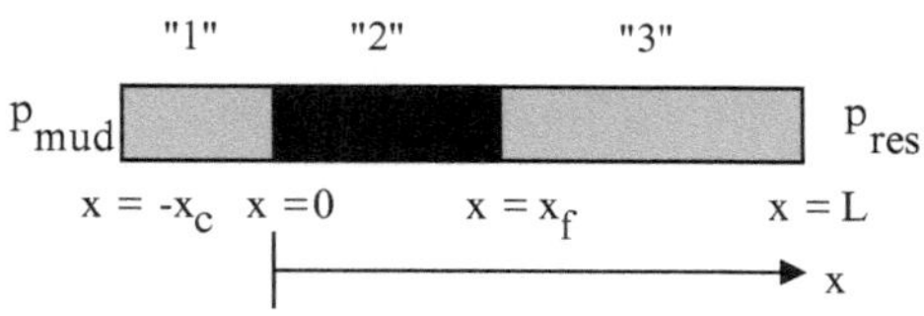

Exact: $x_f(t) = -H + \sqrt{\{H^2 + 2(Hx_{f,o} + \tfrac{1}{2} x_{f,o}^2 + Gt)\}}$

where:

$$G = -\{k_1(p_m - p_r)/(\mu_f \phi_{eff})\} / \{\mu_o k_1/\mu_f k_3 - k_1/k_2 - \phi_{eff} f_s/\{(1-\phi_c)(1-f_s)\}\}$$

$$H = [x_{f,o} \phi_{eff} f_s/\{(1-\phi_c)(1-f_s)\} - \mu_o k_1 L/\mu_f k_3] / \{\mu_o k_1/\mu_f k_3 - k_1/k_2 - \phi_{eff} f_s/\{(1-\phi_c)(1-f_s)\}\}$$

Nomenclature:

- x_c — transient cake thickness
- x_f — transient invasion front
- $x_{f,o}$ — initial displacement (i.e., *spurt*)
- L — lineal core length
- p_m — constant mud pressure P_{mud}
- p_r — constant reservoir pressure P_{res}
- ϕ_{eff} — effective rock porosity
- ϕ_c — mudcake porosity
- k_1 — mudcake permeability to filtrate
- k_2 — rock permeability to filtrate
- k_3 — rock permeability to "oil"
- μ_f — mud filtrate viscosity
- μ_o — viscosity of "oil" or formation fluid
- f_s — mud solid fraction

FIG. 16.4 Exact lineal invasion solution (Chin et al., 1986).

$$\partial^2 p/\partial x^2 + \partial^2 p/\partial y^2 + \partial^2 p/\partial z^2 = 0 \tag{16.1}$$

an elliptic differential equation, while a slightly compressible liquid in the same medium satisfies

$$\partial^2 p/\partial x^2 + \partial^2 p/\partial y^2 + \partial^2 p/\partial z^2 = (\phi \mu c/k)\partial p/\partial t \tag{16.2}$$

In Eqs. (16.1), (16.2), the fluid pressure p(x,y,z,t) is the unknown dependent variable, and x, y, z, and t represent independent variables. The standard symbol ∂ denotes partial derivatives (subscripts will also be used). The quantities ϕ, μ, c, and k denote rock porosity, fluid viscosity, fluid-rock compressibility, and isotropic formation permeability, respectively. Eq. (16.2) is the classical parabolic or heat equation for pressure, so-called because it was first derived and solved in the context of heat transfer (Carslaw and Jaeger, 1946, 1959). On the other hand, for gas flows, under the same assumptions,

$$\partial^2 p^{m+1}/\partial x^2 + \partial^2 p^{m+1}/\partial y^2 + \partial^2 p^{m+1}/\partial z^2 = (\phi \mu c^*/k)\partial p^{m+1}(x,y,z,t)/\partial t \tag{16.3}$$

where m is a nonzero exponent. In Eq. (16.3), the pressure-dependent, "compressibility-like" quantity

$$c^* = m/p(x,y,z,t) \tag{16.4}$$

renders the boundary value problem nonlinear. From a practical viewpoint, this means, say in well test interpretation, that superposition methods do not apply: the sum of individual solutions is itself not a solution. Analytically, nonlinearity implies that the possibilities for closed-form solutions are rare. But our solution methods for compressible flows, even for linear problems, will be numerical anyway. The nonlinear equation as shown in Eq. (16.3), first given by Chin (1993a,b), simply recasts Muskat's exact equation in an analogous form preferable for numerical analysis (Muskat, 1937). The resulting equation is linear-like in appearance and thus allows us to readily adapt working linear numerical schemes to nonlinear problems.

We observe, from classical heat transfer and fluid mechanics, that the constant exponent m describes the thermodynamics of the gas motion. In particular, it is known that

$$\begin{aligned} m &= 1, \text{ for isothermal expansion} \\ &= C_v/C_p, \text{ for adiabatic expansion} \\ &= 0, \text{ for constant volume processes} \\ &= \infty, \text{ for constant pressure processes} \end{aligned} \tag{16.5a,b,c,d}$$

where C_p is the specific heat at constant pressure, and C_v is the specific heat at constant volume. In inhomogeneous, anisotropic media, petroleum *liquids* are governed by

$$\partial(k_x \partial p/\partial x)/\partial x + \partial(k_y \partial p/\partial y)/\partial y + \partial(k_z \partial p/\partial z)/\partial z = \phi \mu c\, \partial p/\partial t \tag{16.6}$$

where k_x, k_y, and k_z are nonuniform permeabilities in the x, y, and z directions. Analogous equations for gases are summarized in Chapter 1.

There is more to formation invasion than Darcy's law $q = -(k/\mu)\ \partial p/\partial x$. Eqs. (16.1)–(16.6) are derived using Darcy's law, a low Reynolds number approximation to the Navier-Stokes momentum equations, in conjunction with a requirement for mass conservation. It is almost never correct to approach simulation by setting, say $q = -(k/\mu)\ \partial p/\partial x = \text{constant}$, to solve a problem, since this does not account for the underlying lineal, radial, or spherical geometry, or for pressure boundary conditions. Yet, this is often done: Darcy fluid mechanics *requires* the solution of pressure boundary value problems.

Problem Formulation

Partial differential equations such as those in Eqs. (16.1)–(16.6) require auxiliary conditions that fix any and all degrees of freedom. Just as the ordinary differential equation

$$d^2p(x)/dx^2 = 0 \tag{16.7a}$$

whose solution

$$p(x) = Ax + B \tag{16.7b}$$

requires two boundary conditions to determine the constants A and B, boundary value problems require analogous boundary conditions, but specified along physical curves. In addition, for problems characterized by obvious time dependences, initial conditions will also be needed. For the most part, the exposition in the remainder of this book, at least for analytical models, requires the reader only to appreciate the fact that the ordinary differential equation

$$d^2p(r)/dr^2 + (1/r)\,dp/dr = 0 \tag{16.8a}$$

for cylindrical radial flow has the solution

$$p(r) = A \log r + B \tag{16.8b}$$

(all logarithms in this book are *natural* logarithms), while the slightly altered "spherical flow" model

$$d^2p(r)/dr^2 + (2/r)\,dp/dr = 0 \tag{16.9a}$$

has the exact solution

$$p(r) = A/r + B \tag{16.9b}$$

These solutions can be easily validated by back-substitution, but we emphasize that the arbitrary constants of integration A and B will vary from problem to problem. Eqs. (16.7)–(16.9) play important roles in constant density invasion problems. Again, Eqs. (16.1)–(16.6) model single-phase flows only, and as indicated, we shall also address miscible flows where viscous diffusion is important and immiscible two-phase flows where capillary pressure and relative permeability cannot be ignored. In order to keep our early discussions elementary, we shall defer the development of these models for now.

Eulerian Versus Lagrangian Description

Equations such as those just given predict pressure as a function of x, y, z, and t. Once pressure solutions are available, the rectangular Darcy velocities are obtained as

$$u(x, y, z, t) = -(k_x/\mu)\, \partial p/\partial x \tag{16.10}$$

$$v(x, y, z, t) = -(k_y/\mu)\, \partial p/\partial y \tag{16.11}$$

$$w(x, y, z, t) = -(k_z/\mu)\, \partial p/\partial z \tag{16.12}$$

The Eulerian velocities in Eqs. (16.10)–(16.12) represent speeds measured at a point in space (x,y,z). This description is useful to reservoir engineers because it provides flow rates at production wells, transient pressures at specific wells for history matching, and pump rates and pressures needed at injectors, among other quantities of interest. If these velocities do not vary with time, the flow is said to be steady. Otherwise, it is transient or unsteady. Thus, a constant velocity, single fluid flow through a linear core is steady, because it appears unchanged from one instant of time to the next, recognizing that numerous fluid elements are actually streaming through the pore spaces. On the other hand, the Eulerian frame of reference is not ideal or convenient for every application. For example, radioactive and chemical tracers are often introduced into injection wells and monitored at production wells in order to study reservoir connectivity and sweep efficiency. This usage addresses the question of where a marked fluid element (or a tagged group of particles) is heading, an objective that requires us to follow the fluid. This is also important in environmental engineering, where the destinations of contaminants, as well as their origins and travel times, are of interest. For these purposes, a Lagrangian description is more suitable. Formation invasion, which deals with traveling fluid fronts, requires Lagrangian solutions to the equations of fluid motion.

Constant Density Versus Compressible Flow

A constant density, or incompressible fluid is a fluid consisting of elements that are not unlike infinitely rigid balls. Any disturbance to a single element is, therefore, instantaneously transmitted throughout the entire field of flow, so that the speed with which information propagates is infinite. Compressible fluids are characterized by elasticity. A fluid element that is disturbed will respond with minute volume changes and finite delay, before passing the disturbance to its neighbors. In borehole annular flows and drillpipe flows, sudden motions manifest themselves as sound waves governed by hyperbolic equations.

In reservoirs supporting Darcy flow, compressibility allows pressure disturbances to slowly diffuse, similar to temperature diffusion in solids. Hence, petroleum engineers often model compressibility using heat equation models.

Steady Versus Unsteady Flow

In elementary reservoir flow analysis, simple single-phase flows are considered, that is, flows containing one and only one fluid species throughout the reservoir medium. For such constant density, incompressible flows, whenever applied pressures remain constant in time, the Eulerian pressure fields are steady and result in steady-state formulations. Only when the effects of compressibility are allowed in a single-phase flow can reservoir transients exist; thus, compressible flows can be both transient and steady. Other transients of importance are introduced in the drilling process. For example, changes in the mud are often accompanied by changes in wellbore pressure and invading fluid viscosity, effects that can (and will) be modeled both analytically and numerically. Not all constant density flows are steady, of course. In two-phase flows of immiscible fluids, for example, in the Eulerian flow obtained by displacing oil with water, the relative saturations of each phase within any particular pore space will vary with time, and then, with position. This type of unsteadiness also exists when the two-phase flow is compressible. For rapid displacements, piston-like, plug, or slug-like flows will result, and the displacement process can be modeled by monitoring the progress of two different single-phase flows separated by a moving interface. (Later we shall explain dimensionlessly what is meant by "rapid" or "slow.") When fresh water displaces saline formation water of like viscosity in a core without mudcake under constant pressure, the flow is steady; but when water displaces oil, or conversely, the flow is unsteady because "total viscosity" changes with time. It is important to understand that different flow limits exist, which will be modeled by different boundary value problem formulations and fluid properties. In single-phase flows, it suffices to consider viscosity, but in two-phase flows, we require concepts like relative permeability and capillary pressure. The flow of a single fluid, which may be steady in an Eulerian description, is unsteady in a Lagrangian model because the individual fluid elements being monitored are always in motion. For this reason, the Eulerian relationships in Eqs. (16.1)–(16.12) provide only a partial solution to the Lagrangian problem. To complete the description, we turn to kinematic considerations that treat moving fronts and interfaces as distinct physical entities.

Incorrect Use of Darcy's Law

Darcy's law states that the local velocity q in a direction s is given by the directional derivative $q = -(k/\mu)\ \partial p/\partial s$, where p is the transient or steady pressure, with k and μ representing permeability and viscosity. Thus, in a lineal flow, we have $q = -(k/\mu)\ \partial p/\partial x$, whereas in a cylindrical or spherical radial flow, we have $q = -(k/\mu)\ \partial p/\partial r$, r being the radial variable. Eqs. (16.10)–(16.12) apply to 3D flows in rectangular or Cartesian coordinates. Darcy's law, a low Reynolds number approximation to the Navier-Stokes equations, does not embody the complete physical description of the invasion process. For example, it does *not* describe mass conservation. Only when the latter requirement is invoked, do we obtain partial differential equations for pressure such as Eqs. (16.1), (16.2), (16.3), or (16.6). These are solved with pressure (or Dirichlet) or flow rate (or Neumann) boundary conditions applied to inlet and outlet faces. Unfortunately, a number of published invasion models actually take $(k/\mu)\ \partial p/\partial x = -q(t)$ as a starting point, where q(t) is specified, leading to an anticipated (but often incorrect) linear variation in pressure. Such approaches do not account for the cylindrical and spherical radial geometric spreading that Eqs. (16.8a), (16.9a) automatically embody. Their results fail to satisfy these equations or their appropriate extensions, for example, it is clear from substitution that $dp/dr = -\mu q(t)/k$ does *not* satisfy $d^2p(r)/dr^2 + (1/r)\ dp/dr = 0$. Still other analyses invoke a universal $\sqrt{t}$ law at the outset, failing to appreciate that this limiting result, correctly derived three decades ago by Outmans (1963), applies to lineal flows only, and then only when cake compaction is insignificant, fluid compressibility is ignored, and formation permeability is high. Because all of this confusion proliferates throughout the invasion literature, this book will list all underlying assumptions used whenever new formulas are derived. We shall state limitations and strengths clearly, and attempt to carefully document all the requisite steps needed to arrive at solutions and numerical algorithms.

MOVING FRONTS AND INTERFACES

The kinematics of moving fronts and interfaces has been studied in different physical contexts for over 200 years. Most notable are the studies of free surfaces in ocean hydrodynamics and vortex sheets in free space (e.g., see Lamb, 1945), and more recently, flame propagation dynamics in combustion analyses. The following derivation, which applies to fluid fronts in porous media, is given in Chin (1993a). Let us consider a moving front or interface located anywhere within a three-dimensional Darcy flow (e.g., any surface marked by red dye), and let $\phi(x,y,z)$ denote the porosity. Furthermore, denote by u, v, and w the Eulerian speed components, and describe our interface by the surface locus of points

$$f(x, y, z, t) = 0 \tag{16.13}$$

An interface, as in Fig. 16.5, is defined by the kinematic property that fluid does not cross it. Hence, the velocity of the fluid normal to the interface must be equal to the velocity of the interface normal to itself. The velocity perpendicular to the surface, from vector algebra, is equal to $-f_t(x, y, z, t)/\sqrt{\left(f_x^2 + f_y^2 + f_z^2\right)}$, while $(uf_x + vf_y + wf_z)/\left\{\phi(x, y, z)\sqrt{\left(f_x^2 + f_y^2 + f_z^2\right)}\right\}$ is the normal velocity of the fluid. The condition that these be equal is therefore given by the equality

$$\partial f(x, y, z, t)/\partial t + (u/\phi)\partial f/\partial x + (v/\phi)\partial f/\partial y + (w/\phi)\partial f/\partial z = 0 \qquad (16.14)$$

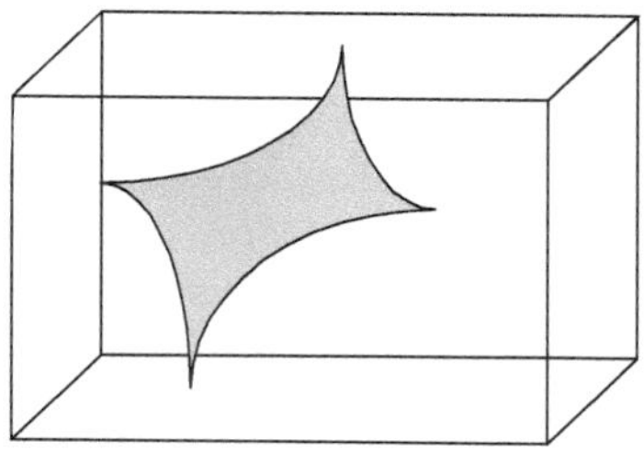

FIG. 16.5 Any surface f(x,y,z,t) = 0 in a reservoir.

Now, from calculus, the total differential df of any function f(x,y,z,t) is

$$df = \partial f/\partial t\, dt + \partial f/\partial x\, dx + \partial f/\partial y\, dy + \partial f/\partial z\, dz \qquad (16.15)$$

Division by dt yields

$$df/dt = \partial f/\partial t + dx/dt\, \partial f/\partial x + dy/dt\, \partial f/\partial y + dz/dt\, \partial f/\partial z \qquad (16.16)$$

Comparison with Eq. (16.14) shows that we can set the so-called substantive, material, or convective derivative df/dt to zero, that is,

$$df/dt = 0 \qquad (16.17)$$

provided we require that

$$dx/dt = u(x, y, z, t)/\phi(x, y, z) \qquad (16.18)$$

$$dy/dt = v(x, y, z, t)/\phi(x, y, z) \qquad (16.19)$$

$$dz/dt = w(x, y, z, t)/\phi(x, y, z) \qquad (16.20)$$

Thus, along the fluid particle trajectories defined by Eqs. (16.18)–(16.20), the function f(x,y,z,t) must be constant, since Eq. (16.17) requires df/dt = 0. This proves that particles on a surface remain on it. Eqs. (16.18)–(16.20) define the fluid fronts and interfaces, but f(x,y,z,t) no longer plays an active role.

Now, the Eulerian velocities u, v, and w in these trajectory equations are determined from a separate and independent host formulation, for example, $\partial^2 p/\partial x^2 + \partial^2 p/\partial y^2 + \partial^2 p/\partial x^2 = (\phi\mu c/k)\ \partial p/\partial t$, or $\partial^2 p/\partial x^2 + \partial^2 p/\partial y^2 + \partial^2 p/\partial x^2 = 0$, or still another flow model. While x, y, and z represent independent variables in the host Eulerian formulation, in the trajectory equations, the same x, y, and z become the dependent variables, with time now being independent. This role reversal, typically resulting in complicated mathematics, is standard in invasion modeling. The trajectory equations form coupled, nonlinear, ordinary (not partial) differential equations. When they are integrated analytically or numerically in time, with all starting positions prescribed, the corresponding particles can be followed as they travel; final results include positions and travel times along the pathlines or streamlines. Also, when u(x,y,z,t), v(x,y,z,t), and w(x,y,z,t) are available, Eqs. (16.18)–(16.20) can be integrated *backward* in time, to determine the origin of a particle or a group of particles.

Observe that, while the solution of $\partial^2 p/\partial x^2 + \partial^2 p/\partial y^2 + \partial^2 p/\partial x^2 = 0$, subject to pressure boundary conditions, does not involve the porosity $\phi(x,y,z)$, the Lagrangian description for the same flow does, as is clear from Eqs. (16.18)–(16.20). That this must be so is intuitively obvious. If we consider the steady-state flow through a kitchen sponge, it is clear that individual fluid particles must travel faster through smaller pore spaces in order to maintain the overall steady flow. Similarly, the traffic speed at the narrowed neck of a steady-state highway flow must exceed that found along multilane stretches. (If it does not, as is usually the case, it is because the flow is unsteady!) The changes in thinking here are critical to invasion modeling, but this philosophy aside, the algebraic manipulations required to produce sought solutions are relatively simple, although at times, quite tedious. For this reason, intermediate steps are retained. The primary products of this book are analytical results and algorithms, *plus* new philosophies and methodologies.

Use of Effective Properties

While the fluid displacement process can be quite complex, for example, as in immiscible two-phase-flow mixing, very often, a piston-like, slug, or plug-like description suffices, at least in offering a qualitative but quantifiable model as the basis for preliminary discussion. For such flows, we can simplify the physical picture and formulate the problem as two idealized single-fluid regions separated by a distinct moving interface or displacement front such as that described above. This boundary is known as a mathematical discontinuity, across which certain physical quantities are conserved and other kinematic requirements enforced. Let us specialize our ideas to the displacement of hydrocarbons by water. The first region, ahead of the front, contains oil or gas together with immobile connate water. Behind the front is the second region, consisting of invading mud filtrate, and immobile residual oil (or gas) left behind the front by displaced hydrocarbons. The symbols denoting connate water and residual oil saturations are S_c and S_{ro} (thus, the initial oil saturation is $1 - S_c$). If the geometric rock porosity is denoted by ϕ, then an effective porosity ϕ_{eff} characterizing the invaded rock can be introduced with the definition $\phi_{eff} = \phi(1 - S_{ro} - S_c)$. This definition can be used for the front trajectories defined by Eqs. (16.18)–(16.20). Again, this provides an approximate description for a rapid displacement process (later, immiscible two-phase flow theory will provide more precise simulations). Note that different definitions of porosity exist in well logging, depending on the type of instrument used for measurement. In this book, by porosity, we mean that associated with the connected pore spaces available for fluid transport, since these pore spaces are the ones implied by the equations governing fluid motion. While the rock under investigation may be uniform and homogeneous, characterized by a single permeability, we shall at times derive our formulas allowing for two permeabilities, so that the results can be used on an ad hoc basis in modeling different permeabilities relative to different formation fluids. This usage is convenient in describing the differences in the flow of water relative to residual oil, versus the movement of oil relative to immobile connate water. This flexibility is consistent with our use of effective porosity and is again offered for convenience only. Finally, while we have emphasized the possibility of two coexisting formation fluids separated by dynamic interfaces, we stress that our results also apply to the case of a single fluid which we may envision as red water displacing blue water. This final example is useful in modeling the displacement of saline formation waters by invading mud filtrate, at least for short initial times, before ionic diffusion smears the separation boundary. It is important to resistivity interpretation.

PROBLEMS AND EXERCISES

1. Eqs. (16.18)–(16.20) represent a system of nonlinearly coupled ordinary differential equations. Review available scientific subroutine libraries and write a program to trace pathlines using available software.
2. Steady liquid flows in homogeneous media satisfy $d^2p/dr^2 + n/r\ dp/dr = 0$ with $n = 0$, 1, and 2 for linear, cylindrical, and spherical flow, respectively. Derive this result from mass conservation considerations, taking into account the geometrical differences between different types of elemental volumes. Then verify the general solutions given in this chapter.

Chapter 17

Static and Dynamic Filtration

Here we introduce the ideas underlying quantitative formation invasion modeling, but restrict ourselves to isotropic Darcy flows dealing with piston-like, slug or plug-like displacements. Miscible and immiscible multiphase flows are considered once simpler techniques have been developed and their limitations are understood. We start with the simplest problems, tracking fluid fronts in cores without mudcakes, progress to mudcake-alone formulations, and finally to problems where the dynamics of the invasion front and the timewise growth of the mudcake are closely coupled. In this last class of problems, the enormous analytical complexities confronting mathematicians are aptly highlighted, complications that occur despite the simplicity of the fluid model used. Hence, we shall be motivated to look for numerical methods that provide greater modeling flexibility, that is, have potential for greater expansion as we attempt to simulate invasion problems that more closely model reality. Computational finite difference methods are introduced in Chapter 20, where they are at first applied to the problems analytically addressed here. However, we extend these algorithms to classes of physical problems where the possibilities for closed-form solution are unlikely. We emphasize that the invasion solutions presented in this book also apply when the flow direction is reversed. For example, they are useful whether drilling overbalanced or underbalanced; they model influx into the wellbore from the formation once the obvious sign change in the pressure differential is made. Dynamic filtration in the borehole, when fluid flows parallel to the hole axis, is also discussed, and the coupling of non-Newtonian annular to Newtonian reservoir flow is developed.

SIMPLE FLOWS WITHOUT MUDCAKE

In this section, we study single-phase flow invasion into rocks, assuming that the influence of mudcake is negligible; the results model, for example, the use of brine or water as the drilling fluid. The formulations for such problems are simple; they highlight the basic differences between the reservoir flow problems covered in undergraduate curricula and the Lagrangian models needed to track moving fronts. Five problems appear in order of increasing complexity.

Quantitative Methods in Reservoir Engineering. http://dx.doi.org/10.1016/B978-0-12-810518-4.00017-7

Homogeneous Liquid in a Uniform Linear Core

The pressure *partial* differential equation governing transient, compressible, lineal, homogeneous, liquid flows having constant properties is $\partial^2 p(x,t)/\partial x^2 = (\phi\mu c/k)\partial p/\partial t$. Here p is pressure, while x and t represent space and time; ϕ, k, μ, and c are rock porosity, rock permeability, fluid viscosity, and net fluid-rock compressibility, respectively. If we assume a constant density, incompressible flow, and ignore the compressibility of the fluid by setting $c=0$, the right side of this equation identically vanishes. Then, the model reduces to the ordinary differential equation $d^2p(x;t)/dx^2=0$ where t is a parameter as opposed to a variable.

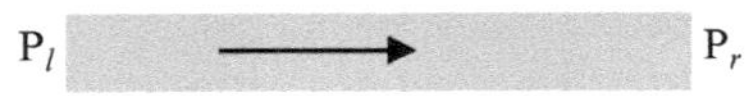

FIG. 17.1 Lineal flow.

The solution to this equation is simply $p(x;t)=Ax+B$. In order to determine the integration constants A and B, boundary conditions for the pressure p(x;t) are required. Let us suppose that the left-side pressure at $x=0$ is P_l, while the right-side pressure at $x=L$ is P_r, as shown in Fig. 17.1. That is, we take $p(0;t)=P_l$ and $p(L;t)=P_r$, which completely determine A and B. If P_l and P_r are constants, then A and B are constants. However, if either or both are functions of time, as may be the case in drilling, then A and B may be functions of time. In this case, the pressure field $p(x;t)=(P_r-P_l)x/L+P_l$ responds immediately to time changes in boundary pressure, since zero compressibility, implying absolute fluid rigidity, requires the instantaneous transmission of information. Henceforth we shall omit t in the argument of pressure whenever we deal with constant density flows, understanding that time dependences *are* parametrically allowed when warranted. The explicit use of t, as in p(x,t), will be reserved for transient compressible flows only.

Now, the fluid velocity q is given by Darcy's law $q=-(k/\mu)dp(x)/dx$, which, in view of our solution, becomes $q=-(k/\mu)(P_r-P_l)/L$. This describes the fluid velocity at a fixed point in space. It is the velocity that an observer fixed to a particular pore space element measures (in our convention, we assume $q>0$ if $P_l>P_r$). For invasion modeling, we are interested in the progress of an initial marked particle(s), and we prefer the alternative Lagrangian description. If we *now* let x denote the tag describing such marked particles, the particle velocity satisfies $dx/dt=q/\phi$ where ϕ is the porosity, as required in Eq. (16.18). This is correct physically, since smaller porosities create faster fronts for the same q, and vice versa. Using our expression for q, we have for the invasion front the ordinary differential equation $dx/dt=-\{k/(\mu\phi)\}(P_r-P_l)/L$ which, for constant porosity, integrates to

$$x(t)=x_O-\{k/(\mu\phi)\}(P_r-P_l)t/L \tag{17.1}$$

where x_0 is the initial marked position. For heterogeneous problems, $\phi = \phi(x)$ and $P_l = P_l(t)$, and the differential equation for x(t) can be integrated in the form $\int \phi(x)\, dx = -\int \{(k/\mu)(P_r - P_l(t))/L\} dt$ using table look-up techniques. Under the assumptions stated in the preceding paragraph, the displacement front in this single-fluid, cake-free, lineal liquid flow example varies, as expected, *linearly* with time. It is physically an uninteresting problem. However, the simplicity of the math allowed us to illustrate concepts introduced in Chapter 16. First, the porosity, which does not appear in Eulerian constant density flow problems, *does* appear in Lagrangian models. Second, the Lagrangian solution cannot be (easily) obtained without solving the Eulerian formulation first. Finally, in progressing from Eulerian to Lagrangian models, the independent variable x literally becomes the dependent variable for the front position. These observations also apply to multidimensional problems.

We can use the solution $x(t) = x_0 - \{k/(\mu\phi)\}(P_r - P_l)\, t/L$ (assuming $x_0 = 0$) to illustrate the basic ideas behind time lapse invasion analysis, that is, we shall take as our *host* model the equation $\{k/(\mu\phi)\}(P_r - P_l)/L = -x(t)/t$. Thus, if the position front x(t) can be monitored or measured as a function of the time t, say using resistivity, radioactive tracer, or Catscan methods, it follows that the quotient x(t)/t yields information about the quantity $\{k/(\mu\phi)\}(P_r - P_l)/L$. Of course, the greater the value of x(t) or t, the smaller the experimental error. This invasion front measurement will provide, at most, the value of the lumped physical quantity $\{k/(\mu\phi)\}(P_r - P_l)/L$. Thus, if any of its single constituent members k, μ, ϕ, P_r, P_l, or L are required, values for the remaining quantities must first be found separately using other means. For example, if the pressure gradient $(P_r - P_l)/L$ and the porosity ϕ is known, then the value of the mobility k/μ is immediately available (but viscosity cannot be determined).

Homogeneous Liquid in a Uniform Radial Flow

Now we repeat the same calculation for cylindrical radial flows. The pressure partial differential equation governing transient, compressible, radial, homogeneous, liquid flows having constant properties is $\partial^2 p(r,t)/\partial r^2 + (1/r)\, \partial p/\partial r = (\phi\mu c/k)\, \partial p/\partial t$. All of our quantities are defined as in the lineal flow, except that the radial coordinate r replaces the lineal coordinate x. If we assume an incompressible fluid with vanishing c, we obtain the differential equation $d^2p(r)/dr^2 + (1/r)\, dp/dr = 0$, whose solution is $p(r) = A \log r + B$. For this radial flow, we impose pressure boundary conditions at the well and outer reservoir radii (that is, at r_{well} and r_{res}) in Fig. 17.2, in the form $p(r_{well}) = P_{well}$ and $p(r_{res}) = P_{res}$. We emphasize that while mudcake effects are not yet included in this formulation, the example itself is not unimportant to formation invasion in real wells. In many shallow wells, muds only slightly thicker than water are used; at other times, wells may be drilled with watery brines that do not produce mudcakes.

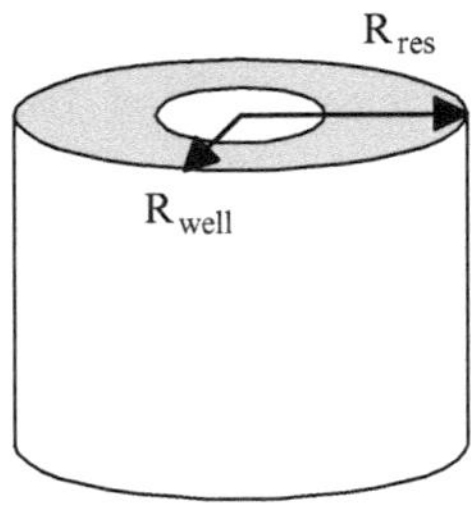

FIG. 17.2 Cylindrical radial flow.

These conditions lead to $P_{well} = A \log r_{well} + B$ and $P_{res} = A \log r_{res} + B$, and subtraction yields $A = (P_{well} - P_{res})/(\log r_{well}/r_{res})$. It is A, and *not* B, that is important when dealing with radial invasion. From $p(r) = A \log r + B$, we find that the radial pressure gradient satisfies $dp(r)/dr = A/r$, so that the Eulerian velocity q satisfies $q(r) = -(k/\mu)\ dp(r)/dr = -Ak/(\mu r)$.

The invasion front r(t) in the Lagrangian description, as in our first example, satisfies $dr/dt = q/\phi$, or $r\ dr = -\{Ak/(\mu\phi)\}dt$. Now consider an initially marked circular ring of tracer particles where $r(t) = r_o$ at $t = 0$. This initial condition leads to an integral in the form $r^2 = r_o^2 - \{2Ak/(\mu\phi)\}t$ so that

$$r(t) = \sqrt{\left[r_o^2 - \{2Akt/(\mu\phi)\}\right]} \tag{17.2}$$

If $P_{well} > P_{res}$ and $r_{well}/r_{res} < 1$, the constant A is negative; for large times t, our solution for r(t) can be approximated by $r(t) \approx \sqrt{\{-2Akt/(\mu\phi)\}}$. Thus, radial front positions will vary like $\sqrt{t}$, even for a uniform liquid in a homogeneous rock without the presence of mudcake. To close this example, we address the meaning of large times. To do this, we rewrite the full solution as $r(t) = [1 - r_o^2\ \mu\phi/\{2Akt\}]^{1/2} \sqrt{\{-2Akt/(\mu\phi)\}}$. Since $\sqrt{(1+\sigma)} \approx 1 + 1/2\ \sigma$ holds for $|\sigma| \ll 1$, it follows that $r(t) \approx \sqrt{\{-2Akt/(\mu\phi)\}}$ only if $|r_o^2\ \mu\phi/\{2Akt\}| \ll 1$. Thus, the meaning of large times must be interpreted *dimensionlessly* in the context of the particular problem itself: it depends on the radial pressure gradient, all characteristic radii, as well as on rock and fluid properties.

By the same token, the use of the term small times must be similarly discussed in the dimensionless context. Here, we rewrite the exact solution for r(t) as $r(t) = r_o\sqrt{[1 - \{2Ak/(\mu\phi r_o^2)\}\ t]}$. Using $\sqrt{(1+\sigma)} \approx 1 + 1/2\ \sigma$, we obtain the approximation $r(t) \approx r_o[1 - \{Ak/(\mu\phi r_o^2)\}t]$. Thus, the front radius varies linearly with t for small times, where small implies $|-\{2Ak/(\mu\phi r_o^2)\}t\ | \ll 1$. Needless to say, other time scales will be introduced by the compressibilities and viscosities of the filtrate, and the displaced reservoir fluid, by the structural characteristics of the mudcake, and so on, in more complicated problems.

Homogeneous Liquid in a Uniform Spherical Domain

The spherical displacement front associated with a point source is useful in studying invasion at the bit. The flow that would idealize Fig. 17.3 possesses spherically symmetry. Cylindrical radial invasion away from the bit (around the drill pipe) is controlled by mudcake buildup. But at the bit, cake does not form, since it is drilled and washed away by the nozzle flow. The equation for compressible, spherical, homogeneous, liquid flows with constant single-phase properties is $\partial^2 p(r,t)/\partial r^2 + (2/r)\ \partial p/\partial r = (\phi\mu c/k)\ \partial p/\partial t$. This differs from the cylindrical model, with a 2/r replacing 1/r, r being the spherical distance from the bit center. For an incompressible liquid, $c = 0$, and we have $d^2p(r)/dr^2 + (2/r)\ dp/dr = 0$.

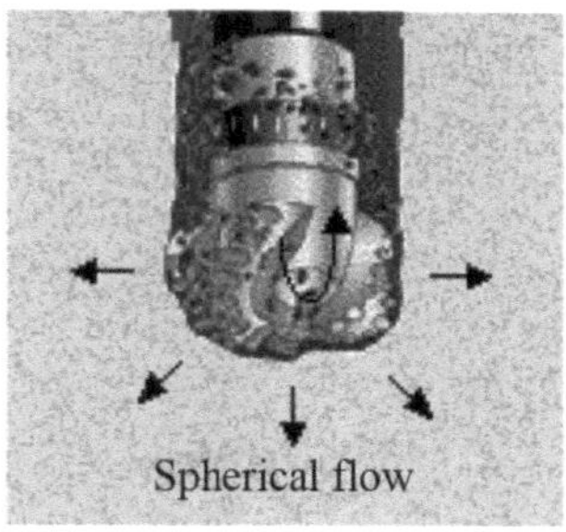

FIG. 17.3 Spherical flow at the drillbit.

The pressure distribution is not logarithmic as in cylindrical flows, since $p(r) = Ar^{-1} + B$. The pressures at the bit radius $r = r_{bit}$ and at an effective radius $r = r_{eff}$ are P_{bit} and P_{pore}, denoting the bit nozzle pressure and the formation pore pressure. That is, we take $p(r_{bit}) = P_{bit}$ and $p(r_{eff}) = P_{pore}$, so $P_{bit} = Ar_{bit}^{-1} + B$ and $P_{pore} = Ar_{eff}^{-1} + B$, or $A = (P_{bit} - P_{pore})/(r_{bit}^{-1} - r_{eff}^{-1})$. Now substitute $dp/dr = -Ar^{-2}$ in Darcy's law $q(r) = -k/(\mu)\ dp(r)/dr = \{Ak/\mu\}r^{-2}$. Since the front velocity satisfies $dr/dt = q/\phi$, we have $dr/dt = \{Ak/(\mu\phi)\}r^{-2}$. If initially $r(t=0) = r_o > 0$, this nonlinear equation can be integrated to give

$$r(t) = \left[r_O^3 + 3Akt/(\mu\phi)\right]^{1/3} \tag{17.3}$$

If $r_o = r_{bit}$, this models invasion from the bit; in general, r_o is the initial radius of any front tracked. For overbalanced drilling, $A > 0$ and the front advances into the rock. If underbalanced, $A < 0$ and the front moves toward the bit. Note that r(t) advances linearly with time over short times, but at large times, its behavior goes like $t^{1/3}$ and *not* $t^{1/2}$. The dimensionless requirements for short and large differ for cylindrical versus spherical flows.

Gas Flow in a Uniform Linear Core

In the first example, we considered single *liquid* flow in a lineal core. Many authors do not distinguish flows of liquids from those of gases. To demonstrate the differences, we now consider a *gas* (still assuming a single homogeneous fluid without mudcake). The equation governing compressible, lineal, homogeneous, *gas* flows with constant properties is $\partial^2 p^{m+1}(x,t)/\partial x^2 = (\phi\mu c^*/k)\partial p^{m+1}/\partial t$ where m + 1 is an exponent (m = 0 for liquids). Here, $c^* = c$ for liquids, while $c^* = m/p(x,t)$ for gases. The constant m refers to thermodynamic limits, for example, m = 1, C_v/C_p, 0, and ∞ for isothermal, adiabatic, constant volume, and constant pressure processes, respectively. (C_v and C_p are specific heats at constant volume and pressure.)

If transient effects are unimportant, this formidable equation reduces to a linear $d^2p^{m+1}(x)/dx^2 = 0$ for $p^{m+1}(x)$ whose solution is $p^{m+1}(x) = Ax + B$. It is $p^{m+1}(x)$, and *not* p(x), that varies linearly in space; therefore, the Darcy velocity $q = -(k/\mu)dp/dx$ will *not* be constant in general. This nonintuitive (but correct) result follows from solving the pressure equation, properly accounting for mass conservation, and not simply assuming constant dp/dx in $q = -(k/\mu)dp/dx$. To determine A and B, boundary conditions are required. As before, the left pressure at x = 0 is P_l, while $P = P_r$ at x = L, so that $p^{m+1}(0) = P_l^{m+1}$ and $p^{m+1}(L) = P_r^{m+1}$, or, $p^{m+1}(x) = (P_r^{m+1} - P_l^{m+1})\,x/L + P_l^{m+1}$, yielding

$$p(x) = \left\{\left(P_r^{m+1} - P_l^{m+1}\right)x/L + P_l^{m+1}\right\}^{1/(m+1)} \tag{17.4a}$$

This does *not* vary linearly with x, except in the liquid limit where m = 0. The isothermal m = 1 case is usually taken in gas well test analysis; other values of m model different thermodynamic processes that may be relevant in a particular reservoir. For the front motion, differentiate pressure to obtain $(m+1)\,p^m(x)dp/dx = \left(P_r^{m+1} - P_l^{m+1}\right)/L$, so that the pressure gradient satisfies

$$dp/dx = \left\{\left(P_r^{m+1} - P_l^{m+1}\right)/((m+1)L)\right\}p^{-m}(x) \tag{17.4b}$$

Therefore,

$$q = -(k/\mu)\,dp(x)/dx = -(k/\mu)\left\{\left(P_r^{m+1} - P_l^{m+1}\right)/((m+1)L)\right\}p^{-m}(x)$$

The front displacement x(t) again satisfies $dx/dt = q/\phi$. Some algebra shows that

$$\begin{aligned}&\left\{L/\left(P_r^{m+1} - P_l^{m+1}\right)\right\}\left\{\left(P_r^{m+1} - P_l^{m+1}\right)x/L + P_l^{m+1}\right\}^{(2m+1)/(m+1)} \\ &\quad = -(kt/(\mu\phi))\left\{\left(P_r^{m+1} - P_l^{m+1}\right)/((m+1)L)\right\} + \text{constant}\end{aligned} \tag{17.5}$$

where the constant is fixed by the initial position of the marker particle. The function x(t) can be obtained by raising each side of Eq. (17.5) to the (m + 1)/(2m + 1)*th* power, and then solving for x as a function of t. Unlike the result for liquids, the front motion is not constant with time. Its motion must vary with time because q needs to vary spatially to conserve mass. The exact variation depends on m. Time lapse analysis for gas reservoirs, therefore, must be handled carefully, taking account of the reservoir's thermodynamic environment. Note that this result does not apply in the presence of multiple fluids, mudcake, or geometric spreading. When these effects, cake compaction, and erosion are important, numerical models are needed.

Flow From a Plane Fracture

The flow of liquids and gases into or out of plane fractures in homogeneous anisotropic media was considered in Chapter 2, which focused on production rates. Reservoir engineers are also interested in *where* produced fluids originate, while stimulation engineers are interested in *how* injected fracture fluids flow. The previous (proppant-free) fracture solution remains valid, but it must be interpreted differently for invasion purposes. It can be used to trace any marked particle (X_o,Y_o) at $t=0$ by integrating a *nonlinearly coupled* pair of ODEs defining the Lagrangian trajectories. Again the flow of liquids in isotropic media is assumed. From the prior integral solution for pressure, we can obtain $\partial p(x,y)/\partial x = \int f(\xi)(x-\xi)/\{(x-\xi)^2+y^2\}d\xi$ and $\partial p(x,y)/\partial y = y \int f(\xi)/\{(x-\xi)^2+y^2\} d\xi$ by partial differentiation. Then, the definitions $dX/dt = -(k/(\mu\phi))\partial P(X,Y)/\partial X$ and $dY/dt = -(k/(\mu\phi))\partial P(X,Y)/\partial Y$, where t is dimensional, lead to

$$dX/dt = -(kP_{ref}/(\mu\phi c))\int f(\xi)\,(x-\xi)/\left\{(x-\xi)^2+y^2\right\} d\xi \tag{17.6a}$$

$$dY/dt = -(kP_{ref}/(\mu\phi c))\, y\int f(\xi)/\left\{(x-\xi)^2+y^2\right\} d\xi \tag{17.6b}$$

These pathline equations suggest the timewise integration given by the recursion formulas in Eqs. (17.7a)–(17.7d). Suppose the dimensionless (x_i,y_i)'s normalized by fracture length c are specified at an initial instant. These values can be used to evaluate Eqs. (17.7a), (17.7b) to yield the instantaneous particle velocity. Then, Eqs. (17.7c), (17.7d) are used to update the position to obtain the final destination (x_f,y_f) at the end of a time step $\Delta t>0$.

$$dx_i/dt = -\left(kP_{ref}/\left(\mu\phi c^2\right)\right)\int f(\xi)\,(x_i-\xi)/\left\{(x_i-\xi)^2+y_i^2\right\} d\xi \tag{17.7a}$$

$$dy_i/dt = -\left(kP_{ref}/\left(\mu\phi c^2\right)\right) y_i\int f(\xi)/\left\{(x_i-\xi)^2+y_i^2\right\} d\xi \tag{17.7b}$$

$$x_f = x_i + (dx_i/dt)\Delta t \tag{17.7c}$$

$$y_f = y_i + (dy_i/dt)\Delta t \tag{17.7d}$$

This applies recursively, starting with any initial value of (X_o/c,Y_o/c), and may continue indefinitely; the finer the integration time step, the greater the physical resolution. More accurate integration schemes exist in the literature, which apply to all the fracture solutions in Chapter 2. Unlike PDE-based finite difference schemes, where convergence and stability depend on the form of the truncation error, and notably the sign of Δt, Eqs. (17.7a)–(17.7d) can be integrated *backwards* in time taking $\Delta t < 0$. This provides the capability of tracing a particle's origin in addition to its destination, an excellent resource in environmental applications where the source of contamination is desired.

FLOWS WITH MOVING BOUNDARIES

Here we progress to flows with nontrivial external and internal moving boundaries. We first consider lineal cake buildup on filter paper, and then we examine the plug flow of two dissimilar liquids in a linear core without mudcake. These two examples set the stage for problems where mudcake growth, formation properties, and invasion front motion are dynamically coupled, which will be treated rigorously in the following section.

Lineal Mudcake Buildup on Filter Paper

In the previous section, we considered formation invasion without the retarding effects of mudcake. In order to understand the physics clearly, we study the problem of *isolated* mudcake growth, as would be obtained in the laboratory lineal flow test setup in Fig. 17.4. We consider a one-dimensional experiment where mud, in essence a suspension of clay particles in water, is allowed to flow through filter paper. Initially, the flow rate is rapid. But as time progresses, solid particles (typically 6–40% by volume for light to heavy muds) such as barite are deposited onto the surface of the paper, forming a mudcake that, in turn, retards the passage of mud filtrate by virtue of the resistance to flow that the cake provides.

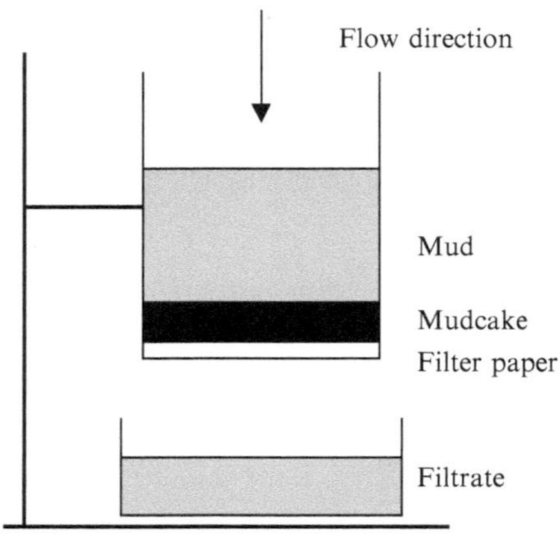

FIG. 17.4 Simple laboratory mudcake buildup.

We therefore expect filtrate volume flow rate and cake growth rate to decrease with time, while filtrate volume and cake thickness continue to increase, but ever more slowly. These qualitative ideas can be formulated precisely, because the problem *is* based on well-defined physical processes. For one, the composition of the homogeneous mud during this filtration does not change: its solid fraction is always constant. Second, the flow *within* the mudcake *is* a Darcy flow and is therefore governed by the equations used by reservoir engineers. The only problem, though, is the presence of a moving boundary, namely, the position interface separating the mudcake from the mud that ultimately passes through it and that continually adds to its thickness. The physical problem, therefore, is a transient process that requires somewhat different mathematics than that taught in partial differential equations courses.

Mudcakes in reality may be compressible, that is, their mechanical properties may vary with applied pressure differential, e.g., as in Fig. 14.7. We shall be able to draw upon reservoir engineering methods for subsidence and formation compaction later. For now, a simple constitutive model for incompressible mudcake buildup, that is, the filtration of a fluid suspension of solid particles by a porous but rigid mudcake, can be constructed from first principles. First, let $x_c(t) > 0$ represent cake thickness as a function of the time, where $x_c = 0$ indicates zero initial thickness. Also, let V_s and V_l denote the volumes of solids and liquids in the mud suspension, and let f_s denote the solid fraction defined by $f_s = V_s/(V_s + V_l)$. Since this does not change throughout the filtration, its time derivative must vanish. If we set $df_s/dt = (V_s + V_l)^{-1} dV_s/dt - V_s(V_s + V_l)^{-2} (dV_s/dt + dV_l/dt) = 0$, we can show that $dV_s = (V_s/V_l)dV_l$. But since, separately, $V_s/V_l = f_s/(1 - f_s)$, it follows that $dV_s = \{f_s/(1 - f_s)\}dV_l$.

This is a conservation of species law for the solid particles making up the mud suspension and does not as yet embody any assumptions related to mudcake buildup. Frequently, we might note, the drilling fluid is thickened or thinned in the process of making hole; if so, the equations derived here should be reworked with $f_s = f_s(t)$ and its corresponding time-dependent pressure drop.

In order to introduce the mudcake dynamics, we observe that the total volume of solids dV_s deposited on an elemental area dA of filter paper during an infinitesimal time dt is $dV_s = (1 - \phi_c)dA\ dx_c$ where ϕ_c is the mudcake porosity. During this time, the volume of the filtrate flowing through our filter paper is $dV_l = |v_n|dA\ dt$ where $|v_n|$ is the Darcy velocity of the filtrate through the cake and past the paper. We now set our two expressions for dV_s equal to form $\{f_s/(1 - f_s)\}dV_l = (1 - \phi_c)dA\ dx_c$, and replace dV_l with $|v_n|dA\ dt$, so that we obtain $\{f_s/(1 - f_s)\}\ |v_n|\ dA\ dt = (1 - \phi_c)\ dA\ dx_c$. The dA's cancel, and we are led to a generic equation governing mudcake growth. In particular, the cake thickness $x_c(t)$ satisfies the ordinary differential equation

$$dx_c(t)/dt = \{f_s/\{(1\text{-}f_s)(1\text{-}\phi_c)\}\}\,|v_n| \tag{17.8a}$$

Now, as in the first example of the previous section, we assume a one-dimensional, constant density, single *liquid* flow. For such flows, the constant Darcy velocity is $(k/\mu)(\Delta p/L)$, where $\Delta p > 0$ is the usual delta p or pressure drop through the core of length L. The corresponding velocity for the present problem is $|v_n| = (k/\mu)(\Delta p/x_c)$ where k is the cake permeability, and μ is the filtrate viscosity. Substitution in Eq. (17.8a) leads to

$$dx_c(t)/dt = \{kf_s\Delta p/\{\mu(1\text{-}f_s)(1\text{-}\phi_c)\}\}/x_c \tag{17.8b}$$

If the mudcake thickness is infinitesimally thin at t=0, with $x_c(0)=0$, Eq. (17.8b) can be integrated, with the result that

$$x_c(t) = \sqrt{[\{2kf_s\Delta p/\{\mu(1\text{-}f_s)(1\text{-}\phi_c)\}\}t]} > 0 \tag{17.9}$$

This demonstrates that cake thickness in a lineal flow grows with time like $\sqrt{t}$. However, it grows ever more slowly, because increasing thickness means increasing resistance to filtrate throughflow, the source of the solid particulates required for mudcake buildup. Consequently, filtrate buildup also slows.

To obtain the filtrate production volume, we combine $dV_l = |v_n|\ dA\ dt$ and $|v_n| = (k/\mu)(\Delta p/x_c)$ to form $dV_l = (k\Delta p dA/\mu)x_c^{-1}dt$. Using Eq. (17.9), $dV_l = (k\Delta p dA/\mu)\ [\{2kf_s\Delta p/\{\mu(1-f_s)(1-\phi_c)\}\}]^{-1/2}(t)^{-1/2}\ dt$. Direct integration, assuming zero filtrate initially, yields

$$\begin{aligned} V_l(t) &= 2(k\Delta p dA/\mu)[\{2kf_s\Delta p/\{\mu(1\text{-}f_s)(1\text{-}\phi_c)\}\}]^{-1/2}(t)^{1/2} \\ &= \sqrt{\{2k\Delta p(1\text{-}f_s)(1\text{-}\phi_c)/(\mu f_s))\}}\sqrt{t}\,dA \end{aligned} \tag{17.10}$$

Chin et al. (1986) and recent industry papers required detailed, tedious laboratory measurement of the cake parameters f_s, ϕ_c, and k. This could pose operational difficulties. It turns out that this procedure is unnecessary: their values can be inferred from the results of simple field-implementable surface filtration experiments discussed later in Chapter 19.

So far, we have encountered two types of $\sqrt{t}$ behavior, first for constant density, radial, single liquid flows without mudcake, and then for lineal mudcake buildup and filtrate production without introducing any underlying rock, just the opposite problem. It turns out that there is still another type of $\sqrt{t}$ behavior, obtained by considering the constant density flows of two sequential fluids through a lineal core without mudcake (treated next). Thus, there are at least three types of $\sqrt{t}$ behavior each governed by different flow parameters or physical processes, and therefore, at least three different $\sqrt{t}$ time scales! Log interpretation, therefore, can be challenging, to say the least.

Plug Flow of Two Liquids in Linear Core Without Cake

Let us consider the Darcy flow through a single lineal core in which one liquid displaces a second in a piston-like, plug, or slug-like manner, as in Fig. 17.5. We assume that the permeability to each fluid is the same, so that a single permeability k suffices. Pressures P_l and P_r are fixed at the left and right sides, with $P_l > P_r$, so that the fluid system flows from left to right. No cake is present. For lineal liquid flows, $\partial^2 p(x,t)/\partial x^2 = (\phi\mu c/k)\ \partial p/\partial t$ describes transient, compressible liquids, with φ, μ, c, and k denoting rock porosity, fluid viscosity, fluid-rock compressibility, and permeability. We address the problem where an invading liquid displaces a preexisting formation liquid, the respective viscosities being μ_1 and μ_2.

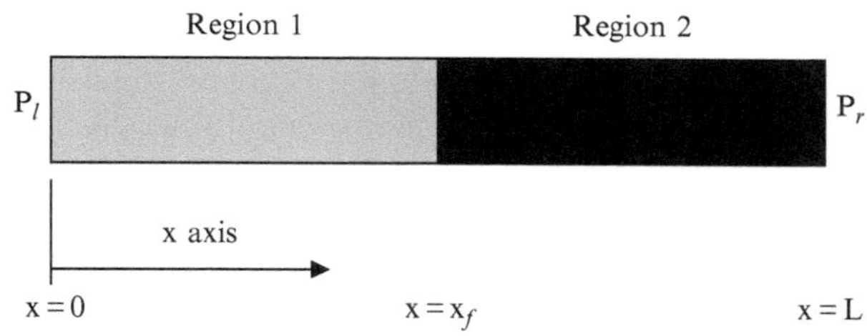

FIG. 17.5 Simple linear flow of two dissimilar fluids.

A second objective of this exercise is the development of mathematical techniques that model internal moving interfaces, for example, the front $x = x_f(t)$ as indicated in Fig. 17.5. For now, though, we may regard the pressure problem as a purely static one. For the incompressible fluids assumed here, the compressibility c vanishes, and the ordinary differential equations for pressure in Layers 1 and 2 become $d^2p_1(x)/dx^2 = 0$ and $d^2p_2(x)/dx^2 = 0$, which have the respective solutions $p_1(x) = Ax + B$ and $p_2(x) = Cx + D$, where A, B, C, and D are integration constants completely determined by the end pressure boundary conditions $p_1(0) = P_l$ and $p_2(L) = P_r$, *and* the interfacial matching conditions at $x = x_f$, $p_1(x_f) = p_2(x_f)$ and $q_1(x_f) = q_2(x_f)$. The pressure continuity equation assumes that interfacial tension effects are negligibly small. Velocity matching, on the other hand, is a kinematic statement stating that the local velocity is single-valued, that is, it takes on one and only one value; the moving interface is convected with this velocity as demonstrated in Chapter 16.

Now, since the $k_1 = k_2 = k$, the Darcy velocities satisfy $q_1 = -(k_1/\mu_1)dp_1(x)/dx = -(k/\mu_1)\,A$ and $q_2 = -(k_2/\mu_2)\,dp_2(x)/dx = -(k/\mu_2)\,C$, so that $A/\mu_1 = C/\mu_2$. This leads to the pressure solution for $0 < x < x_f$,

$$p_1(x) = (\mu_1/\mu_2)(P_r - P_l)x/\{L + x_f(\mu_1/\mu_2 - 1)\} + P_l \qquad (17.11)$$

The pressure solution for $x_f < x < L$ is determined as

$$p_2(x) = (P_r - P_l)(x - L)/\{L + x_f(\mu_1/\mu_2 - 1)\} + P_r \qquad (17.12)$$

The invasion front can be determined, as in all of our preceding examples, by setting $dx_f/dt = q_1/\phi$ assuming that porosity is constant. We now use Eq. (17.11) to obtain $dx_f/dt = -(k/(\phi\mu_1))(\mu_1/\mu_2)(P_r - P_l)/\{L + x_f(\mu_1/\mu_2 - 1)\}$. If we follow the initial marked particle defined by the initial condition $x_f(0) = x_{f,o}$, we obtain the exact integral

$$(\mu_1/\mu_2 - 1)x_f + L = +\Big\{\{(\mu_1/\mu_2 - 1)x_{f,o} + L\}^2 + \{2k(P_l - P_r)/(\phi\mu_2)\}(\mu_1/\mu_2 - 1)\,t\Big\}^{1/2} \qquad (17.13)$$

Depending on the relative values of the μ_1 and μ_2, the displacement front may accelerate or decelerate (detailed calculations will be given in Example 20.1, where this problem is reformulated and solved using finite difference methods). The foregoing analysis is easily reworked to handle time-dependencies in the total differential pressure applied across the core. If $(P_l - P_r)$ is a prescribed function of t, the differential equation should be integrated accordingly, for example, taking $\int(P_r - P_l)\,dt = P_r t - \int P_l(t)dt$. Similar comments apply to situations where $\phi = \phi(x)$. These changes lead to obvious analytical complications, which again motivate the need for numerical models.

COUPLED DYNAMICAL PROBLEMS: MUDCAKE AND FORMATION INTERACTION

Here we derive exact, closed-form, analytical solutions for lineal and radial flows where the growth of the mudcake and the progress of the invasion front are strongly coupled. The first solution was given in Chin et al. (1986), but the radial solution available at the time did not model spurt, and also required numerical analysis. The full solution presented here appears for the first time.

Simultaneous Mudcake Buildup and Filtrate Invasion in a Linear Core (Liquid Flows)

We consider a realistic example where liquid mud filtrate displaces a preexisting formation liquid having a different viscosity. And while this process is ongoing, mudcake thickness is ever-increasing, so that filtrate influx rate is consequently decreasing. All the time, the filtrate-to-formation fluid displacement front moves to the right. In this problem, the dynamics of the mudcake growth are closely coupled to the invasion front motion. In our derivation, there is no assumption that the mudcake is significantly less permeable than the formation, an assumption usually taken to simplify the analysis. The work is exact in this regard, since the relative mobilities between cake, invaded zone, and virgin formation are left as completely free parameters for subsequent evaluation. This important formulation, its solution procedure, and the exact, closed-form, analytical solution for lineal liquid flow were presented in Chin et al. (1986). In the following, we shall reconstruct the steps using the authors' published recipe and reproduce the earlier exact solution.

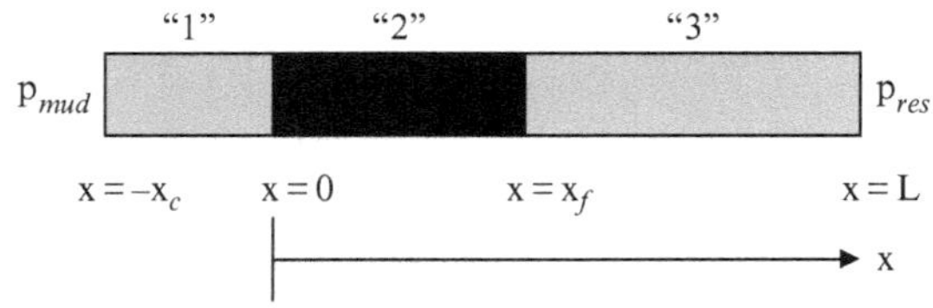

FIG. 17.6 Three-layer lineal flow.

In Fig. 17.6, let Layer 1 denote the mudcake, and Layers 2 and 3, the filtrate-invaded and virgin oil-bearing formations, respectively. The origin $x=0$ is the fixed cake-to-rock interface; also, $x_c>0$ represents the cake thickness, while $x_f>0$ is the displacement front separating invaded from uninvaded rock zones. The transient compressible flow equation assuming constant liquid and rock properties is the classic parabolic partial differential equation, for example, $\partial^2 p_1(x,t)/\partial x^2=(\phi_1\mu_1 c_1/k_1)\ \partial p_1/\partial t$, for Layer 1. If we ignore all compressibilities, in effect considering incompressible liquids with $c=0$, our layered equations reduce to the equations $d^2p_i(x)/dx^2=0$, where $i=1, 2, 3$. These are solved together with the pressure boundary conditions $p_1(-x_c)=p_m$ and $p_3(L)=p_r$, where p_m and p_r represent mud and reservoir pressures. We also invoke interfacial matching conditions for pressure, that is, $p_1(0)=p_2(0)$ and $p_2(x_f)=p_3(x_f)$, plus interfacial matching conditions for velocity, that is, $k_1/\mu_1\ dp_1(0)/dx=k_2/\mu_2\ dp_2(0)/dx$ and $k_2/\mu_2\ dp_2(x_f)/dx=k_3/\mu_3 dp_3(x_f)/dx$.

Note that we have retained three separate permeabilities, k_1, k_2, and k_3 in these equations, as explained in Chapter 16. The k_1 represents, of course, the mudcake permeability. However, while we have but a single rock core, characterized by a single permeability, we shall derive our results with *two* values k_2 and k_3. This flexibility allows us to set $k_2=k_3=k_{rock}$ if desired, or allow them to differ, in order to represent separate permeabilities to filtrate (with residual oil) and oil (in the presence of immobile connate water). This *ad hoc* modeling permits us to mimic two-phase flow relative permeability effects within the framework of a simpler slug displacement approach. We also note that while three viscosities μ_1, μ_2, and μ_3 were explicitly shown for completeness, we in fact assume that $\mu_1=\mu_2=\mu_f$ for the mud filtrate, since the liquid filtrates present in Layers 1 and 2 are identical. Also, we shall later denote $\mu_3=\mu_o$ to represent oil viscosity, that is, the viscosity of the displaced formation fluid. Now, the solutions to our ordinary differential equations for pressure are $p_i(x)=\alpha_i x+\beta_i$, $i=1, 2, 3$. The constants can be determined as

$$\alpha_1 = (p_m\text{-}p_r)/\{(\mu_3 k_1/\mu_2 k_3 - k_1/k_2)x_f - \mu_3 k_1 L/\mu_2 k_3 - x_c\} \qquad (17.14a)$$

$$\beta_1 = p_m + (p_m\text{-}p_r)x_c/\{(\mu_3 k_1/\mu_2 k_3 - k_1/k_2)x_f - \mu_3 k_1 L/\mu_2 k_3 - x_c\} \qquad (17.14b)$$

$$\alpha_2 = (k_1/k_2)(p_m\text{-}p_r)/\{(\mu_3 k_1/\mu_2 k_3 - k_1/k_2)x_f - \mu_3 k_1 L/\mu_2 k_3 - x_c\} \qquad (17.14c)$$

$$\beta_2 = p_m + (p_m\text{-}p_r)x_c/\{(\mu_3 k_1/\mu_2 k_3 - k_1/k_2)x_f - \mu_3 k_1 L/\mu_2 k_3 - x_c\} \qquad (17.14d)$$

$$\alpha_3 = (\mu_3 k_1/\mu_2 k_3)(p_m\text{-}p_r)/\{(\mu_3 k_1/\mu_2 k_3 - k_1/k_2)x_f - \mu_3 k_1 L/\mu_2 k_3 - x_c\} \qquad (17.14e)$$

$$\beta_3 = p_m + (p_m\text{-}p_r)x_c/\{(\mu_3 k_1/\mu_2 k_3 \text{-}k_1/k_2)x_f \text{-}\mu_3 k_1 L/\mu_2 k_3 \text{-}x_c\}$$
$$+x_f\{k_1/k_2\text{-}\mu_3 k_1/\mu_2 k_3\}(p_m\text{-}p_r)/ \tag{17.14f}$$
$$\{(\mu_3 k_1/\mu_2 k_3\text{-}k_1/k_2)x_f\text{-}\mu_3 k_1 L/\mu_2 k_3\text{-}x_c\}$$

Eqs. (17.14a)–(17.14f) completely define the spatial pressure distributions within Layers 1, 2, and 3. However, the solutions to the invasion problem are as yet incomplete because the positions x_c and x_f are unknown time-dependent functions that satisfy other constraints. Consider the mudcake first. Our previous differential equation for cake growth can be evaluated as

$$\begin{aligned} dx_c/dt &= \{f_s/\{(1\text{-}\phi_c)(1\text{-}f_s)\}\}\,|v_n| = \text{-}[f_s/\{(1\text{-}\phi_c)(1\text{-}f_s)\}](k_1/\mu_1) \\ &\times(p_m\text{-}p_r)/\{(\mu_3 k_1/\mu_2 k_3 \text{-} k_1/k_2)x_f \text{-}\mu_3 k_1 L/\mu_2 k_3 \text{-}x_c\} \end{aligned} \tag{17.15}$$

But this cannot be integrated since it depends on the front displacement $x_f(t)$, which satisfies its own dynamic equation. To obtain it, we evaluate the interfacial kinematic condition using the now known Darcy velocity as

$$\begin{aligned} dx_f/dt &= \text{-}\phi_{eff}^{-1}(k_2/\mu_2)\,dp_2(x)/dx \\ &= \text{-}\left(k_1/\mu_2\phi_{eff}\right)(p_m\text{-}p_r)/\{(\mu_3 k_1/\mu_2 k_3 \text{-}k_1/k_2)x_f \text{-}\mu_3 k_1 L/\mu_2 k_3 \text{-}x_c\} \end{aligned} \tag{17.16}$$

Here ϕ_{eff} denotes the effective porosity that Layer 2 offers if immobile fluids are left behind once the filtrate front passes. This usage provides some degree of flexibility in modeling two-phase flow relative permeability effects within the framework of single-phase flow theory. Still, Eqs. (17.15), (17.16) are coupled; at first, recourse to numerical analysis appears necessary, but this is fortunately not the case. It turns out that exact analytical solutions can be obtained. If we assume the initial condition $x_f\,(t=0)=x_{f,o}>0$ for the mud spurt, and $x_c=0$, until $x_f=x_{f,o}>0$ with $x_c(t)>0$, for $x_f>x_{f,o}$, we obtain the solution of Chin et al. (1986),

$$x_f(t) = \text{-}H + \sqrt{\{H^2 + 2(Hx_{f,O} + ½x_{f,O}{}^2 + Gt)\}} \tag{17.17}$$

where

$$\begin{aligned} G = \text{-}&\left\{k_1(p_m\text{-}p_r)/\left(\mu_f\phi_{eff}\right)\right\}/ \\ &\left\{\mu_O k_1/\mu_f k_3 \text{-} k_1/k_2\text{-}\phi_{eff}\,f_s/\{(1\text{-}\phi_c)(1\text{-}f_s)\}\right\} \end{aligned} \tag{17.18}$$

$$\begin{aligned} H = &\left[x_{f,O}\phi_{eff}f_s/\{(1\text{-}\phi_c)(1\text{-}f_s)\}\text{-}\mu_O k_1 L/\mu_f k_3\right]/ \\ &\left\{\mu_O k_1/\mu_f k_3 \text{-} k_1/k_2\text{-}\phi_{eff}f_s/\{(1\text{-}\phi_c)(1\text{-}f_s)\}\right\} \end{aligned} \tag{17.19}$$

Eqs. (17.17)–(17.19) completely describe the progress of the invasion front, as it is affected by filtrate and reservoir liquid viscosities, and mudcake properties and growth. The corresponding equation for mudcake growth is

$$x_c(t) = \left[\phi_{eff} f_s / \{(1-\phi_c)(1-f_s)\}\right] \left(x_f - x_{f,O}\right) \tag{17.20}$$

for which $dx_c/dx_f = \phi_{eff} f_s / \{(1-\phi_c)(1-f_s)\} > 0$. This states that x_f increases if x_c increases; it is interesting that the proportionality factor depends on geometrical parameters only, and not on transport variables like viscosity and permeability. We emphasize that, in general, pure $\sqrt{t}$ behavior is not always obtained, although it does appear in the limit of very large t. The reader, following our earlier Taylor series exercise, should determine the exact dimensionless meaning of large time when pure $\sqrt{t}$ behavior is found. Finally, note that

$$x_f(t) - x_{f,O} = \phi_{eff}^{-1} \sqrt{\left\{2k_1(1-\phi_c)(1-f_s)(p_m-p_r)t/\left(\mu_f f_s\right)\right\}} \tag{17.21}$$

is obtained in the limit when the mobility in the rock core greatly exceeds that of the mudcake. This is the restrictive limit typically considered in the literature; again, our solutions do not invoke any limiting assumptions about relative cake-to-formation mobilities. Finally, we emphasize that these results require us to characterize the mudcake by three independent parameters, namely, the solid fraction f_s, the porosity ϕ_c, and the cake permeability k. The work in Chin et al. (1986) and in recent industry studies requires such empirical inputs and elaborate laboratory. It turns out that all of this is unnecessary, and that a simple lumped parameter defined by convenient lineal filtrate tests on standard filter paper is all that is required. These ideas are pursued in Chapter 19.

Simultaneous Mudcake Buildup and Filtrate Invasion in a Radial Geometry (Liquid Flows)

Here, we shall reconsider the simultaneous mudcake buildup and filtrate invasion problem just discussed, but we shall use realistic radial coordinates. Note that the exact *linear* flow solution in Chin et al. (1986) includes the all-important effect of mud spurt. But while that paper alluded to progress towards a radial solution, the work at that time could not account for any spurt at all because of mathematical complexities and, furthermore, turned to numerical solution as a last resort. Thus, a useful solution was not available, and any applications to time lapse analysis would await further progress. Since then, the result of some significant efforts have led to a closed-form solution. The resulting solution and derivation are described in detail here. This availability, together with the simple recipe for mudcake properties alluded to, brings time lapse analysis closer to reality.

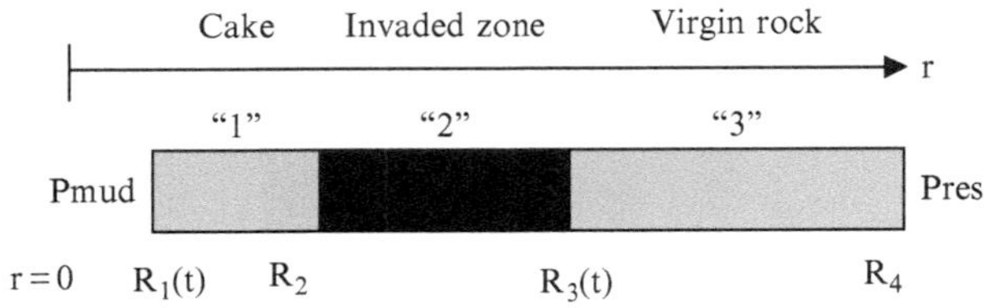

FIG. 17.7 Three-layer radial flow ("mud" denoted by subscript m, reservoir "res" denoted by subscript r).

We consider a realistic example where an incompressible liquid mud filtrate displaces a preexisting incompressible formation *liquid* having a different viscosity (gas displacement is discussed in Chapter 20). Such fluids, flowing in homogeneous, isotropic media, satisfy Laplace's equation for pressure. And while this process is ongoing, mudcake thickness is ever-increasing, so that filtrate influx rate is consequently decreasing; all the time, the filtrate-to-formation fluid displacement front moves to the right. In this problem, as in our lineal one, the dynamics of the mudcake growth are closely coupled to the invasion front motion. In our derivation, there is *no* assumption that the mudcake is significantly less permeable than the formation, an assumption usually taken to simplify the analysis. Also, $\sqrt{t}$ behavior is not presumed at the outset; doing so would be wrong. The work is exact in this regard, since the relative mobilities between cake, invaded zone, and virgin formation are left as completely free parameters for subsequent evaluation.

In Fig. 17.7, let Layer 1 denote the mudcake, and Layers 2 and 3, the filtrate-invaded and virgin oil-bearing formations, respectively. In this axisymmetric problem, the origin $r=0$ is the borehole centerline. Here, $r=R_2$ represents the fixed cake-to-rock interface; R_2 is an absolute constant equal to the borehole radius. Note that $r=R_1(t)$ represents the time-varying radial position of the mud-to-mudcake interface, while $R_3(t)$ denotes the time-dependent invasion front position. Finally, $r=R_4$ represents a fixed effective radius where the reservoir pore pressure p_r is specified. The driving pressure differential is $(p_m - p_r)$, where p_m is the pressure in the borehole. The transient compressible flow equations for constant liquid and rock properties are of standard parabolic form, for example, $\partial^2 p_1(r,t)/\partial r^2 + (1/r)\ \partial p_1/\partial r = (\phi_1 \mu_1 c_1/k_1)\ \partial p_1/\partial t$ for Layer 1. But since we are ignoring all fluid compressibilities, in effect considering constant density liquids with $c=0$, our equations reduce to the differential equations $d^2p_i(r)/dr^2 + (1/r)dp_i/dr = 0$, $i=1, 2, 3$, which have the solutions $p_i(r) = \alpha_i \log r + \beta_i$, $i=1, 2, 3$.

As in our earlier example, the integration constants can be determined from the end pressure boundary conditions $p_1(R_1) = p_m$ and $p_3(R_4) = p_r$. Also, we shall require the interfacial matching conditions $p_1(R_2) = p_2(R_2)$ and $p_2(R_3) = p_3(R_3)$ for pressure, and $(k_1/\mu_1)\ dp_1(R_2)/dr = (k_2/\mu_2)\ dp_2(R_2)/dr$ and $(k_2/\mu_2)\ dp_2(R_3)/dr = (k_3/\mu_3) dp_3(R_3)/dr$ for velocity. Note that we have retained three separate permeabilities, namely, k_1, k_2, and k_3 in these equations. The k_1 represents, of course, the mudcake permeability. However, while we have but a single radial rock core, characterized by a single permeability, we shall derive our results with *two* values k_2 and k_3. This flexibility allows us to set $k_2 = k_3 = k_{rock}$ if desired, or allow them to differ, in order to represent separate permeabilities to filtrate and oil. Note that we have also retained three viscosities μ_1, μ_2, and μ_3, even though the same liquid $\mu_1 = \mu_2$ flows through Layers 1 and 2 (in the previous example, we assumed that $\mu_1 = \mu_2$). This *ad hoc* modeling, consistent with our introduction of ϕ_{eff} earlier, permits us to *mimic* two-phase flow relative permeability effects within the framework of a simpler slug displacement approach (see Chapter 16). The six integration constants are easily found, using elementary algebra, as

$$\alpha_1 = (k_2/\mu_2)(p_r\text{-}p_m)/\log\left[(R_2/R_1)^{k_2/\mu_2}(R_3/R_2)^{k_1/\mu_1}(R_4/R_3)^{k_1k_2\mu_3/\mu_1\mu_2k_3}\right] \quad (17.22a)$$

$$\beta_1 = p_m \text{-} \alpha_1 \log R_1 \quad (17.22b)$$

$$\alpha_2 = (k_1\mu_2/\mu_1k_2)\alpha_1 \quad (17.22c)$$

$$\beta_2 = p_m + \alpha_1 \log(R_2/R_1) \text{-} \alpha_2 \log R_2 \quad (17.22d)$$

$$\alpha_3 = (\mu_3k_1/\mu_1k_3)\alpha_1 \quad (17.22e)$$

$$\beta_3 = p_m + \alpha_1 \log(R_2/R_1) + \alpha_2 \log(R_3/R_2) \text{-} \alpha_3 \log R_3 \quad (17.22f)$$

where all logarithms are natural logarithms. It appears that we have defined the spatial pressure distributions within Layers 1, 2, and 3. However, the solutions to the invasion problem are incomplete because the position fronts $R_1(t)$ and $R_3(t)$ are unknown functions of t. As before, equations for cake growth and displacement front motion must be postulated. For mudcake growth, we have

$$\begin{aligned} \text{-}\,dR_1/dt &= \{f_s/\{(1\text{-}\phi_c)(1\text{-}f_s)\}\}\,|v_n| \\ &= [f_s/\{(1\text{-}\phi_c)(1\text{-}f_s)\}]\,(k_1/\mu_1)\,dp_1/dr \\ &= [f_s/\{(1\text{-}\phi_c)(1\text{-}f_s)\}]\,(k_1/\mu_1)\alpha_1/r \\ &= [f_s/\{(1\text{-}\phi_c)(1\text{-}f_s)\}]\,(k_1/\mu_1)\alpha_1(R_1,R_3)/R_1 \end{aligned} \quad (17.23)$$

The analogous equation for displacement front motion is obtained from

$$\begin{aligned} dR_3/dt &= \text{-}\left\{k_2/\left(\mu_2\phi_{eff}\right)\right\}dp_2/dr \\ &= \text{-}\left\{k_2/\left(\mu_2\phi_{eff}\right)\right\}\alpha_2/r \\ &= \text{-}\left\{k_2/\left(\mu_2\phi_{eff}\right)\right\}\alpha_2(R_1,R_3)/R_3 \end{aligned} \quad (17.24)$$

These nonlinear ordinary differential equations, as in the lineal case, are coupled. But again, it is possible to integrate them in closed, analytical form for general initial conditions. If we assume that $R_3 = R_{spurt} \geq R_2$, when $R_1 = R_2$ (i.e., no cake) at $t = 0$, we find that the displacement front history $R_3(t)$ satisfies

$$
\begin{aligned}
&(k_1R_2^2/\mu_1)[\tfrac{1}{2}(R_3/R_2)^2 \log(R_3/R_2) - \tfrac{1}{4}(R_3/R_2)^2 \\
&\quad - \tfrac{1}{2}(R_{spurt}/R_2)^2 \log(R_{spurt}/R_2) + \tfrac{1}{4}(R_{spurt}/R_2)^2] \\
&+ (k_1k_2\mu_3R_4^2/\mu_1\mu_2k_3)[\tfrac{1}{2}(R_{spurt}/R_4)^2 \log(R_{spurt}/R_4) - \tfrac{1}{4}(R_{spurt}/R_4)^2 \\
&\quad - \tfrac{1}{2}(R_3/R_4)^2 \log(R_3/R_4) + \tfrac{1}{4}(R_3/R_4)^2] \\
&+ \left(k_2R_2^2/4\mu_2\phi_{eff}\right)\{(1-\phi_c)(1-f_s)/f_s\}\times \\
&\quad \Big[\log\Big\{1 + f_s\phi_{eff}\Big\{(R_{spurt}/R_2)^2 - (R_3/R_2)^2\Big\}/\{(1-\phi_c)(1-f_s)\}\Big\} \\
&\quad - f_s\phi_{eff}\Big\{(R_{spurt}/R_2)^2 - (R_3/R_2)^2\Big\}/\{(1-\phi_c)(1-f_s)\} \\
&\quad + f_s\phi_{eff}\Big\{(R_{spurt}/R_2)^2 - (R_3/R_2)^2\Big\}/\{(1-\phi_c)(1-f_s)\}\times \\
&\quad \log\Big\{1 + f_s\phi_{eff}\Big\{(R_{spurt}/R_2)^2 - (R_3/R_2)^2\Big\}/\{(1-\phi_c)(1-f_s)\}\Big\}\Big] = \\
&\quad = \Big\{k_1k_2(p_m - p_r)/\left(\mu_1\mu_2\phi_{eff}\right)\Big\}\,t
\end{aligned}
\tag{17.25}
$$

which does *not*, we emphasize, in general follow $\sqrt{t}$ behavior (e.g., see Outmans, 1963). This exact formula is particularly useful in MWD logging applications where the extent of formation invasion needs to be estimated prior to taking measurements. Eq. (17.25) can be solved by assuming values for R_3 and calculating the corresponding times. The associated cake radius function $R_1(t)$ is then obtained from

$$R_1^2 = R_2^2 + \left(R_{spurt}^2 - R_3^2\right)\left(f_s\phi_{eff}\right)/\{(1-\phi_c)(1-f_s)\} \tag{17.26}$$

It is also possible to show that

$$dR_1^2/dR_3^2 = -[f_s/\{(1-\phi_c)(1-f_s)\}]\phi_{eff} < 0 \tag{17.27}$$

This equation indicates that as our filtration front advances, with R_3^2 increasing, the radius (squared) R_1^2 decreases. This decrease, following the schematic shown in Fig. 17.7, indicates that mudcake thickness is consistently growing. Eq. (17.27) is a Lagrangian mass conservation law that is independent of transport parameters such as permeability and viscosity.

Unlike the lineal cake problem studied earlier where, in principle, the mudcake can increase indefinitely in thickness over time, the maximum radial thickness that can be achieved in this radial example is defined by $R_1(t_{max}) = 0$, and occurs at $t = t_{max}$. At this time, all fluid motions cease, at least within the framework of the piston-like displacements studied in this chapter, and molecular diffusion then becomes the dominant physical player. In order to determine the maximal radial displacement $R_{3,max}$ and its corresponding time scale t_{max}, we set $R_1(t)$ to zero in Eq. (17.26), to obtain

$$R_{3,max} = \sqrt{\left[R_{spurt}^2 + \left\{(1\text{-}\phi_c)(1\text{-}f_s)/\left(f_s\phi_{eff}\right)\right\}R_2^2\right]} \tag{17.28}$$

Then t_{max} is obtained by substituting $R_{3,max}$ into Eq. (17.25), that is,

$$\begin{aligned}
&(k_1R_2^2/\mu_1)[\tfrac{1}{2}(R_{3,max}/R_2)^2 \log (R_{3,max}/R_2) - \tfrac{1}{4}(R_{3,max}/R_2)^2 \\
&\quad - \tfrac{1}{2}(R_{spurt}/R_2)^2 \log (R_{spurt}/R_2) + \tfrac{1}{4}(R_{spurt}/R_2)^2] \\
&+ (k_1k_2\mu_3R_4^2/\mu_1\mu_2k_3)[\tfrac{1}{2}(R_{spurt}/R_4)^2 \log (R_{spurt}/R_4) - \tfrac{1}{4}(R_{spurt}/R_4)^2 \\
&\quad - \tfrac{1}{2}(R_{3,max}/R_4)^2 \log (R_{3,max}/R_4) + \tfrac{1}{4}(R_{3,max}/R_4)^2] \\
&+ \left(k_2R_2^2/4\mu_2\phi_{eff}\right)\{(1\text{-}\phi_c)(1\text{-}f_s)/f_s\}\times \\
&\quad \Big[\log\Big\{1 + f_s\phi_{eff}\Big\{(R_{spurt}/R_2)^2 - (R_{3,max}/R_2)^2\Big\}/\{(1\text{-}\phi_c)(1\text{-}f_s)\}\Big\} \\
&\quad - f_s\phi_{eff}\Big\{(R_{spurt}/R_2)^2 - (R_{3,max}/R_2)^2\Big\}/\{(1\text{-}\phi_c)(1\text{-}f_s)\} \\
&\quad + f_s\phi_{eff}\Big\{(R_{spurt}/R_2)^2 - (R_{3,max}/R_2)^2\Big\}/\{(1\text{-}\phi_c)(1\text{-}f_s)\}\times \\
&\quad \log\{1 + f_s\phi_{eff}\Big\{(R_{spurt}/R_2)^2 - (R_{3,max}/R_2)^2\Big\}/\{(1\text{-}\phi_c)(1\text{-}f_s)\}\Big] = \\
&\quad = \left\{k_1k_2(p_m - p_r)/\left(\mu_1\mu_2\phi_{eff}\right)\right\} t_{max}
\end{aligned} \tag{17.29}$$

In reality, hole plugging is limited by borehole flow erosion, an essential element of the dynamic filtration process, a process that will be discussed and modeled in detail shortly.

Fluid Compressibility

Here we examine the effects of fluid compressibility on invasion. This should not be confused with mudcake and rock compressibility, which represent different physical phenomena. We consider a simple lineal flow example, which will be followed in Chapter 18 by an important and sophisticated formation tester solution useful in modern formation evaluation. We reconsider the liquid and gaseous lineal flows treated earlier, but this time, include the transient effects due to fluid compressibility in a homogeneous core without mudcake. The pressure P(x,t) now depends on both x and t. The relevant geometry is shown in Fig. 17.8, where the left- and right-side pressure boundary conditions are $P(0,t) = P_l$ and $P(L,t) = P_r$, and L is the core length. Let us assume that initially, $P(x,0) = P_o$. We study liquids first, and then reformulate and solve the problem for gases.

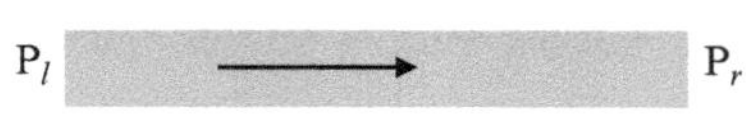

FIG. 17.8 Lineal flow.

For compressible liquids, the partial differential equation governing pressure is $\partial^2 P(x,t)/\partial x^2 = (\phi\mu c/k)\ \partial P/\partial t$ where ϕ, μ, c, and k are porosity, viscosity, compressibility, and permeability. The auxiliary pressure conditions are $P(0,t) = P_l$, $P(L,t) = P_r$, and $P(x,0) = P_o$. This formulation is identical to that of the classic initial and boundary value problem in heat transfer for a rod with prescribed end temperatures and arbitrary initial temperature (Carlsaw and Jaeger, 1946; Tychonov and Samarski, 1964). In reservoir applications, we typically have $P_o = P_r$, but we shall leave the formulation general, since the result may be useful in special experimental situations. This can be solved in closed form using separation of variables and Fourier series (Hildebrand, 1948), but we shall not reproduce the standard derivation. The exact solution is

$$\begin{aligned} P(x,t) = (P_r - P_l)x/L + P_l + (2/\pi)\sum(1/n)[P_o - P_l + (P_r - P_o)(-1)^n] \\ \times \exp\left[-\pi^2 n^2 kt/\left(L^2\phi\mu c\right)\right] \sin n\pi x/L \end{aligned} \tag{17.30}$$

where a summation from $n = 1$ to ∞ is understood. The first line gives the steady-state response; the second is the transient compressible response.

The largest transient contribution to Eq. (17.30) arises from the $n = 1$ term, which has the amplitude decay factor $\exp[-\pi^2 kt/(L^2\phi\mu c)]$. Only when $\pi^2 kt/(L^2\phi\mu c) \rightarrow \infty$, that is, $t \gg L^2\phi\mu c/(\pi^2 k)$ does the effect of compressibility and initial conditions, through the amplitude factor $[P_o - P_l + (P_r - P_o)(-1)^n]$ with $n = 1$, vanish. If we consider the steady solution $P(x,t) = (P_r - P_l)x/L + P_l$ only, the front satisfies $dx/dt = -(k/\phi\mu)\ \partial p(x,t)/\partial x = -(k/\phi\mu)(P_r - P_l)/L$. Its time scale is determined from the quotient $L/(dx/dt)$, that is, $L^2\phi\mu/\{k(P_l - P_r)\}$. Compressibility introduces a time scale proportional to $L^2\phi\mu c/(\pi^2 k)$.

The partial differential equation governing compressible, lineal, homogeneous, gaseous flows having constant properties is nonlinear, satisfying $\partial^2 P^{m+1}(x,t)/\partial x^2 = (\phi\mu c^*/k)\partial P^{m+1}/\partial t$ where the terms have been defined earlier. While the initial and boundary value problem for the function P^{m+1} in

$$\partial^2 P^{m+1}(x,t)/\partial x^2 = \{\phi\mu m/(Pk)\}\ \partial P^{m+1}/\partial t \tag{17.31a}$$

$$P^{m+1}(0,t) = P_l^{\ m+1} \tag{17.31b}$$

$$P^{m+1}(L,t) = P_r^{\ m+1} \tag{17.31c}$$

$$P^{m+1}(x,0) = P_o^{\ m+1} \tag{17.31d}$$

superficially resembles the linear one for compressible liquids, with P replaced by P^{m+1}, the two formulations are different because the constant coefficient $\phi\mu c/k$ in the liquid formulation is replaced by the *function* $\phi\mu m/(Pk)$, which depends on the solution $P(x,t)$. The liquid problem is *linear*, with the sum of individual solutions itself being a solution, rendering superposition using Fourier series possible (well test procedures similarly use superposition techniques). But the latter formulation, because of the pressure-dependence, is nonlinear, and closed-form solution is not possible except for the simplest problems. Nonetheless, we can develop some idea of the time scales that arise on account of compressibility if we approximate the nonlinear coefficient by the constant $\phi\mu m/(P_{avg}k)$, where $P_r < P_{avg} < P_l$, and if we additionally assume that P_o lies in the same range. Then, comparison of the two formulations leads us to infer a formal solution satisfying

$$\begin{aligned} P^{m+1}(x,t) \approx & \left(P_r{}^{m+1} - P_l^{m+1}\right)x/L + P_l^{m+1} \\ & + (2/\pi)\sum(1/n)\left[\left(P_o{}^{m+1} - P_l^{m+1} + \left(P_r{}^{m+1} - P_o{}^{m+1}\right)(-1)^n\right]\right. \\ & \times \exp\left[-\pi^2 n^2 kt P_{avg}/L^2\phi\mu m\right] \sin n\pi x/L \end{aligned} \tag{17.32}$$

which is a very crude approximation to the actual solution. But this formal procedure does provide some indication about the time scales governing transient decay. If we now raise each side of the above to the $1/(m+1)th$ power in order to solve for the transient pressure P(x,t), we would expect a term like exp $[-\pi^2 ktP_{avg}/\{L^2\phi\mu m(m+1)\}]$ to emerge from the algebra. The time scale suggested by this $n=1$ term is quite different from that for linear liquids, and illustrates, through the constant m, the role of reservoir thermodynamics. Compressibility can be important for gas flows, but even for liquids, where our closed-form pressure solutions are relatively simple, the consequences related to front motion are difficult to determine.

Consider, for example, liquid flows. Since lineal front trajectories satisfy $dx/dt = -(k/\phi\mu)\partial P/\partial x$, we have, using Eq. (17.30),

$$\begin{aligned} dx/dt = & -(k/\phi\mu)\, \partial P(x,t)/\partial x \\ = & -(k/\phi\mu)\, \{(P_r - P_l)/L \\ & + (2/L)\sum[P_o - P_l + (P_r - P_o)(-1)^n] \exp\left[-\pi^2 n^2 kt/\left(L^2\phi\mu c\right)\right] \cos n\pi x/L\} \end{aligned} \tag{17.33}$$

We examine a marked fluid element located at the inlet $x=0$ initially. Then, the cos $n\pi x/L$ term in Eq. (17.33) becomes unity, and if we retain only the leading $n=1$ contribution, we have the approximation

$$\begin{aligned} dx \approx & -(k/\phi\mu)\, \{(P_r - P_l)/L \\ & + (2/L)[P_o - P_l - (P_r - P_o)] \exp\left[-\pi^2 kt/\left(L^2\phi\mu c\right)\right]\}\, dt \end{aligned} \tag{17.34}$$

for which

$$\begin{aligned} x \approx & -(k/\phi\mu)\, \{(P_r - P_l)t/L \\ & - 2(2P_o - P_l - P_r)\left[L\phi\mu c/\left(\pi^2 k\right)\right] \left(-1 + \exp\left[-\pi^2 kt/\left(L^2\phi\mu c\right)\right]\right)\} \end{aligned} \tag{17.35}$$

This solution satisfies $x(0)=0$. Therefore, for small times, the effect of compressibility, assuming $P_r = P_o$, will be large or small accordingly as the product $(P_l - P_r)[L\phi\mu c/(\pi^2 k)]$ is large or small.

We emphasize that the exact, closed-form, analytical results obtained in this chapter reveal nontrivial dependences of our filtration front and cake growth formulas on numerous groups of parameters that may have been anticipated from dimensional analysis or dynamic similitude. In the mudcake radial flow analysis, our flagship problem, the derivation did not make any assumptions regarding the relative mobilities in the three different layered flows, and so, are completely general within the framework of their formulation. But the results are restricted to constant density liquid filtrates and liquid formation fluids, that is, not gases, and then, to incompressible mudcakes only. In general, we have found that $\sqrt{t}$ behavior is more the exception than the rule; example numerical calculations will be given later. Under these restrictions, once R_1 and R_3 are known as functions of time, we can evaluate the derived pressure formulas to provide complete spatial pressure distributions at any desired instant in time. In modern MWD and time lapse analysis applications, pressure distributions are of lesser practical interest than the values of formation porosity, permeability, and mobility themselves. Of course, pressures and spatial pressure gradients *are* important in fluid production, that is, the reverse problem where the value of $p_m(t)$ necessary to produce at a prescribed flow rate is important.

DYNAMIC FILTRATION AND BOREHOLE FLOW RHEOLOGY

Here, we shall introduce dynamic filtration analytically and numerically, using recently developed methods in borehole and pipeline flow modeling. These are discussed extensively in the books *Borehole Flow Modeling in Horizontal, Deviated, and Vertical Wells* (Chin, 1992a) and *Computational Rheology for Pipeline and Annular Flow* (Chin, 2001a). However, these references do not develop the reservoir focus needed in coupling borehole flows to the Darcy flows in the rock formation. Dynamic filtration is important to invasion because the erosive effects associated with annular mud flow limit the plugging of the borehole that was predicted on the basis of radial static filtration theory. This in turn affects the flow rate into or out of the reservoir. Once the mudcake thickness reaches equilibrium, the dynamical problem in the borehole becomes steady state. Of course, the flow within the reservoir need not be steady. For example, invasion may continue at a constant rate; the saturations in immiscible two-phase flow may continue to redistribute unsteadily in time, and their dynamically coupled pressures will evolve along with them.

Many drilling muds are formed from mixtures of water or oil and weighting materials. These mixtures, unlike the underlying Newtonian water or oil fluids, are non-Newtonian in character. When formation invasion occurs, solids are left behind at the sandface to create mudcake, while the underlying Newtonian filtrate enters the formation. Therefore, two fluid flow problems arise: Newtonian flow in the reservoir, the subject of this book, and non-Newtonian flow in the borehole annulus, the focus of this section. When fluid in a nonflowing well enters the formation, static filtration occurs, while dynamic filtration occurs when the well is flowing. In the latter, fluid motion is found both parallel and perpendicular to the borehole axis. Borehole annular flows are challenging to model because the fluid is non-Newtonian, or put differently, the rheological model involves a nonlinear relationship between the shear stress and the shear rate. Rheological models for drilling muds commonly used are power law, Bingham plastic, and Herschel-Bulkley. For such nonlinear (as opposed to linear Newtonian) models, for example, doubling the pressure gradient does not double the flow rate, thus making non-Newtonian fluids more difficult and less intuitive to deal with. These mud models are simple in that they are just non-Newtonian. If the base fluid had contained polymers, then both borehole and reservoir flows would be non-Newtonian, and more specialized analysis methods would apply. These are beyond the scope of this book. For more discussion on non-Newtonian flow modeling, refer to the author's book references cited earlier.

Erosion due to Shear Stress

The problem of dynamic filtration, that is, the limiting of mudcake growth due to annular flow erosion, is well known to the invasion literature. But despite its prominence over the decades, no analytical models have appeared, presumably because the complicated erosion process is empirically grounded. And it is becoming more so. Fordham et al. (1991), for example, introduce the concept of a critical invasion rate beyond which mudcakes will not form, while Holditch and Dewan (1991) allude to a mysterious adhesion fraction that presumably controls the buildup process. Whatever the terminology, mere definitions only acknowledge the importance of erosion, but do not contribute to the solution of the physical problem. However, key insights are emerging, and apparently converging on a promising line of research that highlights the importance of viscous shear stress. For example, recent service company efforts described in Fordham et al. (1991) indicate that hydraulic *shear stress* at the mudcake surface limits cake growth. Unfortunately, the paper does not construct the needed model. Interestingly, this same shear stress principle for bed erosion was first proposed in 1990 in a series of *Offshore Magazine* articles by this author, where the principal application was cuttings bed erosion and hole cleaning in highly deviated wells. In these articles, the correlation parameter of physical significance for bed erosion and particulate transport was shown to be the viscous stress obtained at the bed surface. (The work is summarized in *Borehole Flow Modeling* and *Computational Rheology*, where additional physical and modeling concepts are developed along with empirical validation.) Successful correlations were obtained with data obtained from controlled laboratory tests using water-base muds, and from field operations using all-oil muds and invert emulsions. We shall discuss these results in detail later, but for now, we note that the role of bed (or cake) yield shear stress is clearly important to any predictive model for equilibrium thickness. Several analytical models for equilibrium cake thickness based on this shear stress criterion will be developed.

We shall discuss Newtonian flows first, primarily because the simpler analysis enables us to develop the mathematical concepts with clarity. Nonrotating annular flows are considered initially; then, the modifications needed to extend our results to rotating flows and to fluids with increased solids content are presented. Two physical asymptotes are noted, namely, an early time limit where radial filtration into the formation is predominant, and a late time process where a steady-state annular flow satisfying no-slip velocity boundary conditions prevails. Then, we rework our Newtonian flow criteria for power law fluid models that more accurately characterize real drilling muds, first using exact numerical methods based on the classic solution of Fredrickson and Bird (1958), that do not render simplifying geometrical assumptions, and second, using approximate "narrow annulus" methods that allow us to handle rotating drillpipes easily. These approaches apply to concentric annuli only. Finally, following a review of the principal ideas in *Computational Rheology*, special techniques applicable to highly eccentric holes containing fluids having different types of mud rheology are outlined for further development.

Dynamic Filtration in Newtonian Fluids

While borehole annular flows are rarely Newtonian (e.g., fluids such as water or air, where viscous stress is linearly proportional to the rate of strain), many drilling fluids *are* thin and briny, and at times, simply water. Thus, for analysis purposes, the study of Newtonian flows is more than academic. Furthermore, the mathematical simplicity that it offers sheds some insight into the parameters that influence the value of equilibrium cake thickness in the presence of erosive annular flow. Whether our annular flow is Newtonian or power law, concentric or eccentric, it is important to consider two underlying asymptotic fluid-dynamical models. The first applies during small times when borehole fluid enters the formation radially as filtrate, decelerating with time, while the second deals with large times, when invasion rates are so slow that we essentially have classical no-slip velocity boundary conditions. We shall first consider the small time limit, assuming that the drill pipe does not rotate.

For initial times, the mudcake thickness is small compared to the borehole radius, and the near-planar problem can be formulated in simple rectangular coordinates. We have, in particular, the two-dimensional viscous flow shown in Fig. 17.9, where a more or less uniform oncoming annular fluid flows parallel to the cake surface, only to be withdrawn into the formation with a vertical filtration velocity $v_0(t)$. The Navier-Stokes momentum equations governing constant density, viscous, unsteady, Newtonian flow are Eqs. (17.36), (17.37), while the corresponding requirement for mass conservation is given by Eq. (17.38),

$$\rho\left(u_t + uu_x + vu_y\right) = -p_x + \mu\left(u_{xx} + u_{yy}\right) \tag{17.36}$$

$$\rho\left(v_t + uv_x + vv_y\right) = -p_y + \mu\left(v_{xx} + v_{yy}\right) \tag{17.37}$$

$$u_x + v_y = 0 \tag{17.38}$$

Here, u(x,y,t) and v(x,y,t) are Eulerian velocities in the x and y directions, t is time, and p(x,y,t) represents the pressure; also, ρ is mass density and μ is viscosity.

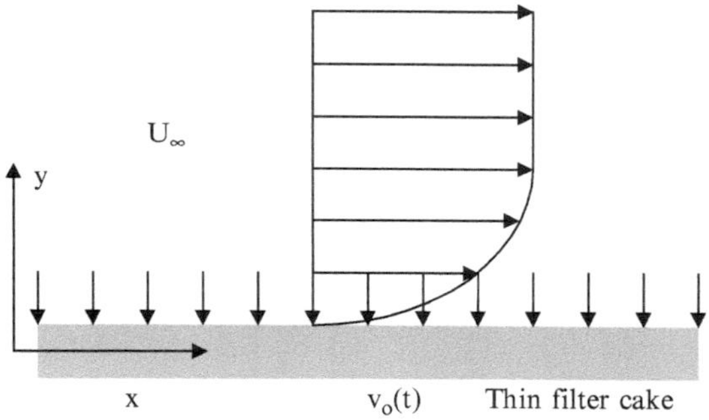

FIG. 17.9 Planar flow past thin mudcake at early times.

We note that Eqs. (17.36)–(17.38) apply in the borehole and do not apply to flows in porous media. Similarly, Darcy's law (that is, $u = -(k/\mu)p_x$, $v = -(k/\mu)p_y$, and $p_{xx} + p_{yy} = 0$, say) applies in the reservoir, and not in the borehole. For the problem in Fig. 17.9, we shall augment Eqs. (17.36)–(17.38) with the velocity boundary conditions in Eqs. (17.39a), (17.39b).

$$u(x, \infty, t) = U_\infty \text{ (a constant)} \tag{17.39a}$$

$$v(x, 0, t) = v_o(t) < 0 \tag{17.39b}$$

Here, $v(x,y,t) = v_o(t)$ is the prescribed filtration rate, for example, as determined from the $\sqrt{t}$ law or its radial extensions. While we have defined U_∞ as the annular velocity parallel to the hole axis in a nonrotating flow, we can more generally take it as the maximum velocity (when both axial and circumferential speeds are present) in flows with rotation, providing increased modeling flexibility. The closed-form solution to this problem is

$$u(x, y, t) = U_\infty \left[1\text{-}\exp\{v_o(t)y/\nu\}\right] \tag{17.40}$$

$$v(x, y, t) = v_o(t) \tag{17.41}$$

$$p(x, y, t) = -\rho v_{o,t}(t)y \tag{17.42}$$

where ν is the kinematic viscosity μ/ρ. Substitution of Eqs. (17.40), (17.41) for u and v into Eq. (17.38) shows that the latter is identically satisfied. Substitution of Eqs. (17.40)–(17.42) into Eq. (17.37) shows that Eq. (17.42) provides the (radial) pressure gradient required in the annulus to produce the filtration speed $v_o(t)$. Finally, substitution of Eqs. (17.40)–(17.42) into Eq. (17.36) shows that our assumptions *do* produce solutions, so long as

$$v_{o,t}(t)\, y \approx 0 \tag{17.43}$$

This is certainly the case, since the filtrate acceleration $dv_o(t)/dt$ is small and y is negligible at the cake surface. In the limit when $v_o(t)$ is constant, Eqs. (17.40)–(17.42) reduce to the classic boundary layer solution for asymptotic suction profiles (Schlichting, 1968). The shear stress at the cake surface $y=0$ is easily determined and is found to be

$$\tau = \mu u_y = -\rho U_\infty v_o(t) \tag{17.44}$$

which is importantly independent of the viscosity μ. Thus, during the initial periods of cake buildup, when the cake has not yet undergone significant compaction, the shear stress at the exposed surface of the mudcake is determined only by the density and mean speed of the annular flow and by the filtration rate. If the shear stress given in Eq. (17.44) exceeds the yield stress of the newly formed cake (or, possibly, some type of gel strength that is important in the interstitial pore spaces of the evolving mudcake), then mudcake buildup is terminated, and equilibrium borehole conditions are reached; filtration into the reservoir, of course, continues.

An asymptotic solution for large times can be developed. For large times, the annular radial geometry and the details of the fully developed laminar or turbulent velocity profile become important. Let us consider, for simplicity, the concentric flow cross-section shown in Fig. 17.10. Again, we assume that the drillpipe does not rotate, and for now, we introduce the constant axial pressure gradient dp/dz where z is directed along the axis of the wellbore. When mudcake growth has built up sufficiently and stabilized, filtration rates (in the radial direction) are extremely small, and the annular flow velocity parallel to the wellbore axis satisfies no-slip velocity boundary conditions at the cake surface R_c. (This is in addition to no-slip enforcement at the pipe radius R_p.) The rectangular coordinates assumed in Fig. 17.9, of course, no longer apply, as the cylindrical radial nature of the flow dominates. Fortunately, an exact solution to this problem exists. From *Borehole Flow Modeling*, the axial velocity $v_z(r)$ parallel to the wellbore axis assuming laminar flow is

$$\begin{aligned} v_z(r) = {} & (4\mu)^{-1}(dp/dz) \\ & \left[r^2 - R_p^2 + \left\{\left(R_c^2 - R_p^2\right)/\log\left(R_p/R_c\right)\right\}\log\left(r/R_p\right)\right] \end{aligned} \tag{17.45}$$

which satisfies the boundary conditions $v_z(R_p) = v_z(R_c) = 0$. The total annular volume flow rate Q pumped is obtained by integrating $v_z(r)$ over concentric areal rings, with incremental areas $2\pi r\, dr$, leading to

$$Q = \int_{R_p}^{R_c} v_z(r)\, 2\pi r\, dr$$

$$= (\pi/2\mu)(dp/dz)\left[{}^1\!/\!_4 \left(R_c^2 - R_p^2\right)^2 - \left\{R_p{}^2 \left(R_c^2 - R_p^2\right)/\log\left(R_c/R_p\right)\right\} \left\{ {}^1\!/\!_2 \left(R_c/R_p\right)^2 \log\left(R_c/R_p\right) - {}^1\!/\!_4 \left(R_c/R_p\right)^2 + {}^1\!/\!_4 \right\}\right]$$

(17.46a,b)

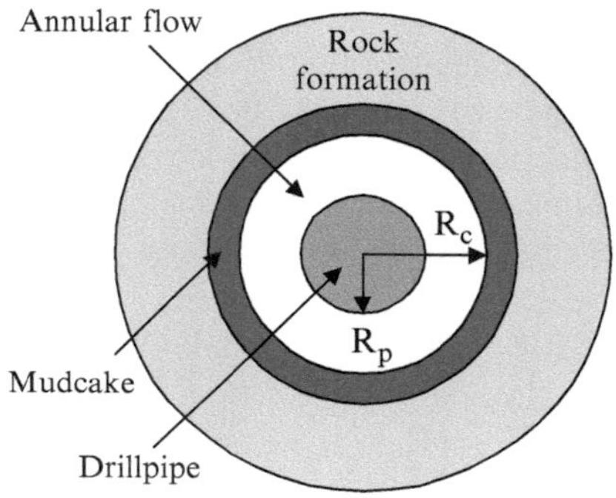

FIG. 17.10 Cross-section of cylindrical flow domain.

Note that the viscous shear stress at the outer mudcake boundary is, using Eq. (17.45), found from

$$\tau_{(Rc)} = \mu(dv_z/dr)(R_c)$$
$$= {}^1\!/\!_4\{2R_c + \left\{\left(R_c^2 - R_p^2\right)/\left(R_c \log\left(R_p/R_c\right)\right)\right\}(dp/dz) \quad (17.47)$$

Thus, the axial pressure gradient satisfies

$$dp/dz = 4\tau_{(Rc)}/\{2R_c + \left\{\left(R_c^2 - R_p^2\right)/\left(R_c \log\left(R_p/R_c\right)\right)\right\} \quad (17.48)$$

If we substitute Eq. (17.48) into Eq. (17.46b), we find that

$$Q = 2\pi\mu^{-1}\left[\tau_{(Rc)}/\{2R_c + \left\{\left(R_c^2 - R_p^2\right)/\left(R_c \log\left(R_p/R_c\right)\right)\right\}\right] \times$$
$$\left[{}^1\!/\!_4 \left(R_c^2 - R_p^2\right)^2 - \left\{R_p{}^2 \left(R_c^2 - R_p^2\right)/\log\left(R_c/R_p\right)\right\} \quad (17.49)$$
$$\left\{ {}^1\!/\!_2 \left(R_c/R_p\right)^2 \log\left(R_c/R_p\right) - {}^1\!/\!_4 \left(R_c/R_p\right)^2 + {}^1\!/\!_4 \right\}\right]$$

expressing Q as a function of the cake radius R_c, the pipe radius R_p, the viscosity μ, *and* the shear stress $\tau_{(Rc)}$ at the cake surface. When Q, R_p, and μ are given, and the yield shear stress τ_{yield} is known empirically from a dynamic erosion test, Eq. (17.49) provides a nonlinear transcendental equation for the equilibrium mudcake radius $R_{c,eq}$, that is,

$$Q = 2\pi\mu^{-1}\left[\tau_{yield}/\{2R_{c,eq} + \left\{\left(R_{c,eq}{}^2 - R_p^2\right)/\left(R_{c,eq}\log\left(R_p/R_{c,eq}\right)\right)\right\}\right] \times$$
$$\left[\tfrac{1}{4}\left(R_{c,eq}{}^2 - R_p^2\right)^2 - \left\{R_p{}^2\left(R_{c,eq}{}^2 - R_p^2\right)/\log\left(R_{c,eq}/R_p\right)\right\}\right.$$
$$\left.\left\{\tfrac{1}{2}\left(R_{c,eq}/R_p\right)^2\log\left(R_{c,eq}/R_p\right) - \tfrac{1}{4}\left(R_{c,eq}/R_p\right)^2 + \tfrac{1}{4}\right\}\right] \tag{17.50}$$

This equation for $R_{c,eq}$ cannot be simply solved, but since $R_{well} > R_{c,eq} > R_p$ where R_{well} is the well radius, it is trivial from a numerical point of view to test for *all* possible radii that may possibly fulfill our geometric constraint. In this manner, the equilibrium mudcake radius can be determined straightforwardly. A useful solution procedure requires our rewriting Eq. (17.49) in the form

$$\tau_{(Rc)} = (Q\mu/2\pi)\{2R_c + \left\{\left(R_c^2 - R_p^2\right)/\left(R_c\log\left(R_p/R_c\right)\right)\right\}/$$
$$\left[\tfrac{1}{4}\left(R_c^2 - R_p^2\right)^2 - \left\{R_p{}^2\left(R_c^2 - R_p^2\right)/\log\left(R_c/R_p\right)\right\}\right. \tag{17.51}$$
$$\left.\left\{\tfrac{1}{2}\left(R_c/R_p\right)^2\log\left(R_c/R_p\right) - \tfrac{1}{4}\left(R_c/R_p\right)^2 + \tfrac{1}{4}\right\}\right]$$

which is independent of the fluid density ρ. For any known set of values Q, μ, and R_p, the cake radius R_c can be varied to produce a sequence of cake wall shear stresses $\tau_{(Rc)}$. If the mudcake yield shear stress is known, then the equilibrium cake radius can be obtained immediately by tabular reference.

Fig. 17.11 displays the results of one such calculation, assuming water as the drilling fluid, flowing at 400 gpm. We also take a drillpipe radius of 0.2 ft. The well radius shown in the following input is not relevant to the calculations, as it does not appear in the host equation (only the annular space between the pipe and cake radii is dynamically significant); it is used only to provide some indication of cake thickness. The calculations correctly show how shear stress increases as the cake *radius* (that is, annular size) decreases. For example, if the mudcake yield stress is known to be 0.0001 psi from an independent erosion test, the tabulated results indicate that the corresponding cake radius (from the origin at the center of the drillpipe) is 0.27 ft. This limit, the result of fluid erosion, prevents the complete hole plugging obtained earlier, which assumes static filtration only. Fig. 17.11 also provides some indication of typical shear stress magnitudes obtained under the stated no-slip assumptions, here being $O(10^{-4})$ psi. It is also of interest to determine the values predicted using Eq. (17.44). We again take a 400 gpm flow of water in the annular region between 0.2 and 0.4 ft radii; this yields an average oncoming free stream speed U_∞ of 2.34 ft/s. If we assume a filtration rate v_o of 1 ft/day, the surface shear stress is found to be 3.65×10^{-7} psi, which appears somewhat small. But for small times, the yield

```
INPUT PARAMETER SUMMARY:
Total volume flow rate (gpm): .4000E+03
Viscosity, drilling mud (cp): .1000E+01
Radius of drill pipe  (feet): .2000E+00
Radius  of bore hole  (feet): .4000E+00

Rcake =  .3900E+00 ft, Thickn =  .1000E-01 ft, Shear =  .1070E-04 psi
Rcake =  .3800E+00 ft, Thickn =  .2000E-01 ft, Shear =  .1216E-04 psi
Rcake =  .3700E+00 ft, Thickn =  .3000E-01 ft, Shear =  .1391E-04 psi
Rcake =  .3600E+00 ft, Thickn =  .4000E-01 ft, Shear =  .1603E-04 psi
Rcake =  .3500E+00 ft, Thickn =  .5000E-01 ft, Shear =  .1864E-04 psi
Rcake =  .3400E+00 ft, Thickn =  .6000E-01 ft, Shear =  .2186E-04 psi
Rcake =  .3300E+00 ft, Thickn =  .7000E-01 ft, Shear =  .2593E-04 psi
Rcake =  .3200E+00 ft, Thickn =  .8000E-01 ft, Shear =  .3113E-04 psi
Rcake =  .3100E+00 ft, Thickn =  .9000E-01 ft, Shear =  .3792E-04 psi
Rcake =  .3000E+00 ft, Thickn =  .1000E+00 ft, Shear =  .4700E-04 psi
Rcake =  .2900E+00 ft, Thickn =  .1100E+00 ft, Shear =  .5947E-04 psi
Rcake =  .2800E+00 ft, Thickn =  .1200E+00 ft, Shear =  .7720E-04 psi
Rcake =  .2700E+00 ft, Thickn =  .1300E+00 ft, Shear =  .1035E-03 psi
Rcake =  .2600E+00 ft, Thickn =  .1400E+00 ft, Shear =  .1447E-03 psi
Rcake =  .2500E+00 ft, Thickn =  .1500E+00 ft, Shear =  .2141E-03 psi
Rcake =  .2400E+00 ft, Thickn =  .1600E+00 ft, Shear =  .3443E-03 psi
Rcake =  .2300E+00 ft, Thickn =  .1700E+00 ft, Shear =  .6303E-03 psi
Rcake =  .2200E+00 ft, Thickn =  .1800E+00 ft, Shear =  .1462E-02 psi
Rcake =  .2100E+00 ft, Thickn =  .1900E+00 ft, Shear =  .6033E-02 psi
```

FIG. 17.11 Mudcake surface shear stress.

strength of the cake may be substantially less than at large times, when the irreversible effects of cake compaction have set in.

Thus, we *might* modify the cake growth recipes offered earlier as follows. Once the transient cake radius $R_c(t)$ determined on a static filtration basis equals the $R_{c,eq}$ as calculated, we can restrict further mudcake growth by fixing cake position at the $R_{c,eq}$ radius. (In reality, this erosive shear stress acts *continuously!*) Of course, invasion will continue into the formation but at a slower, typically unsteady rate dictated by a three-layer model, where the only moving boundary is the displacement front separating two dissimilar fluids.

The yield stress τ_{yield} of the mudcake in question may depend on numerous factors, among them, differential pressure, solids content, viscosifier type, chemical composition, and temperature. The critical invasion rate proposed by Fordham et al. (1991) is simply the lowest volume flow rate Q that will, for a particular mud and drillpipe size, produce viscous shear stresses at the mudcake surface in excess of τ_{yield}. Note that viscous stress and total volume flow rate, in a Newtonian flow, are linearly related; this direct proportionality, as we shall see, disappears in the case of power law fluids.

Modifications for Drillpipe Rotation

So far, we have considered nonrotating drillpipe only. With pipe rotation, two additional physical effects are introduced. First, nonzero azimuthal velocity will alter the stress state acting at the surface of the mud, and second, the rotating flow will induce a centripetal pressure gradient acting in the radial direction that contributes to additional filtration. The dimensionless parameter dictating the importance of rotation is simply the tangential rotation speed at the drillpipe surface divided by the mean axial flow speed in the annulus. The axial velocity parallel to the borehole velocity was given by Eq. (17.45), that is, the exact annular flow solution

$$v_z(r) = (4\mu)^{-1}(dp/dz) \left[r^2 - R_p{}^2 + \left\{\left(R_c^2 - R_p^2\right)/\log\left(R_p/R_c\right)\right\}\log\left(r/R_p\right)\right] \quad (17.45)$$

for nonrotating drillpipe. A solution to the rotating pipe problem without axial flow is available from classical fluid mechanics (Schlichting, 1968). For this problem, the exact solution to the Navier-Stokes equations, written in radial cylindrical coordinates, can be obtained as

$$v_\theta(r) = \omega\left(R_c^2 - R_p{}^2\right)^{-1}\{R_p{}^2R_c^2/r - rR_p{}^2) \quad (17.52)$$

This circumferential velocity vanishes at the outer (cake) radius $r = R_c$, satisfying no-slip conditions. At the drillpipe $r = R_p$, the tangential speed

$$v_\theta(R_p) = \omega\left(R_c^2 - R_p{}^2\right)^{-1}\{R_pR_c^2 - R_p{}^3) = \omega R_p \quad (17.53)$$

corresponds to that for a rotation rate of ω. While Eqs. (17.45), (17.52) individually satisfy the *nonlinear* Navier-Stokes equations, it is not obvious that both solutions can be *linearly* superposed to produce a dynamically correct solution. But this is so because fortuitous simplifications introduced by our concentric geometry cause the nonlinear convective terms to vanish identically! Thus, the vector sum due to Eqs. (17.45), (17.52) can be simply added to produce the total velocity field, as is rigorously demonstrated in *Borehole Flow Modeling*. We emphasize, though, that this superposition does not apply to eccentric Newtonian flows because the convective terms do not disappear, nor does it apply to concentric or eccentric non-Newtonian flows. Now, the circumferential stress $\tau_{(\theta)}$ can be obtained by evaluating

$$\begin{aligned}\tau_{(\theta)}(r) &= \mu r\, d\{v_\theta(r)/r\}/dr \\ &= -2\mu\omega R_p{}^2R_c^2/\{r^2\left(R_c^2 - R_p{}^2\right)\}\end{aligned} \quad (17.54)$$

so that, at the mudcake surface, $\tau_{(\theta)}(r)$ assumes the value

$$\tau_{(\theta),Rc} = -2\mu\omega R_p{}^2R_c^2/\{R_c^2\left(R_c^2 - R_p{}^2\right)\} \quad (17.55)$$

It follows that the total mudcake surface stress satisfies

$$\tau_{total} = \sqrt{\left\{\tau_{(Rc)}{}^2 + \tau_{(\theta),Rc}{}^2\right\}} \quad (17.56)$$

Then, the sequence of numerical steps leading to the construction of the table in Fig. 17.11 can be repeated with Eq. (17.56), with the additional pipe rotation rate ω entering the calculations. In addition to shear stress modifications, drillpipe rotation alters the radial pressure gradient acting at the surface of the mudcake, thereby affecting static filtration somewhat. This effect can be calculated from the centripetal acceleration formula $dp/dr = \rho\, v_\theta(r)^2/r$, where ρ represents the mass density of the mud.

Effect of Solids Concentration

In drilling fluids, fine solids are added to increase mud weight, for the purposes of formation pressure control. These solids are held in suspension using special viscosifiers, and the resultant fluids, it turns out, behave characteristically in non-Newtonian fashion. Whenever possible, the exact rheological properties of the drilling fluid in question should be measured using accepted laboratory techniques. It is of historical interest, however, to cite a viscosity correction formula attributed to Einstein, derived in 1906 in connection with his studies of Brownian motion (e.g., see Landau and Lifshitz, 1959) for an English-language derivation.

A fluid in which numerous fine particles are suspended can be regarded as a homogeneous medium if we are concerned with phenomena having much larger characteristic lengths. Let c represent the concentration of the suspension, that is, the number of particles per unit volume; also, let μ_0 and μ denote, respectively, the original viscosity and the effective viscosity of the modified fluid. If the particles can be approximated as small spheres having a radius R, then Einstein found that $\mu \approx \mu_o(1 + 10/3\pi R^3 c)$ to leading order. Of course, not all drilling fluids behave homogeneously, since density stratification invariably arises when drilling is interrupted during routine operations. *Computational Rheology* describes the consequences in detail. Among the most problematic and interesting physically are the recirculating vortex zones that form in the annulus that impede cuttings transport. Analytical solutions, to include relevant dimensionless parameters, are presented in the book. (These were validated in lab experiments conducted at M-I Drilling Fluids in Houston.)

Turbulent Versus Laminar Flow

The subject of turbulent versus laminar flow is important operationally, but flow turbulence, despite the advances made during the past decades, is still empirically grounded and represents a research discipline in its own right. We shall not extend our annular flow work to turbulent regimes, except note that the ideas discussed so far (and next for power law flows) *can* be applied to appropriate curve-fitted velocity profiles.

CONCENTRIC POWER LAW FLOWS WITHOUT PIPE ROTATION

In Newtonian fluids such as water and air, the shear stress τ is linearly proportional to the rate of strain; for the preceding example, the rate of strain is $dv_z(r)/dr$, and we can write $\tau = \mu dv_z(r)/dr$ where the *constant* of proportionality

μ is the viscosity. Most drilling fluids do not behave like Newtonian fluids, and the study of rheology focuses on the stress behavior of different fluids acting at different shear rates. (Note that the filtrated fluid entering the formation, namely water, *is* Newtonian.) One popular model is the power law fluid. In the simplest case, its constitutive equation is taken in the form

$$\tau = K\,(dv_z(r)/dr)^n \tag{17.57}$$

where the fluid exponent n and the consistency factor K (not to be confused with the Darcy flow permeability) are constants that characterize the fluid itself. For $n=1$, the consistency factor reduces to the Newtonian viscosity μ; in general, the units of K depend on the value of n. (Both n and K can be determined from viscometer measurements using standard laboratory techniques.)

It is important that n and K are constant properties characterizing the fluid and that they remain unchanged regardless of the flow problem. The apparent viscosity of the flow, however, will vary throughout the cross-section of the flow geometry and additionally varies with the pressure gradient, or equivalently, the total flow rate. In other words, the apparent viscosity of a power law flow varies from problem to problem, whereas n and K do not. This fact is not appreciated in drilling engineering. Surface viscometer values for fluid parameters having questionable scientific merit often find routine field usage. Thus, it is not surprising that, at least in cuttings transport analyses, they cannot be correlated with measurable events such as hole cleaning efficiency.

There are other classes of fluids, such as Herschel-Bulkley fluids and Bingham plastics, that follow different stress-strain relationships, which are sometimes useful in different drilling and cementing applications. For a discussion on three-dimensional effects and a rigorous analysis of the stress tensor, the reader should refer to *Computational Rheology*. For now, we shall continue our discussion of mudcake shear stress, but turn our attention to power law fluids. The governing partial differential equations of motion, even for simple relationships of the form given in Eq. (17.57), are nonlinear and therefore rarely amenable to simple mathematical solution. For example, the axial velocity $v_z(r)$ in our cylindrical radial flow satisfies

$$r^{-1}\, d\left\{K(dv_z/dr)^{n-1} r\, dv_z(r)/dr\right\}/dr = dp/dz \tag{17.58}$$

which, despite its simple appearance, is difficult to solve because it is nonlinear. An exact annular flow solution, however, *is* available for nonrotating drillpipes. Thus, in principle, a formula analogous to Eq. (17.51), which relates mudcake edge shear stress, total volume flow rate, pipe radius, and fluid properties, is available. The problem of concentric, nonrotating, annular flow was solved using numerical methods in Fredrickson and Bird (1958). If R_i and R_o are inner and outer radii, where ΔP is a pressure drop, L is a characteristic length, and Q is the annular volume flow rate, these authors show that

$$R_o\Delta P/(2L) = K\left[(2n+1)Q/\left\{n\pi R_o{}^3 Y(1\text{-}R_i/R_o)^{(2n+1)/n}\right\}\right]^n \tag{17.59}$$

while the shear stress at the *outer* wall $r = R_o$ is given by

$$\tau_o = \left(1\text{-}\lambda^2\right) R_o\Delta P/(2L) \tag{17.60}$$

Y and λ in Eqs. (17.59), (17.60), known in chemical engineering as the Fredrickson-Bird Y and λ functions, respectively, depend on n and R_i/R_o only. A condensed tabulation of their results appears in Figs. 17.12 and 17.13.

	R_i/R_o						
n	0.01	0.1	0.2	0.4	0.6	0.8	0.9
1.00	0.6051	0.5908	0.6237	0.7094	0.8034	0.9008	0.9502
0.50	0.6929	0.6270	0.6445	0.7179	0.8064	0.9015	0.9504
0.33	0.7468	0.6547	0.6612	0.7246	0.8081	0.9022	0.9506
0.20	0.8064	0.6924	0.6838	0.7342	0.8128	0.9032	0.9510
0.10	0.8673	0.7367	0.7130	0.7462	0.8124	0.9054	0.9519

FIG. 17.12 Fredrickson-Bird Y function (condensed).

	R_i/R_o						
n	0.01	0.1	0.2	0.4	0.6	0.8	0.9
1.00	0.3295	0.4637	0.5461	0.6770	0.7915	0.8981	0.9495
0.50	0.2318	0.4192	0.5189	0.6655	0.7872	0.8972	0.9493
0.33	0.1817	0.3932	0.5030	0.6587	0.7847	0.8967	0.9492
0.20	0.1503	0.3712	0.4856	0.6509	0.7818	0.8960	0.9491
0.10	0.1237	0.3442	0.4687	0.6429	0.7784	0.8953	0.9489

FIG. 17.13 Fredrickson-Bird λ function (condensed).

If we now eliminate $R_o\Delta P/(2L)$ between Eqs. (17.59), (17.60), we obtain the required result

$$\tau_o = \left(1\text{-}\lambda^2\right)K\left[(2n+1)Q/\left\{n\pi R_o{}^3 Y(1-R_i/R_o)^{(2n+1)/n}\right\}\right]^n \tag{17.61}$$

which relates mudcake edge shear stress, volume flow rate, pipe radius, and fluid properties. In the notation to this chapter, Eq. (17.61) can be rewritten as

$$\tau_{(Rc)} = \left(1\text{-}\lambda^2\right)K\left[(2n+1)Q/\left\{n\pi R_c{}^3 Y\left(1-R_p/R_c\right)^{(2n+1)/n}\right\}\right]^n \tag{17.62}$$

For any particular pair of n and R_p/R_c values, the corresponding Y and λ functions can be obtained from Figs. 17.12 and 17.13. Then, the remainder of the right side of Eq. (17.62) can be evaluated using n, K, R_c, and the prescribed annular volume flow rate Q. The equilibrium mudcake thickness is defined by the condition $\tau_{(Rc)} = \tau_{yield}$ as before, and the procedure for the critical invasion rate discussed earlier carries through unchanged.

CONCENTRIC POWER LAW FLOWS WITH DRILLPIPE ROTATION

The Fredrickson-Bird solution for power law fluids is *exact* in that no geometrical or dynamical simplifications are made. However, the analogous solution to the more general problem which includes drillpipe rotation cannot be as concisely expressed, even numerically using tables, because nonlinearities render computed solutions highly iterative. In order to obtain some qualitative idea about the effects of rotation, it is necessary to invoke a narrow annulus assumption. On doing so, it is possible to obtain explicit closed-form solutions for the axial and circumferential velocities and viscous stresses. In this section, using results obtained in *Computational Rheology*, we extend the procedures developed in our earlier discussions, and we again attempt to relate equilibrium cake radius to prescribed yield stress. As before, we let R_c and R_p denote mudcake and drillpipe radii; also, n and K represent power law coefficients, dp/dz is the axial borehole pressure gradient, and $\omega < 0$, by convention, is the rotation rate. In the cited text, the solution to the general concentric rotating flow problem is derived in its entirety. In applying those results, adopting the notation of the present chapter, let us introduce the constants

$$E_1 = -1/8\,(R_c + R_p)^2\, dp/dz \tag{17.63}$$

$$E_2 = K\{\omega/(R_p - R_c)\}^n \{(R_p + R_c)/2\}^{n+2} \tag{17.64}$$

Then, the velocity $v_z(r)$ parallel to the hole axis is given by

$$\begin{aligned} v_z(r) = {} & \{(r + R_p)/2\}^2 \left\{E_1 + ½((r + R_p)/2)^2 dp/dz\right\}/E_2 \\ & \times (E_2/K)^{1/n} \left[((r + R_p)/2)^{(2n+4)/(n-1)} + ((r + R_p)/2)^{(4n+2)/(n-1)} \right. \\ & \left. \left\{\left(E_1 + ½((r + R_p)/2)^2\, dp/dz\right) / E_2\right\}^2\right]^{(1-n)/2n} (r - R_p) \end{aligned} \tag{17.65}$$

while the circumferential speed $v_\theta(r)$ satisfies $v_\theta(r) = r\,\Omega(r)$ where the *angular velocity* variable $\Omega(r)$ is determined from

$$\begin{aligned} \Omega(r) = {} & (E_2/K)^{1/n} (r - R_c) \left[((r + R_c)/2)^{(2n+4)/(n-1)} \right. \\ & \left. + ((r + R_c)/2)^{(4n+2)/(n-1)} \left\{\left(E_1 + (r + R_c)^2/8\, dp/dz\right)/E_2\right\}^2\right]^{(1-n)/2n} \end{aligned} \tag{17.66}$$

Their corresponding viscous stresses S_{rz} and $S_{r\theta}$, using the notation in *Computational Rheology*, can be calculated from

$$S_{rz} = S_{zr} = \left\{0.5\,r - \left(R_p + R_c\right)^2/(8r)\right\} dp/dz \tag{17.67}$$

$$S_{r\theta} = S_{\theta r} = K\left\{\omega/\left(R_p - R_c\right)\right\}^n \left\{\left(R_p + R_c\right)/2\right\}^{n+2} r^{-2} \tag{17.68}$$

From Eqs. (17.67), (17.68), a total stress S_{total} can be defined in the usual manner as $S_{total} = \sqrt{\left(S_{rz}^{\ 2} + S_{r\theta}^{\ 2}\right)}$. Now, as in our earlier sections, the objective is to eliminate dp/dz in the expression for total annular volume flow rate in favor of S_{total}. This allows us to determine the equilibrium cake radius in terms of a specified mudcake yield stress. To do this, first rewrite the last result in the form $S_{rz}^{\ 2} = S_{total}^{\ 2} - S_{r\theta}^{\ 2}$. Substitution of Eqs. (17.67), (17.68) leads to

$$\begin{aligned}\left\{0.5\,R_c - \left(R_p + R_c\right)^2/(8R_c)\right\}^2 (dp/dz)^2 &= S_{total}^{\ 2} \\ &- K^2\left\{\omega/\left(R_p - R_c\right)\right\}^{2n} \left\{\left(R_p + R_c\right)/2\right\}^{2n+4} R_c^{\ -4}\end{aligned} \tag{17.69}$$

where we have evaluated r at the cake boundary $r = R_c$. Therefore, simple division shows that the axial pressure gradient equals

$$dp/dz = \sqrt{\begin{aligned}&\left[\left\{S_{total}^{\ 2} - K^2\left\{\omega/\left(R_p - R_c\right)\right\}^{2n}\left\{\left(R_p + R_c\right)/2\right\}^{2n+4} R_c^{-4}\right\}\right. \\ &\left./\left\{0.5\,R_c - \left(R_p + R_c\right)^2/(8r)\right\}^2\right]\end{aligned}} \tag{17.70}$$

Thus, Eq. (17.65), which expresses $v_z(r)$ in terms of the pressure gradient dp/dz, can be rewritten in terms of the total stress S_{total} acting at the mudcake surface. As before, the volume flow rate Q can be explicitly found using

$$Q = \int_{R_p}^{R_c} v_z(r)\, 2\pi r\, dr = Q\left(n, K, \omega, R_p, R_c, S_{total}\right) \tag{17.71}$$

If we fix the magnitude of Q in Eq. (17.71), then for any prescribed values of the parameters n, K, ω, and R_p, we can tabulate R_c as a function of S_{total}, or if desired, S_{total} as a function of R_c. For any mudcake yield stress $S_{yield} = S_{total}$, we can therefore obtain the equilibrium mudcake radius $R_{c,eq}$ unambiguously. If we next repeat these calculations for a range of Q*s*, then the minimum value of Q that produces sufficient shear stress so as to erode the mudcake then represents the critical invasion rate.

FORMATION INVASION AT EQUILIBRIUM MUDCAKE THICKNESS

Earlier we considered the three-layer radial invasion problem consisting of mudcake, flushed zone, and uninvaded zone, and obtained a solution for coupled mudcake growth and displacement front motion. Here we revisit that problem, but now assume that the mudcake no longer grows in time because it has reached dynamic equilibrium. The problem nonetheless consists of three layers, these being, again, the mudcake, the flushed zone, and the uninvaded zone. A number of papers refer to a classic formula obtained by solving three coupled pressure equations, each taking the form $d^2p/dr^2 + 1/r\ dp/dr = 0$, as *the* invasion model (e.g., see Muskat, 1937) used as the basis for further development. However, that formula strictly applies to the concentric radial flow of a single fluid through three layers of *nonmoving rock* having different permeabilities. The formula does *not* apply when one of the internal boundaries is moving; for such problems, the pressure boundary value problem as cited is incomplete, as we have noted earlier in this chapter, since the interface equations at the moving boundary *must* be included in the formulation.

Again, the earlier radial mudcake example deals with *two* moving boundaries, namely, the mud-to-mudcake interface, and the displacement front separating two possibly dissimilar fluids within the formation. The problem considered in this section is simpler, because the mudcake, having reached dynamic equilibrium, no longer grows. Its thickness, therefore, is to be regarded as statically fixed in time. The reader should return to our earlier derivation to review the basic assumptions and approach. There, the *cake radius* R_1 in the coefficient function $\alpha_2(R_1,R_3)$ was an *unknown function of time* that was to be determined as part of the solution. Here, we treat R_1 as a *constant* that can be regarded as known, once the shear stress criterion discussed earlier is applied. Thus the integration of our radial displacement front equation proceeds more simply. After some algebra, we obtain the exact solution,

$$
\begin{aligned}
&\tfrac{1}{2}(k_2/\mu_2)R_3^2 \log (R_2/R_1) \\
&+(k_1/\mu_1)R_2^2 \left[\tfrac{1}{2}(R_3/R_2)^2 \log (R_3/R_2) - \tfrac{1}{4}(R_3/R_2)^2\right] \\
&-(k_1k_2\mu_3/\mu_1\mu_2k_3)R_4^2 \left[\tfrac{1}{2}(R_3/R_4)^2 \log (R_3/R_4) - \tfrac{1}{4}(R_3/R_4)^2\right] = \\
&= -k_1k_2(p_r-p_m)t/\left(\mu_1\mu_2\phi_{eff}\right) + \text{constant}
\end{aligned}
\tag{17.72}
$$

The constant of integration in Eq. (17.72) is determined from initial conditions as suggested by the equilibrium solution that would be obtained from the static filtration solution as modified by erosive annular effects.

In our discussion of the filtration process, we had taken the simple-minded view that static filtration will continue until the point at which mudcake thickness reaches the equilibrium thickness derived earlier. At that time, cake growth terminates, but front motions continue as determined by Eq. (17.72). This view is approximate and was adopted for discussion only. In reality, the shear stress in the borehole *continuously* acts on the mudcake as it is formed, so that the interactions between mudcake growth, reservoir Darcy flow, and borehole annular flow can be complicated. We do not pretend to solve this more realistic problem, but we do believe that the principal elements of both static and dynamic filtration processes have been satisfactorily identified.

DYNAMIC FILTRATION IN ECCENTRIC BOREHOLES

Let us reiterate some basic ideas on borehole flow modeling. Simply put, mudcake that lines the borehole controls the filtration rate into the formation when reservoir rock is permeable. It is important to understand how mudcake evolves because good time lapse logging requires a knowledge of the time scales governing key physical processes. For example, invasion is ongoing while drilling, and front movement in the Darcy reservoir depends on viscosity and porosity. If we wished to determine, say, their values from resistivity measurements made at two points in time, at what times should the readings be taken? Knowing how mudcake evolves under static and dynamic conditions is crucial, so that readings are not taken so closely that they are indistinguishable, or too far apart that logging tools cannot respond accurately. When formation and cake permeability are comparable, mudcake growth couples strongly with reservoir flow, posing additional uncertainty in modeling. (This problem is numerically solved later in this book.)

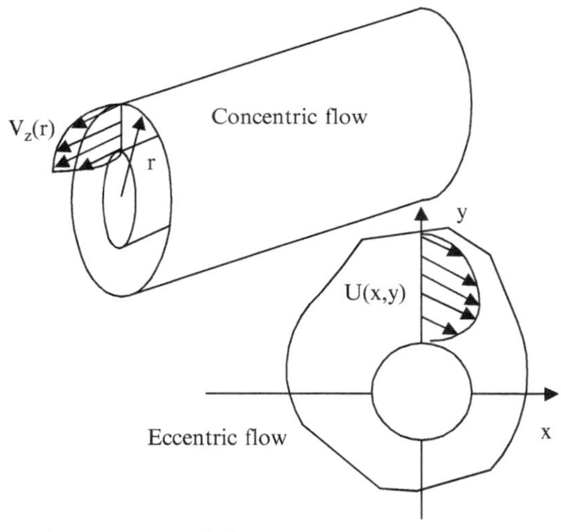

FIG. 17.14 Concentric versus eccentric flows.

Still another problem is borehole geometry. In modern horizontal and multilateral wells, drillpipes rest on the low side of the hole, as shown in Fig. 17.14. Whether the fluids are Newtonian or non-Newtonian, much higher annular velocities are found on the high side. As we have seen, physical quantities like shear stress are also important, in that they affect erosion in dynamic filtration and therefore azimuthally varying filtration rates into the reservoir (Fig. 17.15). The computation of velocity, apparent viscosity, shear rate, and viscous shear stress in eccentric boreholes is necessary for various reasons, for example, pump power requirements and cutting transport efficiency, and accurate solutions are possible using the boundary-conforming, curvilinear mesh systems studied in

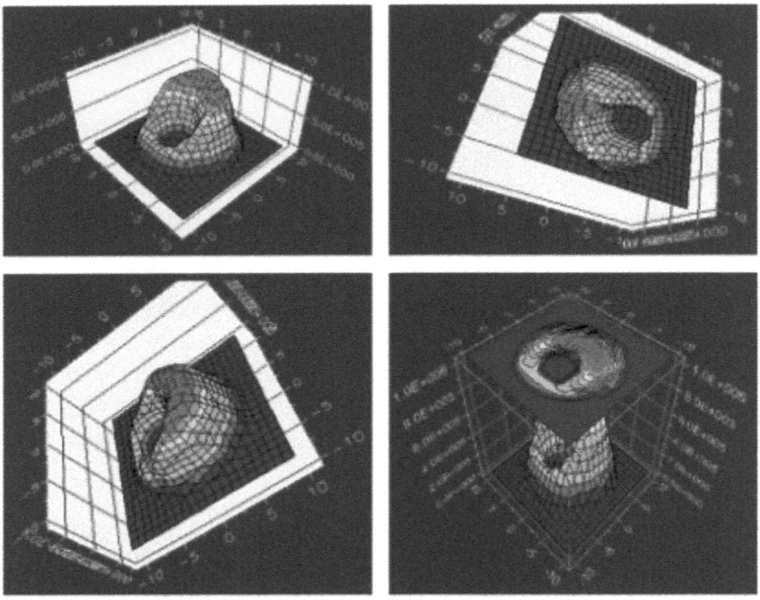

FIG. 17.15 Velocity field in highly eccentric annulus.

Chapters 8 and Chapters 9. Applications of this method are developed in *Computational Rheology for Pipeline and Annular Flow* (Chin, 2001a). Snapshots of typical physical quantities are given in Fig. 17.16.

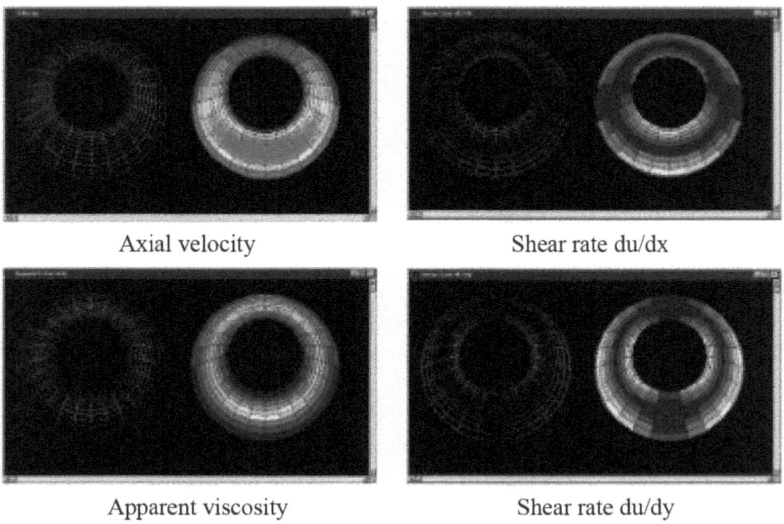

FIG. 17.16 Typical computed quantities.

PROBLEMS AND EXERCISES

1. In this chapter, an implicit solution for the equilibrium thickness of a mudcake was given assuming that its yield shear stress was known and that the fluid was laminar and Newtonian. Extend the solution to (i) turbulent Newtonian, (ii) laminar power law, and (iii) turbulent power law fluids.
2. The time scales over which compressible fluid transients are significant were given in Eqs. (17.30)–(17.35) for lineal flows. Extend these results to radial flows. (Refer to the well test literature for basic solutions.)

Chapter 18

Formation Tester Applications

Formation testers are measurement instruments that retrieve reservoir fluid samples from wells during pauses in drilling operations. When their nozzles are pressed against borehole walls and pump pistons are retracted, petroleum liquids enter flowlines and storage containers. These samples, sometimes analyzed downhole, are later examined at the surface where chemical studies are performed that provide useful information needed for formation evaluation, economic planning and refinery requirements. Importantly, they provide *direct* measurements of properties relevant to production, that is, "Are the candidate fluids *really* hydrocarbons?" "What is the permeability or mobility of the rock?" "Is anisotropy significant?" These questions are not answered indirectly by inferring from resistivity, NMR or sonic logs. Essentially, "What you see is what you get." In this chapter, we outline fluid-dynamical issues that arise in the context of operational and interpretation problems, and address their solutions via math models and software algorithms. Our summaries are comprehensive although far from exhaustive. The subject matter was developed almost exclusively by this author and his colleagues during the past decade, and covered in detailed in published patent, paper and book references to be cited.

BACKGROUND AND ENGINEERING OVERVIEW

Although many formation tester nozzle configurations exist, e.g., number of probes, types of nozzles (pumping versus passive pressure measurement), probe separation and angular orientation, pressure transient mode (steady-state, drawdown-buildup, or oscillatory piston), generic testers are of two varieties. These are "single probe," as in Fig. 18.1A, where a single probe withdraws fluid and simultaneously monitors flowline pressure transients. "Dual probe" devices, as shown in Fig. 18.1B, consist of a pumping probe (which measures source pressure) and a passive pressure measurement probe a fixed axial distance away. In either case, the flowfield is spherical (if isotropic) and ellipsoidal (if transversely isotropic), as opposed to the cylindrical flows encountered in well applications.

Various practical questions arise. During drilling, mud filtrate invades the formation and mixes with hydrocarbon fluids. Depending on mud type, e.g., water-base versus oil, the resulting mixtures can be immiscible or miscible multiphase. The objective behind fluid sampling is clean in situ collection.

Quantitative Methods in Reservoir Engineering. http://dx.doi.org/10.1016/B978-0-12-810518-4.00018-9

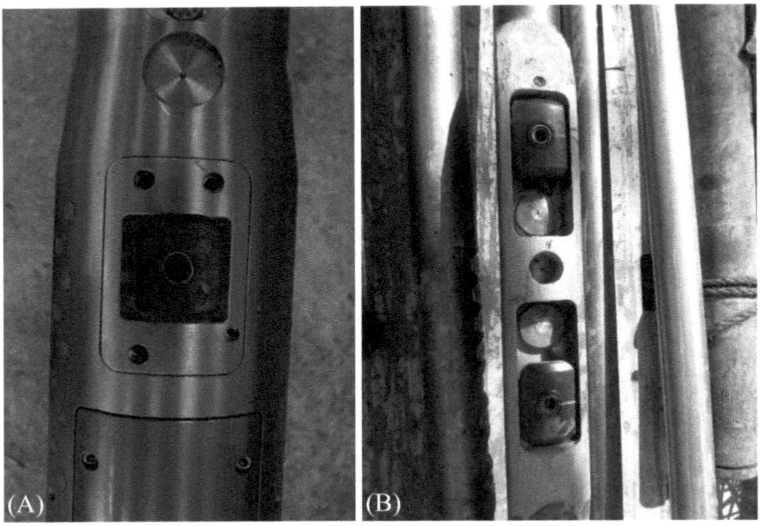

FIG. 18.1 (A) Single-probe formation tester (nozzle embedded in pad). (B) Dual-probe formation tester (pumping plus passive probe).

Thus, operators ask, "How long must pumps be operated in order to obtain true formation fluids and not mud filtrate contaminants?" This depends, of course, on reservoir engineering parameters like relative permeability and capillary pressure, plus dynamic mudcake growth properties that control the rate of filtrate invasion. This multiphase mixing problem, addressed in Chin and Proett (2005) and recently in Chin (2014b), will not be covered in this book, although sample calculated results showing dynamic changes with time will be presented.

However, we shall cover major results in pressure transient analysis and relate these to older and obsolete methods used as recently as two decades ago. In the mid-1960s, workers realized that single probe tools could be used to provide permeability (or more precisely, mobility) during formation tester operations. Essentially, constant flowrate withdrawal continued until steady-state pressures were obtained; then a spherical flow formula connecting rate to pressure drop was used to calculate permeability (or mobility). In the 1990s, service companies developed dual probe tools where pressures were monitored at the pumping or source probe and also a distant observation probe. Here, steady pressure drops were obtained by measurement, but now both k_h and k_v could be obtained (this author extended the well-known Halliburton method to arbitrary dip angles). Anisotropy, of course, is important to hydraulic fracturing, wellbore stability, infill drilling and production planning.

Despite the initial success of these methods, lower and lower permeabilities became more the rule as exploration progressed into the 1990s. These "tight zones" (that is, low permeability, high oil viscosity, or both) proliferated worldwide and steady-state pressure measurements could no longer be obtained economically. Wait times for equilibration often required hours and sometimes days. This introduced not only economic efficiencies, but increased, significantly, the risks of stuck tools lost in the wells. Different measurement procedures were required. An interesting test and interpretation procedure was developed at Halliburton in the late 1990s. An empirical "exponential formula" was used to complement a rapid drawdown-buildup nozzle action that did not require pressure equilibrated data. Permeability predictions, plus pore pressure estimates, which normally required minutes or hours, could now be made satisfactorily in seconds. But why the method worked and how it could be improved, were open questions that could not be answered. For isotropic flows, this author would formulate the complete transient pressure boundary value problem and solve it analytically in closed form. Several interesting results were obtained. First, the "exponential formula" appeared as one special limit of the general solution—hence, its limitations and underlying assumptions, were now understood. And second, because a single formula now described pressure evolution for all time, the ambiguities associated with "early, middle and late time," well testing terms that themselves depend on the unknown permeability or mobility, no longer confronted log analysts.

In 2004, the United States Department of Energy, through its Small Business Innovation Research (SBIR) division, awarded approximately two hundred technology awards in all fields of energy, e.g., nuclear physics, plasma dynamics, air conditioning, green and wind energy, and so on. Four of these addressed fossil fuels, and of these, the author (through his incorporated entity Stratamagnetic Software LLC) won two, both in the area of formation testing. These contracts for high-risk research focused on *Formation Tester Permeability Prediction in Tight Gas Sands* and *Formation Tester Immiscible Flow Response in Horizontally Layered Media*. In the decade that followed, the author and China National Offshore Oil Corporation (CNOOC) developed a meaningful relationship that supported free and open exchange of ideas. These led to two important books Chin et al. (2014b, 2015) published by international scientific publisher John Wiley & Sons where all of the new technology is disclosed for the benefit of the petroleum industry.

These books addressed single and multiphase flow, the latter required in analyzing filtrate contamination and cleaning questions, plus questions that could be answered by examining pressure transients. For instance, "What methods are available to predict permeability (mobility) and pore pressure from short-time data (seconds) obtained in tight zones?" "Is it possible to obtain both k_h and k_v?" How do pump power requirements vary in permeable versus tight zones? "Can ideas and methods developed in resistivity or electromagnetic logging be applied to formation testing?" "Is it possible to design direct noniterative methods to solve 'inverse problems,' that is, without the usual "try and try again" approaches common to the industry?" The two cited books address all of these questions satisfactorily—the present chapter provides an overview into the large body of material now available to the industry.

Mathematical formulations used in pressure transient analysis are of two varieties, namely, "direct" problems where pressure versus time is sought when formation and pumping parameters are given, and "indirect" or "inverse" ones when formation characteristics are required using pressure response as inputs. Understanding the latter requires the development of rigorous direct solutions, and it turns out that four separate formulations govern the direct problem. In the next section, we provide a detailed analysis for "anisotropic media with flow-line storage and skin" effects, to convey an appreciation of the effort needed. Other limits include "isotropic with storage and skin," "anisotropic with storage and no skin" and "isotropic with storage and no skin." The need for separate formulations may not be readily apparent. First consider a mass-spring-damper vibration analogy. The model $m\, d^2y/dt^2 + b\, dy/dt + k\, y = 0$, $y(0) = y_0$ and $dy(0)/dt = y_{t,0}$ is a well-posed second-order differential equation model that satisfies two initial conditions. To study small values of "m," one cannot simply set $m = 0$ since one of the initial conditions cannot be enforced. This analogy occurs with skin effects: the "with skin" model requires a second derivative term not present in the "without skin" approach and the two yield contrasting yet complementary solutions. Furthermore, isotropic models are inherently different from anisotropic ones because the integration volumes are different. Development of inverse models requires an understanding of these subtleties.

ANALYTICAL CLOSED FORM SOLUTION FOR TRANSVERSELY ISOTROPIC FLOW WITH STORAGE AND SKIN EFFECTS

Governing Partial Differential Equation

The anisotropic partial differential equation for weakly compressible liquids in transient flow is

$$k_v\, \partial^2 P/\partial z^2 + k_h \left(\partial^2 P/\partial x^2 + \partial^2 P/\partial y^2\right) = \phi\mu c\, \partial P/\partial t \qquad (18.1)$$

where P is pressure, t is time, and x, y, and z represent ground fixed Cartesian coordinates whose axes are aligned with the principal axes of the permeability tensor. The vertical permeability k_v is taken in the z direction, while horizontal k_h is assumed for both x and y lateral directions. This transversely isotropic model is used to study anisotropy in sedimentary beds.

Eq. (18.1) is very general, but since there are no barriers or other heterogeneities in the flow, we seek more specialized results. As in many transient applications, we shall assume that a uniform initial pressure P_0 exists, whose value is also identical to the farfield pressure at subsequent times. Furthermore, we assume that pressures are constant along ellipsoidal surfaces with $x^2/k_h + y^2/k_h + z^2/k_v = \text{constant}$. This applies even when the source surface (i.e., the tester tool contact nozzle) is not ellipsoidal, provided we are several nozzle diameters

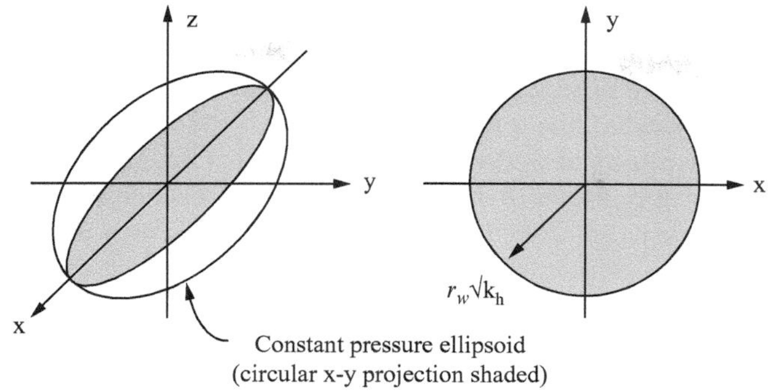

FIG. 18.2 Ellipsoidal flow assumption.

removed. A similar result is known in reservoir engineering; for example, in an areally anisotropic reservoir, constant pressure contours are elliptical far from the well, even if the well is circular and contains fracture imperfections. Fig. 18.2 summarizes several essential geometric elements.

It is therefore convenient to introduce the following dimensionless starred quantities; that is,

$$p^*(r^*,t^*) = \{P(x,y,z,t) - P_0\}/P_{ref} \tag{18.2}$$

$$r^* = \{x^2/k_h + y^2/k_h + z^2/k_v\}^{1/2} \tag{18.3}$$

$$t^* = t/t_{ref} \tag{18.4}$$

where the dimensional constants P_{ref} and t_{ref}, as yet undefined, will be chosen to simplify the math. The choice for pressure allows us to set $p^*(r^*,t^*)$ to zero initially and at the distant boundary for all time. When Eqs. (18.2)–(18.4) are substituted into Eq. (18.1), we obtain, after some algebra, a simple partial differential equation, namely,

$$\partial^2 p^*/\partial r^{*2} + 2/r^*\, \partial p^*/\partial r^* = \phi\mu c/t_{ref}\, \partial p^*/\partial t^* \tag{18.5}$$

that takes on a form identical to the one encountered in isotropic spherical flow! This alone does not allow us to draw mathematical analogies, since boundary conditions must be similarly transformed.

Total Velocity Flux Through Ellipsoidal Surfaces

Two types of physical boundary value problem models can be defined, both of which apply auxiliary conditions at the surface of the ellipsoidal source

$$x^2/k_h + y^2/k_h + z^2/k_v = r^{*}_{w}{}^{2} \tag{18.6}$$

defined by the dimensionless well radius r^*_w. (Idealized "point sources" are not considered here, since they preclude storage modeling.) The first specifies the pressure $p^*(r^*_w, t^*)$ itself, with the object of the solution being the total produced volume flow rate (VFR) taken through the ellipsoidal surface in Eq. (18.6). The second specifies VFR(t), and involves a complicated surface integral to be discussed; the solution, in this case, is the transient well pressure history, with the effects of wellbore storage accounted for. How the ellipsoidal radius r^*_w physically relates to the actual tool nozzle is subtle. Various authors use ad hoc arguments associated with equivalent hydraulic areas, volumes, flow rates, and the like, but these methods cannot be justified and are often incorrect. For our purposes, we observe that r^*_w *can* be rigorously related to the tool response and certain properties of the fluid and the rock formation.

In either formulation, an expression for VFR is required, which can be obtained from simple geometrical considerations. Let Σ represent the entire ellipsoidal surface area in Eq. (18.6) and dS denote the differential surface area on Σ. Then, if **n** is the local unit normal drawn perpendicular to dS, and **q** is the Darcy velocity vector, we simply have

$$\mathrm{VFR}(t) = -\int_{\Sigma} \mathbf{q} \bullet \mathbf{n}\, dS \tag{18.7}$$

The evaluation of this expression, however, requires some care, since **q**, **n**, and dS must each be correctly described. It turns out that it is most convenient to evaluate the local flux $\mathbf{q}\bullet\mathbf{n}$ using original x, y, and z ground-based coordinates. Now, the Darcy expression for velocity is just

$$\mathbf{q} = -k_h/\mu\, \partial p/\partial x\, \mathbf{i} - k_h/\mu\, \partial p/\partial y\, \mathbf{j} - k_v/\mu\, \partial p/\partial z\, \mathbf{k} \tag{18.8}$$

where **i**, **j**, and **k** are unit vectors in the x, y, and z directions. If the surface Σ is represented in the form

$$F(x,y,z;r^*_w) = x^2/k_h + y^2/k_h + z^2/k_v - r^{*2}_w = 0 \tag{18.9}$$

then $\mathbf{n} = (\partial F/\partial x\,\mathbf{i} + \partial F/\partial y\,\mathbf{j} + \partial F/\partial z\,\mathbf{k})/\sqrt{\{(\partial F/\partial x)^2 + (\partial F/\partial y)^2 + (\partial F/\partial z)^2\}}$, which gives

$$\mathbf{n} = (x/k_h\,\mathbf{i} + y/k_h\,\mathbf{j} + z/k_v\,\mathbf{k})/\sqrt{(x^2/k_h^2 + y^2/k_h^2 + z^2/k_v^2)} \tag{18.10}$$

It follows, from Eqs. (18.8), (18.9), that

$$\mathbf{q}\bullet\mathbf{n} = -(1/\mu)\,(x\,\partial p/\partial x + y\,\partial p/\partial y + z\,\partial p/\partial z)\,/ \\ \sqrt{\{x^2/k_h^2 + y^2/k_h^2 + k_v^{-1}(r^{*2}_w - x^2/k_h - y^2/k_h)\}} \tag{18.11}$$

Next, dS is formally constructed by integrating the incremental surface area $dS = \sqrt{\{1 + (\partial z/\partial x)^2 + (\partial z/\partial y)^2\}}\ dx\ dy$ over the projection (e.g., refer to Fig. 18.2) of Σ on the x-y plane, denoted R(x,y), where $z = z(x,y)$ is solved using Eq. (18.9). Let us consider, for example, the equation of a typical ellipsoid in the upper half-plane $z > 0$; that is,

$$z(x,y) = +\,k_v^{1/2}\sqrt{(r^{*2}_w - x^2/k_h - y^2/k_h)} \tag{18.12}$$

Then, it follows that

$$dS = k_v^{1/2} \sqrt{\{x^2/k_h^2 + y^2/k_h^2 + k_v^{-1}(r^{*}{}_w{}^2 - x^2/k_h - y^2/k_h)\}}\, dx\, dy / \sqrt{(r^{*}{}_w{}^2 - x^2/k_h - y^2/k_h)} \tag{18.13}$$

We combine Eqs. (18.11), (18.13) to obtain

$$\mathbf{q} \bullet \mathbf{n}\, dS = -\left(k_v^{1/2}/\mu\right)(x\,\partial p/\partial x + y\,\partial p/\partial y + z\,\partial p/\partial z)\, dx\, dy / \sqrt{(r^{*}{}_w{}^2 - x^2/k_h - y^2/k_h)} \tag{18.14}$$

Now, the quantity x $\partial p/\partial x + y\ \partial p/\partial y + z\ \partial p/\partial z$ can be re-expressed as $P_{ref}\, r^{*}{}_w\, (\partial p^*/\partial r^*)_w$ and is a function of r^* only. As such, it must be constant on $r^* = r^{*}{}_w$. Thus, this quantity can be moved across any integral taken over Eq. (18.11). Since we have restricted ourselves to $z > 0$, the $VFR^+(t)$ is simply

$$\begin{aligned} VFR^+(t) &= -\int_{z>0} \mathbf{q} \bullet \mathbf{n}\, dS \\ &= \left(P_{ref}\, k_v^{1/2}\, k_h/\mu\right) r^{*}{}_w\, (\partial p^*/\partial r^*)_w \iint_R d\xi\, d\eta / \sqrt{(r^{*}{}_w{}^2 - \xi^2 - \eta^2)} \end{aligned} \tag{18.15}$$

where R is the areal projection of the ellipsoid on the x-y plane. To simplify the integration, we introduce a polar transformation of coordinates $\xi = \rho \cos\theta$ and $\eta = \rho \sin\theta$. In the second double integral below, the integration is performed over $0 \le \theta \le 2\pi$ and $0 \le \rho \le r^{*}{}_w$. This leads to

$$\iint_R d\xi\, d\eta / \sqrt{(r^{*}{}_w{}^2 - \xi^2 - \eta^2)} = \iint \rho\, d\rho\, d\theta / \sqrt{(r^{*}{}_w{}^2 - \rho^2)} = 2\pi r^{*}{}_w \tag{18.16}$$

The total VFR accounting for both upper and lower halves of the ellipsoid is just twice the value of $VFR^+(t)$, so that

$$VFR(t) = \left(4\pi r^{*}{}_w{}^2 P_{ref}\, k_v^{1/2} k_h/\mu\right)(\partial p^*/\partial r^*)_w \tag{18.17}$$

which is, we emphasize, expressed in terms of the dimensionless pressure gradient $(\partial p^*/\partial r^*)_w$. This result differs from the conventional one for spherical isotropic flow, where $VFR(t) = 4\pi R_w^2\, k/\mu\, \partial p/\partial r$, which appears as a product of the surface area $4\pi R_w^2$ and an isotropic velocity $k/\mu\ \partial p/\partial r$ constant in all directions. This difference arises because $\partial p/\partial r$ is dimensional whereas $\partial p^*/\partial r^*$ is not, and should not cause confusion. Again, $k_v^{1/2}\, k_h$ is not directly related to any effective permeability, since k_v and k_h have not been completely scaled out; for example, they appear in the definition of r^*. The expression for total volume from rate in Eq. (18.17) again applies to general deviated wells in transversely isotropic media. Two natural boundary value problems can be defined for the ellipsoidal model considered here, namely, the usual pressure and flow rate formulations.

Pressure Boundary Value Problem

In this formulation, a physical pressure $P_{ref}p^*_w(t^*) + P_0$ is prescribed along the ellipsoidal surface defined by Eq. (18.6), and the following boundary value problem is solved.

$$\partial^2 p^*/\partial r^{*2} + 2/r^* \, \partial p^*/\partial r^* = \phi\mu c/t_{ref} \, \partial p^*/\partial t^* \tag{18.18a}$$

$$p^*(r^*, 0) = 0 \tag{18.18b}$$

$$p^*(r^* \rightarrow \infty, t^*) = 0 \tag{18.18c}$$

$$p^*(r^*_w, t^*) = p^*_w(t^*) \tag{18.18d}$$

Once the solution for $p^*(r^*,t^*)$ is obtained, Eq. (18.17) can be evaluated to produce the dimensional VFR. For simplicity, the time scale t_{ref} can be selected as $\phi\mu c$ so that $\phi\mu c/t_{ref} = 1$. This problem can be solved by Laplace transforms. For example, when p^*_w is constant, the solution is

$$p^*(r^*, t^*) = (p^*_w)\,(r_w^*/r^*)\,\mathrm{erfc}\left[1/2\,(r^*/r_w^* - 1)/\sqrt{\left(t^*/r_w^{*2}\right)}\right] \tag{18.18e}$$

At any point in time t^*, the derivative $\partial p^*/\partial r^*$ can be determined and the VFR easily calculated. If p^*_w varies in time, Laplace transform methods can be used to derive a closed-form superposition integral based on the elementary solution in Eq. (18.18e). In this book, the pressure boundary value problem is less significant than the one for flow rate. We now turn to that formulation.

Volume Flow Rate Problem Without Skin Effects

In this formulation, Eqs. (18.18a)–(18.18c) apply, but the total volumetric production rate Q(t) of the ellipsoidal source is specified instead of the pressure in Eq. (18.18d). This rate is *not* generally equal to the Darcy VFR(t) unless wellbore storage effects completely vanish. The physical volumetric balance equation requires us to consider the more general statement $VFR(t) - VC\,\partial p/\partial t = Q(t)$, where V is the storage volume and C is the compressibility of the wellbore fluid. For convenience, we shall take a dimensional production rate Q(t) in the form

$$Q(t) = Q_0 F(t^*) \tag{18.19}$$

where the constant Q_0 has dimensions of VFR and the prescribed function $F(t^*)$ is dimensionless. Then, our problem takes the form

$$\partial^2 p^*/\partial r^{*2} + 2/r^* \, \partial p^*/\partial r^* = \phi\mu c/t_{ref} \, \partial p^*/\partial t^* \tag{18.20a}$$

$$p^*(r^*, 0) = 0 \tag{18.20b}$$

$$p^*(r^* \rightarrow \infty, t^*) = 0 \tag{18.20c}$$

$$(4\pi r^*{}_w{}^2 P_{ref} k_v{}^{1/2} k_h/\mu)(\partial p^*/\partial r^*)_w - VCP_{ref}/t_{ref}\, \partial p^*/\partial t^* = Q_0 F(t^*) \quad (18.20d)$$

Flow Rate Problem With Skin

Eq. (18.20d) provides the boundary condition for the flow rate problem *without* skin effects, and an exact solution can be obtained in closed analytical form (Proett and Chin, 1998). However, it is possible to obtain an exact solution for the more difficult problem including skin effects (Proett and Chin, 2000). Since this more general solution is available, we shall not discuss the skin-free model in this book. Perhaps the greatest difficulty in formulating the problem correctly lies in the form of the skin model used. Conventionally, the *ad hoc* skin model $p_w = p - SR_w\, \partial p/\partial r$ is invoked in well testing and justified empirically without understanding the assumptions and limitations behind it. We shall derive it and show that it arises from velocity continuity (that is, mass conservation approximately) at the planar interface separating the damaged zone and the formation. The derivation shows that it holds only for lineal flows; then, the remainder of this section extends the conventional skin model to encompass multidimensional anisotropic flows.

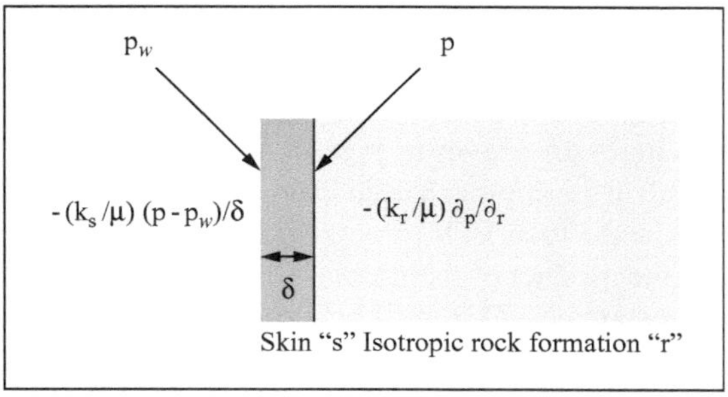

FIG. 18.3 Lineal isotropic flow.

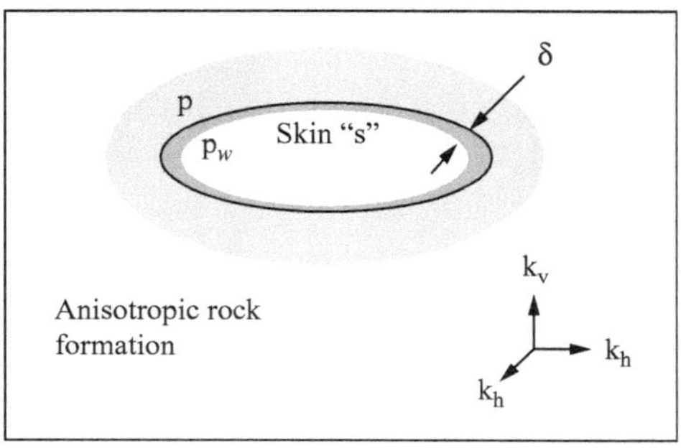

FIG. 18.4 Ellipsoidal anisotropic flow.

In Fig. 18.3, a thin layer of skin of thickness δ is shown adjacent to a much larger rock formation or sandface. The Darcy velocity in the formation is simply $-(k_r/\mu)\ \partial p/\partial r$, where k_r is the formation permeability and μ is the fluid viscosity; this must be identical to the Darcy velocity within the skin, which can be approximated by $-(k_s/\mu)\ (p-p_w)/\delta$, where k_s is the permeability of the skin. Equating the two leads to $p_w = p - SR_w\ \partial p/\partial r$ where $S = \delta k_r/(R_w k_s)$. This derivation is useful for two reasons. First, it provides a direct relationship connecting S to skin thickness and damaged zone permeability. Second, it states that $p_w = p - SR_w\ \partial p/\partial r$ is much more than an empirical model describing skin effects, with an assumed dependence on spatial derivative. Because it actually appears as the consequence of mass conservation, the underlying principle can, in effect, be extended to handle more complicated problems.

We now derive the extended model. Before we proceed, let us state the underlying physical assumptions. We assume that the skin effect under consideration arises from mudcake or invasion damage and also that the permeability of the skin is much less than that of the formation. When this is so, formation properties affect skin or cake growth only minutely; that is, the permeability of the formation, despite its anisotropic character, can be neglected in determining mudcake growth or in characterizing skin. Thus, the cake does not see the formation, even when k_h and k_v are very different; it is therefore appropriate to assume a skin permeability k_s that is isotropic in nature, even when the underlying formation is anisotropic. We consistently assume a skin thickness δ that is uniform areally. Again, since rock properties are unimportant to leading order, there is no reason for δ *not* to be anything but constant around the ellipsoidal skin-to-formation interface shown in Fig. 18.4.

We now generalize the approach used in deriving $p_w = p - SR_w\ \partial p/\partial r$. Within our layer of thin skin, the total VFR through the ellipsoidal surface interface can be represented as the product of the normal Darcy velocity $-(k_s/\mu)\ \{(p-p_w)/\delta\}$ and the total surface area Σ. This must be equal to the total integral of normal velocity $\mathbf{q} \bullet \mathbf{n}$ over dS taken in the formation, where dS represents an incremental surface area of Σ.

$$-(k_s/\mu)\,\{(p-p_w)/\delta\}\,\Sigma = \int_{\Sigma} \mathbf{q}\cdot\mathbf{n}\,dS \tag{18.21}$$

The surface integral in Eq. (18.21) was computed earlier. From Eqs. (18.15), (18.17), we find that the right side of Eq. (18.21) now takes the form $-(4\pi r^*{}_w{}^2\, p_{ref}\, k_v{}^{1/2}\, k_h/\mu)\,(\partial p^*/\partial r^*)_w$; that is,

$$(k_s/\mu)\,\{(p-p_w)/\delta\}\,\Sigma = \left(4\pi r^*{}_w{}^2\, p_{ref}\, k_v{}^{1/2}\, k_h/\mu\right)(\partial p^*/\partial r^*)_w \tag{18.22}$$

Thus, we obtain

$$p_w = p - \left\{4\pi r^*{}_w{}^2\, p_{ref}\, k_v{}^{1/2}\, k_h\,\delta/(k_s\,\Sigma)\right\}(\partial p^*/\partial r^*)_w \tag{18.23}$$

which generalizes the $p_w = p - (\delta k_r/k_s)\ \partial p/\partial r$ obtained for linear isotropic flows. Now, recall that nearfield boundary conditions are applied on the surface of the ellipsoidal source $x^2/k_h + y^2/k_h + z^2/k_v = r^{*}_w{}^2$. Expressions for the surface area Σ in Eq. (18.23) are available in the mathematics literature and depend on the relative values of vertical and horizontal permeabilities.

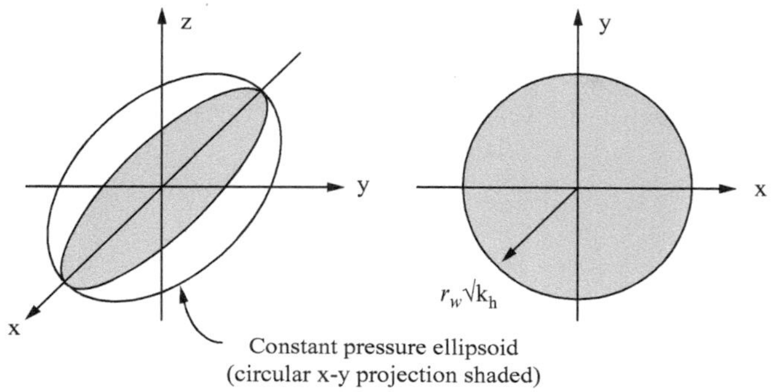

FIG. 18.5 Ellipsoidal flow assumption.

If, as in the usual case, the permeabilities satisfy $k_h > k_v$, so that the ellipsoid represents an *oblate* spheroid, it can be shown that

$$\Sigma = 2\pi k_h\, r^{*}_w{}^2 + \pi k_v\, r^{*}_w{}^2\, \varepsilon^{-1}\, ln\{(1+\varepsilon)/(1-\varepsilon)\} \tag{18.24a}$$

$$\varepsilon = \sqrt{(1 - k_v/k_h)} \tag{18.24b}$$

On the other hand, if the permeabilities satisfy $k_v > k_h$, so that the ellipsoid represents a *prolate* spheroid, it can be shown that

$$\Sigma = 2\pi k_h\, r^{*}_w{}^2 + 2\pi r^{*}_w{}^2 \left\{\sqrt{(k_h k_v)}\right\}\varepsilon^{-1} \arcsin \varepsilon \tag{18.25a}$$

$$\varepsilon = \sqrt{(1 - k_h/k_v)} \tag{18.25b}$$

In the limit $k_h \rightarrow k_v$, the dimensionless parameter ε approaches zero; use of L'Hospital's Rule shows that the areas in Eqs. (18.24a), (18.25a) reduce to the $4\pi R_w^2$ result anticipated for isotropic flows. Ellipsoidal flow assumptions are described schematically in Fig. 18.5.

General Flow Rate Problem Formulation

If we now return to Eqs. (18.20a)–(18.20d) and review the basic formulation, it is obvious that only Eq. (18.20d) needs to be changed to model skin. If we rewrite Eq. (18.20d) in the form $(4\pi r^{*}_w{}^2\, P_{ref}\, k_v{}^{1/2}\, k_h/\mu)$ $(\partial p^{*}/\partial r^{*})_w - VC\partial p/\partial t = Q_0\, F(t^{*})$, it is clear that $\partial p/\partial t$ must be replaced by $\partial p_w/\partial t$, to differentiate between the pressure inside the sandface and that in the well. Doing so, we obtained the extended law

$$(4\pi r^{*}_w{}^2\, P_{ref}\, k_v{}^{1/2}\, k_h/\mu)\,(\partial p^{*}/\partial r^{*})_w - VC\, \partial p_w/\partial t = Q_0\, F(t^{*}) \tag{18.26}$$

Combining Eqs. (18.23), (18.26), we obtain

$$\left(4\pi r^*{}_w{}^2\, P_{ref}\, k_v{}^{1/2}\, k_h/\mu\right)\,(\partial p^*/\partial r^*)_w - VCP_{ref}/t_{ref}\,\partial p^*/\partial t^* + \left\{4\pi r^*{}_w{}^2\, VCP_{ref}\, k_v{}^{1/2}\, k_h\,\delta/(k_s\,\Sigma t_{ref})\right\}\,(\partial^2 p^*/\partial t^*\partial r^*)_w = Q_0\, F(t^*) \tag{18.27}$$

Two transformations map the anisotropic Cartesian equations into spherically symmetric form. If we apply Eqs. (18.2), (18.3), (18.4) and Eqs. (18.28a)–(18.28c), (18.29a)–(18.29c), that is,

$$p^*(r^*, t^*) = p(r, t) \tag{18.28a}$$

$$r = ar^* \tag{18.28b}$$

$$t = t^* \tag{18.28c}$$

$$a = 4\pi\phi c k_v{}^{1/2}\, k_h\, r^*{}_w{}^2\,/(VC) \tag{18.29a}$$

$$t_{ref} = \mu V^2 C^2/\left(16\pi^2\,\phi c\, k_v\, k_h{}^2\, r^*{}_w{}^4\right) \tag{18.29b}$$

$$P_{ref} = \mu Q_0 VC/\left(16\pi^2\,\phi c k_v\, k_h{}^2\, r^*{}_w{}^4\right) \tag{18.29c}$$

to Eq. (18.27), where a is dimensionless, we straightforwardly obtain

$$\partial^2 p/\partial r^2 + 2/r\,\partial p/\partial r = \partial p/\partial t \tag{18.30a}$$

$$p(r, t=0) = 0 \tag{18.30b}$$

$$p(r \to \infty, t) = 0 \tag{18.30c}$$

$$\partial p(r_w, t)/\partial r - \partial p/\partial t + \xi\,\partial^2 p/\partial r\partial t = F(t) \tag{18.30d}$$

where

$$r_w = ar^*{}_w = 4\pi\phi c k_v{}^{1/2}\, k_h\, r^*{}_w{}^3\,/(VC) \tag{18.30e}$$

$$\xi = 16\pi^2\,\phi c k_v\, k_h{}^2\, r^*{}_w{}^4\,\delta/(k_s\,\Sigma VC) > 0 \tag{18.30f}$$

Note that we did not select $t_{ref} = \phi\mu c$, as we had done for our previous pressure boundary value problem. Instead, we chose t_{ref} and P_{ref} judiciously, so that the resulting normalized problem is solved once and only once, thereby eliminating the need to generate numerous type curves, as is typical in well testing.

General Solution

An exact, closed-form, analytical solution to Eq. (18.30) can be derived using transform methods. If $\boldsymbol{p}(r,s)$ denotes the Laplace transform of $p(r,t)$, standard manipulations show that

$$\boldsymbol{p}(r, s) = -\mathbf{F}(s)\, r_w{}^2\, \exp\left\{(r_w - r)\sqrt{s}\right\}/\left[r\left\{r_w s + (1+\eta s)\left(1 + r_w\sqrt{s}\right)\right\}\right] \tag{18.37}$$

Consider constant rate buildup or drawdown and set F(t) = 1, so that **F**(s) = 1/s. Evaluating at the effective well radius r_w, we have

$$\boldsymbol{p}(r_w, \mathrm{s}) = -r_w / \left[\mathrm{s}\left\{r_w \mathrm{s} + (1+\eta \mathrm{s})\left(1 + r_w \sqrt{\mathrm{s}}\right)\right\}\right] \quad (18.38)$$

Eq. (18.38) cannot be inverted by straightforward table lookup. However, a closed-form solution *can* be obtained through the use of some algebra. Let us first rewrite Eq. (18.38) in the form

$$\boldsymbol{p}(r_w, \mathrm{s}) = -1/\left\{\eta \mathrm{s}\left(\mathrm{s}^{3/2} + \mathrm{a}_1 \mathrm{s} + \mathrm{a}_2 \mathrm{s}^{1/2} + \mathrm{a}_3\right)\right\} \quad (18.39a)$$

where

$$\mathrm{a}_1 = (r_w + \eta)/(r_w \eta) \quad (18.39b)$$

$$\mathrm{a}_2 = 1/\eta \quad (18.39c)$$

$$\mathrm{a}_3 = 1/(r_w \eta) \quad (18.39d)$$

are real coefficients. Then we can cast Eq. (18.39a) in the form

$$\boldsymbol{p}(r_w, \mathrm{s}) = -1/\left\{\eta \mathrm{s}\left(\mathrm{s}^{1/2} - \mathrm{x}_1\right)\left(\mathrm{s}^{1/2} - \mathrm{x}_2\right)\left(\mathrm{s}^{1/2} - \mathrm{x}_3\right)\right\} \quad (18.40)$$

where x_1, x_2, and x_3 are the roots of the cubic polynomial equation

$$\mathrm{s}^{3/2} + \mathrm{a}_1 \mathrm{s} + \mathrm{a}_2 \mathrm{s}^{1/2} + \mathrm{a}_3 = 0 \quad (18.41)$$

for the quantity $s^{1/2}$. The required roots can be easily expressed in terms of a_1, a_2, and a_3. Let us introduce the auxiliary quantities

$$\mathrm{Q} = \left(3\mathrm{a}_2 - \mathrm{a}_1^2\right)/9 \quad (18.42a)$$

$$\mathrm{R} = \left(9\mathrm{a}_1 \mathrm{a}_2 - 27\mathrm{a}_3 - 2\mathrm{a}_1^3\right)/54 \quad (18.42b)$$

and then define

$$\mathrm{D} = \mathrm{Q}^3 + \mathrm{R}^2 \quad (18.42c)$$

$$\mathrm{S} = \left(\mathrm{R} + \sqrt{\mathrm{D}}\right)^{1/3} \quad (18.42d)$$

$$\mathrm{T} = \left(\mathrm{R} - \sqrt{\mathrm{D}}\right)^{1/3} \quad (18.42e)$$

Then, it can be shown that

$$\mathrm{x}_1 = (\mathrm{S} + \mathrm{T}) - \mathrm{a}_1/3 \quad (18.43a)$$

$$\mathrm{x}_2 = -(\mathrm{S} + \mathrm{T})/2 - \mathrm{a}_1/3 + i^{1/2}\sqrt{3}\,(\mathrm{S} - \mathrm{T}) \quad (18.43b)$$

$$\mathrm{x}_3 = -(\mathrm{S} + \mathrm{T})/2 - \mathrm{a}_1/3 - i^{1/2}\sqrt{3}\,(\mathrm{S} - \mathrm{T}) \quad (18.43c)$$

Using partial fraction expansions, we can rewrite Eq. (18.40) in the form

$$\begin{aligned} p(r_w, s) = & -[1/\{\eta(x_1-x_2)(x_1-x_3)\}]\left[1/\left\{s\left(s^{1/2}-x_1\right)\right\}\right] \\ & -[1/\{\eta(x_2-x_1)(x_2-x_3)\}]\left[1/\left\{s\left(s^{1/2}-x_2\right)\right\}\right] \\ & -[1/\{\eta(x_3-x_1)(x_3-x_2)\}]\left[1/\left\{s\left(s^{1/2}-x_3\right)\right\}\right] \end{aligned} \quad (18.44)$$

Next observe that the inverse of the Laplace transform $1/\{s(s^{1/2}+a)\}$ is simply $\{1-\exp(a^2t)\,\text{erfc}(a\sqrt{t})\}/a$. Thus, the dimensionless Darcy pressure satisfies

$$\begin{aligned} p(r_w, t) = & \left\{1-\exp\left(x_1{}^2t\right)\text{erfc}\left(-x_1\sqrt{t}\right)\right\}/\{\eta x_1\,(x_1-x_2)(x_1-x_3)\} \\ & + \left\{1-\exp\left(x_2{}^2t\right)\text{erfc}\left(-x_2\sqrt{t}\right)\right\}/\{\eta x_2\,(x_2-x_1)(x_2-x_3)\} \\ & + \left\{1-\exp\left(x_3{}^2t\right)\text{erfc}\left(-x_3\sqrt{t}\right)\right\}/\{\eta x_3\,(x_3-x_1)(x_3-x_2)\} \end{aligned} \quad (18.45)$$

This expression for $p(r_w,t)$ applies *in the formation* only, because it does not yet include the pressure drop through the skin. (It is used for the first term on the right of Eq. 18.23, which shows that the pressure at the well requires a contribution related to $(\partial p^*/\partial r^*)_w$ that can be obtained from Eq. 18.37.) The above model shows how transient invasion problems for compressible liquids in anisotropic formations can be reduced to simple ones using geometric transformations. Far from an academic exercise, the solution represents the backbone of an interpretation method for a modern multi-probed formation tester tool that has proven successful in the field (Chin et al., 2014b, 2015). Extensions to the preceding compressible fluid model that include formation heterogeneities and miscible and immiscible fluids are described in these books.

NEW METHODS FOR RAPID MOBILITY AND PORE PRESSURE PREDICTION

Fast and direct measurements for pore pressure and mobility are crucial to accurate formation evaluation, economic analysis and drilling safety. Real-time predictions can be obtained from formation tester pressure transient analysis. However, in tight sand and heavy oil environments, this is hampered by conventional approaches which require steady-state pressure responses that may require hours of wait time. These imply prohibitive offshore costs and increase the risks of stuck and lost tools; the methods derive from decades-old formulas intended for high-mobility formations which equilibrate rapidly. New math models are introduced that, now assuming high levels of pressure diffusion, require just seconds to yield acceptable pore pressures, and vertical and horizontal permeabilities. These augment traditional methods, providing suites of pressure evaluation tools covering all permeabilities; moreover, they do not require hardware changes to existing tools aside from downhole firmware modifications. For dual-probe tools, we develop analogies to resistivity logging: pump pistons are oscillated sinusoidally, and pressure phase and amplitude changes monitored at observation probes are interpreted using Darcy's law. Second, parallels to sonic logging are considered: pistons are allowed to impact formations suddenly, and signal arrival times measured at observation probes are converted to mobility. Third, for dual and single-probe systems, rapid "drawdown alone" and "drawdown-buildup" approaches that do not use exponential, transcendental or complicated error functions, but instead, efficient rational polynomial expansions, are described. Detailed examples and validations demonstrate the power and versatility behind the new methods for real-time and job planning applications. Fast calculations support increased downhole data usage, thus enhancing real-time capabilities; they free up resources needed to support other MWD/LWD functions.

INTRODUCTION

Formation testers, originally developed for fluid collection, are now widely used in reservoir description, providing detailed near-borehole information related to permeability, mobility, anisotropy and pore pressure. These objectives are met using pressure transient analysis methods which interpret unsteady flowline changes with specially designed mathematical models and operational test procedures.

Different hardware designs are available. A cross-section of tools recently developed by China National Offshore Oil Corporation and its oilfield service subsidiary is shown in the accompanying figure. The single-probe tool, seen top-left in Fig. 18.6A, is preferred in Measurement and Logging-While-Drilling (M/LWD) by virtue of its simple and robust mechanical design. In pressure testing applications, known as "formation testing while drilling" or FTWD, real-time pore pressure and "spherical mobility" (that is, $k_h^{2/3}k_v^{1/3}$) are inferred from short-time drawdown-buildup tests. Interpretation models account for flowline storage and high flow resistance effects that are important at early times. However, individual horizontal and vertical mobilities (or permeabilities, if viscosity is known) cannot be obtained because the datasets are limited.

The dual-probe tester, in the middle photograph, is often used in wireline applications. Ideally, constant flow rate fluid extraction at the pump leads to steady-state profiles with unchanging pressure drops at both source and observation probes. Their values are used to predict horizontal and vertical mobilities; however, in tight or low mobility zones, long wait times often exceeding hours prove to be expensive and conducive to tool-sticking and economic loss. At the end of this chapter, a fast (patent pending) method requiring just seconds of pressure data is described that incorporates phase delay concepts from electromagnetic logging.

Focused probes, at the upper right of Fig. 18.6A and close-up at the bottom, are relatively new and introduce flow features important to both pressure testing and sampling and collection. Fluid is withdrawn through two orifices, the first, an outer circular ring, and the second, a probe similar to that in single-probe tools. These flows are driven by separate hydraulic systems. The ring flow provides a cylindrical barrier to filtrate entering through the mudcake; the center probe, which now "sees" earlier clean-up, provides contact with in situ fluids more quickly. Thus, collected samples are cleaner and more representative

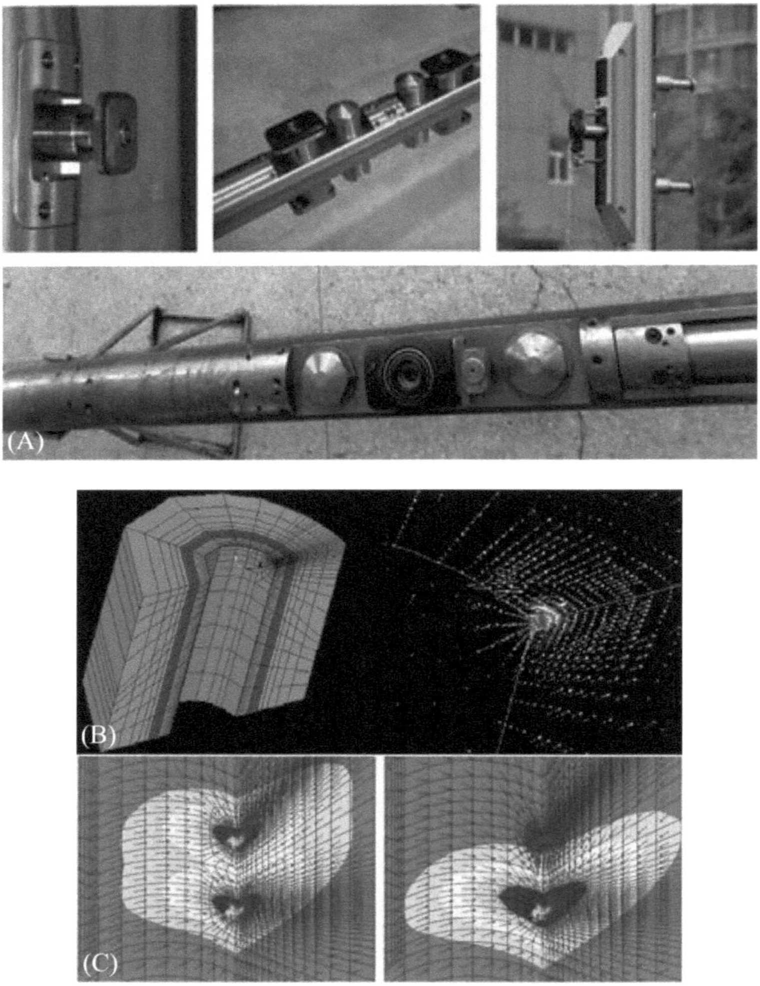

FIG. 18.6 (A) Single, dual and focused probe designs. (B) Finite element model, flow vectors near probe. (C) Pressure contours for drawdown with two source probes (*left*) and one (*right*).

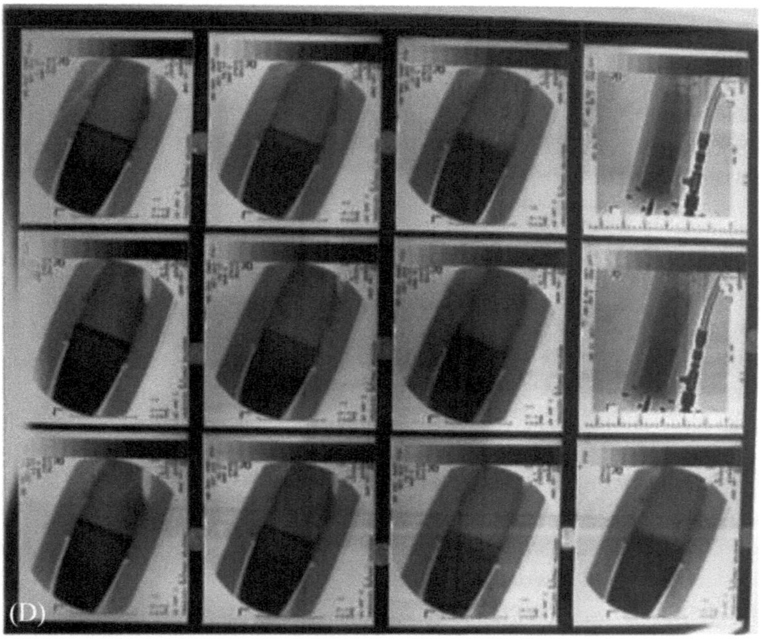

FIG. 18.6, CONT'D (D) Catscan results (flow from top to bottom, darkening lines at center indicate cake growth in time).

of the reservoir; and in pressure testing, predicted mobilities "k/μ" characterize the reservoir fluid since the viscosity μ represents that of oil more and not a time-varying mixture of filtrate and oil.

While pressure testing objectives are clear, the procedures employed leave room for significant improvement. Hardware design is often driven by the availability of sophisticated models. While good models certainly imply good interpretation, they also motivate innovative design changes. Our focus on efficient analysis methods for early time interpretation in low mobility applications continually drives us to re-examine old and new tool design approaches.

Overview

We describe a comprehensive set of integrated formation testing forward and inverse analysis tools developed for wireline and "while drilling" (FTWD) applications in hardware design and pressure transient interpretation. The methods, based on rigorous Darcy flow formulations, are solved analytically in closed form whenever possible and cross-checked in different limits to ensure physical consistency and accuracy.

The transient problem for formation tester liquid pressure response in anisotropic media with flowline storage and skin at arbitrary dip, earlier solved in exact, closed analytical form assuming ellipsoidal sources (using complex complementary error functions), is used to derive exact solutions to several inverse problems where permeabilities are sought when dip angle and source and observation probe pressure drops are given.

First, the zero-skin forward solution is evaluated in the steady-state limit for constant rate pumping. Explicit inverse formulas are derived for all horizontal and vertical permeabilities and dip angles. With pressure drops computed at various dip angles from the forward simulation, derived formulas are used to predict both assumed permeabilities, demonstrating their utility in field interpretation. Neglect of dip angle can lead to significant errors in anisotropy prediction. Moreover, multi-valued inverse solutions exist: for a given set of pressure drops, three permeability pairs are found which require resolution from additional logging data.

Second, the "with-skin" forward solution is evaluated at steady-state for constant rate pumping to develop formulas relating source and observation probe pressure drop, vertical and horizontal permeabilities and skin factor. An algorithm giving possible solutions for both permeabilities and skin at any dip angle when both pressure drops are known is derived. Because only two pressure data points are assumed, additional logging information is needed to render a unique determination.

Third, short-duration "pulse interactions" at the observation probe are used to determine anisotropy (analogous to the pulses emanating from acoustic tools). These are strongest and most advantageous at low permeabilities where diffusion predominates. Short pulses, high in frequency content, provide detailed information. Multi-pulse wave-trains with different flow rates, pulse durations and separations enable multiple fast test suites at the rigsite without requiring new hardware. They are accurate, economical and reduce tool sticking risks in tight zones.

Fourth, "phase delay" approaches for permeability prediction analogous to electromagnetic logging methods are described. Sinusoidal pressure transients are created at the pumping probe. Their amplitudes and phases are measured at one or more observation probes and interpreted using Darcy analysis models. As with pulse interaction methods, phase delay approaches allow short-duration tests that are economical, safe and characterized by high signal-to-noise ratios.

Fifth, a full three-dimensional horizontal well model for single-probe, dual-probe, dual-packer and elongated pad tools with real mandrels in layered media is given, with computations showing effects of azimuth and bed boundary on pressure response and their implications on permeability prediction. While *source models* require dual-probe data for inverse application, *full 3D models* can be used with single-probe FTWD tools (measuring azimuthal pressures) to provide clues related to permeability, anisotropy and bed thickness.

Finally, real-time FTWD pore pressure and mobility prediction is discussed. Such problems, key to drilling safety and rapid economic evaluation, involve transient data distorted by flowline storage effects. Accurate predictions are possible using a minimum of pressure data. We develop rational polynomial expansion methods that do not require exponential, real or complex complementary error functions, and moreover, do not use regression or least-squares smoothing filters that introduce diffusive assumptions beyond those implicit in Darcy's laws. Rapid analysis frees microprocessor resources for other important control and interpretation functions needed during drilling.

FOUNDATION MODEL SUMMARIES

In formation tester pressure transient analysis, two types of applications arise, namely, "forward modeling," in which source and observation probe responses are sought when fluid, formation and tool parameters are specified, and "inverse modeling," in which k_h and k_v permeabilities (and possibly pore pressure) are required when all other parameters are given. Forward models have been developed to a high degree of sophistication. Almost twenty years ago, Proett and Chin (1996) published the first full three-dimensional finite element analysis assuming realistic borehole environments with pad, probe, mandrel, flowline storage and bedding plane effects. These are reviewed in the books due to Chin et al. (2014b, 2015), which also cite extensions to include coupled dynamic mudcake growth and supercharge corrections.

The work of Chin and Proett (2005) provided early finite difference models that included multiphase miscible and immiscible effects. In downhole sampling, the time required to pump until clean in situ fluids are obtained is important. This time scale differs from that used for pressure transient interpretation. Fig. **18.6**B and C from earlier Proett and Chin references illustrate capabilities that have been used to design new tools and interpret transient data obtained in complicated environments. Fig. **18.6**D shows prior Catscan experiments in which dynamic mudcake growth is measured in linear cores with different permeabilities. This is important to supercharge and contamination corrections.

While fully three-dimensional models are important in their own right, they do require complicated inputs, numerical simulation expertise, not to mention sophisticated and expensive computing environments that host high-overhead software and long calculations. Moreover, they do not easily reveal explicit dependencies on parameters like mobility and storage.

Thus, fast methods that retain the basic elements of the physics are desirable, particularly for field office use and real-time downhole analysis. A number of simpler formulations are possible. These range from elementary point source models, which unfortunately "blow up" at the "$r = 0$" origin, to finite radius models, which apply flowline storage and skin boundary conditions at spherical source surfaces (ellipsoidal in transversely isotropic flow), utilized in this chapter.

This chapter also introduces new physical concepts in pressure transient interpretation using innovative math models described in Chin et al. (2014a). So that the ideas are clearly explained, math details are omitted in favor of examples, although shorter summaries are offered. For readers interested in analytical and numerical details, technical reports are available upon request. Several new capabilities and modules are available and, for convenience, are referred to by a "FT" designation with "FT" referring to formation testing.

Module FT-00

This deals with exact transient liquid response in homogeneous anisotropic media. It solves for the unsteady Darcy pressure field about an ellipsoidal source surface immersed in a transversely isotropic infinite homogeneous medium allowing full skin effect and flowline storage boundary conditions. This "nonzero radius source" model is more powerful than limited point source approximations because it handles nearfield boundary conditions without becoming singular at the origin as do point source models. We emphasize that cylindrical borehole effects and drilling fluid invasion are not directly incorporated. The earlier exact analytical solution, posed in terms of the complex complementary error function erfc(z), applied to a single drawdown or buildup and was limited to zero dip angles. To be effectively used for job planning and inverse applications, such as those considered here, improvements were needed.

(1) The erfc function in most scientific software libraries does not converge for certain ranges of complex arguments, and unfortunately, those associated with flowline storage effects on the order of those encountered in real formation testing tools (storage distorts pressure transients and its understanding is important to pressure and mobility interpretation). A new, fast and accurate subroutine is used which converges for a much wider range of complex arguments. Split-second response with fourteen digit accuracy guarantees robust and stable numerical performance.

(2) While the original work formally applied to pressures at all points, that is, all observation points in addition to source points, in reality, the former could not be computed because products of very large and very small numbers were involved. This equivalently meant that time superpositions of different flowrates (necessary to model multirate pumping) could not be computed accurately. Here, instead of dealing with erfc directly, a function of erfc, namely "$\exp(-z^2)\ \mathrm{erfc}(-i\ z)$" where $z = x + iy$, is considered, which avoids the foregoing complications. Thus, pressures can be computed at any observation location, and also, multirate pumping applications can be modeled. With this extended capability available, the original theory was extended to include all nonzero dip angles.

(3) The model applies to liquids only and not to nonlinearly responding gases (gas pumping introduces thermodynamics and high compressibilities). Work in this area is briefly described later under FT-06.

(4) The model applies to an ellipsoidal source immersed in a transversely isotropic homogeneous medium (reducing to a spherical source in isotropic media). This means that there is no cylindrical borehole and there are no layers or barriers. By "skin effect," we imply conventional damage mechanisms that would be found on the surface of the assumed ellipsoidal source. This is physically related to the formation damage seen in drilling mud invasion through cylindrical holes.

(5) Module FT-00 is discussed here because it is used to provide exact pressure data needed for inverse pore pressure and permeability prediction interpretation. Just as important, this high level formulation provides the foundation for our first inverse module FT-01 (solving for permeability and anisotropy exactly) when steady dual probe pressure data is available. FT-01 is based on rigorous math. The equations for FT-00 are evaluated in analytical closed form for asymptotically large times to develop formulas for pressure response that are independent of porosity, fluid compressibility and flowline volume. These formulas are inverted to derive governing algebraic equations for k_h, k_v, and k_h/k_v which can be solved exactly in terms of source and observation probe pressure drops.

Thus, solutions to the inverse problem, which in the literature often imply running forward simulators repeatedly, is actually solved explicitly and exactly using noniterative methods. FT-01 assumes the availability of steady-state dual probe pressure data. This is often not the case with low permeability formations. For such problems, a short-time "pulse interaction" method is developed which importantly uses FT-00 as the primary history matching tool. The pulse interaction method, as will be explained, is most effective in tight formations because high levels of diffusion cause individual pump pulses to interact strongly. This interaction is strongly dependent on formation anisotropy. The pulse interaction method is later complemented by "phase delay" and "rapid drawdown-buildup" analyses. These methods taken together cover a broad range of applications.

Module FT-01

We offer additional details related to the new steady-state inverse method. Although the pressure response in a liquid is linear and superposition applies, the asymptotic evaluation of that expression at large time yields nontrivial dependencies in permeability that lead to different (but mathematically consistent and correct) equations for k_h, k_v and the anisotropy k_h/k_v. In fact, cubic equations of the form $k_h^3+(\ )\ k_h+(\ )=0$, $k_v^{3/2}+(\ )\ k_v+(\ )=0$ and

$(k_h/k_v)+(\ )\ (k_h/k_v)^{1/3}+(\ )=0$ are obtained where ( ) represents various lumped parameters depending on pressure drops, fluid and formation properties. Because k_h generally exceeds k_v, its solution is more reliable with less noise contamination.

As is known from algebra, the polynomial k_h equation may have three real roots, or it may have one real and two complex conjugate roots. Only positive real solutions for k_h are physically meaningful—if several real positive roots are found, other logging data will be needed for a unique determination. On the other hand, negative permeabilities are not meaningful. Small imaginary parts, however, do not rule out the usefulness of roots (with positive real parts) since these typically arise from the use of un-equilibrated pressure data or data inconsistent with Darcy's equations, e.g., pad slippage, transducer calibration and thermal effects.

Module FT-02

Gas pumping, thermodynamic input and flow nonlinearities are addressed in a new simulation module, which applies to forward and inverse applications for single-phase gas flows. Qualitative descriptions and software input screens are similar to those of FT-01 and for brevity are not discussed here. Such applications are increasingly becoming important in the development of unconventional gas resources. The present module provides more accurate characterization.

Module FT-03

FT-00 and FT-01 derive from the same high-level mathematical source formulation, and the forward and inverse solutions used here are "exact" in the sense that they follow from closed form analytical solutions. However, this does not mean that they are exact in a physical sense. Source solutions possess spherical symmetries (ellipsoidal in anisotropic media) that, while elegant theoretically, are not representative of real tools—that is, pad nozzles mounted on solid mandrels.

Consider a real tester, say a single-probe FTWD tool, logging a horizontal well in transversely isotropic media. A top-mounted pressure transducer will "see" essentially k_v while one that is side-mounted will see basically k_h (actually, complicated functions of the two apply at any specific azimuthal angle). It is clear that pressure drops obtained at two different angles can be used to determine the k_h and k_v provided a 3D model is available to perform the required history matching.

This capability presently exists: as a drillstring torques and un-torques, it winds and unwinds, taking it through a range of twist angles. We emphasize that, in contrast to conventional dual-probe interpretation methods assuming axially displaced pressure measurements, it is possible to determine k_h and k_v with a single-probe tool provided a 3D algorithm is used for azimuthal interpretation. Module FT-03 provides this capability, allowing convenient representation for pad and packer sources, arbitrary azimuthal and axial placement for multiple probes and modeling of bed effects.

Other modules, FT-02 and FT-04 to 10, are discussed in Chin *et al* (2014a, 2015) addressing, e.g., nonlinear gas responses and thermodynamic effects, highly transient (as opposed to constant or piecewise-constant multirate) pumping, two phase flow, and so on.

We digress temporarily from our basic model overview and introduce new permeability prediction methods by way of examples before continuing. We begin by presenting the capabilities of the new forward simulator FT-00.

Forward Model Applications, Module FT-00

We summarize the required input parameters by reproducing the software screen shown in Fig. 18.7A. Several blocks are apparent, namely, "Fluid and Formation Parameters," "Tool Properties," and "Pumping Schedule."

Fig. 18.7A shows input parameters for a multi-rate pump schedule with mixed production and injection with both long and short time durations. This supports constant rate pumping, pulse interaction and phase delay modeling. Simulations are extremely fast and typically require at most seconds. The pump schedule used appears in Fig. 18.7B. Source and observation probe transient pressure responses are given in Fig. 18.7C and D. Observe the rapid equilibration in source pressure and close correlation between it and flow rate (that is, compare Fig. 18.7B and C). At the observation probe, as is evident from Fig. 18.7D, slower equilibration and smearing due to diffusion are found. In general, the lower the permeability, the greater the diffusion.

This diffusion is both bad and good. It is "bad" when steady pressure drops are required for input into steady flow models for permeability prediction. However, it is "good" when specially designed transient interpretation approaches are available. For low permeability formations, the dynamical interaction between short duration pressure pulses is strong and highly dependent on anisotropy. Also, phase delays are more apparent and therefore useful in low mobility applications.

Inverse Model Applications, Module FT-01

In the following run, the fluid, formation and tool parameters of Fig. 18.7A are retained, except that the dip angle is changed from 0 to 45 degrees and a constant 10 cc/s pump rate is assumed for all time. Source (left) and observation probe (right) pressure transient responses are shown in Fig. 18.8A—again, observe how diffusion slows the equilibration to steady-state just 15 cm away from the source.

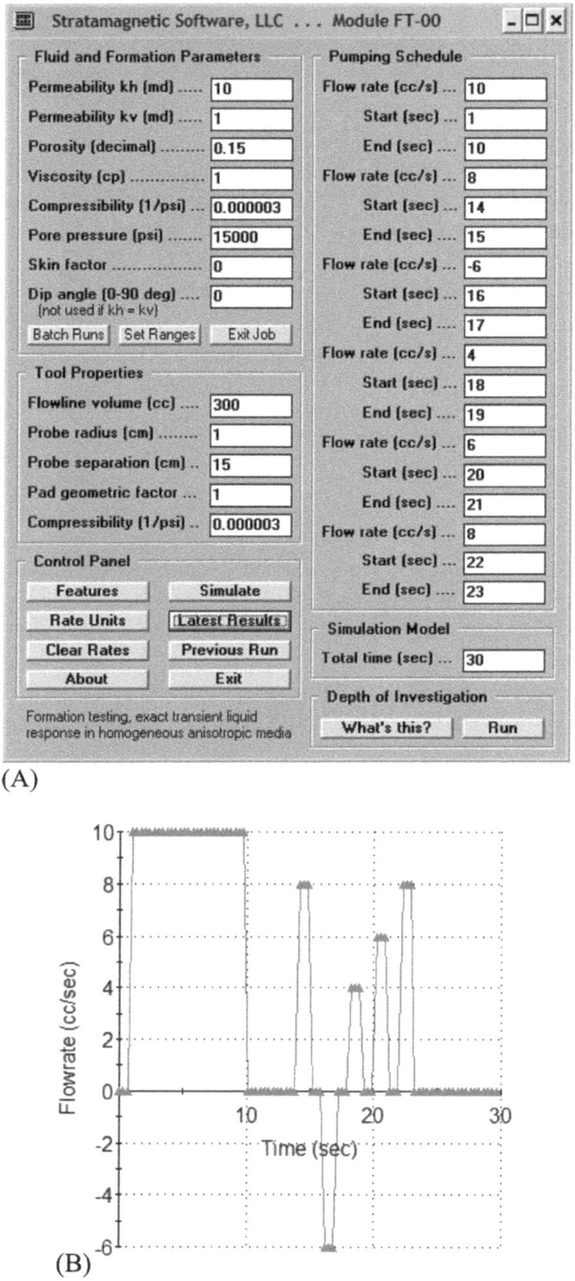

FIG. 18.7 (A) Forward simulation assumptions. (B) Pumpout schedule, volume flow rate.

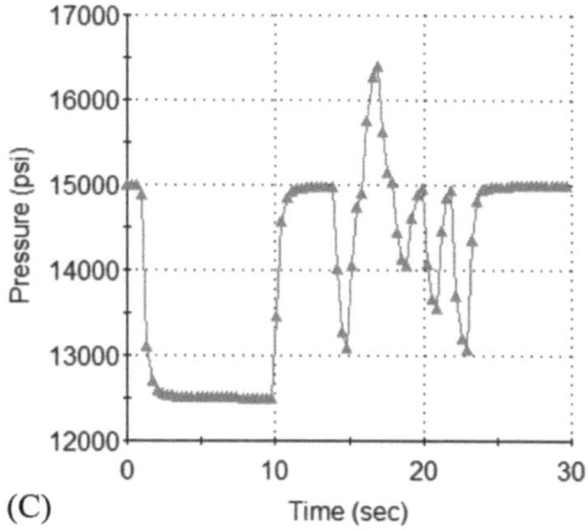

(C)

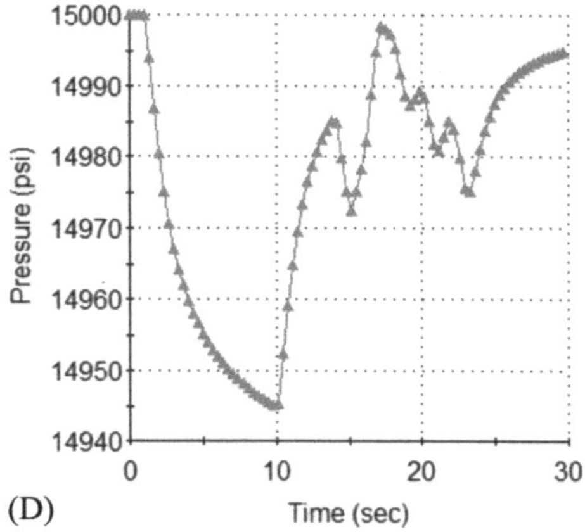

(D)

FIG. 18.7, CONT'D (C) Source probe pressure. (D) Observation probe pressure.

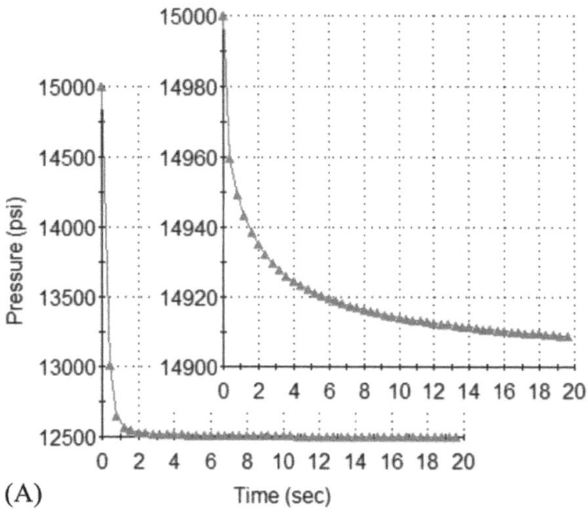

(A)

FIG. 18.8 (A) Source (*bottom*) and observation probe (*top*) pressure responses.

(Continued)

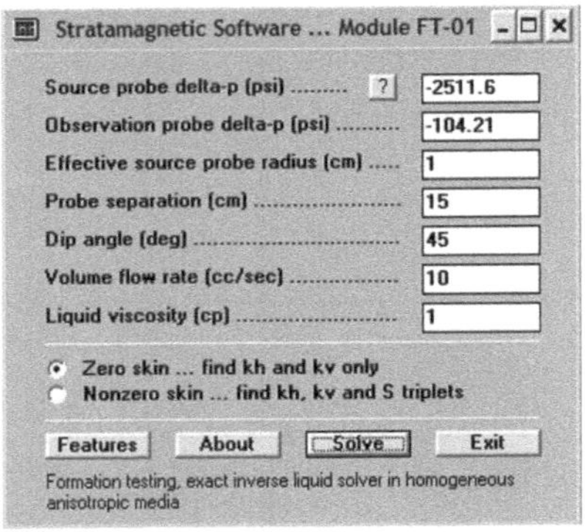

(B)

FIG. 18.8, CONT'D (B) Inverse steady-state solver.

The computed Δp's (probe *minus* a dynamically unimportant pore pressure) are

Time (s)	Δp_{source} (psi)	Δp_{observ} (psi)
0.100E+02	-0.24948E+04	-0.85789E+02
0.200E+02	-0.25001E+04	-0.91263E+02
0.500E+02	-0.25045E+04	-0.96208E+02
0.100E+05	-0.25116E+04	-0.10421E+03

We now consider the inverse problem and assume that pressure pair data (above table) are obtained from a dual probe tool. The input screen in Fig. 18.7A assumes that the skin coefficient is S = 0; thus, the assumptions in the software screen for FT-01 in Fig. 18.8B consistently assume zero skin. First examine our 10,000 s (3 h) data. Exact calculation shows three possible solutions, namely,

```
Tentative permeabilities (md) ...

Complex KH root # 1:  -10.97 +   0.00 i, KV:   0.83
Complex KH root # 2:   10.07 +   0.00 i, KV:   0.99
Complex KH root # 3:    0.91 +   0.00 i, KV: 121.96
```

In this case, two of the roots are easily ruled out; the first k_h is negative, while the third k_h is substantially less than k_v. The remaining k_h and k_v results, at 10.07 and 0.99 md, are almost identical to the assumed 10 and 1 md in the forward simulation creating the data. The method reproduces assumed permeability data exactly in this calculation with nonvanishing dip. This success is a nontrivial event. FT-00 solves a fully transient model (via a complex complementary error function formulation with flowline storage and dip angle) while FT-01 solves an analytically derived polynomial equation valid only at steady-state. Agreement and consistency between the two approaches ensures correct mathematics, physics and software logic. This large time validation case thus provides a demanding test of both models.

In field applications, one might use unsteady data that is not consistent with the math model due to cost considerations and risks of tool sticking. Using 10-s pressure data in this case gives $k_h = 12.49$ md and $k_v = 0.65$ md. Twenty-second data yields 11.68 and 0.74, while fifty-second data leads to 11.02 and 0.83—all acceptable, relative to the 10 and 1 md assumed in FT-00. This accuracy is possible because the formation is relatively permeable. If in the screen of Fig. 18.8B we had checked "nonzero skin," a different less restrictive mathematical model with increased degrees of freedom is used. In the present calculation, the algorithm would return a list of possible solutions, that is, (k_h, k_v, S) triplets, together with the corresponding spherical permeability k_s listed at the far right of the table below (the zero skin solution obtained above is highlighted in red).

kh (md)	kv (md)	S	ks (md)
7.00	8.00	0.62	7.32
7.00	9.00	0.63	7.61
7.00	10.00	0.64	7.88
8.00	5.00	0.52	6.84
8.00	6.00	0.56	7.27
8.00	7.00	0.59	7.65
9.00	3.00	0.36	6.24
9.00	4.00	0.44	6.87
10.00	1.00	0.01	4.64
10.00	2.00	0.21	5.85

Effects of Dip Angle

The effects of dip angle are well known physically. For example, at zero dip, the tester "sees" k_h from all directions, while at 90 degrees, it "sees" k_h from left and right, but k_v from top and bottom (actually, a complicated function of both applies at each azimuthal angle). Since $k_h > k_v$, the measured pressure drop in a vertical well is less than that for one that is horizontal or deviated. As an example, first consider the forward simulation in Fig. 18.9A with zero dip angle and a pump rate fixed at 10 cc/s for all time.

We vary dip from 0 to 90 degrees with other parameters unchanged. Source pressure responses for all runs are identical since they depend on spherical permeability k_s only, as shown in Fig. 18.9B. But, as expected, transient observation probe responses (numbered by dip angle) vary in both magnitude and shape as seen from Fig. 18.9C. For the $k_h = 10$ md, $k_v = 1$ md example here, pressure drops vary over a 200 psi range as dip angles increase. Again, these results are exact.

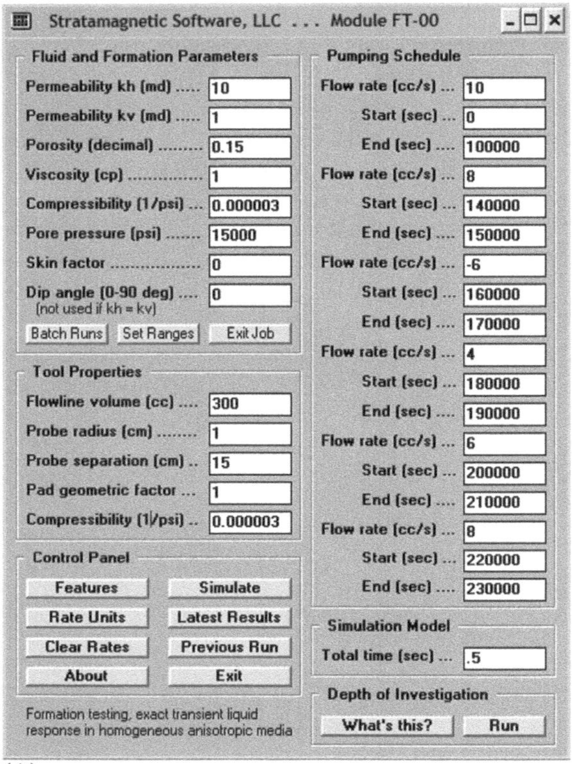

(A)

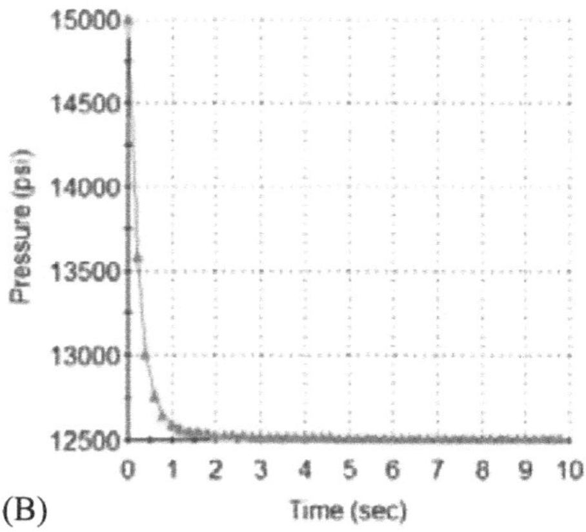

(B)

FIG. 18.9 (A) Constant rate pumping example. (B) Source probe response.

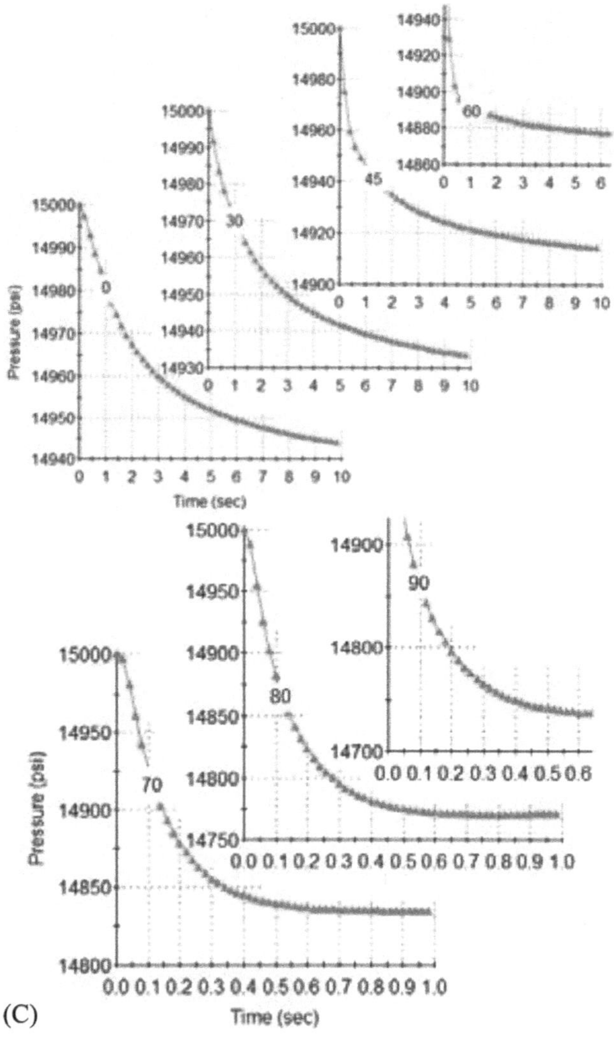

(C)

FIG. 18.9, CONT'D (C) Observation probe response versus dip angle.

We can view the foregoing conclusions from an inverse perspective. Suppose a deviated well were drilled at 45 degrees dip. The 45 degrees dip, zero skin forward simulation gives large time (100,000 s) pressure drops of 2512 psi at the source and 104.6 psi at the distant probe; the corresponding inverse calculation gives a consistently accurate $k_h = 10.02$ md and $k_v = 1.00$ md (in agreement with permeabilities assumed in the forward

FT-00 simulation). But what if, for this measured pressure pair, the exact FT-01 inverse solver were not available? If the conventional industry-standard formula (implicitly assuming zero dip) were used, one would instead calculate $k_h = 7.43$ md and $k_v = 1.81$ md with a k_v/k_h of 0.244 versus an exact value of 0.1. Such large errors imply grave production and economic planning consequences. In the table below, inverse calculations for permeability using FT-01 are performed with the above pressures through a range of dip angles to show the significance of hole deviation.

Dip	kh	kv	kv/kh
0	7.43	1.81	0.244
30	8.35	1.44	0.172
45	10.02	1.00	0.100
60	14.13	0.50	0.035
70	20.88	0.23	0.011
80	41.89	0.06	0.001
90	424.89	0.00	0.000

Inverse Pulse Interaction Approach Using FT-00

The above inverse approach requires fully equilibrated steady pressure drop data at source and observation probes. In a high permeability environment, this is not severe; as seen earlier, 20 s data may well suffice under certain conditions if data integrity is not an issue. In "tight" formations, however, steady observation probe responses may not be possible for hours or days. Even if rig costs were not a concern, the risks of tool sticking are—thus, one seeks permeability prediction methods that respond to earlier time dynamic data.

The use of steady formulas for permeability interpretation is an artificial limitation used only to render the mathematics tractable. As noted, steady conditions are usually achievable in higher permeability formations so that such models are sometimes useful.

In low permeability zones, field experience and exact calculations (using FT-00) show that source probe responses equilibrate very rapidly. Since they depend only on the spherical permeability k_s (and not k_h or k_v individually), the value of k_s inferred from the source probe pressure drop is an accurate one for interpretation purposes. The conventional steady flow permeability formula $k_s = Q\mu / \{4\pi R_w(P_0 - P_s)\}$ can be used which, again, only constrains the relationship between k_h and k_v (here, Q is VFR, μ is viscosity, R_w is effective probe radius, and "$P_0 - P_s$" is the source probe pressure drop). In order to quantify k_h and k_v, additional information is required. Unlike the method underlying FT-01, we shall not draw upon steady-state pressure data obtained at the observation probe.

In the illustrative example below, a k_s of 4.642 md (corresponding to our earlier $k_h = 10$ md, $k_v = 1$ md and $S = 0$ case) is fixed throughout and simulations are performed with different combinations of k_h and k_v. Again, source probe results for the three runs are identical, but observation probe pressure transients are discernible from each other. Fig. **18.10**A and B are clearly different—the former is highly smeared while the pulses in the latter remain distinct; peak pressure drops (from printed FT-00 output not shown) are 19 psi for Fig. **18.10**A and 159 psi for Fig. **18.10**B. Again, these differences are seen from early time transient behavior.

Observation probe pressure transient waveform shapes in Fig. **18.10**B and C are similar, at least on a normalized basis. However, they are very different in magnitude. From printed FT-00 output, peak pressure drops are 159 and 787 psi, respectively. Clear differences in observation probe characteristics suggest that permeability contrasts can be effectively examined using short duration pulse interference rather than long time steady-state drawdown. The dynamical interactions are strongly dependent on anisotropy.

Under what circumstances is our "pulse interaction method" expected to perform well? Interestingly, the lower the permeability, the better the accuracy—a counter-intuitive situation at odds with experience from steady state methods. The explanation is simple: at low mobilities, diffusion predominates, so that dynamical interactions between short pulses with high frequency content are strongest. Thus high signal-to-noise ratios are achieved for history matching.

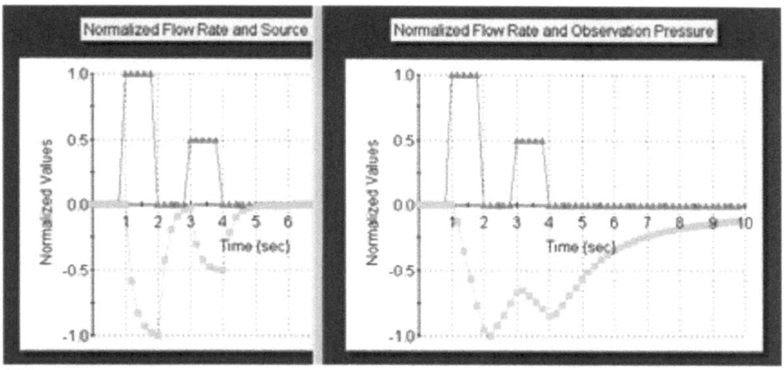

(A)

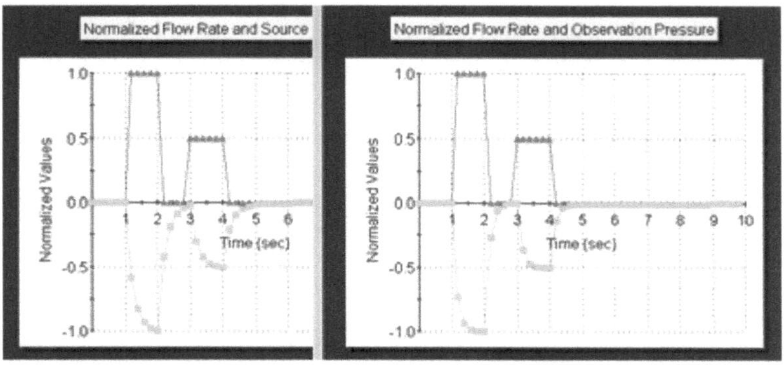

(B)

FIG. 18.10 (A) $k_h = 10$ md, $k_v = 1$ md (that is, $k_h > k_v$). (B) $k_h = k_v = 4.642$ md (that is, isotropic).

(Continued)

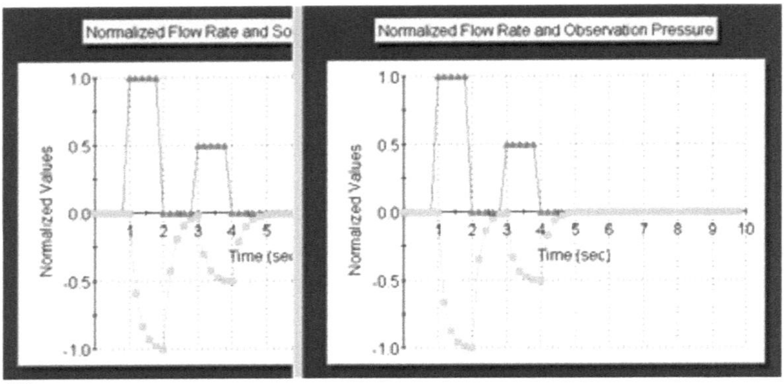

(C)

FIG. 18.10, CONT'D (C) $k_h = 1$ md, $k_v = 100.0$ md (that is, $k_h < k_v$).

Interference effects are most pronounced at low perms when diffusion is dominant, precisely the field condition associated with long wait times, high rig costs and increased risk of tool sticking. In contrast, large time pressure responses associated with constant rate drawdown methods only produce small pressure drops which may not be accurately measured. We emphasize that, at earlier times, the effects of porosity, compressibility and flowline volume do appear, so that calculations using different pulse types with varied durations, amplitudes and time separations is advisable.

Additional research is presently directed at optimizing the pulse sequence used, e.g., evaluating different combinations of pulse amplitude, width, separation and number. It is important to emphasize that no new hardware is required for pulse interaction analysis. Our FT-00 may be used in infinite homogeneous media, but in applications where bedding plane effects are important, the FT-03 simulator discussed later applies.

Because we are evaluating flow differences associated with diffusion itself, it is important that the host math model does not introduce numerical diffusion effects related to truncation errors. These effects, referred to as "artificial viscosity," are prominent with finite difference and finite element simulators, even when second, up to fourth order, schemes are employed. Thus, exact analytical models such as FT-00 should be used to interpret pulse interactions.

Computational Notes

We observe that short time data contain porosity, compressibility and flowline effects. The exact solution in FT-00 is expressed in terms of the complex complementary error function erfc(z) where z=x+iy is a complex number (x and y are related to fluid, formation and tool parameters and here do not represent the spatial coordinates).

In typical function evaluations, e.g., those offered in scientific program libraries, solutions are only possible for a small range of small imaginary values for small real values of the argument near zero, which unfortunately limits the usefulness of our exact solutions (for large real arguments near forty, solutions for all imaginary values are possible). This limitation is most serious for tool applications with flowline volumes such as those typically encountered in logging.

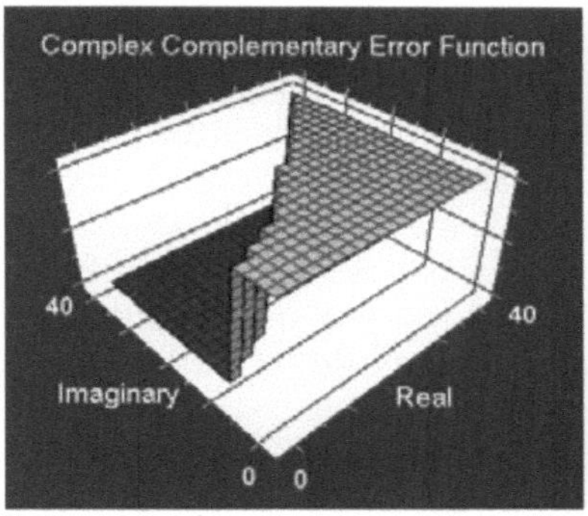

(A)

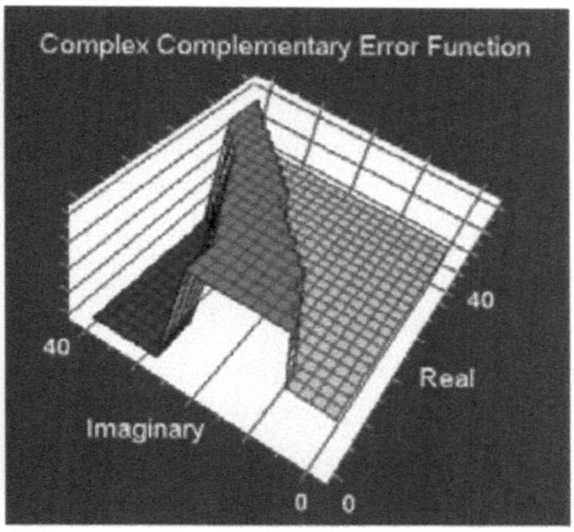

(B)

FIG. 18.11 (A) Standard erfc(z) evaluation for x>0, y>0. (B) Improved erfc function evaluation.

Fig. 18.11A shows Quadrant 1 z ranges where x>0, y>0 for which erfc(x+*i*y) can and cannot be computed by standard (IMSL) algorithms. The lower left (blue) zone shows where solutions cannot be computed due to numerical overflow, while the elevated right (green) zone highlights where solutions can be found. The blue zone contains parameters where flowline volume effects are important. Improvements are now given.

For our approach to be useful, more robust numerical evaluations are now possible to enable a much broader range of arguments for successful calculation, as shown in Fig. 18.11B. The far left (blue) zone still represents solutions that cannot be computed; fortunately, the corresponding arguments do not represent parameters often used in practical formation testing. The far right (green) domain indicates solutions possible using conventional approaches, while the new and sizable middle (red) zone depicts an increased range of newly available computations for erfc(z). Accuracy and speed are emphasized in the method, e.g., hundreds of evaluations per second on typical personal computers. In this discussion, we focused on a zero skin example; space limitations preclude similar discussions on nonzero skin results. However, the basic ideas and results for pulse interaction methods remain unchanged.

Source Model Limitations

The word "exact" often used in the literature (including the author's publications) conveys a sense of confidence and accuracy that may not be deserved. With respect to the present chapter, "exact" refers to closed form analytical solutions for forward and inverse models—a significant endeavor—but which applies only to approximate source formulations of the tester. While the source model used, which applies boundary conditions on an ellipsoidal surface (with nonzero minor and major radii, and which does not "blow up" as in point source models) is useful in this regard, the model nevertheless is approximate in a physical sense.

In other words, the model "sees" a type of spherical symmetry when, in fact, real formation tester nozzles mounted on real mandrels "see" largely ahead and are shielded from behind by solid material. Thus, a real tool in an anisotropic formation will measure different pressure responses as the transducer is rotated azimuthally about the tool axis. A full three-dimensional model accounting for actual tool geometry plus bedding plane effects could be used as a history matching tool to determine k_h and k_v from pressure data collected at different angles. Such data is possible in field practice. Since the drillstring torques and un-torques during operation, it winds and unwinds, offering the opportunity to record pressures at different azimuths.

Whereas earlier examples emphasized the use of dual probe tools, the above ideas apply to single-probe tools such as those in simple wireline and FTWD applications. The basic idea is simple: two pressure measurements are need to provide two permeabilities. These can be taken from axially or azimuthally displaced probes, or both, as in three-probe tools. This leads to the question, "How can we construct a simulator capable of addressing these inverse and job planning needs?"

Full Three-Dimensional Model

Module FT-03 solves the transient Darcy partial differential equation for transversely isotropic media on a boundary-conforming curvilinear mesh in the cross-plane intersecting the tool. High resolution is provided about the borehole using "glove tight, wrap-around" inner meshes while those in the farfield conform to bed boundaries. An algebraically expanding mesh with an origin centered at the pad nozzle or oval pad, or at the center of a packer, provides good resolution axially. The computational domain appears in Fig. 18.12A. The transient equation is solved by finite difference, forward time marching using an Alternating-Direction-Implicit (ADI) scheme for fast speed and numerical stability.

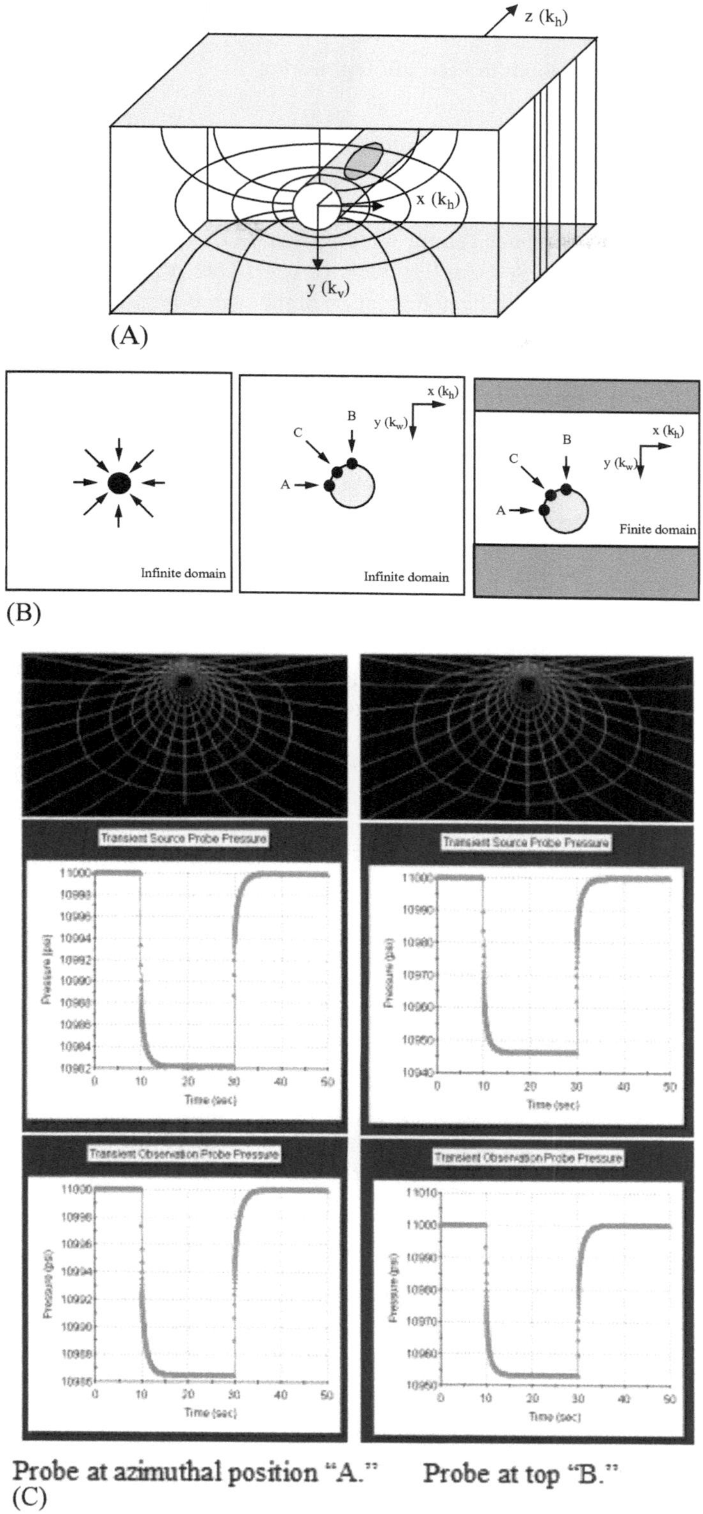

FIG. 18.12 (A) Three-dimensional computational mesh. (B) Borehole orientation. (C) Azimuthal pressure response in layered media.

We focus on key simulation results and consider the diagrams in Fig. 18.12B. The left two show a horizontally oriented borehole in a homogeneous transversely isotropic formation, while the right one shows the same hole bounded by upper and lower planes (along which pressure or no-flow conditions may be prescribed). Again, A "sees" essentially k_h, B sees primarily k_v, while the general point C sees a complicated function of the two. If pressure measurements at any two of A, B or C are available, one can determine both permeabilities by history-matching using a three-dimensional simulator such as FT-03.

For our purposes, we refer to Fig. 18.12C for a low perm run with typical field parameters. The source probe Δp at A is 18 psi (because the k_h it "sees" is high) while that at B is 54 psi (the k_v it sees is low). This sizable difference is detected at the source probe and can be used to estimate anisotropy. We emphasize that a single-probe tool can be used to determine k_h and k_v provided azimuthal measurements are taken. Fig. 18.12C shows observation probe results, assuming that source and observation probes are in line axially (this is not required in general simulations). The effects of bedding plane proximity and centralization are subtle and general conclusions cannot be made. However, the simulator accounts for these as they affect azimuthal transducer placement.

ADVANCED MODEL SUMMARIES

Phase Delay Analysis, Module FT-04

The pulse interaction method, most effective at low perms, is closely related to our dynamical model for "phase delay" analysis. In electromagnetic logging, wave amplitude differences and phase delays between transmitters and receivers are used to infer resistivity anisotropy using Maxwell's equations. A simple formation testing analogy exists. If the pump piston is operated periodically and pressure amplitude differences and time delays are recorded between source and observation probes, then k_h and k_v can be obtained from history matching using a 3D Darcy simulator. Permeability formulas as functions of amplitude and phase delay are given in Chin et al. (2014b, 2015) for simple formations. In general, the lower the permeability, the higher the diffusion and the lengthier the phase delay. Larger delays mean more accurate time measurement and hence better predictions.

The model has since been extended to layered anisotropic media for dipping tool applications, as shown in Fig. 18.13A. This complication means that simple formulas are not available and the interpretation must be pursued numerically.

Fig. 18.13B shows typical transient pressure responses calculated at source (red) and observation (green) probes using the FT-04 simulator of Fig. 18.13A. The large amplitude decreases are typical of low permeability rocks, but small pressure levels at the observation probe may be difficult to measure accurately. On the other hand, time delays are easier and more precise. Phase delays can be calculated using any number of models, e.g., our idealized source model FT-00, the "real tool" simulator FT-03, or the "source in layered media" model in FT-04.

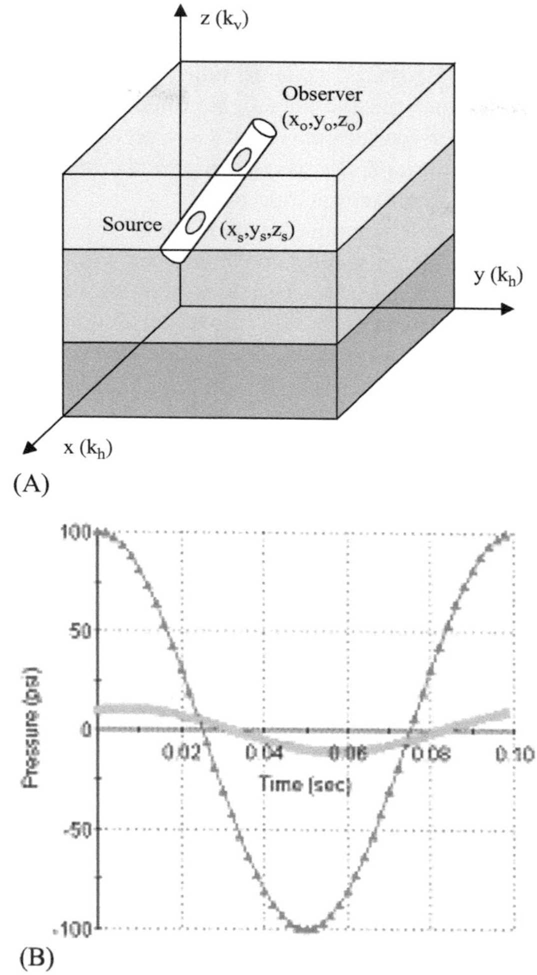

FIG. 18.13 (A) Layered anisotropic media with dipping tool. (B) Pressure response.

Drawdown-Buildup, Module FT-PTA-DDBU

In deep offshore wells where pressure limits separating borehole invasion and fracture are narrow, accurate real-time pore pressure and mobility prediction are essential to safety. Exact models like FT-00 provide forward analyses, calculating transient pressures when fluid, formation and tool properties are given.

FTWD or "pressure-while-drilling" (PWD) requires fast inverse methods that provide pore pressure for safety objectives and gradient analysis for fluid identification, plus mobility analysis for economic and production planning.

These predictions are needed in real-time during drilling. Equations that support FTWD objectives can be derived from the exact, closed form, analytical solution underlying FT-00. Because "while drilling" data are typically early time (due to cost and stuck tool considerations), are often obtained in low mobility applications, or both, derived formulas used must account for highly transient behavior and flowline storage effects, which can distort and mask mobility trends residing in drawdown and buildup data.

Existing methods use exponential, real or complex complementary error functions, complicated integrals, and so on, combined with regression methods for interpretation. For instance, "least squares" fits are often employed, and while reasonable, they introduce arbitrary smoothing assumptions beyond those in Darcy's diffusion laws. It is essential that models invoke no additional flow approximations.

Our inverse solution is derived using rational polynomial expansions that reduce the computational overhead associated with transcendental functions. This allows more data processing to be completed in a given time frame, thus freeing valuable microprocessor resources for other important interpretation and downhole control functions. Closed form analytical solutions are derived and used, and numerical regression and Chi-square methods are never employed.

Two key models have been developed for "drawdown only" and "drawdown-buildup" applications. Both methods apply to transient pressure data distorted by flowline storage effects, and subset models (which run substantially faster for mobility prediction) are also available when steady-state pressures can be found for higher mobility or long test time applications.

Unlike conventional models, ours only require three pressure-time data points along either drawdown or buildup curve, e.g., as shown in Fig. **18.14**A, plus auxiliary data related to test setup. For brevity only, we present examples using buildup data, noting that detailed write-ups are available upon request.

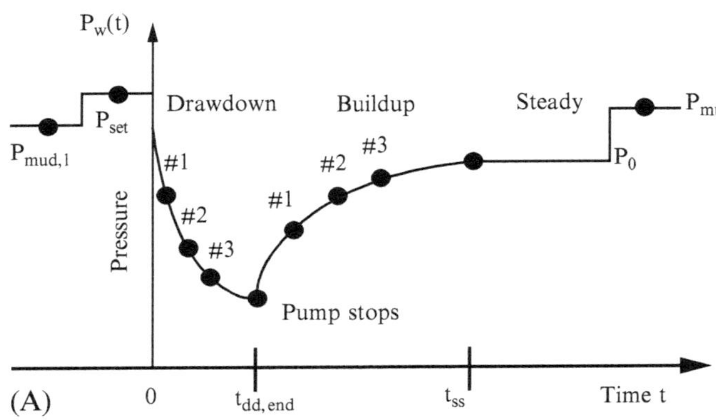

FIG. 18.14 (A) Nomenclature for pressure transient analysis.

In Fig. 18.14B, we use our exact FT-00 forward solver to create the source probe pressure transient data shown. From the input screen, the pore pressure is 10,000 psi while the mobility is 0.1 md/cp.

Fig. 18.14C displays inputs for FT-PTA-DDBU with pressures taken at 10, 15, and 20 s. Rapid calculation gives 9951 psi and 0.11 md/cp, close to exact values. Sensitivity analyses using other time data points yield only small variations. In Fig. 18.14D, we create pressure data for two pretests having different flow rates.

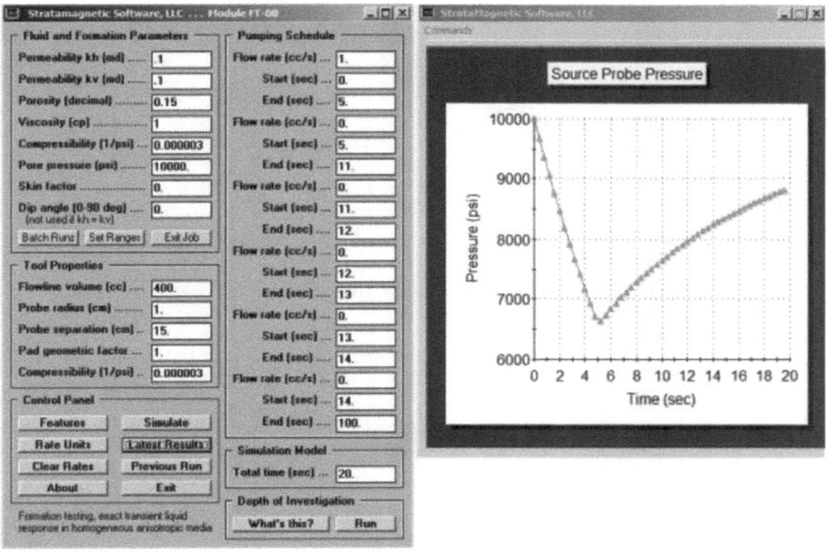

(B)

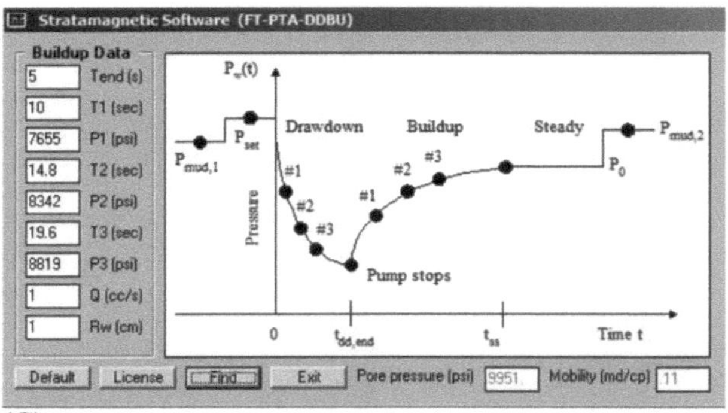

(C)

FIG. 18.14, CONT'D (B) Exact FT-00 forward simulation from single pretest (note large flow-line volume assumed). (C) Predicted pore pressure and mobility.

(Continued)

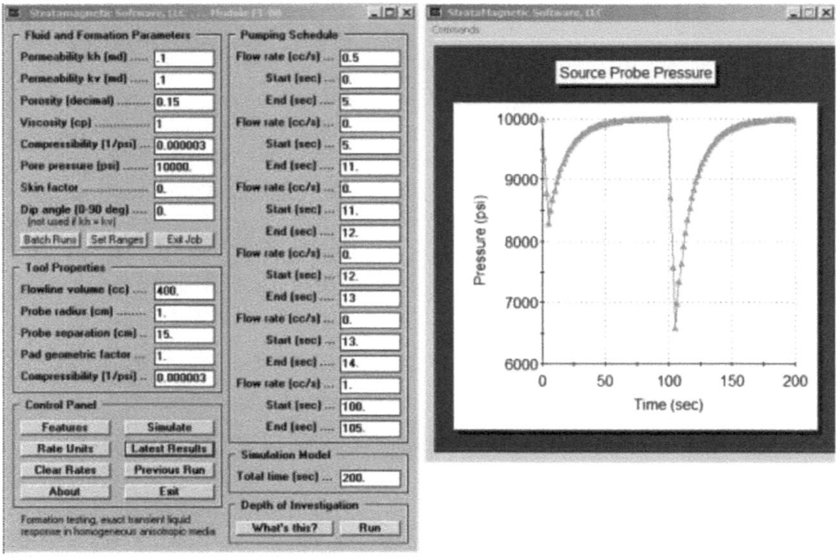

(D)

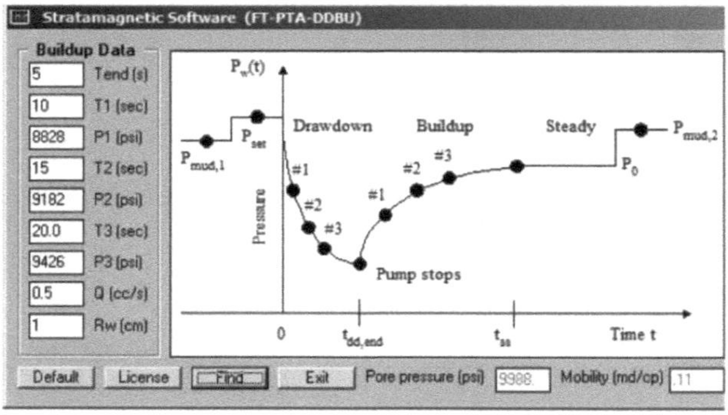

(E)

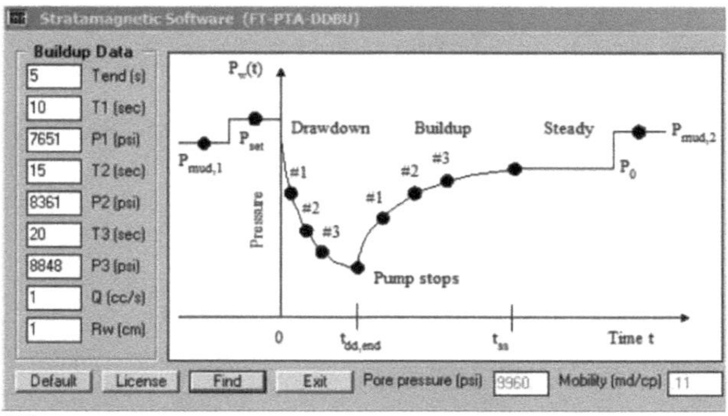

(F)

FIG. 18.14, CONT'D (D) Exact FT-00 forward simulation for two sequential pretests. (E) Predictions (first pretest). (F) Predictions (second pretest).

Fig. 18.14E and F offer predictions of 9988 psi and 0.11 md/cp, and 9960 and 0.11 md/cp, using first and second pretest data, values very close to assumed numbers. Again, the solutions execute rapidly, are analytical and easily programmed, e.g., 10–15 lines of source code depending on the host compiler language. Also, because no iterations are involved, downhole computer microprocessor resources are utilized efficiently and available for other interpretation and control functions. Algorithms have been completed and validated for "drawdown only" as well as buildup test cycles.

Real Pumping, Module FT-06

In real-world applications, flow-rate versus time functions are never constant or piece-wise constant. Flow rates invariably ramp-up and down, with such effects being more dramatic with gas pumping. These features must be modeled numerically and analytical methods are not possible. Fig. 18.15A shows the interface for a liquid-gas simulator which supports flow-rate functions that may be, for instance, trapezoidal as in Fig. 18.15B unlike the piecewise rectangular functions of Fig. 18.7B. Source and observation probe pressures are also given.

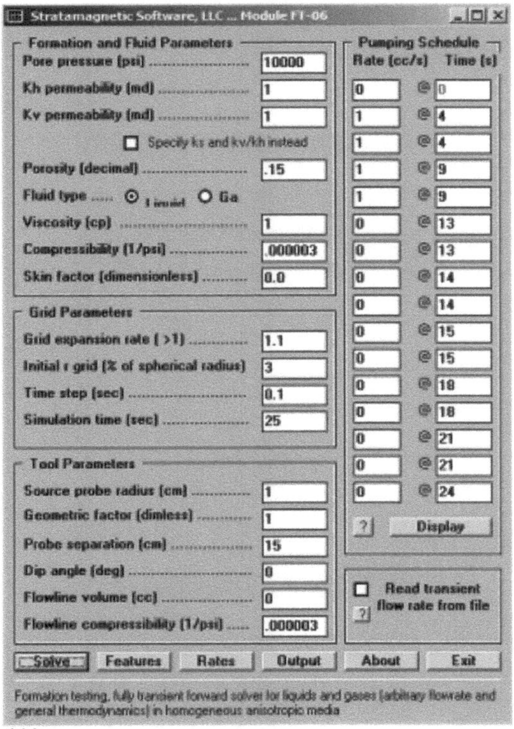

(A)

FIG. 18.15 (A) FT-06 liquid-gas simulator inputs.

(Continued)

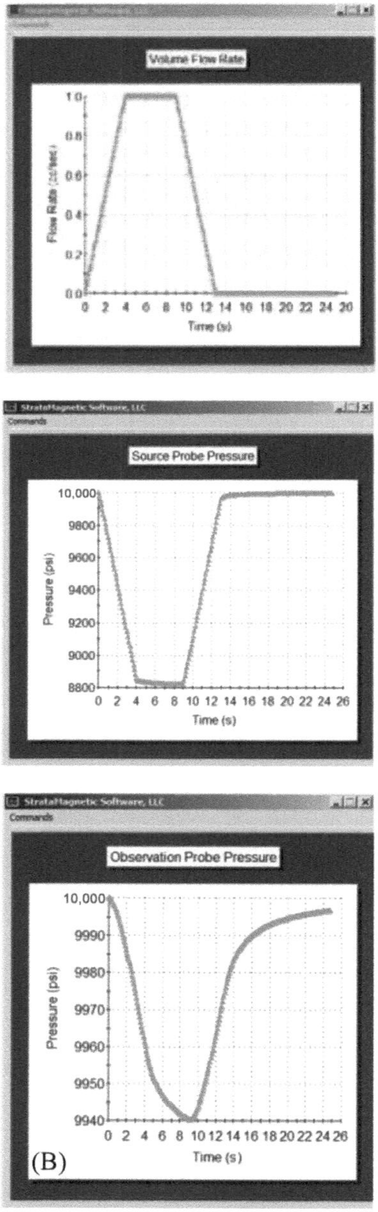

FIG. 18.15, CONT'D (B) FT-06 pump rate and pressure solutions.

Deviated Well Interpretation for Both k_h and k_v

Expressions for phase in deviated wells can be shown to be considerably simpler than those for amplitude. In fact, if the phase θ_L at the observation probe distance L from the center of the source probe is known, then one can demonstrate that

$$\theta_L \approx L\sqrt{[\{\phi\mu c\omega/2\}\{\sin^2\delta/k_h + \cos^2\delta/k_v\}]} \tag{18.46a}$$

thus providing, for the dip angle δ assumed, one equation of two needed to find both k_h and k_v. There are two approaches that can be taken if both horizontal and vertical permeabilities are desired, emphasizing that the two apply to low mobility applications with data taken at early times.

In the first, suppose we have a thick uniform layer where the formation tester resides in a drillstring that is turning or changing direction. Then, Eq. (18.46a) can be evaluated at two different dip angles δ_a and δ_b so that we have two linearly independent equations

$$\theta_{L,a} \approx L\sqrt{[\{\phi\mu c\omega/2\}\{\sin^2\delta_a/k_h + \cos^2\delta_a/k_v\}]} \tag{18.46b}$$

$$\theta_{L,b} \approx L\sqrt{[\{\phi\mu c\omega/2\}\{\sin^2\delta_b/k_h + \cos^2\delta_b/k_v\}]} \tag{18.46c}$$

Then, Eqs. (18.46b), (18.46c) provide two coupled equations in the two unknown permeabilities k_h and k_v which can be directly solved. The result, from simple algebra, is

$$k_h = \left(L^2\phi\mu c\omega/2\right)\left(\sin^2\delta_a\cos^2\delta_b - \sin^2\delta_b\cos^2\delta_a\right)/\left(\theta_a^2\cos^2\delta_b - \theta_b^2\cos^2\delta_a\right) \tag{18.46d}$$

$$k_v = \left(L^2\phi\mu c\omega/2\right)\left(\sin^2\delta_a\cos^2\delta_b - \sin^2\delta_b\cos^2\delta_a\right)/\left(\theta_b^2\sin^2\delta_a - \theta_a^2\sin^2\delta_b\right) \tag{18.46e}$$

We emphasize that Eqs. (18.46d), (18.46e) represent two equations obtained at different dip angles and that the layer is assumed to be thick enough to allow measurements to be taken as the drillstring changes direction. As a check, setting $\delta_a = 0$ degree and $\delta_b = 90$ degrees, and conversely, lead to the simpler formulas derived earlier.

Of course, in thin geological layers, it is not physically possible from a drilling perspective to have a second equation at the same logging point as in Eqs. (18.46b), (18.46c) for different dip angles—thus, having "two equations in two unknowns" as indicated above cannot be done. However, we can still obtain vertical and horizontal permeabilities as follows. Earlier in this chapter, we introduced a single-probe, drawdown-buildup method for low mobility applications which used early-time pressure transient data—we shall refer to this as our "DDBU" method. We use this together with the phase delay approach.

In this second approach, we return to Eq. (18.46a), which provides a relationship connecting k_h and k_v for measured values of the phase θ_L and dip angle δ. To determine horizontal and vertical permeabilities uniquely, a second constraint is needed. This is provided by DDBU models in which the spherical permeability k_s is predicted from early time pressure transient data, so that we know $k_s = k_h^{2/3} k_v^{1/3}$. Combining this with Eq. (18.46a) leads to relationships for the horizontal and vertical permeabilities k_h and k_v, namely,

$$\cos^2\delta\, k_h^3 - \left\{2\theta_L^2 k_s^3 / \left(L^2\phi\mu c\omega\right)\right\} k_h + \sin^2\delta\, k_s^3 = 0 \tag{18.46f}$$

$$\tan^2\delta\, k_v^{3/2} - \left\{2\theta_L^2 k_s^{3/2} / \left(L^2\phi\mu c\omega \cos^2\delta\right)\right\} k_v + k_s^{3/2} = 0 \tag{18.46g}$$

$$\tan^2\delta\, \lambda - \left\{2\theta_L^2 k_s / \left(L^2\phi\mu c\omega\ \cos^2\delta\right)\right\} \lambda^{2/3} + 1 = 0 \tag{18.46h}$$

where

$$\lambda = k_v / k_h \tag{18.46i}$$

represents the so-called "anisotropy" or anisotropy ratio. The foregoing are cubic polynomial equations for k_h, $k_v^{1/2}$ and $\lambda^{1/3}$ which can be solved exactly from standard algebraic formulas. Different equations apply in different situations. In some situations, it may be preferable to work first with k_h if it is typically larger, as it would provide greater numerical accuracy. Once Eq. (18.46f) is solved, the vertical permeability can be obtained from $k_s = k_h^{2/3} k_v^{1/3}$ or Eq. (18.46a). Similar comments apply to Eqs. (18.46g), (18.46h). We importantly emphasize that in obtaining k_s using our drawdown-buildup methods, we also predict the pore pressure from early time data.

Also note that deviated well applications are associated with multiple permeability solutions for the same set of input parameters. Cubic equations provide two types of solutions, namely, (i) all real roots (some of which may be negative), and (ii) complex roots (with one real and one conjugate pair). Input data contaminated by noise, e.g., loose pads, sand in nozzle throats, bubbly solutions, and so on, may lead to false imaginary parts, and calculated solutions must be interpreted with care. We emphasize that this method requires a dual-probe tool because phase delay measurements are required.

In contrast to the dual-probe, steady-state pressure drop method described in Chapter 6 of the formation testing book of Chin et al. (2014b), which in practice requires higher mobilities, the present method is ideally suited to low mobilities (steady-states in low permeability formations may require hours of wait time for equilibration). DDBU was designed for low mobility applications. And in the present phase delay approach, the lower the permeability, the larger the time delay and easier the measurement is made. Thus, the lower the mobility, the better the signal-to-noise ratio, and the more accurate the predictions for both permeabilities.

MULTIPHASE FLOW CLEANING SIMULATOR

For this last section, we introduce multiphase flow analysis and present computed displays of pressure and oil saturation (or formation fluid concentration) fields offered periodically in time. Consider Fig. 18.16A, which contains two field plots, with pressure at the left and concentration at the right for a miscible run (for immiscible runs, concentration plots are replaced by oil saturations). For each of the diagrams in Fig. 18.16A, the left vertical side corresponds to the sandface at the borehole wall—the right side corresponds to the radial far-field. The top and bottom horizontal lines coincide with the top and bottom of the reservoir. Thus, these cross-sections display calculated solutions in the r-z plane for the axisymmetric formulation considered in Chin et al. (2015).

The left pressure plot is uniform vertically, indicating identical pressure profiles at all z stations. The red at the left represents high mud pressure, relative to the lower blue formation pressure at the right. The right concentration plot again indicates a purely radial flow without z variations. Invading blue mud is displacing red formation fluid. The multicolored zones between blue and red in either case represent events in the diffusive mixing zone. Cylindrical radial invasion occurs while drilling. Sometimes the invasion time is short—at other times, it can exceed a day. For long invasion times, it is not necessary to simulate extraordinarily long—we equivalently model the invasion associated with a higher permeability mud for a shorter time. Equivalence formulas are given in Chin (1995, 2002). The relatively short times in Fig. 18.16A and B mimic 24-h invasion. Note how the effects of mud pressure and filtrate invasion are deeper at "1 min" than at "0.33 s," as expected physically.

At some point in time designated by the user, the formation tester starts pumping—it can extract fluid from or inject fluid into the reservoir according to a multi-rate schedule. The left pressure plot in Fig. 18.16C shows the effects of fluid withdrawal, that is, typical fluid sampling performed, in this case, by a single centered nozzle (in general, single, dual and straddle packer probes are permitted, which can be located arbitrarily along the sandface). The left side of the pressure plot in Fig. 18.16C shows a blue-green area associated with the low pressure at the nozzle. Above and below this zone are red colored pressures which indicate higher pressures associated with supercharging—that is, as the nozzle withdraws fluid, high pressure mud invades the formation through the mudcake. Not shown in the pressure plot are reverse flow streamlines that would mark filtrate pumping by the formation tester nozzle.

The right plot in Fig. 18.16C displays the corresponding concentration profile. The blue zone represents the mud filtrate that has penetrated the formation—it is now deeper than that shown in Fig. 18.16B. Fig. 18.16D and E illustrate similar phenomena at later times. Again, note the high supercharge pressures above and below the nozzle, indicating continuing filtrate invasion while the tester nozzle attempts to extract a clean sample. Whether or not this is possible for the input parameters assumed is one question the

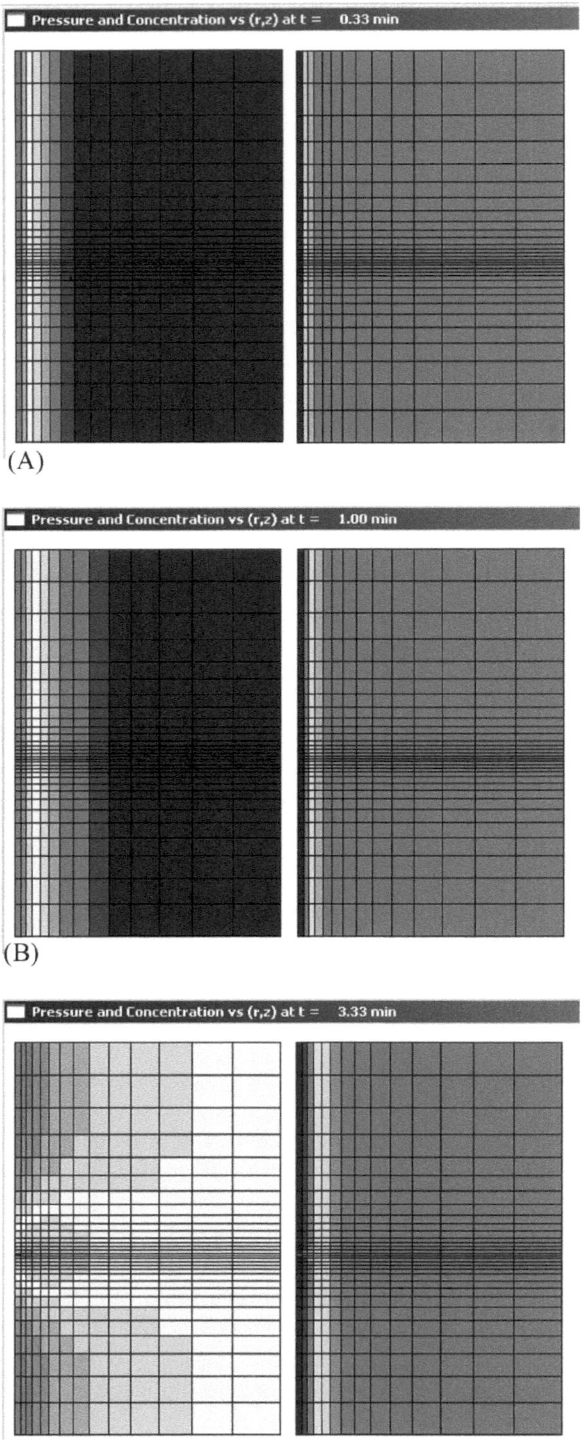

FIG. 18.16 (A) Pressure-concentration profiles, 0.33 s. (B) Pressure-concentration profiles, 1.00 min. (C) Pressure-concentration profiles, 3.33 min.

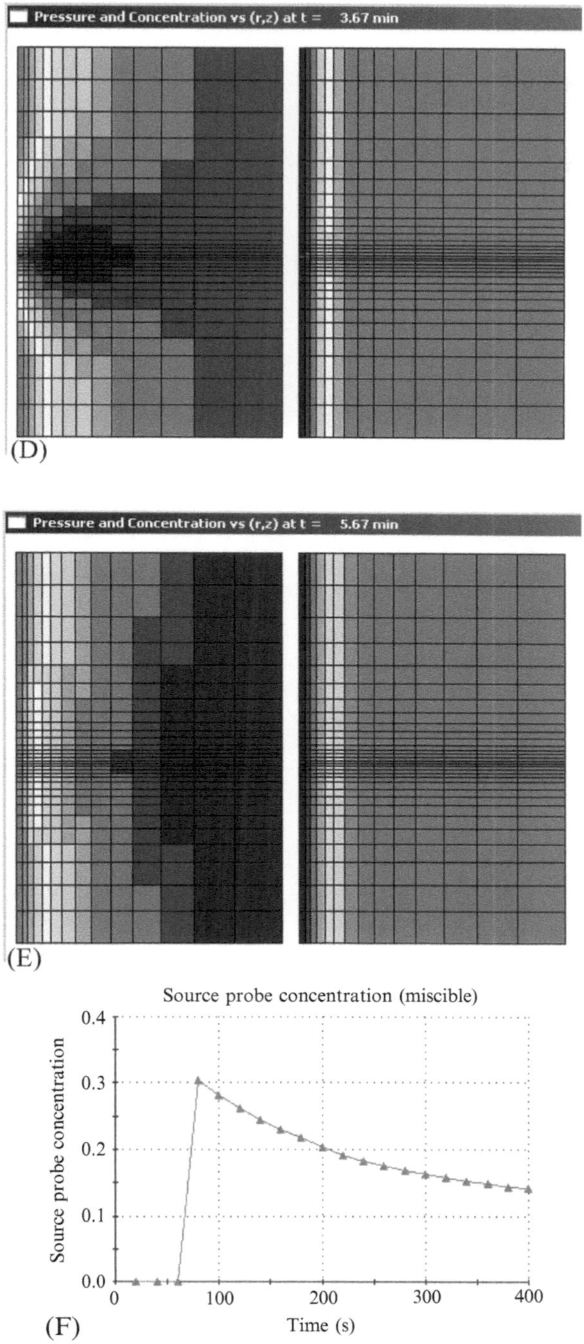

FIG. 18.16, CONT'D (D) Pressure-concentration profiles, 3.67 min. (E) Pressure-concentration profiles, 5.67 min. (F) Formation fluid concentration at source probe.

(Continued)

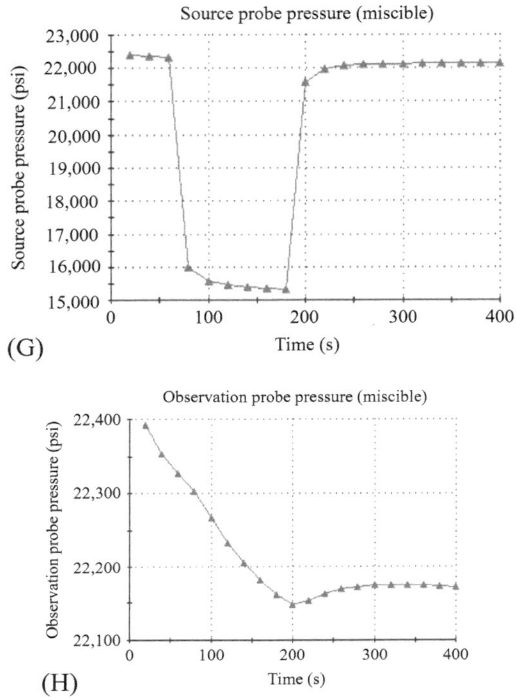

FIG. 18.16, CONT'D (G) Source probe pressure transient history. (H) Observation probe pressure transient history.

simulation addresses. There are several related objectives. Is a clean sample possible? If so, how long must the formation be pumped? If not, how might mud properties and weight to changed? The time scale for adequate clean-up is different from that for pressure transient interpretation. How long must be tool stay in place to ensure good pressure data for permeability and anisotropy prediction—without risking a stuck tool? Good pressure data for permeability prediction, of course, can be obtained even when mud filtrate has not been flushed. Thus, for tools that do not collect samples (e.g., formation-testing-while-drilling tools), the job planning simulator can be used to study pressure transients—while, for wireline formation testers, the simulator serves dual clean-up and pressure transient objectives.

The left plot of Fig. 18.16C indicates probe presence because its low (blue-green) pressures contrast strongly with the high (red) ones due to supercharging. At the right, a small green zone associated with the probe is embedded in the blue filtrate—this small zone is not red because the fluid is still contaminated. The variable meshes used allow high resolution simulations near the probe.

At the end of the simulation—a point in time defined by the user, the pressure-concentration screens shown above may be played back in the software's "movie mode." This playback feature enhances the field engineer's physical intuition about the formation under consideration. In addition, three line graphs automatically appear that conveniently summarize computed results at source and observation probes. Fig. 18.16F displays the concentration of formation fluid at the source probe. For the input parameters assumed in the present simulation, the results are not encouraging (we shall show more optimistic results later). At first, the concentration is zero because the nozzle is not pumping. Once pumping commences, the formation fluid concentration increases to a maximum of 0.3 or 30%—not quite the 90–95% that is deemed adequate by petrophysicists.

There may be several reasons for low dilute levels. Fig. 18.16A–E shows high levels of supercharge pressure while the nozzle is pumping—our nozzle may, in fact, be pumping filtrate just entering the formation. But even if not, it is possible that rapid diffusion between filtrate and formation fluid—encouraged by the particular pore structure in the rock—mixes the two quickly in a detrimental manner. One purpose of the job planning simulator is to identify the reasons for sample contamination and to recommend fixes.

Fig. 18.16G displays the pressure drawdown and buildup calculated for the input parameters assumed. Pressure transient predictions are always checked against exact analytical results to ensure that spatial and time grid parameters yield accurate results. In our case, the complex complementary error function solution of Proett and Chin (2000) was used. This solution assumes a pure spherical (ellipsoidal) source model, so that fluid invasion at the borehole wall is not modeled. Assumed meshes were therefore calibrated against exact results for a perfectly sealed borehole.

That our job planning simulator models both fluid mixing and pressure transient analysis (due to fluid compressibilities) is desirable because it is convenient—but more important, both objectives are accomplished using the same software, thus reducing the uncertainties associated with using multiple simulators. In addition, because the same equations as those in reservoir engineering are used here, the same software can be used for reservoir engineering production predictions by sealing the borehole, expanding the flow domain to field scale, and lowering the borehole pressure to those characteristic of production scenarios.

Finally, as shown in Fig. 18.16H, the software also produces pressure transient outputs at any observation probe location defined by the user. We emphasize that it is not necessary to have an observation probe in the tool to use the software model—that is, the software can be used with single-nozzle tool FTWD applications to ascertain depth of investigation or penetration. Of course, in dual probe applications, the use of pressure drops available at source and observation probes enables prediction of horizontal and vertical permeability. For the present simulation, it is of interest to note the high level of diffusion in

Fig. 18.16H relative to that in Fig. 18.16G. Such indicators are useful in ascertaining the probability of success that might be achieved in so-called "mini-DST" applications. The observation probe can be chosen at any distance from the source probe, e.g., seven inches or ten feet.

Closing Remarks

We have introduced new ideas in formation tester pressure modeling and permeability interpretation. The forward and inverse methods apply to all service company tools, e.g., wireline or "while drilling," in single, dual-probe or packer operation modes. In addition, multiphase flow models important to pumping times needed for clean in situ fluid sampling have been developed. The methods described have been cross-checked and validated in different physical limits to ensure accuracy and consistency. Specific attention was focused on attractive user interface design (with automated color and line graphics) so that convenient field use with minimal training is possible. For further details, the reader is referred to the recent books of Chin et al. (2014b, 2015).

PROBLEMS AND EXERCISES

1. The skin model derived for anisotropic formations assumed the displacement of a formation liquid by a liquid filtrate. Extend the model so that a liquid displaces formation gases having exponent m.
2. Formation invasion during drilling is described by the cylindrical flow models in Chapter 17. Give practical estimates for total invaded fluid volume for typical wells at various stages in the drilling process. The formation tester model in this chapter assumes ellipsoidal flow. How would you use this solution to estimate pump-out time required to remove invaded mud filtrate prior to reservoir fluid sampling? Write a simple program, noting that the roots in Eqs. (18.43a)–(18.43c) are complex.
3. The write-up beneath Eq. (18.45) explained the need for $(\partial p^*/\partial r^*)_w$. Calculate it from Eq. (18.37) to provide the complete well solution.
4. The Eulerian solution in this chapter describes flow at a point. Using it, derive an ordinary differential equation for dr/dt describing the progress of an invasion front in compressible spherical flow. Integrate this result numerically in time for suitable initial conditions. Explain your results.
5. Different physical time scales are found in formation testing practice. For pressure transient testing, with respect to permeability prediction, how do steady-state methods differ from early-time drawdown-buildup approaches? And in multiphase flow clean-up analysis, where invaded mud filtrate is to be removed prior to in situ fluid collection, how do required time scales differ between miscible and immiscible flows?

Chapter 19

Analytical Methods for Time Lapse Well Logging Analysis

Here we continue development of formation invasion models, and present experimental results as reported in Chin et al. (1986) to support our simulation efforts. Importantly, we introduce practical uses for the derived formulas describing coupled mudcake growth and displacement front motion, and in particular, we develop the physical principles and reservoir flow bases underlying time lapse analysis. This work is pursued within the framework of the plug-like displacement models treated so far. Extensions of these inverse prediction methods for miscible and immiscible multiphase flow effects will be considered in Chapter 21. There, powerful methods are developed first to undiffuse fronts that have diffusely smeared and distorted geometrically, in order to recover the original sharp step profiles so that the inverse plug flow models in this chapter apply. Also, methods to unshock saturation fronts developed from immiscible water-oil flows, in order to recover the original smooth flows, are developed so that they can be analyzed accurately for formation information.

EXPERIMENTAL MODEL VALIDATION

"*Formation Evaluation Using Repeated MWD Logging Measurements*" by Chin et al. (1986) summarized a multiyear effort aimed at assessing the viability of time lapse analysis. The exact unsteady front equation given in that paper was simplified assuming high rock-to-cake permeability ratios for further evaluation, thus leaving rock porosity as the sole formation parameter. This being so, porosity could be solved for in terms of displacement front location, time, and mudcake properties. The objective of the work lay, in part, in determining the accuracy of the porosity thus obtained, in comparison to known core-measured values and values available from other types of porosity logs.

Static Filtration Test Procedure

In order to understand the static filtration process, Catscan measurements of flows in radial and lineal cores containing growing mudcakes were obtained over periods of hours, with pressure and mud weight systematically varied in sequential tests. Photographs of the plastic, translucent, radial, and lineal flow

Quantitative Methods in Reservoir Engineering. http://dx.doi.org/10.1016/B978-0-12-810518-4.00019-0

test fixtures appear in Chapter 14. Gel-like mud-to-cake boundaries, sometimes poorly defined due to weak density contrasts, a result of barite present in both mud and mudcake, were enhanced for visualization by adding lightening agents to the mud. The salt water in the Berea sandstone cores tested contrasted well with the fresh water filtrate used, so that special visualization methods were not necessary for the flow internal to the rock. A database of information for mudcake thickness, lineal and radial displacement front position versus time, for different mudcakes and differential pressures, was obtained over a period of months. Cake density and compaction were also measured as a function of distance from the rock and monitored in time (e.g., see Chapter 14). The mudcake model used in the original work and here, described in Collins (1961), is a well-known buildup model for cement slurries. Cake thicknesses obtained by Catscan measurement agreed with predictions, so that the model is also potentially useful in differential sticking.

Dynamic Filtration Testing

Dynamic filtration was studied using a closed-circuit recirculating flow loop. The test section consisted of a foot-long annular core of Berea sandstone, with a 2-in. inner diameter, through which mud flowed. Under differential pressures ranging from 50 to 150 psi, a portion of this recirculating flow is lost as mud filtrate, passing radially into the core and into a collection tank open to a pressure-regulated chamber. The dimensions of the one-foot annular test section were selected to reduce end effects, thus allowing pure radial flow at the center of the core. The test fixture permitted independent control of the differential pressure across the rock and cake, the absolute pressure in the loop, and the mean flow speed. This speed was monitored ultrasonically, while constant system temperatures were controlled by heat exchangers. Pressures were regulated by an accumulator, and bubbles introduced by the replenishing mud supply that replaced lost filtrate were removed by a mechanical separator. The test apparatus was also size-constrained to allow convenient Catscan recording while flowing. The preliminary reported results suggested that mudcake thickness ultimately remains constant with time under laminar flow conditions, showing that surface erosion due to shear stress *does* result in dynamic equilibrium. For turbulent flows, the cakes formed in the tests eroded, possibly because the low-pressure differentials used did not sufficiently compact the mudcake. Only limited data was obtained in these tests, and general conclusions were not drawn.

Measurement of Mudcake Properties

The mudcake model used required independent lab measurements for permeability, porosity, and solid fraction. This implied the need for tedious, time-consuming tasks involving weighing, drying, sorting, and so on, procedures not unlike those reported later by other authors (e.g., Holditch and Dewan, 1991;

Dewan and Chenevert, 1993). The inaccuracies present in such tests pose hurdles to practical field implementation, since any formation predictions obtained would only be as accurate as the mudcake properties. Much of the early effort addressed sensitivities of predicted results as they depended on cake measurements. At the time, no solution to this problem was found, but it turns out that dynamically equivalent information *can* be obtained by a single measurement for filtrate volume and cake thickness at a single large time. This is explained later.

Formation Evaluation From Invasion Data

At the time of the experiments, a closed-form radial front solution as a function of mudcake properties, filtrate and formation fluid viscosity, mud and pore pressure, rock porosity and permeability, and spurt loss was not available. Thus, the precise conditions under which rock permeability, pore pressure, and oil viscosity can be predicted from front data could not be determined. Nor was the form of the ultimate methodology that would host such calculations known: the exact functional relationships were lacking. Thus, the original work focused on rock porosity only, since its role is obvious: when mudcake controls the net filtration rate, the invasion front depends only on porosity, a simple geometric volume variable. The early work, in this sense, evaluated more the uncertainties due to errors in mudcake characterization than it did the formation. To determine these sensitivities, the exact porosities of the Berea sandstone cores were independently measured by direct core analysis methods and were found to vary from 22% to 24%. Fig. 19.1, reproduced from Chin et al. (1986), shows predicted porosities as a function of time after initial radial invasion. Errors at small times are due to two independent effects. The first is poor mudcake definition. The second arises from the neglect of spurt in the derived radial flow formulas, or more precisely, the incorrect assumption of zero spurt. The formulas used, in order to account for the large initial invasion due to spurt loss, responded by predicting porosities that are abnormally low, in the 10% range. This effect corrects for itself over time, since the limited volume of spurt becomes unimportant with time, as the radial front expands geometrically. At least in the runs reported, the time scale required for this correction is about 1 h. If an exact radial flow solution had been available, and spurt loss could be estimated, the waiting time could have been reduced to 10 min.

Time (min)	Porosity (%)	Time (min)	Porosity (%)
1.2	10	49.1	22
3.9	14	64.1	22
9.0	17	81.0	23
16.1	20	100	23
25.6	21	121	23
36.1	21	144	23

FIG. 19.1 Radial flow test, 15 ppg mud, $\Delta p = 150$ psi.

As noted, unsteady cake growth and invasion fronts in time were monitored and captured in Catscan images. In Chapter 14, for example, a sequence of linear flow slides is displayed, obtained over several hours. A radial flow slide showing the central test cross-section containing a ring of mudcake, together with a circular front moving into rock core, is also given. Lineal and radial flow results and predictions, for the 9–15 ppg water-base muds used, proved very repeatable.

Field Applications

Repeat resistivity data obtained from several MWD logging runs taken in the Woodbine Sand in Quitman, Texas, was analyzed using standard multilayer electromagnetic simulations. These calculations determined invasion front radii given resistivity time histories. Then, porosity could be determined using inverse fluid methods, with the procedure repeated at different depths. Typical permeabilities and porosities in the water sands were 200 md and 25%. Resistivity measurements were obtained at 30 min, 1 day, and 31 days. Samples of the 9 ppg water-base mud were retained for laboratory mudcake evaluation. The multilayer electromagnetic code was specialized to four layers, comprising the tool, the mud, and invaded and virgin rock, respectively. A radius R_i separated the latter two zones. This radius was obtained iteratively, with the correct value being the one that reproduces the known tool reading when the logging resistivities R_T and R_{xo} are prescribed. The original paper gave logs calculated at several depth intervals, displayed side by side with the corresponding neutron and density porosity logs (e.g., see Chapter 14). The effective or invasion porosity logs were consistent with neutron porosity and density tool measurements, duplicating qualitative and quantitative features. Exact agreement was not expected, since the latter vertically averaged portions of their signals and were taken at rapid speeds; the high vertical resolution resistivity tool used took instantaneous formation snapshots and yielded accurate readings in this sense. Our porosity refers to *connected pores* that provide conduits for fluid flow.

During the field tests, the classical $\sqrt{t}$ displacement law, strictly valid for lineal invasion, broke down after several days due to significant formation influx, making (simplified) radial flow modeling mandatory. Again, the early work did not address formation properties other than porosity, noting only that quantitative results *may* be possible for hydrocarbon viscosity, formation porosity, and permeability. The principal difficulties with these properties, it was realized, were subtle. In formations with permeabilities greater than a few millidarcies, mudcakes form rapidly and control the invasion process within minutes. As was the case with the work summarized above, the invasion front then depends largely on porosity.

In order to determine hydrocarbon viscosity, rock permeability, and mobility ratio, it is clear that the mobility in the cake must be comparable to that of the formation, in order to create nontrivial dynamical coupling between the two flows, from which the information needed for use in inverse models can be derived or inferred. Loosely speaking, permeable formations should be probed using permeable mudcakes, while impermeable formations require comparably impermeable cakes. In this sense, permeability prediction stands the greatest chance for success when it is extremely low in value. Still, the matter concerning the time separation required between successive resistivity readings needs to be clarified. And finally, how the mudcake properties determined in *lineal* laboratory flow tests *without* underlying impermeable rock are to be used in *radial* time lapse analysis must be resolved. These questions are addressed next, where our exact radial mudcake invasion solution is taken as the basis for time lapse analysis.

CHARACTERIZING MUDCAKE PROPERTIES

The invasion modeling results of Chapter 17 required us to characterize the mudcake by three independent parameters, namely, the solid fraction f_s, the porosity ϕ_c, and the cake permeability k. The theoretical work in Chin et al. (1986) and Collins (1961) required such empirical inputs, and elaborate laboratory procedures were developed to support the volume and Darcy flow resistance measurements needed. It turns out that all of this is unnecessary, if we apply the philosophy underlying time lapse analysis to mudcake properties prediction, using simple lineal filtrate tests performed at the surface, flowing through standard filter paper without underlying rock. The key idea lies in the fact that the foregoing parameters, for incompressible cakes anyway, only affect filtration by way of the two *lumped parameters* μ/k and $f_s/\{(1-f_s)(1-\phi_c)\}$, where μ is the filtrate viscosity.

Simple Extrapolation of Mudcake Properties

In our study of *lineal* cake buildup on filter paper, we found that the mudcake thickness can be written as $x_c(t) = \sqrt{\{2kf_s\Delta p/\{\mu(1-f_s)(1-\phi_c)\}\}}\ \sqrt{t}$. For simplicity, consider a collection vessel having the same area dA as the cross-section of the core sample. (For lineal flows, the complete area A can be substituted in place of dA.) Then, the *filtrate height* $h(t) = V_l(t)/dA$ of the liquid column is simply $h(t) = V_l(t)/dA = \sqrt{\{2k\Delta p(1-f_s)(1-\phi_c)/(\mu f_s)\}}\ \sqrt{t}$. Dividing the first equation by the second,

$$f_s/\{(1-f_s)(1-\phi_c)\} = x_c(t)/h(t) \tag{19.1}$$

while the square of the equation for h(t) yields

$$\mu/k = 2\Delta p\{t/h^2(t)\}(1-f_s)(1-\phi_c)/f_s \tag{19.2}$$

Thus, if $x_c(t)$ and h(t) are both known at some time t, the lumped quantities $f_s/\{(1-f_s)(1-\phi_c)\}$ and μ/k are completely determined. (Our definition of the filtrate height h(t) excludes mud spurt contributions.)

We emphasize that k and $f_s/\{(1 - f_s)(1 - \phi_c)\}$ are material or constitutive *constants* intrinsic to the particular mudcake. The latter is a dimensionless number that depends only upon the packing arrangement of the solid particles making up the mudcake, which in turn depends on the instantaneous pressure gradient and the shearing effects of dynamic filtration, if present. These constants are not unlike others used in engineering analysis, for example, the viscosity of a lubricant or the yield stress of a steel test sample. This being so, their values can be obtained from the simple lineal buildup test just described, and are applied to more general cylindrical radial or spherical flow formulas derived for problems where mudcake and formation interaction are not weak.

The question of mudcake permeability often arises in assessing formation damage, which, for example, manifests itself through reduced production in reservoir engineering or by way of skin effects during transient well tests. Many researchers address this problem by forcing clean water through isolated mudcake (retrieved from filtration vessels) under pressure, thus ensuring a controlled test where the mudcake no longer grows; permeability is calculated by knowing the differential pressure, the cross-sectional area, the filtrate volume, and the water viscosity. This is the standard laboratory procedure used to determine rock core permeability, but its application to mudcake analysis is inconvenient, since it is laborious, and more often than not results in cake damage and tearing. This procedure can be circumvented if we observe that Eqs. (19.1), (19.2) imply that the cake permeability k takes the value

$$k = \mu h(t) x_c(t)/(2t\Delta p) \tag{19.3}$$

which is completely determined using data from the foregoing test. Therefore, it is clear that separate flow tests for mudcake permeability prediction are unnecessary, since the test just described provides the needed information. In order to reduce the experimental error associated with mudcake characterization, the sample time t should be sufficiently large that errors due to initially nonuniform mudcake definition are minimized. This implies a wait of 30–60 min; in fact, a sequence of measurements corresponding to larger and larger wait times might be useful, to be terminated only when calculated results for mudcake properties converge to stable values. It is assumed, of course, that appropriate high temperature and pressure filtration vessels are used whenever necessary to model mudcake growth in deep holes. Experimentally, it has been observed that the mud-to-mudcake interface may be unclear and gel-like at times, thus introducing error into time lapse analysis. It may well be that special muds with easy-to-determine cake thicknesses will need to be formulated if inverse applications are to be successful.

Radial Mudcake Growth on Cylindrical Filter Paper

Many authors presume the universality of $\sqrt{t}$ mudcake-filtration behavior at large times; this may sometimes be valid in lineal flows. However, as we have seen from our general radial mudcake flow results in Chapter 17, this assumption can be wrought with danger. The exact nature of mudcake growth is not only important to interpretation: cake thickness is a useful indicator for both formation damage and probability of differential sticking. While $\sqrt{t}$ behavior provides a "back of the envelope" guess, problems can arise when cake buildup is obviously radial, for example, when mudcake thickness is a substantial fraction of the hole radius, and in newer slimholes, where the buildup process may be uncertain. In this example, we shall investigate the growth of mudcake in a radial flow vessel formed by thin cylindrically formed filter paper, as shown in Fig. 19.2. Although it is possible to study this problem as a formal limit of our three-layer solution, it is instructive to reconsider its formulation from first principles. As shown earlier, the governing ordinary differential equation for an incompressible, isotropic, homogeneous, cylindrical radial Darcy flow is $d^2p(r)/dr^2 + (1/r)\ dp/dr = 0$. Then, the general solution to this equation takes the form $p(r) = A \log r + B$.

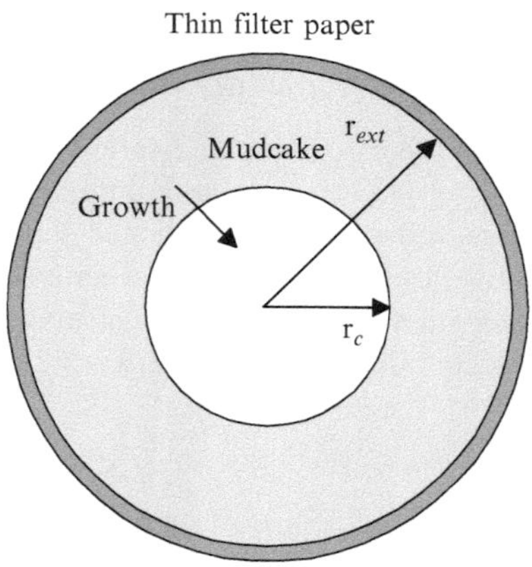

FIG. 19.2 Radial mudcake growth on filter paper.

For this radial flow, we impose the mud pressure P_m at the edge of the growing cake interface and the external pressure P_{ext} at the circularly wrapped filter paper. Thus, the differential pressure acting on the cake ring is $P_m - P_{ext}$. Our boundary conditions are $p(r_c) = P_m$ and $p(r_{ext}) = P_{ext}$. If we now substitute these into the general solution, we find that the integration constant A satisfies $A = (P_m - P_{ext})/(\log r_c/r_{ext})$. The differential equation derived in Chapter 17 for mudcake buildup, for the coordinates used, takes the form

$$\begin{aligned} dr_c(t)/dt &= -\{f_s/\{(1-f_s)(1-\phi_c)\}\}\,|v_n| \\ &= +\{f_s/\{(1-f_s)(1-\phi_c)\}\}\,(k/\mu)\,dp(r_c)/dr \qquad (19.4) \\ &= +\{f_s/\{(1-f_s)(1-\phi_c)\}\}\,(k/\mu)\,A/r_c \end{aligned}$$

where A is again a function of log r_c. This nonlinear ordinary differential equation can be integrated in exact closed form. To fix the constant of integration, we assume that no mudcake exists at $t = 0$; that is, the cake radius is the same as that shown in Fig. 19.2, with $r_c(t = 0) = r_{ext}$. Then, we obtain the following exact implicit solution for radial cake growth as a function of time,

$$\begin{aligned} &\tfrac{1}{2}(r_c/r_{ext})^2 \log (r_c/r_{ext}) - \tfrac{1}{4}(r_c/r_{ext})^2 + 1/4 \\ &= \{kf_s(P_m - P_{ext})/\{\mu(1 - f_s)(1 - \phi_c)r_{ext}{}^2\}\}\,t \end{aligned} \tag{19.5}$$

In deriving Eq. (19.5), we assumed that P_m is constant. If it is instead a function of time, the integral $\int P_m(t)\,dt$ will simply appear in place of $P_m t$.

Now consider the conditions under which this general result reduces to the lineal $\sqrt{t}$ law. This is accomplished by introducing $r_c = r_{ext} - \Delta r$, with $\Delta r > 0$; that is, $r_c/r_{ext} = 1 - \Delta r/r_{ext} = 1 - \delta$ and $\delta > 0$. Then, we expand the above left-hand side *LHS* in Taylor series for small δ, so that $LHS = \tfrac{1}{2}(1 - \delta)^2 \log (1 - \delta) - \tfrac{1}{4}(1 - \delta)^2 + \tfrac{1}{4} \approx \tfrac{1}{2}\delta^2 = \tfrac{1}{2}(\Delta r/r_{ext})^2$. Substitution into Eq. (19.5) and cancellation of common terms yield the cake thickness

$$\Delta r \approx \sqrt{[\{2kf_s(P_m - P_{ext})/\{\mu(1 - f_s)(1 - \phi_c)\}\}\,t]} > 0 \tag{19.6}$$

in agreement with lineal theory. Some indication of the extent to which $\tfrac{1}{2}(1 - \delta)^2 \log (1 - \delta) - \tfrac{1}{4}(1 - \delta)^2 + \tfrac{1}{4}$ can be approximated by $\tfrac{1}{2}\delta^2$ is found by tabulating these functions versus δ, noting that $\delta = \Delta r/r_{ext}$. This is done in Fig. 19.3. The results show that the $\sqrt{t}$ law is satisfactory for $\Delta r/r_{ext} < 0.20$. This applies to radial and lineal mudcake buildup on resistance-free filter paper only and does *not* apply to cake buildup on formations having comparable mobilities.

$\Delta r/r_{ext}$	LHS (exact)	Lineal
.0500	.0012	.0013
.1000	.0048	.0050
.1500	.0107	.0113
.2000	.0186	.0200
.2500	.0285	.0313
.3000	.0401	.0450
.3500	.0534	.0613
.4000	.0681	.0800
.4500	.0840	.1013
.5000	.1009	.1250
.5500	.1185	.1513
.6000	.1367	.1800
.6500	.1551	.2113
.7000	.1733	.2450
.7500	.1911	.2813
.8000	.2078	.3200
.8500	.2230	.3613
.9000	.2360	.4050
.9500	.2456	.4513

FIG. 19.3 Radial versus lineal mudcake theory.

Although thick mudcakes with large values of $\Delta r/r_{ext} > 0.20$ may be uncommon, at least in conventional drilling, with the lineal approximation found to be quite applicable, it may well be that thicker mudcakes are actually desired for accurate time lapse analysis applications. This is so because both the mudcake characterization tests discussed above, and the cruder, direct measurements for k, f_s, and ϕ_c alluded to earlier, ultimately require cake thickness measurements in some form or another. Having a thicker cake to measure, ideally formed from solids with good textural qualities that ensure discernible mud-to-cake boundaries, barring the risks of stuck pipe, of course, reduces the level of experimental uncertainty. Finally, note that a time scale of interest in drilling is the time required for cake to completely plug the well, under the assumption of static filtration. The required formula is useful in evaluating experimental muds drilled in slimholes. When the hole is plugged, we obtain $r_c = 0$. Then, substitution in Eq. (19.5) yields the simple relationship $1/4 = \{kf_s(P_m - P_{ext})/\{\mu(1 - f_s)(1 - \phi_c)r_{ext}^2\}\}\ t$. The time to plug is

$$t_{plug} = \mu(1 - f_s)(1 - \phi_c)r_{ext}^2/\{4kf_s(P_m - P_{ext})\} \tag{19.7}$$

This provides an estimate of the time scale over which plugging may become important, and may be useful operationally in stuck pipe considerations. Again, Eqs. (19.5), (19.7) appear as a result of exact radial flow theory.

POROSITY, PERMEABILITY, OIL VISCOSITY AND PORE PRESSURE DETERMINATION

Here we develop time lapse analysis using three models in the order of increasing complexity. In the first, we address porosity prediction, when mudcake controls the overall invasion. In the second, we consider fluid invasion without the presence of mudcake, and we determine pore pressure, formation permeability, and hydrocarbon viscosity. In the third, the same formation properties are considered, except that the complicating effects of mudcake are not neglected. Numerical examples are given which illustrate the basic ideas.

Simple Porosity Determination

In wells where mudcake controls the overall flow into the formation, and where $\Delta r/r_{ext} < 0.20$ is satisfied, a lineal mudcake model suffices. This being so, we unwrap the cake layer adhering to our wellbore and view the buildup process as a lineal one satisfying the $\sqrt{t}$ law. But the invasion into the formation, of course, is highly radial: in this farfield, the effects of borehole geometry and streamline divergence must be considered in order to conserve mass. Now consider a well with a radius r_{well} and an axial borehole length L, for which the surface area dA of the unwrapped mudcake layer is $2\pi r_{well}L$. Using results from Chapter 17, the total filtrate volume passing through the mudcake at time $t = t^*$ is

$$\begin{aligned} V_l(t^*) &= \sqrt{\{2k\Delta p(1 - f_s)(1 - \phi_c)/(\mu f_s)\}}\ \sqrt{t^*}\, dA \\ &= 2\pi r_{well} L \sqrt{[\{2k\Delta p(1 - f_s)(1 - \phi_c)/(\mu f_s)\}t^*]} \end{aligned} \tag{19.8}$$

For incompressible flow, this equals the formation volume available for filtrate storage; that is, $\pi(r_f^2 - r_{well}^2)L\phi_{eff}$, where ϕ_{eff} is the effective porosity. Thus, we can solve for the effective porosity as

$$\phi_{eff} = 2r_{well}\left\{r_f(t^*)^2 - r_{well}^2\right\}^{-1} \sqrt{[\{2k\Delta p(1 - f_s)(1 - \phi_c)/(\mu f_s)\}t^*]} \qquad (19.9)$$

Ideally, the right side of Eq. (19.9) will be independent of the time t*, but in reality, one anticipates larger measurement errors for small times because cake thicknesses are not yet well defined (e.g., the discussion for Fig. 19.1).

Radial Invasion Without Mudcake

In the preceding example, we showed how formation porosity can be calculated from purely geometric considerations when mudcake controls the flow rate into the reservoir. Sometimes the opposite limit applies: in shallow holes and in special drilling applications, watery brines are sometimes used as the circulating fluid. We shall develop the theory for such problems, and then demonstrate how formation properties can be straightforwardly predicted from time lapse analysis.

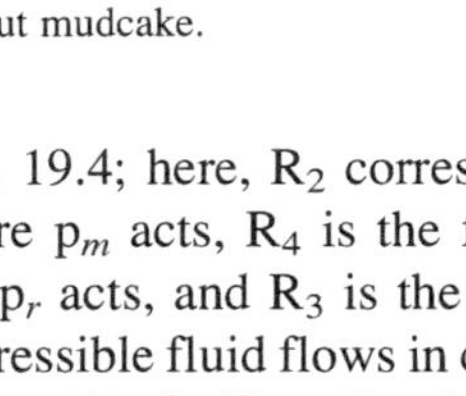

FIG. 19.4 Radial invasion without mudcake.

Let us now refer to Fig. 19.4; here, R_2 corresponds to the fixed borehole radius where a mud pressure p_m acts, R_4 is the fixed effective radius where the reservoir pore pressure p_r acts, and R_3 is the moving invasion front. The complete equation for compressible fluid flows in cylindrical radial coordinates takes the form $\partial^2 p(r,t)/\partial r^2 + (1/r)\ \partial p/\partial r = (\phi\mu c/k)\partial p/\partial t$. We shall deal with constant density flows, so that $d^2p_i(r)/dr^2 + (1/r)dp_i/dr = 0$, $i = 1, 2$. Note that the subscript 1 refers to $R_2 < r < R_3$, while the 2 refers to $R_3 < r < R_4$. The solutions to our ordinary differential equations are $p_1(r) = A \log r + B$ and $p_2(r) = C \log r + D$, where A, B, C, and D are determined by the boundary and matching conditions. These are $p_1(R_2) = p_m$, $p_2(R_4) = p_r$, $p_1(R_3) = p_2(R_3)$, and $(k_1/\mu_1)\ dp_1(R_3)/dr = (k_2/\mu_2)\ dp_2(R_3)/dr$, as in Chapter 17.

In our derivation, we allow the possibility of unequal permeabilities, so that our results can *mimic* relative permeability effects in two-phase immiscible flow. One can show that $A = (p_r - p_m)/\log\{(R_3/R_2)(R_3/R_4)^{-k1\mu2/k2\mu1}\}$, $B = p_m - A \log R_2$, $C = (k_1\mu_2/k_2\mu_1)A$, and $D = p_r - C \log R_4$. Then, the differential equation for radial front motion is as usual found from $dR_3(t)/dt = -\{k_1/(\mu_1\phi)\} \times dp_1/dr = -\{k_1/(\mu_1\phi)\}\ A/R_3$ where A depends on R_3. This nonlinear ordinary differential equation can be integrated exactly in closed analytical form. If we assume a pressure drop that is constant in time, together with the initial condition $R_3(0) = R_2$ (that is, we assume that the radial invasion front initially coincides with the wellbore radius), it follows that

$$\left[(p_m - p_r)\,t/\left(\mu_1 R_2{}^2\right)\right](k_1/\phi) +$$
$$\left\{\tfrac{1}{2}(R_3/R_2)^2 \log(R_3/R_4) - \tfrac{1}{4}(R_3/R_2)^2 - \tfrac{1}{2}\log(R_2/R_4) + \tfrac{1}{4}\right\}(k_1\mu_2/k_2\mu_1)$$
$$= \tfrac{1}{2}(R_3/R_2)^2 \log(R_3/R_2) - \tfrac{1}{4}(R_3/R_2)^2 + \tfrac{1}{4} \qquad (19.10)$$

for the radial front $R_3(t)$. This result could have been obtained as a limit of the three-layer radial solution in Chapter 17, but its self-contained derivation from first principles is instructive and useful in its own right.

We now develop the quantitative basis for time lapse analysis, within the framework of the plug-flow displacement model just discussed, for rock permeability, hydrocarbon viscosity, and pore pressure determination. In *Problem 1*, we consider simultaneous reservoir permeability and hydrocarbon viscosity prediction, while in *Problem 2*, we shall add to these unknowns the formation pore pressure. To fix ideas, we set $k_1 = k_2 = k$ in Eq. (19.10), which is the situation of interest to reservoir engineers. If we multiply Eq. (19.10) by the mud viscosity μ_1, we obtain the fundamental host equation

$$\left[(p_m - p_r)\,t/R_2{}^2\right](k/\phi) +$$
$$\left\{\tfrac{1}{2}(R_3/R_2)^2 \log(R_3/R_4) - \tfrac{1}{4}(R_3/R_2)^2 - \tfrac{1}{2}\log(R_2/R_4) + \tfrac{1}{4}\right\}\mu_2 \qquad (19.11)$$
$$= \mu_1\left\{\tfrac{1}{2}(R_3/R_2)^2 \log(R_3/R_2) - \tfrac{1}{4}(R_3/R_2)^2 + \tfrac{1}{4}\right\}$$

Problem 1. Let us assume that the front position $r = R_3(t)$ is known at two instants in time, say $r = R_3{}^*$ at $t = t^*$, and $r = R_3{}^{**}$ at $t = t^{**}$, for example, as determined from multilayer electromagnetic analysis, as in Chin et al. (1986). Since the values of p_m, p_r, R_2, R_4, and μ_1 are known, we can evaluate the Eq. (19.11) twice, using our data obtained at two points in time, to yield

$$\left[(p_m - p_r)\,t^*/R_2{}^2\right](k/\phi) +$$
$$\left\{\tfrac{1}{2}(R_3{}^*/R_2)^2 \log(R_3{}^*/R_4) - \tfrac{1}{4}(R_3{}^*/R_2)^2 - \tfrac{1}{2}\log(R_2/R_4) + \tfrac{1}{4}\right\}\mu_2 \qquad (19.12a)$$
$$= \mu_1\left\{\tfrac{1}{2}(R_3{}^*/R_2)^2 \log(R_3{}^*/R_2) - \tfrac{1}{4}(R_3{}^*/R_2)^2 + \tfrac{1}{4}\right\}$$

and

$$\left[(p_m - p_r)\, t^{**}/R_2^{\,2}\right](k/\phi) +$$
$$\left\{\tfrac{1}{2}(R_3^{**}/R_2)^2 \log (R_3^{**}/R_4) - \tfrac{1}{4}(R_3^{**}/R_2)^2 - \tfrac{1}{2}\log (R_2/R_4) + \tfrac{1}{4}\right\}\mu_2$$
$$= \mu_1 \left\{\tfrac{1}{2}(R_3^{**}/R_2)^2 \log (R_3^{**}/R_2) - \tfrac{1}{4}(R_3^{**}/R_2)^2 + \tfrac{1}{4}\right\} \qquad (19.12b)$$

In shorthand, letting *RHS* denote right-hand-side quantities, Eqs. (19.12a), (19.12b) become

$$[\;]^*(k/\phi) + \{\;\}^*\mu_2 = RHS^* \qquad (19.13a)$$

$$[\;]^{**}(k/\phi) + \{\;\}^{**}\mu_2 = RHS^{**} \qquad (19.13b)$$

Thus, we have two *linear* equations in the unknowns k/ϕ, a useful lithology indicator related to the well-known Leverett J-function, and the viscosity μ_2. This simple 2×2 system can be solved using elementary algebra. If the porosity of the formation were known from a separate logging measurement, or from late-time-invasion based porosity extrapolation, then these equations would yield solutions for formation permeability and hydrocarbon viscosity.

In order to test our inverse time lapse ideas, let us first generate synthetic front displacement data versus time by assuming appropriate formation and fluid properties for our forward simulation. In Eq. (19.11), the radius R_3 is varied parametrically, and the corresponding invasion time t is computed. The results of a *forward* simulation are shown, where the parameters have been selected for illustrative purposes only (for brevity, only partial numerical results are given).

```
INPUT PARAMETER SUMMARY:
Rock core permeability (darcies):  .1000E-02
Rock core porosity (decimal nbr):  .2000E+00
Viscosity of invading fluid (cp):  .1000E+01
Viscosity, displaced  fluid (cp):  .2000E+01
Pressure at well boundary   (psi):  .1000E+04
Pressure, effective radius  (psi):  .9000E+03
Radius  of the  well bore (feet):  .2000E+00
Reservoir, effective radius  (ft):  .2000E+01
Maximum allowed  number of hours:  .1000E+03

T =  .0000E+00 sec, Rf =  .2000E+00 ft
T =  .2978E+04 sec, Rf =  .3000E+00 ft
T =  .6830E+04 sec, Rf =  .4000E+00 ft
T =  .1148E+05 sec, Rf =  .5000E+00 ft
T =  .1685E+05 sec, Rf =  .6000E+00 ft
T =  .2292E+05 sec, Rf =  .7000E+00 ft
T =  .2962E+05 sec, Rf =  .8000E+00 ft
T =  .3693E+05 sec, Rf =  .9000E+00 ft
T =  .4481E+05 sec, Rf =  .1000E+01 ft
T =  .5323E+05 sec, Rf =  .1100E+01 ft
T =  .6217E+05 sec, Rf =  .1200E+01 ft
T =  .7161E+05 sec, Rf =  .1300E+01 ft
T =  .8151E+05 sec, Rf =  .1400E+01 ft
T =  .9187E+05 sec, Rf =  .1500E+01 ft
T =  .1027E+06 sec, Rf =  .1600E+01 ft
T =  .1139E+06 sec, Rf =  .1700E+01 ft
T =  .1255E+06 sec, Rf =  .1800E+01 ft
T =  .1375E+06 sec, Rf =  .1900E+01 ft
T =  .1498E+06 sec, Rf =  .2000E+01 ft
```

We now apply the *inverse* method developed earlier, and in particular, assume the input parameters given below,

```
INPUT PARAMETER SUMMARY:
Cake-rock "delta pressure" (psi): .1000E+03
Rock core porosity (decimal nbr): .2000E+00
Viscosity of  mud  filtrate (cp): .1000E+01
Radius  of the  well bore (feet): .2000E+00
Reservoir, effective radius (ft): .2000E+01
```

In a field situation, the foregoing (**bold**) inputs would represent best guesses. We next list the results of three separate calculations; additional best guesses for radial invasion front position versus time are shown in **bold** print, whereas predicted formation properties are shown in *italicized* type.

```
TIME LAPSE ANALYSIS PREDICTIONS:

Trial No. 1:
Time of the 1st data point (sec): .6830E+04
Radius  of invasion front (feet): .4000E+00
Time of the 2nd data point (sec): .2962E+05
Radius  of invasion front (feet): .8000E+00
Formation permeability (darcies): .1000E-02
Viscosity, formation  fluid (cp): .2000E+01

Trial No. 2:
Time of the 1st data point (sec): .2962E+05
Radius  of invasion front (feet): .8000E+00
Time of the 2nd data point (sec): .6217E+05
Radius  of invasion front (feet): .1200E+01
Formation permeability (darcies): .1000E-02
Viscosity, formation  fluid (cp): .2000E+01

Trial No. 3:
Time of the 1st data point (sec): .6830E+04
Radius  of invasion front (feet): .4000E+00
Time of the 2nd data point (sec): .6217E+05
Radius  of invasion front (feet): .1200E+01
Formation permeability (darcies): .1000E-02
Viscosity, formation  fluid (cp): .2000E+01
```

Repeated runs for this example indicate that predictions for permeability and formation fluid viscosity are very stable with respect to errors in the input data. In fact, fluid viscosity remained stable to very large changes in assumed parameters, although we have not yet pinpointed the exact reasons for this fortunate circumstance.

Problem 2. Next, suppose that the pore pressure p_r was unknown and additionally desired. Here, we shall rewrite the fundamental result of Eq. (19.11) in the form

$$\begin{aligned} &\left[p_m t/R_2^{\,2}\right]\left(k/\phi\right) + \left(-t/R_2^{\,2}\right)(p_r k/\phi) \\ &\quad + \left\{ \tfrac{1}{2}(R_3/R_2)^2 \log(R_3/R_4) - \tfrac{1}{4}(R_3/R_2)^2 - \tfrac{1}{2}\log(R_2/R_4) + \tfrac{1}{4} \right\}\mu_2 \\ &\quad = \mu_1 \left\{ \tfrac{1}{2}(R_3/R_2)^2 \log(R_3/R_2) - \tfrac{1}{4}(R_3/R_2)^2 + \tfrac{1}{4} \right\} \end{aligned} \tag{19.14}$$

Evaluation of Eq. (19.14) at three instances in time, say t^*, t^{**}, and t^{***}, now yields a 3×3 system of algebraic equations, in particular,

$$[\ ]^*(k/\phi) + (\ )^*\,(p_r k/\phi) + \{\ \}^*\,\mu_2 = RHS^* \tag{19.15a}$$

$$[\ ]^{**}(k/\phi) + (\ )^{**}\,(p_r k/\phi) + \{\ \}^{**}\,\mu_2 = RHS^{**} \tag{19.15b}$$

$$[\ ]^{***}(k/\phi) + (\ )^{***}\,(p_r k/\phi) + \{\ \}^{***}\,\mu_2 = RHS^{***} \tag{19.15c}$$

that is again easily solved using elementary algebra. These three linear equations completely determine the three unknowns k/ϕ, $p_r k/\phi$, and μ_2. Once the values of k/ϕ and $p_r k/\phi$ are known, the pore pressure p_r can be obtained by simple elimination. Then, k/ϕ, p_r, and μ_2 are immediately available.

We emphasize that the times t*, t**, and t*** and their corresponding radii $R_3(t^*)$, $R_3(t^{**})$, and $R_3(t^{***})$ must be chosen so that the resulting simultaneous equations are not ill-conditioned, in the linear algebra sense. If any of the equations are too nearly identical, because the invasion data points are taken too closely in time, the determinant of the coefficient matrix will likely vanish and yield indeterminate or inaccurate solutions. For example, the solution to $x+y=4$ and $x+1.01\,y=4$, while mathematically unique, is unlikely to be physically useful because the result is unstable. One way to ensure correct conditioning is to suddenly change the mud pressure $p_m(t)$. But severe decreases or increases in pressure may lead to dangerous underbalanced drilling or undesired formation fracture, effects that outweigh the need for real-time formation information.

Time Lapse Analysis Using General Muds

Now, we consider the complete radial flow invasion problem modeled in Chapter 17, where general mudcake and formation interaction is allowed. This model studied dynamically coupled mud filtrate invasion, simultaneous water or oil displacement, and time-dependent mudcake buildup. In that application, we had derived the closed-form solution in Eq. (17.25) for the radial invasion front position as a function of differential pressure, mudcake, rock, displaced liquid properties, and time. We shall use that solution as a host time lapse analysis model equation in a manner motivated by the foregoing inverse results. It will be convenient to first rewrite Eq. (17.25) in the form

$$
\begin{aligned}
&\Big[\Big\{R_2^2(1-\phi_c)(1-\phi_s)/4\mu_2\phi_{eff}f_s\Big\}\\
&\quad\times\Big\{\log\Big\{1+f_s\phi_{eff}\Big\{(R_{spurt}/R_2)^2-(R_3/R_2)^2\Big\}/\{(1-\phi_c)(1-f_s)\}\Big\}\\
&\quad-f_s\phi_{eff}\Big\{(R_{spurt}/R_2)^2-(R_3/R_2)^2\Big\}/\{(1-\phi_c)(1-f_s)\}\\
&\quad+f_s\phi_{eff}\Big\{(R_{spurt}/R_2)^2-(R_3/R_2)^2\Big\}/\{(1-\phi_c)(1-f_s)\}\\
&\quad\times\log\Big\{1+f_s\phi_{eff}\Big\{(R_{spurt}/R_2)^2-(R_3/R_2)^2\Big\}/\{(1-\phi_c)(1-f_s)\}\Big\}\Big\}\\
&\quad-k_1(p_m-p_r)t/(\mu_1\mu_2\phi_{eff})\Big]\,k_2\\
&+\Big\{\big(k_1k_2R_4^2/\mu_1\mu_2k_3\big)\Big[\tfrac{1}{2}(R_{spurt}/R_4)^2\log(R_{spurt}/R_4)-\tfrac{1}{4}(R_{spurt}/R_4)^2\\
&\qquad-\tfrac{1}{2}(R_3/R_4)^2\log(R_3/R_4)+\tfrac{1}{4}(R_3/R_4)^2\Big]\Big\}\mu_3\\
&=\Big(\big(-k_1R_2^2/\mu_1\big)\Big[\tfrac{1}{2}(R_3/R_2)^2\log(R_3/R_2)-\tfrac{1}{4}(R_3/R_2)^2\\
&\qquad-\tfrac{1}{2}(R_{spurt}/R_2)^2\log(R_{spurt}/R_2)+\tfrac{1}{4}(R_{spurt}/R_2)^2\Big]\Big)
\end{aligned}
\tag{19.16}
$$

In deriving Eq. (17.25), we assumed $k_2 \neq k_3$ and $\mu_1 \neq \mu_2$. This is not so in applications. Thus, we simplify and write Eq. (19.16) in a more meaningful form, setting $k_2 = k_3 = k_r$, $k_1 = k_c$, $\mu_1 = \mu_2 = \mu_m$, and $\mu_3 = \mu_o$ where k_r is rock permeability, k_c is cake permeability, μ_m is filtrate viscosity, and μ_o is "oil" or displaced liquid viscosity. With this change, Eq. (19.16) becomes

$$
\begin{aligned}
&\Big[\{R_2^{\,2}(1-\phi_c)(1-f_s)/4\mu_m\phi_{eff}f_s\} \\
&\quad \times \Big\{ \log \Big\{1 + f_s\phi_{eff}\Big\{(R_{spurt}/R_2)^2 - (R_3/R_2)^2\Big\}/\{(1-\phi_c)(1-f_s)\}\Big\} \\
&\quad - f_s\phi_{eff}\Big\{(R_{spurt}/R_2)^2 - (R_3/R_2)^2\Big\}/\{(1-\phi_c)(1-f_s)\} \\
&\quad + f_s\phi_{eff}\Big\{(R_{spurt}/R_2)^2 - (R_3/R_2)^2\Big\}/\{(1-\phi_c)(1-f_s)\} \\
&\quad \times \log \Big\{1 + f_s\phi_{eff}\Big\{(R_{spurt}/R_2)^2 - (R_3/R_2)^2\Big\}/\{(1-\phi_c)(1-f_s)\}\Big\}\Big\} \\
&\quad - k_c(p_m - p_r)t/(\mu_m^{\,2}\phi_{eff})\Big]\, k_r \\
&\quad + \Big\{\left(k_cR_4^{\,2}/\mu_m^{\,2}\right)[\tfrac{1}{2}(R_{spurt}/R_4)^2 \log(R_{spurt}/R_4) - \tfrac{1}{4}(R_{spurt}/R_4)^2 \\
&\qquad\qquad - \tfrac{1}{2}(R_3/R_4)^2 \log(R_3/R_4) + \tfrac{1}{4}(R_3/R_4)^2]\Big\}\,\mu_o \\
&= \Big((-k_cR_2^{\,2}/\mu_m)[\tfrac{1}{2}(R_3/R_2)^2 \log(R_3/R_2) - \tfrac{1}{4}(R_3/R_2)^2 \\
&\qquad\qquad - \tfrac{1}{2}(R_{spurt}/R_2)^2 \log(R_{spurt}/R_2) + \tfrac{1}{4}(R_{spurt}/R_2)^2]\Big)
\end{aligned}
\tag{19.17}
$$

As in the previous example, we shall consider two specific time lapse analysis formulations. In the first, we assume that the applied differential pressure is known, and we seek formation permeability and hydrocarbon viscosity only. In the second, we attempt to determine reservoir pore pressure, formation permeability, and hydrocarbon viscosity simultaneously.

Problem 1. Let us examine the physical quantities within the bold brackets [ ], { }, and ( ) of Eq. (19.17). First, our (lumped) mudcake parameters will be regarded as known, since they can be obtained by the simple surface filtrate test defined earlier. That is, the parameter $f_s/\{(1-f_s)(1-\phi_c)\} = x_c(t^*)/h(t^*)$, the cake permeability $k_c = \mu_m h(t^*)x_c(t^*)/(2\Delta p\ t^*)$, the Δp pressure differential used in the filtrate test, and the mud filtrate viscosity μ_m are available from well site measurements. The borehole and effective radii R_2 and R_4 are also considered known, as is the initial spurt radius R_{spurt}. (This becomes less significant with time, as its effect on total invasion depth decreases, and need not accurately specified.) Finally, the effective porosity ϕ_{eff} of the formation can be determined from the large time test in the first example or can be assumed as known from other log measurements, while the pressure differential $(p_m - p_r)$ through the cake and formation is assumed as given. Thus, all of the quantities within our bold brackets are known parameters, with the exception of the time t and its invasion depth $R_3(t)$. As before we evaluate Eq. (19.17) using invasion data from two instances in time, say the radius $R_3(t^*)$ at time t^*, and the radius $R_3(t^{**})$ at t^{**}. The two numerical instances of Eq. (19.17) are

$$[\,]^{*}k_r + \{\,\}^{*}\mu_o = (\,)^{*} \tag{19.18a}$$

$$[\,]^{**}k_r + \{\,\}^{**}\mu_o = (\,)^{**} \tag{19.18b}$$

which provide a 2 × 2 system of algebraic equations for the formation permeability k_r and the hydrocarbon viscosity μ_o. Again, our earlier comments on ill-conditioned equations apply; this means, in practice, that t* and t** cannot be too close together or too far apart.

Problem 2. For this second problem, where the formation pore pressure p_r is regarded as an additional unknown, we rewrite the host invasion equation in a form that separates out the effects of pore pressure, namely,

$$\begin{aligned}
&\Big[\{R_2^2(1-\phi_c)(1-f_s)/4\mu_m\phi_{eff}f_s\}\\
&\quad\times\Big\{\log\Big\{1+f_s\phi_{eff}\Big\{(R_{spurt}/R_2)^2-(R_3/R_2)^2\Big\}/\{(1-\phi_c)(1-f_s)\}\Big\}\\
&\quad -f_s\phi_{eff}\Big\{(R_{spurt}/R_2)^2-(R_3/R_2)^2\Big\}/\{(1-\phi_c)(1-f_s)\}\\
&\quad +f_s\phi_{eff}\Big\{(R_{spurt}/R_2)^2-(R_3/R_2)^2\Big\}/\{(1-\phi_c)(1-f_s)\}\\
&\quad\times\log\Big\{1+f_s\phi_{eff}\Big\{(R_{spurt}/R_2)^2-(R_3/R_2)^2\Big\}/\{(1-\phi_c)(1-f_s)\}\Big\}\Big\}\\
&\quad -k_c p_m t/(\mu_m^2\phi_{eff})\Big]\,k_r\\
&+\Big[\Big[k_c t/(\mu_m^2\phi_{eff})\Big]\Big]\,p_r k_r\\
&+\Big\{(k_c R_4^2/\mu_m^2)\,[\tfrac{1}{2}(R_{spurt}/R_4)^2\log(R_{spurt}/R_4)-\tfrac{1}{4}(R_{spurt}/R_4)^2\\
&\qquad -\tfrac{1}{2}(R_3/R_4)^2\log(R_3/R_4)+\tfrac{1}{4}(R_3/R_4)^2]\Big\}\,\mu_o\\
&=\Big((-k_c R_2^2/\mu_m)\,[\tfrac{1}{2}(R_3/R_2)^2\log(R_3/R_2)-\tfrac{1}{4}(R_3/R_2)^2\\
&\qquad -\tfrac{1}{2}(R_{spurt}/R_2)^2\log(R_{spurt}/R_2)+\tfrac{1}{4}(R_{spurt}/R_2)^2]\Big)
\end{aligned} \tag{19.19}$$

As in Problem 1, we evaluate Eq. (19.19) using invasion radii data from three different instances in time, say $R_3(t^*)$ at time t*, $R_3(t^{**})$ at t**, and the radius $R_3(t^{***})$ at t***. Therefore, the three instances of Eq. (19.19) are

$$[\,]^{*}k_r + [[\,]]^{*}p_r k_r + \{\,\}^{*}\mu_o = (\,)^{*} \tag{19.20a}$$

$$[\,]^{**}k_r + [[\,]]^{**}p_r k_r + \{\,\}^{**}\mu_o = (\,)^{**} \tag{19.20b}$$

$$[\,]^{***}k_r + [[\,]]^{***}p_r k_r + \{\,\}^{***}\mu_o = (\,)^{***} \tag{19.20c}$$

which provide a 3 × 3 system of algebraic equations for the formation permeability k_r, the product $p_r k_r$, and the hydrocarbon viscosity μ_o. (Once k_r and $p_r k_r$ are known, p_r can be deduced.) As before, simple determinant inversion methods from elementary algebra can be used. And again, our earlier comments on ill-conditioned equations and on the stability of calculated formation parameters apply; this means, in practice, that the times t*, t**, and t*** cannot be taken too closely together. (More precisely, displacement fronts must not be spaced too closely together, in order that differences in dynamical effects clearly manifest themselves.)

EXAMPLES OF TIME LAPSE ANALYSIS

While we have demonstrated how quantities of interest, such as permeability, porosity, hydrocarbon viscosity, and pore pressure, can be uniquely obtained, at least from invasion depth data satisfying our equations for piston-like fluid displacement, the actual problem is far from solved even for the simple fluid dynamics model considered here. For one, the tacit assumption that invasion depths can be accurately inferred from resistivity readings is not entirely correct; invasion radii are presently extrapolated from resistivity charts that usually assume concentric layered resistivities, which are at best simplified approximations. And second, since tool response and data interpretation introduce additional uncertainties, not to mention unknown three-dimensional geological effects in the wellbore, time lapse analysis is likely to remain an iterative, subjective, and qualitative process in the near future. With these disclaimers said and done, we now demonstrate via numerical examples how formation parameters might be determined from front radii in actual field runs.

Formation Permeability and Hydrocarbon Viscosity

In this and the following example, we shall first use the exact forward invasion simulation model given by Eq. (17.25) to compute dynamically coupled mudcake growth and radial displacement front motion, where the mud filtrate displaces a more viscous formation fluid. We shall compute radial front position and mudcake boundary as a function of time, and subsequently, using this front information, we shall attempt the backward inversion process where we extract formation permeability and hydrocarbon viscosity values. In other words, we shall generate *synthetic* invasion front displacement data and invert the computed data in order to recover the original formation and fluid properties. This philosophical approach is well known in geophysics, where synthetic P-wave data for known geological structures is created by computer detonations and received surface signals are deconvolved in order to determine the prescribed geology. This validates both forward and backward simulations, demonstrating that the mathematics is at least correct and consistent. Clearly, this does not guarantee success in the field, since the predicted values should ideally be stable with respect to uncertainties in the input data.

The input and output results of a typical radial flow forward simulation are displayed in Fig. 19.5 (bold print denotes input quantities). We have assumed a 1 md, 20% porous rock, and a mudcake having 0.001 md permeability, 10% porosity, and 30% solid fraction. The mud filtrate is taken as water, with a viscosity of 1 cp, and the formation oil is assumed to be 2 cp viscous. Here, the well pressure is taken as 100 psi, and the formation pore pressure is assumed to be 0 psi, acting at wellbore and effective radii of 0.5 and 10 ft, respectively. (Only *differences* in pressure are important for this example.) The numerical calculations show that the borehole completely plugs with mudcake in 17,920,000 s (that is, 4978 h, or 207 days), at which point the invasion front radius terminates at 1.727 ft.

```
INPUT PARAMETER SUMMARY:
Rock core permeability (darcies):  .1000E-02
Rock core porosity (decimal nbr):  .2000E+00
Mud cake permeability  (darcies):  .1000E-05
Mud cake porosity  (decimal nbr):  .1000E+00
Mud solid fraction (decimal nbr):  .3000E+00
Viscosity of invading fluid (cp):  .1000E+01
Viscosity, displaced  fluid (cp):  .2000E+01
Pressure at well boundary  (psi):  .1000E+03
Pressure, effective radius (psi):  .0000E+00
Radius  of the  well bore (feet):  .5000E+00
Reservoir, effective radius (ft):  .1000E+02
Rspurt > Rwell radius @ t=0 (ft):  .6000E+00
Maximum allowed  number of hours:  .1000E+07

T =  .9521E+00 sec, Rf =  .6000E+00 ft, Rc =  .5000E+00 ft
T =  .3242E+05 sec, Rf =  .7000E+00 ft, Rc =  .4875E+00 ft
T =  .1270E+06 sec, Rf =  .8000E+00 ft, Rc =  .4726E+00 ft
T =  .3132E+06 sec, Rf =  .9000E+00 ft, Rc =  .4551E+00 ft
T =  .6284E+06 sec, Rf =  .1000E+01 ft, Rc =  .4348E+00 ft
T =  .1121E+07 sec, Rf =  .1100E+01 ft, Rc =  .4112E+00 ft
T =  .1856E+07 sec, Rf =  .1200E+01 ft, Rc =  .3836E+00 ft
T =  .2921E+07 sec, Rf =  .1300E+01 ft, Rc =  .3512E+00 ft
T =  .4445E+07 sec, Rf =  .1400E+01 ft, Rc =  .3124E+00 ft
T =  .6630E+07 sec, Rf =  .1500E+01 ft, Rc =  .2646E+00 ft
T =  .9862E+07 sec, Rf =  .1600E+01 ft, Rc =  .2012E+00 ft
T =  .1525E+08 sec, Rf =  .1700E+01 ft, Rc =  .9510E-01 ft
T =  .1565E+08 sec, Rf =  .1705E+01 ft, Rc =  .8616E-01 ft
T =  .1607E+08 sec, Rf =  .1710E+01 ft, Rc =  .7614E-01 ft
T =  .1653E+08 sec, Rf =  .1715E+01 ft, Rc =  .6454E-01 ft
T =  .1704E+08 sec, Rf =  .1720E+01 ft, Rc =  .5030E-01 ft
T =  .1763E+08 sec, Rf =  .1725E+01 ft, Rc =  .2982E-01 ft
T =  .1776E+08 sec, Rf =  .1726E+01 ft, Rc =  .2368E-01 ft
T =  .1792E+08 sec, Rf =  .1727E+01 ft, Rc =  .1523E-01 ft
Borehole plugged by mudcake ... run terminated.
```

FIG. 19.5 Numerical results, forward invasion simulation.

Now let us apply the time lapse analysis methodology outlined in Eqs. (19.17), (19.18a), and (19.18b). We shall assume the properties shown in bold print in the following calculations, and then *determine formation permeability and hydrocarbon viscosity using invasion radii information taken at two different points in time*, as computed in Fig. 19.5. We shall attempt this three separate times, in order to demonstrate the utility of the approach.

Fig. 19.6 shows that, in the first attempt, we obtained 0.9573 md and 1.911 cp; in the second and third attempts, we have 1.059 md and 2.131 cp, and 1.016 md and 2.033 cp, respectively. These values compare favorably with the assumed 1 md and 2 cp shown in Fig. 19.5. The disagreement arises because only four decimal places of information are used from Fig. 19.5. Again,

```
INPUT PARAMETER SUMMARY:
Cake-rock "delta pressure" (psi): .1000E+03
Rock core porosity (decimal nbr): .2000E+00
Mud cake permeability  (darcies): .1000E-05
Mud cake porosity  (decimal nbr): .1000E+00
Mud solid fraction (decimal nbr): .3000E+00
Viscosity of  mud  filtrate (cp): .1000E+01
Radius  of the  well bore (feet): .5000E+00
Reservoir, effective radius (ft): .1000E+02
Rspurt > Rwell radius @ t=0 (ft): .6000E+00

TIME LAPSE ANALYSIS PREDICTIONS:

Trial No. 1:
Time of the 1st data point (sec): .3242E+05
Radius  of invasion front (feet): .7000E+00
Time of the 2nd data point (sec): .3132E+06
Radius  of invasion front (feet): .9000E+00
Formation permeability (darcies): .9573E-03
Viscosity, formation  fluid (cp): .1911E+01

Trial No. 2:
Time of the 1st data point (sec): .3132E+06
Radius  of invasion front (feet): .9000E+00
Time of the 2nd data point (sec): .1856E+07
Radius  of invasion front (feet): .1200E+01
Formation permeability (darcies): .1059E-02
Viscosity, formation  fluid (cp): .2131E+01

Trial No. 3:
Time of the 1st data point (sec): .1856E+07
Radius  of invasion front (feet): .1200E+01
Time of the 2nd data point (sec): .3242E+05
Radius  of invasion front (feet): .7000E+00
Formation permeability (darcies): .1016E-02
Viscosity, formation  fluid (cp): .2033E+01
```

FIG. 19.6 Numerical results, inverse invasion simulation.

sensitivity studies must be performed to show that known values of formation properties remain stable to slight errors in input mudcake assumptions. When performing time lapse analysis in the presence of mudcake, significant differences between mudcake and formation mobility heighten this sensitivity. Only when the two are comparable, for example, as in the case where mudcake builds on likewise low permeability rock, can such predictions prove robust, repeatable, and accurate.

Pore Pressure, Rock Permeability, and Fluid Viscosity

In this example, we shall rerun the forward simulation exercise just performed, except that we shall replace the pressure inputs

```
Pressure at well boundary  (psi): .1000E+03
Pressure, effective radius (psi): .0000E+00
```

by

```
Pressure at well boundary  (psi): .5000E+03
Pressure, effective radius (psi): .4000E+03
```

Since the differential pressure (of 100 psi) in both cases remains the same, we would expect the same displacement front and cake buildup history. As Fig. 19.7 shows, we do.

```
INPUT PARAMETER SUMMARY:
Rock core permeability (darcies): .1000E-02
Rock core porosity (decimal nbr): .2000E+00
Mud cake permeability  (darcies): .1000E-05
Mud cake porosity  (decimal nbr): .1000E+00
Mud solid fraction (decimal nbr): .3000E+00
Viscosity of invading fluid (cp): .1000E+01
Viscosity, displaced  fluid (cp): .2000E+01
Pressure at well boundary  (psi): .5000E+03
Pressure, effective radius (psi): .4000E+03
Radius  of the  well bore (feet): .5000E+00
Reservoir, effective radius (ft): .1000E+02
Rspurt > Rwell radius @ t=0 (ft): .6000E+00
Maximum allowed  number of hours: .1000E+06

T =  .9521E+00 sec, Rf =  .6000E+00 ft, Rc =  .5000E+00 ft
T =  .3242E+05 sec, Rf =  .7000E+00 ft, Rc =  .4875E+00 ft
T =  .1270E+06 sec, Rf =  .8000E+00 ft, Rc =  .4726E+00 ft
T =  .3132E+06 sec, Rf =  .9000E+00 ft, Rc =  .4551E+00 ft
T =  .6284E+06 sec, Rf =  .1000E+01 ft, Rc =  .4348E+00 ft
T =  .1121E+07 sec, Rf =  .1100E+01 ft, Rc =  .4112E+00 ft
T =  .1856E+07 sec, Rf =  .1200E+01 ft, Rc =  .3836E+00 ft
T =  .2921E+07 sec, Rf =  .1300E+01 ft, Rc =  .3512E+00 ft
T =  .4445E+07 sec, Rf =  .1400E+01 ft, Rc =  .3124E+00 ft
T =  .6630E+07 sec, Rf =  .1500E+01 ft, Rc =  .2646E+00 ft
T =  .9862E+07 sec, Rf =  .1600E+01 ft, Rc =  .2012E+00 ft
T =  .1525E+08 sec, Rf =  .1700E+01 ft, Rc =  .9510E-01 ft
T =  .1565E+08 sec, Rf =  .1705E+01 ft, Rc =  .8616E-01 ft
T =  .1607E+08 sec, Rf =  .1710E+01 ft, Rc =  .7614E-01 ft
T =  .1653E+08 sec, Rf =  .1715E+01 ft, Rc =  .6454E-01 ft
T =  .1704E+08 sec, Rf =  .1720E+01 ft, Rc =  .5030E-01 ft
T =  .1763E+08 sec, Rf =  .1725E+01 ft, Rc =  .2982E-01 ft
T =  .1776E+08 sec, Rf =  .1726E+01 ft, Rc =  .2368E-01 ft
T =  .1792E+08 sec, Rf =  .1727E+01 ft, Rc =  .1523E-01 ft
Borehole plugged by mudcake ... run terminated.
```

FIG. 19.7 Numerical results, forward invasion simulation.

We wish to illustrate the use of the inverse time lapse analysis model inferred by Eqs. (19.19), (19.20a), (19.20b), and (19.20c). Once we input the known information found in the Input Parameter Summary printed in Fig. 19.8, including the borehole pressure, we shall attempt two predictions for *simultaneous pore pressure, formation permeability, and hydrocarbon viscosity*, using two different sets of time-dependent front displacement data. Unlike the inverse example in Fig. 19.7, however, each set of data now consists of three readings, and not two, because of the additional unknown introduced.

```
INPUT PARAMETER SUMMARY:
Mud pressure  in bore hole (psi): .5000E+03
Rock core porosity (decimal nbr): .2000E+00
Mud cake permeability  (darcies): .1000E-05
Mud cake porosity  (decimal nbr): .1000E+00
Mud solid fraction (decimal nbr): .3000E+00
Viscosity of  mud  filtrate (cp): .1000E+01
Radius  of the  well bore (feet): .5000E+00
Reservoir, effective radius (ft): .1000E+02
Rspurt > Rwell radius @ t=0 (ft): .6000E+00

TIME LAPSE ANALYSIS PREDICTIONS:

Trial No. 1:
Time of the 1st data point (sec): .3242E+05
Radius  of invasion front (feet): .7000E+00
Time of the 2nd data point (sec): .3132E+06
Radius  of invasion front (feet): .9000E+00
Time of the 3rd data point (sec): .1856E+07
Radius  of invasion front (feet): .1200E+01
Formation permeability (darcies): .8404E-03
Viscosity, formation  fluid (cp): .1670E+01
Pore pressure in reservoir (psi): .3999E+03

Trial No. 2:
Time of the 1st data point (sec): .3242E+05
Radius  of invasion front (feet): .7000E+00
Time of the 2nd data point (sec): .1856E+07
Radius  of invasion front (feet): .1200E+01
Time of the 3rd data point (sec): .1525E+08
Radius  of invasion front (feet): .1700E+01
Formation permeability (darcies): .1008E-02
Viscosity, formation  fluid (cp): .2017E+01
Pore pressure in reservoir (psi): .4000E+03
```

FIG. 19.8 Numerical results, inverse invasion simulation.

Observe that the calculated values of permeability, 0.8404 md and 1.008 md, agree well with the assumed 1 md; the calculated hydrocarbon viscosities, 1.670 cp and 2.017 cp, agree with the assumed 2 cp; and finally, the calculated pore pressures, 399.9 psi and 400.0 psi, agree with the assumed 400 psi from Fig. 19.6. Because the only error in the inversion process considered here is truncation error in our three-decimal-place accurate assumptions, the calculated results here and in the previous example provide some indication of computational sensitivities. The preceding results show that, at least in this limited study, pore pressure can be accurately obtained from time lapse analysis. Continuing research in sensitivity analysis will be required if time lapse analysis is to be successful. More than likely, those predicted parameters that prove to be unstable should be obtained by other logging means; such measurements can augment the capabilities developed here. We have developed time lapse analysis methods that are contingent upon the existence of sharp fronts and transitions. Log analysts who deal with resistivity interpretation and tornado charts, however, have rightly criticized those obvious deficiencies that arise in the modeling of resistivity variations using multilayer step and even straight-line ramped profiles. Radii used in the former are arbitrarily selected by eye, while the latter ramped profiles do not resemble real diffused ones with smoothed corners. Later, using ideas borrowed from seismic migration, where a parabolized wave equation is used to image underground formations, we address resistivity migration. An arbitrarily smeared, transient concentration profile is undiffused or migrated backward in time to produce the original step discontinuity. The distinct front radius obtained can be used with time lapse analysis formulas derived here for piston-like flows. (The method is tested using computer generated synthetic lineal and radial flow data.) Similarly, we can unshock saturation discontinuities in two-phase immiscible flow, to recover the original smooth flows for further study; this is demonstrated numerically.

PROBLEMS AND EXERCISES

1. Different fluid transients exist in the reservoir, for instance: (i) unsteady effects arising from fluid compressibility, (ii) time-dependent effects due to changing proportions of oil and water in two-phase, constant density, immiscible flow, and (iii) transient effects due to miscible mixing in incompressible flow. For successful time lapse logging, the correct fluid-dynamical process must be identified in order to use the appropriate interpretation model. Define such a strategy, assuming that you have access to data from other logging instruments.
2. Using the exact solution for transient ellipsoidal flow of compressible liquids in transversely isotropic media derived in Chapter 18, develop an interpretation method to determine k_h and k_v assuming that pressure histories at the source probe and another observation probe are available. For typical formation tester pumping rates, what is the optimum "transmitter to receiver" probe separation for maximum pressure resolution?
3. In the time lapse logging examples considered in this chapter, distinct fluid front positions are assumed to be known as a function of time. In practice, these distances would be inferred from resistivity readings made at different points in time, for example, while drilling, while tripping, and thirty days later. To what extent is this possible in the near future? What types of resistivity tools are available, and when are they used? What are their depth of investigation and vertical resolution capabilities? Resistivity tools actually measure volumetrically averaged quantities rather than properties at a point.

What modeling and computational issues are involved when translating average quantities to point values?

4. In electromagnetic logging, phase delays between transmitter and receiver are used to predict resistivity. Starting with the diffusion model implied by Eq. (18.5), show that, analogously, time delays between an oscillating formation tester piston and an observation probe can be used to predict permeability (Proett and Chin, 2000).

Chapter 20

Complex Invasion Problems: Numerical Modeling

In this chapter, we introduce numerical methods for the solution of complicated invasion problems, and in particular, we use modern finite difference equation modeling. We develop the basic ideas from first principles, initially for steady-state problems and then for problems with moving boundaries. Our discussions, mathematical, numerical, and physical, are self-contained and presented in an easy-to-read manner. Numerical analogies corresponding to the constant density flow analytical models given in Chapter 17 are derived first, coded in Fortran, explained, and executed. Then, computed results are given to illustrate the simulations and to demonstrate their physical correctness. These models include linear and radial incompressible flows, with and without mudcake. Once our basic approach to moving boundary value problems is understood, the numerical modeling is extended to include other real-world effects. These include transients that arise from fluid compressibility, gas displacement by liquids, and mudcake compressibility and compaction. We continue the discussion of piston, slug, or plug-like displacements initiated in Chapter 17 for single-phase flows. The numerical concepts developed in the course of this modeling are generalized to miscible and immiscible flows in Chapter 21. The work in this book on invasion dynamics and numerical simulation, fully self-contained, appears in few petroleum publications. While the modeling concepts used are powerful, the computer implementation is reasonably straightforward. Minor prerequisites include course work in elementary calculus and undergraduate petroleum reservoir flow analysis.

FINITE DIFFERENCE MODELING

Exact analytical solutions to practical engineering problems are rare, and recourse to numerical solutions is often necessary. Finite element, boundary integral (a.k.a., panel), and finite difference methods have been successfully used to solve complicated engineering problems. Recently, new finite difference technologies have been introduced to the petroleum industry. The work of Chin (1992a,b, 2001a,b) applies these methods to annular borehole flow and pipeline modeling, while the approaches of Chin (1993a,b) introduce rigorous modeling concepts to reservoir flow simulation. Chin (1994) applies

Quantitative Methods in Reservoir Engineering. http://dx.doi.org/10.1016/B978-0-12-810518-4.00020-7

finite difference methods to wave propagation problems such as drillstring vibrations, MWD telemetry, and swab-surge. In this chapter, we extend the finite difference techniques introduced in Chapter 7 to more difficult reservoir flow problems. These extremely powerful methods, which can be mastered with a minimum amount of higher math, in fact require no more than a background in simple calculus. Thus, we develop the fundamental ideas, and rapidly progress to state-of-the-art algorithms for steady and transient invasion problems.

Basic Formulas

Let us consider a differentiable function f(x) at three consecutive equidistant locations x_{i-1}, x_i, and x_{i+1}, where i - 1, i, and i + 1 are indexing parameters. Here, we shall assume that all grids are uniformly separated by the constant grid block distance Δx. Now, it is clear from Fig. 20.1 that the first derivative at an intermediate point A between x_{i-1} and x_i is

$$df(x_A)/dx = (x_i - x_{i-1})/\Delta x \tag{20.1}$$

while the first derivative at an intermediate point B between x_i and x_{i+1} is

$$df(x_B)/dx = (x_{i+1} - x_i)/\Delta x \tag{20.2}$$

Hence, the second derivative of f(x) at x_i satisfies

$$d^2f(x_i)/dx^2 = \{df(x_B)/dx - df(x_A)/dx\}/\Delta x \tag{20.3}$$

or

$$d^2f(x_i)/dx^2 = \{f_{i-1} - 2f_i + f_{i+1}\}/(\Delta x)^2 + O(\Delta x)^2 \tag{20.4}$$

Taylor series analysis shows that Eq. (20.4) is second-order accurate in Δx. The $O(\Delta x)^2$ notation describes the order of the truncation error. If Δx is *small*, then $O(\Delta x)^2$ may be regarded as *very small*. Likewise, it is known that

$$df(x_i)/dx = \{f_{i+1} - f_{i-1}\}/(2\Delta x) + O(\Delta x)^2 \tag{20.5}$$

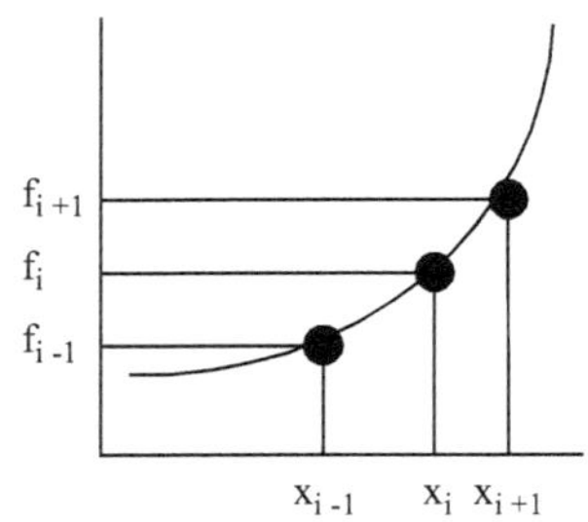

FIG. 20.1 Finite difference discretizations.

is second-order accurate. Eqs. (20.4) and (20.5) are central difference representations for the respective quantities at x_i because they involve left and right quantities at x_{i-1} and x_{i+1}. Note that the backward difference formula

$$d^2f(x_i)/dx^2 = \{f_i - 2f_{i-1} + f_{i-2}\}/(\Delta x)^2 + O(\Delta x) \quad (20.6)$$

for the second derivative is *not* incorrect. But it is not as accurate as the central difference formula, since it turns out to be first-order accurate, the error being only *somewhat small*. Similar comments apply to the forward differencing

$$d^2f(x_i)/dx^2 = \{f_i - 2f_{i+1} + f_{i+2}\}/(\Delta x)^2 + O(\Delta x) \quad (20.7)$$

Alternative representations for the first derivative are the first-order accurate backward and forward difference formulas

$$df(x_i)/dx = \{f_i - f_{i-1}\}/\Delta x + O(\Delta x) \quad (20.8)$$

$$df(x_i)/dx = \{f_{i+1} - f_i\}/\Delta x + O(\Delta x) \quad (20.9)$$

Despite their lesser accuracy, backward and forward difference formulas are often used for practical reasons. For example, they are applied at the boundaries of computational domains. At such boundaries, central difference formulas (e.g., Eqs. 20.4 and 20.5) require values of i that are outside the domain, and hence, undefined. Although high-order accurate backward and forward difference formulas are available, their use often forces simple matrix structures into numerical forms that are not suitable for efficient inversion.

Model Constant Density Flow Analysis

The basic ideas behind the numerical solution of differential equations are reviewed using

$$d^2p(x)/dx^2 = 0 \quad (20.10)$$

whose solution $p(x) = Ax + B$ is determined by two side constraints. Suppose we supplement Eq. (20.10) with the left and right boundary conditions

$$p(0) = P_l \quad (20.11)$$

$$p(x = L) = P_r \quad (20.12)$$

The steady-state pressure solution, applicable to constant density, lineal, liquid flows in a homogeneous core, is

$$p(x) = (P_r - P_l)x/L + P_l \quad (20.13)$$

Suppose that we wish to solve Eq. (20.10) numerically. We introduce along the x-axis the indexes $i = 1, 2, 3, \ldots, i_{max}-1, i_{max}$, where $i = 1$ and i_{max} correspond to the left- and right-side core ends $x = 0$ and $x = L$ (e.g., see Fig. 20.1). With this convention, the constant width grid block size Δx used takes the value

$\Delta x = L/(i_{max} - 1)$. Now, at any position x_i (or simply i), the second derivative in Eq. (20.10) can be approximated using Eq. (20.4), that is,

$$d^2p(x_i)/dx^2 = \{p_{i-1} - 2p_i + p_{i+1}\}/(\Delta x)^2 + O(\Delta x)^2 = 0 \qquad (20.14)$$

so that the finite difference model for our differential equation becomes

$$p_{i-1} - 2p_i + p_{i+1} = 0 \qquad (20.15)$$

The pressures $p_1, p_2, \ldots,$ and p_{imax} at the nodes $i = 1, 2, \ldots,$ and i_{max} are determined by writing Eq. (20.15) for each internal node $i = 2, 3, \ldots,$ and i_{max-1}. This yields $i_{max} - 2$ linear algebraic equations, two short of the number of unknowns i_{max}. The two additional required equations are obtained from boundary conditions, in this case, Eqs. (20.11) and (20.12); in particular, we write $p(0) = P_l$ and $p(L) = P_r$ in the form $p_1 = P_l$ and $p_{imax} = P_r$. To illustrate this, consider the simple case of five nodes (that is, four grid blocks), taking $i_{max} = 5$ and with the grid size $\Delta x = L/(i_{max} - 1) = L/4$. We therefore have

$$\begin{array}{ll} & p_1 = P_l \\ i = 2: & p_1 - 2p_2 + p_3 = 0 \\ i = 3: & p_2 - 2p_3 + p_4 = 0 \\ i = 4: & p_3 - 2p_4 + p_5 = 0 \\ & p_5 = P_r \end{array} \qquad (20.16a,b,c,d,e)$$

Eqs. (20.16a,b,c,d,e) constitute five equations in five unknowns and easily yield to solution, using standard (but tedious) determinant or Gaussian elimination methods from elementary algebra. We could stop here, but we take the solution of Eqs. (20.16a,b,c,d,e) one step further in order to develop efficient solution techniques. The simplicity seen here suggests that we can rewrite the system shown in Eqs. (20.16a,b,c,d,e) in the matrix or linear algebra form

$$\begin{vmatrix} 1 & 0 & & & \\ 1 & -2 & 1 & & \\ & 1 & -2 & 1 & \\ & & 1 & -2 & 1 \\ & & & 0 & 1 \end{vmatrix} \begin{vmatrix} p_1 \\ p_2 \\ p_3 \\ p_4 \\ p_5 \end{vmatrix} = \begin{vmatrix} P_l \\ 0 \\ 0 \\ 0 \\ P_r \end{vmatrix} \qquad (20.17)$$

The left-side coefficient matrix multiplying the unknown vector **p** is said to be banded because its elements fall within diagonal bands. The product shown equals the nonzero right side in Eq. (20.17), which contains the delta-p pressure drop ($P_l - P_r$) that drives the Darcy flow. This delta-p, applied across the entire core, mathematically manifests itself by controlling the top and bottom rows of the governing tridiagonal matrix equation.

It is also interesting to note that our use of central differences physically implies that the pressure at each and every point depends on its left and right neighbors, so that coupled equations necessarily appear. This is not true in certain supersonic flow problems in high-speed aerodynamics, governed by hyberbolic PDEs, where the time-like properties of some space variables may in fact require the use of backward differences! Also observe that the coefficient matrix in Eq. (20.17) is sparse (or empty), with each equation containing at most three unknowns. If each equation had approached i_{max} number of unknowns, the coefficient matrix would have been said to be full. Furthermore, note that our banded matrix possesses a simple tridiagonal (or three-diagonal) structure that is amenable to rapid solution. We shall not review tridiagonal solvers here. They are standard in linear algebra, and we simply note that Eq. (20.17) represents a special instance of

$$\begin{vmatrix} B_1 & C_1 & & & \\ A_2 & B_2 & C_2 & & \\ & A_3 & B_3 & C_3 & \\ & & \cdots\cdots & & \\ & & & & \\ & A_{imax-1} & B_{imax-1} & C_{imax-1} & \\ & & A_{imax} & B_{imax} & \end{vmatrix} \begin{vmatrix} V_1 \\ V_2 \\ V_3 \\ \cdot\cdot \\ \\ V_{imax-1} \\ V_{imax} \end{vmatrix} = \begin{vmatrix} W_1 \\ W_2 \\ W_3 \\ \cdots \\ \\ W_{imax-1} \\ W_{imax} \end{vmatrix} \qquad (20.18)$$

for the unknown vector **V** which, when programmed in the Fortran language, is easily solved by a call to the subroutine TRIDI in Fig. 20.2.

```
      SUBROUTINE TRIDI(A,B,C,V,W,N)
      DIMENSION A(1000), B(1000), C(1000), V(1000), W(1000)
      A(N)  = A(N)/B(N)
      W(N)  = W(N)/B(N)
      DO 100   I = 2,N
      II = -I+N+2
      BN = 1./(B(II-1)-A(II)*C(II-1))
      A(II-1)  = A(II-1)*BN
      W(II-1)  =  (W(II-1)-C(II-1)*W(II))*BN
100   CONTINUE
      V(1)  = W(1)
      DO 200   I = 2,N
      V(I)  = W(I)-A(I)*V(I-1)
200   CONTINUE
      RETURN
      END
```

FIG. 20.2 Tridiagonal equation solver.

Thus, once the coefficient matrixes **A**, **B**, **C**, and **W** are defined in the main body of the computer program, with $B_1 = 1$, $C_1 = 0$, $W_1 = P_l$; $A_2 = A_3 = A_4 = 1$. $B_2 = B_3 = B_4 = -2$, $C_2 = C_3 = C_4 = 1$, $W_2 = W_3 = W_4 = 0$, and, finally, $A_5 = 0$, $B_5 = 1$, $W_5 = P_r$, the statement CALL TRIDI(A,B,C,P,W,5) will solve and store the pressure solution in the elements of **P**. For machine purposes, we shall typically initialize memory using the dummies A(1) = 99 and C(IMAX) = 99, noting that these values do not affect the solution. In general, the internal coefficients are easily defined using the code fragment,

```
      DO 200  I=2,IMAXM1
      A(I) =  1.
      B(I) = -2.
      C(I) =  1.
      W(I) =  0.
200   CONTINUE
```

which is followed by the subroutine call to the tridiagonal matrix solver (in our Fortran, IMAXM1 denotes IMAX-1). In this chapter, we shall study how the engine in the above Fortran will change from problem to problem. For $d^2p(x)/dx^2 = 0$, the exact linear pressure variation will always be obtained for any choice of grid number; unfortunately, this is not so with more complicated equations and formulations. The reader who is not familiar with Fortran should program, execute, and understand this simple example. The ensuing programs in this and Chapter 21 build upon this and are slightly more complicated.

Transient Compressible Flow Modeling

The governing equation for transient, compressible, single-phase, *liquid* flows through homogeneous cores is given by the classical heat equation

$$\partial^2 p(x,t)/\partial x^2 = (\phi\mu c/k)\,\partial p/\partial t \qquad (20.19)$$

Eq. (20.19) provides a useful vehicle for introducing basic ideas and for testing difference schemes for use in forward simulation or time-marching, that is, in modeling events as they evolve in time for given parameters and auxiliary conditions. As before, we shall solve Eq. (20.19) by approximating it with algebraic equations at the nodes formed by a net of coordinate lines, but now, the time coordinate must also be discretized at uniform time intervals. Hence, we deal with numerical solutions in the x-t plane. We replace our space-time continuum with independent variables formed by a discrete set of spatial points $x_i = i\,\Delta x$, where $i = 1, 2, 3, \ldots, i_{max}$, and a discrete set of time points $t_n = n\,\Delta t$, where $n = 1, 2, \ldots$ and so on. We shall represent the function p(x,t) as $P_{i,n}$. We expect that, at any time t_n, the function $P_{i,n}$ at any point x_i will be influenced by its left and right neighbors, so that the central difference formula

$$p_{xx}(x_i, t_n) = \left(P_{i+1,n} - 2P_{i,n} + P_{i-1,n}\right)/(\Delta x)^2 \qquad (20.20)$$

holds. Central differences, however, cannot be used for time derivatives. Since causality requires that events must depend on past and not future history, backward differences apply. Thus, following Eq. (20.8), we must write

$$p_t(x_i, t_n) = \left(P_{i,n} - P_{i,n-1}\right)/\Delta t \qquad (20.21)$$

Then, substitution of Eqs. (20.20) and (20.21) in Eq. (20.19) shows that a difference approximation to the governing partial differential equation is

$$(P_{i+1,n} - 2P_{i,n} + P_{i-1,n})/(\Delta x)^2 = (\phi\mu c/k)\,(P_{i,n} - P_{i,n-1})/\Delta t \tag{20.22}$$

which is $O\{(\Delta x)^2\}$ correct in space but only $O(\Delta t)$ correct in time. Now, we can rewrite Eq. (20.22) in the form $P_{i+1,n} - 2P_{i,n} + P_{i-1,n} = \{\phi\mu c(\Delta x)^2/(k\Delta t)\}(P_{i,n} - P_{i,n-1})$, so that

$$\begin{aligned} P_{i-1,n} - \left[2 + \left\{\phi\mu c(\Delta x)^2/(k\Delta t)\right\}\right] P_{i,n} + P_{i+1,n} \\ = -\left\{\phi\mu c(\Delta x)^2/(k\Delta t)\right\} P_{i,n-1} \end{aligned} \tag{20.23}$$

But Eq. (20.23) for the t_n solution is identical to Eq. (20.15), that is, to $p_{i-1} - 2p_i + p_{i+1} = 0$, except in two minor respects. The 2 in the simpler finite difference equation is replaced by $2 + \{\phi\mu c(\Delta x)^2/(k\Delta t)\}$, while the right-side 0 is replaced by the term $-\{\phi\mu c(\Delta x)^2/(k\Delta t)\}P_{i,n-1}$ assumed to be available from the computed solution in just one earlier time step. For $n = 2$, the $P_{i,2-1}$ or $P_{i,1}$ solution is simply the prescribed initial condition p(x,0). The t_n level solution is obtained as in our foregoing example; that is, Eq. (20.23) is written for each of the internal nodes $i = 2, 3, \ldots, i_{max} - 1$. Left- and right-side boundary conditions are introduced to supplement the resulting incomplete set of algebraic equations. The tridiagonal subroutine is used to solve for the t_n level solution as a function of space. Once this solution is available, it is used to evaluate the right side of Eq. (20.23), and the left side is solved once more in a recursive manner to produce pressure solutions at the subsequent time step.

We emphasize that, in Eq. (20.22) for the time t_n, both $\partial^2 p/\partial x^2$ and $\partial p/\partial t$ are evaluated at the nth time level. This leads to our use of matrix solvers, since all of the resulting nodal equations are algebraically coupled. Finite difference schemes that require matrix inversion are known as implicit schemes. On the other hand, if we had approximated the derivative $\partial^2 p/\partial x^2$ at the earlier (n - 1)th time step, we would have obtained

$$(P_{i+1,n-1} - 2P_{i,n-1} + P_{i-1,n-1})/(\Delta x)^2 = (\phi\mu c/k)(P_{i,n} - P_{i,n-1})/\Delta t \tag{20.24}$$

It is clear from Eq. (20.24) that $P_{i,n}$ can be solved for explicitly by hand in terms of $P_{i-1,n-1}$, $P_{i,n-1}$, and $P_{i+1,n-1}$, thus making matrix inversion unnecessary. Then, $P_{i,n}$ can be updated for every internal i index directly, using a simple calculator. Such explicit schemes, useful when computing machines were uncommon, are less stable than implicit ones, but exceptions can be found.

Numerical Stability

To researchers and practitioners alike, nothing strikes greater fear about simulation than numerical instabilities. Computational instabilities manifest themselves through unrealistic oscillations in pressure buildup or drawdown curves, in unexpected wiggly spatial pressure distributions, and in $O(10^{10}$ psi) overflow messages. How can instabilities be avoided in the development process? One useful tool is the von Neumann stability test. Numerical analysts employ stability tests to evaluate candidate algorithms before embarking on resource-consuming programming efforts. We shall study stability in detail later, but for now, we consider the model heat equation $u_t = u_{xx}$ for $u = u(x,t)$. Let us presume that a discretized u can be approximated by $v(x_i,t_n)$, where $v_{i,n}$ satisfies the explicit model $(v_{i,n+1} - v_{i,n})/\Delta t = (v_{i-1,n} - 2v_{i,n} + v_{i+1,n})/(\Delta x)^2$, where Δt and Δx are time and spatial increments.

How useful is this obvious difference approximation? To obtain some mathematical insight, let us separate variables, and consider an elementary Fourier wave component $v_{i,n} = \psi(t)e^{j\beta x}$, where $j = \sqrt{-1}$. Substitution then yields $\{\psi(t+\Delta t) e^{j\beta x} - \psi(t)e^{j\beta x}\}/\Delta t = \psi(t)[e^{j\beta(x-\Delta x)} - 2e^{j\beta x} + e^{j\beta(x+\Delta x)}]/(\Delta x)^2$. Thus, $\psi(t+\Delta t) = \psi(t) (1 - 4\lambda \sin^2 \beta \Delta x/2)$, where $\lambda = \Delta t/(\Delta x)^2$. And since $\psi(0) = 1$, we have the solution $\psi(t) = (1 - 4\lambda \sin^2 \beta \Delta x/2)^{t/\Delta t}$. For stability, $\psi(t)$ must remain bounded (or finite) as Δt, and thus Δx, approaches zero. This requires that the absolute value $|1 - 4\lambda \sin^2 \beta \Delta x/2| < 1$, thereby establishing definite requirements connecting Δx and Δt. We need not have solved for $\psi(t)$, of course. For example, we could have defined an amplification factor $\alpha = |\psi(t+\Delta t)/\psi(t)|$ from the original equation and determined that $\alpha = |1-4\lambda \sin^2 \beta \Delta x/2| < 1$, leading to the same requirement. Also observe that for large values of $\lambda = \Delta t/(\Delta x)^2$, the time-marching scheme becomes unstable, that is, the explicit method is conditionally stable. Later in this chapter, an absolutely or unconditionally stable implicit scheme is devised for the heat equation for cylindrical and spherical radial coordinates, which reduces to Eq. (20.23) in the lineal limit. We shall prove its von Neumann stability at that time.

Convergence

In our differencing of u(x,t), we denoted its numerical representation by $v_{i,n}$. That u may not, in fact, tend to v, is suggested by this usage. In computational fluid dynamics, the exact functional form of the formally small truncation error is all-important because it determines the type of passive higher-order derivative term that controls the structure of the solution. This determines how well computed solutions actually mirror those of the given partial differential equation. This point was developed in Chapter 13: without evaluating the kinds of derivatives characterizing the neglected terms, whose diffusive versus dispersive effects always remain with the computed solution, the extent to which an "obvious" difference scheme actually models a differential equation cannot be ascertained.

It is also important that the tridiagonal structure in Eq. (20.23) is diagonally dominant; that is, the absolute value of the middle diagonal coefficient, being $2+\{\phi\mu c(\Delta x)^2/(k\Delta t)\} > 2 = 1 + 1$, exceeds the sum of the (unity) coefficients of the side diagonals. This property lends itself to numerical stability, meaning that iterative solutions are not likely to blow up as a result of truncation and round-off errors. This does not guarantee that the computed solutions are correct, but it does buttress the accepted (but questionable) philosophy that any solution is better than no solution. As should be clear from Eq. (20.23), only one additional time level of the solution needs to be stored at any given point, so that two levels of information are required in total. Thus, the Fortran associated with our scheme can be written using two-dimensioned scalar arrays PN(1000) and PNM1(1000) only, representing $P_{i,n}$ and $P_{i,n-1}$, where the Fortran dimension of 1000 might signify 1000 closely spaced nodes. At the end of each time step, we copy PN into PNM1 and repeatedly apply the time-recursive procedure until termination.

It is not necessary (or advisable) to have computer RAM memory allocated for a complete field P(1000,500), say, representing 1000 nodes, and 500 time steps. Intermediate results, such as displacement front location, mudcake thickness, and pressure distributions, can be written to output files for subsequent post-processing and display. Also note that the coefficients **A**, **B**, and **C** need not be recomputed for subsequent time steps, since they are constants defined once and for all. The matrix solver TRIDI in Fig. 20.2 *will* destroy **A**, **B**, **C**, and **W** at the end of each inversion, so that they require redefinition prior to each integration. (Other solvers are available which retain their input values at the expense of increased memory requirements.)

Multiple Physical Time and Space Scales

In Chapter 17, we considered transient front motions whose time scales depend on the relative viscosities of invading and displaced fluids. In addition to these time scales, there now exist additional ones associated with the presence of multiple fluids having different compressibilities. In using computer programs such as those derived here and similar programs available in the industry, it is important to recognize that whether or not calculated solutions capture all the physics associated with these time scales will depend on the filtering effects of grids used, that is, on Δx, Δt, and their ratio. Unfortunately, there are no obvious answers, and it is the engineering evaluation of particular computed solutions as they relate to real-world problems that poses the greatest challenge. In this book, we only demonstrate how algorithms and programs are constructed. We shall *not* delve into grid sensitivity studies and similar validation work, as our goals and objectives are strictly tutorial. Furthermore, our choices for parameters are motivated by simplicity and comparative purposes only, and are not intended to be representative of any particular oil reservoir. With these preliminary remarks completed, let us introduce the subject of numerical invasion simulation with a sequence of examples designed to cover a broad range of physical problems. Independent formulation parameters encompass (i) lineal, cylindrical, and spherical flow domains, (ii) constant density and compressible flow, (iii) possibly dissimilar fluids in formations, (iv) gas versus liquid problems, and finally, (v) the presence of mudcakes with or without compaction.

EXAMPLE 20.1. LINEAL LIQUID DISPLACEMENT WITHOUT MUDCAKE

We have shown how $d^2p(x)/dx^2 = 0$ is easily solved. We now return to an early example for the piston-like Darcy displacement of two constant density *liquids* with different viscosities in a homogeneous lineal core of given permeability k. The transient displacement depends on the relative proportions of fluid initially present and on which portions of the core (i.e., upstream or downstream) they occupy. Now $d^2p(x)/dx^2 = 0$ applies to constant density liquids, but parametric time dependence in the solution is permissible. In this problem, since two liquids are present, two such equations are needed,

$$d^2p_i(x)/dx^2 = 0, i = 1, 2 \tag{20.25}$$

for the first (left) and second (right) sections. For numerical purposes, it will be convenient to define an unknown, upper-case solution vector P(x) by

$$\begin{aligned} P(x) &= p_1(x), \quad 0 < x < x_f \\ &= p_2(x), \quad x_f < x < L \end{aligned} \tag{20.26}$$

where $x = x_f(t)$ represents the position of the unsteady moving front. The boundary value problem for $d^2P(x)/dx^2 = 0$ satisfies the left- and right-side pressure boundary conditions

$$p_1(0) = P_l \tag{20.27a}$$

$$p_2(L) = P_r \tag{20.27b}$$

which are easily programmed as demonstrated earlier. Now, the difference equation corresponding to $d^2P(x)/dx^2 = 0$ at $x = x_f$ does not apply, since the differential equation description of motion breaks down at the boundary separating two distinct fluids where pressure gradients need not be continuous. We therefore replace that equation with an alternative statement that encompasses the requirements posed by the interfacial matching conditions

$$p_1(x_f) = p_2(x_f) \tag{20.28a}$$

$$q_1(x_f) = q_2(x_f) \tag{20.28b}$$

This can be done in any number of ways, but the best choice is a technique that can be carried over to transient compressible flows without modification and that allows us to retain the diagonally dominant features of the time-marching scheme derived earlier. The final result is easily derived.

First, Eq. (20.28b) requires $-(k_1/\mu_1)dp_1(x_f)/dx = -(k_2/\mu_2)dp_2(x_f)/dx$ as a consequence of Darcy's law $q = -(k/\mu)dp(x)/dx$. But since $k_1 = k_2$, this statement simplifies to $(1/\mu_1)dp_1(x_f)/dx = (1/\mu_2)dp_2(x_f)/dx$. Now, we shall denote by *if-* and *if+* the spatial locations infinitesimally close to the left and to the right of the front $x = x_f$, which is itself indexed by $i = if$. (Note that this index satisfies $if = x_f/\Delta x + 1$ in our nodal convention.) Then, in Section 1, we can approximate the pressure gradient $dp_1(x_f)/dx$ using backward differences, while in Section 2, we can apply forward differences (again, differentiation through the interface itself is forbidden since the pressure gradient changes suddenly). This leads to $(1/\mu_1)(p_{if-} - p_{if-1})/\Delta x = (1/\mu_2)(p_{if+1} - p_{if+})/\Delta x$, but since Δx cancels, $(1/\mu_1)(p_{if-} - p_{if-1}) = (1/\mu_2)(p_{if+1} - p_{if+})$. Assuming that surface tension is unimportant,

Eq. (20.28a), which calls for pressure continuity, requires that $p_{if-} = p_{if+}$ or simply p_{if}. Thus, at the interface,

$$(1/\mu_1)\, p_{if-1} - (1/\mu_1 + 1/\mu_2) p_{if} + (1/\mu_2)\, p_{if+1} = 0 \qquad (20.29)$$

applies. However, unlike the difference approximation to the differential equation, which is second-order accurate, our use of backward and forward differences in deriving Eq. (20.29) renders it only $O(\Delta x)$ accurate. In deriving Eq. (20.29), we emphasize that we have used the same mesh size to the left and to the right of the front. This is physically permissible if the two viscosities are comparable, but clearly incorrect if they are not; later, in modeling mudcake flows, we shall find that significant mobility contrasts existing in the problem demand dual mesh systems.

It is interesting, however, to observe that we can rewrite Eq. (20.29) as $p_{if-1} - (1+\mu_1/\mu_2)p_{if} + (\mu_1/\mu_2)p_{if+1} = 0$. In the single-fluid problem where $\mu_1 = \mu_2$, this matching condition reduces to $p_{if-1} - 2p_{if} + p_{if+1} = 0$, which is identical to Eq. (20.15) for the exact differential equation. This fortuitous situation does not apply to compressible transient flows or radial flows. This completes our discussion for the solution of Eqs. (20.25)–(20.28b) for the spatial pressure distribution, which assumes that the front location x_f is prescribed. But the front *does* move with time, and our formulation needs to accommodate this fact. The physical problem is an initial value problem, a transient formulation in which an interface, initially located at $x = x_{f,o}$ moves with time—even though Eq. (20.25) governing time-dependent pressure does not contain time derivatives!

We can solve this unsteady problem by first producing the pressure distribution as just outlined, then updating the front location $x = x_f$, and subsequently, repeating this process recursively, as required. The update formula is obtained from the kinematic requirement that

$$\begin{aligned} dx_f/dt &= u/\phi = -(k/\mu_1\phi)\, dp_1/dx \\ &= -(k/\mu_1\phi)\left(p_{if} - p_{if-1}\right)/\Delta x \end{aligned} \qquad (20.30)$$

in Section 1. This kinematic statement, formally derived in Chapter 16, was used extensively in the analytical invasion modeling pursued in Chapter 17. If we evaluate the right side of Eq. (20.30) with the pressure solution just obtained, denoting existing solutions for p and x_f as *old*, then the *new* x_f is obtained by approximating Eq. (20.30) as

$$\left(x_{f,new} - x_{f,old}\right)/\Delta t = -(k/\mu_1\phi)\left(p_{if} - p_{if-1}\right)_{old}/\Delta x \qquad (20.31)$$

or

$$x_{f,new} = x_{f,old} - \{k\Delta t/(\mu_1\phi\Delta x)\}\left(p_{if} - p_{if-1}\right)_{old} \qquad (20.32)$$

With this new front position available, we again solve for the pressure, followed by a front update, and so on. In Fig. 20.3, the Fortran listing showing the

```
.
C     INITIAL SETUP
      IMAX = XCORE/DX +1
      IMAXM1 = IMAX-1
      IFRONT = XFRONT/DX +1
      .
      N = 0
      T = 0.
      NSTOP = 0
      MINDEX = 1
      TIME(1) = 0.
      XPLOT(1) = XFRONT
C
C     START TIME INTEGRATION
      DO 300  N=1,NMAX
      T = T+DT
      DO 200  I=2,IMAXM1
      A(I) =  1.
      B(I) = -2.
      C(I) =  1.
      W(I) =  0.
 200  CONTINUE
      A(1) = 99.
      B(1) = 1.
      C(1) = 0.
      W(1) = PLEFT

      A(IMAX) = 0.
      B(IMAX) = 1.
      C(IMAX) = 99.
      W(IMAX) = PRIGHT
      IF(VISCIN.EQ.VISCDP) GO TO 240
      A(IFRONT) =  1./VISCL
      B(IFRONT) = -1./VISCL -1./VISCR
      C(IFRONT) =  1./VISCR
      W(IFRONT) =  0.
240   CALL TRIDI(A,B,C,VECTOR,W,IMAX)
      DO 250  I=1,IMAX
      P(I) = VECTOR(I)
250   CONTINUE
      PGRAD = (P(IFRONT)-P(IFRONT-1))/DX
      XFRONT = XFRONT - (K*DT/(PHI*VISCL))*PGRAD
      IFRONT = XFRONT/DX +1
      IF(XFRONT.GE.XMAX.OR.XFRONT.LE.XMIN) NSTOP=1
      .
      .
      WRITE(*,280) N,T,XFRONT,IFRONT
280   FORMAT(1X,'T(',I4,')= ',E8.3,' sec, Xf= ',E8.3,' ft, I= ',I3)
      MINDEX = MINDEX+1
      TIME(MINDEX) = T
      XPLOT(MINDEX) = XFRONT
300   CONTINUE
400   WRITE(*,10)
      CALL GRFIX(XPLOT,TIME,MINDEX)
      STOP
      END
```

(A)

FIG. 20.3 (A) Fortran source code (Example 20.1).

```
   INPUT PARAMETER SUMMARY:
   Rock core permeability (darcies): .100E+00
   Rock core porosity (decimal nbr): .200E+00
   Viscosity of invading fluid (cp): .100E+02
   Viscosity, displaced  fluid (cp): .100E+01
   Pressure at left boundary  (psi): .100E+03
   Pressure at right boundary (psi): .000E+00
   Length of rock core sample  (ft): .100E+01
   Initial "xfront" position (feet): .500E+00
   Integration space step size (ft): .200E-02
   Integration time step size (sec): .100E+01
   Maximum allowed  number of steps: .200E+04

Time (sec)  Position (ft)
  .000E+00    .500E+00     |             *
  .600E+02    .539E+00     |              *
  .120E+03    .576E+00     |               *
  .180E+03    .611E+00     |                *
  .240E+03    .644E+00     |                 *
  .300E+03    .676E+00     |                  *
  .360E+03    .706E+00     |                   *
  .420E+03    .736E+00     |                    *
  .480E+03    .764E+00     |                     *
  .540E+03    .792E+00     |                      *
  .600E+03    .818E+00     |                      *
  .660E+03    .844E+00     |                       *
  .720E+03    .870E+00     |                        *
  .780E+03    .895E+00     |                         *
  .840E+03    .919E+00     |                         *
  .900E+03    .942E+00     |                          *
  .960E+03    .965E+00     |                           *
  .102E+04    .988E+00     |                            *
```

(B)

```
   INPUT PARAMETER SUMMARY:
   Rock core permeability (darcies): .100E+00
   Rock core porosity (decimal nbr): .200E+00
   Viscosity of invading fluid (cp): .100E+01
   Viscosity, displaced  fluid (cp): .100E+02
   Pressure at left boundary  (psi): .100E+03
   Pressure at right boundary (psi): .000E+00
   Length of rock core sample  (ft): .100E+01
   Initial "xfront" position (feet): .500E+00
   Integration space step size (ft): .200E-02
   Integration time step size (sec): .100E+01
   Maximum allowed  number of steps: .200E+04

Time (sec)  Position (ft)
                           ------------------------------
  .000E+00    .500E+00     |             *
  .600E+02    .542E+00     |               *
  .120E+03    .586E+00     |                *
  .180E+03    .635E+00     |                  *
  .240E+03    .690E+00     |                    *
  .300E+03    .753E+00     |                      *
  .360E+03    .830E+00     |                        *
  .420E+03    .938E+00     |                            *
```

(C)

FIG. 20.3, CONT'D (B) Numerical results (Example 20.1). (C) Numerical results (Example 20.1).

structural components of the recursive algorithm is given. The front matching conditions and position updating logic are shown in bold print. Details related to dimension statements, interactive input queries, print statements, and so on, are omitted for brevity. Only those salient features that relate to the algorithm are replicated. Note that the Fortran statement `IFRONT = XFRONT/DX +1`, because `IFRONT` is a Fortran *integer* variable, will discard the fractional part of the right-side division. This means that the algorithm will not move `IFRONT` from one time step to the next unless it has advanced sufficiently. In this sense, the scheme is not truly boundary conforming; however, it is easily modified at the expense of programming complexity. Small meshes, in general, should be used in modeling invasion front motions.

We shall consider *two* computational limits that demonstrate the physics of piston-like fluid displacement, as well as the correctness of the program. For the first example, consider the simulation input and solution in Fig. 20.3B. Note the high viscosity of the invading fluid relative to that of the displaced fluid. The plot and tabulated results correctly show that the front *decelerates* with time. This is so because fluid of increased viscosity displaces and replaces fluid having lower viscosity, with the relative proportion of the former increasing with time, as the low viscosity fluid is forced out the right side of the core. Hence, continual slowdown is anticipated and is indeed obtained. In our second example, we reverse the role of the two fluids and allow a less viscous fluid to displace one having much higher viscosity. As the latter is forced through the core and emptied, fluid having lower viscosity replaces it, so that it naturally *accelerates* through the core. Again, our computed results are physically correct; also note the differences in the time scales of the two problems.

It is clear that our calculations produce results that make physical sense. Of course, in the present problem where an analytical solution is available, there is no need to resort to numerical methods. But the solution is useful because it allows us to study the effects of grid selection, that is, the role of Δx and Δt in affecting computed solutions. We emphasize that the above calculations provide the *time scales* characteristic of the displacement flows. Both fronts start at the midpoint of the core, and both simulations terminate near the end of the core. Their total transit times are obviously different. These time scales, as our earlier closed-form solution

$$(\mu_1/\mu_2-1)x_f + L = +\{\{(\mu_1/\mu_2-1)x_{f,o} + L\}^2 + \{2k(P_l - P_r)/(\phi\mu_2)\}(\mu_1/\mu_2-1)\,t\}^{1/2} \quad (17.13)$$

shows, depend on numerous parameters, combined in well-defined groups. For example, both $(\mu_1/\mu_2 - 1)$ and $2k(P_l - P_r)t/(\phi\mu_2)$ are individually important.

The power of well-formulated numerical models lies, of course, in their potential for simple extension. For example, if the left- and right-side boundary pressures PLEFT and PRIGHT are to be prescribed functions of time, these constants are easily replaced by Fortran function statements. Likewise, time dependences in the left side invading fluid viscosity VISCL are readily incorporated. These generalizations are not unusual to actual drilling situations. Changes in mud weight, which alter borehole pressure, are used for formation control; these changes are effected by varying both solids and viscosifier content. Finally, some notes on the computational efficiency of the scheme are in order. Using an old 1990s vintage Pentium PC, 1000 time steps requires approximately two seconds for a 500 grid block problem, all the time printing intermediate solutions to the screen. (This is the slowest part of the process and can be omitted for increased speed). The compiled code, dimensioned for a maximum of 1000 grid blocks, requires 40,000 bytes of RAM memory. By contrast, canned finite element simulators designed to solve general 3D problems, by contrast, can require orders-of-magnitude more computing times for the same number of steps.

EXAMPLE 20.2. CYLINDRICAL RADIAL LIQUID DISPLACEMENT WITHOUT CAKE

We now rework the preceding problem and alter the formulation so that it handles cylindrical radial flows. Thus, we replace Eq. (20.10) (that is, $d^2p(x)/dx^2 = 0$) by Laplace's equation in cylindrical radial flows,

$$d^2p(r)/dr^2 + (1/r)\,dp(r)/dr = 0 \tag{20.33}$$

The required changes are minor. Using Eq. (20.14), we find that a simple change of notation gives $d^2p(r_i)/dr^2 = \{p_{i-1} - 2p_i + p_{i+1}\}/(\Delta r)^2 + O(\Delta r)^2$. Similarly, from Eq. (20.5), $dp(r_i)/dr = \{p_{i+1} - p_{i-1}\}/(2\Delta r) + O(\Delta r)^2$. We shall define the radial variable r by $r = R_{well} + (i-1)\Delta r$ so that $i = 1$ corresponds to the left boundary of the computational grid. Then, substitution in Eq. (20.33) and minor rearrangement lead to

$$\begin{aligned}[1 - ½\Delta r/\{R_{well} + (i\text{-}1)\Delta r\}]\,p_{i-1} - 2\,p_i \\ + [1 + ½\Delta r/\{R_{well} + (i\text{-}1)\Delta r\}]\,p_{i+1} = 0\end{aligned} \tag{20.34}$$

Recall that the matrix coefficients **A**, **B**, **C**, and **W** of the finite difference equation for the lineal flow model $d^2p(x)/dx^2 = 0$, extracted from the simple formula $[1]\ p_{i-1} - 2\ p_i + [1]\ p_{i+1} = 0$, were defined by the code fragment

```
      DO 200  I=2,IMAXM1
      A(I)  =  1.
      B(I)  = -2.
      C(I)  =  1.
      W(I)  =  0.
200   CONTINUE
```

Comparison with Eq. (20.34) shows that the only required change needed to model fully radial flow effects is a correction $\pm\frac{1}{2}\Delta r/\{R_{well}+(i-1)\Delta r\}$ to the **C** and **A** matrix coefficients. That is, we replace the preceding code with

```
      DO 200  I=2,IMAXM1
      CORRECT = 0.5*DX/(WELRAD + (I-1)*DX)
      A(I) =  1. - CORRECT
      B(I) = -2.
      C(I) =  1. + CORRECT
      W(I) =  0.
200   CONTINUE
```

Of course, there will be additional input and output nomenclature changes, calling for wellbore and farfield radii, starting front radii, and so on. For readability, we have retained DX to represent the radial mesh length Δr, in order to limit the number of typographical changes; WELRAD represents the wellbore radius. The Fortran source code for this example, appearing in Fig. 20.4A, uses the same front matching logic as does lineal flows.

We shall consider *two* computational limits that demonstrate the physics of radial displacement flows, as well as the correctness of the computer program. For the first example, we assume simulation input parameters that are identical

```
C     INITIAL SETUP
      IMAX = (XCORE-WELRAD)/DX +1
      IMAXM1 = IMAX-1
      IFRONT = (XFRONT-WELRAD)/DX +1
      .
      N = 0
      T = 0.
      NSTOP = 0
      MINDEX=1
      TIME(1) = 0.
      XPLOT(1) = XFRONT
C
C     START TIME INTEGRATION
      DO 300  N=1,NMAX
      T = T+DT
      DO 200  I=2,IMAXM1
      CORRECT = 0.5*DX/(WELRAD + (I-1)*DX)
      A(I) =  1. - CORRECT
      B(I) = -2.
      C(I) =  1. + CORRECT
      W(I) =  0.
 200  CONTINUE
      A(1) = 99.
      B(1) = 1.
      C(1) = 0.
      W(1) = PLEFT
```

(A)

FIG. 20.4 (A) Fortran source code (Example 20.2).

```
      A(IMAX) = 0.
      B(IMAX) = 1.
      C(IMAX) = 99.
      W(IMAX) = PRIGHT
      IF(VISCIN.EQ.VISCDP) GO TO 240
      A(IFRONT) =  1./VISCL
      B(IFRONT) = -1./VISCL -1./VISCR
      C(IFRONT) =  1./VISCR
      W(IFRONT) =  0.
240   CALL TRIDI(A,B,C,VECTOR,W,IMAX)
      DO 250  I=1,IMAX
      P(I) = VECTOR(I)
250   CONTINUE
      PGRAD = (P(IFRONT)-P(IFRONT-1))/DX
      XFRONT = XFRONT - (K*DT/(PHI*VISCL))*PGRAD
      IFRONT = (XFRONT-WELRAD)/DX +1
      .
      WRITE(*,280) N,T,XFRONT,IFRONT
280   FORMAT(1X,'T(',I4,')= ',E8.3,' sec, Rf= ',E10.5,' ft,I= ',I3)
      MINDEX = MINDEX+1
      TIME(MINDEX) = T
      XPLOT(MINDEX) = XFRONT
300   CONTINUE
400   WRITE(*,10)
      CALL GRFIX(XPLOT,TIME,MINDEX)
      STOP
      END
```

```
INPUT PARAMETER SUMMARY:
Rock core permeability (darcies): .100E+00
Rock core porosity (decimal nbr): .200E+00
Viscosity of invading fluid (cp): .100E+01
Viscosity, displaced  fluid (cp): .100E+02
Pressure at well boundary  (psi): .100E+03
Pressure, effective radius (psi): .000E+00
Radius   of   the   bore   hole (ft): .100E+03
Reservoir effective radius  (ft): .101E+03
Initial "Rfront" position (feet): .101E+03 (i.e., 100.5)
Integration space step size (ft): .200E-02
Integration time step size (sec): .100E+01
Maximum allowed  number of steps: .200E+04
Number spatial DR grids selected: .500E+03

COMPUTED RESULTS:
T(   0)= .000E+00 sec, Rf= .10050E+03 ft, I= 250
T(  60)= .600E+02 sec, Rf= .10054E+03 ft, I= 271
T( 120)= .120E+03 sec, Rf= .10059E+03 ft, I= 294
T( 180)= .180E+03 sec, Rf= .10064E+03 ft, I= 318
T( 240)= .240E+03 sec, Rf= .10069E+03 ft, I= 346
T( 300)= .300E+03 sec, Rf= .10075E+03 ft, I= 377
T( 360)= .360E+03 sec, Rf= .10083E+03 ft, I= 416
T( 420)= .420E+03 sec, Rf= .10094E+03 ft, I= 470
```

(B)

FIG. 20.4, CONT'D (B) Numerical results (Example 20.2).

(Continued)

```
INPUT PARAMETER SUMMARY:
Rock core permeability (darcies): .100E+00
Rock core porosity (decimal nbr): .200E+00
Viscosity of invading fluid (cp): .100E+01
Viscosity, displaced  fluid (cp): .100E+02
Pressure at well boundary  (psi): .100E+03
Pressure, effective radius (psi): .000E+00
Radius  of  the  bore  hole (ft): .100E+00
Reservoir effective radius  (ft): .110E+01
Initial "Rfront" position (feet): .600E+00
Integration space step size (ft): .200E-02
Integration time step size (sec): .100E+01
Maximum allowed  number of steps: .200E+04
Number spatial DR grids selected: .500E+03

Time (sec)  Position (ft)
                             ______________________________
  .000E+00    .600E+00       |               *
  .600E+02    .647E+00       |                *
  .120E+03    .695E+00       |                  *
  .180E+03    .745E+00       |                   *
  .240E+03    .796E+00       |                     *
  .300E+03    .849E+00       |                      *
  .360E+03    .906E+00       |                        *
  .420E+03    .967E+00       |                          *
  .480E+03    .104E+01       |                            *
```

(C)

FIG. 20.4, CONT'D (C) Numerical results (Example 20.2).

to those of the first run in Example 20.1, plus wellbore and farfield radii of 100 and 101 ft, so that the net radial extent of 1 ft equals the core length of the previous example. This large radius allows the program to mimic purely lineal flows; we shall compare the computed results with those obtained for exact lineal flow. For such large radii, the effect of the radial terms should be insignificant. If so, then the computed radial front positions should be identical to those in Fig. 20.3C. The two-decimal-place bold numbers in Fig. 20.4B, when compared to their three-decimal place counterparts in Fig. 20.3C, demonstrate that exactly the same water-to-oil displacement results are obtained as we expected. This provides a useful computing and programming check.

Next, we consider a physical situation where the geometric effects of radial spreading must be important, and accordingly we select a small slimhole radius of 0.1 ft and a farfield radius of 1.1 ft. These choices therefore fix the length of the core to 1 ft. Again, we initialize our front position to the center of the core sample. Computed results demonstrate important geometric effects. From t = 360 to 420 s, the radial front has advanced from r = 0.906 to 0.967 ft, for a total extent of 0.061 ft. If we refer to Fig. 20.3C for the lineal result, in the same time period, the front has advanced from x = 0.830 to 0.938 ft, for a total of 0.108 ft. The decrease in distance obtained in the radial case is clearly the result of geometric spreading, and the twofold change indicates that such effects can be significant for small-diameter holes. These changes are all-important to resistivity interpretation and modeling.

EXAMPLE 20.3. SPHERICAL RADIAL LIQUID DISPLACEMENT WITHOUT CAKE

Now let us rework the preceding cylindrical radial problem, and alter the analytical and numerical formulations so that they handle spherical radial flows. Such formulations model invasion at the drillbit and also point fluid influx into formation testers at small times. We shall replace the governing equation for cylindrical radial flows, namely, $d^2p(r)/dr^2 + (1/r)dp(r)/dr = 0$ in Eq. (20.33), by the spherical flow equation

$$d^2p(r)/dr^2 + (2/r)\,dp(r)/dr = 0 \tag{20.35}$$

Again, we are restricted to constant density flows in homogeneous rocks. The required changes are minor, since we have merely substituted a "2/r" variable coefficient in favor of 1/r. Instead of Eq. (20.34), we have

$$\begin{aligned}&[1 - \Delta r/\{R_{well} + (i\text{-}1)\Delta r\}]\,p_{i-1} - 2\,p_i\\ &\quad + [1 + \Delta r/\{R_{well} + (i\text{-}1)\Delta r\}]\,p_{i+1} = 0\end{aligned} \tag{20.36}$$

The code fragment

```
      DO 200  I=2,IMAXM1
      CORRECT = 0.5*DX/(WELRAD + (I-1)*DX)
      A(I)  =  1. - CORRECT
      B(I)  = -2.
      C(I)  =  1. + CORRECT
      W(I)  =  0.
200   CONTINUE
```

appearing in the cylindrical radial program requires only a one-line change in order to implement Eq. (20.36), so that instead we have

```
      DO 200  I=2,IMAXM1
      CORRECT = DX/(WELRAD + (I-1)*DX)
      A(I)  =  1. - CORRECT
      B(I)  = -2.
      C(I)  =  1. + CORRECT
      W(I)  =  0.
200   CONTINUE
```

As before, there are obvious input and output nomenclature changes, calling for bit and farfield radii, starting front radii and so on. (Again, for readability, we have retained DX to represent the radial mesh length Δr.) The source code is similar to that in Fig. 20.4A, except for the single line change just described. In order to demonstrate the differences between cylindrical and spherical radial flows, we have assumed parameters identical to those in the second run of Example 20.2. At $t = 480$ s, the *cylindrical* radial position is 1.04 ft, whereas at the same instant, the spherical radial position is 0.852 ft, which is significantly less. As the calculated results in Fig. 20.5A and B show, the spherical

```
INPUT PARAMETER SUMMARY:
Rock core permeability (darcies): .100E+00
Rock core porosity (decimal nbr): .200E+00
Viscosity of invading fluid (cp): .100E+01
Viscosity, displaced  fluid (cp): .100E+02
Pressure at "bit" boundary (psi): .100E+03
Pressure, effective radius (psi): .000E+00
Radius  at  the  drill  bit (ft): .100E+00
Reservoir effective radius  (ft): .110E+01
Initial "Rfront" position (feet): .600E+00
Integration space step size (ft): .200E-02
Integration time step size (sec): .100E+01
Maximum allowed  number of steps: .200E+04
Number spatial DR grids selected: .500E+03
```

(A)

```
Time (sec)  Position (ft)
                                   ______________________________
  .000E+00    .600E+00     |              *
  .600E+02    .637E+00     |               *
  .120E+03    .673E+00     |                *
  .180E+03    .706E+00     |                 *
  .240E+03    .738E+00     |                  *
  .300E+03    .768E+00     |                   *
  .360E+03    .797E+00     |                    *
  .420E+03    .825E+00     |                    *
  .480E+03    .852E+00     |                     *
  .540E+03    .878E+00     |                      *
  .600E+03    .904E+00     |                       *
  .660E+03    .928E+00     |                       *
  .720E+03    .952E+00     |                        *
  .780E+03    .975E+00     |                         *
  .840E+03    .998E+00     |                         *
  .900E+03    .102E+01     |                          *
  .960E+03    .104E+01     |                          *
  .102E+04    .106E+01     |                           *
  .108E+04    .108E+01     |                            *
```

(B)

FIG. 20.5 (A) Numerical results (Example 20.3). (B) Numerical results (Example 20.3).

front requires more time to reach the farfield boundary defined by the effective radius r = 1.1 ft. Its acceleration is less than that in the previous example as a result of increased geometric spreading.

EXAMPLE 20.4. LINEAL LIQUID DISPLACEMENT WITHOUT MUDCAKE, INCLUDING COMPRESSIBLE FLOW TRANSIENTS

In this example, we shall revisit Example 20.1 but include the additional effect of nonvanishing fluid compressibility. This being the case, $d^2p/dx^2 = 0$ is no longer the governing equation. Instead, the governing partial differential equation is the *heat equation*

$$\partial^2 p(x,t)/\partial x^2 = (\phi\mu c/k)\, \partial p/\partial t \tag{20.19}$$

which requires *initial conditions* for spatial pressure distribution in addition to those for front position. Its finite difference approximation, as derived earlier, takes the form

$$\begin{aligned} &P_{i-1,n} - \left[2 + \left\{\phi\mu c(\Delta x)^2/(k\Delta t)\right\}\right] P_{i,n} + P_{i+1,n} \\ &= -\left\{\phi\mu c(\Delta x)^2/(k\Delta t)\right\} P_{i,n-1} \end{aligned} \tag{20.23}$$

instead of the simpler equation

$$(1)\,P_{i-1} - 2\,P_i + (1)\,P_{i+1} = 0 \tag{20.37}$$

derived for $d^2p/dx^2 = 0$. The finite difference program of Example 20.1 can be modified to handle transients due to fluid compressibility. First, the right side of Eq. (20.23) indicates that pressure information from one earlier time step is required before the tridiagonal equations can be solved. Thus, an initial condition is required, so that the program user must enter an initial pressure.

When a new formation is penetrated, the initial pressure will always be equal to the reservoir pore pressure. However, in this book and in the code, we shall leave this input completely general, if only for code flexibility and the possibility that the program will be used in special experimental situations. Once the pressure field in space is obtained for a particular time step, it must be copied into the pressure array for the earlier pressure before pressures can be recursively advanced and integrated in time. The bookkeeping of an earlier time pressure array means that an additional Fortran dimension statement, plus more allocated memory, will be required. Aside from new input statements required for fluid compressibilities, we shall need to modify the matrix coefficients **B** and **W** as required by Eqs. (20.23) and (20.37). That is, the -2 of Eq. (20.37) is to be replaced by the term $-2 - \{\phi\mu c(\Delta x)^2/(k\Delta t)\}$ of Eq. (20.23), while the 0 of Eq. (20.37) is now to be replaced by the $-\{\phi\mu c(\Delta x)^2/(k\Delta t)\}P_{i,n-1}$ of Eq. (20.23). (Note that the former change increases numerical stability by increasing diagonal dominance.) The interfacial velocity matching conditions derived in Example 20.1 do not change. But the meaning of the product μc must be understood: it is different on either side of the front, which again moves from time step to time step. The required Fortran source code changes are shown in bold print in Fig. 20.6A. The array $P_{i,n-1}$ is denoted by PNM1, and the initial pressure is PINIT. In order to determine the transient effects of fluid compressibility, we reconsider one of the data sets used in Example 20.1, where water displaces an oil with ten times the viscosity. The corresponding compressibilities are taken as 3×10^{-6}/psi and 50×10^{-6}/psi, while the initial pressure was assumed to be equal to the right-side reservoir pressure. If we compare computed results, which now include time scales related to fluid compressibilities and moving fronts, with those in Example 20.1 (see Fig. 20.3C), we find that in the present run, the effect of compressible flow transients on displacement front position with time is minimal. One Fortran subtlety deserves elaboration. The 200 do-loop defines *two* separate difference equations for the flows left and right of the front, but W(I) = -TERM*PNM1(I) refers to a *single* pressure. So long as the front does not move more than one mesh in a time step, errors due to copying water pressure as oil pressure, or conversely, do not exist (e.g., refer to the 260 loop); pressure continuity assures that both blocks contain equal pressures.

```
            .
            .
C           INITIAL SETUP
            IMAX = XCORE/DX +1
            IMAXM1 = IMAX-1
            IFRONT = XFRONT/DX +1
            N = 0
            T = 0.
            DO 100  I=1,IMAX
            PNM1(I) = PINIT
 100        CONTINUE
            NSTOP = 0
            MINDEX=1
            TIME(1) = 0.
            XPLOT(1) = XFRONT
C
C           START TIME INTEGRATION
            DO 300  N=1,NMAX
            T = T+DT
            DO 200  I=2,IMAXM1
            IF(I.LT.IFRONT) COMP = COMPL
            IF(I.GE.IFRONT) COMP = COMPR
            IF(I.LT.IFRONT) VISC = VISCL
            IF(I.GE.IFRONT) VISC = VISCR
            TERM = PHI*VISC*COMP*DX*DX/(K*DT)
            A(I) =  1.
            B(I) = -2.-TERM
            C(I) =  1.
            W(I) =  -TERM*PNM1(I)
 200        CONTINUE
            A(1) = 99.
            B(1) = 1.
            C(1) = 0.
            W(1) = PLEFT
```

(A)

FIG. 20.6 (A) Fortran source code (Example 20.4).

```
      A(IMAX) = 0.
      B(IMAX) = 1.
      C(IMAX) = 99.
      W(IMAX) = PRIGHT
      IF(VISCIN.EQ.VISCDP) GO TO 240
      A(IFRONT) =  1./VISCL
      B(IFRONT) = -1./VISCL -1./VISCR
      C(IFRONT) =  1./VISCR
      W(IFRONT) =  0.
240   CALL TRIDI(A,B,C,VECTOR,W,IMAX)
      DO 250  I=1,IMAX
      P(I) = VECTOR(I)
250   CONTINUE
      PGRAD = (P(IFRONT)-P(IFRONT-1))/DX
      XFRONT = XFRONT - (K*DT/(PHI*VISCL))*PGRAD
      IFRONT = XFRONT/DX +1
      DO 260  I=1,IMAX
      PNM1(I) = P(I)
260   CONTINUE
      WRITE(*,280) N,T,XFRONT,IFRONT
280   FORMAT(1X,'T(',I4,')= ',E8.3,' sec, Xf= ',E8.3,' ft, I= ',I3)
      MINDEX = MINDEX+1
      TIME(MINDEX) = T
      XPLOT(MINDEX) = XFRONT
300   CONTINUE
400   WRITE(*,10)
      WRITE(4,10)
      CALL GRFIX(XPLOT,TIME,MINDEX)
      STOP
      END
```

```
INPUT PARAMETER SUMMARY:
Rock core permeability (darcies): .100E+00
Rock core porosity (decimal nbr): .200E+00
Viscosity of invading fluid (cp): .100E+01
Viscosity, displaced  fluid (cp): .100E+02
Compr ... invading fluid (1/psi): .300E-05
Compr .. displaced fluid (1/psi): .500E-04
Pressure at left boundary  (psi): .100E+03
Pressure at right boundary (psi): .000E+00
Pressure, initial time t=0 (psi): .000E+00
Length of rock core sample  (ft): .100E+01
Initial "xfront" position (feet): .500E+00
Integration space step size (ft): .200E-02
Integration time step size (sec): .100E+01
Maximum allowed  number of steps: .200E+04
Number spatial DX grids selected: .500E+03
```

```
Time (sec)  Position (ft)
                              ______________________________
 .000E+00    .500E+00     |              *
 .600E+02    .542E+00     |                *
 .120E+03    .587E+00     |                 *
 .180E+03    .637E+00     |                   *
 .240E+03    .692E+00     |                    *
 .300E+03    .755E+00     |                      *
 .360E+03    .833E+00     |                         *
 .420E+03    .944E+00     |                             *
```

(B)

FIG. 20.6, CONT'D (B) Numerical results (Example 20.4).

EXAMPLE 20.5. VON NEUMANN STABILITY OF IMPLICIT TIME SCHEMES

The *implicit* time scheme in Example 20.4 turns out to be stable numerically, and it is of interest to examine its von Neumann characteristics for a wider class of transient flow formulations. In particular, let us consider those encompassing lineal, cylindrical, and spherical radial limits, that is

$$\partial^2 p(x,t)/\partial x^2 = (\phi\mu c/k)\,\partial p/\partial t \tag{20.38}$$

$$\partial^2 p/\partial r^2 + 1/r\,\partial p/\partial r = (\phi\mu c/k)\,\partial p/\partial t \tag{20.39}$$

$$\partial^2 p/\partial r^2 + 2/r\,\partial p/\partial r = (\phi\mu c/k)\,\partial p/\partial t \tag{20.40}$$

and specifically examine

$$\partial^2 p/\partial r^2 + N/r\,\partial p/\partial r = (\phi\mu c/k)\,\partial p/\partial t \tag{20.41}$$

where N = 0, 1, or 2 accordingly as the flow domain is lineal, cylindrical, or spherical. (Other values for N are related to nonconventional fractal descriptions that have been the subject of recent reservoir description studies.) We shall now difference Eq. (20.41) as suggested by Eq. (20.22), and approximate $\partial p/\partial r$ by the central difference formula $(P_{i+1,n} - P_{i-1,n})/(2\Delta r)$, while the reciprocal 1/r is evaluated at the center point i. This leads to

$$\begin{aligned} &(P_{i-1,n} - 2P_{i,n} + P_{i+1,n})/(\Delta r)^2 \\ &+ (N/r_i)(P_{i+1,n} - P_{i-1,n})/(2\Delta r) = (\phi\mu c/k)(P_{i,n} - P_{i,n-1})/\Delta t \end{aligned} \tag{20.42}$$

or

$$\begin{aligned} &\{1 - N\Delta r/(2r_i)\}P_{i-1,n} - \left\{2 + \phi\mu c(\Delta r)^2/(k\Delta t)\right\}P_{i,n} \\ &+ \{1 + N\Delta r/(2r_i)\}P_{i+1,n} = -\left\{\phi\mu c(\Delta r)^2/(k\Delta t)\right\}P_{i,n-1} \end{aligned} \tag{20.43}$$

which immediately shows how the lineal flow algorithm given in the foregoing example can be modified to handle cylindrical radial and spherical flow effects. (That is, we now have the generalized matrix coefficients $\mathbf{A} = A_i = 1 - N\Delta r/(2r_i)$ and $\mathbf{C} = C_i = 1 + N\Delta r/(2r_i)$ instead of unit coefficients.) This represents the only required change. In order to determine its numerical stability, we shall examine *Fourier wave components* having the form

$$P_{i,n} = \zeta^n e^{j\beta(i\Delta r)} \tag{20.44}$$

where $j = \sqrt{-1}$, β is a disturbance wavenumber (e.g., see Chin, 1994 for more detailed discussion) and ζ represents the amplification factor introduced earlier. Substitution in Eq. (20.43) gives

$$\zeta = 1/\left[1 + \left\{4k\Delta t/\left(\phi\mu c(\Delta r)^2\right)\right\}\sin^2\beta\Delta r/2 - jkN\Delta t/(\phi\mu c r_i\Delta r)\right] \tag{20.45}$$

For stability, we require that $|\zeta| < 1$. This is possible provided

$$\left[1 + \left\{4k\Delta t/\left(\phi\mu c(\Delta r)^2\right)\right\} \sin^2 \beta\Delta r/2\right]^2 + \left[kN\Delta t/(\phi\mu c r_i \Delta r)\right]^2 > 1 \quad (20.46)$$

Since $\sin^2 \beta\Delta r/2 > 0$, the inequality always holds if $\Delta t > 0$, thus guaranteeing stability. (Of course, mesh sizes must be kept small in order to reduce truncation errors and to ensure convergence to solutions of the PDE.) Unlike the conditionally stable explicit scheme studied earlier, this implicit scheme, which requires only tridiagonal matrix inversion, is unconditionally stable. We have tacitly assumed a positive time step $\Delta t > 0$ in arriving at this stability, which is the usual case. But in Chapter 21, we shall introduce reverse time integration where we have $\Delta t < 0$. For such applications, the stability requirements are altered, and the nature of the numerical truncation errors changes.

EXAMPLE 20.6. GAS DISPLACEMENT BY LIQUID IN LINEAL CORE WITHOUT MUDCAKE, INCLUDING COMPRESSIBLE FLOW TRANSIENTS

The piston-like displacement of formation gas by liquid filtrate, even without the complicating presence of mudcake, poses very difficult mathematical obstacles to solution. (More accurate two-phase immiscible flow modeling is pursued in Chapter 21.) To this author's knowledge, the problem has not been correctly solved in the literature, despite its importance in studying flows in tight gas sands. Many investigators simply assume

$$\partial^2 p(x,t)/\partial x^2 = (\phi\mu c/k)\, \partial p/\partial t \quad (20.47)$$

which applies to *liquids* only, also applies to gases, with the appropriate value of c. (Note that $c_{water} \approx 0.000003\ psi^{-1}$, whereas gas values, highly dependent on pressure, may be several hundred times this.) But in fact, as we have noted, the relevant equation for *gases* is

$$\partial^2 p^{m+1}(x,t)/\partial x^2 = \{\phi\mu m/(kp)\}\, \partial p^{m+1}/\partial t \quad (20.48)$$

with m being Muskat's thermodynamic exponent. (Eqs. 20.47 and 20.48 both assume lineal, isotropic flow.) Note that $m = 1$ for isothermal problems, whereas for adiabatic flows, $m = C_v/C_p \approx 0.7$ in the case of many gases. Let us demonstrate the nature of the complexities by examining the elementary case of *incompressible* gas displacement by *incompressible* liquids. Then, we shall proceed directly to a formulation that models the general displacement of gas by liquid, with moving fronts and nonnegligible transient compressibility effects. This study will highlight the importance of numerical methods, and we shall also, drawing on the von Neumann stability results of Example 20.5, demonstrate how seemingly unrelated pieces of information can provide insight into designing stable, robust, computational algorithms.

Incompressible Problem

For reference purposes, we shall consider the flow domain shown in Fig. 20.7. In the incompressible limit, Eq. (20.47) reduces to $d^2p_1(x)/dx^2=0$, and Eq. (20.48) becomes $d^2p_2^{m+1}(x)/dx^2 = 0$. We have introduced the 1 and 2 subscripts to denote the left-side liquid and right-side gas flows, respectively; these subscripts also remind us that these flows satisfy very different ordinary differential equations. These second-order equations admit the solutions $p_1(x)=Ax+B$ and $p_2^{m+1}(x) = Cx + D$. For our gas flow, it is important to understand that it is not the pressure $p_2(x)$ that varies linearly, but the function $p_2^{m+1}(x)$. Now, the solution to $p_1(x)$ satisfying the condition $p_1(0)=P_L$ is $p_1(x)=Ax+P_L$, while the solution to $p_2^{m+1}(x)$ satisfying the right-side pressure boundary condition $p_2(L)=P_R$ is $p_2^{m+1}(x) = C(x\text{-}L) + p_R^{m+1}$. (Again, L is the length of the core.) So far, A and C are unknown, but they are, in principle, fixed by invoking the continuity of pressure and velocity at $x=x_f$. Since $p_2(x) = \{C(x\text{-}L) + p_R^{m+1}\}^{1/(m+1)}$, the continuity of pressure requires that we have $Ax_f + P_L = \{C(x_f\text{-}L) + p_R^{m+1}\}^{1/(m+1)}$. Next, we evaluate the derivative $dp_2(x)/dx = \{C/(m+1)\}\{C(x\text{-}L) + p_R^{m+1}\}^{-m/(m+1)}$. Thus, continuity of velocity requires $(1/\mu_1)dp_1(x_f)/dx=(1/\mu_2)dp_2(x_f)/dx$, or $A/\mu_1 = (1/\mu_2)\{C/(m+1)\}\{C(x_f\text{-}L) + p_R^{m+1}\}^{-m/(m+1)}$, since permeability is uniform throughout. In summary, we solve

$$Ax_f + P_L = \{C(x_f\text{-}L) + p_R^{m+1}\}^{1/(m+1)} \tag{20.49a}$$

$$A/\mu_1 = (1/\mu_2)\{C/(m+1)\}\{C(x_f\text{-}L) + p_R^{m+1}\}^{-m/(m+1)} \tag{20.49b}$$

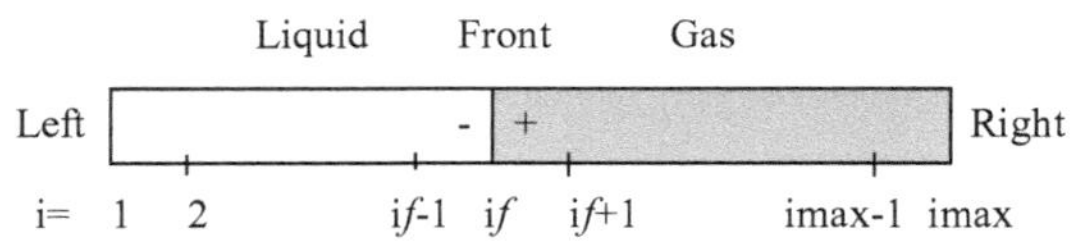

FIG. 20.7 Gas displacement by liquid.

analytically. It is clear that A can be eliminated between Eqs. (20.49a) and (20.49b), but this leaves an intractable nonlinear equation for C. Even if explicit expressions for A and C are obtained, the integration of the displacement front equation $dx_f/dt=-k/(\mu\phi)dp_1(x_f)/dx=-kA/(\mu\phi)$ leads to complexities. These worsen when transient effects due to compressibility must be modeled. Thus, we are motivated to formulate the problem numerically, drawing on the success of Example 20.4 and the stability information obtained in Example 20.5.

Transient, Compressible Problem

The finite differencing required to model Eq. (20.47) has been discussed, and in Example 20.4, we in fact considered displacements by dissimilar *liquids* having different viscosities and compressibilities. Again, the transients that arise are of two types, namely, the usual compressible ones found in well testing and those due to front motions that depend on mobility contrasts. Eq. (20.48), given its similarity to Eq. (20.47), can be differenced in a like manner, provided we observe that the right-side coefficient $\phi\mu m/(kp)$, or $\phi\mu c^*/k$ in our earlier notation, is *not* constant but dependent on p(x,t), which continuously evolves in time. (Numerically, this pressure can be evaluated at the previous time step, at any instant in the forward time integration.) Let us recall that Example 20.4 was successfully solved by approximating Eq. (20.47) using

$$\begin{aligned} &P_{i-1,n} - \left[2 + \left\{\phi\mu c(\Delta x)^2/(k\Delta t)\right\}\right] P_{i,n} + P_{i+1,n} \\ &= -\left\{\phi\mu c(\Delta x)^2/(k\Delta t)\right\} P_{i,n-1} \end{aligned} \tag{20.23}$$

This equation still applies to the left of the moving front in Fig. 20.7, where the invading *liquid* resides. To the right of the front, however, Eq. (20.48) applies. Since implicit finite difference equations of the form given in Eq. (20.42) are unconditionally stable, we attempt to difference Eq. (20.48) in such a way as to take advantage of that stability. To do this, we observe that our $\partial^2 p^{m+1}(x,t)/\partial x^2 = \{\phi\mu m/(kp)\}\ \partial p^{m+1}/\partial t$ can be expanded as

$$\partial^2 p(x,t)/\partial x^2 + (m/p)\,(\partial p/\partial x)^2 = \{\phi\mu m/(kp)\}\partial p/\partial t \tag{20.50}$$

If we difference all old terms exactly as before, and approximate our new ones with the $O(\Delta x)^2$ accurate formulas

$$\begin{aligned} (m/p)(\partial p/\partial x)^2 = &\ (m/P_{i,n-1})\{(P_{i+1,n-1} - P_{i-1,n-1})/(2\Delta x)\} \\ &\times \{(P_{i+1,n} - P_{i-1,n})/(2\Delta x)\} \end{aligned} \tag{20.51}$$

$$\phi\mu m/(kp) = \phi\mu m/(kP_{i,n-1}) \tag{20.52}$$

we obtain

$$\begin{aligned} &\{1 - m(\Delta x)(\partial p/\partial x)_{i,n-1}/(2P_{i,n-1})\}P_{i-1,n} \\ &- \left[2 + \left\{\phi\mu m(\Delta x)^2/(kP_{i,n-1}\Delta t)\right\}\right] P_{i,n} \\ &+ \{1 + m(\Delta x)(\partial p/\partial x)_{i,n-1}/(2P_{i,n-1})\}P_{i+1,n} \\ &= -\phi\mu m(\Delta x)^2/(k\Delta t) \end{aligned} \tag{20.53}$$

Thus, the Fortran source code developed in Example 20.4 to model displacement by dissimilar liquids can be easily modified to handle gas displacement by liquids, if to the right of the front, we instead apply Eq. (20.53). The front matching condition

$$(1/\mu_1)\, p_{if\text{-}1} - (1/\mu_1 + 1/\mu_2) p_{if} + (1/\mu_2)\, p_{if+1} = 0 \tag{20.29}$$

still applies at each time step; again, it embodies pressure and velocity continuity, and is not related to fluid compressibility. In the source code modifications shown in Fig. 20.8A, the Muskat exponent m is denoted EM. The numerical

```
      .
      .
C     START TIME INTEGRATION
      DO 300  N=1,NMAX
      T = T+DT
      DO 200  I=2,IMAXM1
      IF(I.LT.IFRONT) A(I) = 1.
      IF(I.LT.IFRONT) B(I) =-2.-PHI*VISCL*COMPL*DX*DX/(K*DT)
      IF(I.LT.IFRONT) C(I) = 1.
      IF(I.LT.IFRONT) W(I) =-(PHI*VISCL*COMPL*DX*DX/(K*DT))*PNM1(I)
      IF(I.GE.IFRONT) DPDX =  (PNM1(I+1)-PNM1(I-1))/(2.*DX)
      IF(I.GE.IFRONT) A(I) = 1.  -EM*DX*DPDX/(2.*PNM1(I))
      IF(I.GE.IFRONT) B(I) =-2.-PHI*VISCR*EM*DX*DX/(K*DT*PNM1(I))
      IF(I.GE.IFRONT) C(I) = 1.  +EM*DX*DPDX/(2.*PNM1(I))
      IF(I.GE.IFRONT) W(I) =-(PHI*VISCR*EM*DX*DX/(K*DT))
 200  CONTINUE
      A(1) = 99.
      B(1) = 1.
      C(1) = 0.
      W(1) = PLEFT
      A(IMAX) = 0.
      B(IMAX) = 1.
      C(IMAX) = 99.
      W(IMAX) = PRIGHT
      IF(VISCIN.EQ.VISCDP) GO TO 240
```

(A)

```
      A(IFRONT) =  1./VISCL
      B(IFRONT) = -1./VISCL -1./VISCR
      C(IFRONT) =  1./VISCR
      W(IFRONT) =  0.
 240  CALL TRIDI(A,B,C,VECTOR,W,IMAX)
      DO 250  I=1,IMAX
      P(I) = VECTOR(I)
 250  CONTINUE
      PGRAD = (P(IFRONT)-P(IFRONT-1))/DX
      XFRONT = XFRONT -  (K*DT/(PHI*VISCL))*PGRAD
      IFRONT = XFRONT/DX +1
      DO 260  I=1,IMAX
      PNM1(I) = P(I)
 260  CONTINUE
      .
```

FIG. 20.8 (A) Fortran source code (Example 20.6).

```
INPUT PARAMETER SUMMARY:
Rock core permeability (darcies): .100E-02
Rock core porosity (decimal nbr): .100E+00
Viscosity, invading  liquid (cp): .100E+01
Viscosity of  displaced gas (cp): .200E-01
Compr .. invading liquid (1/psi): .300E-05
Muskat m exponent of gas (real#): .700E+00
Pressure at left boundary  (psi): .200E+03
Pressure at right boundary (psi): .100E+03
Pressure, initial time t=0 (psi): .100E+03
Length of rock core sample  (ft): .100E+01
Initial "xfront" position (feet): .200E+00
Integration space step size (ft): .100E-01
Integration time step size (sec): .100E+01
Maximum allowed  number of steps: .100E+05
Number spatial DX grids selected: .101E+03

  Time (sec)  Position (ft)
                             ______________________________
   .000E+00    .200E+00     |    *
   .600E+02    .221E+00     |    *
   .120E+03    .239E+00     |     *
   .180E+03    .256E+00     |      *
   .240E+03    .272E+00     |      *
   .300E+03    .287E+00     |       *
   .360E+03    .302E+00     |       *
   .420E+03    .316E+00     |       *
   .480E+03    .329E+00     |        *
   .540E+03    .342E+00     |        *
   .600E+03    .354E+00     |         *
   .660E+03    .366E+00     |         *
   .720E+03    .378E+00     |         *
   .780E+03    .389E+00     |          *
   .840E+03    .400E+00     |          *
   .900E+03    .411E+00     |          *
   .960E+03    .421E+00     |           *
   .102E+04    .431E+00     |           *
   .
   .
   .192E+04    .563E+00     |               *
   .198E+04    .571E+00     |               *
   .204E+04    .578E+00     |                *
   .210E+04    .586E+00     |                *
   .216E+04    .593E+00     |                *
   .222E+04    .601E+00     |                *
   .228E+04    .608E+00     |                 *
   .234E+04    .615E+00     |                 *
   .240E+04    .622E+00     |                 *
   .246E+04    .629E+00     |                 *
   .252E+04    .636E+00     |                 *
   .258E+04    .643E+00     |                  *
   .264E+04    .650E+00     |                  *
   .270E+04    .656E+00     |                  *
```

(B)

FIG. 20.8, CONT'D (B) Numerical results (Example 20.6).

(Continued)

```
INPUT PARAMETER SUMMARY:
Rock core permeability (darcies): .100E-02
Rock core porosity (decimal nbr): .500E-01
Viscosity, invading  liquid (cp): .100E+01
Viscosity of  displaced gas (cp): .200E-01
Compr .. invading liquid (1/psi): .300E-05
Muskat m exponent of gas (real#): .700E+00
Pressure at left boundary  (psi): .200E+03
Pressure at right boundary (psi): .100E+03
Pressure, initial time t=0 (psi): .100E+03
Length of rock core sample  (ft): .100E+01
Initial "xfront" position (feet): .200E+00
Integration space step size (ft): .100E-01
Integration time step size (sec): .100E+01
Maximum allowed  number of steps: .100E+05
Number spatial DX grids selected: .101E+03
```

```
 Time (sec)  Position (ft)
                            ________________________________
  .000E+00    .200E+00     |    *
  .600E+02    .239E+00     |     *
  .120E+03    .272E+00     |      *
  .180E+03    .302E+00     |       *
  .240E+03    .329E+00     |        *
  .300E+03    .354E+00     |         *
  .360E+03    .378E+00     |         *
  .420E+03    .400E+00     |          *
  .480E+03    .421E+00     |           *
  .540E+03    .441E+00     |           *
  .600E+03    .461E+00     |            *
  .660E+03    .479E+00     |             *
  .720E+03    .497E+00     |             *
  .
  .
  .204E+04    .796E+00     |                       *
  .210E+04    .807E+00     |                        *
  .216E+04    .818E+00     |                        *
  .222E+04    .829E+00     |                        *
  .228E+04    .839E+00     |                         *
  .234E+04    .850E+00     |                         *
  .240E+04    .860E+00     |                         *
  .246E+04    .870E+00     |                          *
  .252E+04    .880E+00     |                          *
  .258E+04    .890E+00     |                          *
  .264E+04    .900E+00     |                           *
  .270E+04    .910E+00     |                           *
  .276E+04    .920E+00     |                           *
  .282E+04    .929E+00     |                            *
  .288E+04    .939E+00     |                            *
  .294E+04    .948E+00     |                            *
  .300E+04    .957E+00     |                             *
```

(C)

FIG. 20.8, CONT'D (C) Numerical results (Example 20.6).

results displayed in Figs. 20.8B and C are obtained for two different values of porosity, with all other parameters otherwise held fixed.

One Fortran subtlety deserves elaboration. The 200 do-loop defines *two* separate difference equations for the flows left and right of the front, but the pressure updating in the 260 do-loop refers to a *single* pressure. So long as the front does not move more than one mesh in a time step, errors due to copying liquid pressure as gas pressure, or conversely, do not exist, assuming small capillary pressures. Pressure continuity assures that both blocks will contain identical pressures.

EXAMPLE 20.7. SIMULTANEOUS MUDCAKE BUILDUP AND DISPLACEMENT FRONT MOTION FOR INCOMPRESSIBLE LIQUID FLOWS

In this last exercise, we reconsider the problem of dynamically coupled invasion front motion and mudcake growth in lineal flow; this was studied analytically in Chapter 17, where it was solved in closed form, but we shall approach its solution numerically. This is pursued for several reasons. First, we wish to demonstrate how problems with moving boundaries and disparate space *scales* (characterizing mudcake and rock) are formulated and solved with finite differences. Second, computational methods are ultimately needed because they are more convenient when cake compaction, time-dependent applied pressures, and formation heterogeneities are required. Because the present problem can be described analytically, we at least possess a tool with which to evaluate the quality of more approximate solution methods. In the foregoing examples, we emphasized how the effects of transients due to fluid compressibility, and the nonlinear effects of gas displacement by liquids, can be numerically modeled. For the present, we return to simple incompressible flows of liquids to illustrate the main ideas, so that we need not address the complicating, but nonetheless straightforward, effects. Here, we have instead two moving boundaries: the displacement front within the rock and the surface of the mudcake, which moves in such a way as to increase cake thickness with time. Thus, analytical and computational changes to our schemes are required. In addition, as we have noted, disparate space scales enter the numerical formulation in a subtle way: mudcakes are thin relative to the distance that the filtrate penetrates the formation. The problem domain is shown in Fig. 20.9A.

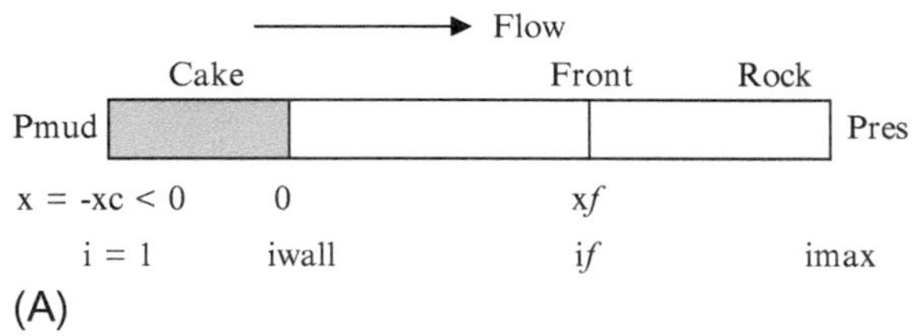

FIG. 20.9 (A) Three-layer lineal flow problem.

(Continued)

For simplicity, we assume that in the cake and rock, the permeabilities k_c and k_r are constant, although they can be different constants. Therefore, whether we start with $d(k_c\ dp/dx)/dx = 0$ or $d(k_r\ dp/dx)/dx = 0$, the permeabilities factor out, leaving

$$d^2p(x)/dx^2 = 0 \tag{20.10}$$

in either case. Now, we can approximate Eq. (20.10) with the central difference formula used earlier, namely,

$$d^2p(x_i)/dx^2 = \{p_{i-1} - 2p_i + p_{i+1}\}/(\Delta x)^2 + O(\Delta x)^2 = 0 \tag{20.14}$$

Our combined mudcake-growth and displacement-front-movement problem, with its clearly disparate length scales, is not unlike boundary layer or shock layer type flows in classical fluid mechanics. That is, the cake is extremely thin, while the scale of the front motion is orders of magnitude larger: any attempt to characterize both flows using the same physical measures of length is likely to result in inaccuracy. Therefore, we would like to select Δx, as usual, for the rock, but Δx_c for the mudcake, with $\Delta x_c \ll \Delta x$. Fortunately, this does not lead to numerical complexity, since the grid length completely drops out when applying Eq. (20.14) to Eq. (20.10). (This is not the case with radial flows, or transient compressible flows, where minor changes are needed.)

Thus, Eq. (20.15), which we have reproduced as follows, applies throughout the entire domain of flow in the general case where we have both different permeabilities and grid sizes.

$$p_{i-1} - 2p_i + p_{i+1} = 0 \tag{20.15}$$

This does not mean that changes won't be needed; they are needed, and they will be discussed shortly. But for now, the pressures $p_1, p_2, \ldots,$ and p_{imax} at the nodes $i = 1, 2, \ldots,$ and i_{max} are determined as in our earlier examples by writing Eq. (20.15) for each of the *internal nodes* $i = 2, 3, \ldots,$ and i_{max-1} (with two notable exceptions, to be discussed). This yields $i_{max} - 2$ number of linear equations, two short of the number of unknowns i_{max}. The two additional required equations are obtained from boundary conditions; in this case,

$$p(-x_c) = P_{mud},\ x_c > 0 \tag{20.54}$$

$$p(L) = P_{res} \tag{20.55}$$

for the mud and the farfield reservoir. That is, we assume that $p_1 = P_{mud}$ and $p_{imax} = P_{res}$ where L is the core length. This leads to the coupled equations

$$\begin{array}{lll}
 & p_1 & = P_{mud} \\
i=2: & p_1 - 2p_2 + p_3 & = 0 \\
i=3: & p_2 - 2p_3 + p_4 & = 0 \\
i=4: & p_3 - 2p_4 + p_5 & = 0 \\
\cdot & & \\
i = i\mathit{wall} & & \\
\cdot & & \\
i = i\mathit{front} \text{ or } i\mathit{f}: & & \\
\cdot & & \\
i = imax\text{-}2: & p_{imax-3} - 2p_{imax-2} + p_{imax-1} & = 0 \\
i = imax\text{-}1: & p_{imax-2} - 2p_{imax-1} + p_{imax} & = 0 \\
 & p_{imax} & = P_{res}
\end{array} \tag{20.56}$$

or

$$\begin{vmatrix} 1 & 0 & & & \\ 1 & -2 & 1 & & \\ & 1 & -2 & 1 & \\ \ldots & & & & \\ \ldots & & & & \\ \ldots & & & & \\ \ldots & & & & \\ & 1 & -2 & 1 & \\ & & 1 & -2 & 1 \\ & & & 0 & 1 \end{vmatrix} \begin{vmatrix} p_1 \\ p_2 \\ p_3 \\ . \\ . \\ . \\ . \\ p_{imax-2} \\ p_{imax-1} \\ p_{imax} \end{vmatrix} \begin{matrix} = \\ = \\ = \\ = \\ = \\ = \\ = \\ = \\ = \\ = \end{matrix} \begin{vmatrix} P_{mud} \\ 0 \\ 0 \\ .. \\ .. \\ .. \\ .. \\ 0 \\ 0 \\ P_{res} \end{vmatrix} \tag{20.57}$$

just as we had obtained for simple liquid flows. The crucial differences, however, arise from the matching conditions that need to be enforced at the mud-to-mudcake and displacement front interfaces. Let the subscripts c and r denote cake and rock properties, while mf and o denote mud filtrate and formation oil or displaced fluid. At the front interface separating invading from displaced fluids, velocity continuity requires that $-(k_r/\mu_{mf})dp_r(x_f-)/dx$ to the left of the front equal the velocity $-(k_r/\mu_o)dp_r(x_f+)/dx$ just to the right.

Matching Conditions at Displacement Front

Since rock permeability cancels, we have $(1/\mu_{mf})dp(x_f-)/dx = (1/\mu_o)dp(x_f+)/dx$. Now we shall denote by *if*- and *if*+ the spatial locations infinitesimally close to the left and right of the front $x = x_f$, which is itself indexed by $i = if$. Then, we can approximate the pressure gradient $dp(x_f-)/dx$ using backward differences, whereas the gradient $dp(x_f+)/dx$ can be modeled using forward differences. (Again, differentiation through the interface itself is forbidden since the pressure gradient in general changes suddenly.) This process leads to $(1/\mu_{mf})(p_{if-} - p_{if-1})/\Delta x = (1/\mu_o)(p_{if+1} - p_{if+})/\Delta x$, or $(1/\mu_{mf})(p_{if-} - p_{if-1}) = (1/\mu_o)(p_{if+1} - p_{if+})$. Now, since surface tension is unimportant, pressure continuity requires that $p_{if-} = p_{if+}$ or simply p_{if}. Thus, at the interface, the matching condition

$$(1/\mu_{mf})\, p_{if-1} - (1/\mu_{mf} + 1/\mu_o)p_{if} + (1/\mu_o)\, p_{if+1} = 0 \tag{20.58}$$

applies, and straightforward changes are made to Eqs. (20.56) and (20.57) corresponding to the row defined by i = ifront. Unlike the central difference approximation, which is second-order accurate, our use of backward and forward differences in deriving Eq. (20.58) renders it only $O(\Delta x)$ accurate.

Matching Conditions at the Cake-to-Rock Interface

It is tempting to invoke similar arguments at the index $i = i_{wall}$ representing the cake-to-rock interface, modifying Eq. (20.58) in the obvious manner to account for differences between mudcake and rock permeabilities. This would lead to

$$k_c\, p_{iwall\text{-}1} - (k_c + k_r)\, p_{iwall} + k_r\, p_{iwall+1} = 0 \qquad (20.59)$$

In Eq. (20.59), viscosity drops out identically, since the same filtrate flows through the mudcake as through the flushed zone in the rock. However, any attempt to use Eq. (20.59) would produce gross numerical error and poor physical resolution in the mudcake, since identical grid sizes Δx are implicit in its derivation. Also, the fact that Eq. (20.59) is not as numerically stable as $p_{i-1} - 2p_i + p_{i+1} = 0$, say, would lead to inaccuracies if our algorithm were extended to transient compressible flows. Thus, we need to return to basics and consider the more general statement

$$-(k_c/\mu_{mf})\, dp(x_{wall}-)/dx = -(k_r/\mu_{mf})\, dp(x_{wall}+)/dx \qquad (20.60)$$

Since physical length scales in the mudcake are much smaller than those characterizing the rock, we wish to use the mesh Δx_s in the cake, and the usual Δx in the rock, such that $\Delta x_s \ll \Delta x$, in constructing our backward and forward differences. This leads to the matching condition

$$(k_c/\Delta x_s)\, p_{iwall\text{-}1} - (k_c/\Delta x_s + k_r/\Delta x)\, p_{iwall} + (k_r/\Delta x)\, p_{iwall+1} = 0 \qquad (20.61)$$

since $p_{iwall-} = p_{iwall+}$. This difference equation, if we choose $k_c \propto \Delta x_s$ and $k_r \propto \Delta x_s$, leads directly to our desired $p_{iwall\text{-}1} - 2p_{iwall} + p_{iwall+1} = 0$! Thus, we use Eq. (20.61) in Eqs. (20.56) and (20.57) to separate the finite difference equation blocks for the cake and flushed zone flows; this matching condition applies at the matrix row corresponding to $i = iwall$.

Coding Modifications

Eqs. (20.58) and (20.61) represent, theoretically, the most significant modifications, but other equally important details must be addressed. At the end of each time step, we advance the displacement front using Eqs. (20.30)–(20.32), as in Example 20.1. In the notation of the present example, we have

$$x_{f,new} = x_{f,old} - \{k_r \Delta t/(\mu_{mf} \phi \Delta x)\}(p_{if} - p_{if\text{-}1})_{old} \qquad (20.62)$$

where the p refers to the pressure in the flushed zone. The mud-to-mudcake boundary $x = x_c(t)$ is updated using our earlier mudcake growth formula, modified to accommodate the sign convention of Fig. 20.9A, that is,

$$dx_c/dt = -\{f_s/\{(1\text{-}f_s)(1\text{-}\phi_c)\}\}|v_n| < 0 \qquad (20.63)$$

where $|v_n|$ is proportional to the Darcy velocity $(k_c/\mu_{mf})dp(x_c)/dx$ at the cake surface. The mudcake counterpart to Eq. (20.62) is therefore

$$x_{c,new} = x_{c,old} + \{f_s/\{(1\text{-}f_s)(1\text{-}\phi_c)\}\}\{k_c \Delta t/(\mu_{mf} \Delta x)\}(p_2 - p_1)_{old} \qquad (20.64)$$

```
      .
C     Mudcake properties can be entered as shown, but lumped data
C     from the  filtration  test  in Chapter 4 is more convenient.
      WRITE(*,36)
 36   FORMAT(' Mud cake  permeability (darcies):   ',$)
      READ(*,32) KCAKE
      WRITE(*,37)
 37   FORMAT(' Mud cake porosity  (decimal nbr):   ',$)
      READ(*,32) PHIMUD
      WRITE(*,38)
 38   FORMAT(' Mud solid fraction (decimal nbr):   ',$)
      READ(*,32) FS
      .
C     INITIAL SETUP
      IWALL = 3
      IMAX   = XCORE/DX + IWALL
      IMAXM1 = IMAX-1
      IFRONT = XFRONT/DX + IWALL
      .
      N = 0
      T = 0.
      XCAKE = 0.
      .
C     START TIME INTEGRATION
      DO 300  N=1,NMAX
      T = T+DT
      DO 200  I=2,IMAXM1
      A(I) =  1.
      B(I) = -2.
      C(I) =  1.
      W(I) =  0.
 200  CONTINUE
      A(1) = 99.
      B(1) = 1.
      C(1) = 0.
      W(1) = PLEFT
      A(IMAX) = 0.
      B(IMAX) = 1.
      C(IMAX) = 99.
      W(IMAX) = PRIGHT
      IF(VISCIN.EQ.VISCDP) GO TO 240
      A(IFRONT) =  1./VISCL
      B(IFRONT) = -1./VISCL -1./VISCR
      C(IFRONT) =  1./VISCR
      W(IFRONT) =  0.
 240  A(IWALL) =  KC/DXCAKE
      B(IWALL) = -KC/DXCAKE -K/DX
      C(IWALL) =  K/DX
      W(IWALL) =  0.
      CALL TRIDI(A,B,C,VECTOR,W,IMAX)
      DO 250  I=1,IMAX
      P(I) = VECTOR(I)
 250  CONTINUE
      PGRAD  = (P(IFRONT)-P(IFRONT-1))/DX
      XFRONT = XFRONT -  (K*DT/(PHI*VISCL))*PGRAD
      PGRADC = (P(2)-P(1))/DXCAKE
      XCAKE  = XCAKE+(FS/((1.-PHIMUD)
     1          *(1.-FS)))*(KC/VISCL)*PGRADC*DT
```

(B)

FIG. 20.9, CONT'D (B) Fortran source code (Example 20.7).

(Continued)

```
          IWALL   = -XCAKE/DXCAKE + 3
          IFRONT  = XFRONT/DX + IWALL
          IMAX    = XCORE/DX + IWALL
          .
          WRITE(*,280) N,T,XFRONT,IFRONT,XCAKE,IWALL
          MINDEX = MINDEX+1
          TIME(MINDEX) = T
          XPLOT(MINDEX) = XFRONT
          XC(MINDEX) = -XCAKE
300       CONTINUE
          .
```

```
INPUT PARAMETER SUMMARY:
Rock core permeability (darcies): .100E+01
Rock core porosity (decimal nbr): .100E+00
Mud cake permeability  (darcies): .100E-02
Mud cake porosity  (decimal nbr): .100E+00
Mud solid fraction (decimal nbr): .100E+00
Viscosity of invading fluid (cp): .100E+01
Viscosity, displaced  fluid (cp): .100E+01
Pressure at left boundary  (psi): .100E+03
Pressure at right boundary (psi): .000E+00
Length of rock core sample  (ft): .100E+01
Initial "xfront" position (feet): .100E+00
DX grid size in rock sample (ft): .200E-02
DX grid size in the mudcake (ft): .200E-03
Integration time step size (sec): .100E+00
Maximum allowed  number of steps: .100E+04
```

```
INVASION FRONT POSITION VERSUS TIME:
Time (sec)  Position (ft)
                              ______________________________
  .000E+00    .100E+00        | *
  .600E+01    .283E+00        |      *
  .120E+02    .388E+00        |          *
  .180E+02    .470E+00        |            *
  .240E+02    .540E+00        |              *
  .300E+02    .602E+00        |                *
  .360E+02    .659E+00        |                  *
  .420E+02    .710E+00        |                    *
  .480E+02    .759E+00        |                     *
  .540E+02    .804E+00        |                       *
  .600E+02    .847E+00        |                        *
  .660E+02    .888E+00        |                         *
  .720E+02    .927E+00        |                          *
  .780E+02    .965E+00        |                            *

MUD CAKE THICKNESS VERSUS TIME:
Time (sec)  Position (ft)
                              ______________________________
  .000E+00    .000E+00        |
  .600E+01    .226E-02        |    *
  .120E+02    .356E-02        |       *
  .180E+02    .457E-02        |          *
  .240E+02    .544E-02        |             *
  .300E+02    .620E-02        |               *
  .360E+02    .690E-02        |                 *
  .420E+02    .753E-02        |                   *
  .480E+02    .813E-02        |                    *
  .540E+02    .869E-02        |                      *
  .600E+02    .922E-02        |                       *
  .660E+02    .973E-02        |                         *
  .720E+02    .102E-01        |                          *
  .780E+02    .107E-01        |                            *
```

(C)

FIG. 20.9, CONT'D (C) Numerical results (Example 20.7).

Numerically, as Eq. (20.56) suggests, three separate matrix regimes (separated by two matching conditions) are required. Initially, the i_{wall} index shown in Fig. 20.9A must at least equal three; in this minimal setup, $i=1$ handles the left mud pressure boundary condition, $i=3$ handles Eq. (20.61), while $i=2$ would correspond to a single finite difference equation $p_{i-1}-2p_i+p_{i+1}=0$ written for the index $i=2$. More initial mudcake grids, of course, would lead to inaccuracy, since mudcake thickness is vanishingly small at initial times (unless, of course, mesh sizes were significantly decreased). The suggested value of three allows us to grow our cake outward as time advances. Finally, observe that we shall dynamically adjust our meshes at each time step. The leading index $i=1$ is always assigned to the moving mud-to-mudcake boundary. Then we take the mudcake-to-rock interface at $i_{wall}=|x_c|/\Delta x+3$ (so that $i_{wall}=3$ if $x_c=0$), and additionally, $i_f=x_f/\Delta x+i_{wall}$. Also, while Eq. (20.63) explicitly requires that the mudcake properties f_s, ϕ_c, and k_c be available, we understand from Chapter 19 that they can equivalently be replaced by lumped parameters obtained from the surface filtration test developed there. Relevant portions of the Fortran source code that we designed to implement the foregoing changes are shown in Fig. 20.9B, where we have added descriptive comment statements as needed. Key notes corresponding to ideas we have emphasized so far are shown in bold type for emphasis. Finally, typical computed results are displayed in Fig. 20.9C.

Modeling Formation Heterogeneities

Rock heterogeneities such as internal filter cake, or damaged zones, are easily modeled by allowing k_r to vary with x. If so, the differential equation $d^2p(x)/dx^2=0$ no longer applies, as it is derived for constant permeabilities only. Instead, we must consider

$$d(k_r\, dp/dx)/dx = 0 \tag{20.65}$$

$$k_r(x)\, d^2p(x)/dx^2 + (dk_r/dx)\, dp/dx = 0 \tag{20.66}$$

Then, the central difference approximation

$$k_r(x_i)\,\{p_{i-1}-2p_i+p_{i+1}\}/(\Delta x)^2 + \{(k_{r,i+1}-k_{r,i-1})/(2\Delta x)\}\{(p_{i+1}-p_{i-1})/(2\Delta x)\} = 0 \tag{20.67}$$

leads to

$$\{p_{i-1}-2p_i+p_{i+1}\} + \{(k_{r,i+1}-k_{r,i-1})/(4k_r(x_i))\}(p_{i+1}-p_{i-1}) = 0 \tag{20.68}$$

or

$$[1-\{(k_{r,i+1}-k_{r,i-1})/(4k_r(x_i))\}]\,p_{i-1}-2p_i + [1+\{(k_{r,i+1}-k_{r,i-1})/(4k_r(x_i))\}]\,p_{i+1} = 0 \tag{20.69}$$

Thus, the only required change when $k_r(x)$ is an explicitly prescribed function of x is the replacement of $p_{i-1} - 2p_i + p_{i+1} = 0$ by Eq. (20.69). When $k_r(p)$ is a function of p, which additionally depends on x, the physics changes, and the algorithm modifications discussed next are required.

Mudcake Compaction and Compressibility

Mudcake compaction, meaning pressure-dependent permeability and porosity, is easily handled. For example, if $k_c = k_c(p)$, the governing pressure equation in the mudcake

$$d(k_c\, dp/dx)/dx = 0 \tag{20.70}$$

becomes

$$k_c(p)\, d^2p(x)/dx^2 + (dk_c/dp)\,(dp/dx)^2 = 0 \tag{20.71}$$

Following the lead of earlier examples, Eq. (20.71) can be linearized about pressure values obtained from one earlier time step, so that we can write

$$k_c(p_{old})\, d^2p(x)/dx^2 + \{(dk_c/dp)(dp/dx)\}_{old}\,(dp/dx) = 0 \tag{20.72}$$

This is exactly Eq. (20.66), with $k_r(x)$ replaced by $k_c(p_{old})$ and (dk_r/dx) replaced by $\{(dk_c/dp)(dp/dx)\}_{old}$. Thus, an equation analogous to Eq. (20.69) is easily obtained. The function $k_c(p)$ and the function $\phi_c(p)$ in Eq. (20.64) could be hard-coded into the main program, or declared as subroutines or statement functions, as desired. Finally, mudcake compressibility transients are easily modeled using the ideas developed in Example 20.4. For such problems, instead of the nonlinear ODE $d(k_c dp/dx)/dx = 0$, we would instead solve the nonlinear parabolic equation

$$\partial\{(k_c(p)/\mu)\, \partial p(x,t)\}/\partial x = \partial\{\phi_c(p)\mu c(p)\, p(x,t)\}/\partial t \tag{20.73}$$

where we have introduced the pressure-dependent compressibility c(p). The required Fortran changes are left to the interested reader.

Modeling Borehole Activity

We have developed the foregoing example in detail assuming lineal flows, but as we have shown, the extension to cylindrical radial flows requires but two lines of Fortran changes. These are specifically redefinitions for the **A** and **C** matrix coefficients as indicated in Example 20.2. We also observe that mud weight is often increased or decreased during drilling and that this weight change is accomplished by adding or removing solid particulates (e.g., *barite*), and modifying viscosifiers (e.g., *bentonite*). The mud pressure PLEFT and the invading filtrate viscosity VISCIN can be redefined as general time-dependent Fortran statement functions. This allows modeling flexibility that cannot be achieved with exact analytical solutions. Finally, note that the erosive effects of dynamic filtration can be used to limit radial mudcake growth to an equilibrium value, by introducing if-then programming logic (e.g., do *not* update x_c if $x_c > x_{c,equil}$).

We have selected a cross-section of examples, ranging from constant density, two-fluid flows without mudcake to flows with transients due to compressibility to problems with mudcake thickening with time. Naturally, other combinations of problems with lineal, radial, and spherical geometries, single or multiple fluids in formations, compressible mudcake, general transient effects, and so on, can be modeled by combining appropriate pieces of theory and source code. Finally, this author warns prospective users of canned computational fluid mechanics software of likely formulation errors. In an environment driven by high-resolution graphics and user-friendly screen interfaces, it is important to understand precisely which equations are solved and the methodology employed. The highly specialized problems typical of formation invasion applications are unlikely to be pre-programmed in commercial solvers; users should direct technical questions to development and not sales staff.

PROBLEMS AND EXERCISES

1. The skin model derived in Chapter 18 for wellbore damage does not apply when the damaged zone is extensive. For such problems, two fully coupled partial differential models (i.e., two heat equations) must be solved simultaneously. Formulate this problem for cylindrical radial flows and solve it numerically. Evaluate the extent to which the skin model applies (or does not apply) in typical well testing applications.
2. The three-layer invasion model for piston-like displacements was solved analytically in Chapter 17 for lineal and radial problems. Solve these same problems numerically, and compare their solutions with the exact ones.

Chapter 21

Forward and Inverse Multiphase Flow Modeling

In this final chapter, we present new ideas on immiscible and miscible flow modeling with respect to formation invasion and time lapse analysis. In particular, we first study forward simulation methods, where the evolution of an initial state dynamically in time is considered. Then, we focus on inverse time lapse analysis applications that attempt to uncover formation evaluation information from data collected by well logging instruments. Whereas our earlier models assume piston-like flows associated with discontinuous step changes in fluid properties, the forward and inverse, miscible and immiscible flow models here are generally smeared by diffusion, stretched by geometric spreading, and characterized by steep saturation shock fronts. We pose, and importantly solve, what we call the resistivity migration problem, wherein the distinct fronts from which prescribed smeared profiles evolve are recovered by reverse diffusion using methods similar to the parabolized wave methods in seismic migration. We also show how the sharp saturation discontinuities obtained in immiscible water-oil flows can be unshocked in order to recover the original smooth saturation distributions for further information processing. The work in this chapter is *not* a tutorial on multiphase flow, although all derivations do proceed from first principles. This book assumes some exposure to reservoir flow analysis, for example, to concepts such as Darcy's law, miscible flow, relative permeability, and capillary pressure, but it does not require any exposure to the research literature or any experience in numerical simulation.

Problem hierarchies. We first discuss immiscible two-phase flows in the Buckley-Leverett limit of zero capillary pressure, and we provide exact, analytical, closed-form solutions for early-time, near-well invasion problems, which can be modeled by a planar flow. Also, since mud filtration rates are at their largest, the effects of capillary pressure can be ignored. For the problem in which saturation shocks form, shock-fitting is used to obtain the correct physical solution. Then, we turn our attention to miscible flows, where the competing effects of convection and diffusion are important. (This model can be used to refine the water-phase description obtained in the immiscible discussion.) Here new closed form solutions are given, and numerical models are developed. Using these models, the basic ideas behind resistivity migration and undiffusion are introduced using lineal and radial flow examples.

Quantitative Methods in Reservoir Engineering. http://dx.doi.org/10.1016/B978-0-12-810518-4.00021-9

With these discussions completed, we proceed to two more difficult problems. First, we consider deep, late-time invasion, when filtration rates are likely to be the smallest; here, the effects of radial geometric divergence and capillary pressure cannot be ignored. A two-phase flow model is formulated which assumes that a highly impermeable mudcake controls the filtration rate into the flushed zone. This is solved numerically for a range of parameters that alter the ratio of inertial to capillary forces from very low to very high, in the latter case, showing how shock formation as suggested by the Buckley-Leverett limit of zero capillary pressure is recovered. We show how nearly discontinuous saturation solutions can be reversed or *unshocked* using a high-order accurate numerical finite difference model. Second, the immiscible flow problem where mud filtrate invades a radial core is considered, but the usual assumption that a highly impermeable mudcake controls the flow rate into the core is not invoked. Thus, the model applies to mud filtrate invasion into very low permeability hydrocarbon zones with mixed water and oil. This combined analytical and computational model is developed using ideas obtained by integrating our two-phase flow formulation with the numerical mudcake growth model designed in Chapter 20. Finally, because much of two-phase flow modeling, by virtue of its inherent mathematical difficulties, is necessarily numerical, we refer the reader to our prior discussion in Chapter 13 on artificial viscosity, numerical diffusion, and convergence to correct solutions.

IMMISCIBLE BUCKLEY-LEVERETT LINEAL FLOWS WITHOUT CAPILLARY PRESSURE

In this section, we shall study the immiscible, constant density flow through a homogeneous lineal core where the effects of capillary pressure are insignificant. In particular, we shall derive exact, analytical, closed form solutions for the forward modeling problem for a single core. These solutions include those for saturation, pressure, and shock front velocity, for arbitrary relative permeability and fractional flow functions. We shall determine what formations properties can be inferred, assuming the existence of a propagating front, when the front velocity is known. The Darcy velocities are

$$q_w = -(k_w/\mu_w)\partial P_w/\partial x \qquad (21.1)$$

$$q_{nw} = -(k_{nw}/\mu_{nw})\partial P_{nw}/\partial x \qquad (21.2)$$

where μ_w and μ_{nw} are viscosities, and k_w and k_{nw} are relative permeabilities, the subscripts w and nw here denoting wetting and nonwetting phases. For mathematical simplicity, we assume zero capillary pressures P_c, so that

$$P_{nw}\text{-}P_w = P_c = 0 \qquad (21.3)$$

For water injection problems, this assumes that the displacement is fast (or, inertia dominated), so that surface tension can be neglected; however, when water breakthrough occurs, the assumption breaks down locally. In formation invasion, this zero capillary pressure assumption may be valid during the early periods of invasion near the well, when high filtrate influx rates are possible, as the resistance offered by mudcakes is minimal. For slow flows, capillary pressure is important; but generally, fast and slow must be characterized dimensionlessly in the context of the model. Since $P_{nw} = P_w$ holds, the pressure gradient terms in Eqs. (21.1), (21.2) are identical. If we divide Eq. (21.2) by Eq. (21.1), these cancel and we obtain

$$q_{nw} = (k_{nw}\mu_w / k_w \mu_{nw}) q_w \tag{21.4}$$

At this point, we invoke mass conservation, and assume for simplicity a constant density, incompressible flow. Then, it follows that

$$\partial q_w / \partial x = -\phi \partial S_w / \partial t \tag{21.5}$$

$$\partial q_{nw} / \partial x = -\phi \partial S_{nw} / \partial t \tag{21.6}$$

where S_w and S_{nw} are the wetting and non-wetting saturations. Since the fluid is incompressible, these saturations must sum to unity; that is,

$$S_w + S_{nw} = 1 \tag{21.7}$$

Then, upon adding Eqs. (21.5), (21.6), and simplifying with Eq. (21.7), it follows that

$$\partial (q_w + q_{nw}) / \partial x = 0 \tag{21.8}$$

Thus, we conclude that a one-dimensional, lineal, constant density flow without capillary pressure admits the general total velocity integral

$$q_w + q_{nw} = q(t) \tag{21.9}$$

where an arbitrary functional dependence on time is permitted. We have not yet stated what q(t) is, or how it is to be determined; this crucial issue is discussed in detail later. It is convenient to define the fractional flow function f_w for the wetting phase by the quotient

$$f_w = q_w / q \tag{21.10}$$

Then, for the nonwetting phase, we obtain

$$f_{nw} = q_{nw} / q = (q - q_w) / q = 1 - f_w \tag{21.11}$$

where we used Eq. (21.9). Eqs. (21.10), (21.11) can be rewritten as

$$q_w = q\, f_w \tag{21.12}$$

$$q_{nw} = q(1 - f_w) \tag{21.13}$$

Substituting into Eq. (21.4), the function q(t) drops out, so that

$$1 - f_w = (k_{nw}\mu_w / k_w \mu_{nw}) f_w \tag{21.14}$$

$$f_w(S_w, \mu_w / \mu_{nw}) = 1 / \{1 + (k_{nw}\mu_w / k_w \mu_{nw})\} \tag{21.15}$$

The function $f_w(S_w,\mu_w/\mu_{nw})$ in Eq. (21.15), we emphasize, is a function of the *constant* viscosity ratio μ_w/μ_{nw} and the saturation *function* S_w itself. According to Eq. (21.12), q_w must likewise be a function of S_w. Thus, we can rewrite Eq. (21.5) with the more informed nomenclature

$$\begin{aligned} \partial S_w/\partial t &= -\phi^{-1}\partial q_w/\partial x \\ &= -\phi^{-1} q\, \partial f_w(S_w,\mu_w/\mu_{nw})/\partial x \\ &= -\phi^{-1} q\, df_w(S_w,\mu_w/\mu_{nw})/S_w\, \partial S_w/\partial x \end{aligned} \tag{21.16}$$

or

$$\partial S_w/\partial t + \{q(t)/\phi\}\, df_w(S_w,\mu_w/\mu_{nw})/dS_w\, \partial S_w/\partial x = 0 \tag{21.17}$$

Eq. (21.17) is a first-order nonlinear partial differential equation for the saturation $S_w(x,t)$. Its general solution can be easily constructed using concepts from elementary calculus. The total differential dS_w for the function $S_w(x,t)$ can be written in the form

$$dS_w = \partial S_w/\partial t\, dt + \partial S_w/\partial x\, dx \tag{21.18}$$

If we divide Eq. (21.18) by dt, we find that

$$dS_w/dt = \partial S_w/\partial t + dx/dt\, \partial S_w \tag{21.19}$$

Comparison with Eq. (21.17) shows that we can certainly set

$$dS_w/dt = 0 \tag{21.20}$$

provided

$$dx/dt = \{q(t)/\phi\}\, df_w(S_w,\mu_w/\mu_{nw})/dS_w \tag{21.21}$$

Eq. (21.20) states that the saturation S_w is constant along a trajectory whose speed is defined by Eq. (21.21). (This constant may vary from trajectory to trajectory.) In two-phase immiscible flows, we conclude that it is the characteristic velocity $dx/dt = \{q(t)/\phi\}\, df_w(S_w,\mu_w/\mu_{nw})/dS_w$ that is important, and not the simple $dx/dt = q(t)/\phi$ obtained for single-phase flow. But when shocks form, it turns out that Eq. (21.39) applies.

Example Boundary Value Problems

If the filtration rate q(t) is a constant, say q_O, Eq. (21.21) takes the form

$$dx/dt = \{q_O/\phi\}\, df_w(S_w,\mu_w/\mu_{nw})/dS_w \tag{21.22}$$

Since the derivative $df_w(S_w,\mu_w/\mu_{nw})/dS_w$ is also constant along trajectories (as a result of Eq. 21.20), depending only on the arguments S_w and μ_w/μ_{nw}, it follows that Eq. (21.22) can be integrated in the form

$$x - \{(q_O/\phi) df_w(S_w,\mu_w/\mu_{nw})/dS_w\}\, t = \text{constant} \tag{21.23}$$

That S_w is constant when x - { ...} t is constant can be expressed as

$$S_w(x,t) = G(x - \{(q_O/\phi)df_w(S_w,\mu_w/\mu_{nw})/dS_w\}t) \tag{21.24}$$

where G is a general function. Note that the method by which we arrived at Eq. (21.24) is known as the method of characteristics (Hildebrand, 1948).

General Initial Value Problem

We now explore the meaning of Eq. (21.24). Let us set t=0 in Eq. (21.24). Then, we obtain

$$S_w(x,0) = G(x) \tag{21.25}$$

In other words, the general saturation solution to Eq. (21.17) for constant $q(t) = q_o$ satisfying the initial condition $S_w(x,0) = G(x)$, where G is a prescribed initial function, is *exactly* given by Eq. (21.24)!

Thus, it is clear that the finite difference numerical solutions offered by some authors are not really necessary because problems without capillary pressure can be solved analytically. Actually, such computational solutions are more damaging than useful because the artificial viscosity and numerical diffusion introduced by truncation and round-off error smear certain singularities (or, infinities) that appear as exact consequences of Eq. (21.17). Such numerical diffusion, we emphasize, appears as a result of finite difference and finite element schemes only, and can be completely avoided using the more labor-intensive method of characteristics. For a review of these ideas, refer to Chapter 13. As we shall show later, capillary pressure effects become important when singularities appear; modeling these correctly is crucial to correct strength and shock position prediction.

To examine how these singularities arise in the solution of Eq. (21.17), take partial derivatives of Eq. (21.24) with respect to x, so that

$$\partial S_w(x,t)/\partial x = \{G'\}\{1 - t(q_O/\phi)d^2f_w/dS_w^2\, \partial S_w(x,t)/\partial x\} \tag{21.26}$$

Solving for $\partial S_w(x,t)/\partial x$, we obtain

$$\partial S_w(x,t)/\partial x = G'/\{1 + t(q_O/\phi)(G')d^2f_w/dS_w^2\} \tag{21.27}$$

Now, the fractional flow function $f_w(S_w,\mu_w/\mu_{nw})$ is usually obtained from laboratory measurement and is to be considered as prescribed for the purposes of analysis. Let us focus our attention on the denominator of Eq. (21.27). If it remains positive, then the spatial derivative $\partial S_w(x,t)/\partial x$, is well-behaved for all time. If, however, $(q_O/\phi)(G')d^2f_w/dS_w^2 < 0$, then it follows that the denominator vanishes in the finite breakthrough time given by

$$t_{breakthrough} = -\phi/\{q_O\, G'\, d^2f_w/S_w^2\} \tag{21.28}$$

at which point the spatial derivative of saturation $\partial S_w(x,t)/\partial x \to \infty$ becomes singular, approaching infinity, increasing without bound. In reservoir engineering, this is known by various terms including water breakthrough, shocks, or saturation discontinuities. Since S_w undergoes rapid change, it is also said to be multivalued, or doublevalued. Whether or not this discontinuity exists in reality cannot be determined within the scope of our zero capillary pressure analysis. When saturation gradients become large, the capillary forces that we have neglected may become important, and cannot be excluded a priori in any analysis. When infinite saturation gradients form, as they have formed here, low-order theory breaks down, and recourse to a model that offers finer physical resolution is required.

General Boundary Value Problem for Infinite Core

Note that the argument of G{ } appearing in the solution of Eq. (21.24) takes the general form $x - \{(q_O/\phi)df_w(S_w\mu_w/\mu_{nw})/dS_w\}t$. There is nothing sacred about this expression, and we could have multiplied it by two, five, or $-\phi/(q_O\, df_w/dS_w)$. With the last choice, we can rewrite Eq. (21.24) as

$$S_w(x,t) = H\{t - \phi x/(q_O\, df_w/dS_w)\} \tag{21.29}$$

If we set $x=0$ throughout in Eq. (21.29), we find that $S_w(0,t) = H(t)$. Thus, the saturation solution to Eq. (21.17) satisfying the boundary condition $S_w(0,t) = H(t)$, where H is a prescribed function, is given by Eq. (21.29).

Variable q(t)

If the filtration rate q(t) is a general function of time, we return to Eq. (21.21) and rewrite it in the differential form

$$dx = \{q(t)/\phi\}df_w(S_w,\mu_w/\mu_{nw})/dS_w\, dt \tag{21.30}$$

Since Eq. (21.20) states that S_w is still constant along a trajectory, the term $df_w(S_w,\mu_w/\mu_{nw})/dS_w$ is likewise constant. Thus, the integral of Eq. (21.30) is simply

$$x - \phi^{-1}df_w(S_w,\mu_w/\mu_{nw})/dS_w \int q(t)\, dt = \text{constant} \tag{21.31}$$

where $\int q(t)dt$ denotes the indefinite integral (e.g., $\int q_O\, dt = q_O t$ is obtained for our constant rate problem). Following a line of reasoning similar to that leading to Eq. (21.24), since S_w is constant whenever the left side of Eq. (21.31) is constant, we have the equivalent functional statement

$$S_w(x,t) = G\left(x - \phi^{-1}df_w(S_w,\mu_w/\mu_{nw})/dS_w \int q(t)\, dt\right) \tag{21.32}$$

Eq. (21.32) is the general saturation solution for time-dependent q(t). If the integrated function $\int q(t)\, dt$ vanishes for $t=0$, this solution satisfies the initial condition specified by Eq. (21.25). If the function does not vanish, some minor algebraic manipulation is required to obtain the correct format.

Mudcake-Dominated Invasion

So far, we have not stated how the velocity q(t), possibly transient, is determined. If we assume that the flow at the inlet to our lineal core is controlled by mudcake, as is often the case, the fluid dynamics within the core will be unimportant in determining q(t). (This assumption is removed in our last example.) Then, the general mudcake model in Chapter 17 for single-phase filtrate flows provides the required q(t). In fact,

$$x_f(t) = \phi_{eff}^{-1} \sqrt{\left\{2k_1(1-\phi_C)(1-f_s)(p_m-p_r)t/\left(\mu_f f_s\right)\right\}} \tag{21.33}$$

when the effect of spurt and the presence of the formation are neglected. The fluid influx rate q(t) through the mudcake is therefore given by

$$q(t) = \phi_{eff} dx_f(t)/dt = \tfrac{1}{2}\, t^{-1/2} \sqrt{\left\{2k_1(1-\phi_C)(1-f_s)(p_m-p_r)/\left(\mu_f f_s\right)\right\}} \tag{21.34}$$

which can be substituted in the nonlinear saturation equation

$$\partial S_w/\partial t + \{q(t)/\phi\}\, df_w(S_w, \mu_w/\mu_{nw})/dS_w\, \partial S_w/\partial x = 0 \tag{21.35}$$

This can be integrated straightforwardly using the method of characteristics. So long as singularities and saturation fronts do not form, saturations obtained as a function of space and time will be smooth, and shocks will not appear.

Shock Velocity

We shall consider the problem that arises when saturation shocks *do* form. (Problems with smooth but rapidly varying properties are addressed in our capillary pressure analysis.) In order to discuss saturation discontinuities and steep gradients, we must complete the formulation by specifying initial and boundary conditions. We assume that at $t=0$, our core is held at the constant water saturation S_w^i throughout, where the italicized *i* denotes initial conditions. At the left boundary $x=0$, where fluid influx occurs, we assume that the water saturation is fixed at a constant value S_w^l where the italicized *l* denotes left. (Normally, this value is unity for water filtrates, but it may differ for certain water-oil muds.) That is, we take

$$S_w(x,0) = S_w^i \tag{21.36}$$

$$S_w(0,t) = S_w^l \tag{21.37}$$

As discussed, we can expect shockwaves and steep saturation discontinuities to form in time, depending on the exact form and values of our fractional flow functions and initial conditions. We shall assume that the particular functions do lead to piston-like shock formation very close to the borehole. The shock boundary value problem just stated *can* be solved in closed form, and, in fact, is the petroleum engineering analog of the classic nonlinear signaling problem ($\rho_t + c(\rho)\,\rho_x = 0$, $\rho = \rho_o$ for $x > 0$, $t = 0$, and $\rho = g(t)$ for $t > 0$, $x = 0$) discussed in the wave mechanics book of Whitham (1974).

We shall not rederive the mathematics, but will draw on Whitham's results only. For brevity, define for convenience the function

$$Q(S_w) = \{q(t)/\phi\}\, df_w(S_w, \mu_w/\mu_{nw})/dS_w \tag{21.38}$$

where q(t) is given in Eq. (21.34). It turns out that the shock propagates with a shock speed equal to

$$V_{shock} = \{Q_w(S_w^l) - Q_w(S_w^i)\}/(S_w^l - S_w^i) \tag{21.39}$$

If the injection rate q(t), the core porosity ϕ, and the speed of the front V_{shock} separating saturations S_w^l from S_w^i are known, then since S_w^l is available at the inlet of the core, Eqs. (21.38), (21.39) yield information relating the initial formation saturation $S_w{}^i$ to the fractional flow derivative $df_w(S_w, \mu_w/\mu_{nw})/dS_w$. Eq. (21.15) shows that the fractional flow function satisfies $f_w(S_w \mu_w/\mu_{nw}) = 1/\{1 + (k_{nw}\mu_w/k_w\mu_{nw})\}$. Thus, if additional lithology information is available about the form of the relative permeability functions, the viscosity ratio μ_w/μ_{nw} can be extracted, thus yielding μ_{nw}. We emphasize that this solution for the nonlinear saturation problem does *not* apply to the linear single-phase flow where red water displaces blue water.

Pressure Solution

Now we derive the solution for the corresponding transient pressure field. Let us substitute Eqs. (21.1), (21.2) (that is, Darcy's laws $q_w = -(k_w/\mu_w)\partial P_w/\partial x$ and $q_{nw} = -(k_{nw}/\mu_{nw})\partial P_{nw}/\partial x$) into Eq. (21.9) (or $q_w + q_{nw} = q(t)$). Also, from Eq. (21.3), we find that $P_{nw} = P_w$. Thus, we obtain the governing pressure equation

$$\{(k_w(S_w)/\mu_w) + (k_{nw}(S_w)/\mu_{nw})\}\partial P_w/\partial x = -q(t) \tag{21.40}$$

so that the *pressure gradient* satisfies

$$\partial P_w/\partial x = -q(t)/\{(k_w(S_w)/\mu_w) + (k_{nw}(S_w)/\mu_{nw})\} \tag{21.41}$$

Since the saturation function $S_w(x,t)$, following Whitham's solution to the signaling problem is a simple step function in the x direction whose hump moves at the shock velocity, we conclude that the *pressure gradient* in Eq. (21.41) takes on either of two constant values, depending on whether S_w equals $S_w{}^i$ or $S_w{}^l$ locally. Thus, on either side of the shock front, we have different but linear pressure variations with space, when time is held fixed. This situation is shown in Fig. 21.1. At the shock front itself, the requirement that pressure be continuous and single-valued, a consequence of our zero capillary pressure assumption, is itself sufficient to uniquely define the time-varying pressure distribution across the entire core.

Now we outline the computational procedure. At the left of the core, the saturation specification $S_w{}^l$ completely determines the value of the linear variation $\partial P_w(S_w{}^l)/\partial x$, following the arguments of the preceding paragraph. Since the exact value of pressure P^l is assumed to be known at $x=0$ (that is, the interface between the rock core and the mudcake), knowledge of the constant rate of change of pressure throughout completely defines the pressure variation starting at $x=0$. Unlike reservoir engineering problems, we are not posing a pressure problem for the core in order to calculate flow rate; our flow rate is completely prescribed by the mudcake. In this problem, saturation constraints fix both pressure gradients, which in turn fix the right-side pressure. The radial flow extension of this procedure leads to an estimate for reservoir pore pressure.

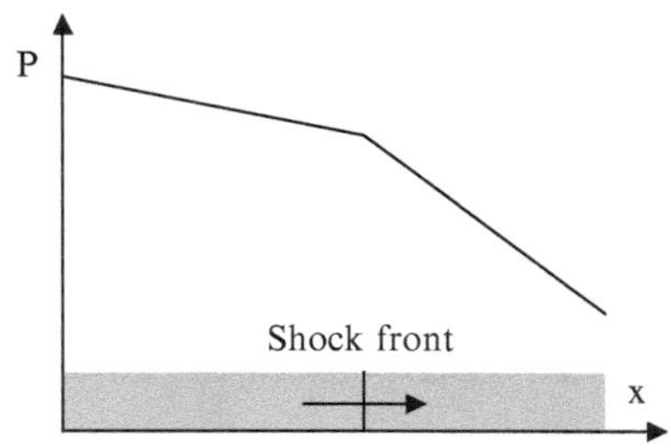

FIG. 21.1 Pressure in lineal core.

In finite length core flows *without* mudcakes, it *is* appropriate to specify both the left and right pressures P^l and P^r, and determine the corresponding q(t). Since q(t) is now unknown, the shock velocity cannot be written down a priori, so that the manner in which the step solution for saturation propagates is uncertain. Strong nonlinear coupling between the pressure and saturation equations is obtained, and iterative numerical solutions are required, which will be discussed later. Before embarking on radial flows with capillary pressure, we turn to multiphase flows of miscible fluids, where diffusive processes predominate.

MOLECULAR DIFFUSION IN FLUID FLOWS

Fluid flows need not be purely homogeneous, as in single-phase flows, nor need they be definable by clearly discernible differences in properties, as in multiphase immiscible flows. For simplicity, let us consider *mixtures* having two components only; the composition of the mixture is described by the concentration C, defined as the ratio of the mass of one component to the total mass of the fluid in a given volume element. With the passage of time, this concentration changes in two ways. When there is macroscopic motion of the fluid, mechanical mixing of the flow results; if we ignore thermal conduction and internal friction, this change is thermodynamically reversible and does not result in energy dissipation. But a change in composition will also occur by the molecular transfer of the components from one part of the fluid to another. The equalization of concentration by this direct change of composition is called diffusion. Diffusion is an irreversible process; like thermal conduction and internal viscous friction, it is one of the sources of energy dissipation in fluid mixtures. If we denote by ρ the total density of the fluid, the equation of mass continuity for the total mass of the fluid is, as before,

$$\partial\rho/\partial t + \nabla \bullet (\rho \mathbf{q}) = 0 \tag{21.42}$$

where $\mathbf{q}$ is the velocity vector and ∇ denotes the gradient operator from vector calculus. The corresponding momentum, or Darcy equations, remain unchanged. In the absence of diffusion, the composition of any given fluid element would remain unchanged as it moved about. That is, the total derivative dC/dt would be zero, so that $dC/dt = \phi\, \partial C/\partial t + \mathbf{q} \bullet \nabla C = 0$. This can be written, using Eq. (21.42) for mass continuity, in the form

$$\phi\, \partial(\rho C)/\partial t + \nabla \bullet (\rho \mathbf{q} C) = 0 \tag{21.43}$$

as a continuity equation for a component of the mixture. But when diffusion occurs, besides the flux $\rho\mathbf{q}C$ of the component under investigation, there is another flux which results in the transfer of the components even when the fluid mass as a whole is at rest. The general concentration equation describing both mass transport and diffusion takes the form (Peaceman, 1977)

$$\phi\, \partial C/\partial t + \mathbf{q} \bullet \nabla C = \kappa \nabla^2 C \tag{21.44}$$

where κ is the diffusivity coefficient. In radial cylindrical coordinates, Eq. (21.44) can be written as

$$\phi\, \partial C/\partial t + v_{(r)}\, \partial C/\partial r = \kappa \left\{ \partial^2 C/\partial r^2 + (1/r)\, \partial C/\partial r \right\} \tag{21.45}$$

where $v_{(r)}$ is the underlying radial Darcy velocity, for example, as obtained in Chapters 17 and 20, or

$$\phi\, \partial C/\partial t + \left(v_{(r)} - \kappa/r\right) \partial C/\partial r = \kappa\, \partial^2 C/\partial r^2 \tag{21.46}$$

What might be a typical value of κ? Peaceman and Rachford (1962), for example, assumed a value of $\kappa = 10^{-3}$ cm^2/s. This corresponded to an experimental situation where oil was flooded by solvent of equal density, from a thin rectangular channel in Lucite packed with uniform Ottawa sand. We shall discuss Eq. (21.46) in more detail later, but for now, it is useful to consider lineal flows for which motivating exact analytical solutions are available.

Exact Lineal Flow Solutions

For one-dimensional *lineal* flows, the convective-diffusion equation for a *constant* velocity U takes the form

$$\phi\, \partial C/\partial t + U\, \partial C/\partial x = \kappa\, \partial^2 C/\partial x^2 \tag{21.47}$$

Let us assume that at $t=0$, the concentration varies linearly with x in the form $C_0 + \alpha\, x$, whereas at the inlet boundary $x=0$, the concentration is imposed in the form $C_1 + \beta\, t$. While the linear variations appear somewhat limiting, they can be generally interpreted as first-order Taylor series representations to more general initial and boundary conditions. In mathematical form,

$$C(x > 0, 0) = C_0 + \alpha\, x \tag{21.48}$$

$$C(0, t > 0) = C_1 + \beta\, t \tag{21.49}$$

The exact solution to this initial-boundary value problem is straightforwardly obtained using Laplace transforms and can be shown to be

$$\begin{aligned} C(x,t) = {} & C_0 + \alpha\,(x - Ut/\phi) \\ & + \tfrac{1}{2}(C_1 - C_0)\left\{\mathrm{erfc}\,\tfrac{1}{2}(x - Ut/\phi)/(\kappa t)^{1/2} + e^{Ux/\kappa}\,\mathrm{erfc}\,\tfrac{1}{2}(x + Ut/\phi)/(\kappa t)^{1/2}\right\} \\ & + \{(\beta + \alpha U/\phi)(2U/\phi)\}\left\{(x + Ut/\phi)e^{Ux/\kappa}\,\mathrm{erfc}\,\tfrac{1}{2}(x + Ut/\phi)/(\kappa t)^{1/2}\right. \\ & \left. - (x - Ut/\phi)\,\mathrm{erfc}\,\tfrac{1}{2}(x - Ut/\phi)/(\kappa t)^{1/2}\right\} \end{aligned} \tag{21.50}$$

where erfc denotes the complementary error function. These solutions show that, in a coordinate system moving with the speed U, the width of the transition zone increases and smears with time (Marle, 1981). Several limits of Eq. (21.50) immediately come to mind. If $\alpha = \beta = 0$,

$$\begin{aligned} C(x,t) = {} & C_0 + \tfrac{1}{2}(C_1 - C_0) \\ & \times \left\{\mathrm{erfc}\,\tfrac{1}{2}(x - Ut/\phi)/(\kappa t)^{1/2} + e^{Ux/\kappa}\,\mathrm{erfc}\,\tfrac{1}{2}(x + Ut/\phi)/(\kappa t)^{1/2}\right\} \end{aligned} \tag{21.51}$$

If, in addition, $U=0$,

$$C(x,t) = C_0 + \tfrac{1}{2}(C_1 - C_0)\left\{\mathrm{erfc}\,\tfrac{1}{2}x/(\kappa t)^{1/2} + \mathrm{erfc}\,\tfrac{1}{2}x/(\kappa t)^{1/2}\right\} \tag{21.52}$$

This solution, at least in lineal flows, describes the large-time behavior in problems with thick mudcakes that effectively shut off the influx of filtrate.

Numerical Analysis

The numerical formulation for the heat-like equation $\phi\,\partial C/\partial t + U\,\partial C/\partial x = \kappa\,\partial^2 C/\partial x^2$ given in Eq. (21.47) proceeds in the same manner as that for Eq. (20.19), or $\partial^2 p(x,t)/\partial x^2 = (\phi\mu c/k)\,\partial p/\partial t$, since the former can be written as

$$\partial^2 C/\partial x^2 = \phi/\kappa\,\partial C/\partial t \tag{21.53}$$

in the U = 0 limit. In this limit, Eqs. (20.22), (20.23) apply without change. If we replace P in Eq. (20.23) by C, and $\phi\mu c/k$ by ϕ/κ, we have

$$\begin{aligned} C_{i-1,n} - \left[2 + \left\{\phi(\Delta x)^2/(\kappa\Delta t)\right\}\right] C_{i,n} + C_{i+1,n} \\ = -\left\{\phi(\Delta x)^2/(\kappa\Delta t)\right\} C_{i,n-1} \end{aligned} \tag{21.54}$$

Then, the algorithm and Fortran implementation developed for compressible transient flows applies without change. In the limit when U does not vanish, we write the governing PDE in the form $\partial^2 C/\partial x^2 = \phi/\kappa\,\partial C/\partial t + U/\kappa\,\partial C/\partial x$, or

$$\partial^2 C/\partial x^2 - U/\kappa\,\partial C/\partial x = \phi/\kappa\,\partial C/\partial t \tag{21.55}$$

Applying central differences to all spatial derivatives and backward differences to the first-order time derivative, we have

$$\begin{aligned} (C_{i-1,n} - 2C_{i,n} + C_{i+1,n})/\Delta x^2 - (U/\kappa)(C_{i+1,n} - C_{i-1,n})/(2\Delta x) \\ = \phi/\kappa(C_{i,n} - C_{i,n-1})/\Delta t \end{aligned} \tag{21.56}$$

or

$$\begin{aligned} C_{i-1,n} - 2C_{i,n} + C_{i+1,n} - \left(U\Delta x^2/\kappa\right)(C_{i+1,n} - C_{i-1,n})/(2\Delta x) \\ = \left\{\left(\phi\Delta x^2\right)/(\kappa\Delta t)\right\}(C_{i,n} - C_{i,n-1}) \end{aligned} \tag{21.57}$$

Thus, we again have the familiar tridiagonal difference equation

$$\begin{aligned} &[1 + U\Delta x/2\kappa]\, C_{i-1,n} \\ &- \left[2 + \left(\phi\Delta x^2\right)/(\kappa\Delta t)\right] C_{i,n} \\ &+ [1 - U\Delta x/2\kappa]\, C_{i+1,n} = -\left(\phi\Delta x^2\right)/(\kappa\Delta t)\}\, C_{i,n-1} \end{aligned} \tag{21.58}$$

which bears superficial resemblance to our pressure equation for radial flows. Peaceman and Rachford (1962) discuss this model in their investigation of miscible reservoir flow modeling. Also, Lantz (1971) offers very enlightening discussions on numerical diffusion, and in particular examines the types of numerical diffusion and truncation error that arise in different kinds of discretization schemes. For example, instead of the central differencing used in Eq. (21.56) for the first derivative, we might have assumed

$$\partial C/\partial x \approx (U/\kappa)(C_{i+1,n}-C_{i,n})/\Delta x \tag{21.59a}$$

$$\partial C/\partial x \approx (U/\kappa)(C_{i,n}-C_{i-1,n})/\Delta x \tag{21.59b}$$

$$\partial C/\partial x \approx (U/\kappa)(C_{i+1,n-1}-C_{i,n-1})/\Delta x \tag{21.59c}$$

or

$$\partial C/\partial x \approx (U/\kappa)(C_{i,n-1}-C_{i-1,n-1})/\Delta x \tag{21.59d}$$

We caution that issues beyond accuracy are involved. As noted in Chapter 13, the computed diffusivity is not the physical diffusivity κ, but a combination of that plus numerical diffusion due to truncation errors.

Diffusion in Cake-Dominated Flows

Close to the well, immiscible flows containing propagating saturation discontinuities may exist. But very often, flows are obtained that do not contain shocks. These include immiscible flows with and without capillary pressure, and miscible flows governed by highly diffusive processes, where discontinuities never form.

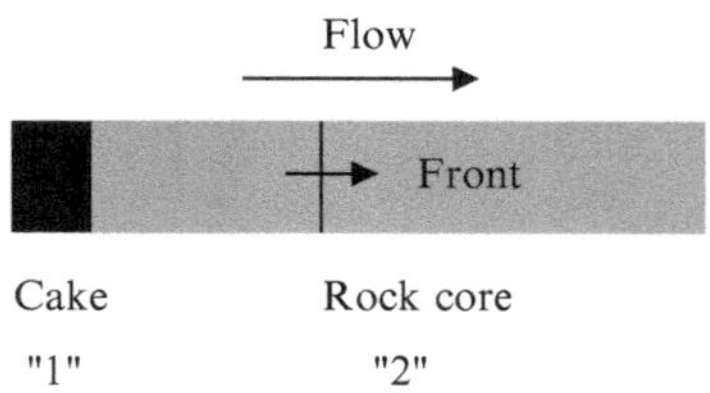

FIG. 21.2 Diffusive front motion.

For purely diffusive flows, sharp (fresh versus saline water resistivity) discontinuities always smear in time. The dynamics of such flows are very important in log interpretation. For this class of problems, the speed of the fresh-to-saline water interface slows appreciably once the mudcake establishes itself at the borehole walls, as we have demonstrated in Chapter 17. This is especially true in the case of radial flows, where geometric spreading significantly slows the front. For such problems, the speed of the underlying flow U can be neglected after some time, when diffusion predominates. The problem is shown in Fig. 21.2.

Resistivity Migration

Let us suppose that the ultimate electromagnetic wave resistivity tool were available and capable of determining the exact, continuous, or even discontinuous variation of electrical properties in the formation as a function of the radial coordinate r in a concentric problem. (*Resistivity* and *concentration* are used interchangeably, since they are related through logging tool measurements.) In order to use the piston-like displacement results assumed in Chapter 19 for time lapse analysis, a front having a distinct constant radius would *have* to be inferred from a generally continuous distribution of resistivities. Typically, this is done in any of several ways: by eye, by arithmetic, geometric, or harmonic averaging, or by using the improved method of Chin et al. (1986) as discussed in Chapter 19, all of which are ad hoc. Actually, a simple and exact solution to this problem is possible. What we wish to do, at any particular instant in time, given a smeared concentration profile that will generally vary with radial position, is to extrapolate that profile back to time $t=0$ when the front is truly discontinuous. This problem formulation appears incredible, since diffusion is physically irreversible. For example, in heat transfer, the effect of an instantaneous point heat source is a diffusion width that grows with time; the diffused temperature distribution *never* evolves backward to become a point source. However, while physical diffusion is irreversible, the computational process isn't. It turns out that we can undiffuse a smeared front using reverse diffusion and recover original sharp transitions by marching backward in time using a host diffusion equation. Of course, the initial profile must be sufficiently transient, since a steady-state profile is obviously devoid of historical content. Such migration methods are used in seismic imaging and geophysics. In particular, wave-equation-based methods, introduced by Claerbout (1985a,b) at M.I.T. and Stanford, and formalized by the multiple scale analyses in Chin (1994), lead to a parabolized wave equation which is just the heat equation in disguise.

By applying these methods to our smeared concentrations, we can recover any sharp discontinuities, if they in fact existed. In doing so, we obtain the location of the radial front for use in the plug-flow time lapse analysis equations developed and used in Chapters 17 and 19. In addition to this front position, we can uncover the time scale of the reverse diffusion process as a byproduct of the reverse time integration. The key idea is simple: differential equations of evolution *do* exist, and their application to deconvolution is not at all unusual. There are some problems, however. Since the *end starting conditions* are likely to be complicated functions of space, determined at discrete points, the reverse diffusion must be accomplished numerically in time. But finite difference methods produce truncation and round-off errors that are associated with their own thermodynamic irreversibility and entropy production. Thus, the scheme has to be designed so that it is perfectly reversible in order to be usable for time lapse analysis purposes. This is accomplished by retaining the next highest order finite difference contributions neglected in Chapter 20.

Lineal Diffusion and Undiffusion Examples

For simplicity, consider the fresh-to-saline water invasion problem, where mudcake forms and grows at the inlet entrance. At first, mud filtrate motions are extremely rapid, and fluid movements dominate the convection-diffusion process. However, as mudcake forms, the influx of filtrate decreases rapidly with time, and eventually, diffusion dominates the dynamics. For simplicity, we first study lineal flows where the effects of radial geometric spreading are unimportant. In our examples, because fluid convection is negligible, we consider $\kappa\, \partial^2 C/\partial x^2 = \phi\, \partial C/\partial t$. For numerical purposes, we fix the left-side ($x = 1$) concentration at $C = 10\%$, while the right ($x = 11$) is held at $C = 90\%$. For visual clarity, all concentrations to the left side of $x = 6$ are initially 10%, while those values to the right are 90%.

There are several objectives for pursuing the test cases described here. For one, if the initial value problem when time reversal starts has progressed to steady-state, straight-line conditions, it is clear that all transient information will have been lost and that no amount of reverse diffusion will return the steady-state system to its initial step profile. (The steady-state solution is obtained by solving $d^2C/dx^2 = 0$, taking the straight line joining C values at the left and right boundaries.) The degree of smear and its percentage approach to steady state are therefore important research questions. Second, we need to determine if the method is applicable to radial flows, if it proves successful for lineal ones. This objective is important because any spatial distribution of concentration obtained radially is a consequence of both diffusion and geometric spreading. Geometric spreading worsens the undiffusion process because diffusion effects are less clear. The method *must* account for both mechanisms if the initial step profile is to be recovered properly. In the following results, we deemphasize the values of the numerical inputs themselves; note that ten one-foot grid blocks were selected, with 500 time steps taken forward, then followed by 500 taken backwards. The real parameters of computational significance, of course, are the dimensionless ones that affect truncation errors. Solutions are both tabulated and plotted using a simple ASCII text plotter; the wiggles in our plotter are due to character spacing and font control issues and not instability, as tabulated results clearly show. Observe the strong initial discontinuity in the C(x,t) profile used. The bottom solution in Fig. 21.3A represents the final spatial profile obtained before we reverse integrate in time. The profile is smeared, almost to the point where a straight-line steady solution is obtained. Carefully study the reverse diffusion results in Fig. 21.3B.

Despite truncation errors after 1000 time steps, the last tabulation-plot in Fig. 21.3B shows that we have recaptured the step initial condition in three ways: we (1) obtained the exact left-to-right concentration values of 10% and 90%, (2) correctly imaged the transition boundary between the $x = 5$ to 6 ft nodes, and (3) extracted the two solutions just quoted using exactly the same number of backward time steps as we did forward time steps. In time lapse

```
Concentration vs distance @ time .5000E+00 sec.
  Position (ft)    C%
                             -------------------------------
   .100E+01    .100E+02      | *
   .200E+01    .100E+02      | *
   .300E+01    .100E+02      | *
   .400E+01    .100E+02      | *
   .500E+01    .103E+02      | *
   .600E+01    .897E+02      |                            *
   .700E+01    .900E+02      |                            *
   .800E+01    .900E+02      |                            *
   .900E+01    .900E+02      |                             *
   .100E+02    .900E+02      |                            *
   .110E+02    .900E+02      |                             *

Concentration vs distance @ time .1000E+02 sec.
  Position (ft)    C%
                             -------------------------------
   .100E+01    .100E+02      | *
   .200E+01    .100E+02      | *
   .300E+01    .100E+02      | *
   .400E+01    .102E+02      | *
   .500E+01    .153E+02      |   *
   .600E+01    .847E+02      |                           *
   .700E+01    .898E+02      |                            *
   .800E+01    .900E+02      |                            *
   .900E+01    .900E+02      |                            *
   .100E+02    .900E+02      |                            *
   .110E+02    .900E+02      |                             *

Concentration vs distance @ time .1000E+03 sec.
  Position (ft)    C%
                             -------------------------------
   .100E+01    .100E+02      | *
   .200E+01    .103E+02      | *
   .300E+01    .117E+02      | *
   .400E+01    .176E+02      |   *
   .500E+01    .351E+02      |         *
   .600E+01    .649E+02      |                    *
   .700E+01    .824E+02      |                          *
   .800E+01    .883E+02      |                            *
   .900E+01    .897E+02      |                            *
   .100E+02    .900E+02      |                            *
   .110E+02    .900E+02      |                             *

Concentration vs distance @ time .2495E+03 sec.
  Position (ft)    C%
                             -------------------------------
   .100E+01    .100E+02      | *
   .200E+01    .124E+02      |  *
   .300E+01    .172E+02      |   *
   .400E+01    .265E+02      |      *
   .500E+01    .413E+02      |           *
   .600E+01    .587E+02      |                  *
   .700E+01    .735E+02      |                       *
   .800E+01    .828E+02      |                          *
   .900E+01    .874E+02      |                            *
   .100E+02    .892E+02      |                            *
   .110E+02    .900E+02      |                             *
```

(A)

FIG. 21.3 (A) A diffusing lineal flow.

```
Concentration vs distance @ time .2000E+03 sec.
  Position (ft)    C%
                            ______________________________
   .100E+01    .100E+02     |  *
   .200E+01    .116E+02     |  *
   .300E+01    .154E+02     |    *
   .400E+01    .243E+02     |       *
   .500E+01    .401E+02     |            *
   .600E+01    .599E+02     |                  *
   .700E+01    .757E+02     |                        *
   .800E+01    .846E+02     |                           *
   .900E+01    .883E+02     |                            *
   .100E+02    .896E+02     |                            *
   .110E+02    .900E+02     |                             *

Concentration vs distance @ time .1000E+03 sec.
  Position (ft)    C%
                            ______________________________
   .100E+01    .100E+02     |  *
   .200E+01    .103E+02     |  *
   .300E+01    .117E+02     |  *
   .400E+01    .176E+02     |    *
   .500E+01    .350E+02     |          *
   .600E+01    .650E+02     |                    *
   .700E+01    .824E+02     |                          *
   .800E+01    .883E+02     |                            *
   .900E+01    .897E+02     |                            *
   .100E+02    .899E+02     |                            *
   .110E+02    .900E+02     |                             *

Concentration vs distance @ time .8000E+01 sec.
  Position (ft)    C%
                            ______________________________
   .100E+01    .100E+02     |  *
   .200E+01    .999E+01     |  *
   .300E+01    .100E+02     |  *
   .400E+01    .105E+02     |  *
   .500E+01    .130E+02     |   *
   .600E+01    .870E+02     |                           *
   .700E+01    .895E+02     |                            *
   .800E+01    .899E+02     |                            *
   .900E+01    .900E+02     |                             *
   .100E+02    .900E+02     |                            *
   .110E+02    .900E+02     |                            *

Concentration vs distance @ time .0000E+00 sec.
  Position (ft)    C%
                            ______________________________
   .100E+01    .100E+02     |  *
   .200E+01    .999E+01     |  *
   .300E+01    .100E+02     |  *
   .400E+01    .106E+02     |  *
   .500E+01    .833E+01     |*
   .600E+01    .917E+02     |                             *
   .700E+01    .894E+02     |                            *
   .800E+01    .900E+02     |                            *
   .900E+01    .900E+02     |                            *
   .100E+02    .900E+02     |                            *
   .110E+02    .900E+02     |                            *
```

(B)

FIG. 21.3, CONT'D (B) An "undiffusing" lineal flow.

analysis, the front position obtained in the last plot might be used as input to the piston-displacement formulas derived in Chapters 7 and 19. Similar results for radial flows can be obtained.

Radial Diffusion and Undiffusion Examples

Here, we repeat the foregoing problem with the same computational parameters, except that the lineal equation $\kappa\ \partial^2 C/\partial x^2 = \phi\ \partial C/\partial t$ is replaced by its cylindrical radial counterpart, namely, $\kappa\ (\partial^2 C/\partial r^2 + 1/r\ \partial C/\partial r) = \phi\ \partial C/\partial t$. We introduce strongly divergent radial effects by assuming a small borehole radius of 0.25 ft relative to our one-foot grid blocks. Again, the difference scheme is integrated 500 time steps, at which point the smeared and geometrically distorted concentration profile is undiffused in time for an identical 500 time steps. Once more, our computed results suggest that smeared resistivity profiles can be successfully deconvolved to produce the original sharp front. The last display of Fig. 21.4A represents the final radial profile obtained before reverse time integration begins. The time-reversed computations are shown in Fig. 21.4B.

As before, we recaptured the exact initial step concentration profile, to include concentration values, location of the discontinuity, and total time to undiffuse. While the starting radial concentration profile is substantially smeared, and significantly different from the lineal flow solution obtained at this point in Fig. 21.3A, we again successfully undiffuse our starting flow.

IMMISCIBLE RADIAL FLOWS WITH CAPILLARY PRESSURE AND PRESCRIBED MUDCAKE GROWTH

In this section, we shall consider immiscible radial flows with capillary pressure and prescribed mudcake growth. In particular, we shall derive the relevant governing equations, develop the numerical finite difference algorithm and the Fortran implementation, and proceed to demonstrate the computational model in both forward and inverse modes.

Governing Saturation Equation

Let us now repeat the lineal flow derivation given earlier but include the effects of radial geometric spreading and nonvanishing capillary pressure. Again, analogous Darcy laws apply, namely,

$$q_w = -(k_w/\mu_w)\ \partial P_w/\partial r \qquad (21.60)$$

$$q_{nw} = -(k_{nw}/\mu_{nw})\ \partial P_{nw}/\partial r \qquad (21.61)$$

Unlike flows in rectangular systems, the mass continuity equations in cylindrical radial coordinates take the form

```
Concentration vs distance @ time .5000E+00 sec.
  Position (ft)      C%
                                 ______________________________
  .100E+01    .100E+02          | *
  .200E+01    .100E+02          | *
  .300E+01    .100E+02          | *
  .400E+01    .100E+02          | *
  .500E+01    .103E+02          | *
  .600E+01    .897E+02          |                           *
  .700E+01    .900E+02          |                           *
  .800E+01    .900E+02          |                           *
  .900E+01    .900E+02          |                            *
  .100E+02    .900E+02          |                           *
  .110E+02    .900E+02          |                            *
Concentration vs distance @ time .1000E+01 sec.
  Position (ft)      C%
                                 ______________________________
  .100E+01    .100E+02          | *
  .200E+01    .100E+02          | *
  .300E+01    .100E+02          | *
  .400E+01    .100E+02          | *
  .500E+01    .107E+02          | *
  .600E+01    .895E+02          |                           *
  .700E+01    .900E+02          |                           *
  .800E+01    .900E+02          |                           *
  .900E+01    .900E+02          |                            *
  .100E+02    .900E+02          |                           *
  .110E+02    .900E+02          |                            *
Concentration vs distance @ time .1000E+02 sec.
  Position (ft)      C%
                                 ______________________________
  .100E+01    .100E+02          | *
  .200E+01    .100E+02          | *
  .300E+01    .100E+02          | *
  .400E+01    .103E+02          | *
  .500E+01    .159E+02          |   *
  .600E+01    .853E+02          |                          *
  .700E+01    .898E+02          |                           *
  .800E+01    .900E+02          |                           *
  .900E+01    .900E+02          |                           *
  .100E+02    .900E+02          |                           *
  .110E+02    .900E+02          |                            *
Concentration vs distance @ time .1000E+03 sec.
  Position (ft)      C%
                                 ______________________________
  .100E+01    .100E+02          | *
  .200E+01    .107E+02          | *
  .300E+01    .128E+02          |  *
  .400E+01    .202E+02          |    *
  .500E+01    .393E+02          |           *
  .600E+01    .685E+02          |                    *
  .700E+01    .839E+02          |                         *
  .800E+01    .887E+02          |                           *
  .900E+01    .898E+02          |                           *
  .100E+02    .900E+02          |                           *
  .110E+02    .900E+02          |                            *
Concentration vs distance @ time .2000E+03 sec.
  Position (ft)      C%
                                 ______________________________
  .100E+01    .100E+02          | *
  .200E+01    .141E+02          |  *
  .300E+01    .194E+02          |    *
  .400E+01    .299E+02          |       *
  .500E+01    .466E+02          |             *
  .600E+01    .652E+02          |                   *
  .700E+01    .789E+02          |                        *
  .800E+01    .860E+02          |                          *
  .900E+01    .888E+02          |                           *
  .100E+02    .897E+02          |                           *
  .110E+02    .900E+02          |                            *
Concentration vs distance @ time .2495E+03 sec.
  Position (ft)      C%
                                 ______________________________
  .100E+01    .100E+02          | *
  .200E+01    .163E+02          |   *
  .300E+01    .226E+02          |     *
  .400E+01    .334E+02          |         *
  .500E+01    .486E+02          |              *
  .600E+01    .648E+02          |                   *
  .700E+01    .774E+02          |                       *
  .800E+01    .848E+02          |                          *
  .900E+01    .882E+02          |                           *
  .100E+02    .895E+02          |                           *
  .110E+02    .900E+02          |                            *
```

(A)

FIG. 21.4 (A) A diffusing radial flow.

(Continued)

```
Concentration vs distance @ time .2000E+03 sec.
   Position (ft)     C%
                              ______________________________
    .100E+01    .100E+02      |  *
    .200E+01    .141E+02      |   *
    .300E+01    .194E+02      |     *
    .400E+01    .299E+02      |        *
    .500E+01    .466E+02      |              *
    .600E+01    .653E+02      |                    *
    .700E+01    .789E+02      |                         *
    .800E+01    .860E+02      |                           *
    .900E+01    .888E+02      |                            *
    .100E+02    .897E+02      |                            *
    .110E+02    .900E+02      |                             *

Concentration vs distance @ time .5000E+02 sec.
   Position (ft)     C%
                              ______________________________
    .100E+01    .100E+02      |  *
    .200E+01    .101E+02      |  *
    .300E+01    .107E+02      |  *
    .400E+01    .141E+02      |   *
    .500E+01    .303E+02      |         *
    .600E+01    .745E+02      |                       *
    .700E+01    .875E+02      |                            *
    .800E+01    .897E+02      |                            *
    .900E+01    .900E+02      |                            *
    .100E+02    .900E+02      |                            *
    .110E+02    .900E+02      |                             *

Concentration vs distance @ time .4000E+01 sec.
   Position (ft)     C%
                              ______________________________
    .100E+01    .100E+02      |  *
    .200E+01    .100E+02      |  *
    .300E+01    .999E+01      |  *
    .400E+01    .107E+02      |  *
    .500E+01    .109E+02      |  *
    .600E+01    .894E+02      |                            *
    .700E+01    .896E+02      |                            *
    .800E+01    .900E+02      |                            *
    .900E+01    .900E+02      |                             *
    .100E+02    .900E+02      |                            *
    .110E+02    .900E+02      |                            *

Concentration vs distance @ time .0000E+00 sec.
   Position (ft)     C%
                              ______________________________
    .100E+01    .100E+02      |  *
    .200E+01    .100E+02      |  *
    .300E+01    .996E+01      |  *
    .400E+01    .108E+02      |  *
    .500E+01    .815E+01      | *
    .600E+01    .916E+02      |                             *
    .700E+01    .895E+02      |                            *
    .800E+01    .900E+02      |                            *
    .900E+01    .900E+02      |                            *
    .100E+02    .900E+02      |                            *
    .110E+02    .900E+02      |                            *
```

(B)

FIG. 21.4, CONT'D (B) An "undiffusing" radial flow.

$$\partial q_w/\partial r + q_w/r = -\phi\, \partial S_w/\partial t \tag{21.62}$$

$$\partial q_{nw}/\partial r + q_{nw}/r = -\phi\, \partial S_{nw}/\partial t \tag{21.63}$$

If we add Eqs. (21.62), (21.63), and observe that

$$S_w + S_{nw} = 1 \tag{21.64}$$

is constant for incompressible flows, it follows that

$$r\, \partial(q_w + q_{nw})/\partial r + (q_w + q_{nw}) = 0 \tag{21.65}$$

or, equivalently, $\{r\,(q_w+q_{nw})\}_r=0$, so that

$$r\,(q_w+q_{nw}) = Q(t) \tag{21.66}$$

Here the function Q(t), having dimensions of length squared per unit time (not to be confused with volume flow rate), is determined by its value at the wellbore sandface. In particular, since only mud filtrate is obtained there, we have

$$Q(t) = R_{well}\, q(t) \tag{21.67}$$

where R_{well} is the radius of the borehole and q(t) is the velocity through the mudcake obtained on a lineal flow basis, given by the expression derived earlier in this chapter, namely,

$$q(t) = \phi_{eff}\, dx_f(t)/dt = \tfrac{1}{2}\, t^{-\frac{1}{2}} \sqrt{\left\{2k_1(1-\phi_C)(1-f_s)(p_m-p_r)/(\mu_f f_s)\right\}} \tag{21.34}$$

A means for handling the square root singularity at $t=0$ is given later. Note that another choice of q(t), for thick mudcakes, is found in the radial cake growth formula derived in Chapter 19. At this point, it is convenient to introduce the capillary pressure function P_c and write it as a function of the water saturation S_w, taking

$$P_c(S_w) = P_{nw} - P_w \tag{21.68}$$

Then, the nonwetting velocity in Eq. (21.61) can be written in the form $q_{nw} = -(k_{nw}/\mu_{nw})\ \partial P_{nw}/\partial r = -(k_{nw}/\mu_{nw})\ \partial(P_c+P_w)/\partial r$. If we substitute this and Eq. (21.60) into Eq. (21.66), we obtain

$$r(k_w/\mu_w + k_{nw}/\mu_{nw})\,\partial P_w/\partial r + r(k_{nw}/\mu_{nw})\,\partial P_c/\partial r = -Q(t) \tag{21.69}$$

or, more precisely,

$$r(k_w/\mu_w + k_{nw}/\mu_{nw})\,\partial P_w/\partial r + r(k_{nw}/\mu_{nw})\,P_c'(S_w)\partial S_w/\partial r = -Q(t) \tag{21.70}$$

This yields

$$\partial P_w/\partial r = -\{Q(t) + r(k_{nw}/\mu_{nw})P_c'(S_w)\partial S_w/\partial r\}/\{r(k_w/\mu_w + k_{nw}/\mu_{nw})\} \tag{21.71}$$

so that Eq. (21.60) becomes

$$q_w = (k_w/\mu_w)\{Q(t) + r(k_{nw}/\mu_{nw})P_c'(S_w)\partial S_w/\partial r\}/\{r(k_w/\mu_w + k_{nw}/\mu_{nw})\} \tag{21.72}$$

If we combine Eqs. (21.72), (21.62), that is, $\partial q_w/\partial r + q_w/r = -\phi\ \partial S_w/\partial t$, we have

$$-\phi\,\partial S_w/\partial t = (\partial/\partial r + 1/r) \tag{21.73}$$

$$(k_w/\mu_w)\{Q(t) + r(k_{nw}/\mu_{nw})P_c'(S_w)\partial S_w/\partial r\}/\{r(k_w/\mu_w + k_{nw}/\mu_{nw})\}$$

where it is understood that the relative permeabilities k_w and k_{nw} are both prescribed functions of S_w. This is the nonlinear governing equation for water saturation. Once S_w is known, the oil saturation S_{nw} can be obtained using Eq. (21.64) as $S_{nw} = 1 - S_w$. In order to simplify notation, let us reintroduce the fractional flow function first used in Eq. (21.15), namely,

$$F(S_w) = 1/\{1 + \mu_w k_{nw}/\mu_{nw} k_w\} \tag{21.74}$$

and, in addition, the function

$$G(S_w) = \{k_{nw}/\mu_{nw}\} F(S_w) P_c'(S_w) \tag{21.75}$$

Then, Eq. (21.73) can be expressed succinctly in the form

$$\begin{aligned} &-\phi\, \partial S_w/\partial t - \{Q(t)F'(S_w) + G(S_w)\}/r\, \partial S_w/\partial r \\ &= G'(S_w)(\partial S_w/\partial r)^2 + G(S_w)\partial^2 S_w/\partial r^2 \end{aligned} \tag{21.76}$$

Numerical Analysis

Eq. (21.76) is conveniently solved, again using finite difference time-marching schemes. We always central difference our first and second-order space derivatives, while backward differencing in time, with respect to the nodal point (r_i, t_n). Furthermore, we shall evaluate all nonlinear saturation-dependent coefficients at their previous values in time. This leads to

$$\begin{aligned} &[1 - (Q_n F'_{i,n-1} + G_{i,n-1})\Delta r/(2G_{i,n-1} r_i) \\ &-G'_{i,n-1}\, (\partial S_w/\partial r)_{i,n-1} \Delta r/(2G_{i,n-1})]\, SW_{i-1,n} \\ &+\left[-2 + \phi(\Delta r)^2/(G_{i,n-1}\Delta t)\right] SW_{i,n} \\ &+[1 + (Q_n F'_{i,n-1} + G_{i,n-1})\Delta r/(2G_{i,n-1} r_i) \\ &+G'_{i,n-1}\, (\partial S_w/\partial r)_{i,n-1} \Delta r/(2G_{i,n-1})]\, SW_{i+1,n} \\ &= +\, \phi(\Delta r)^2 SW_{i,n-1}/(G_{i,n-1}\Delta t) \end{aligned} \tag{21.77}$$

which importantly assumes tridiagonal form for rapid matrix inversion while maintaining $O(\Delta x)^2$ accuracy in space. Note that $r_i = R_{well} + (i-1)\Delta r$. Straightforward von Neumann analysis shows that the time-dependent scheme implied by Eq. (21.77) is conditionally stable, with the exact time step limitations depending on the form of the relative permeability and capillary pressure functions. Following the rules established in Chapter 20, we write Eq. (21.77) for the internal nodes i = 2, 3, ..., imax-1, and augment the resulting system of linearized equations with the mud filtrate boundary condition $SW_{1,n} = SW^l = 1$ (for 100% water saturation) and the saturation $SW_{imax,n} = SW^r < 1$ at a distant effective radius. To start the time-marching calculations, the right side of Eq. (21.77) is assumed as $SW_{i,n-1} = SW^r < 1$ for the very first value of the time index n. In this discussion, SW^r also represents the initial uniform water saturation in the reservoir. Once the left side of Eq. (21.77) is inverted using the tridiagonal matrix solver TRIDI, $SW_{i,n}$ is copied into $SW_{i,n-1}$ on the right side, and the calculations are continued recursively.

```
      .
C     START RECURSIVE TIME INTEGRATION
      DO 300  N=1,NMAX
      T = T+DT
      THOURS = T/3600.
      DO 200  I=2,IMAXM1
      RI = WELRAD+(I-1)*DR
      SW = SNM1(I)
      DSDR =(SNM1(I+1)-SNM1(I-1))/(2.*DR)
      TERM1=((Q(T)*FP(SW)+G(SW))*DR)/(2.*G(SW)*RI)
      TERM2=  DR*DR*PHI/(G(SW)*DT)
      TERM3= (GP(SW)*DR/G(SW))*DSDR/2.
      A(I) =  1.- TERM1-TERM3
      B(I) = -2.+ TERM2
      C(I) =  1.+ TERM1+TERM3
      W(I) =  TERM2*SNM1(I)
 200  CONTINUE
      A(1) = 99.
      B(1) = 1.
      C(1) = 0.
      W(1) = SL
      A(IMAX) = 0.
      B(IMAX) = 1.
      C(IMAX) = 99.
      W(IMAX) = SR
      CALL TRIDI(A,B,C,VECTOR,W,IMAX)
      DO 250  I=1,IMAX
      S(I) = VECTOR(I)
 250  CONTINUE
      DO 260  I=1,IMAX
      SNM1(I) = S(I)
 260  CONTINUE
      CALL GRFIX(S,XPLOT,IMAX)
 300  CONTINUE
      .
```

FIG. 21.5 Nonlinear saturation solver.

Fortran Implementation

Eq. (21.77) is easily programmed in Fortran. Because the implicit scheme is second-order accurate in space, thus rigidly enforcing the diffusive character of the capillary pressure effects assumed in this formulation, we do not obtain the oscillations at saturation shocks or the saturation overshoots having $S_w > 1$ often cited. The exact Fortran producing the results shown later is displayed in Fig. 21.5 and in several function statements given later. For convenience, the saturation derivatives $F'(S_w)$ and $G'(S_w)$ are denoted FP and GP (P indicates *prime* for derivatives).

Typical Calculations

In this section, we shall perform a suite of validation runs designed to demonstrate the stability and physical correctness of the two-phase flow algorithm. In the calculations, a borehole radius of 0.2 ft and an effective reservoir radius of 2 ft are assumed. The water saturation at the borehole sandface is assumed to be unity, since it consists entirely of water-base mud filtrate. At the farfield boundary or effective radius, the water saturation is taken as 0.10. (This is also assumed to be the initial reservoir water saturation.) In addition, we discretize the radial coordinate using 0.1 ft grids, assume time steps of 0.001 s, and take the porosity of the rock as 20%. Note that for the 20 grid block mesh used, 1000 time steps requires approximately one second on typical personal computers. Multiphase flow properties are conveniently defined in Fortran function statements. In our calculations, the relative permeability curves and fractional flow functions are specified in the following code fragment

```
FUNCTION F(SW)
REAL KDARCY,KABS,KW,KNW
KDARCY = 0.001
KABS = KDARCY*0.00000001/(12.*12.*2.54*2.54)
KW = KABS * SW**2.
KNW = KABS*(SW-1.)**2.
VISCIN = 1.
VISCDP = 2.
VISCL  = 0.0000211*VISCIN
VISCR  = 0.0000211*VISCDP
F = 1. +VISCL*KNW/(VISCR*KW)
F = 1./F
RETURN
END
```

In the preceding calculations, an absolute permeability of 0.001 Darcies is assumed for the formation, and the wetting and nonwetting relative permeability functions, defined in terms of the water saturation S_w, are taken in the form $k_w = S_w^2$ and $k_{nw} = (S_w\text{-}1)^2$ for simplicity. Our water and oil viscosities are taken as 1 and 2 cp, respectively. The fractional flow function just defined is independent of the absolute permeability, of course, and depends only on the viscosity ratio. The function $G(S_w)$ is similarly defined by

```
FUNCTION G(SW)
REAL KDARCY,KABS,KNW
KDARCY = 0.001
KABS = KDARCY*0.00000001/(12.*12.*2.54*2.54)
KNW = KABS*(SW-1.)**2.
VISCDP = 2.
VISCR  = 0.0000211*VISCDP
G = KNW*F(SW)*PCP(SW)/VISCR
RETURN
END
```

while the capillary pressure function is defined by $P_c = 35$ (1-Sw) psi, again for simplicity, through the function block

```
FUNCTION PC(SW)
PC = 1.-SW
PC = 144.*35.*PC
RETURN
END
```

Derivatives of P_c, F, and G with respect to water saturation can be easily taken by introducing function statements that define the differentiation process. We now discuss typical calculations, designed to test the properties of the scheme, such as saturation overshoots, unstable oscillations, and so on. We shall find that the algorithm given is physically consistent. For example, it will not yield water saturations that exceed unity or fall below zero; thus, oil will not be created or destroyed, at least not in an obvious manner. The fully implicit scheme, unlike the explicit schemes used in many commercial IMPES models (to be discussed), does not produce numerical oscillations at the head of the shock. But instabilities do arise when the saturation shock reflects back upstream from the fictitious i = imax effective radius boundary; these instabilities, however, are irrelevant to our simulations. Finally, when mud filtrate is completely shut off, the water-oil saturation front *never* moves and must remain stationary—a trait not shared by several commercial simulators because of numerical roundoff.

Let us now discuss specific calculations. In this very first example, we set our mud filtration invasion rate identically to zero, using the function statement

```
FUNCTION Q(T)
Q = 0.
RETURN
END
```

The partial results shown in Fig. 21.6A for the near-well nodes indicate that the water front correctly stays absolutely static, with the remainder of the computational domain remaining unperturbed, despite the 1000 steps taken in time.

Again, we note that Q is not the volume flow rate, but the product of well radius and radial Darcy velocity at the sandface. For the assumed radius of 0.2 ft, a typical velocity may be assumed as 0.1 ft/h, so that Q = (0.2 ft) (0.1 ft/h) = 0.0000055 ft^2/s in the units used. We determine if the calculated invasion rates are physically reasonable, and in the process, we examine the stability of the algorithm. In Fig. 21.6B, 50,000 time steps of 0.001 s are taken, requiring 1 min of computations, and sample early and late time results are given. Truncation error is negligible in this stable scheme.

```
Water saturation at time (hrs): .167E-04
Position node Water Sat
                                  ___________________________________-
  .100E+01    .100E+01          |                                    *
  .200E+01    .100E+00          |  *
  .300E+01    .100E+00          |  *
  .400E+01    .100E+00          |  *
  .500E+01    .100E+00          |  *
  .600E+01    .100E+00          |  *
  .700E+01    .100E+00          |  *

Water saturation at time (hrs): .150E-03
Position node Water Sat
                                  ___________________________________-
  .100E+01    .100E+01          |                                    *
  .200E+01    .100E+00          |  *
  .300E+01    .100E+00          |  *
  .400E+01    .100E+00          |  *
  .500E+01    .100E+00          |  *
  .600E+01    .100E+00          |  *
  .700E+01    .100E+00          |  *

Water saturation at time (hrs): .267E-03
Position node Water Sat
                                  ___________________________________-
  .100E+01    .100E+01          |                                    *
  .200E+01    .100E+00          |  *
  .300E+01    .100E+00          |  *
  .400E+01    .100E+00          |  *
  .500E+01    .100E+00          |  *
  .600E+01    .100E+00          |  *
  .700E+01    .100E+00          |  *
```

(A)

FIG. 21.6 (A) Zero mud filtrate influx.

(Continued)

Questions of large versus small Q are only meaningful dimensionlessly, when the effects of inertia are measured against those due to capillary pressure. Since the functional form of these changes from problem to problem, because relative permeability and capillary pressure curves will often vary substantially, a parameter as simple or as elegant as the Reynolds number in elementary Newtonian fluid mechanics is not in general available. But, fortunately, we can understand the stability properties of the numerical scheme by examining different parameter limits of the present problem. It is clear from the two foregoing runs that inertia is not important, since little of the fluid is actually moving. In this next example, we assume the comparatively large constant value of $Q = 1$, to simulate water breakthrough known to reservoir engineers.

Large Q*s* model rapid influxes of injected water and should result in sharp saturation discontinuities; for such problems, there is little smearing at the shock due to capillary pressure. This is not to say that capillary pressure is unimportant: it is, because of the singular role it plays in defining the correct saturation discontinuity. (The shock-fitting used in the Buckley-Leverett solution process is unnecessary in the present high-order formulation.) For the $Q = 1$, 2, and 3 calculations shown in the following figures, 3000 time steps of 0.001 s each were used. In Figs. 21.6C–E, note how the effects of radial geometrical spreading are captured in the gently sloping curve, while steep

```
Water saturation at time (hrs): .167E-04
Position node Water Sat
                             _______________________________
  .100E+01    .100E+01      |                              *
  .200E+01    .100E+00      | *
  .300E+01    .100E+00      | *
  .400E+01    .100E+00      | *
  .500E+01    .100E+00      | *
  .
  .
  .150E+02    .100E+00      | *
  .160E+02    .100E+00      | *
  .170E+02    .100E+00      | *
  .180E+02    .100E+00      | *

Water saturation at time (hrs): .903E-02
Position node Water Sat
                             _______________________________
  .100E+01    .100E+01      |                              *
  .200E+01    .110E+00      | *
  .300E+01    .100E+00      | *
  .400E+01    .100E+00      | *
  .500E+01    .100E+00      | *
  .
  .
  .150E+02    .100E+00      | *
  .160E+02    .100E+00      | *
  .170E+02    .100E+00      | *
  .180E+02    .100E+00      | *

Water saturation at time (hrs): .126E-01
Position node Water Sat
                             _______________________________
  .100E+01    .100E+01      |                              *
  .200E+01    .114E+00      | *
  .300E+01    .100E+00      | *
  .400E+01    .100E+00      | *
  .500E+01    .100E+00      | *
  .600E+01    .100E+00      | *
  .700E+01    .100E+00      | *
  .800E+01    .100E+00      | *
  .900E+01    .100E+00      | *
  .100E+02    .100E+00      | *
  .110E+02    .100E+00      | *
  .120E+02    .100E+00      | *
  .130E+02    .100E+00      | *
  .140E+02    .100E+00      | *
  .150E+02    .100E+00      | *
  .160E+02    .100E+00      | *
  .170E+02    .100E+00      | *
  .180E+02    .100E+00      | *
```

(B)

FIG. 21.6, CONT'D (B) Very slow constant injection rate.

(Continued)

saturation gradients are computed as sudden changes. Also note that the calculations shown are extremely stable and that no numerical oscillations appear in the results. Moreover, we never obtain any water saturations that exceed unity in our $O(\Delta x^2)$ accurate implicit scheme. However, we have found that instabilities will arise after the shock reaches the farfield computational boundary and reflects. By this time, the calculations have no physical meaning, so that the existence of this instability is not germane to our applications.

```
Water saturation at time (hrs): .167E-04
Position node Water Sat

                              ------------------------------
  .100E+01    .100E+01      |                            *
  .200E+01    .938E+00      |                          *
  .300E+01    .877E+00      |                        *
  .400E+01    .827E+00      |                      *
  .500E+01    .789E+00      |                     *
  .600E+01    .758E+00      |                    *
  .700E+01    .664E+00      |                 *
  .800E+01    .278E+00      |      *
  .900E+01    .116E+00      | *
  .100E+02    .101E+00      | *
  .110E+02    .100E+00      | *
  .120E+02    .100E+00      | *
  .130E+02    .100E+00      | *
  .140E+02    .100E+00      | *
  .150E+02    .100E+00      | *
  .160E+02    .100E+00      | *
  .170E+02    .100E+00      | *
  .180E+02    .100E+00      | *

Water saturation at time (hrs): .500E-04
Position node Water Sat

                              ------------------------------
  .100E+01    .100E+01      |                            *
  .200E+01    .975E+00      |                           *
  .300E+01    .945E+00      |                          *
  .400E+01    .913E+00      |                         *
  .500E+01    .881E+00      |                        *
  .600E+01    .849E+00      |                       *
  .700E+01    .819E+00      |                      *
  .800E+01    .794E+00      |                     *
  .900E+01    .774E+00      |                     *
  .100E+02    .760E+00      |                    *
  .110E+02    .751E+00      |                    *
  .120E+02    .739E+00      |                    *
  .130E+02    .682E+00      |                  *
  .140E+02    .372E+00      |         *
  .150E+02    .130E+00      | *
  .160E+02    .102E+00      | *
  .170E+02    .100E+00      | *
  .180E+02    .100E+00      | *

Water saturation at time (hrs): .667E-04
Position node Water Sat

                              ------------------------------
  .100E+01    .100E+01      |                            *
  .200E+01    .981E+00      |                           *
  .300E+01    .957E+00      |                          *
  .400E+01    .931E+00      |                         *
  .500E+01    .903E+00      |                         *
  .600E+01    .875E+00      |                        *
  .700E+01    .848E+00      |                       *
  .800E+01    .822E+00      |                      *
  .900E+01    .798E+00      |                     *
  .100E+02    .779E+00      |                     *
  .110E+02    .764E+00      |                    *
  .120E+02    .754E+00      |                    *
  .130E+02    .747E+00      |                    *
  .140E+02    .739E+00      |                    *
  .150E+02    .708E+00      |                   *
  .160E+02    .507E+00      |             *
  .170E+02    .164E+00      |  *
  .180E+02    .100E+00      | *
```

(C)

FIG. 21.6, CONT'D (C) Q = 1, constant rate, high inertia flow.

```
Water saturation at time (hrs): .167E-04
Position node Water Sat

                           -------------------------------
  .100E+01    .100E+01     |                             *
  .200E+01    .964E+00     |                           *
  .300E+01    .925E+00     |                          *
  .400E+01    .884E+00     |                         *
  .500E+01    .846E+00     |                        *
  .600E+01    .811E+00     |                       *
  .700E+01    .784E+00     |                      *
  .800E+01    .764E+00     |                     *
  .900E+01    .749E+00     |                     *
  .100E+02    .716E+00     |                    *
  .110E+02    .510E+00     |              *
  .120E+02    .165E+00     |   *
  .130E+02    .106E+00     |  *
  .140E+02    .100E+00     |  *
  .150E+02    .100E+00     |  *
  .160E+02    .100E+00     |  *
  .170E+02    .100E+00     |  *
  .180E+02    .100E+00     |  *
```

```
Water saturation at time (hrs): .333E-04
Position node Water Sat

                           -------------------------------
  .100E+01    .100E+01     |                             *
  .200E+01    .981E+00     |                            *
  .300E+01    .957E+00     |                           *
  .400E+01    .931E+00     |                          *
  .500E+01    .903E+00     |                          *
  .600E+01    .876E+00     |                         *
  .700E+01    .848E+00     |                        *
  .800E+01    .822E+00     |                       *
  .900E+01    .798E+00     |                      *
  .100E+02    .778E+00     |                      *
  .110E+02    .762E+00     |                     *
  .120E+02    .752E+00     |                     *
  .130E+02    .745E+00     |                     *
  .140E+02    .737E+00     |                     *
  .150E+02    .709E+00     |                    *
  .160E+02    .520E+00     |              *
  .170E+02    .172E+00     |    *
  .180E+02    .100E+00     |  *
```

(D)

FIG. 21.6, CONT'D (D) Q=2, constant rate, high inertia flow.

```
Water saturation at time (hrs): .167E-04
Position node Water Sat

                           -------------------------------
  .100E+01    .100E+01     |                             *
  .200E+01    .976E+00     |                            *
  .300E+01    .947E+00     |                           *
  .400E+01    .916E+00     |                          *
  .500E+01    .883E+00     |                         *
  .600E+01    .851E+00     |                        *
  .700E+01    .821E+00     |                       *
  .800E+01    .794E+00     |                      *
  .900E+01    .772E+00     |                      *
  .100E+02    .756E+00     |                     *
  .110E+02    .746E+00     |                     *
  .120E+02    .735E+00     |                     *
  .130E+02    .688E+00     |                   *
  .140E+02    .396E+00     |          *
  .150E+02    .138E+00     |   *
  .160E+02    .103E+00     |  *
  .170E+02    .100E+00     |  *
  .180E+02    .100E+00     |  *
```

(E)

FIG. 21.6, CONT'D (E) Q=3, constant rate, high inertia flow.

Mudcake-Dominated Flows

Now we consider time-dependent influx flows of the kind created by real mudcakes. Earlier we found that the invasion speed at $t=0$ was infinite, behaving like $t^{-1/2}$. Such singularities, if implemented exactly, would cause instabilities in finite difference schemes. Fortunately, we can circumvent this difficulty without introducing any artificial devices, by considering the effects of nonzero mud spurt. From Chapter 17, the filtration thickness x(t) in a lineal flow varies like $dx/dt=\alpha/x$, where α is a constant. If $x(0)=x_{spurt}$, it follows that $x(t)=\sqrt{\left(2\alpha t + x_{spurt}^2\right)}$; then, the speeds $dx/dt=\alpha/\sqrt{\left(2\alpha t + x_{spurt}^2\right)}$ and $q(t)=\phi\ dx/dt=\alpha\phi/\sqrt{\left(2\alpha t + x_{spurt}^2\right)}$ are *never* infinite. This lineal model is used because the controlling mudcake forms lineally; a radial model can, of course, be substituted in slimhole applications. The spurt model is implemented by the Fortran function definition (Fig. 21.7).

```
      FUNCTION Q(T)
C     MUDCAKE MODEL, ALPHA = 1.
      PHI = 0.2
      WELRAD = 0.2
      SPURT =0.1
      SPURT2 = SPURT**2
      ALPHA = 1.
      Q = WELRAD*ALPHA*PHI/SQRT(SPURT2+2.*ALPHA*T)
      RETURN
      END
```

FIG. 21.7 Mudcake-dominated invasion.

In the sequence of snapshots in Fig. 21.8A–C, the formation and movement of the saturation shocks are shown for high, very high, and very slow invasion rates, all using 0.001 s time steps. Again, complete stability is obtained, without numerical saturation oscillations.

Unshocking a Saturation Discontinuity

In time lapse analysis, we may detect a moving saturation front, and may wish to look within or unscramble the steeply formed flow for additional fluid-dynamical information. Here, resistivity migration means unsteepening the shock, carefully untracing its history, being dynamically consistent with the effects of capillary pressure and nonlinear relative permeability functions. Unlike the miscible flow problem, where the dominant physical process involved the unsmearing of a diffused front, several complications enter the present problem. First, radial spreading again exists. But the high-order derivative term, now related to capillary pressure instead of molecular diffusion, serves dual purposes: it smears the flow throughout the entire flow domain, and it is instrumental in controlling the shock formation. (Shocks, remember, do not exist in the miscible flow problem.)

Water saturation at time (hrs): .167E-04

Position node	Water Sat
.100E+01	.100E+01
.200E+01	.810E+00
.300E+01	.475E+00
.400E+01	.132E+00
.500E+01	.101E+00
.600E+01	.100E+00
.700E+01	.100E+00
.800E+01	.100E+00
.900E+01	.100E+00
.100E+02	.100E+00
.110E+02	.100E+00
.120E+02	.100E+00
.130E+02	.100E+00
.140E+02	.100E+00
.150E+02	.100E+00
.160E+02	.100E+00
.170E+02	.100E+00
.180E+02	.100E+00

Water saturation at time (hrs): .117E-03

Position node	Water Sat
.100E+01	.100E+01
.200E+01	.905E+00
.300E+01	.831E+00
.400E+01	.776E+00
.500E+01	.639E+00
.600E+01	.222E+00
.700E+01	.108E+00
.800E+01	.100E+00
.900E+01	.100E+00
.100E+02	.100E+00
.110E+02	.100E+00
.120E+02	.100E+00
.130E+02	.100E+00
.140E+02	.100E+00
.150E+02	.100E+00
.160E+02	.100E+00
.170E+02	.100E+00
.180E+02	.100E+00

Water saturation at time (hrs): .383E-03

Position node	Water Sat
.100E+01	.100E+01
.200E+01	.939E+00
.300E+01	.879E+00
.400E+01	.829E+00
.500E+01	.793E+00
.600E+01	.764E+00
.700E+01	.688E+00
.800E+01	.340E+00
.900E+01	.121E+00
.100E+02	.101E+00
.110E+02	.100E+00
.120E+02	.100E+00
.130E+02	.100E+00
.140E+02	.100E+00
.150E+02	.100E+00
.160E+02	.100E+00
.170E+02	.100E+00
.180E+02	.100E+00

Water saturation at time (hrs): .833E-03

Position node	Water Sat
.100E+01	.100E+01
.200E+01	.956E+00
.300E+01	.908E+00
.400E+01	.863E+00
.500E+01	.823E+00
.600E+01	.792E+00
.700E+01	.771E+00
.800E+01	.750E+00
.900E+01	.671E+00
.100E+02	.312E+00
.110E+02	.119E+00
.120E+02	.101E+00
.130E+02	.100E+00
.140E+02	.100E+00
.150E+02	.100E+00
.160E+02	.100E+00
.170E+02	.100E+00
.180E+02	.100E+00

(A)

FIG. 21.8 (A) High filtration rate mudcake model ($\alpha = 1$).

(Continued)

```
Water saturation at time (hrs): .167E-04
Position node Water Sat
                            ________________________________
  .100E+01    .100E+01      |                              *
  .200E+01    .891E+00      |                          *
  .300E+01    .813E+00      |                       *
  .400E+01    .726E+00      |                    *
  .500E+01    .385E+00      |          *
  .600E+01    .125E+00      | *
  .700E+01    .101E+00      | *
  .800E+01    .100E+00      | *
  .900E+01    .100E+00      | *
  .100E+02    .100E+00      | *
  .110E+02    .100E+00      | *
  .120E+02    .100E+00      | *
  .130E+02    .100E+00      | *
  .140E+02    .100E+00      | *
  .150E+02    .100E+00      | *
  .160E+02    .100E+00      | *
  .170E+02    .100E+00      | *
  .180E+02    .100E+00      | *
```

```
Water saturation at time (hrs): .150E-03
Position node Water Sat
                            ________________________________
  .100E+01    .100E+01      |                              *
  .200E+01    .953E+00      |                           *
  .300E+01    .904E+00      |                          *
  .400E+01    .857E+00      |                        *
  .500E+01    .817E+00      |                       *
  .600E+01    .788E+00      |                      *
  .700E+01    .767E+00      |                      *
  .800E+01    .738E+00      |                     *
  .900E+01    .595E+00      |                *
  .100E+02    .202E+00      |     *
  .110E+02    .108E+00      | *
  .120E+02    .100E+00      | *
  .130E+02    .100E+00      | *
  .140E+02    .100E+00      | *
  .150E+02    .100E+00      | *
  .160E+02    .100E+00      | *
  .170E+02    .100E+00      | *
  .180E+02    .100E+00      | *
```

```
Water saturation at time (hrs): .833E-03
Position node Water Sat
                            ________________________________
  .100E+01    .100E+01      |                              *
  .200E+01    .978E+00      |                            *
  .300E+01    .952E+00      |                           *
  .400E+01    .924E+00      |                          *
  .500E+01    .894E+00      |                         *
  .600E+01    .864E+00      |                        *
  .700E+01    .836E+00      |                        *
  .800E+01    .810E+00      |                       *
  .900E+01    .788E+00      |                      *
  .100E+02    .772E+00      |                      *
  .110E+02    .760E+00      |                     *
  .120E+02    .752E+00      |                     *
  .130E+02    .743E+00      |                     *
  .140E+02    .711E+00      |                    *
  .150E+02    .511E+00      |              *
  .160E+02    .162E+00      |  *
  .170E+02    .105E+00      | *
  .180E+02    .100E+00      | *
```

(B)

FIG. 21.8, CONT'D (B) Very high filtration rate mudcake model ($\alpha = 5$).

```
Water saturation at time (hrs): .167E-04
Position node Water Sat
                              _______________________________
  .100E+01     .100E+01     |                              *
  .200E+01     .101E+00     | *
  .300E+01     .100E+00     | *
  .400E+01     .100E+00     | *
  .500E+01     .100E+00     | *
  .600E+01     .100E+00     | *
  .700E+01     .100E+00     | *
  .800E+01     .100E+00     | *
  .900E+01     .100E+00     | *
  .100E+02     .100E+00     | *
  .110E+02     .100E+00     | *
  .120E+02     .100E+00     | *
  .130E+02     .100E+00     | *
  .140E+02     .100E+00     | *
  .150E+02     .100E+00     | *
  .160E+02     .100E+00     | *
  .170E+02     .100E+00     | *
  .180E+02     .100E+00     | *

Water saturation at time (hrs): .120E-02
Position node Water Sat
                              _______________________________
  .100E+01     .100E+01     |                              *
  .200E+01     .186E+00     |    *
  .300E+01     .102E+00     | *
  .400E+01     .100E+00     | *
  .500E+01     .100E+00     | *
  .600E+01     .100E+00     | *
  .700E+01     .100E+00     | *
  .800E+01     .100E+00     | *
  .900E+01     .100E+00     | *
  .100E+02     .100E+00     | *
  .110E+02     .100E+00     | *
  .120E+02     .100E+00     | *
  .130E+02     .100E+00     | *
  .140E+02     .100E+00     | *
  .150E+02     .100E+00     | *
  .160E+02     .100E+00     | *
  .170E+02     .100E+00     | *
  .180E+02     .100E+00     | *

Water saturation at time (hrs): .167E-02
Position node Water Sat
                              _______________________________
  .100E+01     .100E+01     |                              *
  .200E+01     .231E+00     |     *
  .300E+01     .104E+00     | *
  .400E+01     .100E+00     | *
  .500E+01     .100E+00     | *
  .600E+01     .100E+00     | *
  .700E+01     .100E+00     | *
  .800E+01     .100E+00     | *
  .900E+01     .100E+00     | *
  .100E+02     .100E+00     | *
  .110E+02     .100E+00     | *
  .120E+02     .100E+00     | *
  .130E+02     .100E+00     | *
  .140E+02     .100E+00     | *
  .150E+02     .100E+00     | *
  .160E+02     .100E+00     | *
  .170E+02     .100E+00     | *
  .180E+02     .100E+00     | *
```

(C)

FIG. 21.8, CONT'D (C) Very slow filtration rate model ($\alpha = 0.001$).

```
      .
C     START TIME INTEGRATION
      DO 300   N=1,NMAX
      IF(N.LT.2000) T = T+DT
      IF(N.GE.2000) T = T-DT
      THOURS = T/3600.
      DO 200   I=2,IMAXM1
      RI = WELRAD+(I-1)*DR
      SW = SNM1(I)
      DSDR =(SNM1(I+1)-SNM1(I-1))/(2.*DR)
      IF(N.LT.2000) TERM1=((Q(T)*FP(SW)+G(SW))*DR)/(2.*G(SW)*RI)
      IF(N.GE.2000) TERM1=((-Q(T)*FP(SW)+G(SW))*DR)/(2.*G(SW)*RI)
      TERM2=  DR*DR*PHI/(G(SW)*DT)
      TERM3=  (GP(SW)*DR/G(SW))*DSDR/2.
      A(I)  =   1.- TERM1-TERM3
      B(I)  = -2.+ TERM2
      C(I)  =   1.+ TERM1+TERM3
      W(I)  =   TERM2*SNM1(I)
200   CONTINUE
      .
300   CONTINUE
```

FIG. 21.9 Unshocking a steep gradient.

Can we undo all of these two-phase flow effects? The answer appears to be a definitive, "Yes." To evaluate this numerical reversibility, we execute the program for *2000* time steps, assuming $\alpha = 1$, and then reverse the direction of time as well as the direction of filtrate movement, as shown in Fig. 21.9 by the bold print modifications to our earlier source code. Forward simulation results are given in Fig. 21.10A, while successfully migrated, or unshocked results, are shown in Fig. 21.10B. The potential applications of this important capability are vast indeed and are under investigation.

IMMISCIBLE FLOWS WITH CAPILLARY PRESSURE AND DYNAMICALLY COUPLED MUDCAKE GROWTH

In the foregoing formulation, we assumed that q(t) was available from a knowledge of mudcake properties, and we solved for the resulting two-phase flow in the rock. Of course, this is not generally the case. Consider the limit in which no mudcake forms on the rock: here the time-dependent flow through the rock is determined by the saturations and pressures at the inlet and outlet boundaries of the problem. For the cake-free problem just described, the PDEs governing saturation and pressure are nonlinearly coupled, and the time-dependent flow rate through the core must be determined iteratively. This is also the case when the mobility in the mudcake is comparable to that of the formation. But there is a complication. For such problems, this cake growth must be additionally determined as part of the solution; it does not alone dictate the filtrate influx but depends strongly on two-phase flow details in the rock.

```
Water saturation at time (hrs): .167E-04
Position node Water Sat

                           ______________________________
  .100E+01    .100E+01     |                             *
  .200E+01    .810E+00     |                       *
  .300E+01    .475E+00     |             *
  .400E+01    .132E+00     | *
  .500E+01    .101E+00     | *
  .600E+01    .100E+00     | *
  .700E+01    .100E+00     | *
  .800E+01    .100E+00     | *
  .900E+01    .100E+00     | *
  .100E+02    .100E+00     | *
  .110E+02    .100E+00     | *
  .120E+02    .100E+00     | *
  .130E+02    .100E+00     | *
  .140E+02    .100E+00     | *
  .150E+02    .100E+00     | *
  .160E+02    .100E+00     | *

Water saturation at time (hrs): .117E-03
Position node Water Sat

                           ______________________________
  .100E+01    .100E+01     |                             *
  .200E+01    .905E+00     |                         *
  .300E+01    .831E+00     |                       *
  .400E+01    .776E+00     |                      *
  .500E+01    .639E+00     |                  *
  .600E+01    .222E+00     |     *
  .700E+01    .108E+00     | *
  .800E+01    .100E+00     | *
  .900E+01    .100E+00     | *
  .100E+02    .100E+00     | *
  .110E+02    .100E+00     | *
  .120E+02    .100E+00     | *
  .130E+02    .100E+00     | *
  .140E+02    .100E+00     | *
  .150E+02    .100E+00     | *
  .160E+02    .100E+00     | *


Water saturation at time (hrs): .350E-03
Position node Water Sat

                           ______________________________
  .100E+01    .100E+01     |                             *
  .200E+01    .937E+00     |                           *
  .300E+01    .876E+00     |                         *
  .400E+01    .825E+00     |                       *
  .500E+01    .790E+00     |                      *
  .600E+01    .758E+00     |                     *
  .700E+01    .651E+00     |                  *
  .800E+01    .257E+00     |      *
  .900E+01    .112E+00     | *
  .100E+02    .101E+00     | *
  .110E+02    .100E+00     | *
  .120E+02    .100E+00     | *
  .130E+02    .100E+00     | *
  .140E+02    .100E+00     | *
  .150E+02    .100E+00     | *
  .160E+02    .100E+00     | *
  .170E+02    .100E+00     | *
  .180E+02    .100E+00     | *
```

(A)

FIG. 21.10 (A) Forward shock formation.

(Continued)

```
Water saturation at time (hrs): .550E-03
Position node Water Sat
                            -------------------------------
  .100E+01    .100E+01      |                             *
  .200E+01    .947E+00      |                           *
  .300E+01    .893E+00      |                         *
  .400E+01    .845E+00      |                        *
  .500E+01    .806E+00      |                       *
  .600E+01    .779E+00      |                      *
  .700E+01    .753E+00      |                     *
  .800E+01    .658E+00      |                  *
  .900E+01    .277E+00      |      *
  .100E+02    .115E+00      | *
  .110E+02    .101E+00      | *
  .120E+02    .100E+00      | *
  .130E+02    .100E+00      | *
  .140E+02    .100E+00      | *
  .150E+02    .100E+00      | *
  .160E+02    .100E+00      | *
  .170E+02    .100E+00      | *
  .180E+02    .100E+00      | *

Water saturation at time (hrs): .211E-03
Position node Water Sat
                            -------------------------------
  .100E+01    .100E+01      |                             *
  .200E+01    .923E+00      |                          *
  .300E+01    .855E+00      |                        *
  .400E+01    .806E+00      |                       *
  .500E+01    .765E+00      |                     *
  .600E+01    .645E+00      |                  *
  .700E+01    .240E+00      |     *
  .800E+01    .110E+00      | *
  .900E+01    .101E+00      | *
  .100E+02    .100E+00      | *
  .110E+02    .100E+00      | *
  .120E+02    .100E+00      | *
  .130E+02    .100E+00      | *
  .140E+02    .100E+00      | *
  .150E+02    .100E+00      | *
  .160E+02    .100E+00      | *
  .170E+02    .100E+00      | *
  .180E+02    .100E+00      | *

Water saturation at time (hrs): .106E-04
Position node Water Sat
                            -------------------------------
  .100E+01    .100E+01      |                             *
  .200E+01    .747E+00      |                     *
  .300E+01    .255E+00      |     *
  .400E+01    .108E+00      | *
  .500E+01    .101E+00      | *
  .600E+01    .100E+00      | *
  .700E+01    .100E+00      | *
  .800E+01    .100E+00      | *
  .900E+01    .100E+00      | *
  .100E+02    .100E+00      | *
  .110E+02    .100E+00      | *
  .120E+02    .100E+00      | *
  .130E+02    .100E+00      | *
  .140E+02    .100E+00      | *
  .150E+02    .100E+00      | *
  .160E+02    .100E+00      | *
  .170E+02    .100E+00      | *
  .180E+02    .100E+00      | *
```

(B)

FIG. 21.10, CONT'D (B) Backward shock migration.

Flows Without Mudcakes

In order to solve the latter, it is instructive to formulate and discuss the former one without mudcake first. In doing so, we derive the complete set of two-phase flow equations required later, and we demonstrate some essential ideas. Let us recall that we had determined

$$r\,(k_w/\mu_w + k_{nw}/\mu_{nw})\,\partial P_w/\partial r + r\,(k_{nw}/\mu_{nw})\,P_c{}'(S_w)\partial S_w/\partial r = -Q(t) \quad (21.70)$$

In the previous section, Q(t) was assumed to be known; this being the case, the derived saturation equation could be solved independently of the pressure equation, so that a pressure differential equation was not required. Now, we expect that any derived governing pressure equation must reduce to an anticipated $\partial^2 P_w/\partial r^2 + (1/r)\partial P_w/\partial r = 0$ in the single-phase flow limit. This can be accomplished by differentiating Eq. (21.70) with respect to the radial coordinate; this differentiation eliminates the explicit appearance of Q(t) which is, again, unknown. Since $k_w = k_w(S_w)$ and $k_{nw} = k_{nw}(S_w)$, straightforward manipulations show that we can write the desired equation as

$$\begin{aligned}
&\partial^2 P_w/\partial r^2 + [1/r + \{(k_w{}'/\mu_w + k_{nw}{}'/\mu_{nw})/\,(k_w/\mu_w + k_{nw}/\mu_{nw})\}\partial S_w/\partial r]\\
&\times \partial P_w/\partial r = -\big[\big(\partial^2 S_w/\partial r^2 + 1/r\,\partial S_w/\partial r\big)\,(\mu_w k_{nw}/\mu_{nw} k_w)\,P_c{}'(S_w)\\
&+\big\{(\mu_w k_{nw}{}'/\mu_{nw} k_w)P_c{}'(S_w) + (\mu_w k_{nw}/\mu_{nw} k_w)P_c{}''(S_w)\big\}\\
&\times(\partial S_w/\partial r)^2\big]/\big[1 + (\mu_w k_{nw}/\mu_{nw} k_w)\big]
\end{aligned} \quad (21.78)$$

Now, there exist two dependent variables in the present problem, namely, pressure and saturation. Pressure is governed by Eq. (21.78), while saturation satisfies Eq. (21.73),

$$\begin{aligned}
&-\phi\,\partial S_w/\partial t = (\partial/\partial r + 1/r)\\
&(k_w/\mu_w)\{Q(t) + r(k_{nw}/\mu_{nw})P_c{}'(S_w)\partial S_w/\partial r\}/\{r(k_w/\mu_w + k_{nw}/\mu_{nw})\}
\end{aligned} \quad (21.73)$$

where Q(t), now *not* a prescribed function, merely stands for the functional combination

$$Q(t) = -r(k_w/\mu_w + k_{nw}/\mu_{nw})\partial P_w/\partial r - r(k_{nw}/\mu_{nw})\,P_c{}'(S_w)\partial S_w/\partial r \quad (21.79)$$

as is clear from Eq. (21.70). If the initial spatial distributions for P_w and S_w are prescribed, a reasonable numerical solution process might solve Eqs. (21.78), (21.73) sequentially for a time step, before proceeding to the next. We shall, in fact, adopt this procedure. The solution procedure for saturation has been discussed and was implemented using Eq. (21.77). We shall retain that procedure for the present problem. For the pressure solution, in order to simplify our nomenclature, we recast Eq. (21.78) in the form

$$\partial^2 P_w/\partial r^2 + COEF\ \partial P_w/\partial r = RHS \tag{21.80}$$

where *COEF* and *RHS* denote the coefficient and right-hand-side terms. Then, adopting the central difference approximation $(PW_{i-1} - 2\,PW_i + PW_{i+1})/\Delta r^2 + COEF_i\ (PW_{i+1} - PW_{i-1})/(2\Delta r) = RHS_i$, we rewrite Eq. (21.80) as

$$(1 - COEF_i\ \Delta r/2)\ PW_{i-1} - 2\,PW_i + (1 + COEF_i\ \Delta r/2)\ PW_{i+1} = RHS_i\ \Delta r^2 \tag{21.81}$$

Insofar as the tridiagonal solver TRIDI is concerned, the coefficients **A**, **B**, **C**, and **W** take the form $A_i = (1 - COEF_i\ \Delta r/2)$, $B_i = -2$, $C_i = (1 + COEF_i\ \Delta r/2)$, and $W_i = RHS_i\ \Delta r^2$ for the internal nodes i = 2, 3, ..., imax-1. In addition, A(1) = 99, B(1) = 1, C(1) = 0, W(1) = P_{left}, and A(IMAX) = 0, B(IMAX) = 1, C(IMAX) = 99, W(IMAX) = P_{right}, where P_{left} and P_{right} denote the prescribed pressures at the inlet and outlet boundaries.

Observe that *COEF* and *RHS* will always be evaluated by second-order accurate central differences in space at the previous time step. Also, a starting initial pressure distribution is required that is analogous to our initial condition for saturation. Selected portions of the Fortran source code engine required to implement this algorithm are given in Fig. 21.11. They are obtained by simple modification of our earlier program designed to solve two-phase flows when the flow rate is a prescribed function of time. Finally, observe that we do *not* use the *outlet saturation boundary condition* of Collins (1961) because our outlet is a fictitious computational boundary that is *internal* to the reservoir.

Note that "Multiple Factors That Influence Wireline Formation Tester Pressure Measurements and Fluid Contact Estimates," by M.A. Proett, W.C. Chin, M. Manohar, R. Sigal, and J. Wu, SPE Paper 71566, presented at the 2001 SPE Annual Technical Conference and Exhibition in New Orleans, Louisiana, Sep. 30–Oct. 3, 2001, extends the work in this chapter to higher order, ensuring that mass is accurately conserved at strong saturation discontinuities. For further information or a complimentary copy of the paper, the reader should write or contact the author directly at wilsonchin@aol.com.

In the following calculations, two tabulations are shown per frozen instant in time, the first for spatial saturation distribution and the second for the corresponding pressure. (The pressure and time units shown are not germane to our discussion, since they were chosen to replicate an entire range of weak to strong inertia-to-capillary force effects.) The tabulated solution sets appear in Fig. 21.12A–C.

```
      .
      .
C     INITIALIZATION
      T = 0.
      DO 100  I=1,IMAX
      SNM1(I) = SZERO
      XPLOT(I) = WELRAD+(I-1)*DR
      P(I) = PINIT
 100  CONTINUE
C
C     START TIME INTEGRATION
      DO 300  N=1,NMAX
      T = T+DT
      THOURS = T/3600.
C
C     PRESSURE EQUATION
      DO 150  I=2,IMAXM1
      RI = WELRAD+(I-1)*DR
      SW = SNM1(I)
      DSDR   = (SNM1(I+1)-SNM1(I-1))/(2.*DR)
      DSDR2  = DSDR**2.
      D2SDR2 = (SNM1(I-1)-2.*SNM1(I)+SNM1(I+1))/(DR*DR)
      DEL2S  =  D2SDR2+(1./RI)*DSDR
      COEF = 1./RI
     1 +((KWP(SW)/VISCL+KNWP(SW)/VISCR)/
     2   (KW(SW) /VISCL+KNW(SW) /VISCR))*DSDR
      RHS = PCP(SW)*DEL2S*(VISCL*KNW(SW)/(VISCR*KW(SW)))
     1       +DSDR2*VISCL*KNWP(SW)*PCP(SW)/(VISCR*KW(SW))
     2       +DSDR2*PCPP(SW)*(VISCL*KNW(SW)/(VISCR*KW(SW)))
      RHS = -RHS*F(SW)
      A(I) = 1.-COEF*DR/2.
      B(I) = -2.
      C(I) = 1.+COEF*DR/2.
      W(I) = RHS*DR*DR
 150  CONTINUE
      A(1) = 99.
      B(1) = 1.
      C(1) = 0.
      W(1) = PLEFT
      A(IMAX) = 0.
      B(IMAX) = 1.
      C(IMAX) = 99.
      W(IMAX) = PRIGHT
      CALL TRIDI(A,B,C,VECTOR,W,IMAX)
      DO 160  I=1,IMAX
      P(I) = VECTOR(I)
 160  CONTINUE
C
C     SATURATION EQUATION
      DO 200  I=2,IMAXM1
      RI = WELRAD+(I-1)*DR
      SW = SNM1(I)
      DSDR = (SNM1(I+1)-SNM1(I-1))/(2.*DR)
      DPDR = (P(I+1)-P(I-1))/(2.*DR)
      Q = -RI*(KW(SW)/VISCL+KNW(SW)/VISCR)*DPDR
     1    -RI*(KNW(SW)/VISCR)*PCP(SW)*DSDR
      TERM1=((Q*FP(SW)+G(SW))*DR)/(2.*G(SW)*RI)
      TERM2=  DR*DR*PHI/(G(SW)*DT)
      TERM3= (GP(SW)*DR/G(SW))*DSDR/2.
      A(I) =  1.- TERM1-TERM3
      B(I) = -2.+ TERM2
```

FIG. 21.11 Implicit pressure—implicit saturation solver.

(Continued)

```
      C(I) =  1.+ TERM1+TERM3
      W(I) =  TERM2*SNM1(I)
 200  CONTINUE
      A(1) = 99.
      B(1) = 1.
      C(1) = 0.
      W(1) = SL
      A(IMAX) = 0.
      B(IMAX) = 1.
      C(IMAX) = 99.
      W(IMAX) = SR
      CALL TRIDI(A,B,C,VECTOR,W,IMAX)
      DO 250  I=1,IMAX
      S(I) = VECTOR(I)
 250  CONTINUE
      DO 260  I=1,IMAX
      SNM1(I) = S(I)
 260  CONTINUE
      IF(MOD(N,60).NE.0) GO TO 300
      WRITE(*,10)
      WRITE(4,10)
      WRITE(*,280) THOURS
      WRITE(4,280) THOURS
 280  FORMAT('   Water saturation at time (hrs):' E9.3)
      CALL GRFIX(S,XPLOT,IMAX,1)
      WRITE(*,281) THOURS
      WRITE(4,281) THOURS
 281  FORMAT('   Pressure versus r @ time (hrs):' E9.3)
      CALL GRFIX(P,XPLOT,IMAX,2)
 300  CONTINUE
      .
      .
      STOP
      END
C
      FUNCTION F(SW)
      REAL KDARCY,KABS,KW,KNW
      KDARCY = 0.001
      KABS = KDARCY*0.00000001/(12.*12.*2.54*2.54)
      KW = KABS * SW**2.
      KNW = KABS*(SW-1.)**2.
      VISCIN = 1.
      VISCDP = 2.
      VISCL  = 0.0000211*VISCIN
      VISCR  = 0.0000211*VISCDP
      F = 1. +VISCL*KNW/(VISCR*KW)
      F = 1./F
      RETURN
      END
```

FIG. 21.11, CONT'D

```
Water saturation at time (hrs): .167E-03
Position (ft) Water Sat
                              ______________________________
  .200E+00     .100E+01      |                             *
  .300E+00     .807E+00      |                       *
  .400E+00     .388E+00      |          *
  .500E+00     .120E+00      | *
  .600E+00     .101E+00      | *
  .700E+00     .100E+00      | *
  .800E+00     .100E+00      | *
  .900E+00     .100E+00      | *
  .100E+01     .100E+00      | *
  .110E+01     .100E+00      | *
  .120E+01     .100E+00      | *
  .130E+01     .100E+00      | *
  .140E+01     .100E+00      | *
  .150E+01     .100E+00      | *
  .160E+01     .100E+00      | *
  .170E+01     .100E+00      | *
  .180E+01     .100E+00      | *
  .190E+01     .100E+00      | *

Pressure versus r @ time (hrs): .167E-03
Position (ft)  Pressure
                              ______________________________
  .200E+00     .144E+09      |                             *
  .300E+00     .129E+09      |                         *
  .400E+00     .108E+09      |                     *
  .500E+00     .888E+08      |                 *
  .600E+00     .765E+08      |              *
  .700E+00     .663E+08      |            *
  .800E+00     .574E+08      |          *
  .900E+00     .496E+08      |         *
  .100E+01     .426E+08      |       *
  .110E+01     .363E+08      |      *
  .120E+01     .305E+08      |     *
  .130E+01     .252E+08      |    *
  .140E+01     .203E+08      |   *
  .150E+01     .157E+08      |  *
  .160E+01     .114E+08      | *
  .170E+01     .738E+07      *
  .180E+01     .359E+07      |
  .190E+01     .000E+00      |
```

(A)

FIG. 21.12 (A) Early time saturation and pressure.

(Continued)

The early time saturation solution shown in Fig. 21.12A indicates that inertia effects are not yet strong. This is clear, since reference to our source code shows that we have initialized our pressure field to a constant value throughout, so that the flow is initially stagnant. At t = 0+, a sudden applied pressure differential is introduced (that is, PLEFT - PRIGHT > 0), and fluid movement commences. However, the saturation shock has not formed, and the flow is controlled by capillary pressure. Note how the computed pressure shows a mild slope discontinuity, not unlike that presumed in Chapter 17.

Fig. 21.12B illustrates the start of saturation shock formation, an event not unlike the piston-like displacements assumed early in this book. It is interesting to observe that *immiscible two-phase flow theory will predict piston-like fronts when they exist, but when they do not, will produce smooth flows*. Thus, immiscible flow theory is more general and more powerful. However, it suffers from several practical disadvantages. Calculations are almost always numerical and

```
Water saturation at time (hrs): .667E-03
Position (ft) Water Sat

 .200E+00    .100E+01
 .300E+00    .919E+00
 .400E+00    .854E+00
 .500E+00    .811E+00
 .600E+00    .756E+00
 .700E+00    .491E+00
 .800E+00    .149E+00
 .900E+00    .103E+00
 .100E+01    .100E+00
 .110E+01    .100E+00
 .120E+01    .100E+00
 .130E+01    .100E+00
 .140E+01    .100E+00
 .150E+01    .100E+00
 .160E+01    .100E+00
 .170E+01    .100E+00
 .180E+01    .100E+00
 .190E+01    .100E+00
```

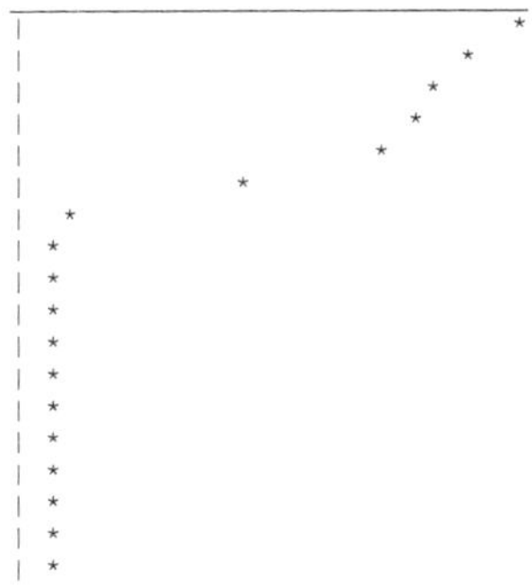

```
Pressure versus r @ time (hrs): .667E-03
Position (ft)  Pressure

 .200E+00    .144E+09
 .300E+00    .129E+09
 .400E+00    .117E+09
 .500E+00    .106E+09
 .600E+00    .962E+08
 .700E+00    .845E+08
 .800E+00    .697E+08
 .900E+00    .599E+08
 .100E+01    .514E+08
 .110E+01    .438E+08
 .120E+01    .368E+08
 .130E+01    .304E+08
 .140E+01    .245E+08
 .150E+01    .189E+08
 .160E+01    .138E+08
 .170E+01    .891E+07
 .180E+01    .433E+07
 .190E+01    .000E+00
```

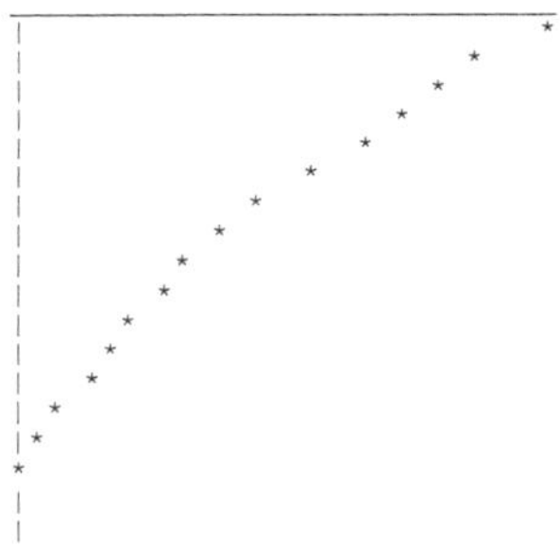

(B)

FIG. 21.12, CONT'D (B) Intermediate time saturation and pressure.

(Continued)

produce little intuitive insight; also, the relative permeability and capillary pressure functions that are required may not be known accurately.

Finally, note that while the saturation profiles in Fig. 21.12A–C have evolved significantly from the beginning to the end of the calculations, our pressure profiles have remained more or less invariant with time. This indicates the existence of two global time scales in the problem, one governing pressure and the other governing saturation. Also, while the pressure gradient profile is mildly discontinuous, the saturation profile is strongly discontinuous.

The invariance of the pressure solution with time is not unexpected, although it is not always obtained. Since-steady state *all-water* and steady-state *all-oil* pressure distributions are identical for a fixed pressure differential, one might expect that all of the intervening mixed fluid pressure states will not deviate far from the profile obtained on a single-phase flow basis. The latter solution can therefore be used to initialize the pressure solver for rapid convergence. The

```
Water saturation at time (hrs): .283E-02

Position (ft) Water Sat

                           -------------------------------
  .200E+00    .100E+01     |                             *
  .300E+00    .978E+00     |                            *
  .400E+00    .951E+00     |                           *
  .500E+00    .921E+00     |                          *
  .600E+00    .891E+00     |                         *
  .700E+00    .861E+00     |                        *
  .800E+00    .834E+00     |                        *
  .900E+00    .811E+00     |                       *
  .100E+01    .793E+00     |                      *
  .110E+01    .780E+00     |                      *
  .120E+01    .772E+00     |                      *
  .130E+01    .765E+00     |                     *
  .140E+01    .755E+00     |                     *
  .150E+01    .691E+00     |                   *
  .160E+01    .359E+00     |         *
  .170E+01    .127E+00     |  *
  .180E+01    .102E+00     |  *
  .190E+01    .100E+00     |  *
Pressure versus r @ time (hrs): .283E-02
Position (ft)   Pressure

                           -------------------------------
  .200E+00    .144E+09     |                             *
  .300E+00    .125E+09     |                         *
  .400E+00    .111E+09     |                      *
  .500E+00    .100E+09     |                   *
  .600E+00    .899E+08     |                 *
  .700E+00    .809E+08     |               *
  .800E+00    .726E+08     |              *
  .900E+00    .648E+08     |           *
  .100E+01    .576E+08     |          *
  .110E+01    .508E+08     |         *
  .120E+01    .445E+08     |        *
  .130E+01    .386E+08     |       *
  .140E+01    .331E+08     |     *
  .150E+01    .275E+08     |    *
  .160E+01    .196E+08     |   *
  .170E+01    .117E+08     |*
  .180E+01    .562E+07     *
  .190E+01    .000E+00     |
```

(C)

FIG. 21.12, CONT'D (C) Late time saturation and pressure.

converged solution would contain the propagating slope discontinuities required at the water-oil interface. Intuitive arguments such as this, when plausible, can motivate more efficient numerical schemes for research purposes. It is important to observe that the transient saturation equation may be either parabolic or hyberbolic in nature (e.g., see Hildebrand, 1948), depending on the importance of the capillary pressure term relative to the convection term. The form of the equation given in Eq. (21.76) illustrates this distinction very clearly. When capillary pressure is important, the $G(S_w)\ \partial^2 S_w/\partial r^2$ term *must* be retained, so that

$$-\phi\, \partial S_w/\partial t - \{Q(t)F'(S_w) + G(S_w)\}/r\, \partial S_w/\partial r = G'(S_w)(\partial S_w/\partial r)^2 + G(S_w)\, \partial^2 S_w/\partial r^2 \qquad (21.76)$$

is heat-like. The equation $\partial S_w/\partial t \propto \partial^2 S_w/\partial r^2$ is clearly diffusive, and it is not unlike the pressure diffusion equation used in transient compressible well test simulation (see Chapter 20). But when inertia is more important, the second derivative term $\partial^2 S_w/\partial r^2$ can be neglected, at least until shocks form. With this term neglected, Eq. (21.76) reduces to the first-order wave equation

$$-\phi\, \partial S_w/\partial t - \{Q(t)F'(S_w) + G(S_w)\}/r\, \partial S_w/\partial r = G'(S_w)(\partial S_w/\partial r)^2 \quad (21.82)$$

which is the radial form of the Buckley-Leverett equation studied earlier for lineal flows. Whether or not the saturation equation is parabolic or hyperbolic, the pressure equation

$$\partial^2 P_w/\partial r^2 + COEF\, \partial P_w/\partial r = RHS \quad (21.80)$$

is always *elliptic-like* and time-independent, at least to the extent that the variables *COEF* and *RHS* are evaluated at the previous time step. In any event, both governing equations, Eqs. (21.76), (21.80), contain second-order spatial derivative terms and are associated with well-defined boundary value problems and boundary conditions.

These formulations were solved using second-order accurate implicit schemes in the work just presented; that is, our approach was implicit pressure, implicit saturation. This is in contrast to the popular implicit pressure, *explicit* saturation codes used in the industry, which are only conditionally stable. (The von Neumann stability of both implicit and explicit schemes was considered in Chapter 20.) This so-called IMPES scheme, in addition to its stability problems, yields undesirable saturation oscillations and overshoots that are often fixed by upstream (that is, backward) differencing of spatial derivatives. But this solution actually introduces more problems than it fixes. As Lantz (1971) demonstrates, this stabilizes the numerical problem, at the expense of adding artificial viscosity by way of the truncation terms. Thus, the physically meaningful diffusion coefficient G in the $G(S_w)\ \partial^2 S_w/\partial r^2$ term of Eq. (21.76) is no longer the only diffusion in the problem: a numerical diffusion comparable in size to G is introduced that contaminates the computed solution. This has the effect of misplacing the position of the saturation shock and miscalculating the magnitude of the saturation discontinuity. These problems are well known and solved in the aerospace industry, where they arise in high-speed wing design. Mathematical problems should be addressed within the context of the equation itself. However, the basic issues (Chin, 1993a) are still overlooked by petroleum investigators overly concerned with field agreement.

Modeling Mudcake Coupling

Now that we understand immiscible two-phase flow formulations, both analytically and numerically, we address the problem where an additional mudcake Darcy flow appears at the inlet to our radial geometry. This flow satisfies its own pressure differential equation and is characterized by a *moving mud-to-mudcake boundary* and a *fixed mudcake-to-rock interface*. The problem is shown in Fig. 21.13, where x applies to both lineal and radial flows. In order

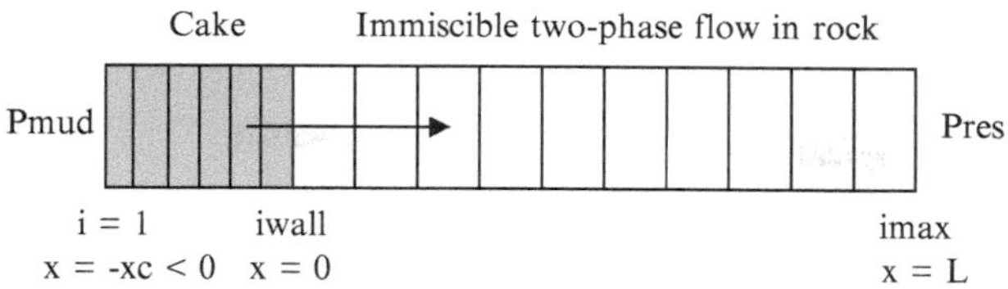

FIG. 21.13 Two-layer mudcake-rock, immiscible flow model.

to solve this coupled problem, the algorithms developed in Example 20.7 of Chapter 20 and the immiscible flow problem just completed must be coupled.

Let us first review the mudcake formulation developed earlier. Again, the flow in the mudcake is assumed to be single-phase, and because compressibility is neglected, the pressure distribution P(x,t) satisfies

$$d(k_c\, dP/dx)/dx = 0 \tag{21.83}$$

where the mudcake absolute permeability k_c may be prescribed as a function of x, or given as a function of P, or taken as a constant for simplicity. We choose the latter for convenience, so that the simple ordinary differential equation

$$d^2P/dx^2 = 0 \tag{21.84}$$

applies. Note that P(x,t) will depend parametrically on time, even though there are no time-dependent derivatives in Eq. (21.84), because a moving boundary will be allowed. We also assume that the mud filtrate and the formation water are identical, so that only two fluids need to be modeled. Other formulations are possible but will not be treated here. If oil base muds are used, three separate fluids must be accounted for, namely, the oil filtrate, the formation hydrocarbon, and the connate water. If two different waters (e.g., fresh versus saline) are present, gravity effects may have to be accounted for. And if combined water-oil muds are considered, the mudcake flow formulation is necessarily two-phase as it is in the formation. These formulations add to numerical complexity and do not introduce new ideas.

Now *how* do we couple Eq. (21.84) for the single-phase flow in the *growing* mudcake to Eqs. (21.78), (21.73), (21.79) describing the two-phase immiscible flow in the rock? It is clear that the *grid expansion method* used in Example 20.7 and suggested in Fig. 21.13 cannot be used: the number of nodes increases with time as the cake thickens, but the saturation solution $S_{i,n}$ requires information at earlier nonexistent spatial nodes. An elementary solution to the problem, fortunately, is available, and requires us first to *transform the boundary value problem for the cake into a boundary condition for the rock flow.* Since Eq. (21.84) applies, where x actually refers to the radial coordinate, the exact solution $P = Ar + B$ applies. Then, the simple solution $P = A(r - R_{cake}) + P_{left}$ satisfies $P = P_{left}$ at $r = R_{cake}$. (Here, P_{left} is the borehole mud pressure, acting on the exposed face of the mud cake located at $r = R_{cake}$.) The pressure at the mudcake-to-rock interface is given by the expression $P = A(R_{well} - R_{cake}) + P_{left}$, where $r = R_{well}$ is the wellbore radius without cake. The fluid velocity at the cake-to-rock interface is k_{cake} dP/dr or $k_{cake}A$. This must be equal to the Darcy velocity $k_{rock}(PW_{2,n}-PW_{1,n})/\Delta r$ evaluated from the two-phase flow solution. Setting the two equal, that is, $k_{cake} A = k_{rock}(PW_{2,n}-PW_{1,n})/\Delta r$, and noting that pressure continuity requires that $PW_{1,n} = A(R_{well} - R_{cake}) + P_{left}$ lead to the fact that

$$\begin{aligned}&[k_{cake}\Delta r + k_{rock}(R_{well} - R_{cake})]\, PW_{1,n}\\ &-k_{rock}(R_{well} - R_{cake})PW_{2,n} = k_{cake}P_{left}\Delta r\end{aligned} \tag{21.85}$$

where we have eliminated the constant A, and PW is the wetting phase pressure. Since the mudcake-to-rock interface is completely saturated with water, the permeability k_{rock} is exactly the absolute permeability.

Unchanging Mudcake Thickness

In dynamic filtration, the mudcake ceases to grow once equilibrium conditions are achieved in the borehole (see Chapter 18). This invasion is modeled by a cake thickness that is a *prescribed constant*, which *does not* vary with time. Then, the *only* algorithmic change to the Fortran code in the foregoing section requires us to replace

```
A(1)  = 99.
B(1)  = 1.
C(1)  = 0.
W(1)  = PLEFT
```

by

```
KCAKE = 0.001
KC = KCAKE*0.00000001/(12.*12.*2.54*2.54)
RCAKE = 0.01/12.
.
.
A(1)  = 99.
B(1)  = KC*DR + K*(WELRAD-RCAKE)
C(1)  = -K*(WELRAD-RCAKE)
W(1)  = KC*PLEFT*DR
```

Typical saturation and pressures in Fig. 21.14A–C for early, intermediate, and late times illustrate shock formation and propagation. The parameters were selected to cover the entire range of inertial-to-capillary force ratios.

Transient Mudcake Growth

When transient mudcake growth *is* allowed, for example, as in static filtration or nonequilibrium dynamic filtration, conceptual but simple coding changes are required. For thin mudcake-to-borehole radii ratios, the *lineal* cake growth model

$$dx_c/dt = -\{f_s/\{(1-f_s)(1-\phi_c)\}\}|v_n| < 0 \tag{20.63}$$

applies, where $|v_n|$ is proportional to the Darcy velocity $(k_c/\mu_{mf})\, dp(x_c)/dx$ at the cake surface. Note that Eq. (20.62) describing displacement fronts in the rock is

```
Water saturation at time (hrs): .167E-03
Position (ft) Water Sat

                           ------------------------------
  .200E+00    .100E+01     |                            *
  .300E+00    .715E+00     |                   *
  .400E+00    .213E+00     |    *
  .500E+00    .105E+00     | *
  .600E+00    .100E+00     | *
  .700E+00    .100E+00     | *
  .800E+00    .100E+00     | *
  .900E+00    .100E+00     | *
  .100E+01    .100E+00     | *
  .110E+01    .100E+00     | *
  .120E+01    .100E+00     | *
  .130E+01    .100E+00     | *
  .140E+01    .100E+00     | *
  .150E+01    .100E+00     | *

Pressure versus r @ time (hrs): .167E-03
Position (ft)  Pressure

                           ------------------------------
  .200E+00    .117E+09     |                            *
  .300E+00    .103E+09     |                        *
  .400E+00    .807E+08     |                  *
  .500E+00    .682E+08     |               *
  .600E+00    .589E+08     |             *
  .700E+00    .510E+08     |           *
  .800E+00    .442E+08     |         *
  .900E+00    .382E+08     |       *
  .100E+01    .328E+08     |      *
  .110E+01    .279E+08     |     *
  .120E+01    .235E+08     |    *
  .130E+01    .194E+08     |  *
  .140E+01    .156E+08     |  *
  .150E+01    .121E+08     | *
  .160E+01    .878E+07     |*
  .170E+01    .568E+07     *
  .180E+01    .276E+07     |
  .190E+01    .000E+00     |
```

(A)

```
Water saturation at time (hrs): .100E-02
Position (ft) Water Sat

                           ------------------------------
  .200E+00    .100E+01     |                            *
  .300E+00    .929E+00     |                         *
  .400E+00    .867E+00     |                        *
  .500E+00    .823E+00     |                      *
  .600E+00    .792E+00     |                     *
  .700E+00    .720E+00     |                   *
  .800E+00    .374E+00     |         *
  .900E+00    .125E+00     | *
  .100E+01    .101E+00     | *
  .110E+01    .100E+00     | *
  .120E+01    .100E+00     | *
  .130E+01    .100E+00     | *
  .140E+01    .100E+00     | *
  .150E+01    .100E+00     | *
  .160E+01    .100E+00     | *
  .170E+01    .100E+00     | *
  .180E+01    .100E+00     | *
  .190E+01    .100E+00     | *

Pressure versus r @ time (hrs): .100E-02
Position (ft)  Pressure

                           ------------------------------
  .200E+00    .119E+09     |                            *
  .300E+00    .106E+09     |                        *
  .400E+00    .954E+08     |                      *
  .500E+00    .863E+08     |                   *
  .600E+00    .782E+08     |                 *
  .700E+00    .706E+08     |               *
  .800E+00    .603E+08     |             *
  .900E+00    .503E+08     |          *
  .100E+01    .431E+08     |        *
  .110E+01    .367E+08     |       *
  .120E+01    .308E+08     |     *
  .130E+01    .255E+08     |    *
  .140E+01    .205E+08     |   *
  .150E+01    .159E+08     |  *
  .160E+01    .115E+08     |*
  .170E+01    .747E+07     *
  .180E+01    .363E+07     |
  .190E+01    .000E+00     |
```

(B)

FIG. 21.14 (A) Early time solution. (B) Intermediate time solution.

(Continued)

```
Water saturation at time (hrs): .267E-02
Position (ft) Water Sat
                           -------------------------------
  .200E+00    .100E+01     |                            *
  .300E+00    .971E+00     |                           *
  .400E+00    .936E+00     |                          *
  .500E+00    .900E+00     |                         *
  .600E+00    .866E+00     |                       *
  .700E+00    .835E+00     |                       *
  .800E+00    .809E+00     |                      *
  .900E+00    .791E+00     |                     *
  .100E+01    .779E+00     |                     *
  .110E+01    .770E+00     |                     *
  .120E+01    .752E+00     |                    *
  .130E+01    .634E+00     |                 *
  .140E+01    .249E+00     |     *
  .150E+01    .112E+00     | *
  .160E+01    .101E+00     | *
  .170E+01    .100E+00     | *
  .180E+01    .100E+00     | *
  .190E+01    .100E+00     | *

Pressure versus r @ time (hrs): .267E-02
Position (ft)  Pressure
                           -------------------------------
  .200E+00    .116E+09     |                            *
  .300E+00    .102E+09     |                        *
  .400E+00    .907E+08     |                     *
  .500E+00    .816E+08     |                   *
  .600E+00    .736E+08     |                 *
  .700E+00    .664E+08     |               *
  .800E+00    .597E+08     |             *
  .900E+00    .536E+08     |           *
  .100E+01    .479E+08     |          *
  .110E+01    .426E+08     |         *
  .120E+01    .377E+08     |       *
  .130E+01    .324E+08     |      *
  .140E+01    .244E+08     |    *
  .150E+01    .183E+08     |  *
  .160E+01    .133E+08     | *
  .170E+01    .861E+07     |*
  .180E+01    .418E+07     *
  .190E+01    .000E+00     |
```

(C)

FIG. 21.14, CONT'D (C) Late time solution.

not used here, since saturation discontinuities are allowed to form naturally in immiscible flows, if they exist. Eq. (20.63) is approximated by

$$x_{c,new} = x_{c,old} + \{f_s/\{(1\text{-}f_s)(1\text{-}\phi_c)\}\}\{k_c\Delta t/(\mu_{mf}\Delta x)\}(p_2\text{-}p_1)_{old} \tag{20.64}$$

where $(p_2 - p_1)_{old}/\Delta x$ represents the pressure gradient in the cake. But our mudcake pressure solution $P = Ar + B$ shows that $dP/dr = A$ does not depend on position, and that at any instant, it is a constant that does not change through the cake. This being the case, its value can be extrapolated from the velocity matching interfacial condition $k_{cake}\ dP/dr = k_{rock}(PW_{2,n}\text{-}PW_{1,n})/\Delta r$, that is,

$$dP/dr = (k_{rock}/k_{cake})(PW_{2,n}\text{-}PW_{1,n})/\Delta r \tag{21.86}$$

Thus, the only required addition to the Fortran immediately preceding is the following update logic in boldface type.

```
C     INITIAL SETUP
      RCAKE = WELRAD
      KCAKE = 0.001
      KC = KCAKE*0.00000001/(12.*12.*2.54*2.54)
      FS=0.2
      PHIMUD=0.2
      .
      .
C     Update cake position immediately after pressure integration.
      RATIO  = K/KC
      PGRADC = RATIO*(P(2)-P(1))/DR
      RCAKE  = RCAKE+(FS/((1.-PHIMUD)*(1.-FS)))*(KC/VISCL)*PGRADC*DT
      .
      .
      A(1)  = 99.
      B(1)  = KC*DR + K*(WELRAD-RCAKE)
      C(1)  = -K*(WELRAD-RCAKE)
      W(1)  = KC*PLEFT*DR
```

The uppermost line represents the mudcake initial condition; that is, at time t = 0, the surface of the infinitesimally thin cake coincides with the borehole radius. In Fig. 21.15A–C, the computational parameters are identical to those in Fig. 21.14A–C, except that the cake grows from zero thickness, as opposed to being fixed at 0.01 in. for all time. Since the mudcake considered in Fig. 21.15 is typically thinner than that in Fig. 21.14 for any instant in time,

```
Water saturation at time (hrs): .167E-03
Position (ft) Water Sat
                                  ______________________________
  .200E+00     .100E+01          |                             *
  .300E+00     .803E+00          |                       *
  .400E+00     .373E+00          |             *
  .500E+00     .118E+00          | *
  .600E+00     .101E+00          | *
  .700E+00     .100E+00          | *
  .800E+00     .100E+00          | *
  .900E+00     .100E+00          | *
  .100E+01     .100E+00          | *
  .110E+01     .100E+00          | *
  .120E+01     .100E+00          | *
  .130E+01     .100E+00          | *
  .140E+01     .100E+00          | *
  .150E+01     .100E+00          | *
  .160E+01     .100E+00          | *
  .170E+01     .100E+00          | *
  .180E+01     .100E+00          | *
  .190E+01     .100E+00          | *

Pressure versus r @ time (hrs): .167E-03
Position (ft)  Pressure
                                  ______________________________
  .200E+00     .141E+09          |                             *
  .300E+00     .125E+09          |                         *
  .400E+00     .104E+09          |                     *
  .500E+00     .862E+08          |                 *
  .600E+00     .743E+08          |              *
  .700E+00     .644E+08          |            *
  .800E+00     .558E+08          |          *
  .900E+00     .482E+08          |         *
  .100E+01     .414E+08          |       *
  .110E+01     .352E+08          |      *
  .120E+01     .296E+08          |     *
  .130E+01     .245E+08          |    *
  .140E+01     .197E+08          |  *
  .150E+01     .152E+08          | *
  .160E+01     .111E+08          |*
  .170E+01     .717E+07          *
  .180E+01     .349E+07          |
  .190E+01     .000E+00          |
```

(A)

FIG. 21.15 (A) Early time solution.

(Continued)

we expect greater relative invasion. In fact, we do observe increased water saturation and deeper penetration of the saturation shock into the rock.

Another interesting observation concerns pressure drops computed at different points in the radial core sample. In the normalized units selected, $P_{left} = 0.144 \times 10^9$ was assumed at the borehole edge of the mudcake, while $P_{right} = 0$ was taken at the far right effective radius. Fig. 21.15C shows that a pressure of 0.113×10^9 was obtained at the mudcake-to-rock interface. *In this calculation, the rock and not the mudcake supports the greatest portion of the total pressure drop.* The computations pursued here, in loose terms, model invasion in very tight zones and in problems having highly permeable cakes.

```
Water saturation at time (hrs): .100E-02
Position (ft) Water Sat

                             ---------------------------------
  .200E+00     .100E+01      |                               *
  .300E+00     .937E+00      |                             *
  .400E+00     .879E+00      |                          *
  .500E+00     .833E+00      |                        *
  .600E+00     .803E+00      |                        *
  .700E+00     .774E+00      |                       *
  .800E+00     .650E+00      |                   *
  .900E+00     .249E+00      |      *
  .100E+01     .111E+00      |  *
  .110E+01     .101E+00      |  *
  .120E+01     .100E+00      |  *
  .130E+01     .100E+00      |  *
  .140E+01     .100E+00      |  *
  .150E+01     .100E+00      |  *
  .160E+01     .100E+00      |  *
  .170E+01     .100E+00      |  *
  .180E+01     .100E+00      |  *
  .190E+01     .100E+00      |  *

Pressure versus r @ time (hrs): .100E-02
Position (ft)  Pressure

                             ---------------------------------
  .200E+00     .131E+09      |                               *
  .300E+00     .117E+09      |                          *
  .400E+00     .105E+09      |                        *
  .500E+00     .947E+08      |                     *
  .600E+00     .857E+08      |                   *
  .700E+00     .776E+08      |                 *
  .800E+00     .693E+08      |               *
  .900E+00     .570E+08      |            *
  .100E+01     .481E+08      |          *
  .110E+01     .409E+08      |        *
  .120E+01     .344E+08      |      *
  .130E+01     .284E+08      |     *
  .140E+01     .228E+08      |    *
  .150E+01     .177E+08      |   *
  .160E+01     .129E+08      |*
  .170E+01     .832E+07      *
  .180E+01     .404E+07      |
  .190E+01     .000E+00      |
```

(B)

FIG. 21.15, CONT'D (B) Intermediate time solution.

(Continued)

We emphasize that we have obtained stable numerical results, without saturation overshoots and local oscillations, all using second-order accurate spatial central differencing, without having to introduce special upwind operators. The methods developed are stable and require minimal computing since they are based on tridiagonal equations. Several subtle aspects of numerical simulation as they affect miscible diffusion and immiscible saturation shock formation are discussed in Chapter 13.

```
Water saturation at time (hrs): .267E-02
Position (ft) Water Sat
                           _______________________________
  .200E+00    .100E+01     |                             *
  .300E+00    .973E+00     |                            *
  .400E+00    .941E+00     |                           *
  .500E+00    .907E+00     |                          *
  .600E+00    .874E+00     |                         *
  .700E+00    .843E+00     |                        *
  .800E+00    .816E+00     |                      *
  .900E+00    .796E+00     |                     *
  .100E+01    .783E+00     |                     *
  .110E+01    .773E+00     |                     *
  .120E+01    .764E+00     |                    *
  .130E+01    .734E+00     |                    *
  .140E+01    .525E+00     |             *
  .150E+01    .170E+00     |   *
  .160E+01    .105E+00     | *
  .170E+01    .100E+00     | *
  .180E+01    .100E+00     | *
  .190E+01    .100E+00     | *

Pressure versus r @ time (hrs): .267E-02
Position (ft)  Pressure
                           _______________________________
  .200E+00    .113E+09     |                             *
  .300E+00    .987E+08     |                         *
  .400E+00    .882E+08     |                      *
  .500E+00    .794E+08     |                    *
  .600E+00    .717E+08     |                  *
  .700E+00    .647E+08     |                *
  .800E+00    .583E+08     |              *
  .900E+00    .523E+08     |            *
  .100E+01    .468E+08     |           *
  .110E+01    .417E+08     |         *
  .120E+01    .369E+08     |       *
  .130E+01    .323E+08     |      *
  .140E+01    .269E+08     |     *
  .150E+01    .192E+08     |   *
  .160E+01    .137E+08     | *
  .170E+01    .885E+07     |*
  .180E+01    .430E+07     *
  .190E+01    .000E+00     |
```

(C)

FIG. 21.15, CONT'D (C) Late time solution.

General Immiscible Flow Model

Earlier we showed how a first-order nonlinear equation arises in immiscible two-phase flow. We derived Eq. (21.17), that is, $\partial S_w/\partial t + \{q(t)/\phi\}\, df_w(S_w,\mu_w/\mu_{nw})/dS_w \;\; \partial S_w/\partial x = 0$ for saturation in one-dimensional systems, and indicated that it applied to high-rate invasion problems where capillary pressure could be ignored. This equation was accurate at least until the appearance of saturation shocks and steep flow gradients. Then, the low-order description breaks down locally, but it could still be used provided we introduce a shock that satisfies certain externally imposed constraints that fall outside the scope of the simple formulation. In fact, mass conservation requires us to take the shock velocity in the form given by Eq. (21.39), namely, $V_{shock} = \{Q_w(S_w^l) - Q_w(S_w^i)\}/(S_w^l - S_w^i)$. But the patched solution is incomplete, since the *structure* and *thickness* of the shock cannot be resurrected.

In order to recover the complete details of the flow, recourse to the high-order partial differential equation *with* capillary pressure is necessary. In radial flow, the required Eq. (21.76) shows that the more detailed physical model is $-\phi\, \partial S_w/\partial t - \{Q(t)\, F'(S_w) + G(S_w)\}/r \;\; \partial S_w/\partial r = G'(S_w)(\partial S_w/\partial r)^2 + G(S_w)\;\; \partial^2 S_w/\partial r^2$. The $G(S_w)\;\; \partial^2 S_w/\partial r^2$ term is all-important, as we have seen (e.g., refer to

Fig. 21.6C) because it produces the shock structure naturally; also, it will affect the propagation speed somewhat, and the shock speed so obtained will differ from the V_{shock} given here. In addition, this second-order derivative completely determines the particular flux that is conserved across shocks and implicitly contains the entropy condition that dictates the manner in which shocks form. The key idea, we emphasize, is the crucial role that the high-order derivative term plays: it may be negligible for a while, but it must be correctly accounted for at the shock because it is large. This being the case, it is imperative that the correct high-order terms be modeled and that the included terms remain free of undesirable numerical diffusion.

In this chapter, the coupling of dynamic mudcake growth to immiscible fluid flow was studied as a purely radial problem. Idealizations were undertaken in order to extend the diffusion ideas first presented in Chapter 13 to broader problems involving two-phase flow. In practical applications, many physical mechanisms are simultaneously at work in the formation, for example, reservoir heterogeneities, axial variations, and miscible mixing. At the same time, the auxiliary conditions that apply to well logging tools are far from simple. Consider formation-testing-while-drilling. During drilling, two-phase flow invasion takes place while the mudcake builds; this establishes initial conditions that apply once fluid sampling commences. When pretest samples are taken, mudcake is first removed at the piston, and complicated three-dimensional boundary conditions that model skin and flow line storage effects must be used (e.g., see Chapter 18 for an elementary discussion on tester modeling). This operational procedure can be simulated in detail using the building blocks described in this book. A comprehensive numerical model has been developed for tester applications. It can predict (1) the pumping times required to purge the near-well formation of mud filtrate before uncontaminated petroleum fluids are accessible, (2) the tool power requirements associated with such pumping processes, and (3) continually refined formation evaluation parameters based on compressible and incompressible fluid flow pressure transients. This modeling effort is reported in "Sample Quality Prediction with Integrated Oil and Water-Based Mud Invasion Modeling," *SPE Paper No. 77964, SPE Asia Pacific Oil & Gas Conference and Exhibition (APOGCE)*, Oct. 2002, Melbourne, Australia by M. Proett, D. Belanger, M. Manohar, and the present author.

PROBLEMS AND EXERCISES

1. Select several available immiscible two-phase flow simulators, and define conditions that would lead to water breakthrough in finite time. Assume different capillary pressure functions. How are breakthrough times and locations affected? Is mass conserved across the saturation discontinuity? Rerun your problem sets with capillary pressure identically zero and compare results.

2. For the one-dimensional immiscible flow formulation considered here, write a program comparing backward, central, and forward difference approximations and their effects on mass conservation. What general conclusions can you draw?

Cumulative References

Aguilera, R., 1980. Naturally Fractured Reservoirs. PennWell Publishing Company, Tulsa.

Aguilera, R., Artindale, J.S., Cordell, G.M., Ng, M.C., Nicholl, G.W., Runions, G.A., 1991. Horizontal Wells—Formation Evaluation, Drilling, and Production. Gulf Publishing, Houston.

Allen, D., Auzerais, F., Dussan, E., Goode, P., Ramakrishnan, T.S., Schwartz, L., Wilkinson, D., Fordham, E., Hammond, P., Williams, R., 1991. Invasion revisited. Schlumberger Oilfield Rev., 10–23.

Allen, D.F., Jacobsen, S.J., 1987. Resistivity profiling with a multi-frequency induction Sonde. In: SPWLA 28th Annual Logging Symposium, London, June 29–July 2.

Allen, M.B., Pinder, G.F., 1982. The convergence of upstream collocation in the Buckley-Leverett problem. SPE Paper No. 10978, In: 57th Annual Fall Technical Conference and Exhibition of the Society of Petroleum Engineers, New Orleans, La., September 26–29.

Ames, W.F., 1977. Numerical Methods for Partial Differential Equations. Academic Press, New York.

Ashley, H., Landahl, M.T., 1965. Aerodynamics of Wings and Bodies. Addison-Wesley, Reading, MA.

Aziz, K., Settari, A., 1979. Petroleum Reservoir Simulation. Applied Science Publishers, London.

Batchelor, G.K., 1970. An Introduction to Fluid Dynamics. Cambridge University Press, London.

Bear, J., 1972. Dynamics of Fluids in Porous Media. American Elsevier Publishing, Inc., New York.

Bisplinghoff, R.L., Ashley, H., Halfman, R., 1955. Aeroelasticity. Addison-Wesley, Reading, MA.

Broussard, S., 1989. The annulus effect. Schlumberger Tech. Rev. 37 (1), 41–47.

Cardwell, W.T., 1959. The meaning of the triple value in noncapillary Buckley-Leverett theory. Pet. Trans. AIME 216, 271–276.

Carnahan, B., Luther, H.A., Wilkes, J.O., 1969. Applied Numerical Methods. John Wiley & Sons, New York.

Carrier, G.F., Krook, M., Pearson, C.E., 1966. Functions of a Complex Variable. McGraw-Hill, New York.

Carslaw, H.S., Jaeger, J.C., 1946, 1959. Conduction of Heat in Solids. Oxford University Press, London.

Chappelear, J.E., Williamson, A.S., 1981. Representing wells in numerical reservoir simulation: Part 2–Implementation. Soc. Petrol. Eng. J., 339–344.

Chen-Charpentier, B., Kojouharov, H.V., 2001. Modeling of Subsurface Biobarrier Formation. In: Erickson, L.E. (Ed.), Proceedings of the 2000 Conference on Hazardous Waste Research. Great Plains/Rocky Mountain Hazardous Substance Research Center, Manhattan, KS, pp. 228–237.

Cherry, J.A., Freeze, R.A., 1979. Groundwater. Prentice-Hall, Englewood Cliffs, NJ.

Chin, W.C., 1977. Numerical solution for viscous transonic flow. AIAA J., 1360–1362.

Chin, W.C., 1978a. Pseudo-transonic equation with a diffusion term. AIAA J., 87–88.

Chin, W.C., 1978b. Algorithm for inviscid flow using the viscous transonic equation. AIAA J., 848–849.

Chin, W.C., 1978c. Type-independent solutions for mixed compressible flows. AIAA J., 854–856.

Chin, W.C., 1978d. Aerodynamics of Wings and Bodies I, II and III, Course Notes. Boeing Commercial Airplane Company, Seattle.

Chin, W.C., 1979. On the design of thin subsonic airfoils. ASME J. Appl. Mech., 6–8.

Chin, W.C., 1981. Direct approach to aerodynamic design problems. ASME J. Appl. Mech., 721–726.

Chin, W.C., 1984. Thin airfoil theory for planar inviscid shear flow. ASME J. Appl. Mech., 19–26.

Chin, W.C., 1988. Simulating horizontal well fracture flow. Offshore, 49–52.

Chin, W.C., 1992a. Borehole Flow Modeling in Horizontal, Deviated and Vertical Wells. Gulf Publishing, Houston.

Chin, W.C., 1992b. Petrocalc 14: Horizontal and Vertical Annular Flow Modeling, Petroleum Engineering Software for the IBM PC and Compatibles. Gulf Publishing, Houston.

Chin, W.C., 1993a. Modern Reservoir Flow and Well Transient Analysis. Gulf Publishing, Houston.

Chin, W.C., 1993b. 3D/SIM: 3D Petroleum Reservoir Simulation for Vertical, Horizontal, and Deviated Wells, Petroleum Engineering Software for the IBM PC and Compatibles. Gulf Publishing, Houston.

Chin, W.C., 1994. Wave Propagation in Petroleum Engineering, with Applications to Drillstring Vibrations, Measurement-While-Drilling, Swab-Surge and Geophysics. Gulf Publishing, Houston.

Chin, W.C., 1995. Formation Invasion, With Applications to Measurement-While-Drilling, Time Lapse Analysis and Formation Damage. Gulf Publishing, Houston.

Chin, W.C., 2000. General three-dimensional electromagnetic model for nondipolar transmitters in layered anisotropic media with dip. Well Logging Technol. J. 24, 262–278.

Chin, W.C., 2001a. Computational Rheology for Pipeline and Annular Flow. Butterworth-Heinemann, Boston, MA.

Chin, W.C., 2001b. RheoSim 2.0: Advanced Rheological Flow Simulator. Butterworth-Heinemann, Boston, MA.

Chin, W.C., 2002. Quantitative Methods in Reservoir Engineering, first ed. Elsevier Science, Woburn, MA.

Chin, W.C., 2008. Formation Testing Pressure Transient and Contamination Analysis. E&P Press, Houston.

Chin, W.C., 2012a. FTWD-Processing-Algorithms-V56a. Stratamagnetic Software Internal Report. July.

Chin, W.C., 2012b. Managed Pressure Drilling: Modeling, Strategy and Planning. Elsevier Science, Woburn, MA.

Chin, W.C., 2013. Formation tester flow analysis in anisotropic media with flowline storage and skin at arbitrary dip. Well Logging Technol. J. 37, 1–12.

Chin, W.C., 2014a. Electromagnetic Well Logging: Models for MWD/LWD Interpretation and Tool Design. John Wiley & Sons, Hoboken, NJ (a).

Chin, W.C., 2014b. Wave Propagation in Drilling, Well Logging and Reservoir Applications. John Wiley & Sons, Hoboken, NJ.

Chin, W.C., 2015a. Formation Evaluation Using Phase Shift Periodic Pressure Pulse Testing in Anisotropic Media. United States Provisional Patent Application No. 62172120, June 7.

Chin, W.C., 2015b. Multiple Drawdown Pressure Transient Analysis for Low Mobility Formation Testing. United States Provisional Patent Application No. 62175549, June 15.

Chin, W.C., 2016a. Quantitative Methods in Reservoir Engineering, Second Edition—With New Topics in Formation Testing and Multilateral Well Flow Analysis. Elsevier Science, Woburn, MA.

Chin, W.C., 2016b. Reservoir Engineering in Modern Oilfields: Vertical, Deviated, Horizontal and Multilateral Well Systems. John Wiley & Sons, Hoboken, NJ.

Chin, W.C., 2016c. Resistivity Modeling: Propagation, Laterolog and Micro-Pad Analysis. John Wiley & Sons, Hoboken, NJ.

Chin, W.C., Proett, M.A., 1997. Formation Evaluation Using Phase Shift Periodic Pressure Pulse Testing. United States Patent No. 5,672,819 issued Sept. 30.

Chin, W.C., Proett, M.A., 2005. Formation tester immiscible and miscible flow modeling for job planning applications. In: SPWLA 46th Annual Logging Symposium, New Orleans, Louisiana, June 26–29.

Chin, W.C., Rizzetta, D.P., 1979. Airfoil design in subcritical and supercritical flow. ASME J. Appl. Mech., 761–766.

Chin, W.C., Su, Y., Sheng, L., Li, L., Bian, H., Shi, R., 2014a. Measurement-While-Drilling Signal Analysis, Optimization and Design. John Wiley & Sons, Hoboken, NJ.

Chin, W.C., Suresh, A., Holbrook, P., Affleck, L., Robertson, H., 1986. Formation evaluation using repeated MWD logging measurements. Paper No. U, In: SPWLA 27th Annual Logging Symposium, Houston, TX, June 9–13.

Chin, W.C., Zhou, Y., Feng, Y., Yu, Q., 2015. Formation Testing: Low Mobility Pressure Transient Analysis. John Wiley & Sons, Hoboken, NJ.

Chin, W.C., Zhou, Y., Feng, Y., Yu, Q., Zhao, L., 2014b. Formation Testing Pressure Transient and Contamination Analysis. John Wiley & Sons, Hoboken, NJ.

Chin, W.C., Zhuang, X., 2007. Formation tester inverse permeability interpretation for liquids in anisotropic media with flowline storage and skin at arbitrary dip. In: SPWLA 48th Annual Logging Symposium, Austin, Texas, June 3–6.

Churchill, R.V., 1960. Complex Variables and Applications. McGraw-Hill, New York.

Claerbout, J.F., 1985a. Fundamentals of Geophysical Data Processing. Blackwell Scientific Publishers, Oxford.

Claerbout, J.F., 1985b. Imaging the Earth's Interior. Blackwell Scientific Publishers, Oxford.

Cobern, M.E., Nuckols, E.B., 1985. Application of MWD resistivity relogs to evaluation of formation invasion. In: SPWLA 26th Annual Logging Symposium, June 17–20.

Cole, J.D., 1949. Problems in Transonic Flow (Ph.D. Thesis). California Institute of Technology, Pasadena, CA.

Cole, J.D., 1968. Perturbation Methods in Applied Mathematics. Blaisdell Publishing, Massachusetts.

Collins, R.E., 1961. Flow of Fluids Through Porous Materials. Reinhold Publishing, New York.

Courant, R., Friedrichs, K.O., 1948. Supersonic Flow and Shock Waves. Springer-Verlag, Berlin.

Craig, F.F., 1971. The Reservoir Engineering Aspects of Waterflooding, SPE Monograph Series. Society of Petroleum Engineers, Dallas.

Dahlquist, G., Bjorck, A., 1974. Numerical Methods. Prentice-Hall, Englewood Cliffs, NJ.

Dake, L.P., 1978. Fundamentals of Reservoir Engineering. Elsevier Scientific Publishing, Amsterdam.

Dewan, J.T., Chenevert, M.E., 1993. Mudcake buildup and invasion in low permeability formations: application to permeability determination by measurement while drilling, Paper NN. In: SPWLA 34th Annual Logging Symposium, June 13–16.

Dewan, J.T., Holditch, S.A., 1992. Radial Response Functions for Borehole Logging Tools. Topical Report, Contract No. 5089-260-1861, Gas Research Institute.

Ding, D.Y., Wu, Y.S., Farah, N., Wang, C., Bourbiaus, B., 2014. Numerical simulation of low permeability unconventional gas reservoirs. In: SPE Paper 167711, SPE/EAGE European Unconventional Conference and Exhibition, Vienna, Austria, February 25–27.

Doll, H.G., 1955. Filtrate invasion in highly permeable sands. Pet. Eng., B-53–B-66.

Douglas, J., Blair, P.M., Wagner, R.J., 1958. Calculation of linear waterflood behavior including the effects of capillary pressure. Pet. Trans. AIME 213, 96–102.

Douglas, J., Peaceman, D.W., Rachford, H.H., 1959. A method for calculating multi-dimensional immiscible displacement. Pet. Trans. AIME 216, 297–306.

Durlofsky, L.J., Aziz, K., 2004. Advanced Techniques for Reservoir Simulation and Modeling of Nonconventional Wells. Final Report, DOE Award DE-AC26-99BC15213, Department of Petroleum Engineering, Stanford University.

Durst, D.G., Vento, M.C., 2012. Unconventional shale play selective fracturing using multilateral technology. In: SPE Paper 151989-MS, SPE/EAGE European Unconventional Resources Conference and Exhibition, Vienna, Austria, March 20–22.

Economides, M.J., Nolte, K.G., 1987. Reservoir Stimulation. Schlumberger Educational Services, Houston.

Estrada, R., Kanwal, R.P., 1987. The Carleman type singular integral equations. SIAM Rev. 29 (2), 263–290.

Fayers, F.J., Sheldon, J.W., 1959. The effect of capillary pressure and gravity on two-phase fluid flow in a porous medium. Pet. Trans. AIME 216, 147–155.

Fordham, E.J., Allen, D.F., Ladva, H.K.J., Alderman, N.J., 1991. The principle of a critical invasion rate and its implications for log interpretation. SPE Paper No. 22539, In: 66th Annual Technical Conference and Exhibition of the Society of Petroleum Engineers, Dallas, TX, October 6–9.

Fraser, L., Williamson, D., Haydel, S., 1994. MMH fluids reduce formation damage in horizontal wells. Pet. Eng. Int., 44–49.

Fredrickson, A.G., Bird, R.B., 1958. Non-Newtonian flow in annuli. Ind. Eng. Chem. 50, 347.

Gakhov, F.D., 1966. Boundary Value Problems. Pergamon Press, London.

Garabedian, P.R., 1964. Partial Differential Equations. John Wiley & Sons, New York.

Gerald, C.F., 1980. Applied Numerical Analysis. Addison-Wesley, Reading, MA.

Gondouin, M., Heim, A., 1964. Experimentally determined resistivity profiles in invaded water and oil sands for linear flows. J. Pet. Technol., 337–348.

Gradshteyn, I.S., Ryzhik, I.M., 1965. Table of Integrals, Series, and Products. Academic Press, New York.

Heckman, T., Olsen, G., Scott, K., Seiller, B., Simpson, M., Blasingame, T., 2013. Best practices for reserves estimation in conventional reservoirs—present and future considerations. In: SPE Unconventional Resources Conference, The Woodlands, Texas, April 10–12.

Hildebrand, F.B., 1948. Advanced Calculus for Applications. Prentice-Hall, Englewood Cliffs, NJ.

Hildebrand, F.B., 1965. Methods of Applied Mathematics. Prentice-Hall, Englewood Cliffs, NJ.

Holditch, S.A., Dewan, J.T., 1991. The Evaluation of Formation Permeability Using Time Lapse Logging Measurements During and After Drilling. Annual Report, Contract No. 5089-260-1861, Gas Research Institute.

Hovanessian, S.A., Fayers, F.J., 1961. Linear water flood with gravity and capillary effects. Soc. Petrol. Eng. J., 32–36.

Jameson, A., 1975. Numerical computation of transonic flows with shock waves. In: Oswatitsch, K., Rues, D. (Eds.), Symposium Transsonicum II, Proceedings, September 8–13. International Union of Theoretical and Applied Mechanics, Gottingen.

Kober, H., 1957. Dictionary of Conformal Representations. Dover Publications, New York.

Lamb, H., 1945. Hydrodynamics. Dover Press, New York.

Landau, L.D., Lifshitz, E.M., 1959. Fluid Mechanics. Pergamon Press, London.

Lane, H.S., 1993. Numerical simulation of mud filtrate invasion and dissipation, Paper D. In: SPWLA 34th Annual Logging Symposium, June 13–16.

Lantz, R.B., 1971. Quantitative evaluation of numerical diffusion (truncation error). Soc. Petrol. Eng. J., 315–320.

Lapidus, L., Pinder, G.F., 1982. Numerical Solution of Partial Differential Equations in Science and Engineering. John Wiley & Sons, New York.

Latil, M., 1980. Enhanced Oil Recovery. Gulf Publishing, Houston.

Lee, E.H., Fayers, F.J., 1959. The use of the method of characteristics in determining boundary conditions for problems in reservoir analysis. Pet. Trans. AIME 216, 284–289.

Lee, S.H., Milliken, W.J., 1993. The productivity index of an inclined well in finite difference reservoir simulation. Paper No. SPE-25247-MS, In: SPE Symposium on Reservoir Simulation, New Orleans, LA, February 28 to March 3.

Liepmann, H.W., Roshko, A., 1957. Elements of Gasdynamics. John Wiley & Sons, New York.

Lighthill, M.J., 1958. Fourier Analysis and Generalised Functions. Cambridge University Press, London.

Lighthill, M.J., 1945. A Mathematical Method of Cascade Design. Memo No. 2104, British Aeronautical Research Council Reports.

Marle, C.M., 1981. Multiphase Flow in Porous Media. Gulf Publishing, Houston.

Matthews, C.S., Russell, D.G., 1967. Pressure Buildup and Flow Tests in Wells, SPE Monograph Series. Society of Petroleum Engineers, Dallas.

McEwen, C.R., 1959. A numerical solution of the linear displacement equation with capillary pressure. Pet. Trans. AIME 216, 412–415.

Messenger, J.U., 1981. Lost Circulation. PennWell Books, Tulsa, OK.

Mikhlin, S.G., 1965. Multidimensional Singular Integrals and Integral Equations. Pergamon Press, London.

Milne-Thomson, L.M., 1940. Hydrodynamical images. Proc. Camb. Philos. Soc. 36, 246–247.

Milne-Thomson, L.M., 1958. Theoretical Aerodynamics. Macmillan & Co., London.

Milne-Thomson, L.M., 1968. Theoretical Hydrodynamics. Macmillan Co., New York.

Moretti, G., Salas, M.D., 1972. Numerical analysis of viscous one-dimensional flows. In: Numerical Methods in Fluid Dynamics: AGARD Lecture Series No. 48. Advisory Group for Aerospace Research and Development, North Atlantic Treaty Organization, von Karman Institute, Rhode-Saint-Genese, Belgium.

Muskat, M., 1937. Flow of Homogeneous Fluids Through Porous Media. McGraw-Hill, New York.

Muskat, M., 1949. Physical Principles of Oil Production. McGraw-Hill, New York.

Muskhelishvili, N.I., 1953. Singular Integral Equations. P. Noordhoff N.V, Holland.

Nayfeh, A., 1973. Perturbation Methods. John Wiley & Sons, New York.

Oates, G.C., 1978. The Aerothermodynamics of Aircraft Gas Turbine Engines. Technical Report AFAPL-TR-78-52, Air Force Aero Propulsion Laboratory.

Outmans, H.D., 1963. Mechanics of static and dynamic filtration in the borehole. Soc. Petrol. Eng. J., 236–244.

Peaceman, D.W., 1977. Fundamentals of Numerical Reservoir Simulation. Elsevier Scientific Publishing, Amsterdam.

Peaceman, D.W., 1978. Interpretation of well-block pressures in numerical reservoir simulation. Soc. Petrol. Eng. J., 183–194.

Peaceman, D.W., 1983. Interpretation of well-block pressures in numerical reservoir simulation with nonsquare grid blocks and anisotropic permeability. Soc. Petrol. Eng. J., 531–543.

Peaceman, D.W., Rachford, H.H., 1962. Numerical calculation of multidimensional miscible displacement. Soc. Petrol. Eng. J., 327–339.

Phelps, G.D., Stewart, G., Peden, J.M., 1984. The analysis of the invaded zone characteristics and their influence on wireline log and well-test interpretation. SPE Paper No. 13287, In: 59th Annual Technical Conference and Exhibition, Houston, TX, September 16–19.

Proett, M.A., Belanger, D., Manohar, M., Chin, W.C., 2002. Sample quality prediction with integrated oil and water-based mud invasion modeling. SPE Paper No. 77964, In: SPE Asia Pacific Oil & Gas Conference and Exhibition (APOGCE), October 2002, Melbourne, Australia.

Proett, M.A., Chin, W.C., 1996. Supercharge pressure compensation with new wireline formation testing method. In: 37th Annual SPWLA Symposium, New Orleans, Louisiana, June 16–19.

Proett, M.A., Chin, W.C., 1998. Exact spherical flow solution with storage for early-time test interpretation. SPE J. Pet. Technol. 50, 34–36.

Proett, M.A., Chin, W.C., 2000. Advanced permeability and anisotropy measurements while testing and sampling in real-time using a dual probe formation tester. SPE Paper No. 64650, In: Seventh International Oil & Gas Conference and Exhibition, November 2000, Beijing, China.

Proett, M.A., Chin, W.C., Manohar, M., Sigal, R., Wu, J., 2001. Multiple factors that influence wireline formation tester pressure measurements and fluid contact estimates. SPE Paper No. 71566, In: 2001 SPE Annual Technical Conference and Exhibition, New Orleans, LA, September 30 to October 3.

Richtmyer, R.D., Morton, K.W., 1957. Difference Methods for Initial Value Problems. Interscience Publishers, New York.

Roache, P.J., 1972. Computational Fluid Dynamics. Hermosa Publishers, Albuquerque, NM.

Saad, M.A., 1966. Thermodynamics for Engineers. Prentice-Hall, Englewood Cliffs, NJ.

Sabet, M.A., 1991. Well Test Analysis. Gulf Publishing, Houston.

Scheidegger, A.E., 1957. The Physics of Flow Through Porous Media. University of Toronto Press, Toronto.

Schlichting, H., 1968. Boundary Layer Theory. McGraw-Hill, New York.

Scholz, N., 1977. Aerodynamics of Cascades. Advisory Group for Aerospace Research and Development (AGARD), North Atlantic Treaty Organization (NATO), Neuilly sur Seine, France.

Semmelbeck, M.E., Holditch, S.A., 1988. The effects of mud-filtrate invasion on the interpretation of induction logs. SPE Form. Eval., 386–392.

Sharpe, H.N., Anderson, D.A., 1991. Orthogonal curvilinear grid generation with preset internal boundaries for reservoir simulation. Paper No. 21235, In: Eleventh SPE Symposium on Reservoir Simulation, Anaheim, CA, February 17–20.

Sheldon, J.W., Zondek, B., Cardwell, W.T., 1959. One-dimensional, incompressible, noncapillary, two-phase fluid flow in a porous medium. Pet. Trans. AIME 216, 290–296.

Sichel, M., 1966. The effect of longitudinal viscosity on the flow at a nozzle throat. J. Fluid Mech., 769–786.

Slattery, J.C., 1981. Momentum, Energy, and Mass Transfer in Continua. Robert E. Krieger Publishing Company, New York.

Spiegel, M.R., 1964. Schaum's Outline Series: Complex Variables. McGraw-Hill, New York.

Spreiter, J.R., 1950. The Aerodynamic Forces on Slender Plane and Cruciform Wing and Body Combinations. NACA Report No. 962.

Stakgold, I., 1968. Boundary Value Problems of Mathematical Physics, vol. II Macmillan Company, New York.

Streeter, V.L., 1961. Handbook of Fluid Dynamics. McGraw-Hill, New York.

Streltsova, T.D., 1988. Well Testing in Heterogeneous Formations. John Wiley & Sons, New York.

Tamamidis, P., Assanis, D.N., 1991. Generation of orthogonal grids with control of spacing. J. Comput. Phys. 94, 437–453.

Thomas, G.B., 1960. Calculus and Analytic Geometry. Addison-Wesley, Reading, MA.

Thomas, G.W., 1982a. Principles of Hydrocarbon Reservoir Simulation. International Human Resources Development Corporation, Boston.

Thomas, P.D., 1982b. Composite three-dimensional grids generated by elliptic systems. AIAA J., 1195–1202.

Thomas, P.D., Middlecoff, J.F., 1980. Direct control of the grid point distribution in meshes generated by elliptic equations. AIAA J., 652–656.

Thompson, J.F., 1978. Numerical Solution of Flow Problems Using Body-Fitted Coordinate Systems. Lecture Series 1978-4. von Karman Institute for Fluid Dynamics, Brussels, Belgium.

Thompson, J.F., 1984. Grid generation techniques in computational fluid dynamics. AIAA J., 1505–1523.

Thompson, J.F., Warsi, Z.U.A., Mastin, C.W., 1985. Numerical Grid Generation. Elsevier Science Publishing, New York.

Thwaites, B., 1960. Incompressible Aerodynamics. Oxford Press, Oxford.

Tobola, D.P., Holditch, S.A., 1989. Determination of reservoir permeability from repeated induction logging. SPE Paper No. 19606, In: 64th Annual Technical Conference and Exhibition of the Society of Petroleum Engineers, San Antonio, Texas, October 8–11.

Tychonov, A.N., Samarski, A.A., 1964. Partial Differential Equations of Mathematical Physics, vol. I Holden-Day, San Francisco.

Tychonov, A.N., Samarski, A.A., 1967. Partial Differential Equations of Mathematical Physics, vol. II Holden-Day, San Francisco.

van Dyke, M.D., 1956. Second-Order Subsonic Airfoil Theory Including Edge Effects. NACA Report No. 1274, National Advisory Committee for Aeronautics.

van Dyke, M., 1964. Perturbation Methods in Fluid Mechanics. Academic Press, New York.

van Everdingen, A.F., Hurst, W., 1949. The application of the Laplace transformation to flow problems in reservoirs. Trans. AIME 186, 305–324.

van Golf-Racht, T.D., 1982. Fundamentals of Fractured Reservoir Engineering. Elsevier Scientific Publishing, Amsterdam.

van Poollen, H.K., Breitenbach, E.A., Thurnau, D.H., 1968. Treatment of individual wells and grids in reservoir modeling. Soc. Petrol. Eng. J., 341–346.

Weinig, F.S., 1964. Theory of two-dimensional flow through cascades. In: Hawthorne, W.R. (Ed.), High Speed Aerodynamics and Jet Propulsion, Vol. X: Aerodynamics of Turbines and Compressors. Princeton University Press, Princeton, NJ.

Wesseling, P., 1992. An Introduction to Multigrid Methods. John Wiley & Sons, Chichester.

White, J.W., 1982. General mapping procedure for variable area duct acoustics. AIAA J., 880–884.

Whitham, G.B., 1974. Linear and Nonlinear Waves. John Wiley & Sons, New York.

Williamson, A.S., Chappelear, J.E., 1981. Representing wells in numerical reservoir simulation: Part 1–Theory. Soc. Petrol. Eng. J., 323–338.

Wolfsteiner, C., Durlofsky, L.J., Aziz, K., 2003. Calculation of well index for nonconventional wells on arbitrary grids. Comput. Geosci. 7, 61–82.

Woods, L.C., 1961. The Theory of Subsonic Plane Flow. Cambridge University Press, Cambridge.

Yih, C.S., 1969. Fluid Mechanics. McGraw-Hill, New York.

Zarnowski, R., Hoff, D., 1991. A finite difference scheme for the Navier–Stokes equations of one-dimensional, isentropic, compressible flow. SIAM J. Numer. Anal., 78–112.

Zhou, Y., Hao, Z., Feng, Y., Yu, Q., Chin, W.C., 2014a. Formation testing: new methods for rapid mobility and pore pressure prediction. Paper No. OTC 24890-MS, In: Offshore Technology Conference Asia, Kuala Lumpur, Malaysia, March 25–28.

Zhou, Y., Zhao, L., Feng, Y., Yu, Q., Chin, W.C., 2014b. Formation testing: new methods for rapid mobility and pore pressure prediction. Paper No. IPTC 17214-MS, In: International Petroleum Technology Conference, Doha, Qatar, January 20–22.

Index

Note: Page numbers followed by *f* indicate figures.

D

E

F

G

H

I

N

O

P

Q

R

S

T

U

V